Signals and Communication Technology

More information about this series at http://www.springer.com/series/4748

Wolfgang Utschick
Editor

Communications in Interference Limited Networks

Springer

Editor
Wolfgang Utschick
Technische Universität München
Munich
Germany

ISSN 1860-4862 ISSN 1860-4870 (electronic)
Signals and Communication Technology
ISBN 978-3-319-22439-8 ISBN 978-3-319-22440-4 (eBook)
DOI 10.1007/978-3-319-22440-4

Library of Congress Control Number: 2015950002

Springer Cham Heidelberg New York Dordrecht London

Printed on acid-free paper

Springer International Publishing AG Switzerland is part of Springer Science+Business Media (www.springer.com)

To Ralf Kötter

Preface

The classical approaches to cope with interference in wireless communication systems are based on the principle of interference avoidance and minimization. This is classically achieved through proper network planning as well as various orthogonal medium access techniques. While network planning is largely concerned with the geographic situation of a given network, relying on interference control through attenuation of wireless signals, medium access techniques exploit orthogonality in the signals themselves to keep different users separate. However, as the demands on modern wireless communication systems increase dramatically, interference still stands to become the limiting performance factor. Orthogonality, either induced by time, frequency, or space, cannot be achieved perfectly but can, at best, offer to mitigate interference. In addition to this imperfect interference control, it is known that orthogonal access techniques are suboptimal with respect to the fundamentally achievable rates in a network. In contrast to and beyond orthogonal access techniques, centralized and decentralized methods have been proposed to avoid and handle interference as well as approaches that resolve interference constructively. In fact, an exiting insight into the operation of networks is that it may be beneficial, in terms of an overall throughput, to actively create and manage interference. Thus, when handled properly, "mixing" of data in networks becomes a useful tool for operation rather than a nuisance, as it has been treated traditionally. The development of mobile, robust, ubiquitous, reliable, and instantaneous communication being a driving and enabling factor of an information centric economy, the understanding, mitigation, and exploitation of interference in networks must be seen as a centrally important task.

This book exemplarily introduces the fundamentals of various approaches to aim at these objectives. It summarizes and extends the collective scientific work that has been conducted under the umbrella of a priority program by the same name that has been supported by the Deutsche Forschungsgesellschaft (DFG) from 2009 to 2015. The chapters in this book are related to a wide range of challenges and topics, among which are ad hoc networks, analog computation in networks, cellular networks, cooperative communications, device-to-device communications, evidence-based computation models, interference alignment, interference limited networks, interference management, mesh networks, multicasting in networks, multiple-relay networks, network coding, physical layer security, robust communications, satellite networks, and many more. With this book the authors provide an extensive and substantial overview of current research in a buoyant technological area for future wireless communication systems.

The idea of this book was born during a series of meetings over the span of the DFG priority program *Communications in Interference Limited Networks* that was started with a joint initiative of a group of researchers in 2009. My special thanks go to Holger Boche (Technische Universität München), Robert Fischer (Universität Ulm), Peter Höher (Christian-Albrechts-Universität zu Kiel), Anja Klein (Technische Universität Darmstadt), Volker Kühn (Universität Rostock), Berthold Lankl (Universität der Bundeswehr München), and Erich Lutz (Deutsches Zentrum für Luft- und Raumfahrt). My special thanks also to my colleagues in the coordination group, Johannes Huber (Friedrich-Alexander-Universität Erlangen-Nürnberg), and Ralf Kötter (Technische Universität München). In particular, I want to thank Ralf Kötter (2009†) for all the fun during preparing the proposal of the program and the unforgettable time with him at the Technische Universität München. It was a great tragedy that Ralf could not participate in the program any more.

Finally, I want to thank all authors and co-authors who contributed to this book, and Christoph Baumann, Gayathri Umashankar, and Mathini Manoj from Springer International Publishing AG, and last but not least, Thomas Wiese, Technische Universität München, who did an incredible job of supporting the collection and harmonization of all material.

Munich, Germany
July 2015

Wolfgang Utschick

Contents

1 **Coding Techniques for Transmitting Packets Through Complex Communication Networks** 1
Martin Bossert, Vladimir Sidorenko and Antonia Wachter-Zeh

2 **Modulo-Type Precoding for Networks** 31
Robert F.H. Fischer, Michael Cyran, Sebastian Stern and Johannes B. Huber

3 **Enabling the Multi-User Generalized Degrees of Freedom in Cellular Interference Networks with Multi-User Coding** 53
Rick Fritschek and Gerhard Wunder

4 **The Information-Theoretic Constant-Gap Optimality of Treating Interference as Noise in Interference Networks** 75
Soheil Gherekhloo, Anas Chaaban and Aydin Sezgin

5 **Interference-Aware Analog Computation over the Wireless Channel: Fundamentals and Strategies** 97
Mario Goldenbaum, Sławomir Stańczak and Holger Boche

6 **Bounds on the Outage Constrained Capacity of the Gaussian Relay Channel** 119
Andreas Gründinger, Lennart Gerdes, Michael Joham and Wolfgang Utschick

7 **Balancing for Interference-Limited Multi-User Satellite Communications** 139
Andreas Gründinger, Michael Joham and Wolfgang Utschick

8 Distributed Channel Selection for Underlay Device-to-Device Communications: A Game-Theoretical Learning Framework 173
Setareh Maghsudi and Sławomir Stańczak

9 Cyclic Interference Alignment Via Polynomial Rings for Multi-user Communication Channels 191
Henning Maier, Johannes Schmitz and Rudolf Mathar

10 Multicast in Networks of Broadcast Channels—Part I: Submodular Models and Optimization 215
Maximilian Riemensberger and Wolfgang Utschick

11 Multicast in Networks of Broadcast Channels—Part II: Representation of Bounds on the Multicast Capacity Region 237
Maximilian Riemensberger and Wolfgang Utschick

12 Arbitrarily Varying Channels—A Model for Robust Communication in the Presence of Unknown Interference 259
Rafael F. Schaefer, Holger Boche and H. Vincent Poor

13 Resource Allocation and Pricing in Non-cooperative Interference Networks with Malicious Users 285
Fei Shen, Anil Kumar Chorppath, Eduard Jorswieck and Holger Boche

14 Interference Alignment Aided by Non-Regenerative Relays 327
Rakash SivaSiva Ganesan, Hussein Al-Shatri, Xiang Li, Anja Klein and Tobias Weber

15 Stochastic Geometry for Analysis of Coordination and Cooperation in Interference-Limited Ad Hoc Networks 347
Andrey Skrebtsov, Guido H. Bruck and Peter Jung

16 Cooperative Diversity Under Spatial Interference Correlation in Wireless Networks 367
Ralph Tanbourgi, Holger Jäkel and Friedrich K. Jondral

17 Physical-Layer Key Generation and Reconciliation 393
Jon Wallace, Rashid Mehmood, Rajesh Sharma, Werner Henkel, Oana Graur, Nazia Islam and Alexandra Filip

18 Physical Layer Cooperation in One-Way Relaying Systems 431
Meng Wu, Frank Ludwig, Dirk Wübben, Armin Dekorsy, Karl-Dirk Kammeyer and Steffen Paul

19 Physical Layer Cooperation in Two-Way Relaying Systems 451
Meng Wu, Frank Ludwig, Dirk Wübben, Armin Dekorsy and Steffen Paul

20 Generalized Multi-Carrier Waveforms in Two-Way Relay Systems 475
Matthias Woltering, Stephan Schedler, Dirk Wübben, Armin Dekorsy and Volker Kühn

21 Linear Multi-Cell Precoding for Throughput Optimization Considering Outage 501
Xin Zhang, Richard Fritzsche, Andreas Festag and Gerhard Fettweis

Chapter 1
Coding Techniques for Transmitting Packets Through Complex Communication Networks

Martin Bossert, Vladimir Sidorenko and Antonia Wachter-Zeh

Abstract Random Linear Network Coding (RLNC) is a technique to disseminate information in a network. Various error scenarios require algebraic code constructions with high error-correcting capability in order to transmit packets reliably through such a network. It was shown that subspace codes, in particular lifted rank-metric codes, are suitable for this purpose, in contrast to Hamming metric in the case of a classical transmission. The mainly used codes are Gabidulin codes. In this contribution, we will introduce Gabidulin codes and describe several error-erasure decoding algorithms. Further, an extension of Gabidulin codes is introduced which allows to decode beyond half the minimum rank distance, the interleaved Gabidulin codes. Further, we will introduce (partial) unit memory codes based on Gabidulin codes. Such convolutional codes are of particular interest in so-called multi-shot transmissions since memory between different transmission is introduced. Finally, we will show a significant difference of Gabidulin and Reed-Solomon codes in case of list decoding. Namely, that the list size can grow exponentially for a decoding radius below the Johnson bound for rank metric codes.

1.1 Introduction

Random linear network coding (RLNC) is a technique for transmitting packets through complex communication networks. In Fig. 1.1, such a network is depicted.

M. Bossert (✉) · A. Wachter-Zeh
Ulm University, Albert-Einstein-Allee 43, 89081 Ulm, Germany
e-mail: martin.bossert@uni-ulm.de

V. Sidorenko
Institute for Information Transmission Problems,
Russian Academy of Sciences, Moscow, Russia
e-mail: vladimir.sidorenko@uni-ulm.de

A. Wachter-Zeh
Technion—Israel Institute of Technology, Haifa, Israel
e-mail: antonia@cs.technion.ac.il

W. Utschick (ed.), *Communications in Interference Limited Networks*,
Signals and Communication Technology, DOI 10.1007/978-3-319-22440-4_1

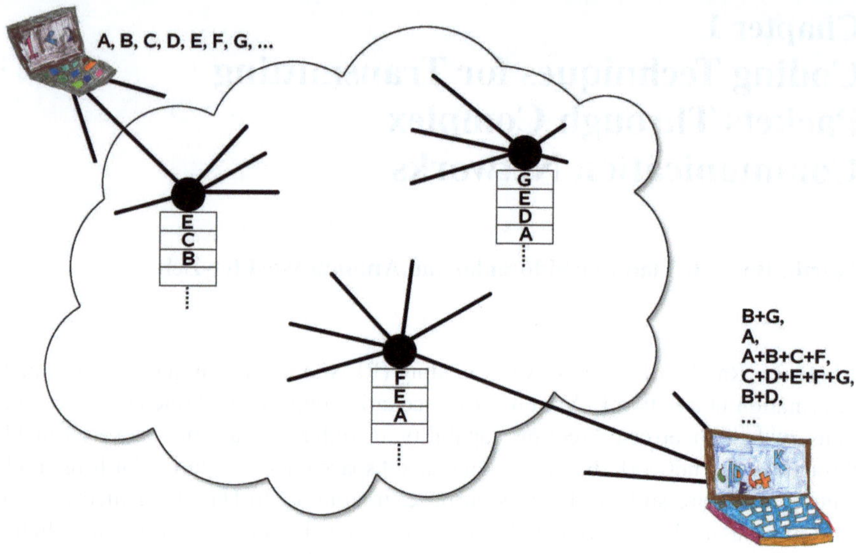

Fig. 1.1 Random linear network [4]

We want to transmit information through a network from a source to a destination. In RLNC, the sent packets are $(n + m)$-dimensional vectors over a finite field[1] $\mathbb{F}_q$ and each node calculates its outgoing packets as random linear $\mathbb{F}_q$-combinations of its incoming packets. The transmission takes place in shots, where in each shot the source inserts n packets $\mathbf{x}_0, \ldots, \mathbf{x}_{n-1}$ (rows of a matrix $\mathbf{X} \in \mathbb{F}_q^{n\times(n+m)}$) into the network and the destination collects $\tilde{n} \geq n$ packets $\mathbf{y}_0, \ldots, \mathbf{y}_{\tilde{n}-1}$ (rows of a matrix $\mathbf{Y} \in \mathbb{F}_q^{\tilde{n}\times(n+m)}$). If the transmission is error-free, the collected packets at the destination are random linear combinations of the transmitted vectors, which implies that with high probability the row spaces of $\mathbf{X}$ and $\mathbf{Y}$ are the same. In [20], the row space of $\mathbf{X}$ is seen as the message. Due to the linear combinations of the network, we get:

$$\mathbf{Y} = \mathbf{A}\mathbf{X},$$

where the matrix $\mathbf{A} \in \mathbb{F}_q^{\tilde{n}\times n}$ describes the transformation of the network. If ℓ nodes perform a random linear combination of incoming packets during one shot, random additive errors within the network can be described as

$$\mathbf{Y} = \mathbf{A}\mathbf{X} + \mathbf{B}\mathbf{Z},$$

where $\mathbf{B} \in \mathbb{F}_q^{\tilde{n}\times\ell}$ represents the linear transformation of the network and the rows of $\mathbf{Z} \in \mathbb{F}_q^{\ell\times(n+m)}$ are the error vectors $\mathbf{z}_0, \ldots, \mathbf{z}_{\tilde{n}-1}$ of the corresponding edge of the

[1] $\mathbb{F}_q$ denotes a finite field with q elements, where q is a prime power.

network with elements from $\mathbb{F}_q$. E.g, $\mathbf{z}_i = 0$ means that on the edge corresponding to the ith row of $\mathbf{B}$ there did not occur any error.

Note that the only property which is not changed by a transmission over an error-free random linear network coding channel is the row space of the transmitted matrix $\mathbf{X}$. Therefore, information is encoded into subspaces instead of specific matrices.

In [36], it is shown that subspace codes and rank-metric codes are closely related. In [36], a construction is given where Gabidulin codes are used for error control with respect to errors and erasures. We transmit a codeword

$$\mathbf{X} = \begin{pmatrix} \mathbf{I}_n & \mathbf{M} \end{pmatrix},$$

where $\mathbf{M}$ is a $(n \times m)$ code matrix of a matrix code $\mathcal{C}_M$ and $\mathbf{I}_n$ is the $n \times n$ identity matrix. In [36], the use of Gabidulin codes in rank-metric is proposed. It follows that

$$\mathbf{Y} = \mathbf{AX} + \mathbf{E}_{out},$$

is received, where $\mathbf{E}_{out}$ is the error matrix. Now, let $\mathbf{S}$ be a matrix which brings $\mathbf{Y}$ to echelon form. Then, we have

$$\mathbf{SY} = \begin{pmatrix} \mathbf{G} & \mathbf{R} \\ 0 & \mathbf{C} \end{pmatrix}.$$

The matrix $\mathbf{R}$ has r rows such that $\mathbf{C}$ consists of $\tilde{n} - r$ rows. The paper [36] extended $\mathbf{R}$ to n rows by inserting specific zero rows to obtain $\hat{\mathbf{R}}$ and showed that $\hat{\mathbf{R}}$ can be decomposed into

$$\begin{aligned} \hat{\mathbf{R}} &= \mathbf{M} + \mathbf{LM} + \mathbf{DC} + \mathbf{E}_{rest} \\ &= \mathbf{M} + \mathbf{E}_{row} + \mathbf{E}_{col} + \mathbf{E}_{rest}, \end{aligned}$$

where $\mathbf{M}$ is the transmitted codeword. The matrix $\mathbf{L}$ is a known $(n \times n)$ matrix such that $\mathbf{LM}$ is a generalized rank row erasure. In [36], this is called an erasure. The $(\tilde{n} - r \times m)$ matrix $\mathbf{C}$ is known as well and $\mathbf{DC}$ is a generalized rank column erasure which is called deviation in [36]. The remaining $\mathbf{E}_{rest}$ is a random rank error. This decomposition will not be taken into consideration since we do not regard erasures. We will only discuss $\mathbf{Y} = \mathbf{AX} + \mathbf{E}_{out}$.

Clearly, the matrix $\mathbf{A}$ can be used to check whether with row operations the identity matrix can be recovered. For simplicity, the case where $\mathbf{A}$ does not have full rank will not be considered in the following. Then, the Gabidulin code must be decoded in order to correct errors. Therefore, in the next section we describe Gabidulin codes and their decoding.

1.2 Gabidulin Codes

While in [5] network coding with error correction is studied for the operator channel, in [16] the benefits of network coding over routing in random networks are discussed. In [35], codes for network coding have been introduced, which are based on rank-metric codes, in particular Gabidulin codes. Since these codes are not commonly known, we will briefly describe them in the following.

The notion rank norm outside of coding theory is known (implicitly) in linear algebra by the well known inequality

$$\mathrm{rk}(A+B) \leq \mathrm{rk}(A) + \mathrm{rk}(B)$$

Explicitly, the rank norm was introduced by Hua [17]. Later, it was introduced to coding theory in [6, 11, 29].

For a given basis of $\mathbb{F}_{q^m}$ over $\mathbb{F}_q$, there exists a vector space isomorphism which maps each vector $\mathbf{x} \in \mathbb{F}_{q^m}^n$ to a matrix $\mathbf{X} \in \mathbb{F}_q^{m \times n}$. Let $\mathrm{rk}(\mathbf{x})$ denote the rank of $\mathbf{X}$ over $\mathbb{F}_q$ and let $\mathcal{R}_q(\mathbf{X})$, $\mathcal{C}_q(\mathbf{X})$ denote the row and column space of $\mathbf{X}$ in $\mathbb{F}_q^n$. We use the notation as vector (e.g. from $\mathbb{F}_{q^m}^n$) or matrix (e.g. from $\mathbb{F}_q^{m \times n}$) equivalently, whatever is more convenient. The *rank distance* of two vectors $\mathbf{x}, \mathbf{y} \in \mathbb{F}_{q^m}$ is defined by

$$\mathrm{d_R}(\mathbf{x}, \mathbf{y}) = \mathrm{rk}(\mathbf{x} - \mathbf{y})$$

and we define the *minimum rank distance* d of a code $\mathcal{C}$ as

$$d = \min\left\{ \mathrm{d_R}(\mathbf{c}_1, \mathbf{c}_2) \ : \ \mathbf{c}_1, \mathbf{c}_2 \in \mathcal{C}, \mathbf{c}_1 \neq \mathbf{c}_2 \right\}.$$

An $\mathcal{C}_{n,M,d}$ code over $\mathbb{F}_{q^m}$ denotes a code (not necessarily linear) of cardinality M and minimum rank distance d. Its codewords are in $\mathbb{F}_{q^m}^n$ or equivalently represented as matrices in $\mathbb{F}_q^{m \times n}$. W.l.o.g. we assume in this paper that $n \leq m$. If this is not the case, we consider the transpose of all matrices such that $n \leq m$ holds. We call n the *length* of such a block code in rank metric over $\mathbb{F}_{q^m}$.

The cardinality M of an $\mathcal{C}_{n,M,d}$ code with $n \leq m$ is limited by a *Singleton*-like upper bound (see [11, 36]):

$$M \leq q^{\min\{n(m-d+1), m(n-d+1)\}} = q^{m(n-d+1)}. \tag{1.1}$$

For linear codes of length $n \leq m$Gabidulin and dimension k, this implies $d \leq n-k+1$. If the cardinality of a code fulfills (1.1) with equality, the code is called a *Maximum Rank Distance* (MRD) code. A linear MRD code over $\mathbb{F}_{q^m}$ of length $n \leq m$, dimension k and minimum rank distance $d = n-k+1$ is denoted by $\mathsf{MRD}[n,k]$ and has cardinality $M = q^{mk}$. In the following we introduce the class of Gabidulin codes which are are MRD codes.

The codes introduced in [11] have length n, dimension k, minimum rank distance $d = n - k + 1$, and are nowadays called Gabidulin codes. Gabidulin codes are constructed over an extension field $\mathbb{F}_{q^m}$ of $\mathbb{F}_q$. A generator matrix is given by:

$$\mathbf{G} = \begin{bmatrix} g_1 & g_2 & \cdots & g_n \\ g_1^{[1]} & g_2^{[1]} & \cdots & g_n^{[1]} \\ g_1^{[2]} & g_2^{[2]} & \cdots & g_n^{[2]} \\ \vdots & \vdots & \ddots & \vdots \\ g_1^{[k-1]} & g_2^{[k-1]} & \cdots & g_n^{[k-1]} \end{bmatrix}, \tag{1.2}$$

where $g_j \in \mathbb{F}_{q^m}$, $j = 1, \ldots, n$, are linearly independent over $\mathbb{F}_q$. The notation $g^{[i]} = g^{q^i}$ denotes the i-th Frobenius power of g.

A parity-check matrix has the form

$$\mathbf{H} = \begin{bmatrix} h_1 & h_2 & \cdots & h_n \\ h_1^{[1]} & h_2^{[1]} & \cdots & h_n^{[1]} \\ \vdots & \vdots & \cdots & \vdots \\ h_1^{[d-2]} & h_2^{[d-2]} & \cdots & h_n^{[d-2]} \end{bmatrix}, \tag{1.3}$$

where $h_j \in \mathbb{F}_{q^m}$, $j = 1, \ldots, n$, are linearly independent over $\mathbb{F}_q$, and

$$\mathbf{G}\mathbf{H}^\top = \mathbf{0}.$$

Gabidulin codes are the rank-metric analogs of Reed-Solomon codes and found many applications including network coding. Interleaving or the direct sum of Gabidulin codes allows both decreasing the redundancy and increasing the error correcting capability for network coding. For Gabidulin codes, we proposed a transform-domain algorithm correcting both errors and erasures [32]. We showed how to generalize this algorithm for interleaved Gabidulin codes. The transform-domain approach allows to simplify derivations and proofs and also simplifies finding the error vector after solving the key equation. Details on these results can be found in [33].

The classical definition of Gabidulin codes requires the concept of linearized polynomials. A linearized polynomial (or q-polynomial) over $\mathbb{F}_{q^m}$ is a polynomial of the form

$$f(x) = \sum_{i=0}^{t} f_i x^{[i]}$$

where $f_i \in \mathbb{F}_{q^m}$. The maximum i for which $f_i \neq 0$ is called the q-degree of $f(x)$ and is denoted by $deg_q f(x)$. The addition of linearized polynomials $f(x)$ and $g(x)$ is defined componentwise. Multiplication uses the symbolic product: $f(x) \circ g(x) = f(g(x))$. The result is a linearized polynomial of q-degree $deg_q f(x) + deg_q g(x)$.

The symbolic product is distributive and associative, but noncommutative. The set of linearized polynomials over $\mathbb{F}_{q^m}$ with the operation of polynomial addition and symbolic multiplication forms a noncommutative ring.

1.2.1 Decoding Using the Linearized Extended Euclidean Algorithm (LEEA)

There exist several efficient algorithms for decoding Gabidulin codes. A syndrome-based bounded minimum distance (BMD) decoding algorithm based on solving a key equation was introduced by Gabidulin [11]. The first step of the decoding process is to compute the syndrome. This can be done by multiplying the received word $\mathbf{r}$ with the transposed of the parity check matrix $\mathbf{H}$. We use the syndrome values $\mathbf{s} = (s_0, \ldots, s_{n-k+1}) := \mathbf{r} \cdot \mathbf{H}^T$ to calculate the syndrome polynomial $s(x) = \sum_{i=0}^{n-k-1} s_i x^i$. The key equation which has to be solved is given by

$$\Omega(x) \equiv \Lambda(s(x)) = \Lambda(x) \circ s(x) \mod x^{[n-k]},\ deg\ \Omega(x) < deg\ \Lambda(x).$$

This key equation can be solved for the error span polynomial $\Lambda(x)$. The direct way to accomplish this task is to solve a linear system of equations based on the key equation:

$$\Omega_i = \sum_{j=0}^{i} \Lambda_j s_{i-j}^{[j]} = \sum_{j=0}^{t} \Lambda_j s_{i-j}^{[j]} = 0, \forall i \in [t, n-k-1-1].$$

This is equivalent to the following homogeneous linear system of equations:

$$\begin{pmatrix} \Omega_t \\ \Omega_{t+1} \\ \vdots \\ \Omega_{n-k-1} \end{pmatrix} = \begin{pmatrix} s_t^{[0]} & s_{t-1}^{[1]} & \ldots & s_0^{[t]} \\ s_{t+1}^{[0]} & s_t^{[1]} & \ldots & s_1^{[t]} \\ \vdots & \vdots & \ddots & \vdots \\ s_{n-k-1}^{[0]} & s_{n-k-2}^{[1]} & \ldots & s_{n-k-1-t}^{[t]} \end{pmatrix} \cdot \begin{pmatrix} \Lambda_0 \\ \Lambda_1 \\ \vdots \\ \Lambda_t \end{pmatrix} = \mathbf{0}. \quad (1.4)$$

If we use Gaussian elimination, the complexity of this step is bounded by $\mathcal{O}(n^3)$ operations in $\mathbb{F}_{q^m}$. However, it is possible to exploit the structure of the syndrome matrix in order to solve the key equation with complexity $\mathcal{O}(n^2)$ in $\mathbb{F}_{q^m}$ by applying the Linearized Extended Euclidean Algorithm (LEEA) which is given in Algorithm 1. The algorithm uses a subroutine RIGHTDIV$(a(x); b(x))$ which finds two polynomials $q(x)$ and $r(x)$ such that

$$a(x) = q(x) \circ b(x) + r(x),$$

and $deg_q r(x) < deg_q b(x)$. The existence of such an algorithm is ensured by the fact that the ring of linearized polynomials with addition $+$ and multiplication $\circ$ is a Euclidean domain.

Algorithm 1:
$r_{out}(x); u_{out}(x); v_{out}(x) \leftarrow \text{RIGHTLEEA}\big(a(x); b(x); d_{stop}\big)$

Input: $a(x); b(x) \in \mathbb{L}_{q^m}[x]$ with $\deg_q a(x) \geq \deg_q b(x)$;
stopping degree d_{stop}

Initialize: $i \leftarrow 1$,
$r^{(-1)}(x) \leftarrow a(x), r^{(0)}(x) \leftarrow b(x)$,
$u^{(-1)}(x) \leftarrow 0, u^{(0)}(x) \leftarrow x^{[0]}$,
$v^{(-1)}(x) \leftarrow x^{[0]}, v^{(0)}(x) \leftarrow 0$

while $\deg_q r^{(i-1)}(x) \geq d_{stop}$ **do**
  $q^{(i)}(x); r^{(i)}(x) \leftarrow \text{RIGHTDIV}\big(r^{(i-1)}(x); r^{(i-2)}(x)\big)$ $u^{(i)}(x) \leftarrow u^{(i-2)}(x) - q^{(i)}(u^{(i-1)}(x))$
  $v^{(i)}(x) \leftarrow v^{(i-2)}(x) - q^{(i)}(v^{(i-1)}(x))$ $i \leftarrow i + 1$

Output: $r_{out}(x) \leftarrow r^{(i-1)}(x); u_{out}(x) \leftarrow u^{(i-1)}(x); v_{out}(x) \leftarrow v^{(i-1)}(x)$

Therefore, we use the syndrome polynomial $s(x)$ and $x^{[n-k]}$ as input to the LEEA which in turn gives us $\Lambda(x)$. Next, we need to find a basis of the root space of $\Lambda(x)$. This is relatively easy due to the structure of linearized polynomials. This basis can be used for determining the error. Therefore it is necessary to solve a system of equations. The overall complexity is in the order $\mathcal{O}(n^2)$ over $\mathbb{F}_{q^m}$. We can further reduce the complexity by applying a Gao-like decoding approach [47]. This algorithm uses an alternative transformed key equation which can be solved by applying the LEEA and outputs directly the linearized evaluation polynomial of the estimated codeword.

1.2.2 *Linearized Shift-Register Synthesis*

Another algorithm that can be applied for solving the key equation efficiently is a generalization of the Berlekamp-Massey algorithm [31]. A problem which is equivalent to the problem of solving the key equation is the problem of linearized shift-register synthesis: Let $s = s_1, s_2, \ldots, s_N$ be a sequence over a field $\mathbb{F}_{q^m}$. Find the smallest integer $l \geq 0$ for which there is a vector $\sigma = (\sigma_1, \ldots, \sigma_l)$ over $\mathbb{F}_{q^m}$ such that

$$s_n = -\sum_{i=1}^{l} \sigma_i s_{n-i}^{q^i} \quad for\ n = l + 1, \ldots, N$$

and find a suitable vector σ which is called the connection vector. Given a connection vector $\sigma = (\sigma_1, \ldots, \sigma_l)$ over $\mathbb{F}_{q^m}$, the corresponding connection q-polynomial is defined by

$$\sigma(x) = \sum_{i=0}^{l} \sigma_i x^{[i]}, \tag{1.5}$$

where $\sigma_0 \triangleq 1$ and $\deg_q \sigma(x) \leq l$. A linearized-feedback shift-register is completely determined by length l and connection vector σ, hence it can be denoted by (l, σ). Given a sequence s and a shift-register (l, σ), let the *discrepancy* $d_n(l, \sigma)$ of a sequence element s_n be defined by

$$d_n(l, \sigma) = \begin{cases} 0 & \text{for } n = 1, \ldots, l, \\ \sum\limits_{i=0}^{l} \sigma_i s_{n-i}^{[i]} & \text{for } n = l+1, \ldots, N. \end{cases} \tag{1.6}$$

A shift-register (l, σ) generates the element s_n, if $d_n(l, \sigma) = 0$. The shift-register generates a sequence if it generates every element of the sequence.

Algorithm 2:
$l; \sigma(x); \sigma'(x); n' \leftarrow$ Richter–Plass(s)

Input: $s = s_1, \ldots, s_N$
Initialize: $l \leftarrow 0, \sigma(x) \leftarrow x$
$\sigma'(x) \leftarrow x, n' \leftarrow 0, d' \leftarrow 1$

for *each n from* 1 *to N* **do**
  $d \leftarrow \sum_{j=0}^{l} \sigma_j s_{n-j}^{[j]}$ **if** $d \neq 0$ **then**
    if $2l \geq n$ **then**
      $\sigma(x) \leftarrow \sigma(x) - d\left(\frac{x}{d'}\right)^{[n-n']} \otimes \sigma'(x)$
    else
      $\tilde{\sigma}(x) \leftarrow \sigma(x)$ $\sigma(x) \leftarrow \sigma(x) - d\left(\frac{x}{d'}\right)^{[n-n']} \otimes \sigma'(x)$ $l \leftarrow n - l$ $\sigma'(x) \leftarrow \tilde{\sigma}(x)$,
      $n' \leftarrow n, d' \leftarrow d$

Output: $l, \sigma(x)$, and $\sigma'(x), n'$

The problem of linearized shift-register synthesis can be tackled by solving a linear system of equations. Applying straightforward approaches like Gaussian elimination have a computational complexity of order $\mathcal{O}(N^3)$. Solving the problem with Algorithm 2 reduces the complexity to order $\mathcal{O}(N^2)$. The algorithm was independently introduced by Paramonov and Tretjakov [26] and by Richter and Plass [28].

Algorithm 2 is an iterative approach. First, the algorithm initializes the setting of the shift-register with $(l, \sigma) = (0, x)$. The for-loop sequentially processes the elements of s. Within the loop, the discrepancy $d_n(l, \sigma)$ is computed using the current shift-register (l, σ). The discrepancy is zero if and only if (l, σ) is able to generate the element s_n. In this case there is no need to change the shift-register. Hence, the algorithm directly continues with the processing of the next element. If the discrepancy is not equal to zero, the shift-register (l, σ) is modified in order to obtain the shortest shift-register $(\hat{l}, \hat{\sigma})$ which is able to generate s_n and all previous elements. The algorithm outputs both, length l and a connection polynomial σ.

The time complexity of this algorithm may be estimated by $\mathcal{O}(lN) \leq \mathcal{O}(N^2)$ operations in $\mathbb{F}_{q^m}$ since the algorithm has to process N elements of the sequence s and the processing of each element requires at most $\mathcal{O}(l) \leq \mathcal{O}(N)$ operations in $\mathbb{F}_{q^m}$.

In addition, the uniqueness of the solution can be checked. For decoding Gabidulin codes up to half the code distance, a decoding failure can be declared when the solution is not unique. Furthermore, it is possible to calculate all solutions which is interesting for decoding Gabidulin codes beyond half the code distance.

1.2.3 Remarks

Fast decoding up to half the minimum rank distance of Gabidulin codes by applying a Gao-like algorithm is explained in [47]. Sidorenko et al. [31] introduces the generalization of the Berlekamp-Massey algorithm and proves its correctness. Furthermore it is explained how to check uniqueness of the solution and how to find all possible solutions.

1.3 Interleaved Gabidulin Codes

Interleaving or the direct sum of ℓ Gabidulin codes with rank distance d can be applied for random network coding. For codes over $\mathbb{F}_{q^m}$, a fast decoding algorithm exists which corrects errors of rank up to $\frac{\ell}{\ell+1}(d-1)$ with high probability. Thus, with interleaving the error correcting radius can be enlarged. The algorithm is based on fast skew-feedback shift-register synthesis and for fixed ℓ it has time complexity $\mathcal{O}(m^3 \log m)$ operations in the base field $\mathbb{F}_q$.

In network coding, rows of a matrix $\mathbf{C} \in \mathbb{F}_q^{m \times n}$ (with a prefix of length m) are sent via the network as packets. In many practical cases the packet length is much more than the number m of packets in the message, so we would like to have $n \gg m$ in this case. In order to increase the parameter n by a factor of $\ell \in \mathbb{N}$, the direct sum of ℓ codewords of the code $\mathcal{C}$ can be used. Given a matrix code $\mathcal{C} \subseteq \mathbb{F}_q^{m \times n}$ and a number ℓ, the ℓ-interleaved code $\mathcal{IC} \subseteq \mathbb{F}_q^{m \times \ell n}$ consists of all $m \times \ell n$ matrices C over $\mathbb{F}_q$ having the following block form

$$\mathbf{C} = \left(\mathbf{C}^{(1)}\ \mathbf{C}^{(2)} \ldots \mathbf{C}^{(\ell)}\right), \mathbf{C}^{(\ell)} \in \mathcal{C}, \tag{1.7}$$

where $l = 1, \ldots, \ell$. The rate and the distance of the interleaved code are the same as for the original code $\mathcal{C}$.

We assume $m = n$ for the interleaved Gabidulin code. Note, that Loidreau and Overbeck in [24, 25] used interleaving of transposed Gabidulin codes.

We describe a decoder for any ℓ-interleaved Gabidulin code $\mathcal{IC} \subseteq \mathbb{F}_q^{m \times \ell n}$ having distance d. This decoder guarantees to correct all error words of rank less than $d/2$ and it corrects with high probability all error words of rank t if

$$t \leq \frac{\ell}{\ell+1}(d-1).$$

The probability $\mathrm{P_f}(t)$ of a decoding failure can be upper bounded by $\mathrm{P_f}(t) < 4/q^m$ for the cases that words of fixed rank are equiprobable. This probability is very small in practice and coincides with the bound from [24, 25] and $\mathrm{P_f}(t)$ practically vanishes if $t < \lfloor \frac{\ell}{\ell+1}(d-1) \rfloor$.

1.3.1 Decoding Interleaved Gabidulin Codes

Consider interleaving of ℓ identical Gabidulin codes $\mathcal{G}(q^m; n, k)$ and index ℓ runs from 1 to ℓ, i.e., $\ell = 1, \ldots, \ell$.

Definition 1.3.1 [32] The ℓ-interleaved Gabidulin code is

$$\mathcal{IG}(q^m, \ell; \ell n, k) = \left\{ \left(\mathbf{c}^{(1)} \ \ldots \ \mathbf{c}^{(\ell)} \right) : \mathbf{c}^{(l)} \in \mathcal{G}(q^m; n, k) \right\}. \tag{1.8}$$

Here $\mathbf{c} = \left(\mathbf{c}^{(1)} \ \ldots \ \mathbf{c}^{(\ell)} \right) \in \mathbb{F}_{q^m}^{\ell n}$ denotes the concatenation of ℓ vectors $c^{(\ell)}$. The codeword $\mathbf{c} \in \mathcal{IG}(q^m, \ell; \ell n, k)$ is transmitted and a vector $\mathbf{r} = \left(\mathbf{r}^{(1)} \ \ldots \ \mathbf{r}^{(\ell)} \right) \in \mathbb{F}_{q^m}^{\ell n}$ is received. The error on the channel is $\mathbf{e} = \left(\mathbf{e}^{(1)} \ \ldots \ \mathbf{e}^{(\ell)} \right) \in \mathbb{F}_{q^m}^{\ell n}$, where $\mathbf{r} = \mathbf{c} + \mathbf{e}$. Let $\mathrm{rk}(\mathbf{e}) = t$ then we can represent the error vector as

$$\mathbf{e} = \mathbf{a}\mathbf{B}, \quad \text{where } \mathbf{a} \in \mathbb{F}_{q^m}^{t}, \ \mathbf{B} \in \mathbb{F}_q^{t \times \ell n},$$

and $\mathrm{rk}(\mathbf{a}) = \mathrm{rk}(\mathbf{B}) = t$. We can write $\mathbf{B}$ as

$$\mathbf{B} = \left(\mathbf{B}^{(1)}, \ldots, \mathbf{B}^{(\ell)} \right), \ \mathbf{B}^{(l)} \in \mathbb{F}_q^{t \times n},$$

where $\mathrm{rk}(\mathbf{B}^{(l)}) \leq t$. Then we have

$$\mathbf{e}^{(l)} = \mathbf{a}\mathbf{B}^{(l)} \tag{1.9}$$

and the syndrome of the l-th code is

$$\mathbf{s}^{(l)} = \mathbf{a}\mathbf{B}^{(l)}\mathbf{H}^T. \tag{1.10}$$

We use the notation $\left(x^{(l)}\right)^{[i]} = x^{(l)[i]}$. let

$$\mathbf{F}^{(l)} = \mathbf{B}^{(l)}\mathbf{H}^T = \begin{pmatrix} f_1^{(l)} & \dots & f_1^{(l)[d-2]} \\ f_2^{(l)} & \dots & f_2^{(l)[d-2]} \\ \vdots & \vdots & \vdots \\ f_t^{(l)} & \dots & f_t^{(l)[d-2]} \end{pmatrix}, \tag{1.11}$$

where

$$f_i^{(l)} = \sum_{j=1}^{n} B_{i,j}^{(l)}\mathbf{h}_j, \quad i = 1, \dots, t. \tag{1.12}$$

Thus we have the following system of equations with unknowns a_i, $f_i^{(l)}$, $i = 1, \dots, t$,

$$\sum_{i=1}^{t} a_i f_i^{(l)[j-1]} = s_j^{(l)}, \quad j = 1, \dots, d-1. \tag{1.13}$$

We introduce again a q-linearized polynomial

$$\sigma(x) = \sum_{i=0}^{t} \sigma_i x^{[i]} \tag{1.14}$$

having all linear combinations of $a_1, \dots, a_t$ as roots and $\sigma_0 = 1$. The key equation is:

$$s_i^{(l)} = -\sum_{j=1}^{t} \sigma_j \left(s_{i-j}^{(l)}\right)^{[j]}, \tag{1.15}$$

for $l = 1, \dots, \ell$ and $i = t+1, \dots, d-1$. To solve this key equation we should solve the problem that given vectors $s^{(1)}, \dots, s^{(\ell)} \in \mathbb{F}_{q^m}^{d-1}$ of length $d-1$, find the minimum integer $t \geq 0$ such that the system of equations has a solution $\sigma(x)$. A solution can be found by solving the key equation for $t = 0, 1, \dots, d-1$. The complexity of this approach is $\mathcal{O}(\ell d^3)$. In the following section we give an efficient solution having complexity $\mathcal{O}(\ell d^2)$. After the key equation is solved we calculate vector a. After a vector a is found, the error words $\mathbf{e}^{(l)}$ can be found separately for every interleaved code.

1.3.2 Remarks

Details on these results can be found in [32]. Further details and bounds can be found in [30]. Further, in [46], an interpolation-based unique decoder as well as a list decoder for interleaved Gabidulin codes were presented.

1.4 Partial Unit Memory Codes

In non-coherent multi-shot RLNC, the unknown and time-variant network is used several times. In order to create dependencies between the different shots, convolutional network codes are used. Any convolutional code can be described as a Partial Unit Memory (PUM) code. PUM codes can be constructed based on block codes. Thus, for PUM codes over rank-metric Gabidulin codes can be used. One construction is based on the parity-check matrix of the code, and another on the generator matrix. The constructions require a modified rank metric, the so-called sum rank metric. For the sum rank metric, the free rank distance, the extended row rank distance and its slope were defined [40]. The construction of PUM codes based on the parity-check matrix of Gabidulin codes achieves the upper bound on the free rank distance.

PUM codes over rank metric can efficiently be decoded when errors, erasures and deviations occur. The decoding complexity is cubic with the length. Moreover, we described how lifting of these codes can be applied for error correction in RLNC [45].

In [22], *Unit Memory* (UM) codes were defined as a special class of rate k/n convolutional codes with memory $m = 1$ and overall constraint length $\nu = k$. In [21], this idea was extended to *Partial Unit Memory* (PUM) codes where the memory of the convolutional code is also $m = 1$, but the overall constraint length is $\nu < k$.

Known results of (P)UM codes based on block codes are e.g. *Reed–Solomon* (RS) [19, 27, 50] or BCH codes [7, 8]. The use of block codes makes an algebraic description of these convolutional codes possible. A convolutional code can be characterized by its *free distance* and the *average linear increase (slope)* of the *extended row distance*. These distance measures determine the error–correcting capability of the convolutional code. In [21, 22, 27, 39] upper bounds for the free (Hamming) distance and the slope of (P)UM codes were derived. There are constructions that achieve the upper bound for the free (Hamming) distance, e.g. [19, 27, 50]. A construction of (P)UM codes based on RS codes with optimal free distance and half the optimal slope was given in [50].

1.4.1 Convolutional Codes

Definition 1.4.1 (*Convolutional Code* [40]) A rate k/n convolutional code $\mathcal{C}$ over $\mathbb{F}$ with memory m is defined by its $k \times n$ generator matrix in polynomial form:

$$\mathbf{G}(D) = \begin{pmatrix} g_{11}(D) & g_{12}(D) & \dots & g_{1n}(D) \\ \vdots & \vdots & \ddots & \vdots \\ g_{k1}(D) & g_{k2}(D) & \dots & g_{kn}(D) \end{pmatrix},$$

where $g_{ij}(D) = g_{ij}^{(0)} + g_{ij}^{(1)} D + g_{ij}^{(2)} D^2 + \dots + g_{ij}^{(m)} D^m$ with $g_{ij}^{(\ell)} \in \mathbb{F}$, $\ell = 0, \dots, m$, for $i = 1, \dots, k$ and $j = 1, \dots, n$.

We denote the *constraint lengths* ν_i, the *memory* m and the *overall constraint length* ν of the convolutional code $\mathcal{C}$ according to [3] as follows.

Definition 1.4.2 (*Memory and Constraint Length* [40]) The constraint length for the i–th input of a polynomial generator matrix $\mathbf{G}(D)$ is

$$\nu_i \stackrel{\text{def}}{=} \max_{1 \le j \le n} \{\deg g_{ij}(D)\}.$$

The memory is the maximum constraint length:

$$m \stackrel{\text{def}}{=} \max_{1 \le i \le k} \{\nu_i\}.$$

The overall constraint length ν is defined as the sum of the constraint lengths ν_i:

$$\nu \stackrel{\text{def}}{=} \sum_{i=1}^{k} \nu_i.$$

A codeword $\mathbf{c}(D) = \mathbf{c}_0 + \mathbf{c}_1 D + \mathbf{c}_2 D^2 + \dots$ of the convolutional code $\mathcal{C}$ is generated by $\mathbf{c}(D) = \mathbf{u}(D) \cdot \mathbf{G}(D)$, where $\mathbf{u}(D) = u_0 + u_1 D + u_2 D^2 + \dots$ is the information.

A polynomial $(n-k) \times n$ parity–check matrix $\mathbf{H}(D)$ of $\mathcal{C}$ is defined such that for every $\mathbf{c}(D) \in \mathcal{C}$, $\mathbf{c}(D) \cdot \mathbf{H}^T(D) = 0$ holds, where $\mathbf{H}(D) = [h_{ij}(D)]_{1 \le i \le (n-k), 1 \le j \le n}$ with $h_{ij}(D) = h_{ij}^{(0)} + h_{ij}^{(1)} D + h_{ij}^{(2)} D^2 + \dots + h_{ij}^{(m_H)} D^{m_H}$, where $h_{ij}^{(\ell)} \in \mathbb{F}$ for $\ell = 0, \dots, m_H$.

As usual, we can represent $\mathbf{G}(D)$ and $\mathbf{H}(D)$ as semi–infinite matrices $\mathbb{G}$ and $\mathbf{H}$ over $\mathbb{F}$ with the submatrices $\mathbf{G}_0, \mathbf{G}_1, \dots \mathbf{G}_m$ and $\mathbf{H}_0, \mathbf{H}_1, \dots, \mathbf{H}_{m_H}$, where $\mathbf{G}_i$, $i = 0, \dots, m$ are $k \times n$–matrices and $\mathbf{H}_i$, $i = 0, \dots, m_H$ are $(n-k) \times n$ matrices. In general, the number of submatrices in $\mathbf{H}$ and $\mathbf{G}$ is not equal, i.e., $m_H \neq m$. We call m_H the *dual memory* of $\mathcal{C}$. If both $\mathbf{G}$ and $\mathbf{H}$ are in minimal basic encoding form, the overall number of memory elements, i.e., the overall constraint length ν is the same in both representations [10].

$\mathbf{c}(D)$ and $\mathbf{u}(D)$ can be represented as causal infinite sequences: $\mathbf{c} = (\mathbf{c}_0\ \mathbf{c}_1\ \mathbf{c}_2 \dots)$ and $\mathbf{u} = (\mathbf{u}_0\ \mathbf{u}_1\ \mathbf{u}_2 \dots)$, where $\mathbf{c}_j \in \mathbb{F}^n$ and $\mathbf{u}_j \in \mathbb{F}^k$ for all j. Then, $\mathbf{c} = \mathbf{u} \cdot \mathbf{G}$ and $\mathbf{c} \cdot \mathbf{H}^T = \mathbf{0}$.

1.4.2 (Partial) Unit Memory Codes

(P)UM codes are a special class of convolutional codes with memory $m = 1$ [21, 22], i.e., the semi–infinite generator matrix $\mathbf{G}$ is given by:

$$\mathbf{G} = \begin{pmatrix} \mathbf{G}_0 & \mathbf{G}_1 & & \\ & \mathbf{G}_0 & \mathbf{G}_1 & \\ & & \ddots & \ddots \end{pmatrix}, \tag{1.16}$$

where $\mathbf{G}_0$ and $\mathbf{G}_1$ are $k \times n$ matrices. For an (n, k) UM code, both matrices have full rank. For an $(n, k \mid k_1)$ PUM code, $\text{rk}(\mathbf{G}_0) = k$ and $\text{rk}(\mathbf{G}_1) = k_1 < k$:

$$\mathbf{G}_0 = \begin{pmatrix} \mathbf{G}_{00} \\ \mathbf{G}_{01} \end{pmatrix}, \quad \mathbf{G}_1 = \begin{pmatrix} \mathbf{G}_{10} \\ \mathbf{0} \end{pmatrix},$$

where $\mathbf{G}_{00}$ and $\mathbf{G}_{10}$ are $k_1 \times n$ matrices and $\mathbf{G}_{01}$ is a $(k - k_1) \times n$–matrix. For both cases, we have the following encoding rule:

$$\mathbf{c}_j = \mathbf{u}_j \cdot \mathbf{G}_0 + \mathbf{u}_{j-1} \cdot \mathbf{G}_1.$$

Note that the overall constraint length for UM codes is $\nu = k$ and for PUM codes $\nu = k_1$, since this is the number of symbol which influence the next block.

There are restrictions on the code rate of (P)UM codes when a certain number of full–rank submatrices $\mathbf{H}_i$ should exist. This full–rank condition, $\text{rk}(\mathbf{H}_i) = n - k$ for all $i = 0, \dots, m_H$ is used in the following construction.

Lemma 1.4.1 (Rate Restriction for UM Codes [40]) *An (n, k) UM code with overall constraint length $\nu = k$ has rate*

$$R = \frac{(n-k) \cdot m_H}{(n-k) \cdot (m_H + 1)}$$

if the parity–check matrix $\mathbf{H}$ in minimal basic encoding form consists of $m_H + 1$ full–rank submatrices $\mathbf{H}_i$ for $m_H \geq 1$.

Lemma 1.4.2 (Rate Restriction for PUM Codes [40]) *Let an $(n, k \mid k_1)$ PUM code with $\nu = k_1 < k$ be given. Its rate is*

$$R = \frac{k}{n} > \frac{m_H}{m_H + 1},$$

if the parity–check matrix $\mathbf{H}$ *in minimal basic encoding form consists of* $m_H + 1$ *submatrices* $\mathbf{H}_i$ *for* $m_H \geq 1$.

If we use the *parity–check* matrix to construct (P)UM codes, the following theorem guarantees that there is always a corresponding generator matrix that defines a (P)UM code.

Theorem 1.4.1 *For each semi–infinite parity–check matrix* $\mathbf{H}$, *where the* $(m_H + 1)$ *submatrices* $\mathbf{H}_i$ *are* $(n-k)\times n$ *matrices of* $\mathrm{rk}(\mathbf{H}_i) = n-k, \forall i$, *and* $R \geq m_H/(m_H+1)$, *there exists a generator matrix* $\mathbf{G}$ *such that* $\mathbf{G} \cdot \mathbf{H}^T = \mathbf{0}$ *and* $\mathbf{G}$ *defines a (P)UM code with* $k_1 \geq (k-1)/2$ [40].

1.4.3 Distance Measures

Definition 1.4.3 (*Sum Rank Weight* [40]) Let a vector $\mathbf{v} \in \mathbb{F}^{\ell n}$ be given and let it be decomposed into subvectors:

$$\mathbf{v} = \left(\mathbf{v}_0 \ \mathbf{v}_1 \ \ldots \ \mathbf{v}_{\ell-1}\right),$$

with $\mathbf{v}_i \in \mathbb{F}^n$ for all i. We define the *sum rank weight* $\mathrm{wt}_{rk}(\mathbf{v})$ as the *sum* of the rank norms of the subvectors:

$$\mathrm{wt}_{\mathrm{rk}}(\mathbf{v}) \stackrel{\mathrm{def}}{=} \sum_{i=0}^{\ell-1} \mathrm{rk}(\mathbf{v}_i), \tag{1.17}$$

for $0 \leq \ell \leq \infty$.

Hence, we define the *sum rank distance* between two sequences $\mathbf{v}^{(1)}, \mathbf{v}^{(2)}$ of length ℓn by

$$\mathrm{d}(\mathbf{v}^{(1)}, \mathbf{v}^{(2)}) \stackrel{\mathrm{def}}{=} \mathrm{wt}_{\mathrm{rk}}(\mathbf{v}^{(1)} - \mathbf{v}^{(2)}) = \sum_{i=0}^{\ell-1} \mathrm{d}_{\mathrm{R}}(\mathbf{v}_i^{(1)}, \mathbf{v}_i^{(2)}). \tag{1.18}$$

The *free rank distance* d_{free} is defined as follows.

Definition 1.4.4 (*Free Rank Distance* [40]) The minimum sum rank distance between any two codewords $\mathbf{c}^{(1)}, \mathbf{c}^{(2)}$ from a convolutional code $\mathcal{C}$ is the *free rank distance* d_{free}:

$$d_{\mathrm{free}} \stackrel{\mathrm{def}}{=} \min_{\mathbf{c}^{(1)} \neq \mathbf{c}^{(2)}} \left\{\mathrm{d}(\mathbf{c}^{(1)}, \mathbf{c}^{(2)})\right\} = \min_{\mathbf{c}^{(1)} \neq \mathbf{c}^{(2)}} \left\{\sum_{i=0}^{\infty} \mathrm{d}_{\mathrm{R}}(\mathbf{c}_i^{(1)}, \mathbf{c}_i^{(2)})\right\}. \tag{1.19}$$

Generally, the error–correcting capability of convolutional codes is determined by *active distances*. In the following, we define the *extended row rank distance* which is an important active distance in the sum rank metric.

Let $\mathcal{C}^r(\ell)$ denote the set of all codewords $\mathbf{c}^{(\ell)}$ corresponding to paths in the minimal code trellis which diverge from the zero state at depth j and return to the zero state for the *first* time after $\ell+1$ branches at depth $j+\ell+1$. W.l.o.g., we assume $j=0$ as we only consider time–invariant convolutional codes.

The extended row rank distance of order ℓ is defined as the minimum sum rank weight of all codewords in $\mathcal{C}^r(\ell)$.

Definition 1.4.5 (*Extended Row Rank Distance* [40]) The extended row rank distance of order $\ell = 1, 2, \ldots$ is defined as

$$d_\ell^r \stackrel{\text{def}}{=} \min_{\mathbf{c} \in \mathcal{C}^r(\ell)} \left\{ \mathrm{wt}_{\mathrm{rk}}(\mathbf{c}) \right\}. \tag{1.20}$$

The minimum of the ℓth order extended row rank distances gives the free rank distance:

$$d_{\text{free}} = \min_\ell \left\{ d_\ell^r \right\}.$$

As for Hamming metric, the extended row rank distance d_ℓ^r can be lower bounded by a linear function $d_\ell^r \geq \max\{\alpha\ell + \beta, d_{\text{free}}\}$ where $\beta \leq d_{\text{free}}$ and α denotes the *slope* (the average linear increase). The slope is an important parameter for determining the error–correcting capability and is defined as follows (analog to [8]).

Definition 1.4.6 (*Slope* [40]) The average linear increase of the extended row rank distance (slope) is defined as

$$\alpha \stackrel{\text{def}}{=} \lim_{\ell \to \infty} \left\{ \frac{d_\ell^r}{\ell} \right\}. \tag{1.21}$$

1.4.4 Upper Bounds

Assume, the codewords of a convolutional code are considered in *Hamming* metric. Let us then denote the ℓth order active row *Hamming* distance by $d_\ell^{H,r}$ and the free *Hamming* distance of this convolutional code by d_{free}^H.

Theorem 1.4.2 (Connection between Distances [40]) *For the free rank/Hamming distance and the extended row rank/Hamming distance, the following holds:*

$$d_{\text{free}} \leq d_{\text{free}}^H, \quad \textit{and} \quad d_\ell^r \leq d_\ell^{H,r}, \quad \ell = 1, 2, \ldots.$$

For the rank norm $\mathrm{rk}(\mathbf{v}_i)$ and the Hamming norm $\mathrm{wt}_H(\mathbf{v}_i)$ of a vector $\mathbf{v}_i$ it holds that: $\mathrm{rk}(\mathbf{v}_i) \leq \mathrm{wt}_H(\mathbf{v}_i)$ and hence also

$$\sum_{i=0}^{\ell-1} \mathrm{rk}(\mathbf{v}_i) \leq \mathrm{wt}_H(\mathbf{v}_0 \ldots \mathbf{v}_{\ell-1}).$$

Consequently, the upper bounds for the free distance and the slope of (P)UM codes based on Hamming metric [27, 39] also hold for (P)UM codes based on the sum rank metric.

Corollary 1.4.1 (Upper Bounds [40]) *For an (n, k) UM code, where $\nu = k$, the free rank distance is upper bounded by:*

$$d_{\text{free}} \leq 2n - k + 1. \tag{1.22}$$

For an $(n, k \mid k_1)$ PUM code, where $\nu = k_1 < k$, the free rank distance is upper bounded by:

$$d_{\text{free}} \leq n - k + \nu + 1. \tag{1.23}$$

For both UM and PUM codes, the average linear increase (slope) is upper bounded by:

$$\alpha \leq n - k. \tag{1.24}$$

1.4.5 Construction

We show now an explicit construction using normal bases.

Definition 1.4.7 (*(P)UM Code Based on Gabidulin Codes* [40]) For some $m_H \geq 1$, let a rate $R = k/n = c \cdot m_H/(c \cdot (m_H + 1))$ UM code or a rate $R = k/n > m_H/(m_H+1)$ PUM code over $\mathbb{F}$ be defined by its semi–infinite parity–check matrix $\mathbf{H}$. Let each submatrix $\mathbf{H}_i$, $i = 0, \ldots, m_H$ be the parity–check matrix of an (n, k) Gabidulin code $\mathcal{G}^{(i)}$:

$$\mathbf{H}_i = \mathcal{V}_{n-k}(\mathbf{h}^{(i)}) = \mathcal{V}_{n-k}\big(h_1^{(i)}\ h_2^{(i)}\ \ldots\ h_n^{(i)}\big),$$

for all $i = 0, \ldots, m_H$. Additionally, let

$$\mathbf{H}^{(c)} \overset{\text{def}}{=} \begin{pmatrix} \mathbf{H}_0 \\ \mathbf{H}_1 \\ \ldots \\ \mathbf{H}_{m_H} \end{pmatrix} \quad \text{define} \quad \mathcal{G}^{(c)}, \tag{1.25}$$

$$\mathbf{H}^{(r(i))} \overset{\text{def}}{=} \big(\mathbf{H}_i\ \mathbf{H}_{i-1}\ \ldots\ \mathbf{H}_0\big) \quad \text{define} \quad \mathcal{G}^{(r_i)}, \tag{1.26}$$

for all $i = 1, \ldots, m_H$. $\mathcal{G}^{(c)}$ is an $(n^{(c)}, k^{(c)})$ Gabidulin code and $\mathcal{G}^{(r_i)}$ is an $(n^{(r_i)}, k^{(r_i)})$ Gabidulin code with

$$n^{(c)} = n, \quad k^{(c)} = n - (m_H + 1)(n - k),$$
$$n^{(r_i)} = (i + 1)n, \quad k^{(r_i)} = i \cdot n + k, \quad i = 1, \ldots, m_H.$$

Hence, not only each submatrix has to define a Gabidulin code, but also the rows and columns of submatrices of $\mathbf{H}$.

Now, we give an explicit construction that fulfills the requirements of Definition 1.4.7. To ensure that (1.25) is fulfilled and $\mathbf{H}^{(c)}$ defines a Gabidulin code, $\mathbf{H}_1$ has to be the continuation of $\mathbf{H}_0$, i.e., $\mathbf{h}^{(1)} = \big((h_1^{(0)})^{[n-k]}, (h_2^{(0)})^{[n-k]}, \dots, (h_n^{(0)})^{[n-k]}\big)$. Also, $\mathbf{h}^{(2)}$ has to be the continuation of $\mathbf{h}^{(1)}$ and so on. Hence,

$$\begin{aligned} \mathbf{H}^{(c)} &= \mathcal{V}_{(m_H+1)(n-k)}\big(\mathbf{h}^{(0)}\big) \\ &= \mathcal{V}_{(m_H+1)(n-k)}\big(h_1^{(0)}\ h_2^{(0)}\ \dots\ h_n^{(0)}\big). \end{aligned} \tag{1.27}$$

In order to fulfill (1.26), we have to ensure that all elements from $\mathbb{F}$ in the set

$$\begin{aligned} \mathcal{H} &\overset{\text{def}}{=} \{h_1^{(0)}, \dots, h_n^{(0)}, h_1^{(1)}, \dots, h_n^{(1)}, \dots, h_1^{(m_H)}, \dots, h_n^{(m_H)}\} \\ &= \{h_1^{(0)}, \dots, h_n^{(0)}, (h_1^{(0)})^{[n-k]}, \dots, (h_n^{(0)})^{[n-k]}, \\ &\quad \dots, (h_1^{(0)})^{[m_H(n-k)]}, \dots, (h_n^{(0)})^{[m_H(n-k)]}\} \end{aligned} \tag{1.28}$$

with $|\mathcal{H}| = (m_H + 1) \cdot n$ are linearly independent over $\mathbb{F}_q$.

To obtain an explicit construction of such a (P)UM code, we can use a *normal basis*. A basis $\mathcal{B} = \{b_0, b_1, \dots, b_{s-1}\}$ of $\mathbb{F}$ over $\mathbb{F}_q$ is a normal basis if $b_i = b^{[i]}$ for all i and $b \in \mathbb{F}$ is called a *normal element*. There is a normal basis for any finite extension field $\mathbb{F}$ [23]. For our construction, we use a normal element b to define

$$\begin{aligned} \mathbf{h}^{(0)} = \big(b^{[0]} \dots b^{[n-k-1]} \big| b^{[(m_H+1)(n-k)]} \dots b^{[(m_H+2)(n-k)-1]} \big| \\ b^{[2(m_H+1)(n-k)]} \dots b^{[2(m_H+2)(n-k)-1]} \big| \dots\big). \end{aligned} \tag{1.29}$$

This $\mathbf{h}^{(0)}$ is used to define $\mathbf{H}^{(c)}$ (1.27) and hence also $\mathbf{H}$ is defined (1.25).

To make sure that also (1.26), (1.28) are fulfilled, we require a certain minimal field size. If $(n-k)$ divides n, then $\mathbf{h}^{(0)}$ can be divided into subvectors, each of length $(n-k)$ (1.29) and the field size has to fulfill $s \geq (m_H + 1) \cdot n$ to ensure that all elements in $\mathcal{H}$ are linearly independent (1.28) and that (1.26) is fulfilled. If $(n-k)$ does not divide n, the last subvector in $\mathbf{h}^{(0)}$ is shorter than $n-k$. Equations (1.28) and (1.26) can be

$$\mathbf{H} = \begin{pmatrix} \mathbf{H}_0 & & \\ \mathbf{H}_1 & \mathbf{H}_0 & \\ & \mathbf{H}_1 & \ddots \\ & & \ddots \end{pmatrix} = \left(\begin{array}{cccccc|cccccc|c} a^{[0]} & a^{[1]} & a^{[4]} & a^{[5]} & a^{[8]} & a^{[9]} & & & & & & & \\ a^{[1]} & a^{[2]} & a^{[5]} & a^{[6]} & a^{[9]} & a^{[10]} & & & & & & & \\ \hline a^{[2]} & a^{[3]} & a^{[6]} & a^{[7]} & a^{[10]} & a^{[11]} & a^{[0]} & a^{[1]} & a^{[4]} & a^{[5]} & a^{[8]} & a^{[9]} & \\ a^{[3]} & a^{[4]} & a^{[7]} & a^{[8]} & a^{[11]} & a^{[12]} & a^{[1]} & a^{[2]} & a^{[5]} & a^{[6]} & a^{[9]} & a^{[10]} & \\ \hline & & & & & & a^{[2]} & a^{[3]} & a^{[6]} & a^{[7]} & a^{[10]} & a^{[11]} & \dots \\ & & & & & & a^{[3]} & a^{[4]} & a^{[7]} & a^{[8]} & a^{[11]} & a^{[12]} & \dots \end{array}\right). \tag{1.30}$$

guaranteed in general if

$$s \geq (m_H + 1) \cdot \left\lceil \frac{n}{n-k} \right\rceil \cdot (n-k). \tag{1.31}$$

This implies the restriction for $(n-k)|n$ as a special case.

The following lemma shows that the parity–check matrix constructed in such a way is in minimal basic form.

Lemma 1.4.3 *Let a (P)UM code based on Gabidulin codes be defined by its parity–check matrix* $\mathbf{H}$ *as in Definition 1.4.7. Then,* $\mathbf{H}(D)$ *is in minimal basic form [40].*

The construction is demonstrated in the following example.

Example 1.4.1 Let us construct an $(6, 4 \mid 2)$ PUM code with $m_H = 1$. Hence, $s \geq 12$ (1.31) and we define the code e.g. over $\mathbb{F}_{2^{12}}$. In this field, there exists a *normal element* a such that $a^{[0]}, a^{[1]}, \dots, a^{[11]}$ are all linearly independent over $\mathbb{F}_2$ (i.e., this is a *normal basis*).

In order to guarantee also the continuation with $n-k=2$, we choose $\mathbf{h}^{(0)}$ as in (1.29) and $\mathbf{h}^{(1)}$ is defined by (1.25), (1.27):

$$\mathbf{h}^{(0)} = \left(a^{[0]}\ a^{[1]}\ a^{[4]}\ a^{[5]}\ a^{[8]}\ a^{[9]}\right),$$
$$\mathbf{h}^{(1)} = \left(a^{[2]}\ a^{[3]}\ a^{[6]}\ a^{[7]}\ a^{[10]}\ a^{[11]}\right).$$

The semi–infinite parity–check matrix $\mathbf{H}$ is given in Eq. (1.30).

As required by (1.25), (1.26) $\mathbf{H}_0, \mathbf{H}_1, \left(\mathbf{H}_0\ \mathbf{H}_1\right)$ and $\begin{pmatrix}\mathbf{H}_0\\ \mathbf{H}_1\end{pmatrix}$ each define Gabidulin codes.

The corresponding generator matrix $\mathbf{G}$ consists of two submatrices $\mathbf{G}_0, \mathbf{G}_1$ which are (4×6)-matrices, where the submatrices $\mathbf{G}_{00}, \mathbf{G}_{01}, \mathbf{G}_{10}$ are (2×6)-matrices, since $k_1 = \nu = m_H(n-k) = 2$. Hence, $\mathbf{H}^T$ defines a PUM code based on Gabidulin codes (to ensure that there exists such a PUM code see also Theorem 1.4.1). Note that the generator submatrices are not necessarily generator matrices of a Gabidulin code.

1.4.6 Calculation of Distances

In order to estimate the extended row rank distance d_ℓ^r (1.20) for the construction of Definition 1.4.7 and dual memory $m_H = 1$, let us consider all paths in the set $\mathcal{C}^{(r)}(\ell)$. As defined in (1.20), d_ℓ^r is the minimum sum rank weight (1.17) of all possible words in $\mathcal{C}^{(r)}(\ell)$.

Theorem 1.4.3 ([40]) *Let an* $(n, k \mid k_1)$ *(P)UM code with* $m_H = 1$ *be given where the submatrices of* $\mathbf{H}$ *define Gabidulin codes as in Definition 1.4.7. Let the rate* $R \geq 1/2$*, then the extended row rank distance* d_ℓ^r *(1.20) is given by*

$$d_1^r = 2(n-k)+1,$$
$$d_\ell^r \geq \left\lceil \frac{\ell+1}{2} \right\rceil \cdot (n-k+1), \quad \ell = 2, 3 \ldots . \tag{1.32}$$

This yields the following results for the free rank distance and the slope of a PUM code based on Gabidulin codes constructed as in Definition 1.4.7 with $m_H = 1$.

Theorem 1.4.4 *For $R > 1/2$, the $(n, k \mid k_1)$ PUM code based on Gabidulin codes with $m_H = 1$ achieves the upper bound of the free rank distance d_{free} and half the optimal slope α [40]:*

$$d_{\text{free}} = 2(n-k)+1 = n-k+\nu+1,$$
$$\alpha = \frac{n-k+1}{2}.$$

1.4.7 Remarks

We considered (P)UM codes based on Gabidulin codes. We defined general distance measures for convolutional codes based on a modified rank metric – the sum rank metric – and derived upper bounds for (P)UM codes based on the sum rank metric. In addition, an explicit construction of (P)UM codes based on Gabidulin codes was given and its free rank distance and slope were calculated for dual memory $m_H = 1$. Our PUM construction achieves the upper bound for the free rank distance and half the optimal slope.

Details on these results can be found in [40, 41]. Some proofs are omitted, but can be found in the long version [40]. In [45], another construction of rank-metric convolutional codes based on the generator matrix was shown. Further, we gave an efficient decoding algorithm for this class of PUM codes and showed explicitly how lifted rank-metric PUM codes can be applied to random linear network coding.

1.5 Bounds for List Decoding

Gabidulin codes are the rank-metric equivalent of Reed–Solomon codes. However, there is an interesting difference since no polynomial-time list decoding algorithm beyond half the minimum distance is known for Gabidulin codes but for RS codes. Bounds on the list size of rank metric codes help to understand whether polynomial-time list decoding is possible or not. Three bounds on the list size were shown in [43]. The first is a lower exponential bound for Gabidulin codes and shows that for Gabidulin codes no polynomial-time list decoding beyond the Johnson radius exists. The second is an exponential upper bound, which holds for any rank metric code of length n and minimum rank distance d. The third bound proves that there exists a rank metric code such that the list size is exponential in the length for any radius greater than half the minimum distance. This implies that there cannot exist a polynomial upper bound depending only on n and d as the Johnson bound for Hamming metric. All three bounds reveal significant differences to codes in Hamming metric.

A list decoding algorithm returns the list of all codewords within distance at most τ from any given word. In Hamming metric, the Johnson upper bound [2, 14, 18] shows for any code in Hamming metric of length n and minimum Hamming distance d_H that the size of this list is polynomial in n when τ is less than the Johnson radius $\tau_J = n - \sqrt{n(n - d_H)}$. Although this fact has been known since the 1960s, a polynomial-time list decoding algorithm for *Reed–Solomon* (RS) codes up to the Johnson radius was found not earlier than 1999 by Guruswami and Sudan [15] as a generalization of the Sudan algorithm [38]. Further, in Hamming metric, it can be shown that there exists a code such that the list size becomes *exponential* in n beyond the Johnson radius [13, 14]. It is not known whether this bound also holds for RS codes.

A lower bound on the maximum list size, which is exponential in the length n of the code, rules out polynomial-time list decoding since already writing down the list has exponential complexity. On the other hand, a polynomial upper bound—as the Johnson bound for Hamming metric—shows that a polynomial-time list decoding algorithm might exist.

In [42] a lower bound for list decoding of Gabidulin codes of length n and minimum rank distance d was given, which shows that the list size can be exponential in n when the radius is at least the Johnson radius $\tau_J = n - \sqrt{n(n - d)}$.

The first bound in [43] is an *exponential upper bound* for any rank metric code and provides no conclusion about polynomial-time list decodability. Remarkably, the second bound shows that there exists a rank metric code over $\mathbb{F}_{q^m}$ of length $n \leq m$ such that the list size is *exponential* in the length n when the decoding radius is greater than *half the minimum distance*. For these codes, hence, no polynomial-time list decoding can exist. The derivations apply connections between constant-rank codes and constant-dimension codes by Gadouleau and Yan [12]. Moreover, our results show that purely as a function of the length n and the minimum rank distance d, there cannot exist a polynomial upper bound similar to the Johnson bound in Hamming metric.

1.5.1 Subspace Codes

A *Grassmannian* of dimension r is the set of all subspaces of $\mathbb{F}_q^n$ of dimension $r \leq n$ and denoted by $\mathcal{G}_q(n, r)$. The cardinality of $\mathcal{G}_q(n, r)$ is given by the Gaussian binomial:

$$\left|\mathcal{G}_q(n, r)\right| = \begin{bmatrix} n \\ r \end{bmatrix} \overset{\text{def}}{=} \prod_{i=0}^{r-1} \frac{q^n - q^i}{q^r - q^i},$$

with the upper and lower bounds (see e.g. [20, Lemma 4])

$$q^{r(n-r)} \leq \begin{bmatrix} n \\ r \end{bmatrix} \leq 4q^{r(n-r)}. \tag{1.33}$$

For two subspaces $\mathcal{U}, \mathcal{V}$ in $\mathbb{F}_q^n$, we denote by $\mathcal{U} + \mathcal{V}$ the smallest subspace containing the union of $\mathcal{U}$ and $\mathcal{V}$. The *subspace distance* between $\mathcal{U}, \mathcal{V}$ in $\mathbb{F}_q^n$ is defined by

$$\begin{aligned} \mathrm{d_S}(\mathcal{U}, \mathcal{V}) &= \dim(\mathcal{U} + \mathcal{V}) - \dim(\mathcal{U} \cap \mathcal{V}) \\ &= 2\dim(\mathcal{U} + \mathcal{V}) - \dim(\mathcal{U}) - \dim(\mathcal{V}). \end{aligned}$$

It can be shown that the subspace distance is a metric [20].

A *subspace code* is a non-empty subset of subspaces of $\mathbb{F}_q^n$ and has minimum subspace distance d_S, when all subspaces in the code have subspace distance at least d_S. The codewords of a subspace code are therefore subspaces. An $\mathsf{CD}(n, \mathrm{d_S}, r)$ code denotes a *constant-dimension code* of dimension r and minimum subspace distance $\mathrm{d_S}$, i.e., it is a special subspace code and is a subset of $\mathcal{G}_q(n, r)$.

1.5.2 Constant-Rank Codes

A special class of rank metric codes are *constant-rank codes*. Such a (not necessarily linear) $\mathsf{CR}(n, d, r)$ code over $\mathbb{F}_{q^m}$ has length $n \leq m$, consists of words in $\mathbb{F}_{q^m}^n$ or equivalently of matrices in $\mathbb{F}_q^{m \times n}$, has minimum rank distance d and each codeword has rank exactly r.

The maximum cardinalities of a constant-dimension and constant-rank code for fixed parameters will be denoted by $|\mathsf{CD}(n, d_S, r)|_{\max}$ and $|\mathsf{CR}(n, d_R, r)|_{\max}$.

Further, $\mathcal{B}_\tau(\mathbf{a})$ denotes a ball of radius τ in rank metric around a word $\mathbf{a} \in \mathbb{F}_{q^m}^n$ and $\mathcal{S}_\tau(\mathbf{a})$ denotes a sphere in rank metric of radius τ around the word $\mathbf{a}$. The cardinality of a sphere of radius τ is the number of $m \times n$ matrices in $\mathbb{F}_q$, which have rank distance *exactly* τ from a word $\mathbf{a}$ and the cardinality of a ball of radius τ is the number of $m \times n$ matrices in $\mathbb{F}_q$, which have rank distance *less than or equal to* τ.

1.5.3 Constant-Dimension and Constant-Rank Codes

In this subsection, we recall known connections between constant-dimension and constant-rank codes by Gadouleau and Yan [12] and generalize some of their results, since we will use them in the next sections for bounding the list size.

Proposition 1.5.1 (Maximum Cardinality, [12]) *For all q and $1 \leq \delta \leq r \leq n \leq m$, the **maximum** cardinality of a $\mathsf{CR}(n, d_R = \delta + r, r)$ constant-rank code over $\mathbb{F}_{q^m}$ is upper bounded by the maximum cardinality of a constant-dimension code as follows:*

$$|\mathsf{CR}(n, d_R = \delta + r, r)|_{\max} \leq |\mathsf{CD}(n, d_S = 2\delta, r)|_{\max} .$$

The following proposition shows explicitly how to construct constant-rank codes out of constant-dimension codes and is a generalization of [12, Proposition 3] to constant-dimension codes of arbitrary cardinalities.

Proposition 1.5.2 (Construction of a Constant-Rank Code [44]) *Let* M *be a* $\mathsf{CD}(m, d_{S,M}, r)$ *and* N *be a* $\mathsf{CD}(n, d_{S,N}, r)$ *constant-dimension code with* $r \leq \min\{n, m\}$ *and cardinalities* $|\mathsf{M}|$ *and* $|\mathsf{N}|$*. Then, there exists a* $\mathsf{CR}(n, d_R, r)$ *constant-rank code* C *of cardinality* $\min\{|\mathsf{M}|, |\mathsf{N}|\}$ *with* $\mathcal{C}_q(\mathsf{C}) \subseteq \mathsf{M}$ *and* $\mathcal{R}_q(\mathsf{C}) \subseteq \mathsf{N}$*. Furthermore,*

$$d_R \geq \frac{1}{2}\, d_{S,M} + \frac{1}{2}\, d_{S,N},$$

1.5.4 Constant-Dimension Codes from Lifted MRD Codes

There are several contributions about constructions of constant-dimension codes and bounds on their cardinality, see e.g. [1, 9, 20, 34, 36, 37, 48, 49]. For our application of constant-dimension codes, the cardinality does not have to be optimal. *Lifted MRD codes* (see [36]) are sufficient and are shown in the following.

Definition 1.5.1 (*Lifting,* [36]) Let the mapping $\mathcal{I} : \mathbb{F}_q^{r \times (n-r)} \to \mathcal{G}_q(n, r)$ be given by $\mathbf{X} \mapsto \mathcal{I}(\mathbf{X}) = \mathcal{R}_q([\mathrm{i}_r\, \mathbf{X}])$, where i_r denotes the $r \times r$ identity matrix. The subspace $\mathcal{I}(\mathbf{X})$ is called **lifting** of the matrix **X**.

If we apply this map on all codewords of a block code C, then the constant-dimension code $\mathcal{I}(\mathsf{C})$ is called lifting of C.

The following two lemmas show lifted MRD codes for some explicit parameters.

Lemma 1.5.1 (Lifted MRD Code, Even Distance, [36]) *Let* d *be an even integer, let* $\tau \geq d/2$ *and* $\tau \leq n - \tau$*. Let a linear* $\mathsf{MRD}[\tau, \tau - d/2 + 1]$ *code* C *over* $\mathbb{F}_{q^{n-\tau}}$ *of length* τ*, minimum rank distance* $d_R = d/2$ *and cardinality* M_R *be given.*

Then, the lifting of the transposed codewords, i.e.,

$$\mathcal{I}(\mathsf{C}^T) \overset{\text{def}}{=} \{\mathcal{I}(\mathbb{C}^T) = \mathcal{R}_q([\mathrm{i}_\tau\ \mathbb{C}^T]) : \mathbb{C} \in \mathsf{C}\}$$

is a $\mathsf{CD}(n, d_S, \tau)$ *constant-dimension code of cardinality* $M_S = M_R = q^{(n-\tau)(\tau-d/2+1)}$*, minimum subspace distance* $d_S = d$ *and lies in the Grassmannian* $\mathcal{G}_q(n, \tau)$*.*

Lemma 1.5.2 (Lifted MRD Code, Odd Distance, [36]) *Let* d *be an odd integer and let* $\tau \geq {}^{(d-1)}/_2 + 1$*. Then,*

- *for* $\tau \leq m - \tau$ *and a linear* $\mathsf{MRD}[\tau, \tau - {}^{(d-1)}/_2 + 1]$ *code* C *over* $\mathbb{F}_{q^{m-\tau}}$*, the lifting* $\mathcal{I}(\mathsf{C}^T)$ *is a* $\mathsf{CD}(m, d_S = d - 1, \tau)$ *constant-dimension code of cardinality* $q^{(m-\tau)(\tau-{}^{(d-1)}/_2+1)}$*,*

- *for* $\tau \leq n-\tau$, $n \leq m$ *and a linear* $\mathsf{MRD}[\tau, \tau - {}^{(d+1)}/_2 + 1]$ *code* C *over* $\mathbb{F}_{q^{n-\tau}}$, *the lifting* $\mathcal{I}(\mathsf{C}^T)$ *is a* $\mathsf{CD}(n, d_S = d+1, \tau)$ *constant-dimension code of cardinality* $q^{(n-\tau)(\tau - {}^{(d+1)}/_2 + 1)} = q^{(n-\tau)(\tau - {}^{(d-1)}/_2)} < q^{(m-\tau)(\tau - {}^{(d-1)}/_2 + 1)}$.

Lifted MRD codes are sufficient for our bounds, although there are constant-dimension codes of higher cardinality, e.g. [9].

We analyze the question of *polynomial-time list decodability* of rank metric codes. Thus, we want to bound the maximum number of codewords in a ball of radius τ around a received word $\mathbf{r}$. This number will be called the maximum *list size* ℓ in the following. The worst-case complexity of a possible list decoding algorithm directly depends on ℓ.

Problem 1.5.1 (*Maximum List Size* [43]) Let C be an $\mathcal{C}_{n,M,d}$ code over $\mathbb{F}_{q^m}$ of length $n \leq m$, cardinality M and minimum rank distance $d_R = d$. Let $\tau < d$. Find a lower and upper bound on the maximum number of codewords ℓ in a ball of rank radius τ around a word $\mathbf{r} = (r_0 \ r_1 \ \dots \ r_{n-1}) \in \mathbb{F}_{q^m}^n$. Hence, find a bound on

$$\ell \overset{\text{def}}{=} \max_{\mathbf{r} \in \mathbb{F}_{q^m}^n} \left\{ \left| \mathsf{C} \cap \mathcal{B}_\tau(\mathbf{r}) \right| \right\}.$$

For an upper bound, we have to show that the bound holds for *any* received word $\mathbf{r}$, whereas for a lower bound it is sufficient to show that there exists (at least) one $\mathbf{r}$ for which this bound on the list size is valid. Further, we denote the list of all codewords of C in the ball of rank radius τ around a given word $\mathbf{r} \in \mathbb{F}_{q^m}^n$ by:

$$\begin{aligned} \mathcal{L} &\overset{\text{def}}{=} \mathsf{C} \cap \mathcal{B}_\tau(\mathbf{r}) \\ &= \left\{ \mathbf{c}_1, \mathbf{c}_2, \dots, \mathbf{c}_{|\mathcal{L}|} \ : \ \mathbf{c}_i \in \mathsf{C} \text{ and } \operatorname{rk}(\mathbf{r} - \mathbf{c}_i) \leq \tau, \ \forall i \right\}, \end{aligned} \tag{1.34}$$

with cardinality $|\mathcal{L}| \leq \ell$.

1.5.5 *Connection Between Constant-Rank Codes and List Size*

As in (1.34), denote the list of codewords when decoding up to $\tau < d$ errors with an $\mathcal{C}_{n,M,d_R=d}$ code C by

$$\begin{aligned} \mathcal{L} = \left\{ \mathbf{c}_1, c_2, \dots, \mathbf{c}_{|\mathcal{L}|} \right\} &= \mathsf{C} \cap \mathcal{B}_\tau(\mathbf{r}) \\ &= \sum_{i=0}^{\tau} \left(\mathsf{C} \cap \mathcal{S}_i(\mathbf{r}) \right), \end{aligned}$$

for some received word $\mathbf{r} \in \mathbb{F}_{q^m}^n$. Consider only the codewords in rank distance *exactly* τ from the received word, i.e.:

$$\{\mathbf{c}_1, \mathbf{c}_2, \ldots, \mathbf{c}_{\overline{\ell}}\} \overset{\text{def}}{=} \mathsf{C} \cap \mathcal{S}_\tau(\mathbf{r}).$$

Clearly, this gives a lower bound on the maximum list size: $\ell \geq \overline{\ell} = |\mathsf{C} \cap \mathcal{S}_\tau(\mathbf{r})|$. Now, consider a translate of all codewords of rank distance exactly τ as follows:

$$\overline{\mathcal{L}} \overset{\text{def}}{=} \{\mathbf{r} - \mathbf{c}_1, \mathbf{r} - \mathbf{c}_2, \ldots, \mathbf{r} - \mathbf{c}_{\overline{\ell}}\}.$$

This set $\overline{\mathcal{L}}$ is a $\mathsf{CR}(n, d_R \geq d, \tau)$ constant-rank code since $\text{rk}(\mathbf{r} - \mathbf{c}_i) = \tau$ for all $i = 1, \ldots, \overline{\ell}$ and its minimum rank distance is at least d, since

$$\text{rk}(\mathbf{r} - \mathbf{c}_i - \mathbf{r} + \mathbf{c}_j) = \text{rk}(\mathbf{c}_i - \mathbf{c}_j) \geq d, \quad \forall i, j, \; i \neq j.$$

The cardinality of this constant-rank code is $\overline{\ell}$ and for $\tau < d$, it is non-linear, since the rank of its codewords is less than its minimum distance.

Hence, a translate of the list of all codewords in rank distance exactly τ from the received word can be interpreted as a constant-rank code, which makes it possible to use bounds on the cardinality of a constant-rank code to obtain bounds on the list size ℓ.

1.5.6 Upper Bound on the List Size

The upper bound presented in this subsection is an upper bound on the list size when decoding rank metric codes and holds for *any* code in rank metric and *any* received word.

Theorem 1.5.1 (Upper Bound on the List Size [44]) *Let* $\lfloor (d-1)/2 \rfloor < \tau < d \leq n \leq m$. *Then, for* ***any*** $\mathcal{C}_{n,M,d}$ *code over* $\mathbb{F}_{q^m}$ *in rank metric, the maximum list size is upper bounded as follows:*

$$\begin{aligned} \ell &\leq 1 + \sum_{t=\lfloor \frac{d-1}{2} \rfloor + 1}^{\tau} \frac{\begin{bmatrix} n \\ 2t+1-d \end{bmatrix}}{\begin{bmatrix} t \\ 2t+1-d \end{bmatrix}} \\ &\leq 1 + 4 \sum_{t=\lfloor \frac{d-1}{2} \rfloor + 1}^{\tau} q^{(2t-d+1)(n-t)}. \end{aligned}$$

Notice that this bound gives (almost) the same upper bound as we showed in [42, Theorem 2], which can slightly be improved if we use better upper bounds for constant-dimension codes, for example the iterated Johnson bound for constant-dimension codes [49, Corollary 3]. However, the Wang–Xing–Safavi-Naini bound provides a nice closed-form expression and is asymptotically tight.

Unfortunately, our upper bound on the list size of rank metric codes is exponential in the length of the code. However, the lower bound of Sect. 1.5.7 will show that any upper bound depending only on the length $n \leq m$ and the minimum rank distance d has to be exponential in $(\tau - \lfloor (d-1)/2 \rfloor)(n-\tau)$, since there exists a rank metric code with such a list size.

1.5.7 *Lower Bound on the List Size*

In this subsection, we prove the most significant difference to codes in Hamming metric: We derive the existence of a rank metric code over $\mathbb{F}_{q^m}$ of length $n \leq m$ with exponential list size for any decoding radius *greater than half the minimum distance*. First, we prove the existence of a certain constant-rank code.

Theorem 1.5.2 (Constant-Rank Code [43]) *Let* $\lfloor (d-1)/2 \rfloor < \tau < d \leq n \leq m$ *and* $\tau \leq n - \tau$. *Then, there exists a* $\mathsf{CR}(n, d_R \geq d, \tau)$ *constant-rank code over* $\mathbb{F}_{q^m}$ *of cardinality* $q^{(n-\tau)(\tau - \lfloor (d-1)/2 \rfloor)}$.

This can now directly be used to show the existence of a rank metric code with exponential list size.

Theorem 1.5.3 (Lower Bound on the List Size [43]) *Let* $\lfloor (d-1)/2 \rfloor < \tau < d \leq n$ *and* $\tau \leq n - \tau$. *Then, there exists an* $\mathcal{C}_{n,M,d_R \geq d}$ *code* C *over* $\mathbb{F}_{q^m}$ *of length* $n \leq m$ *and minimum rank distance* $d_R \geq d$, *and a word* $\mathbf{r} \in \mathbb{F}_{q^m}^n$ *such that*

$$\ell \geq \left| \mathsf{C} \cap \mathcal{B}_\tau(\mathbf{r}) \right| \geq q^{(n-\tau)(\tau - \lfloor (d-1)/2 \rfloor)}. \tag{1.35}$$

This rank metric code C is not a linear code since it has codewords of weight τ, but minimum rank distance d.

For constant code rate $R = k/n$ and constant relative decoding radius τ/n, where $\tau > \lfloor (d-1)/2 \rfloor$, (1.35) gives

$$\ell \geq q^{n^2(1-\tau/n)(\tau/n - 1/2(1-R))} = q^{n^2 \cdot const}.$$

Therefore, the lower bound on list decoding of this $\mathcal{C}_{n,M,d_R \geq d}$ code is exponential in $n \leq m$ for any $\tau > \lfloor (d-1)/2 \rfloor$ and Theorem 1.5.3 shows that there exist codes, where the number of codewords in a rank metric ball around the all-zero word is exponential in n, thereby prohibiting polynomial-time list decoding.

The next corollary shows that the restriction $\tau \leq n - \tau$ does not limit the code rate for which Theorem 1.5.3 shows an exponential behavior of the list size. For the special case of $\tau = \lfloor (d-1)/2 \rfloor + 1$, the condition $\tau \leq n - \tau$ is always fulfilled for even minimum distance since $d \leq n$. For odd minimum $d - 1 \leq n$ has to hold. Note that $d = n$ is a trivial code.

Corollary 1.5.1 (Special Case $\tau = \lfloor (d-1)/2 \rfloor + 1$) *Let* $n \leq m$, $\tau = \lfloor (d-1)/2 \rfloor + 1$ *and* $d \leq n - 1$ *when* d *is odd. Then, there exists a* $\mathcal{C}_{n,M,d_R \geq d}$ *code* C *over* $\mathbb{F}_{q^m}$ *with* $n \leq m$ *and a word* $\mathbf{r} \in \mathbb{F}_{q^m}^n$ *such that* $|\mathsf{C} \cap \mathcal{B}_\tau(\mathbf{r})| \geq q^{(n-\tau)}$ [43].

This corollary hence shows that for any $n \leq m$ and *any code rate*, there exists a rank metric code of rank distance at least d whose list size can be exponential in n.

1.5.8 Interpretation and Conclusion

This section interprets the results from the previous sections and compares them to known bounds on list decoding in Hamming metric (see e.g. [14, Chaps. 4 and 6]).

Theorem 1.5.3 shows that for any $n \leq m$ and d, there is a code over $\mathbb{F}_{q^m}$ of minimum rank distance at least d and a word in $\mathbb{F}_{q^m}^n$ such that there is a ball of any radius $\tau > \lfloor (d-1)/2 \rfloor$, which contains a number of codewords that is exponential in the length n. Hence, for these rank metric codes *no* polynomial-time list decoding algorithm beyond half the minimum distance exists. However, this does not mean that this holds for *any* rank metric code. In particular, the theorem does not provide a conclusion if there exists a *linear* code or even a *Gabidulin* code with this list size. In order to find a *polynomial* upper bound, it will be necessary to use further properties of the code (such as linearity or a concrete weight distribution) in the derivation.

In particular, for Gabidulin codes, there is still an unknown region between half the minimum distance and the Johnson radius since we could only prove that the list size can be exponential beyond the Johnson radius (see [42]).

Further, our lower bound from Theorem 1.5.3 implies that there cannot exist a polynomial upper bound depending only on the length of the code n and minimum rank distance d similar to the Johnson bound in Hamming metric.

These results show surprising differences between codes in Hamming and rank metric. *Any* ball in Hamming metric of radius less than the Johnson radius $\tau_J = n - \sqrt{n(n-d)}$ always contains a *polynomial* number of codewords of *any* code in Hamming metric of length n and minimum Hamming distance d. Moreover, in Hamming metric there exist codes such that the list size is exponential in n if the radius is slightly greater than the Johnson radius [13, 14], whereas in rank metric we proved that this happens directly beyond half the minimum distance.

1.5.9 Remarks

Details on these results can be found in [42–44].

Acknowledgments This work was supported from 2009 to 2013 by the Deutsche Forschungsgemeinschaft (DFG) under grant No. Bo-867/21.

References

1. Bachoc C, Passuello A, Vallentin F (2013) Bounds for projective codes from semidefinite programming. Adv Math Commun 7(2):127–145
2. Bassalygo LA (1965) New upper bounds for error correcting codes. Prob Inf Transm 1(4):41–44
3. Bossert M (1999) Channel coding for telecommunications. Wiley, Chichester
4. Bossert M (2012) Einführung in die Nachrichtentechnik. Oldenburg Verlag
5. Cai N, Yeung RW (2002) Network coding and error correction. In: Proceedings of IEEE Information Theory Workshop, pp 119–122
6. Delsarte P (1978) Bilinear forms over a finite field, with applications to coding theory. J Comb Theory Ser A 25(3):226–241
7. Dettmar U, Shavgulidze S (1992) New optimal partial unit memory codes. Electron Lett 28:1748–1749
8. Dettmar U, Sorger U (1993) New optimal partial unit memory codes based on extended BCH codes. Electron Lett 29(23):2024–2025
9. Etzion T, Silberstein N (2009) Error-correcting codes in projective spaces via rank-metric codes and ferrers diagrams. IEEE Trans Inf Theory 55(7):2909–2919
10. Forney G (1970) Convolutional codes I: algebraic structure. IEEE Trans Inf Theory 16(6):720–738
11. Gabidulin EM (1985) Theory of codes with maximum rank distance. Prob Peredachi Inf 21(1):3–16
12. Gadouleau M, Yan Z (2010) Constant-rank codes and their connection to constant-dimension codes. IEEE Trans Inf Theory 56(7):3207–3216
13. Goldreich O, Rubinfeld R, Sudan M (2000) Learning polynomials with queries: the highly noisy case. SIAM J Discrete Math 13(4). doi:10.1137/S0895480198344540
14. Guruswami V (1999) List decoding of error-correcting codes: winning thesis of the 2002 ACM doctoral dissertation competition. Lecture notes in computer science. Springer, Berlin
15. Guruswami V, Sudan M (1999) Improved decoding of reed-solomon and algebraic-geometry codes. IEEE Trans Inf Theory 45(6):1757–1767
16. Ho T, Koetter R, Medard M, Karger DR, Effros M (2003) The benefits of coding over routing in a randomized setting
17. Hua LK (1951) A theorem on matrices over a field and its applications. Acta Math Sinica 1(2):109–163
18. Johnson S (1962) A new upper bound for error-correcting codes. IRE Trans Inf Theory 8(3):203–207. doi:10.1109/TIT.1962.1057714
19. Justesen J (1993) Bounded distance decoding of unit memory codes. IEEE Trans Inf Theory 39(5):1616–1627
20. Koetter R, Kschischang FR (2008) Coding for errors and erasures in random network coding. IEEE Trans Inf Theory 54(8):3579–3591. doi:10.1109/TIT.2008.926449
21. Lauer GS (1979) Some optimal partial-unit memory codes. IEEE Trans Inf Theory 23(2):240–243
22. Lee LN (1976) Short unit-memory byte-oriented binary convolutional codes having maximal free distance. IEEE Trans Inf Theory :349–352
23. Lidl R, Niederreiter H (1996) Finite fields. Encyclopedia of mathematics and its applications. Cambridge University Press, Cambridge
24. Loidreau P, Overbeck R (2006) Decoding rank errors beyond the error correcting capability, pp 186–190
25. Overbeck R (2006) Decoding interleaved gabidulin codes and ciphertext-security for GPT variants. Preprint
26. Paramonov A, Tretjakov O (1991) An analogue of Berlekamp-Massey algorithm for decoding codes in rank metric. In: Proceedings of MIPT
27. Pollara F, McEliece RJ, Abdel-Ghaffar KAS (1988) Finite-state codes. IEEE Trans Inf Theory 34(5):1083–1089

28. Richter G, Plass S (2004) Error and erasure decoding of rank-codes with a modified Berlekamp-Massey algorithm. In: ITG Fachbericht, pp 203–210
29. Roth RM (1991) Maximum-rank array codes and their application to crisscross error correction. IEEE Trans Inf Theory 37(2):328–336
30. Sidorenko V, Bossert M (2010) Decoding interleaved gabidulin codes and multisequence linearized shift-register synthesis. In: IEEE international symposium on information theory, pp 1148–1152. doi:10.1109/ISIT.2010.5513676
31. Sidorenko V, Richter G, Bossert M (2011) Linearized shift-register synthesis. IEEE Trans Inf Theory 57(9):6025–6032
32. Sidorenko V, Wachter-Zeh A, Chen D (2012) On fast decoding of interleaved gabidulin codes. In: The XIII international symposium—problems of redundancy in information and control systems
33. Sidorenko V, Li W, Chen D (2013) On transform domain decoding of gabidulin codes. In: Accepted for the eigth international workshop on coding and cryptography (WCC 2013)
34. Silberstein N, Etzion T (2011) Enumerative coding for grassmannian space. IEEE Trans Inf Theory 57(1):365–374
35. Silva D, Kschischang FR (2007) Using rank-metric codes for error correction in random network coding. In: IEEE international symposium on information theory, pp 796–800. doi:10.1109/ISIT.2007.4557322
36. Silva D, Kschischang FR, Koetter R (2008) A Rank-Metric Approach to Error Control in Random Network Coding. IEEE Trans Inf Theory 54(9):3951–3967
37. Skachek V (2008) Recursive code construction for random networks. ArXiv preprint arXiv:08063650
38. Sudan M (1997) Decoding of Reed Solomon codes beyond the error-correction bound. J Complex 13(1):180–193. doi:10.1006/jcom.1997.0439
39. Thommesen C, Justesen J (1983) Bounds on distances and error exponents of unit memory codes. IEEE Trans Inf Theory 29(5):637–649
40. Wachter A, Sidorenko V, Bossert M, Zyablov V (2011) On (partial) unit memory codes based on Gabidulin codes. Prob Inf Transm 47(2):38–51
41. Wachter A, Sidorenko V, Bossert M, Zyablov V (2011) Partial unit memory codes based on Gabidulin codes. In: IEEE international symposium on information theory (ISIT 2011)
42. Wachter-Zeh A (2012) Bounds on list decoding Gabidulin codes. In: Thirteenth international workshop on algebraic and combinatorial coding theory (ACCT 2012), pp 329–334
43. Wachter-Zeh A (2013) Bounds on list decoding of rank metric codes. IEEE Trans Inf Theory 59(11):7268–7277
44. Wachter-Zeh A (2013) Bounds on polynomial time list decoding of rank metric codes. In: IEEE international symposium on information theory (ISIT), vol 59(11), pp 7268–7277
45. Wachter-Zeh A, Sidorenko V (2012) Rank metric convolutional codes for random linear network coding. In: IEEE International symposium on network coding (Netcod 2012)
46. Wachter-Zeh A, Zeh A (2014) List and unique error-erasure decoding of interleaved gabidulin codes with interpolation techniques. Des Codes Crypt 73(2):547–570
47. Wachter-Zeh A, Afanassiev V, Sidorenko V (2013) Fast decoding of Gabidulin codes. Des Codes Crypt 66(1):57–73
48. Wang H, Xing C, Safavi-Naini R (2003) Linear authentication codes: bounds and constructions. IEEE Trans Inf Theory 49(4):866–872
49. Xia ST, Fu FW (2009) Johnson type bounds on constant dimension codes. Des Codes Crypt 50(2):163–172
50. Zyablov V, Sidorenko V (1994) On periodic (partial) unit memory codes with maximum free distance. Lect Notes Comput Sci 829:74–79

Chapter 2
Modulo-Type Precoding for Networks

Robert F.H. Fischer, Michael Cyran, Sebastian Stern and Johannes B. Huber

Abstract In this chapter, we address scenarios where the tasks of (modulo-type) precoding for the multiple-input/multiple-output (MIMO) broadcast channel, network coding with its associated finite-field matrix channel, and channel coding meet or complement each other. By enlightening dualities, similarities, and differences between the areas and corresponding schemes, a deeper understanding of their mutual interaction is gained. Moreover, this allows for a transfer of schemes and strategies from one field to another one. Exemplarily, schemes operating at the intersection of complex-valued and finite-field/modulo processing are addressed. First, an overview on modulo-type precoding and its latest version via finite-field preprocessing is given; the connections and specific restrictions of the different approaches are illustrated. The advantages of modulo-type precoding are addressed when additional requirements, such as per-antenna power constraints and a reduced degree of coordination in a network MIMO scenario, are imposed. Finally, the application of precoding to finite-field channels is discussed, either as differential network coding or as selection precoding.

2.1 Introduction

Since two decades, the interest in communication schemes with multiple input and multiple output signals, forming a so-called *multiple-input/multiple-output (MIMO) channel*, has grown enormously. Meanwhile, MIMO techniques are in the standard

R.F.H. Fischer (✉) · S. Stern
Ulm University, Albert-Einstein-Allee 43, 89081 Ulm, Germany
e-mail: robert.fischer@uni-ulm.de

S. Stern
e-mail: sebastian.stern@uni-ulm.de

M. Cyran · J.B. Huber
FAU Erlangen-Nürnberg, Cauerstrasse 7, 91058 Erlangen, Germany
e-mail: michael.cyran@fau.de

J.B. Huber
e-mail: johannes.huber@fau.de

W. Utschick (ed.), *Communications in Interference Limited Networks*,
Signals and Communication Technology, DOI 10.1007/978-3-319-22440-4_2

repertoire of each communication engineer. Not only *multi-antenna* point-to-point systems can be treated under this umbrella term, but the joint consideration of signals is particularly useful in *multi-user schemes*. Via joint processing, interference between users can be dealt with in a constructive way, rather than as unavoidable, noise-like disturbance.

Interference and joint processing can occur/be performed on different layers in the OSI communication model. In this chapter, we address the *physical* and the *network* layer. Specifically, *modulo-type precoding* [19] and *network coding* [15] are considered, which have been subject of immense research activity over the last decade. Even though these techniques pursue similar or dual purposes, up to now interdependencies between the areas have only rarely been investigated or even utilized. In this chapter, we address scenarios where the tasks of (modulo-type) precoding, network coding, and channel coding meet or complement each other. By enlightening dualities, similarities, and differences between the fields and corresponding schemes, a deeper understanding of their mutual interaction is gained. Moreover, this allows for a transfer of schemes and strategies from one field to another one.

First, we present the broadcast channel and the finite-field matrix channel which arises in network coding. Then, similarities and dualities are briefly discussed. In Sect. 2.2, the connection between complex-valued and finite-field channels is depicted for the situation of precoding for the broadcast channel. Even though the channel is real-/complex-valued, the use of finite-field arithmetic for precoding is an interesting new option. The optimization of wireless multi-base-station schemes w.r.t. per-antenna power constraints is considered in Sect. 2.3. The possibility of reducing the required (wired) backhaul traffic employing a decentralized processing with hierarchical coordination among base stations and via finite-field precoding are given. Finally, in Sect. 2.4, precoding is applied to finite-field channels. Differential linear network coding and selection precoding are techniques transferring knowledge from the continuous to the finite-field world, thereby enabling new degrees of freedom in system design.

2.1.1 MIMO Broadcast Channel

A block diagram of the MIMO *broadcast channel (BC)*, describing the downlink transmission from a central *base station (BS)*, equipped with B antennas, to K independent, non-cooperating *user equipments (UEs)* is visualized in Fig. 2.1. The fundamental input/output relation is given by

$$\boldsymbol{y} = \boldsymbol{H}\boldsymbol{x} + \boldsymbol{n} \ , \tag{2.1}$$

where $\boldsymbol{x} = [x_1, \dots, x_B]^\mathsf{T} \in \mathbb{C}^B$ is the vector of transmit symbols, $\boldsymbol{y} = [y_1, \dots, y_K]^\mathsf{T}$ $\in \mathbb{C}^K$ the vector of receive symbols, $\boldsymbol{n} = [n_1, \dots, n_K]^\mathsf{T} \in \mathbb{C}^K$ is the noise vector and

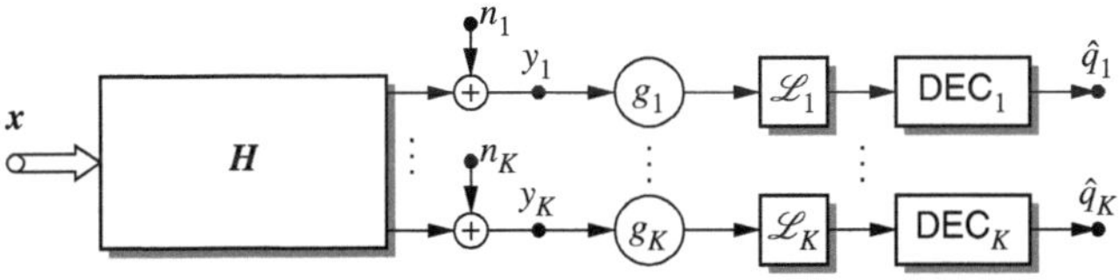

Fig. 2.1 Broadcast channel, i.e., downlink transmission from a central base station with B antennas (joint transmit vector $\boldsymbol{x}$) to the K individual users (scaling g_k, metric calculation $\mathcal{L}_k$, and channel decoding)

$$\boldsymbol{H} = \left[h_{k,b} \right]_{\substack{k=1,\dots,K \\ b=1,\dots,B}} \tag{2.2}$$

is the $K \times B$ channel matrix collecting the channel gains $h_{k,b}$ when assuming flat fading channels.[1] We stick to the standard assumption of a *block-fading channel*, i.e., it is randomly drawn according to some (known) distribution, but remains constant over a transmission burst. The transmit symbols are expected to be zero-mean and with variance $\sigma_x^2 \stackrel{\text{def}}{=} \mathrm{E}\{|x_b|^2\}$, $b = 1, \dots, B$, and the zero-mean white Gaussian noise has variance $\sigma_n^2 \stackrel{\text{def}}{=} \mathrm{E}\{|n_k|^2\}$, $k = 1, \dots, K$. Please note, since the channel is modeled in the *equivalent complex baseband*, all signals and coefficients are complex-valued. Unless otherwise stated, we assume perfect channel knowledge at the joint transmitter.

2.1.2 Finite-Field Matrix Channels

Meanwhile, *finite-field matrix channels*—i.e., MIMO channels where all quantities are drawn from and the arithmetic is carried out over a finite field—emerged in several situations, in particular in *(random) linear network coding ((R)LNC)* [40] or in lattice-coded MIMO systems, where *integer-forcing (IF)* [25, 33] receivers or transmitters are employed (cf. also Sect. 2.2.3). The most general form is the *multiplicative additive matrix channel (MAMC)* [40], depicted in Fig. 2.2. Its input-output relation reads

$$\boldsymbol{Y} = \boldsymbol{A}\boldsymbol{X} + \boldsymbol{E} . \tag{2.3}$$

Here, usually *packets* (or codewords, vectors) of length l (equivalent to l time steps in the BC) are considered and n packets are fed to the network in parallel.[2] Thus, the transmit signal is given as matrix $\boldsymbol{X} \in \mathbb{F}_q^{n \times l}$. The linear channel, modeled by the channel matrix $\boldsymbol{A} \in \mathbb{F}_q^{N \times n}$, multiplicatively distorts $\boldsymbol{X}$ and superimposes an additive

---

[1] If, in addition to the multi-user interference, intersymbol interference (ISI) occurs, the usual way is to apply orthogonal frequency-division multiplexing (OFDM) to deal with the ISI. The MIMO model is then valid per subcarrier.

[2] In terms of RLNC, one channel usage, i.e., the transmission of one transmit and the reception of one receive matrix, is called one *generation*. In terms of the BC, this is a transmission burst.

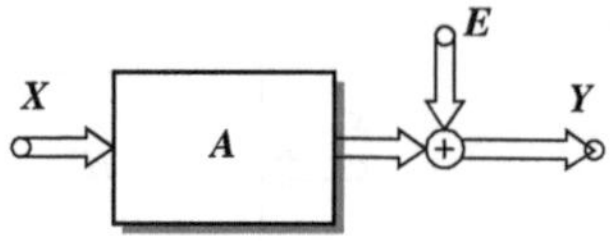

Fig. 2.2 Multiplicative additive matrix channel

error, expressed by an error matrix $\boldsymbol{E} \in \mathbb{F}_q^{N\times l}$. The quantities n and N, with $N \geq n$, denote the numbers of transmit and receive packets, respectively. In RLNC, $\boldsymbol{A}$ is randomly chosen (typically uniformly over all full-column-rank $(N \times n)$ matrices over $\mathbb{F}_q$ [9] and may change after each packet [40].

For some applications it is reasonable to consider two types of degenerated finite-field matrix channels, namely the *multiplicative matrix channel* (MMC) $\boldsymbol{Y} = \boldsymbol{AX}$, and the *additive matrix channel* (AMC) $\boldsymbol{Y} = \boldsymbol{X} + \boldsymbol{E}$.

2.1.3 Analogies and Dualities

In communications and information theory, analogies and dualities are very helpful and powerful tools. A famous example is the duality between *source coding* and *channel coding* [17, 24]. In source coding redundancy is eliminated, while in channel coding redundancy is deliberately added. Via this duality, schemes from the one field can be converted to the other one, cf. [16, 58].

Another prominent example is the *uplink/downlink duality* [28, 38, 46, 47, 56], which states that the *multiple-access channel* and the *broadcast channel* are dual to each other. This not only led to the derivation of the capacity region of the BC [49] but also sparked the design of transmitter-side techniques which are dual to receiver-side approaches, cf. *Tomlinson-Harashima precoding (THP)* as the dual to *decision-feedback equalization (DFE)* (aka successive interference cancellation [20]), or *vector precoding* [35] as the dual to *maximum-likelihood detection.*

Besides such dualities, *analogies* and *similarities* are of major importance. As can be seen from Figs. 2.1 and 2.2 or (2.1) and (2.3), respectively, the MIMO BC and the MAMC are counterparts to each other, existing in two different worlds of arithmetics. Moreover, *precoding (PC)* for the MIMO BC (see Sect. 2.2) and *network coding (NC)* share some general principles, too. In both fields a number of incoming signals are treated jointly, as can be seen in Fig. 2.3. *Modulo-linear combinations* of the signals X_a and X_b (in NC) and of the signal a and the interference f (in PC), respectively, are calculated and processed. However, due to the different fields $\mathbb{F}_q$ and $\mathbb{C}$ the modulo operation is inherent (and given) in the one situation but has to be forced (and can be designed [1]) in the other one. Moreover, the important parameter of (transmit) power, obvious over $\mathbb{C}$, is not existent over $\mathbb{F}_q$ or has to be defined explicitly as a cost (cf. Sect. 2.4).

In the following sections, we exemplarily address schemes operating at the intersection of complex-valued and finite-field/modulo processing. First, an overview on modulo-type precoding and a recent proposal using finite-field preprocessing is

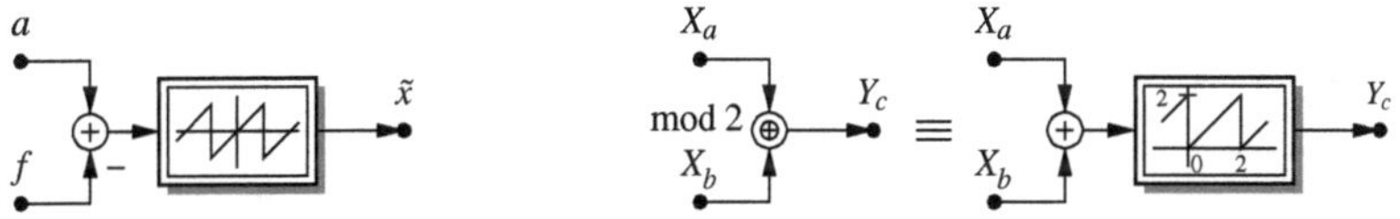

Fig. 2.3 General principles of precoding for the BC (*left*) and (binary) network coding (*right*)

given; the connections and differences of the approaches are illustrated. The advantage of precoding when additional constraints are imposed is addressed. Finally, the application of precoding to finite-field channels is discussed.

2.2 Connection Between Complex-Valued and Finite-Field Channels in Precoding

Precoding for the broadcast channel has now been studied since more than one decade and has undergone significant developments. Still, one of the fundamental bases for the treatment of this scenario is the *uplink-downlink duality* [38, 46, 47, 56]. As a result, schemes developed for the joint reception in a multipoint-to-point scenario are transferred (dualized) to joint transmitter-side processing. However, very recently, a new concept entered the scene: the connection between complex-valued and finite-field calculations. In this section, we briefly review the different variants of preprocessing for the broadcast channel and highlight the respective advantages. In particular, we discuss the combination of channel coding with the respective precoding scheme.

2.2.1 *Conventional Schemes*

The simplest version of precoding for the broadcast channel is *linear preequalization (LPE)*, cf. Fig. 2.4, first row. If the preprocessing matrix $\boldsymbol{W}$ is chosen according to the *zero-forcing (ZF)* criterion, it is simply the (right pseudo) inverse of the channel matrix $\boldsymbol{H}$. In MIMO communications, significant gains can be achieved optimizing the processing according to the *minimum mean-squared error (MMSE)* criterion, trading off transmit power enhancement (the dual to noise enhancement in receiver-side equalization) and residual interference. Since decoupled AWGN channels for the users are created, each user can employ any conventional channel coding scheme (ENC_k) and mapping to signal points ($\mathcal{M}_k$) known from the AWGN channel.

Some improvements can be obtained if *Tomlinson-Harashima Precoding (THP)* is used, see Fig. 2.4, second row. The main idea is to employ *modulo arithmetics* which can be interpreted as representing data by a multiplicity of signal points and selecting the most suited one in a symbol-by-symbol fashion [19]. Via the feedforward matrix

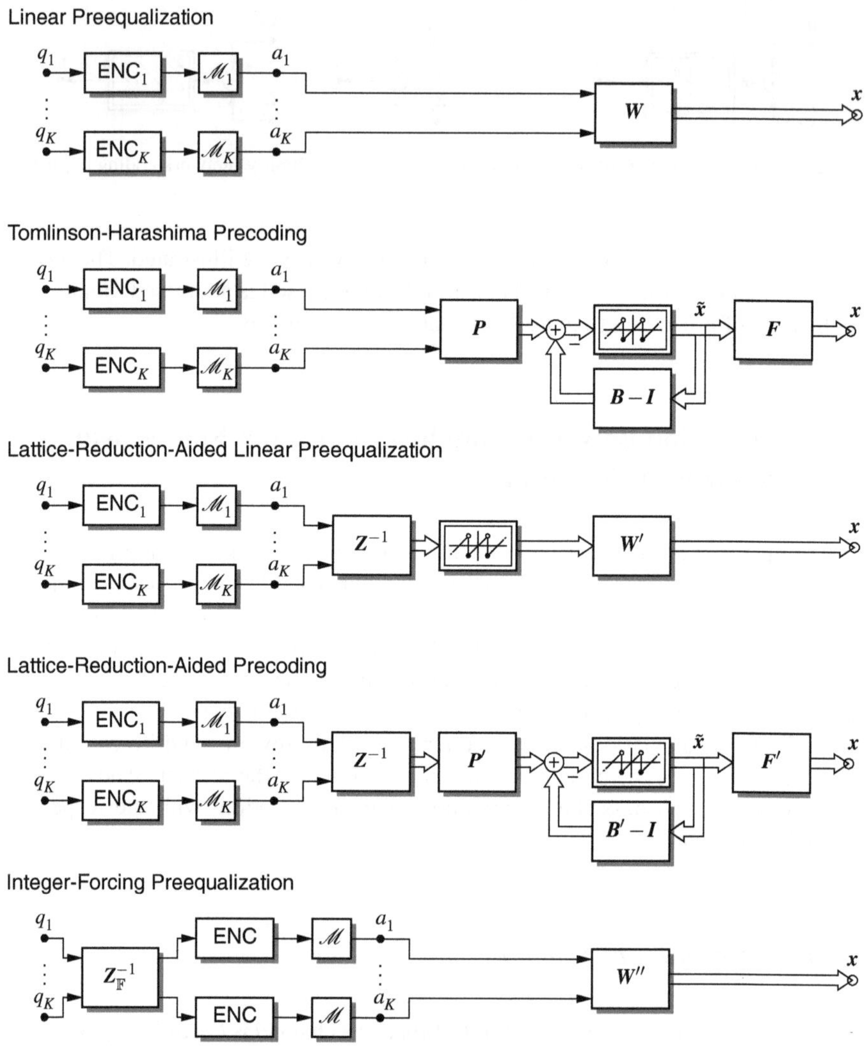

Fig. 2.4 Visualization of the various precoding concepts for the broadcast channel. *Top to bottom* linear preequalization; Tomlinson-Harashima precoding; lattice-reduction-aided linear preequalization; lattice-reduction-aided precoding; integer-forcing (linear) preequalization

$\boldsymbol{F}$, the channel matrix is transformed to lower triangular (spatially causal) form. The transmit symbols are generated successively via the feedback loop (lower triangular, unit main diagonal feedback matrix $\boldsymbol{B}$); the modulo device restricts the amplitude and hence the transmit power. The optimization of the encoding order is described by the permutation matrix $\boldsymbol{P}$. Noteworthy, as in LPE, channel coding can be employed on top of the equalization scheme. However, due to the inherent modulo congruence, the set partition in the coded modulation scheme has to match with the multiple representation, i.e., the partitions also have to be modulo periodic and the receiver-side decoding (metric calculation $\mathcal{L}_k$ and DEC_k in Fig. 2.1) has to take the periodic extension into account [19].

Using uncoded transmission, both schemes, LPE and THP (regardless whether optimized according to the ZF or MMSE criterion), exhibit only a diversity order of one.

2.2.2 *Lattice-Reduction-Aided Schemes*

In 2002/03 the concept of *lattice-reduction-aided (LRA)* equalization has been introduced in [55] and then generalized and transferred to precoding in [41, 51, 52]. The main idea is, not to equalize the channel itself, but to find a more suited description.

When using signal constellations drawn from a regular grid, the signal points at the output of a MIMO channel form (a subset of) a *lattice*. However, each lattice can be described by different bases—for equalization basis vectors as short and as close to orthogonality as possible are highly preferable. The task of finding such a basis is known as *lattice (basis) reduction*. Efficient algorithms for decomposing a matrix $\boldsymbol{H}$ (whose rows span the lattice) according to $\boldsymbol{H} = \boldsymbol{Z} \cdot \boldsymbol{H}_{\mathrm{red}}$ exist [3, 32, 54], where $\boldsymbol{Z}$ is a change-of-basis matrix with integer entries and a unit-magnitude determinant—a so-called *unimodular* matrix—and $\boldsymbol{H}_{\mathrm{red}}$ has desired properties (basically, $\boldsymbol{H}_{\mathrm{red}}^{-1}$ should cause least noise enhancement). Applying this representation in receiver-side equalization, not the transmitted signal constellation is equalized but "only" the *signal grid*. In other words, integer linear combinations of the signal points are detected. These linear combinations are, in a final step, resolved by the inverse change-of-basis matrix, which is unimodular, too.

Utilizing the uplink-downlink duality, *LRA linear preequalization* and *LRA precoding* result, cf. third and fourth row in Fig. 2.4. Integer linear combinations of the (encoded and mapped) data symbols are calculated (unimodular matrix $\boldsymbol{Z}^{-1}$) followed by either linear preequalization or THP of the so-called *reduced channel* given by $\boldsymbol{H}_{\mathrm{red}} = \boldsymbol{Z}^{-1} \cdot \boldsymbol{H}$. It has been shown that LRA equalization/precoding achieves the full diversity order offered by the MIMO channel [44]. Noteworthy, as in the case of conventional schemes, LRA schemes can be optimized according to the ZF or the MMSE criterion, see, e.g., [4, 5, 7, 53].

It is true that LRA equalization so far has been studied almost exclusively for uncoded transmission. This is because channel coding can easily be applied independently of/cascaded with the equalization part. The only restriction, as already present in THP, is that the coded modulation fits to the periodic extension/modulo reduction. Since all data streams are encoded and decoded individually (from a_k to $y_k, k = 1, \ldots, K$, basically independent parallel AWGN channels with periodic continuation of the signal constellation are present) each user may employ an individual code.

2.2.3 Integer-Forcing Schemes

Very recently, a new philosophy for equalization in MIMO schemes, named *integer forcing (IF)*, has been proposed [59]. It originates from *compute-and-forward* relaying schemes [33] and physical-layer network coding [18] and is closely related to LRA equalization with subtile, however essential differences. A precoding version of the integer-forcing strategy for distributed antenna systems, termed *reverse compute-and-forward (RCoF)*, has been presented in [25].

The main difference, when comparing the third (LRA LPE) and last (IF) row in Fig. 2.4, is that the order of channel encoding and equalization of the integer part is reversed. Not the encoded symbols are linearly combined but the information symbols prior to encoding. In doing so, it has to be assumed that in IF the source symbols are drawn from the finite field $\mathbb{F}_p$, where p is a prime. In our view, the restriction to a prime field is the much more important conceptional difference between LRA and IF than that of studying uncoded and coded transmission, respectively; we cannot share the view given in [59, BottomofPage7678]. Indeed, considering uncoded transmission, a signal constellation whose cardinality M is a prime, and employing a so-called linear mapping [18] (in its simplest case the natural mapping of the finite-field elements 0, 1, ..., $p-1$ to the integers 0, 1, ..., $p-1$), both versions are identical.

In LRA equalization integer linear combinations of data symbols/code words in the signal space are decoded. In order to resolve these linear combinations by calculating only integer combinations (over $\mathbb{R}$), the integer matrix (change-of-basis matrix) $\boldsymbol{Z}$ has to be unimodular (see above). In IF this resolving is done over the finite field—an inverse matrix has to exist over $\mathbb{F}_p$, only. In turn, the finite-field equivalent to the integer matrix, $\boldsymbol{Z}_{\mathbb{F}}$, may have any determinant.

A comparison of the respective rows in Fig. 2.4 reveals that in LRA and IF modulo-M integer linear combinations are calculated prior to the linear equalization of the residual (reduced) part of the channel. First, the modulo reduction (equivalence modulo M) relaxes the constraints on the existence of $\boldsymbol{Z}^{-1}$; not an inverse over $\mathbb{R}$ is required but only over $\mathbb{Z}_M$, the integers modulo M. If $M = p$ is a prime we have $\det(\boldsymbol{Z}) \in \mathbb{Z} \setminus M\mathbb{Z}$; in this case (and assuming linear encoding) the operations in LRA and IF, respectively, are completely equivalent as the arithmetic of the prime field $\mathbb{F}_p$ is identical to that of calculating over $\mathbb{Z} \mod p$. Second, having a close look at the operation of LRA LPE, it is revealed that $\boldsymbol{Z}$ may have any determinant unequal to zero. Performing all operations over the integers preserves the discrete nature of the signal constellation; inverting a non-unit determinant matrix $\boldsymbol{Z}$ over the reals destroys the regular structure of the symbols prior to $\boldsymbol{W}'$ but precoding still works. Third, it can be shown that IF even works for the more practical case of M not being a prime, in particular $M = 2^m$, $m \in \mathbb{N}$. Here, $\det(\boldsymbol{Z}) \in 2\mathbb{Z} + 1$ is required. Numerical simulations do not show a clear advantage for one of the strategies.

Noteworthy, by replacing the preequalization matrix $\boldsymbol{W}''$ with a THP structure, an integer-forcing precoding strategy would be possible as well (not shown here; cf. also the receiver-side counterpart [34]).

Dropping the unimodular constraint on $\boldsymbol{Z}$ in the IF strategy has two main disadvantages. First, up to now, no well-performing and efficient strategy on calculating $\boldsymbol{Z}$ with arbitrary determinant is available; see the "full" search in [59], the distributed approach in [25], or again the lattice-reduction strategy in [37]. Despite channel matrices can be constructed where IF provides an infinite gain over LRA equalization (cf. [59, Appendix A]), no gains for i.i.d. Rayleigh fading channels could be observed up to now. Second, IF precoding requires the employment of identical channel codes for the users; in case of different desired rates, the lower-rate codes have to be subcodes of the highest-rate code [25]. This gives less flexibility in the system design than in an LRA scheme. However, in distributed antenna systems (coordinated multipoint, network MIMO, see also Sect. 2.3) IF is of interest. If, as in [25], the residual equalization via $\boldsymbol{W}'' \approx \boldsymbol{H}_{\text{red}}^{-1}$ is not present, only finite-field symbols have to be communicated from the central unit over the backhaul network to the base stations, which perform the encoding of the precombined source words.

To conclude, the IF philosophy has changed the perspective on LRA schemes and has sparked new research eliminating the unnecessary restrictions imposed up to now. However, as LRA schemes are already diversity-optimum, no further immense gains can be expected.

2.2.4 Summary

As a summary, Table 2.1 compares the discussed precoding schemes w.r.t. the equalization task, the utilized degree of freedom, and the constraints on the signal constellations and the codes.

Table 2.1 Comparison of linear preequalization (LPE), Tomlinson-Harashima precoding (THP), LRA preequalization/precoding (LRA), and integer-forcing preequalization (IF) w.r.t. the equalization task, the utilized degree of freedom, and the constraints on the signal constellations and the codes

	Equalization task (ZF case)	Degree of freedom	Constraints on signal constellation	Constraints on codes
LPE	$\boldsymbol{HW} = \boldsymbol{I}$ full equal	–	–	Codes for AWGN channel can be used
THP	$\boldsymbol{PHF} = \boldsymbol{B}$ Lower triangular	Modulo-congruent signal points	Periodic continuation required	Have to work under modulo reduction
LRA	$\boldsymbol{HW}' = \boldsymbol{Z}$ Integer unimodular matrix	Integer linear combinations	Integer lin. comb. and periodic continuation required	Have to work under modulo reduction
IF	$\boldsymbol{HW}'' = \boldsymbol{Z}$ Integer matrix	Integer linear combinations	Integer lin. comb. and periodic continuation required	Same code for UEs; Has to work under modulo reduction

2.3 Precoding for Distributed MIMO

In *distributed MIMO* (also denoted as *network MIMO* or *coordinated multipoint*), e.g., [29], a number of non-collocated *base stations (BSs)* are grouped and jointly communicate with a number of (non-cooperating) *user equipments (UEs)*, distributed over some service area. Assuming perfect coordination of the BSs, a variant of the BC is present and hence all preequalization/precoding strategies (cf. Fig. 2.4) for the MIMO broadcast channel can be employed.

However, since the transmit antennas are not at the same location some important extra considerations have to be taken into account. First, in addition to the usual *sum power constraint (SPC)*, here a *per-antenna power constraint (APC)* is of much more interest; this problem is addressed in Sect. 2.3.1. Second, since the data has to be communicated to/between the BSs, the coordination effort and the respectively induced backhaul traffic has to be taken into account; see Sects. 2.3.2 and 2.3.3.

The most obvious network MIMO downlink scenario, used, e.g., in [6, 11, 12, 42], is illustrated in the left part of Fig. 2.5 (denomination of the symbols as in Fig. 2.4). In our example, $B = 3$ BSs (with sectorized antennas and taken from a larger arrangement of BSs; distance r_{BS}) serve the shaded area, where $K = 3$ single-antenna UEs are located. Since network MIMO is particularly suited for supplying cell-edge regions, UEs close to BSs (distance smaller than $r_{\min}$) may be excluded from joint processing, cf. [6, 11]. The channel matrix $\boldsymbol{H}$ for this scenario can be derived from the geometry and includes the antenna pattern, attenuation, path loss, and shadowing as well as fast fading effects. Details can be found in [6, 13, 14, 42].

In network MIMO, the BSs are usually connected to a central unit (CU) via wired backhaul links enabling the coordinated transmission.

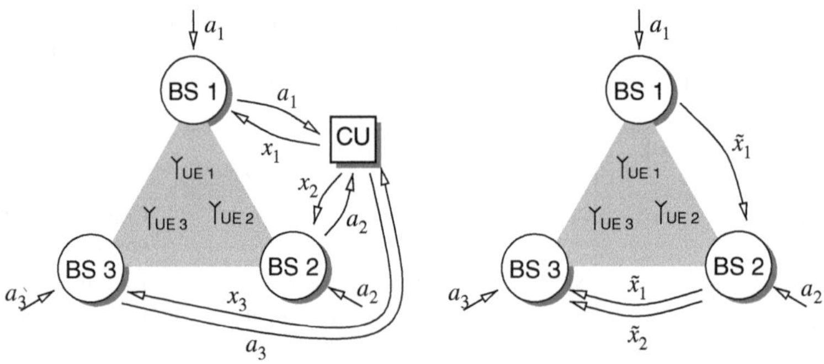

Fig. 2.5 Geometrical network MIMO system model for $B = 3$ and $K = 3$. *Left* full coordination via a central unit (CU); *Right* hierarchical scheme without central instances

2.3.1 Optimization Under Per-Antenna Power Constraints

In the literature, the optimization of LPE subject to a per-antenna power constraint has been given in [45, 57]. Recently, such an additional constraint has been applied to the more advanced precoding versions, cf. [6].

The performance is preferably expressed in terms of the *signal-to-interference-plus-noise ratio (SINR)* of the users. Let the users be renumbered according an optimized processing order (obtained, e.g., from conventional THP and described by a permutation matrix $\boldsymbol{P}$). Then, the SINR of user u in case of THP, cf. Fig. 2.4, second row, is given by

$$\mathrm{SINR}_k \stackrel{\text{def}}{=} \frac{|c_{k,k}|^2}{\frac{\sigma_n^2}{\sigma_a^2} + \sum\limits_{l,\, l>k} |c_{k,l}|^2}\,, \tag{2.4}$$

where $\boldsymbol{C} = [c_{k,l}] = \boldsymbol{HF}$ denotes the *end-to-end cascade* from data to receive symbols. Due to successive encoding in THP, users with lower index (processed first) do not contribute to the interference. This main advantage over linear schemes can be used to "shape" the interference [6].

Knowing the actual channel realization, it is appropriate to adjust the precoding scheme such that the minimum SINR over all users is maximized. Hence, the optimization task (minSINR criterion) for finding the optimum feedforward matrix $\boldsymbol{F}$ reads

$$\underset{\boldsymbol{F}}{\operatorname{argmax}}\ \mathrm{minSINR} \stackrel{\text{def}}{=} \underset{\boldsymbol{F}}{\operatorname{argmax}}\ \min_{k=1,\dots,K} \mathrm{SINR}_k\,, \tag{2.5}$$

taking the SPC $\mathrm{E}\{\boldsymbol{x}^{\mathsf{H}}\boldsymbol{x}\} \le P_{\text{sum}}$ and the APC $\mathrm{E}\{|x_b|^2\} \le P_{\text{per}}$, $b = 1, \dots, B$, into account. After the determination of the feedforward matrix $\boldsymbol{F}$, the gain matrix $\boldsymbol{G} = \mathrm{diag}(g_1, \dots, g_K)$, cf. Fig. 2.1, is adjusted such that $\boldsymbol{GHF}$ has unit main diagonal, and the feedback matrix $\boldsymbol{B}$ is given as the lower triangular part thereof.

This resulting optimization task may be solved by resorting to a *second-order cone program*, cf. [45, Algorithm 2]. The sorting according to the *BLAST algorithm* [21, 23] almost always provides the optimum ordering (permutation matrix $\boldsymbol{P}$) [50].

Figure 2.6 shows the bit error rate (BER) curves[3] assuming uncoded transmission (16-ary QAM constellation with Gray labeling and variance σ_a^2) for the $K \times B = 3 \times 3$ network MIMO system model. The channel parameters are taken from [6, 13, 14]. Either no (gray curves) or a per-antenna power constraint of $P_{\text{per}} = 1.5\,\sigma_a^2$ (black curves) is imposed. In each case a sum power constraint $P_{\text{sum}} = 3\,\sigma_a^2$ is active. LPE and THP are optimized according to the minSINR criterion, see also [45]. However, in case of LPE this optimization does not provide substantial gains over the conventional sum-MSE solution, where, in case of an APC, the feedforward matrix is simply scaled to meet the constraint, cf. [6].

[3] All results in this section are displayed over the ratio of *transmitted energy per information bit* E_{b} and one-sided noise power spectral density N_0', where the average channel attenuation is eliminated. For details on the normalization, see [6].

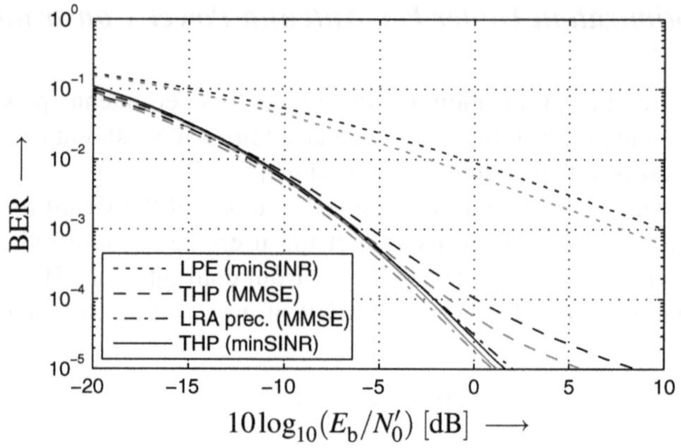

Fig. 2.6 Bit error rate over the ratio of transmitted energy per information bit E_b and equivalent noise power spectral density N_0'. Geometrical channel model with $r_\mathrm{min} = 125$ m. *Dotted* minSINR LPE; *Dashed* MMSE THP; *Dashed-dotted* MMSE LRA precoding. *Solid* minSINR THP. *Gray* only SPC $P_\mathrm{sum} = 3\,\sigma_a^2$ active; *black* additional APC $P_\mathrm{per} = 1.5\,\sigma_a^2$. Inter-site distance $r_\mathrm{BS} = 500$ m

As can be seen from the figure, the APC $P_\mathrm{per} = 1.5\,\sigma_a^2$ causes a loss of about 2.5 dB for LPE. Performance can significantly be improved if conventional MMSE THP is employed, which provides a gain of up to 15 dB over LPE. However, the curves for THP and LPE flatten out to diversity order one. Very good performance can also be obtained using LRA precoding, where the feedforward matrix is also simply scaled to meet the APC [6]. This scheme has full diversity order 3; no flattening of the BER curves occur. In the low BER regime, THP optimized according to the minSINR criterion gives even better results as both power constraints are explicitly taken into account in the optimization. An APC of $P_\mathrm{per} = 1.5\,\sigma_a^2$ does almost not lead to a performance degradation.

In summary, it can be stated that compared to LPE non-linear precoding is much more attractive in the network MIMO scenario with per-antenna power constraint. The individual powers at the antennas can even be restricted down to $P_\mathrm{per} = \sigma_a^2 \ldots 2\,\sigma_a^2$ without too much loss in performance.

2.3.2 Coordination Effort and Hierarchical Precoding

In network MIMO using TH-type precoding, one has to distinguish between two different coordination/joint processing tasks: (i) the precoding matrices $\boldsymbol{B}$ and $\boldsymbol{F}$ together with the optimum permutation $\boldsymbol{P}$ have to be computed, and (ii) given the data symbols (i.e., the data vector $\boldsymbol{a}$), the vector $\boldsymbol{x}$ of transmit symbols has to be calculated.

Assuming a block-fading channel (constant channel matrix $\boldsymbol{H}$ over a transmission burst), the calculation of the matrices has only to be done once per channel realization. In contrast, the calculation of $\boldsymbol{x}$ has to be carried out at each time step. To this end, all data may be communicated to a CU (cf. Fig. 2.5), processed, and then transferred back to the BSs. Assuming that the symbols are initially available at the BSs, this procedure requires the transmission of six complex numbers per time step. The same amount is also required for the decentralized coordination described in [11, 12], where each BS passes its knowledge to the other BSs.

Conceptually, the main part of THP is the successive encoding (feedback structure) of the data symbols. However, due to the feedforward matrix $\boldsymbol{F}$, for the calculation of each transmit symbol x_b all precoded symbols $\tilde{x}_k$ are required. The exchange of these symbols causes the huge backhaul traffic in decentralized coordination. Let β denote the number of complex symbols communicated per time step, we have $\beta = 6$ for the example at hand.

To overcome this problem, in [11, 12] a *hierarchical* distribution of knowledge among the BSs has been proposed. The thereby required backhaul transmission is indicated in the right part of Fig. 2.5 and amounts to only $\beta = 3$. As can be seen, the level of knowledge on the data of the other users increases from BS to BS. Such a successive procedure is only possible if the (usually joint) feedforward processing is modified; it is easy to see that such a structure is reflected in the fact that $\boldsymbol{F}$ has *lower triangular* structure.

Basically, the minSINR optimization briefly discussed above can (with slight modifications) be applied to this setting, too [11]. However, in addition to the $K! = 6$ possible permutations of the users (as in conventional THP), here also the BSs are non-equivalent and an optimized ordering (relabeling) of them has to be done. In [11], a greedy strategy has been presented to preselect a set of suited sortings among all $K!B! = 36$ options. The minSINR optimization is then carried out for these candidates and the sorting leading to the best performance (largest minSINR) is chosen for transmission.

In Fig. 2.7, the BER obtained by numerical simulation is shown for the mentioned hierarchical scheme. Both the results for full search (all sortings tested) and the greedy preselection (six candidates) are shown. For comparison, the results for fully-coordinated schemes are repeated from Fig. 2.6 (APC $P_{\text{per}} = 1.5\,\sigma_a^2$). As can be seen, employing hierarchical precoding (full search), there is only a loss of about 2.5 dB in the low BER regime (BER $\approx 10^{-4}$) when compared to fully joint preprocessing via THP (minSINR criterion). However, compared to LPE, which also requires $\beta = 6$ backhaul transmissions, hierarchical THP shows significantly better results. Hierarchical THP even outperforms fully-coordinated THP optimized according to the sum MSE when considering the very-low-BER range (BER $< 10^{-4}$). The greedy preselection does not cost too much in performance but lowers the initial computation effort significantly.

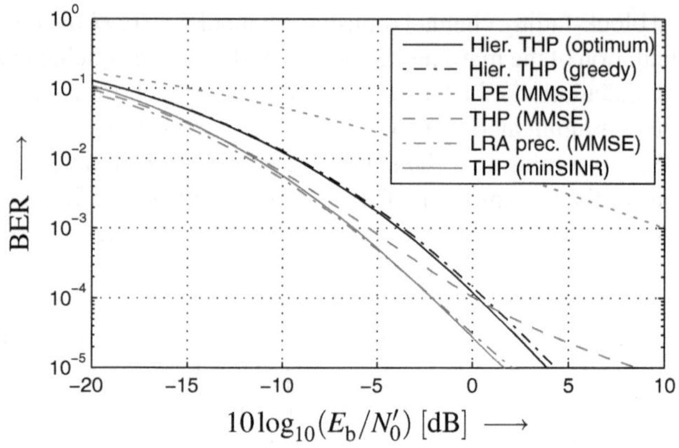

Fig. 2.7 Bit error rate over the ratio of transmitted energy per information bit E_b and equivalent noise power spectral density N_0'. Geometrical channel model with $r_{min} = 125$ m. *Black solid* hierarchical THP with full search; all sortings tested. *Black dashed-dotted* hierarchical THP with greedy strategy; six candidates. *Gray* curves from Fig. 2.6 for comparison. SPC $P_{sum} = 3\,\sigma_a^2$; APC $P_{per} = 1.5\,\sigma_a^2$. Inter-site distance $r_{BS} = 500$ m

2.3.3 *Selection of the Coordination Strategy*

In many applications, a maximum tolerable BER is defined, corresponding to a minimum requirement on the SINR. If a "good" channel realization is present, the required performance may already be achieved without any precoding, hence requiring no backhaul traffic ($\beta = 0$). For "medium" channel instances, hierarchical precoding is often sufficient ($\beta = 3$); only in case of "bad" channel conditions, fully coordinated THP ($\beta = 6$) may have to be used. In other words, the desired performance should be achieved with the smallest amount of joint processing/backhaul traffic as possible.

In [12], a selection algorithm has been proposed, which either selects hierarchical or full THP. To this end, a minimally required SINR, denoted as $\text{minSINR}_{\text{thr}}$, is specified, e.g., based on system requirements or on the statistical distribution of minSINR (cf. (2.5)). As long as the hierarchical scheme achieves this SINR, it is used; otherwise fully coordinated THP is chosen. Additionally, "very bad" channels for which even full THP gives $\text{minSINR} < \text{minSINR}_{\text{out}}$ may be excluded from transmission at all (e.g., via higher protocol layers) and an *outage* may be declared.

Via the choice of $\text{minSINR}_{\text{thr}}$ and $\text{minSINR}_{\text{out}}$ a trade-off between performance and backhaul traffic is enabled. When lowering $\text{minSINR}_{\text{thr}}$ the hierarchical scheme is more often sufficient and the backhaul signaling is reduced. Since the SINR depends on the channel SNR, a trade-off between minimally required SNR (e.g., to guarantee $\text{BER} \leq 10^{-4}$) and backhaul traffic $\text{E}\{\beta\}$ is possible, see Fig. 2.8.

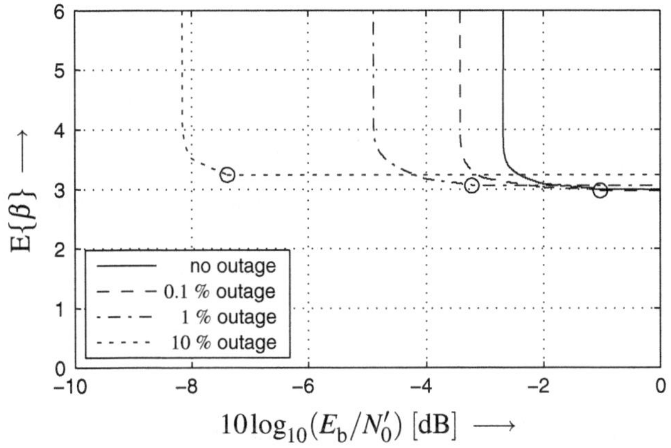

Fig. 2.8 Trade-off between minimum required SNR and average backhaul traffic to obtain a BER of 10^{-4}. 16QAM constellation; APC $P_{\mathrm{B}} = 1.5\,\sigma_a^2$. *Curves* go over into a horizontal branch when $\mathrm{minSINR}_{\mathrm{thr}} < \mathrm{minSINR}_{\mathrm{out}}$ (*marked by circles*)

When setting $\mathrm{minSINR}_{\mathrm{thr}} = \infty$, the selection is disabled and conventional THP (with APC) is chosen leading to $\beta = 6$ but the lowest required SNR. If $\mathrm{minSINR}_{\mathrm{thr}}$ is lowered, the hierarchical scheme is selected more often and $\mathrm{E}\{\beta\}$ decreases but the required SNR increases slightly. Only if $\mathrm{minSINR}_{\mathrm{thr}}$ is very low, the hierarchical scheme is almost always active and $\mathrm{E}\{\beta\} \rightarrow 3$. Allowing outage (0.1, 1, and 10 % in Fig. 2.8), the required SNR can significantly be lowered as the worst-case channels are eliminated. All curves go over into a horizontal branch when $\mathrm{minSINR}_{\mathrm{thr}} < \mathrm{minSINR}_{\mathrm{out}}$. For more details see [12].

2.3.4 Quantization of Precoded Symbols

The schemes discussed so far need to communicate *complex-valued* symbols via the backhaul. In practice, a quantization of these symbols is indispensable. Hence, another degree of freedom—the tradeoff between quantization error/performance and the required data rate (number of bits per symbol)—is enabled.

In this context, precoding based on integer forcing (cf. Sect. 2.2), specifically the *reverse compute-and-forward (RCoF) scheme* in [25] is of special interest. Since precoding is purely done over the finite field, only a few bits are required to communicate such symbols. However, in RCoF further joint preequalization via $\boldsymbol{W}''$ (cf. Fig. 2.4 last row) or via a lower triangular $\boldsymbol{F}$ (cf. Sect. 2.3.2) is not possible. Hence, performance is degraded for lowering the backhaul traffic.

2.4 Precoding for Finite-Field Channels

In Sect. 2.1.3 we discussed the usefulness of dualities and analogies in communications and pointed out that PC and NC are related in some sense. We now substantiate this point of view by presenting two precoding approaches for finite-field matrix channels, which can be seen as counterparts of conventional approaches for real (or complex) valued channels: *differential linear network coding (DLNC)* related to differential encoding, and *selection precoding (SP)* as a counterpart to vector precoding. All definitions of Sect. 2.1.2 apply here again.

2.4.1 Differential Linear Network Coding

The common approach for communicating over the MAMC is the usage of *lifted rank metric codes* [30, 39], which results in a rate loss $L_{\text{lift}} \stackrel{\text{def}}{=} n/l$, where l is the packet length and n is the number of simultaneously transmitted packets (cf. Sect. 2.1.2). This loss arises due to the lifting operation, i.e., transmitting the identity matrix at the beginning of each generation, which can be seen as pilot symbols for channel sounding. In the following, we present an approach that overcomes this drawback.

In case of *differential phase-shift keying (DPSK)* [36], information is transmitted in the phase transition between two consecutive transmit symbols, rather than in the absolute phase. Thus, any constant phase offset caused by the channel (and not known to the receiver) can be eliminated, at the cost of a (slightly) degraded performance compared to coherent transmission. We adopt this idea to the RLNC scenario, which is described by the MAMC (cf. Sect. 2.1.2, working over $\mathbb{F}_q$)

$$\boldsymbol{Y} = \boldsymbol{A}\boldsymbol{X} + \boldsymbol{E}\ . \tag{2.6}$$

Information is not represented directly in the matrix $\boldsymbol{X}$, but in the transition between two consecutive transmit matrices. Such an idea was also present in the field of differential space-time codes [26, 27, 31, 43]. As a consequence, a *constant* multiplicative distortion $\boldsymbol{A}$ is irrelevant to the receiver. To obtain a constant network channel matrix, we force the intermediate nodes to keep their linear coding coefficients constant during the transmission. However, changes in the network channel matrix can occur when intermediate nodes leave or join the network, e.g., due to system failures, low battery, or system reboots.

If nodes leave or join the network between generations, the change of the network channel matrix can be expressed as

$$\boldsymbol{A}_i = \boldsymbol{A}_{i-1} + \Delta\boldsymbol{A}_i\ , \tag{2.7}$$

where $\boldsymbol{A}_i$ denotes the network channel matrix in generation i and $\Delta\boldsymbol{A}_i$ the *channel deviation*. In [8], it has been shown that the rank of the channel deviation is upper bounded by the sum of the *node weights* of the leaving/joining nodes. The node weight of a node ν is defined as $w(\nu) \stackrel{\text{def}}{=} \min\{n, \text{In}\{\nu\}, \text{Out}\{\nu\}\}$, where $\text{In}\{\nu\}$ and $\text{Out}\{\nu\}$ denote the numbers of ν's incoming and outgoing edges, respectively. We call networks with a very small probability of leaving/joining nodes as *slowly-varying networks*.

For brevity let us restrict to square matrices, i.e., $n = N = l$; the generalization to non-square matrices can be found in [8, 10]. For differential modulation, we assume to have a sequence of full rank source words $\boldsymbol{S}_i$. The DLNC transmit symbol in generation i is generated as

$$\boldsymbol{X}_i = \boldsymbol{X}_{i-1} \cdot \boldsymbol{S}_i\,, \tag{2.8}$$

where $\boldsymbol{X}_0 \stackrel{\text{def}}{=} \boldsymbol{I}_n$ ($\boldsymbol{I}_n$: identity matrix of dimension n) is the initialization word.

Differential demodulation at the destination node is done by calculating the product $\boldsymbol{Y}_{i-1}^{+} \cdot \boldsymbol{Y}_i$, where $\boldsymbol{Y}_{i-1}^{+}$ is the pseudoinverse of the previously received matrix. This results in the demodulated symbol $\tilde{\boldsymbol{S}}_i$ [8]

$$\tilde{\boldsymbol{S}}_i \stackrel{\text{def}}{=} \boldsymbol{Y}_{i-1}^{+} \cdot \boldsymbol{Y}_i = \boldsymbol{S}_i + \tilde{\boldsymbol{E}}_i\,, \tag{2.9}$$

where $\tilde{\boldsymbol{E}}_i$ is the *effective error matrix*. Thus, the application of differential modulation and demodulation transforms the MAMC into an AMC with additive error $\tilde{\boldsymbol{E}}_i$, as visualized in Fig. 2.9. The effective error comprises the effects of the differential demodulation process and of the slowly varying network. The former causes the additive error to have (approximately) doubled rank compared to the underlying channel, cf. noncoherent demodulation of DPSK where the error variance is doubled. The latter causes a rarely occurring impulsive rank error, which is equal to the rank of $\Delta\boldsymbol{A}$.

When applying the differential approach, we are restricted to full-rank transmit matrices $\boldsymbol{S}_i$, and thus, suffer a rate loss, well approximated by $L_{\text{dlnc}} = 1/q$, cf. [30]. This loss is negligible as long as q is large, which is a prerequisite in RLNC. More challenging are the circumstances that the additive errors occur in pairs, and that the scheme is sensitive to network topology variations. Due to the resulting effective error structure in DLNC—the doubled additive error, plus the rarely occurring impulsive

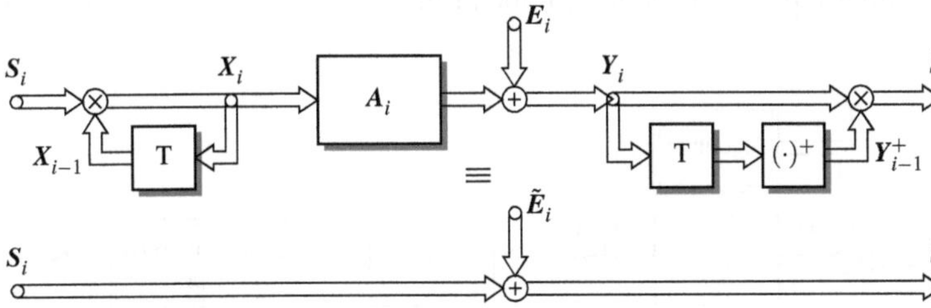

Fig. 2.9 Block diagram of a DLNC scheme, converting the MAMC into an AMC. The "T"-block depicts a unit delay element

error peaks caused by leaving/joining nodes—the commonly applied rank-metric Gabidulin block codes [22] are no longer the best choice. Instead of these block codes, convolutional-type codes are preferable. In [8], it was shown that in particular *partial unit memory* Gabidulin codes [48] match to the present situation. Using them, higher reliabilities and/or higher rates are possible via DLNC compared to the lifting approach in slowly varying network coding applications.

2.4.2 Selection Precoding

In Sect. 2.2 the concept of modulo-type precoding for the MIMO broadcast channel was explained. The main idea in such schemes—THP and its generalization—is that each data symbol is represented by multiple, modulo congruent signal points; from this set the most suited one is selected for transmission. In doing so, the channel may not only be equalized but, additionally, desired properties of the transmit signal can be achieved, i.e., some kind of signal shaping is performed [19]. This principle can be transferred to finite-field channels, leading to a scheme denoted as *selection precoding (SP)* [2].

A prerequisite is the definition of modulo-congruent signal points in finite fields $\mathbb{F}_{2^m}$. This can be done in terms of cosets and the respective *coset decomposition.* Given an additive subgroup $\mathbb{G}_{2^\mu}$ in $\mathbb{F}_{2^m}$, $\mu < m$, the respective cosets are defined as

$$\mathcal{C}^{(\zeta)} \stackrel{\text{def}}{=} \{e_\zeta + a \mid a \in \mathbb{G}_{2^\mu}\}\,, \quad e_\zeta \in \mathbb{F}_{2^m}\,. \tag{2.10}$$

The *coset leaders* e_ζ are then used as data-carrying symbols; all elements within the coset are modulo congruent and represent the same information. Given the data symbols e_ζ any element from $\mathbb{G}_{2^\mu}$ can be added without change of information. Conversely, given any element $x \in \mathbb{F}_{2^m}$ a *finite field modulo operation* can be defined such that the coset leader e_ζ of the corresponding coset, i.e., $x \in \mathcal{C}^{(\zeta)}$, is returned.

The structure of the selection precoding scheme is depicted in Fig. 2.10. For brevity, we restrict ourselves to the MMC and square network channel matrices $\boldsymbol{A}$; the generalization is easily possible. The elements of the source matrix $\boldsymbol{S}$ are drawn from the set of coset leaders e_ζ. A precoding matrix $\boldsymbol{U}$, whose elements are drawn from the subgroup $\mathbb{G}_{2^\mu}$, is added to obtain the *effective data matrix* $\boldsymbol{V} \stackrel{\text{def}}{=} \boldsymbol{S} + \boldsymbol{U}$.

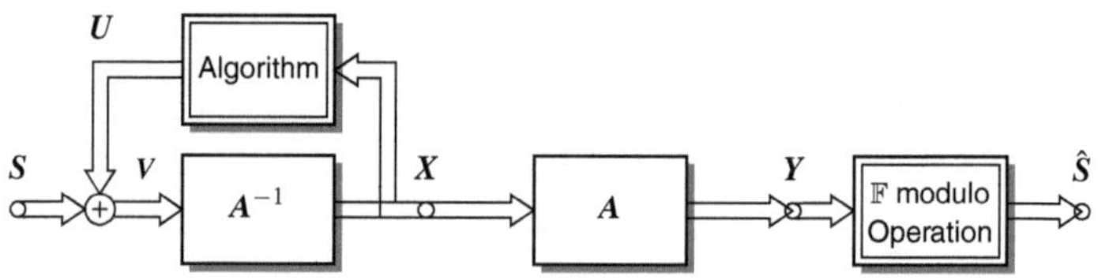

Fig. 2.10 Selection precoding for the multiplicative matrix channel

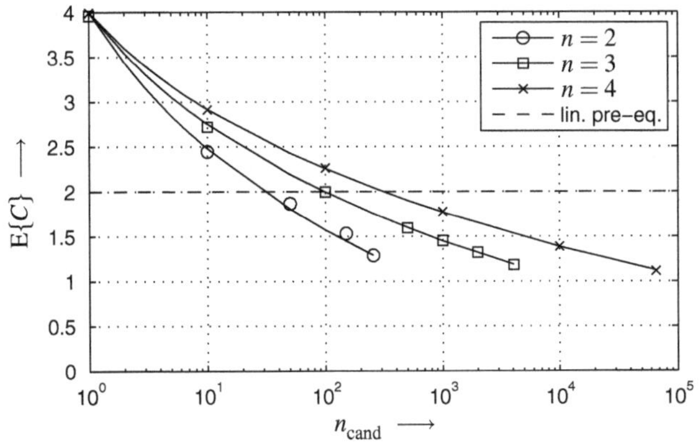

Fig. 2.11 Average cost per symbol over the number of tested candidates for the selection precoding scheme with $n = 2, 3$, and 4. $\mathbb{F}_{2^8}$ and $\mathbb{G}_{16}$, i.e., $m = 8$ and $\mu = 4$

This matrix is precoded via[4] $\boldsymbol{A}^{-1}$, thus, the SP transmit matrix, which is injected into the network, is given as $\boldsymbol{X} = \boldsymbol{A}^{-1}\boldsymbol{V}$. Assuming error-free transmission, the receive matrix is $\boldsymbol{Y} = \boldsymbol{V}$. After an (element-wise) finite-field modulo operation (recovery of the coset leader), the source symbols are perfectly recovered, i.e., $\hat{\boldsymbol{S}} = \boldsymbol{S}$.

In [2], we have shown how to adapt the optimization to the current situation, i.e., how to find the optimum precoding matrix $\boldsymbol{U}_{\text{opt}}$. To this end, an optimization criterion has to be defined. This can be done in terms of a *cost function* $C(\boldsymbol{X})$, which assigns a non-negative, real-valued cost to each finite-field matrix $\boldsymbol{X} \in \mathbb{F}_{2^m}^{n\times n}$. In [2], two examples are given, namely the number of ones (i.e., the Hamming weight) and the number of signal changes in the binary vector representation of the $\mathbb{F}_{2^m}$ elements.[5]

Then, the optimization problem reads

$$\boldsymbol{U}_{\text{opt}} = \underset{\boldsymbol{U}\in\mathbb{G}_{2^\mu}^{n\times\ell}}{\operatorname{argmin}}\ C(\boldsymbol{A}^{-1}(\boldsymbol{S}+\boldsymbol{U}))\ . \tag{2.11}$$

This search may be accomplished in a random fashion, i.e., by randomly generating n_{cand} candidate precoding matrices $\boldsymbol{U}$, and selecting the one, which leads to the lowest transmit cost. Via n_{cand} a performance/complexity trade-off is enabled.

In Fig. 2.11 the average cost $\mathrm{E}\{C\}$ per symbol is plotted over the number n_{cand} of tested candidates per transmission block for $n = 2, 3$, and 4. The coset decomposition of $\mathbb{F}_{2^8}$ is generated with respect to $\mathbb{G}_{16}$. The solid lines represent the analytically derived average cost per symbol, cf. [2]. From the simulations results, which perfectly

[4] Again, we assume the network channel matrix to be a full rank matrix. However, it is known from the conventional setting that non-linear schemes (e.g., THP) can even be used on singular channels where linear schemes fail. This stabilization due to the multiple representation of symbols and the degree of freedom to choose from them is an additional advantage of the present scheme.

[5] Both cost functions have the same distribution for the cost and, thus, lead to the same performance.

match with the analytic solution, we can conclude that employing selection precoding is able to achieve a lower average cost than the reference scheme (dashed lines) at the same rate.

Acknowledgments The work of Robert Fischer and Johannes Huber was supported by the German Research Foundation (DFG) under Grants FI 982/4-1, FI 982/4-2, FI 982/4-3, and HU 634/11-3, respectively.

References

Project-Related Publications

1. Abay Ü, Fischer RFH (2011) Comparison of generalized tomlinson-harashima precoding strategies for the broadcast channel. In: Proceedings of international ITG workshop on smart antennas (WSA), Aachen, Germany, Feb 2011
2. Cyran M, Fischer RFH, Huber JB (2014) Selection precoding for the finite-field multiplicative matrix channel. IEEE Commun Lett 18(2):360–363
3. Fischer RFH (2010) From gram-schmidt orthogonalization via sorting and quantization to lattice reduction. In: Proceedings of joint workshop on coding and communications (JWCC), Santo Stefano Belbo, Italy, Oct 2010
4. Fischer RFH (2011) Efficient Lattice-reduction-aided mmse decision-feedback equalization. In: Proceedings of international conference on acoustics, speech and signal processing (ICASSP), Prag, Czech Republic, May 2011
5. Fischer RFH (2012) Complexity-performance trade-off of algorithms for combined lattice reduction and QR decompositions. Int J Electron Commun (AEÜ) 66(11):871–879
6. Fischer RFH, Heinrichs S (2013) The network MIMO Downlink with per-antenna power constraint: linear preequalization vs. tomlinson-harashima precoding. In: Proceedings of international ITG workshop on smart antennas (WSA), Stuttgart, Germany, Mar 2013
7. Fischer RFH, Windpassinger C, Stierstorfer C, Siegl C, Schenk A, Abay Ü (2011) Lattice-reduction-aided MMSE equalization and the successive estimation of correlated data. Int J Electron Commun (AEÜ) 65(8):688–693
8. Puchinger S, Cyran M, Fischer RFH, Bossert M, Huber JB (2015) Error correction for differential linear network coding in slowly-varying networks. In: Proceedings of international ITG conference on systems, communications and coding (SCC), Hamburg, Germany, Feb 2015
9. Schotsch B, Cyran M, Huber JB, Fischer RFH, Vary P (2015) An upper bound on the outage probability of random linear network codes with known incidence matrices. In: Proceedings of international ITG conference on systems, communications and coding (SCC), Hamburg, Germany, Feb 2015
10. Seidl M, Cyran M, Fischer RFH, Huber JB (2013) A differential encoding approach to random linear network coding. In: Proceedings of international ITG Conference on systems, communications and coding (SCC), Munich, Germany, Jan 2013
11. Stern S, Fischer RFH (2015) Hierarchical precoding for the network MIMO downlink. In: Proceedings of international ITG conference on systems, communications and coding (SCC), Hamburg, Germany, Feb 2015
12. Stern S, Fischer RFH (2015) Selection of the coordination strategy in the network MIMO downlink. In: Proceedings of international ITG workshop on smart antennas (WSA), Ilmenau, Germany, Mar 2015
13. 3GPP (2002) Technical Specification Group Radio Access Network, RF System Scenarios, Technical Report TR 25.996 V3.3.0, 2002

Additional Literature

14. 3GPP (2006) Technical Specification Group Radio Access Network, Physical Layer Aspects for Evolved Universal Terrestrial Radio Access (UTRA), Technical Report TR 25.814 V7.1.0, 2006
15. Ahlswede R, Cai N, Li SYR, Yeung RW (2000) Network information flow. IEEE Trans Inf Theory 46(4):1204–1216
16. Ancheta TC (1976) Syndrome-source-coding and its universal generalization. IEEE Trans Inf Theory 22(4):432–436
17. Caire G, Shamai S, Verdú S (2003) Lossless data compression with error correcting codes. In: Proceedings of the international symposium on information theory 2003, Yokohama, Japan, Jun/Jul 2003
18. Feng C, Silva D, Kschischang FR (2013) An algebraic approach to physical-layer network coding. IEEE Trans Inf Theory 59(11):7576–7596
19. Fischer RFH (2002) Precoding and signal shaping for digital transmission. Wiley, New York
20. Fischer RFH, Windpassinger C, Lampe A, Huber JB (2002) Space-time transmission using tomlinson-harashima precoding. In: Proceedings of the international ITG conference on source and channel coding (SCC), Berlin, Germany, pp. 139–147, Jan 2002
21. Foschini G (1996) Layered space-time architecture for wireless communication in a fading environment when using multiple antennas. Bell Laboratories Tech J 41–59
22. Gabidulin EM (1985) Theory of codes with maximum rank distance. Probl Inf Transm 21(1):1–12
23. Golden G, Foschini G, Valenzuela R, Wolniasky P (1999) Detection algorithm and initial laboratory results using the V-BLAST space-time communication architecture. Electron Lett 14–15
24. Gupta A, Verdú S (2011) Operational duality between lossy compression and channel coding. IEEE Trans Inf Theory 57(6):3171–3179
25. Hong SN, Caire G (2013) Compute-and-forward strategies for cooperative distributed antenna systems. IEEE Trans Inf Theory 59(9):5227–5243
26. Hochwald BM, Sweldens W (2000) Differential unitary space-time modulation. IEEE Trans Commun 48(12):2041–2052
27. Hughes BL (2000) Differential space-time modulation. IEEE Trans Inf Theory 46(7):2567–2578
28. Jindal N, Vishwanath S, Goldsmith AJ (2004) On the duality of gaussian multiple-access and broadcast channels. IEEE Trans Inf Theory 50(5):768–783
29. Karakayali MK, Foschini GJ, Valenzuela RA (2006) Network coordination for spectrally efficient communication in cellular systems. IEEE Wirel Commun 13(4):56–61
30. Kötter R, Kschischang FR (2008) Coding for errors and erasures in random linear network coding. IEEE Trans Inf Theory 54(8):3579–3591
31. Lampe L, Schober R, Fischer RFH (2003) Coded differential space-time modulation for flat fading channels. IEEE Trans Wireless Commun 2(3):582–590
32. Lenstra A, Lenstra H, Lovász L (1982) Factoring polynomials with rational coefficients. Mathematische Annalen 515–534
33. Nazer B, Gastpar M (2011) Compute-and-forward: harnessing interference through structured codes. IEEE Trans Inf Theory 57(10):6463–6486
34. Ordentlich O, Erez E, Nazer B (2013) Successive integer-forcing and its sum-rate optimality. In: Annual allerton conference on communication, control, and computing, Allerton, USA, pp 282–292, Oct 2013
35. Peel CB, Hochwald BM, Swindlehurst AL (2005) A Vector-perturbation technique for near-capacity multiantenna multiuser communication-part i: channel inversion and regularization. IEEE Trans Commun 53(1):195–202
36. Proakis J, Salehi M (2008) Digital communications, 5th edn. McGraw-Hill
37. Sakzad A, Harshan J, Viterbo E (2013) Integer-forcing MIMO linear receivers based on lattice reduction. IEEE Trans Wireless Commun 12(10):4905–4915

38. Schubert M, Boche H (2002) A unifying theory for uplink and downlink multi-user beamforming. In: IEEE International zurich seminar (IZS), Zurich, Switzerland, Feb 2002
39. Silva D, Kschischang FR, Kötter R (2008) A rank-metric approach to error control in random linear network coding. IEEE Trans Inf Theory 54(9):3951–3967
40. Silva D, Kschischang FR, Kötter R (2010) Communication over finite-field matrix channels. IEEE Trans Inf Theory 56(3):1296–1305
41. Stierstorfer C, Fischer RFH (2006) Lattice-reduction-aided tomlinson-harashima precoding for point-to-multipoint transmission. Int J Electron Commun (AEÜ) 60(4):328–330
42. Stierstorfer C, Siegl C, Fischer RFH, Wild T, Hoek C (2010) Network MIMO downlink transmission. In: Proceedings of international OFDM workshop (InOWo), Hamburg, Germany, Sept 2010
43. Tarokh V, Jafarkhani H (2000) A differential detection scheme for transmit diversity. IEEE J Sel Areas Commun 18(7):1169–1174
44. Taherzadeh M, Mobasher A, Khandani AK (2007) LLl reduction achieves the receive diversity in MIMO decoding. IEEE Trans Inf Theory 53(12):4801–4805
45. Tölli A, Codreanu M, Juntti M (2008) Linear multiuser MIMO transceiver design with quality of service and per-antenna power constraints. IEEE Trans Signal Process 56(7):3049–3055
46. Viswanath P, Tse DNC (2003) Sum capacity of the vector gaussian broadcast channel and uplink-downlink duality. IEEE Trans Inf Theory 49(8):1912–1921
47. Vishwanath S, Jindal N, Goldsmith A (2003) Duality, achievable rates, and sum-rate capacity of gaussian MIMO broadcast channels. IEEE Trans Inf Theory 49(10):2658–2668
48. Wachter A, Sidorenko VR, Bossert M, Zyablov VV (2011) On (Partial) unit memory codes based on gabidulin codes. Prob Inf Transm 47(2):117–129
49. Weingarten H, Steinberg Y, Shamai S (2006) The capacity region of the Gaussian multiple-input multiple-output broadcast channel. IEEE Trans Inf Theory 52(9):3936–3964
50. Windpassinger C (2004) Detection and precoding for multiple input multiple output channels. Ph.D. Thesis, Friedrich-Alexander-Universität Erlangen-Nürnberg
51. Windpassinger C, Fischer RFH (2003) Low-complexity near-maximum-likelihood detection and precoding for MIMO systems using lattice reduction. In: IEEE information theory workshop, Paris, France, pp 345–348, Mar/Apr 2003
52. Windpassinger C, Fischer RFH, Huber JB (2004) Lattice-reduction-aided broadcast precoding. IEEE Trans Commun 52(12):2057–2060
53. Wübben D, Böhnke R, Kühn V, Kammeyer KD (2004) Near-maximum-likelihood detection of MIMO systems using MMSE-based lattice reduction. In: IEEE International conference on communications, Paris, France, pp 798–802, Jun 2004
54. Wübben D, Seethaler D, Jalden J, Matz G (2011) Lattice reduction. IEEE Signal Process Mag 28(3):70–91
55. Yao H, Wornell GW (2002) Lattice-reduction-aided detectors for MIMO communication systems. In: IEEE global telecommunications conference, Taipei, Taiwan, pp 424–428, Nov 2002
56. Yu W, Cioffi JM (2004) Sum capacity of Gaussian vector broadcast channels. IEEE Trans Inf Theory 50(9):1875–1892
57. Yu W, Lan T (2007) Transmitter optimization for the multi-antenna downlink with per-antenna power constraints. IEEE Trans Signal Process 55(6):2646–2660
58. Zamir R, Shamai S, Erez U (2002) Nested linear/lattice codes for structured multiterminal binning. IEEE Trans Inf Theory 48(6):1250–1276
59. Zhan J, Nazer B, Erez U, Gastpar M (2014) Integer-forcing linear receivers. IEEE Trans Inf Theory 60(12):7661–7685

Chapter 3
Enabling the Multi-User Generalized Degrees of Freedom in Cellular Interference Networks with Multi-User Coding

Rick Fritschek and Gerhard Wunder

Abstract Recent investigations have shown that multi-user gain can be enabled in frequency-flat, time-invariant, single-antenna interfering multiple access channels. This chapter will provide an overview of results, which lead to these recent key results. We will present conclusions of two deterministic approximation models and discuss the advantages and disadvantages. We start with results which were obtained with the linear deterministic model and show the necessity to use a richer model, the lower triangular deterministic model. Moreover, we shine a light on the connection to the interfering broadcast channel and therefore duality for the linear deterministic setting. We will see, that considerable multi-user gain can be enabled through signal-scale based coding in both deterministic approximations. We show that the multi-user gain of the linear deterministic model can be achieved in the Gaussian model as well, but is not optimal at all points. At the end, we discuss the possible transfer to Gaussian channels and give an outlook on ongoing research into the lower triangular deterministic model, where the upper bound can be reached at all points.

3.1 Introduction

Cellular interference networks are the backbone of the wireless communication world and will handle most of the expected capacity growth due to the smartphone revolution in the upcoming years. However the major share of this growth comes from shrinking the cell size by leveraging new infrastructure and higher bandwidth which is indeed costly for the network operators. Another factor was believed to contribute as well: cooperative multi-point concepts (CoMP). However, these concepts require

R. Fritschek (✉)
Communications and Information Theory Group, Technische Universität Berlin, Einsteinufer 25, 10587 Berlin, Germany
e-mail: rick.fritschek@tu-berlin.de

G. Wunder
Wireless Communication and Networks, Fraunhofer Heinrich–Hertz–Institute, Einsteinufer 37, 10587 Berlin, Germany
e-mail: gerhard.wunder@hhi.fraunhofer.de

W. Utschick (ed.), *Communications in Interference Limited Networks*, Signals and Communication Technology, DOI 10.1007/978-3-319-22440-4_3 channel state information at the transmitters of which the impact has been highly underestimated in the past. Indeed the CoMP gains seen in LTE-A networks are rather disappointing and often such modes are not even switched on. This gives some incentive to look at new coding models which can cope much better with imperfect CSI. Interference is one of the major limiting factors of wireless networks and it scales with SNR, therefore noise becomes less important in high SNR regimes. This fact motivated investigations of interference affected networks to gain more insight into its nature. The simplest model of an interference network, the interference channel (IC), is still unsolved for more than 40 years. However, recent investigations have achieved a major breakthrough. In 2008, Etkin et al. [6] have shown the channel capacity of the Gaussian interference channel (GIC) within a single bit. The gap towards the capacity is independent of the power and noise, and these results are therefore called constant-gap capacity approximations. This landmark paper resulted in a number of follow-up investigations on different models like the k-user IC [5], X-channel [13] and also cellular channels [19]. One of the main factors, resulting in the breakthrough of the GIC, was the introduction of the so-called generalized degrees of freedom (GDoF).

Degrees of freedom are a notion of how the achievable rate behaves in the high SNR regime and are based on the capacity of the single-link additive white Gaussian noise channel (AWGN-Channel). The channel equation for the AWGN-channel is given by

$$y = \sqrt{\text{SNR}}x + z,$$

where $z \in \mathcal{CN}(0, 1)$. The input is constrained by average unit power $\text{E}\{|x|^2\} \leq 1$. The capacity is given as

$$C_{\text{AWGN}} = \log(1 + \text{SNR}), \tag{3.1}$$

and one can see that the capacity scales with log SNR, when SNR gets large. Therefore log SNR is related to the DoF. We can define a notion of degrees of freedom for every other network, based on this behaviour with

$$d_{\text{sym}} := \lim_{\text{SNR}\to\infty} \frac{C_{\text{sym}}(\text{SNR})}{\log \text{SNR}},$$

where d_{sym} defines the degrees of freedom and therefore tells us about how the networks capacity $C_{\text{sym}}(\text{SNR})$ scales in comparison to the single-link AWGN-channel for the high SNR regime. However, degrees of freedom have the drawback, that they are independent of the interference strength in the network. They provide a minimum achievable point over the whole interference regime. DoF is therefore too simple to capture the interference behaviour of the network and cannot yield constant-gap capacity approximations. Etkin et al. defined a more sophisticated notion of DoF, coined GDoF which incorporates the changes in DoF due to the interference strength.

$$d_{\text{sym}}(\alpha) := \lim_{\text{SNR},\text{INR}\to\infty;\ \frac{\log \text{INR}}{\log \text{SNR}}=\alpha} \frac{C_{\text{sym}}(\text{INR},\text{SNR})}{C_{\text{awgn}}(\text{SNR})},$$

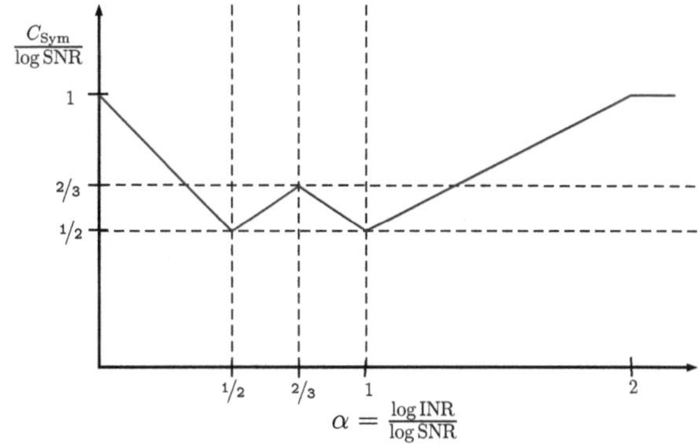

Fig. 3.1 The generalised degrees of freedom for the two-user Gaussian interference channel. The symmetric capacity is illustrated and shows the achievable DoF per link dependent on the interference strength

where $d_{\text{sym}}(\alpha)$ defines the GDoF, dependent on the interference-to-signal ratio α. This interference dependent DoF yields the so-called w-curve, which is depicted in Fig. 3.1. One can see, that there are 5 different regimes, where the first three ($0 \leq \alpha \leq 1$) belong to the weak interference regime, followed by the strong and the very strong interference regime. A detailed description of the different regimes can be found in [6]. The GDoF are an important step in the characterisation of the capacity and can provide a much more detailed view of the interference behaviour of interference affected network models. Still, achieving results for more complex channel models is hard due to the noise properties of the Gaussian channels. A branch of research therefore investigated deterministic approximations of Gaussian channel models. Starting from now, all channels without special mentioning (such as LD-, LTD- or LDM of) will refer to their Gaussian version.

An interesting and successful approximation emerged with the work of Avestimehr et al. [1]. In this work, the so-called linear deterministic model (LDM) was proposed, which is based on the binary expansion of the real input signals. Imagine the example of the point-to-point AWGN channel from above, within the framework of the LDM it can be equivalently written as

$$y = h2^n x + z, \tag{3.2}$$

with a fine channel gain $h \in (1, 2]$ and a coarse channel gain 2^n with $n \in \mathbf{N}$. One can see, that all channel gains greater or equal to one can be expressed in this form. The LDM makes the further assumption that the input signals have a peak power constraint of 1. Moreover, the fine channel gain gets approximated by 1, resulting in the correspondence $n = \lfloor \log \text{SNR} \rfloor$.[1] The bits in the binary expansion are viewed as bit-vectors, and its entries as levels. These bit-vectors get truncated at noise level to approximate the influence of noise, which makes the model deterministic. Moreover, channel gains result in a shift of the bit-vectors. Surprisingly, this rather easy approximation produced results, which are close to the Gaussian channel equivalents. For

[1]For the *real* scalar Gaussian channel, the correspondence would be $n = \lfloor \frac{1}{2} \log \text{SNR} \rfloor$.

some channels even within a constant-gap of the Gaussian channel, e.g. the GIC in [2]. However, certain channels show that the approximation is limited in such a way, that it cannot produce results good enough for constant-gap approximations, e.g. the two-user interfering multiple access channel [9, 10]. Interestingly, a new deterministic model emerged in [14], where the fine channel gain is not approximated by one, but rather incorporated as binary expansion. This new model is called lower triangular deterministic model (LTDM), due to the fine channel gains forming a lower triangular matrix in the channel equations of the model. It is shown in [14], that this model is able to overcome previous problems of the linear deterministic X-channel and yields a constant-gap capacity approximation at the cost of some arbitrary small outage set. One of the key techniques, used in [14] is interference alignment. Interference alignment emerged with the work of [5, 13] and is used to align the interfering signals in some dimension to achieve the optimal DoF in interference networks, enabling dimensions for interference-free transmission. Interference alignment methods can be divided into two classes. One class is vector space alignment methods which use the classical dimensions of time, frequency or multiple antennas to align interference. These methods were successfully applied to several channel models, including for example the X-channel or the k-user interference channel. However, these methods fail for single-antenna, time-invariant and frequency-flat channel models, since these dimensions are needed for alignment. In these cases, the second class of interference alignment, signal-scale alignment, can be used. This class of methods uses the signal scale to align interference. Results include [3, 7, 14, 16–18] where lattice codes were used to align interfering codewords and establish optimal DoF. Most of these results build on intuition from the schemes of corresponding linear deterministic models, since this approximation also provides a direct view and hence insights into the signal scale. In many cases, linear deterministic schemes even provided a precise structure for the corresponding lattice schemes.

3.1.1 *Road to Constant-Gap Sum-Capacity Approximations of Cellular Channels*

As stated above, one of the earliest results for cellular interference channels is in [19], where the interfering multiple access channel (IMAC) and the interfering broadcast channel (BC) were subject of investigation. The IMAC and IBC are the simplest example for cellular channels, where multiple users communicate to base stations or vice versa and are limited by inter-cell interference between the users of different cells. In [19] the DoF (per cell)[2] of the k-user (per cell), G-cell IMAC was shown to be

$$\frac{k}{(\sqrt[G-1]{k}+1)^{G-1}}, \tag{3.3}$$

[2] All of the following DoF and GDoF results in this chapter will be per cell.

which goes to one, as the number of users approach infinity. Moreover, a duality result was proven, such that the same DoF can be achieved by the corresponding IBC. These two results were astonishing, as they state that a large number of users can diminish interference effects and yield interference-free DoF asymptotically. However, they are dependent on time or frequency dimensions and therefore on vector space alignment. Moreover, DoF results are not strong enough to give insights into the interference dependent nature of the channel model, as stated above. Investigating the GDoF was the next logical step in conjunction with the use of the LDM model in hope for a general scheme in a time-invariant frequency-flat setting. Such a scheme could possibly be transferred to the Gaussian model to yield a constant-gap approximation of the capacity. Among the first investigations of LDM cellular channel models is [4], where the multiple access channel interfering with a point-to-point (MAC-P2P) link was investigated. In the weak interference regime, it could be shown, that only for certain channel parameters multi-user gain is present. Moreover, multi-user gain of the GDoF vanishes for the interference regimes above $\alpha = \frac{2}{3}$. Nevertheless, the result (3.3) suggest that in case of an interfering multiple access channel, multi-user gain is still present in the whole regime since it yields $\frac{2}{3}$ DoF as minimum GDoF for the whole interference range. We will therefore focus on the investigations of the deterministic 2-user IMAC in this chapter. We present recent results for the LDM channel models. Moreover, we show the necessity of investigation of the LTDM IMAC (LTD-IMAC) and that a constant-gap sum-capacity approximation of the LTD-IMAC is possible. We also lay out on-going investigations towards the transfer to the Gaussian IMAC.

3.2 The Interfering Multiple Access Channel

The two-user interfering multiple access channel consists of two MAC channels, each one forming a cell. To get a consistent notation, also in view towards the interfering broadcast channel, we use the notation of a_{ik}^j, where j represents the receiver cell, and i and k the transmitter cell and user respectively. We therefore have the two users x_{11} and x_{12} communicating with y^1 and x_{21}, x_{22} with y^2. Hence, the channel equations are given by

$$
\begin{aligned}
y^1 &= h_{11}^1 2^{n_{11}^1} x_{11} + h_{12}^1 2^{n_{12}^1} x_{12} + h_{21}^1 2^{n_{21}^1} x_{21} + h_{22}^1 2^{n_{22}^1} x_{22} + z^1 \\
y^2 &= h_{11}^2 2^{n_{11}^2} x_{11} + h_{12}^2 2^{n_{12}^2} x_{12} + h_{21}^2 2^{n_{21}^2} x_{21} + h_{22}^2 2^{n_{22}^2} x_{22} + z^2,
\end{aligned}
$$

where $z^j \in \mathcal{CN}(0, 1)$ is assumed to be zero mean and unit variance Gaussian noise. Also each transmitted signal has an associated average power constraint $\mathrm{E}\{|x_{ik}^2|\} \leq P$. Note that we use the aforementioned channel gain notation of (3.2).

We define a shift property of the channel, which is

$$\frac{n_{i1}^i}{n_{i2}^i} \neq \frac{n_{i1}^j}{n_{i2}^j} \tag{3.4}$$

for $i, j \in \{1, 2\}$, $i \neq j$, and states that the ratio of the two direct coarse channel gains must be different to the ratio of the coarse interference channel gains in order to activate multi-user gain. It is assumed that $n_{21}^1 = n_{22}^1 =: n_2^1$, $n_{11}^2 = n_{12}^2 =: n_1^2$, stating that the coarse interference strength caused by $\mathbf{x}_{ij}$ at the receivers is the same. Note that the restriction is justified in case when the distance between the two cells is significantly larger than the cell dimension itself. With this assumption, the shift property becomes

$$n_{i1}^i \neq n_{i2}^i, \tag{3.5}$$

which means that the two direct signals cannot have equal coarse channel gain in order to enable multi-user gain in the linear deterministic model.

Moreover, we assume without loss of generality that $n_{11}^1 \geq n_{12}^1$, $n_{21}^2 \geq n_{22}^2$ and the difference between the two coarse channel gains is denoted as $n_{11}^1 - n_{12}^1 =: \Delta_1$ and $n_{21}^2 - n_{22}^2 =: \Delta_2$.

3.3 Linear Deterministic Approximation of the IMAC in the Weak Interference Regime

The linear deterministic model (LDM) models the input signals x_{ik} as bit vectors $\mathbf{x}_{ik}$. This is achieved by a binary expansion of the real input signal. The resulting bits constitute the new bit vector. The positions within the vector will be referred to as levels. To model the signal impairment induced by noise, the bit vectors will be truncated at noise level and only the n most significant bits are received at y^j. This is done by shifting the incoming bit vector for $q - n$ positions $\mathbf{y} = \mathbf{S}^{q-n}\mathbf{x}$, with $q := \max n_{ik}^j$. Where $\mathbf{S}$ is the shift matrix defined as

$$\mathbf{S} = \begin{pmatrix} 0\,0 \cdots 0\,0 \\ 1\,0 \cdots 0\,0 \\ 0\,1 \cdots 0\,0 \\ \vdots\,\vdots\,\ddots\,\vdots\,\vdots \\ 0\,0 \cdots 1\,0 \end{pmatrix}. \tag{3.6}$$

By scaling of the channel gains, we may assume an unit average power constraint in the corresponding Gaussian channel. This unit average power constraint is assumed to be a peak power constraint in the linear deterministic channel [1]. Using a peak power constraint instead of an average constraint reduces the rate within a constant-gap [2]. Superposition at the receivers is modelled via binary addition of the incoming

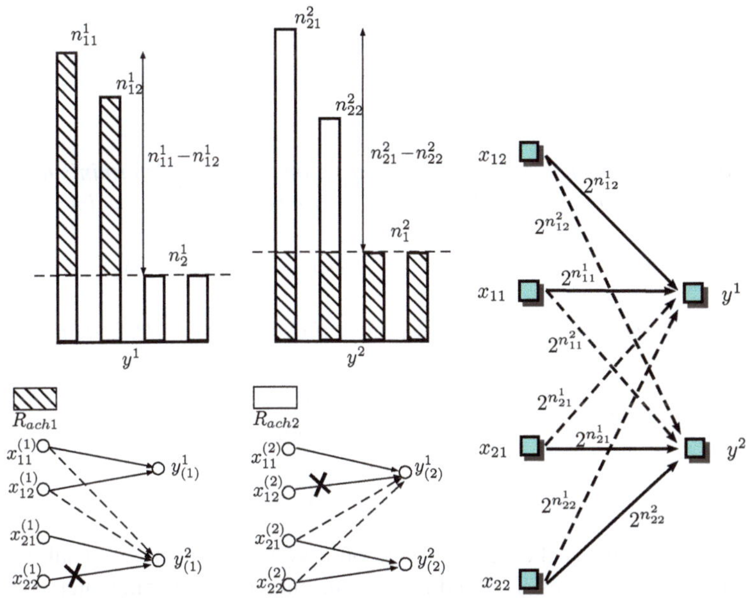

Fig. 3.2 (left) Illustration of the two subsystems. As one can see, letting the second user in each sub-system be silent, results in the one-sided interference MAC-P2P model. Hence, achievable schemes are applicable; (right) Illustration of the LD-IMAC

bit vectors on the individual levels. Carry over is not used to limit the superposition on the specific level where it occurs. Therefore, bit-levels can be used independently. This property enables a form of bit-level alignment, which is used by the following schemes. The channel gain is represented by n_{ik}^j-bit levels which corresponds to $\lceil \log \mathrm{SNR} \rceil$ of the original channel. We therefore approximate the fine channel gain by 1. With this definitions the model can be written as

$$\mathbf{y}^1 = \mathbf{S}^{q-n_{11}^1}\mathbf{x}_{11} \oplus \mathbf{S}^{q-n_{12}^1}\mathbf{x}_{12} \oplus \mathbf{S}^{q-n_{21}^1}\mathbf{x}_{21} \oplus \mathbf{S}^{q-n_{22}^1}\mathbf{x}_{22} \tag{3.7}$$

$$\mathbf{y}^2 = \mathbf{S}^{q-n_{11}^2}\mathbf{x}_{11} \oplus \mathbf{S}^{q-n_{12}^2}\mathbf{x}_{12} \oplus \mathbf{S}^{q-n_{21}^2}\mathbf{x}_{21} \oplus \mathbf{S}^{q-n_{22}^2}\mathbf{x}_{22} \tag{3.8}$$

and is depicted in Fig. 3.2.

3.3.1 Approximate Sum-Capacity

We confine the following investigation of the linear deterministic IMAC to the very weak interference regime. We can demonstrate the drawbacks of the linear deterministic approximation in this regime, without going into the complex schemes in higher interference regimes where both cells are coupled. For the very weak interference case it is assumed that the sum of both interference parts of the signals are below the direct signal level. This is stated in the condition that $n_2^1 + n_1^2 \leq \min(n_{11}^1, n_{21}^2)$. For a symmetric model, this condition becomes: $0 \leq \alpha \leq \frac{1}{2}$, with $\alpha := \frac{n_i}{n_1}$, where $n_i := n_2^1 = n_1^2$ and $n_1 := n_{11}^1 = n_{21}^2$.

Theorem 3.3.1 ([8]) *The achievable sum-rate for the linear deterministic interfering multiple access channel (LD-IMAC) in the weak interference regime, is*

$$R_\Sigma \leq n_{12} + n_{22} - n_2^1 - n_1^2 + \phi(n_2^1, \Delta_1) + \phi(n_1^2, \Delta_2), \tag{3.9}$$

where the function ϕ for $p, q \in \mathbb{N}_0$, following the notation of [4], is defined as

$$\phi(p,q) := \begin{cases} q + \frac{l(p,q)q}{2} & \text{if } l(p,q) \text{ is even,} \\ p - \frac{(l(p,q)-1)q}{2} & \text{if } l(p,q) \text{ is odd,} \end{cases} \tag{3.10}$$

where $l(p,q) := \lfloor \frac{p}{q} \rfloor$ for $q > 0$ and $l(p,0) = 0$.

Proof The proof of the theorem is based on a system split into two sub-systems. Each of these sub-systems $\mathcal{R}_{ach}^{(1)}$ and $\mathcal{R}_{ach}^{(2)}$ represent a MAC channel interfering with a point-to-point link, with one-sided interference, see Fig. 3.2. The channel equations for these two sub-systems are given by

$$\mathbf{y}_1^{(1)} = \mathbf{S}^{q^{(1)}-(n_{11}^1-n_2^1)}\mathbf{x}_{11}^{(1)} \oplus \mathbf{S}^{q^{(1)}-(n_{12}^1-n_2^1)}\mathbf{x}_{12}^{(1)} \tag{3.11}$$
$$\mathbf{y}_2^{(1)} = \mathbf{S}^{q^{(1)}-n_1^2}\mathbf{x}_{11}^{(1)} \oplus \mathbf{S}^{q^{(1)}-n_1^2}\mathbf{x}_{12}^{(1)} \oplus \mathbf{S}^{q^{(1)}-n_1^2}\mathbf{x}_{21}^{(1)} \tag{3.12}$$
$$\oplus\, \mathbf{S}^{q^{(1)}-n_1^2}\mathbf{x}_{22}^{(1)} \tag{3.13}$$

for $\mathcal{R}_{ach}^{(1)}$ and

$$\mathbf{y}_2^{(2)} = \mathbf{S}^{q^{(2)}-(n_{21}^2-n_1^2)}\mathbf{x}_{21}^{(2)} \oplus \mathbf{S}^{q^{(2)}-(n_{22}^2-n_1^2)}\mathbf{x}_{22}^{(2)} \tag{3.14}$$
$$\mathbf{y}_1^{(2)} = \mathbf{S}^{q^{(2)}-n_2^1}\mathbf{x}_{11}^{(2)} \oplus \mathbf{S}^{q^{(2)}-n_2^1}\mathbf{x}_{12}^{(2)} \oplus \mathbf{S}^{q^{(2)}-n_2^1}\mathbf{x}_{21}^{(2)} \tag{3.15}$$
$$\oplus\, \mathbf{S}^{q^{(2)}-n_2^1}\mathbf{x}_{22}^{(2)} \tag{3.16}$$

for $\mathcal{R}_{ach}^{(2)}$. The achievability scheme for the IMAC is an extended version of the scheme already used for the MAC-P2P in [4]. Leaving the weaker direct signal silent results in a channel model similar to the ones used in [4], resulting in the sum-rates

$$R_\Sigma^{(1)} \leq n_1^2 + \zeta^{(1)} + \phi(n_2^1, \Delta_1) \tag{3.17}$$

$$R_\Sigma^{(2)} \leq n_2^1 + \zeta^{(2)} + \phi(n_1^2, \Delta_2) \tag{3.18}$$

for the sub-systems. Where $\zeta^{(1)} := n_{12}^1 - n_2^1 - n_1^2$, $\zeta^{(2)} := n_{22}^2 - n_2^1 - n_1^2$. This result was proven for the MAC-P2P in [4] with use of optimal bit-level assignment and an orthogonal bit-level assignment scheme. The bits are chosen such that a maximum

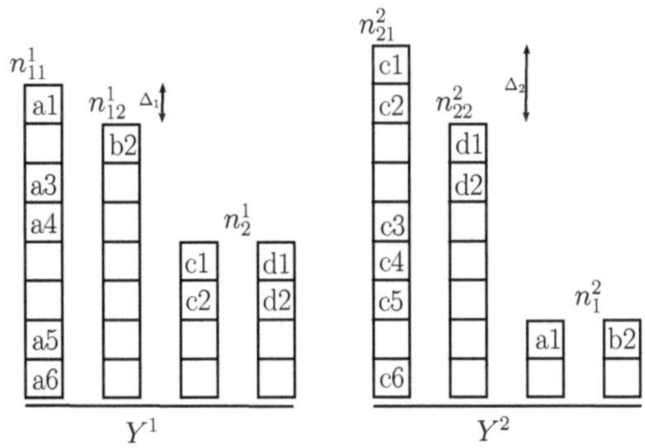

Fig. 3.3 An example for the LD-IMAC scheme is shown. One can see the specific structure of bit-level assignment, which is a possible optimal assignment, reaching the upper bound of 14 bit levels in the specific example

number of information can be transmitted in the common part of the signal to the intended cell, while an interference alignment is achieved at the unintended cell. Finally, the sum rate for the overall system can be obtained by adding the sub-systems: $R_{\Sigma}^{(1)} + R_{\Sigma}^{(2)} = R_{\Sigma}$

$$
\begin{aligned}
R_{\Sigma} &\le n_1^2 + \zeta^{(1)} + \phi(n_2^1, \Delta_1) + n_2^1 + \zeta^{(2)} + \phi(n_1^2, \Delta_2) \\
&= n_{12} + n_{22} - n_2^1 - n_1^2 + \phi(n_2^1, \Delta_1) + \phi(n_1^2, \Delta_2)
\end{aligned}
$$

□

The question is now, if this achievable scheme and the corresponding sum-rate are optimal. We therefore have to investigate upper bounds for the specific model. The following theorem shows these upper bounds for the LD-IMAC.

Theorem 3.3.2 ([8]) *The sum rate for the LD-IMAC system model in the weak interference regime, can be bounded from above by*

$$
R_{\Sigma} \le n_{11}^1 + n_{21}^2 - \frac{n_1^2}{2} - \frac{n_2^1}{2}. \tag{3.19}
$$

Proof A proof for the two symmetrical weak interference cases can be found in [8]. Note that, with these proof techniques, the mixed cases can be shown as well. □

Comparing the upper bounds with the achievable sum-rates shows, that the achievable scheme can reach the upper bound at certain points (see Fig. 3.4). We save a detailed description of these specific points for later on. First of all, we show that these results for the LD-IMAC can be proved for the LD-IBC as well.

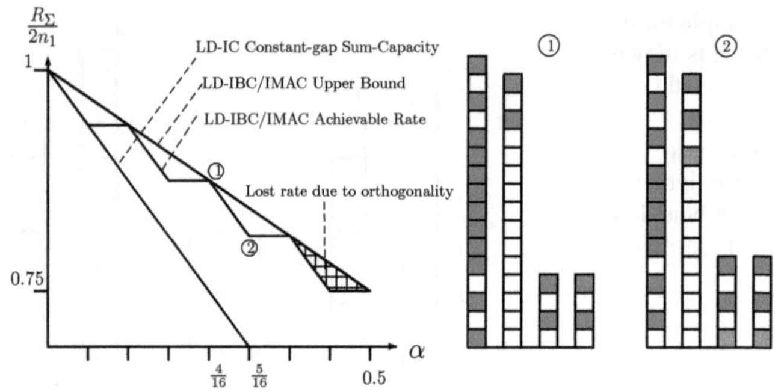

Fig. 3.4 The figure shows the w-curve for the interference range $0 \leq \alpha \leq \frac{1}{2}$ on the left side and the corresponding bit-level assignment scheme on the right side

3.3.2 Duality Between IMAC and IBC

Information-theoretic problems for the broadcast channel (BC) are usually much harder than the corresponding problems for the MAC. But there are certain relations between these models, which can help to solve these BC problems. These relations are usually referred to as *duality*. Duality means that the achievable scheme, upper bound or the capacity of two systems is the same, hence the networks are comparable. For an example, we refer to [11], where a duality result of the capacity region between the Gaussian BC and MAC model was shown. Duality itself can help to solve hard problems, by translating them to simpler dual models, solving them there and translating the solution back. A dual model is usually defined as a network in which the nodes are the same as in the original model, but with reversed directions of transmissions. With this definition, the dual model of the IMAC is the interfering broadcast channel (IBC). The Gaussian interference broadcast channel (IBC) consists of 2 transmitters and 4 receivers. With the notation from above, the transmitter x_1 communicates two messages with y_1^1 and y_2^1, whereas x_2 communicates two messages with y_1^2 and y_2^2, respectively. The IBC channel equations are therefore given by

$$Y_k^j = \sum_{i=1}^{2} h_{ik}^j X_i + Z_k^j \tag{3.20}$$

where $Z_k^j \in \mathcal{CN}(0, 1)$ is again a zero mean and unit variance Gaussian random variable and X has an associated average power constraint $\mathrm{E}\{|X_i|^2\} \leq P$. As in the IMAC case, we use the LDM to approximate the Gaussian channel model. All general LDM assumptions also apply for the IBC. Therefore the channel model can be written as:

$$\mathbf{y}_1^1 = \mathbf{S}^{q-n_{11}^1}\mathbf{x}_1 \oplus \mathbf{S}^{q-n_{21}^1}\mathbf{x}_2 \tag{3.21}$$

$$\mathbf{y}_2^1 = \mathbf{S}^{q-n_{12}^1}\mathbf{x}_1 \oplus \mathbf{S}^{q-n_{22}^1}\mathbf{x}_2 \tag{3.22}$$

$$\mathbf{y}_1^2 = \mathbf{S}^{q-n_{21}^2}\mathbf{x}_2 \oplus \mathbf{S}^{q-n_{11}^2}\mathbf{x}_1 \quad (3.23)$$

$$\mathbf{y}_2^2 = \mathbf{S}^{q-n_{22}^2}\mathbf{x}_2 \oplus \mathbf{S}^{q-n_{12}^2}\mathbf{x}_1. \quad (3.24)$$

The question is now, if the achievable sum-rate and the capacity are the same as in the LD-IMAC.

Theorem 3.3.3 ([8]) *The achievable sum-rate for the linear deterministic interfering broadcast channel (LD-IBC) in the weak interference regime, is*

$$R_\Sigma \leq n_{12}^1 + n_{22}^2 - n_2^1 - n_1^2 + \phi(n_2^1, \Delta_1) + \phi(n_1^2, \Delta_2). \quad (3.25)$$

Proof The proof of the theorem is based on the particular structure of LDM coding schemes. Since these schemes are linear, one can use a result of [15], which states that *any linear deterministic network which is linearly solvable obeys linear reciprocity.* In other words, a scheme can be found for the dual LD-IBC, which results in the same achievable sum-rate as the LD-IMAC. A particular bit-level assignment strategy for the LD-IBC would be to use the LD-IMAC bit-level assignment vectors upside down: $\mathbf{x}_1^{BC} = \mathbf{x}_1^{MAC} \oplus \mathbf{x}_2^{MAC}$, we refer to Fig. 3.5 as an example. □

Moreover, the next theorem shows an upper bound, which is the same as in the LD-IMAC.

Theorem 3.3.4 ([8]) *The sum rate for the LD-IBC system model in the weak interference regime, can be bounded from above by*

$$R_\Sigma \leq n_{11}^1 + n_{21}^2 - \frac{n_1^2}{2} - \frac{n_2^1}{2}. \quad (3.26)$$

Proof A proof for the theorem can be found in [8]. □

These two theorems establish a duality relationship between the LD-IMAC and LD-IBC by means of achievable sum-rate and sum-rate upper bound.

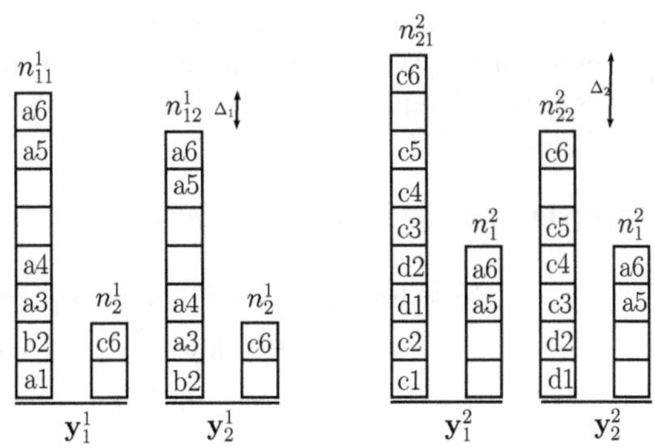

Fig. 3.5 An example for a coding scheme of the LD-IBC system model is shown. The example is the exact dual case to the one in Fig. 3.3. We can see the basic strategy of turning corresponding IMAC bit-level vectors upside down to form a IBC bit-level vector

3.3.3 Transfer to the Gaussian IMAC

As mentioned in the introduction, part of the motivation of the linear deterministic model is to approximate the Gaussian model in a way such that we gain insights about it. These insights can help and maybe directly lead to corresponding schemes for the Gaussian model. The LDM and its achievability schemes only consider the signal scale of the network model to align interference. A natural technique to imitate this property is, therefore, the use of lattice coding. The first examples of a successful application of lattice coding techniques for interference alignment can be found in [3, 18], which investigated the many-to-one and one-to-many GIC, and the three-user GIC, respectively. These techniques were also used in [7, 17]. In this section we will focus on the latter one, which used the lattice techniques to transfer the LD-IMAC/IBC schemes to the Gaussian IMAC/IBC setting. The general lattice coding strategy in view of the deterministic model is to partition the available SNR of the Gaussian model into smaller intervals or *levels*, imitating the LDM bit-level structure. An independent lattice code will then be used for every level. In [3, 7, 18] a result from [12] is used which states that under the rate restriction of

$$R \leq \log\left(\frac{P}{N}\right), \tag{3.27}$$

a lattice code $(\gamma \Lambda_C + v) \cap S$ exists with arbitrary small error probability. This code consists of a lattice $\Lambda_C \in \mathbb{R}^n$, a scaling factor $\gamma \in \mathbb{R}$, a translation $v \in \mathbb{R}^n$ and a spherical shaping region $S \subset \mathbb{R}^n$ with power P per dimension. N denotes the noise variance per dimension. The power partition dictates the available power per level and the noise is given through the decoding strategy. In particular, every level is treated as a point-to-point additive noise channel, such that (3.27) can be applied. Decoding occurs per level starting from the top, while all other levels are treated as Gaussian noise. After the codeword of a level is decoded, it will be subtracted of the rest and decoding proceeds with the next highest level. This procedure is done successively until every codeword is decoded. If a level happens to be the summation of several codewords, only the sum will be decoded. This means that two interfering levels will be treated as one codeword. The sum-rate of lattice coding for the Gaussian IMAC is therefore the sum over all level rates. The rate for each level is dependent on its power and noise. Whereas the noise is a summation over the general noise and the power of all subsequent levels. This results in the following achievable sum-rate for the Gaussian IMAC:

Theorem 3.3.5 ([7]) *The overall achievable sum rate* R_Σ *is*

$$\begin{aligned} R_\Sigma > {} & \log P_1^{\beta_1} - \log P_1^{\alpha_2} + \log P_2^{\beta_2} - \log P_2^{\alpha_1} \\ & + \phi(\log P_1^{\alpha_2}, \log P_1^{(1-\beta_1)}) + \phi(\log P_2^{\alpha_1}, \log P_2^{(1-\beta_2)}) \\ & - 9 - 6\lfloor L_2 \rfloor - 6\lfloor L_1 \rfloor \end{aligned} \tag{3.28}$$

with ϕ defined as in the LDM case. Moreover, $\frac{\log SNR_{i2}}{\log SNR_{i1}} = \beta_i$ and $\frac{\log INR_i^j}{\log SNR_{i1}} = \alpha_j$, for $i, j \in \{0, 1\}, i \neq j$ and $P_i := SNR_{i1}$. These parameters correspond to α, β which are used in the LDM channel model.

Proof The particular coding scheme and proof of the achievable sum-rate can be found in [7]. □

One can see that the achievable rate of the Gaussian channel is within a constant-gap of $2(2 + \log 4) + 6\lfloor L_1 \rfloor + 6\lfloor L_2 \rfloor$ bits of the LDM rate using the correspondence $n_{ik}^j = \lceil \log |h_{ik}^j|^2 P \rceil$. Note that the gap also depends on the scheme structure itself. As the shift between the two direct signals gets smaller, the number of levels will increase. This will also increase L and therefore the gap. A violation of the shift property (3.4) would result in a breakdown of the coding scheme. As we have seen in the previous section, the LDM achievable sum-rate can reach the upper bound at certain points (see Fig. 3.4). The question is, if a similar bound can be proven for the corresponding Gaussian IMAC. In [2], techniques were shown to transform the mutual information of the GIC into the LDM and vice versa, while introducing a constant bound. This method can be used to prove the following theorem.

Theorem 3.3.6 *The sum-rate for the Gaussian IMAC can be upper bounded by the corresponding LD-IMAC upper bound, within a constant number of bits. The Gaussian IMAC is therefore bounded from above by*

$$R_\Sigma \leq n_{11}^1 + n_{21}^2 - \frac{n_1^2}{2} - \frac{n_2^1}{2} + 22.2. \tag{3.29}$$

Proof The proof is an extension of the two-user IC case in [2] and is omitted. □

The theorem can be proven for the Gaussian IBC as well. This establishes a constant-gap sum-capacity result at points, where the interference strength divided by Δ is an even number. At points with an odd ratio, the sum-capacity is within Δ plus the constant-gap. This raises the question, if the upper bound is too big, and in fact should be rather of a step-like shape, or if the achievable scheme of lattice coding is not sufficient enough. As we will see, the latter is the case. The lattice coding scheme inherits its problems directly from the underlying LDM scheme. The LDM scheme cannot reach the upper bound, since it uses exploitation of the orthogonality of the bit-levels. The lower triangular deterministic model brings a possible solution to this problems, because the schemes do not rely on the independence of bit-levels anymore.

3.4 Lower Triangular Deterministic Model

In contrast to the LDM, the LTDM does not approximate the fine channel gain by 1. Instead, the fine channel gain h is also written as a binary expansion resulting in a discrete convolution between the bits of h and x. To model the signal impairment induced by noise, the bits in the convolution with exponent less than zero (at noise level) will be truncated and only the n most significant bits are received at y. The incoming bit vector can be written as $\mathbf{Hx}$, where $\mathbf{H}$ is a lower triangular matrix defined as

$$\mathbf{H} = \begin{pmatrix} 1 & 0 & \cdots & 0 & 0 \\ [h]_1 & 1 & \cdots & 0 & 0 \\ [h]_2 & [h]_1 & \cdots & 0 & 0 \\ \vdots & \vdots & \ddots & \vdots & \vdots \\ [h]_{n-1} & [h]_{n-2} & \cdots & [h]_1 & 1 \end{pmatrix}, \tag{3.30}$$

where $[h]_i$ represents the i-th bit in the binary expansion of h. As in the LDM, superposition at the receivers is modelled via binary addition of the incoming bit vectors. Moreover, the general assumptions of the LDM also apply for the LTDM. We have limited the investigation of the LDM to the weak interference regime, and in particular to $\alpha \leq \frac{1}{2}$. However, for the LTD-IMAC, we want to present the results for the whole interference regime. Since both MAC channels are coupled for $\alpha > \frac{1}{2}$ in a sense, that one cannot separate both achievable schemes, the number of cases would increase. In order to reduce the complexity of the investigation and presenting a clear overview, we assume that $n_{11}^1 = n_{21}^2 := n_1, n_{12}^1 = n_{22}^2 := n_2, n_1^2 = n_2^1 = n_i$. We remark that these assumptions keep the general channel structure intact, and the proofs can be extended to the more general cases as well. Thus, the LTD-IMAC channel model is governed by the following equations

$$\mathbf{y}^1 = \mathbf{H}_{11}^1 \bar{\mathbf{x}}_{11} \oplus \mathbf{H}_{12}^1 \bar{\mathbf{x}}_{12} \oplus \mathbf{H}_2^1 \left(\bar{\mathbf{x}}_{21}^c \oplus \bar{\mathbf{x}}_{22}^c \right) \tag{3.31}$$

$$\mathbf{y}^2 = \mathbf{H}_1^2 (\bar{\mathbf{x}}_{11}^c \oplus \bar{\mathbf{x}}_{12}^c) \oplus \mathbf{H}_2^2 \bar{\mathbf{x}}_{21} \oplus \mathbf{H}_{22}^2 \bar{\mathbf{x}}_{22}, \tag{3.32}$$

where $\mathbf{y}^1, \mathbf{y}^2 \in \mathbb{F}_2^{n_1}$, $\mathbf{H}_{ik}^1, \mathbf{H}_{ik}^2 \in \mathbb{F}_2^{n_1 \times n_1}$, $\mathbf{x}_{ik} \in \mathbb{F}_2^{n_{ik}^j}$ and $\mathbf{x}_{ik}^c \in \mathbb{F}_2^{n_i}$ with $i, j, k \in \{1, 2\}$. Note that for a better distinction between the common and private part, we denote the common part of $\bar{\mathbf{x}}$ as $\bar{\mathbf{x}}^c$. Also $\bar{\mathbf{x}}$ refers to the shifted version of $\mathbf{x}$.

3.4.1 Approximate Sum-Capacity

In order to present a clear and short overview, we further restrict this section to the cases, where full multi-user gain can be reached. This means, that n_2 has to be strong enough to support the gain. Note that there are several cases between the IC sum-rate and the full multi-user gain sum-rate, a full investigation of these cases can be found in [9].

Theorem 3.4.1 *For every $\delta \in (0, 1]$ and $n_1, n_2, n_i \in \mathbb{N}$ such that $n_1 \geq n_2$ and n_2 strong enough to support full multi-user gain, there exists a set $B \subset (1, 2]^{2\times3}$ of Lebesgue measure $\mu(B) \leq \delta$ such that for all channel gains $h_{ik}^j \in (1, 2]^{2\times3} \backslash B$ the achievable sum rate for the IMAC system model is:*

$$R_\Sigma \geq$$

$$\begin{cases} 2(n_1 - \frac{1}{2}n_i) - 2\log(64/\delta) & \textit{for } \alpha \leq \frac{1}{2} \textit{ and } \Delta \leq \frac{1}{2}n_i, \\ n_1 + n_i - 2\log(128/\delta) & \textit{for } \frac{1}{2} \leq \alpha \leq \frac{3}{5} \textit{ and } n_2 \geq n_i + \frac{1}{2}(n_1 - n_i), \\ 2n_1 - \frac{2}{3}n_i - 2\log(128/\delta) & \textit{for } \frac{3}{5} \leq \alpha \leq 1 \textit{ and } n_2 \geq n_1 - \frac{1}{3}n_i, \\ \frac{4}{3}n_i - 2\log(128/\delta) & \textit{for } 1 \leq \alpha \leq \frac{3}{2} \textit{ and } n_2 \geq \frac{1}{3}n_i, \\ 2n_1 - 2\log(128/\delta) & \textit{for } \frac{3}{2} \leq \alpha \leq \infty \textit{ and } n_2 + n_i \leq 2n_1 \end{cases} \quad (3.33)$$

Proof The proof of the theorem depends on the conditions under which the subspaces, spanned by the used bit-levels, are independent. This is dependent on the size of the channel gain matrices which in turn depend on the coarse channel gain of the strongest direct link. If the spanned subspaces are independent, then there exist a unique input for the corresponding output, and the messages can be decoded. In [14] a sufficient condition on the number of bit-levels, i.e. rate, was derived under which independence holds with high probability. This conditions were proved for the LTD-X-channel and since the structure is similar to the LTD-IMAC, we can use the result to get a decoding condition for the LTD-IMAC as well. The following lemma states the decoding criterion.

Lemma 3.4.1 ([14], Lemma 7) *Let* $\delta \in (0, 1]$ *and* $\mathbf{N} \in \mathbb{Z}_+^{2\times 2}$ *such that* $\min\{n_{11}, n_{22}\} \geq \max\{n_{12}, n_{21}\}$. *Assume* $\bar{R}_{11}^P$, $\bar{R}_{11}^C$, $\bar{R}_{12}$, $\bar{R}_{21}$, $\bar{R}_{22}^P$ $\bar{R}_{22}^C \in \mathbb{Z}_+$ *satisfy,*

$$\bar{R}_{11}^C + \max\{\bar{R}_{21}, \bar{R}_{22}^C\} + \bar{R}_{12} + \bar{R}_{11}^P \leq n_{11} - \log\left(\tfrac{32}{\delta}\right) \quad (3.34)$$

$$\max\{\bar{R}_{21}, \bar{R}_{22}^C\} + \bar{R}_{12} + \bar{R}_{11}^P \leq n_{12} - \log\left(\tfrac{32}{\delta}\right) \quad (3.35)$$

$$\bar{R}_{12} + \bar{R}_{11}^P \leq n_{12} + n_{21} - n_{22} \quad (3.36)$$

and

$$\bar{R}_{22}^C + \max\{\bar{R}_{12}, \bar{R}_{11}^C\} + \bar{R}_{21} + \bar{R}_{22}^P \leq n_{22} - \log\left(\tfrac{32}{\delta}\right) \quad (3.37)$$

$$\max\{\bar{R}_{12}, \bar{R}_{11}^C\} + \bar{R}_{21} + \bar{R}_{22}^P \leq n_{21} - \log\left(\tfrac{32}{\delta}\right) \quad (3.38)$$

$$\bar{R}_{21} + \bar{R}_{22}^P \leq n_{12} + n_{21} - n_{11} \quad (3.39)$$

Then, the bit allocation [...] for the (modulated) deterministic X-channel allows successful decoding at both receivers for all channel gains ($g_{mk} \in (1, 2]^{2\times 3}$) *except for a set* $B \subset (1, 2]^{2\times 3}$ *of Lebesgue measure* $\mu(B) \leq \delta$.

Proof Given in [14]. □

We can use the lemma for $\frac{1}{2} \leq \alpha < 1$ by setting

$\bar{R}^C_{11} := R^c_{11}$	$n_{11} := n_1$
$\bar{R}^C_{22} := R^c_{21}$	$n_{12} := n_2$
$\max\{\bar{R}_{21}, \bar{R}^C_{22}\} := R^c_{12}$	$n_{12} + n_{21} - n_{22} := n_i$
$\max\{\bar{R}_{12}, \bar{R}^C_{11}\} := R^c_{22}$	$n_{22} := n_1$
$\bar{R}_{12} := \max\{R^c_{11}, R^c_{12}\}$	$n_{21} := n_2$
$\bar{R}_{21} := \max\{R^c_{21}, R^c_{22}\}$	$n_{12} + n_{21} - n_{11} := n_i$
$\bar{R}^P_{11} := R^{p1}_{11}$	
$\bar{R}^P_{22} := R^{p1}_{22}$	

Due to the special structure in the weak interference case ($\alpha < \frac{1}{2}$), the gap in the lemma will be smaller.

This lemma tells us that for all rates which obey the conditions, the spanned subspaces are independent except for a small set of measure $\mu(B) \leq \delta$. This means that there exists a unique solution and the signals can be decoded.

It remains to choose the rate allocations for the bit vectors, check them against the lemma conditions and show that they achieve the rates in the theorem. We have to choose different rate allocations for the cases

$$\mathrm{I}: \ \alpha \in [0, \tfrac{1}{2}), \ \mathrm{II}: \ \alpha \in [\tfrac{1}{2}, \tfrac{3}{5}), \ \mathrm{III}: \ \alpha \in [\tfrac{3}{5}, 1] \tag{3.40}$$

$$\mathrm{IV}: \ \alpha \in (1, \tfrac{3}{2}], \ \mathrm{V}: \ \alpha \in (\tfrac{3}{2}, \infty). \tag{3.41}$$

Common and private signal parts are indicated with the superscript c and p respectively, such that $\mathbf{x} = [\mathbf{x}^c; \mathbf{x}^p]$. The private parts of the signal can be used to communicate solely to the intended receiver, without affecting the other cell. We dedicate the R^c_{ik} most significant bits, and the R^p_{i1} least significant bits of $\mathbf{x}_{ik}$ to carry information, see Fig. 3.6. Except for $\alpha < \frac{1}{2}$ there is just one private part. We continue with a brief explanation of the different regimes in Fig. 3.9 and the corresponding design rules.

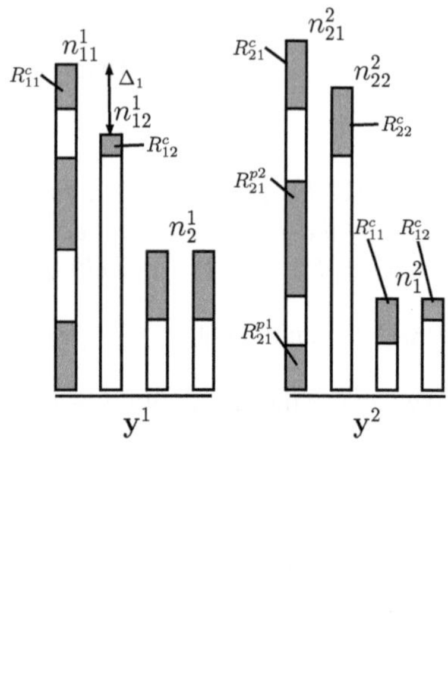

Fig. 3.6 Illustration of the rate allocations and the design rules for the achievable scheme in the range $\alpha < \frac{1}{2}$. At y^1, we have an example for the case: $\frac{1}{2}n_i < \Delta < n_i$ and at y^2 we have an example for the case: $\Delta \leq \frac{1}{2}n_i$

I The first case is the weak interference regime where $\alpha \leq \frac{1}{2}$. In this regime, interference is reduced by half due to the multi-user gain. Design rules are directly inspired by the corresponding LDM investigations. This means that the upper half of the common part of both direct signals is used, as well as the free private part below the aligned interference (see Fig. 3.6). We can therefore align the whole interference in half of its signal scale, at the unintended receiver. If n_i is too low, the scheme needs to be adjusted accordingly, resulting in one of the two sub-cases in [10].

II For $\alpha > \frac{1}{2}$, the previous scheme fails, since both common parts exceed the restrictions of the decoding lemma. The available common part is $n_1 - n_i$ and we use half of it for the direct signals in order to ensure alignment. The private part is still the available signal scale beneath the aligned interference. Figure 3.7 shows the case for n_2 strong enough to support full multi-user gain.

III For $\alpha > \frac{3}{5}$, the private part is not strong enough anymore to support the previous scheme. From now on, the whole private part is used, which is $n_1 - n_i$. Therefore, a rate of n_i is available for the common part and alignment. We need to have space for the two common parts of the direct signal, and the aligned interference of both of them. Therefore, the scheme is to use one-third of the top of every common signal part. As in previous cases, n_2 has to be strong enough to support the full gain. An example for both cases is presented in Fig. 3.8.

IV At $\alpha = 1$, the private part vanishes completely. The whole signal will appear as interference at the unintended receiver. Moreover, for $\alpha > 1$ the interference will be stronger than the direct signals. Design rules for the full multi-user gain case are the same as in the previous regime, except that we do not have a private part. If n_2 is weaker than one-third of the interference, the sum-rate and scheme are governed by a sub-case. For $n_2 = 0$, the model falls back to the IC model.

V For $\alpha \geq \frac{3}{2}$ the interference is strong enough, that it can be completely mitigated, provided that n_2 is strong enough. The full (interference-free) sum-rate can be achieved, by using $\frac{n_1}{2}$ as common part of both users. A weaker n_2 will result in a sub-case with less multi-user gain and for $n_2 = 0$ it will result in a fall back to the IC case.

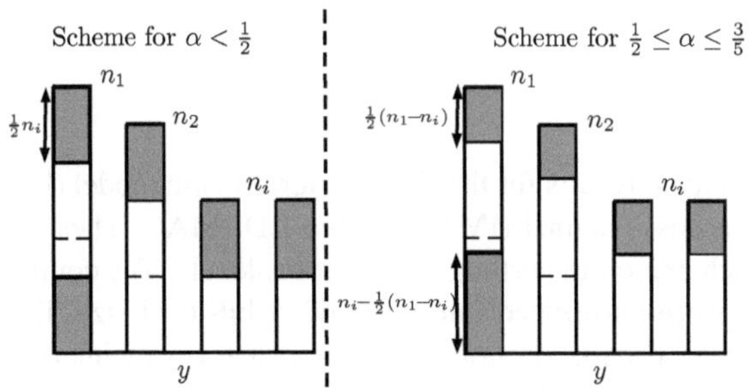

Fig. 3.7 Illustration of the rate allocations and the design rules for the achievable scheme in the range $\frac{1}{2} < \alpha < \frac{3}{5}$ for an exemplary cell with $\alpha = \frac{4}{7}$. With a comparison to the scheme of the previous regime to illustrate the problem with the decoding bound on n_1

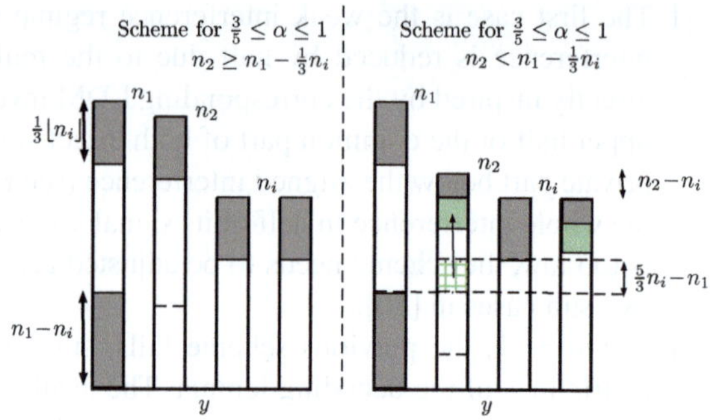

Fig. 3.8 Illustration of the rate allocations for the achievable scheme in the range $\frac{3}{5} \le \alpha \le 1$ for an exemplary cell with $\alpha = \frac{2}{3}$. We show the two achievable schemes for $n_2 \ge n_1 - \frac{1}{3}n_i$ and $n_2 < n_1 - \frac{1}{3}n_i$

Following these design rules, it is shown in [9], that the achievable scheme obeys the modified versions of Lemma 3.4.1 and can reach the sum-rate of Theorem 3.4.1. □

The question is now, how close are those achievable sum-rates to the upper bounds. The next theorem answers this question.

Theorem 3.4.2 *The sum rate for the symmetric LTD-IMAC system model with n_2 strong enough to support full multi-user gain, can be bounded from above by*

$$R_\Sigma \le \begin{cases} 2(n_1 - \frac{1}{2}n_i) & \text{for } 0 \le \alpha \le \frac{1}{2} \text{ and } \Delta \le \frac{1}{2}n_i, \\ 2(\frac{1}{2}n_1 + \frac{1}{2}n_i) & \text{for } \frac{1}{2} \le \alpha \le \frac{3}{5} \text{ and } n_2 \ge n_i + \frac{1}{2}(n_1 - n_i), \\ 2(n_1 - \frac{1}{3}n_i) & \text{for } \frac{3}{5} \le \alpha \le 1 \text{ and } n_2 \ge n_1 - \frac{1}{3}n_i, \\ \frac{4}{3}n_i & \text{for } 1 \le \alpha \le \frac{3}{2} \text{ and } n_2 \ge \frac{1}{3}n_i, \\ 2n_1 & \text{for } \frac{3}{2} \le \alpha \le \infty \text{ and } n_2 + n_i > 2n_1. \end{cases} \quad (3.42)$$

Proof A proof for the bounds can be found in [9]. □

This theorem shows that the achievable schemes and upper bounds coincide within a constant bit-gap. The upper bounds therefore show the constant-gap approximated sum-capacity of the symmetrical IMAC channel.

3.5 Discussion

We have shown recent results for the linear deterministic model (LDM) of the interfering multiple access channel (IMAC) called LD-IMAC. These results are based on schemes which exploit the orthogonality of bit-levels. The common part and also the private part were partitioned into parts of Δ-bit-level size. Due to the special LDM structure, these parts do not overlap and a form of interference alignment on the signal-scale can be achieved. Therefore, interference can be aligned in half of its signal-scale dimensions and multi-user gain can be achieved. Due to the Δ-based alignment structure, the achievable sum-rate is dependent on the ratio of the total interference strength divided by Δ and forms a step-like function (see Fig. 3.4). Moreover, we have shown that an upper bound can be found, which forms a linear function on top of the achievable scheme. The upper bound coincides with the achievable sum-rate at points where the $\frac{n_i}{\Delta}$ ratio is *even*. Therefore, the sum-rate is only optimal at these points. These results can be extended towards the LDM interfering broadcast channel (LD-IBC), which shows an *duality* between both cellular models. This means, that both, the achievable sum-rate and the upper bound, are the same for both models. Since the LDM only approximates the Gaussian IMAC, it is interesting if these results also hold in the original model. We have shown that these LDM schemes can be transferred to the Gaussian IMAC/IBC through layered lattice coding schemes, as indicated in the introduction. Moreover, we presented the result of an extension of a proof in [2] which can show, that the Gaussian upper bound is upper bounded by the corresponding LDM upper bound to within a constant bit-gap. Although this transfers all results of the LD-IMAC/IBC to its Gaussian equivalent, the lattice coding structure is directly based on the underlying LDM structure and it therefore inherits the problems with the step-like achievable sum-rate. As we have seen, this sum-rate is not optimal at all points and the question arises if the upper bound can be even lowered or the LDM-based achievable scheme is not good enough. It turns out, that the problem is related to the LDM. In fact, the LDM approximation of the fine channel gain to 1 is an oversimplification, resulting in a restricted class of achievable schemes. In Sect. 3.4 we show, that the lower triangular deterministic model (LTDM), introduced in [14], solves this problem by incorporating the exact fine channel gain as binary expansion into the model. This enables a broader class of achievable schemes, which are not limited to orthogonal bit-level structures anymore. This broader class of achievable schemes can reach the upper bound of the *LD*-IMAC within a constant bit-gap and an arbitrary small outage set. Moreover, similar techniques as in the LD-IMAC can yield an upper bound for the LTD-IMAC. This upper bound also coincides with the new LTD sum-rate within a constant bit-gap. We have shown these results for the whole interference regime of the LTD-IMAC. Figure 3.9 shows the GDoF for the symmetric LTD-IMAC in comparison to the symmetric LTD-IC. Multi-user gain is present in the whole interference range. Moreover, the lowest DoF point is at $\alpha = 1$ and $\frac{2}{3}$ DoF, which is in accordance with previous results for vector space alignment methods in [19], see Eq. (3.3). Another observation is, that the interference-free DoF can be reached for $\alpha \geq \frac{3}{2}$ in the LTD-IMAC, whereas for the IC-case, α has to be bigger than 2 to reach it. Current research investigates possible duality relations between the LTD-IMAC and LTD-IBC. The scheme structure of the LTD-IMAC suggest, that a transfer to the corresponding LTD-IBC is possible. Moreover, the major goal is to have results for the Gaussian channel. Layered lattice coding schemes, which were used for the LD-IMAC/IBC, are not longer working in the LTDM case, since bit-levels overlap in cells. But the results for the X-channel in [14] suggest, that LTD schemes can provide insights for possible Gaussian schemes. In particular, LTDM bit-level allocation would constraint the input constellation points such that they can be decoded for all channel gains, except an arbitrary small outage set.

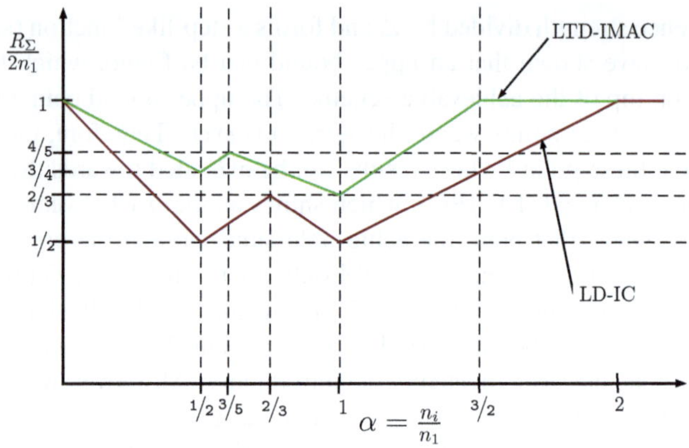

Fig. 3.9 A GDoF comparison between the symmetric LTD-IMAC and the symmetric LD-IC is shown

Acknowledgments The work of Rick Fritschek was supported by the German Research Foundation (DFG) under Grants WU 598/1-3 and WU 598/2-3.

References

1. Avestimehr S, Diggavi SN, Tse DNC (2011) Wireless network information flow: a deterministic approach. IEEE Trans Inf Theory 57(4):1872-1905
2. Bresler G, Tse D (2008) The two-user gaussian interference channel: a deterministic view. Euro Trans Telecommun 19(4):333–354
3. Bresler G, Parekh A, Tse D (2010) The approximate capacity of the many-to-one and one-to-many Gaussian interference channels. IEEE Trans Inf Theory 56(9):4566–4592
4. Bühler J, Wunder G (2011) On interference alignment and the deterministic capacity for cellular channels with weak symmetric cross links. In: IEEE international symposium on information theory proceedings (ISIT), Saint-Petersburg, Russia
5. Cadambe V, Jafar S (2008) Interference alignment and degrees of freedom of the k-user interference channel. IEEE Trans Inf Theory 54(8):3425–3441
6. Etkin R, Tse D, Wang H (2008) Gaussian interference channel capacity to within one bit. IEEE Trans Inf Theory 54(12):5534–5562
7. Fritschek R, Wunder G (2014) Enabling the multi-user generalized degrees of freedom in the gaussian cellular channel. In: Proceedings of the IEEE information theory workshop (ITW), Hobart, Australia
8. Fritschek R, Wunder G (2014) Upper bounds and duality relations of the linear deterministic sum capacity for cellular systems. In: Proceeding of the IEEE international conference on communications (ICC), Sydney, Australia
9. Fritschek R, Wunder G (2015) Constant-gap sum-capacity approximation of the deterministic interfering multiple access channel. In: Proceedings of the IEEE international symposium on information theory (ISIT), Hong Kong, China, to appear
10. Fritschek R, Wunder G (2015) Deterministic imac revisited: constant-gap capacity in the weak interference case. In: Proceeding of the IEEE international conference on communications (ICC), London, UK, to appear

11. Jindal N, Vishwanath S, Goldsmith A (2004) On the duality of Gaussian multiple-access and broadcast channels. IEEE Trans Inf Theory 50(5):768–783
12. Loeliger HA (1997) Averaging bounds for lattices and linear codes. IEEE Trans Inf Theory 43(6)
13. Maddah-Ali M, Motahari A, Khandani A (2008) Communication over mimo x channels: interference alignment, decomposition, and performance analysis. IEEE Trans Inf Theory 54(8):3457–3470
14. Niesen U, Maddah-Ali M (2013) Interference alignment: from degrees of freedom to constant-gap capacity approximations. IEEE Trans Inf Theory 59(8):4855–4888
15. Raja A, Prabhakaran V, Viswanath P (2009) Reciprocity in linear deterministic networks under linear coding. In: IEEE information theory workshop on networking and information theory (ITW), Taormina, Italy
16. Saha S, Berry RA (2011) Sum-capacity of a class of k-user gaussian interference channels within $o(k \log k)$ bits. In: Allerton conference 2011
17. Saha S, Berry RA (2012) Symmetric k-user gaussian interference channels: approximate sum-capacity via deterministic modeling. In: Allerton conference 2012
18. Sridharan S, Jafarian A, Vishwanath S, Jafar SA, Shamai S (2008) A layered lattice coding scheme for a class of three user gaussian interference channels. In: Allerton conference 2008
19. Suh C, Tse D (2008) Interference alignment for cellular networks. In: Proceedings of the allerton conference on communication, control, and computing, Monticello, IL

Chapter 4
The Information-Theoretic Constant-Gap Optimality of Treating Interference as Noise in Interference Networks

Soheil Gherekhloo, Anas Chaaban and Aydin Sezgin

Abstract Treating interference as noise is one of the simplest methods for the management of interference in wireless networks. Despite its simplicity, treating interference as noise (TIN) was shown to be information-theoretically optimal for certain Gaussian interference channels (IC) with very-weak (noisy) interference. In this chapter, we consider cellular networks, such as networks that consists of a point-to-point channel interfering with a multiple access channel (MAC). The sum-capacity of such networks is studied with main focus on the constant-gap optimality of TIN rather than on its exact optimality. It turns out that TIN in its naive variant, where all transmitters are active and receivers use TIN for decoding, is not the best choice for certain networks. In fact, a scheme that combines both time division multiple access and TIN (TDMA-TIN) strictly outperforms the naive TIN scheme. Furthermore, it is shown that in some regimes, TDMA-TIN achieves the sum-capacity within a constant gap for Gaussian networks. Additionally, it is shown that, even for very-weak interference, there are some regimes where a combination of interference alignment with power control and treating interference as noise at the receiver side outperforms TDMA-TIN. As a consequence, on the one hand treating interference as noise in a cellular uplink is approximately optimal in certain regimes. On the other hand, those regimes cannot be simply described by the strength of interference, requiring a careful design of wireless networks.

S. Gherekhloo (✉) · A. Sezgin
Institute for Digital Communication Systems, Ruhr-University Bochum,
44801 Bochum, Germany
e-mail: soheil.gherekhloo@rub.de

A. Sezgin
e-mail: aydin.sezgin@rub.de

A. Chaaban
Department of Electrical Engineering, King Abdullah University of Science and Technology, Thuwal, Saudi Arabia
e-mail: anas.chaaban@kaust.edu.sa

W. Utschick (ed.), *Communications in Interference Limited Networks*,
Signals and Communication Technology, DOI 10.1007/978-3-319-22440-4_4

4.1 Introduction

Recent surveys show that the demand for higher data rates has increased to a level such that the supply offered by the current technology used in wireless networks is insufficient. This is due to the fact (among other issues) that current technology is aiming at keeping the users in a network well separated in utilizing the resources, be it time, code and/or frequency by relying on orthogonal multiple access protocols like time-division multiple access (TDMA), code-division multiple access (CDMA) or frequency-division multiple access (FDMA). However, the trend towards higher demand will increase and even accelerate with the dawn of the internet-of-things (IoT) era.

The new generation of wireless networks approach this challenge in multiple ways. One approach is to reduce the cell sizes of the network by setting up more base-stations, which results in a densification of the network. Yet another approach is to open the access to the resources such that multiple users can have access simultaneously and transmit (or receive) at the same time within the same bandwidth (in-band). This change of paradigm leads to networks, which are more and more affected by the interference within a cell as well as from the interference from neighboring cells. Thus, essentially we are facing a system which is interference limited rather than being noise-limited. As a consequence, new methods and techniques have been designed to deal with the interference, among which there are candidates deployed at the receivers such as successive interference decoding and cancellation [1]. Either as a stand-alone method or combined with a processing at the transmitter such as superposition coding [1], those algorithms are able to improve the performance of the system in some cases (depending on the operating point of the system) significantly. This is especially true for interference alignment, a strategy which is by now an integral part of communication strategies aiming at performance gains in regimes with high signal-to-noise power ratios [2]. One drawback of those schemes is the involved coordination between the nodes which is needed so that the schemes can unfold their full potential. Furthermore, the complexity of those schemes is in general high at both the transmitter and the receiver which precludes their employment in practical systems. Alternatively, one could simply treat the interference as noise. This sounds, also very low in complexity, to be a suboptimal approach. Indeed, this is one of the widespread strategies on how interference is dealt with in practical systems. The question addressed in this chapter is, whether treating interference as noise is a suboptimal approach in general? Or, from the opposite perspective, are there conditions with respect to the communication structure/topology such that the simple scheme of treating interference as noise is optimal? Although very provocative at first, a positive outcome would be very important for the system designer as this would not require a major shift from the current policy. In more details, the system designer would have the insight that in certain cases, the information theoretic optimal strategy on how to deal with the interference is simply to treat it as noise and there is no need to employ more sophisticated (and thus more complex) schemes as they are not providing a better performance. Now, in order to quantize the performance obtained by treating interference as noise, upper bounds are needed for comparison purposes as this would indicate how close (or how far) one can get to obtain optimal performance. This characterization could be done by exact measures whenever they are within reach or by utilizing approximate performance measures as a doable alternative. In this chapter, we consider different signal-to-noise power ratio SNR regimes to allow varying viewpoints on the optimality of treating interference as noise. For this, we introduce several measures such as the channel capacity and the generalized degrees of freedom to quantitatively determine the performance of treating interference as noise.

4.2 Optimality of Treating Interference as Noise

In order to study the optimality of treating interference as noise, let us first consider the smallest possible scenario where interference becomes a major issue. This scenario is the so-called interference channel [3]. The interference channel consists of two transmitter-receiver pairs interfering with each other as shown in Fig. 4.1. The inputs of the channel are random variables X_1 and X_2 with powers P_1 and P_2, and the outputs are:

$$Y_k = h_d X_k + h_{cj} X_j + Z_k,$$

where $j \neq k$, $j, k \in \{1, 2\}$. Here, Z_k is a Gaussian noise with zero mean and unit variance (normalized). The channels $h_{cj} \in \mathbb{R}$ represent interference caused by Tx j and h_d denotes the desired channel.

A fundamental information-theoretic criterion is the capacity of this channel. The capacity is defined as the highest achievable communication rate, measured in bits per transmission. Let us denote the rate of user k by R_k. Thus, the highest sum-rate $R_1 + R_2$ is of great interest since it indicates the overall performance of the whole system. We denote this highest sum-rate by C_Σ.

Fig. 4.1 A 2-user interference channel (IC)

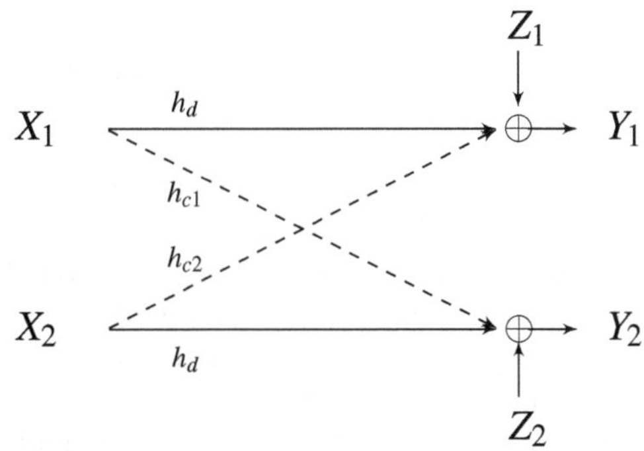

It is known that the capacity of an AWGN point-to-point channel with signal-to-noise ratio SNR is given by

$$C = \frac{1}{2}\log_2(1 + \mathsf{SNR}),$$

which is achievable by random-coding [4]. Now the following question arises: What is the best code for an IC? In particular, what is to be done with the interference in the IC? One might decode the interference, avoid it, or ignore it. The last variant is of practical interest due to its simplicity. By ignoring interference here we mean that it is treated as noise. Thus, each transmitter-receiver pair is treated as an AWGN point-to-point channel with a signal-to-interference-plus-noise ratio of

$$\mathsf{SINR}_k = \frac{\mathsf{SNR}_k}{1 + \mathsf{INR}_k},$$

where $\mathsf{SNR}_k = h_d^2 P_k$, and $\mathsf{INR}_k = h_{cj}^2 P_j$. The achievable rate can thus be written simply as

$$R_k = \frac{1}{2}\log_2(1 + \mathsf{SINR}_k), \quad k = 1, 2. \tag{4.1}$$

Despite its simplicity, the optimality of this scheme has been shown for the IC with what came to be known as "noisy interference". The conditions for noisy interference will be presented later. But first, let us catch the story from its beginning with the work of Etkin et al. [5].

4.2.1 GDOF Perspective

The achievable rate in (4.1) provides a lower bound on the sum-capacity C_Σ given by

$$C_\Sigma \geq \sum_{k=1}^{2} \frac{1}{2}\log_2(1 + \mathsf{SINR}_k) = \underline{C}_\Sigma. \tag{4.2}$$

To examine the tightness of this lower bound, one needs an upper bound for comparison. Several upper bounds on the capacity of the IC existed since the start of the study of this problem. Among these bounds are the cut-set bounds and Kramer's Genie-aided bound [6]. However, both bounds do not coincide with the achievable sum-rate $\underline{C}_\Sigma$. A new upper bound is required.

The main idea of the required bound came from the work in [5] on the approximate capacity of the IC. Let us restrict the attention to the symmetric IC for now. That is, $h_{c1} = h_{c2} = h_c$, and $P_1 = P_2 = P$. In their work, Etkin et al. proved that

$$C_\Sigma \leq \log_2\left(1 + \mathsf{INR} + \frac{\mathsf{SNR}}{1 + \mathsf{INR}}\right), \tag{4.3}$$

where $\mathsf{INR} = h_c^2 P$. The idea of this bound is based on a "genie" giving side information to the receivers. The side information given to receiver k is $h_{ck}X_k + Z_j$. Then the upper bound follows by using some standard information-theoretic steps. Etkin et al. further noticed that if we write $\mathsf{INR} = \rho^\alpha$ and $\mathsf{SNR} = \rho$ with $\alpha > 0$, then the upper bound above approaches

$$\underline{C}_\Sigma = \log_2(1 + \mathsf{SINR})$$

as $\mathsf{SNR} \to \infty$, as long as $\alpha \leq \frac{1}{2}$. This approach was denoted as a generalized degrees-of-freedom (GDoF) approach. The GDoF of the symmetric IC is defined as

$$d(\alpha) = \lim_{\rho \to \infty} \frac{C_\Sigma(\alpha, \rho)}{\frac{1}{2}\log_2(\rho)}.$$

If we compare the GDoF corresponding to the upper bound (4.3) and the GDoF corresponding to the lower bound (4.2) in the regime of interest ($\alpha \leq \frac{1}{2}$), we can see both have a GDoF of

$$2 - 2\alpha.$$

Thus, the treating interference as noise lower bound and the Etkin-Tse-Wang (ETW) upper bound coincide from GDoF point of view.

Back to the asymmetric IC, and the sum-capacity. In this case, treating interference as noise achieves the sum-capacity of the IC within a constant gap (within 1 bit) as long as $\alpha_1 + \alpha_2 \leq 1$, where $\alpha_i = \frac{\log_2(\mathsf{INR}_i)}{\log_2(\rho)}$.

Building on the idea of the upper bound of ETW, the exact sum-capacity of the IC with noisy interference was characterized by [7–9], as we discuss in the following paragraphs.

4.2.2 *Finite SNR Regime*

It turns out the ETW upper bound in (4.3) can be tightened by replacing the ETW genie side-information given to receiver k ($h_{ck}X_k + Z_j$) by $S_k = h_{ck}X_k + W_k$, where W_k is Gaussian noise with variance η_k correlated with Z_k via a correlation coefficient $\rho_k \in [-1, 1]$. Here, η and ρ are free parameters that can be chosen to minimize the upper bound.

The resulting genie information satisfies two properties: Usefulness and smartness. A useful genie is one which allows deriving a closed-form upper bound on the sum-capacity (such as the ETW genie). A smart genie is one which does not

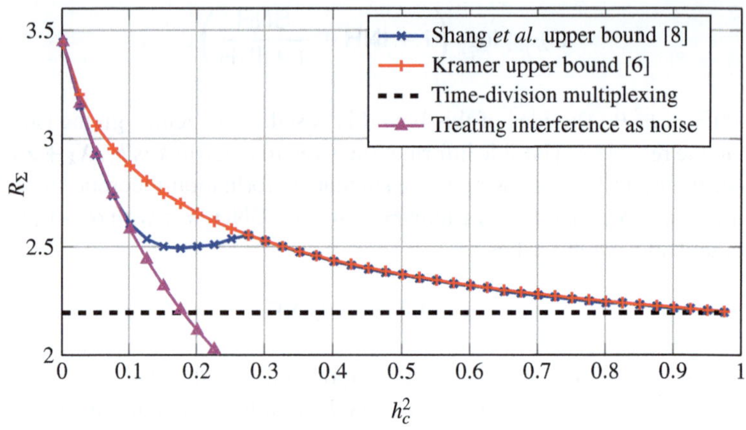

Fig. 4.2 Sum-capacity bounds for a symmetric interference channel with $P = 10$ and $h_d = 1$ versus the interference channel gain. Notice the optimality of treating interference as noise for low h_c^2

increase the capacity of the channel. The modified genie S_k is useful regardless of the parameters of the channel and the genie, and is smart if the IC satisfies

$$|h_{c2}(1 + h_{c1}^2 P_1)| + |h_{c1}(1 + h_{c2}^2 P_2)| < 1 \tag{4.4}$$

for $h_d = 1$.[1] The condition above is known as the noisy interference condition, and the corresponding IC is a noisy IC. The sum-capacity of an IC satisfying this condition is given by

$$C_\Sigma = \sum_{k=1}^{2} \frac{1}{2} \log_2(1 + \mathsf{SINR}_k). \tag{4.5}$$

Figure 4.2 shows upper bounds and lower bounds on the sum-capacity of a symmetric IC versus the interference channel gain h_c^2 with $h_d = 1$. For this channel, the noisy interference regime is given by $|h_c(1 + h_c^2 P)| < 0.5$. As the interference increases beyond this regime, treating interference as noise looses its optimality, and is outperformed by time-division multiplexing. It is worth to note at this point that further upper bounds on the capacity of the interference channel have been derived in [10, 11], which generalize the bounds by Kramer [6] and Shang et al. [8], and are tighter in the range that shows a large gap in Fig. 4.2. Furthermore, the noisy interference condition has also been derived for the K-user interference channel (consisting of K transmitter-receiver pairs) in [12].

[1]Note that we can always assume $h_d = 1$ without loss of generality, by normalization of the IC.

The results so far aim at characterizing the regimes exactly, in which treating interference as noise is capacity optimal. This approach requires deriving upper bounds which are (using the terminology of [7]) both useful and smart, i.e., they are computable in closed form and tight. Only then, we are able to state the exact optimality. The methods for achieving this goal are rather limited, and this is one of the main reasons why the progress on determining the capacity of more general networks is slow. In order to achieve some progress, an alternative method has to be used. This method aims for approximate capacity characterizations, i.e., deriving upper bounds which are within an additive constant gap from the rate achievable by treating interference as noise. In some way, this is following the footsteps of [5], where this was performed for the basic IC. In the following, we report the findings when taking this philosophy to the analysis of interference networks.

4.3 Approximate Optimality

In this section, we study the approximate optimality of treating interference as noise (TIN) for the point-to-point system interfering with a multiple access system (PIMAC). First, we introduce the system model in details.

4.3.1 System Model

The system we consider consists of three transmitters and two receivers as shown in Fig. 4.3. Transmitter 1 (Tx1) and transmitter 3 (Tx3) constitute a MAC with their desired receiver, namely Rx1. The communication over the MAC is interfered by Tx2 which communicates with Rx2.

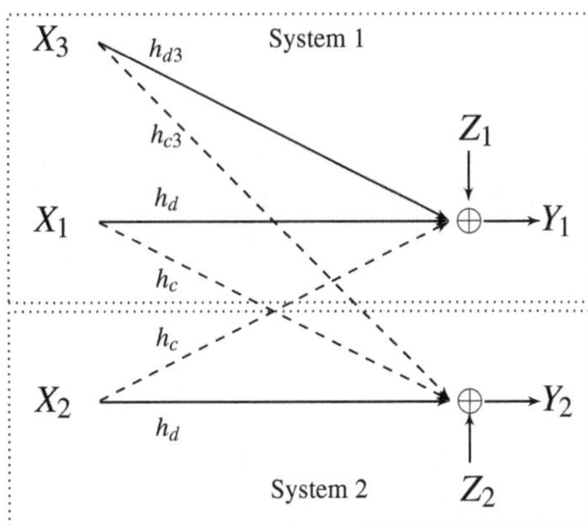

Fig. 4.3 System model of the Gaussian PIMAC

In more details, Tx1 and Tx3 want to communicate with Rx1 and Tx2 wants to communicate with Rx2. To do this, Txi, $i \in \{1, 2, 3\}$, sends a codeword of length n symbols $X_i^n = (X_{i1}, \ldots, X_{in}) \in \mathbb{R}^n$. The transmitters of the PIMAC have an average power constraint P. Hence, the following condition must be satisfied

$$\frac{1}{n}\sum_{t=1}^{n} \mathbb{E}[X_{it}^2] = P_i \leq P. \tag{4.6}$$

After n transmissions, the received signals of the two receivers denoted Y_1^n and Y_2^n can be written as

$$Y_1^n = h_d X_1^n + h_c X_2^n + h_{d3} X_3^n + Z_1^n, \tag{4.7}$$
$$Y_2^n = h_c X_1^n + h_d X_2^n + h_{c3} X_3^n + Z_2^n, \tag{4.8}$$

where Z_j^n, $j \in \{1, 2\}$, are independent sequences representing the additive white Gaussian noise (AWGN) at the receivers, and the constants h_k, $k \in \{d, c, d3, c3\}$ represent the real positive (static) channel coefficients. Then noise sequences are independent and identically distributed Gaussian random variables with zero mean and unit variance. We assume that global channel state information is available at all nodes.

Since we are interested in the interference limited scenario, we assume that all signal-to-noise and interference-to-noise ratios are larger than 1, i.e.,

$$\min\{h_d^2, h_c^2, h_{d3}^2, h_{c3}^2\}P > 1. \tag{4.9}$$

Rx1 decodes $\hat{X}_1^n$ and $\hat{X}_3^n$ from its received signal Y_1^n, and Rx2 decodes $\hat{X}_2^n$ from Y_2^n. The probability of error, achievable rates R_1, R_2, R_3, and capacity region $\mathcal{C}$ are defined in the standard Shannon sense [1]. The sum capacity is the maximum achievable sum-rate which is given by

$$C_\Sigma = \max_{(R_1,R_2,R_3)\in\mathcal{C}} R_\Sigma, \tag{4.10}$$

where $R_\Sigma = \sum_{i=1}^{3} R_i$. In this work, we study an approximation of the capacity for the PIMAC at high SNR, namely the GDoF. To present the definition of the GDoF, we need to define the capacity pre-log parameters (α's). Let α_k be defined as

$$\alpha_k = \frac{\log_2(h_k^2 P)}{\log_2(\rho)}, \text{ where } k \in \{d, c, d3, c3\}, \tag{4.11}$$

where $\frac{1}{2}\log(\rho)$ approximates the capacity of the reference point to point channel with $\rho > 1$. Hence, we can write $h_k^2 P = \rho^{\alpha_k}$. Doing this, the SINR at Rx2 when all Tx's send with full power can be written as

$$\mathsf{SINR}_2 = \frac{\rho^{\alpha_d}}{\rho^{\alpha_c} + \rho^{\alpha_{c3}} + 1}. \tag{4.12}$$

The parameter α_k is an approximate ratio between the capacity of a P2P channel and the reference channel. Then, the GDoF of the PIMAC is defined as

$$d(\boldsymbol{\alpha}) = \lim_{\rho \to \infty} \frac{C_{\Sigma}(\boldsymbol{\alpha}, \rho)}{\frac{1}{2}\log_2 \rho}, \tag{4.13}$$

where $\boldsymbol{\alpha} = (\alpha_d, \alpha_c, \alpha_{d3}, \alpha_{c3})$. In the next section, we present simple ways of dealing with interference in PIMAC.

4.3.2 Simple Interference Management in PIMAC

In this sub-section, we studying the (sub-)optimality of transmission schemes that are not too demanding in terms of computing and decoding complexity. In what follows, we present three simple ways of dealing with interference.

- **Naive-TIN**: In this scheme, all Tx's send their signals simultaneously during the whole transmission with their maximum power. At the receiver side, the strategy is the same as if there is no interference. Hence, the Rx's decode their desired messages while they completely ignore the interference. Doing this, Rx1 decodes X_1^n and X_3^n as in the multiple access channel, and Rx2 decodes X_2^n as in a point-to-point channel. The achievable sum-rate is given by

$$R_{\Sigma,\text{TIN}} \leq \frac{1}{2}\log_2\left(1 + \frac{\rho^{\alpha_d} + \rho^{\alpha_{d3}}}{1 + \rho^{\alpha_c}}\right) + \frac{1}{2}\log_2\left(1 + \frac{\rho^{\alpha_d}}{1 + \rho^{\alpha_c} + \rho^{\alpha_{c3}}}\right). \tag{4.14}$$

  This scheme is the simplest variant of the transmission schemes considered here, since no coordination between Tx's is required and the Rx's do not need to decode the interference. Interestingly, despite of the simplicity of this scheme, it is optimal in some regimes of many networks such as the 2-user IC [5, 7–9], the K-user IC [12], and the X channel [13]. This has been discussed for the 2-user IC in Sect. 4.2.2.
- **Interference avoidance (IA)**: Compared to Naive-TIN, in this scheme, some Tx's stay silent and thus cause no interference to unintended Rx's. This can be done, by avoiding more than one active Tx, unless the active Tx's want to communicate with the same receiver. Applying this scheme, to the PIMAC, we have four cases of choosing the set of active Tx's. These four cases are illustrated in Fig. 4.4. While in the first three cases, we have only one active Tx, in the remaining case, Tx1 and Tx3 communicate simultaneously with Rx1 as in the multiple access channel. Using this scheme, the following sum-rate is achievable

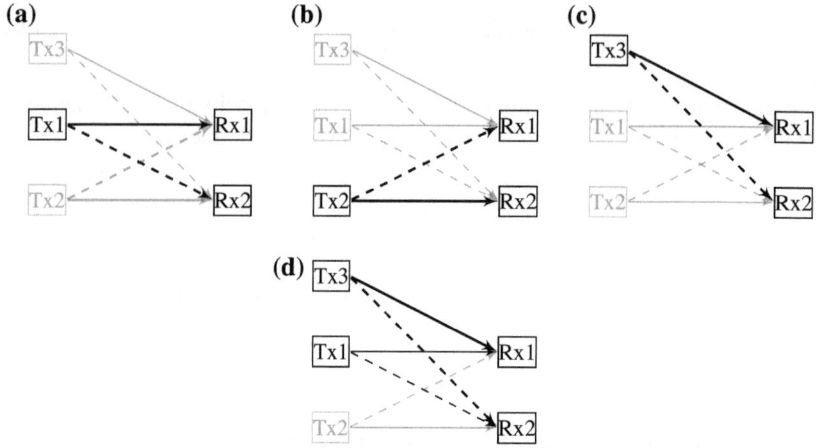

Fig. 4.4 This figure illustrates different cases of interference avoidance. In all figures, the *solid* and the *dashed lines* represent the desired and interference channels, respectively. While in (**a**)–(**c**), we have only one active transmitter, in case (**d**), both MAC users are active. As it is shown, in all four cases, the desired receiver of the active transmitters does not experience any interference

$$R_{\Sigma,\text{IA}} = \max\left\{\frac{1}{2}\log_2\left(1+\rho^{\alpha_d}\right), \frac{1}{2}\log_2\left(1+\rho^{\alpha_{d3}}\right)\right\}. \tag{4.15}$$

- **Interference avoidance with TIN (IA-TIN)** This scheme which resembles a trade of between naive-TIN and interference avoidance, is inspired by the TIN-optimality result for the 2-user IC. As it is well known by now, there is a regime for the 2-user IC in which TIN is optimal [5, 7–9]. In this scheme, we reduce the PIMAC to the 2-user IC by switching either Tx1 or Tx3 off to avoid an additional interference to Rx2. The two active Tx's send their signals simultaneously with full power during the whole transmission. At the receiver side, Rx's decode their desired message while the interference is treated as noise. The two cases of IA-TIN is shown in Fig. 4.5. Using this scheme, the following sum-rate is achievable

$$\begin{aligned} R_{\Sigma,\text{IA-TIN}} = \max\Bigg\{&\frac{1}{2}\log_2\left(1+\frac{\rho^{\alpha_d}}{1+\rho^{\alpha_c}}\right)+\frac{1}{2}\log_2\left(1+\frac{\rho^{\alpha_d}}{1+\rho^{\alpha_c}}\right), \\ &\frac{1}{2}\log_2\left(1+\frac{\rho^{\alpha_{d3}}}{1+\rho^{\alpha_c}}\right)+\frac{1}{2}\log_2\left(1+\frac{\rho^{\alpha_d}}{1+\rho^{\alpha_{c3}}}\right)\Bigg\}. \end{aligned} \tag{4.16}$$

In what follows, we study the optimality of these schemes with respect to the GDoF. To do this, we need upper bounds on the capacity of the PIMAC, which is presented next.

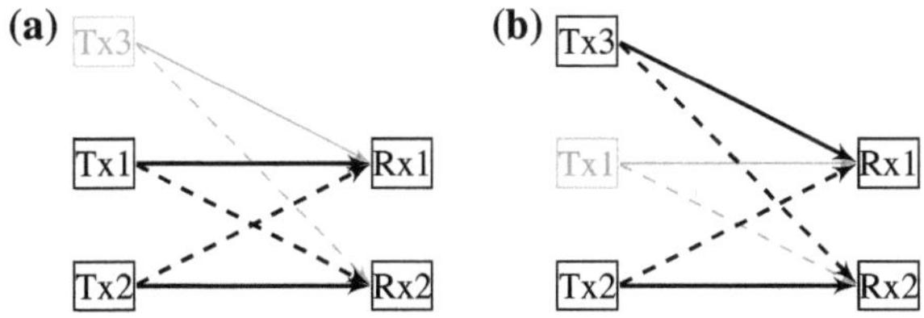

Fig. 4.5 Using IA-TIN, the channel operates as a 2-user IC. The *solid* and the *dashed lines* represent the desired and interference channels, respectively. As it can be seen, while Tx2 is active in both cases, Tx1 and Tx3 which are MAC transmitters are not active simultaneously. Hence, in these cases, PIMAC is reduced to a 2-user IC

4.3.3 Upper Bound for the Capacity of the PIMAC

The sum-capacity of the Gaussian PIMAC is upper bounded by

$$C_\Sigma \leq \frac{1}{2}\log_2\left(1+\rho^{\alpha_c}+\rho^{\alpha_{d3}}+\frac{\rho^{\alpha_d}}{1+\rho^{\alpha_c}}\right)+\frac{1}{2}\log_2\left(1+\rho^{\alpha_c}+\frac{\rho^{\alpha_d}}{1+\rho^{\alpha_c}}\right), \tag{4.17}$$

$$C_\Sigma \leq \frac{1}{2}\log_2\left(1+\rho^{\alpha_d}+\frac{1}{2}\rho^{\alpha_c}+\frac{\rho^{\alpha_{d3}}}{1+\rho^{\alpha_{c3}}}\right)+\frac{1}{2}\log_2\left(1+\rho^{\alpha_{c3}}+\frac{\rho^{\alpha_d}}{1+\rho^{\alpha_c}}\right), \tag{4.18}$$

$$C_\Sigma \leq \frac{1}{2}\log_2\left(1+\rho^{\alpha_c}+\frac{\rho^{\alpha_d}+\rho^{\alpha_{d3}}}{1+\rho^{\alpha_c-\alpha_d}(\rho^{\alpha_d}+\rho^{\alpha_{d3}})}\right) \tag{4.19}$$
$$+\frac{1}{2}\log_2\left(1+\rho^{\alpha_{c3}}+\rho^{\alpha_c}+\frac{\rho^{\alpha_d}}{1+\rho^{\alpha_c}}\right)+1 \quad \text{if} \quad \alpha_{d3}-\alpha_d \leq \alpha_{c3}-2\alpha_c,$$

$$C_\Sigma \leq \frac{1}{2}\log_2\left(1+\rho^{\alpha_c}+\frac{\rho^{\alpha_d}+\rho^{\alpha_{d3}}}{1+\rho^{\alpha_{c3}-\alpha_{d3}}(\rho^{\alpha_d}+\rho^{\alpha_{d3}})}\right) \tag{4.20}$$
$$+\frac{1}{2}\log_2\left(1+\rho^{\alpha_{c3}}+\rho^{\alpha_c}+\frac{\rho^{\alpha_d}}{1+\rho^{\alpha_c}}\right)+1, \quad \text{if} \quad \alpha_d-\alpha_{d3} \leq \alpha_c-2\alpha_{c3}.$$

Details of the proof of the upper bounds are provided in [14]. Shortly, all the upper bounds are established by using the so called genie-aided method, in which some additional well chosen side information is provided to some specific nodes in the system resulting in bounds which are both computable and reasonably tight. In the first and second bounds, we provide as side information the message of one interferer to Rx2. Hence, the resulting PIMAC is more capable than the original one and the capacity of the former is an upper bound for the capacity of the latter. The enhanced PIMAC channel is treated similar to the IC. In order to establish the latter two bounds, we provide a noisy weighted sum of the desired signals to each Rx.

In the following section, the GDoF optimality of the schemes proposed in Sect. 4.3.2 is studied.

4.3.4 GDoF Optimality of Simple Schemes

In this analysis, we focus on regimes where

$$\alpha_c \leq \frac{\alpha_d}{2}. \tag{4.21}$$

This means that in the symmetric 2-user IC with nodes Tx1, Tx2, Rx1, and Rx2, the sum of the powers of the interference signals at both Rx's is less than or equal to the power of the desired received signal at each Rx on logarithmic scale. The reason of this assumption is that by switching Tx3 off, the PIMAC setup is reduced into a symmetric 2-user IC with the desired and interference channel parameters α_d and α_c, respectively. We know from [5] that TIN is GDoF optimal in the 2-user IC as long as the condition in (4.21) is satisfied. Our goal is to study the GDoF optimality of the schemes presented in Sect. 4.3.2 when we have an additional transmitter to the 2-user IC which operates in the noisy interference regime. To do this, we compare the achievable GDoF of these schemes with the upper bounds on the GDoF of the PIMAC. In what follows, first we convert the achievable sum-rates in (4.14)–(4.16) to the achievable GDoF. Next, we derive the upper bounds on the GDoF from (4.17)–(4.20). Finally, the achievable GDoF for each scheme is compared with upper bounds.

Using the definition of the GDoF in (4.13) together with the interference limited constraint in (4.9) and the assumption in (4.21), we convert the achievable sum-rate of TIN given in (4.14) into GDoF as follows

$$\begin{aligned}\underline{d}_{\text{TIN}}(\boldsymbol{\alpha}) &= \lim_{\rho\to\infty} \frac{R_{\Sigma,\text{TIN}}}{\frac{1}{2}\log_2(\rho)} \\ &\geq \lim_{\rho\to\infty} \frac{\log_2\left(\frac{\rho^{\max\{\alpha_d,\alpha_{d3}\}}}{2\rho^{\alpha_c}}\right) + \log_2\left(\frac{\rho^{\alpha_d}}{3(\rho^{\max\{\alpha_c,\alpha_{c3}\}})}\right)}{\log_2(\rho)} \\ &= \max\{\alpha_d, \alpha_{d3}\} - \alpha_c + (\alpha_d - \max\{\alpha_c, \alpha_{c3}\})^+. \end{aligned} \tag{4.22}$$

Similarly, we can write the achievable GDoF of IA and IA-TIN from (4.15) and (4.16), respectively. Doing this we obtain

$$\underline{d}_{\text{IA}}(\boldsymbol{\alpha}) \leq \max\{\alpha_d, \alpha_{d3}\}, \tag{4.23}$$

$$\underline{d}_{\text{IA-TIN}}(\boldsymbol{\alpha}) \leq \max\{2(\alpha_d - \alpha_c), (\alpha_{d3} - \alpha_c)^+ + (\alpha_d - \alpha_{c3})^+\}. \tag{4.24}$$

Next, we convert the upper bounds on the capacity of the PIMAC to GDoF upper bounds. Consider the expression in (4.17). This can be rewritten as

$$C_\Sigma \leq \frac{1}{2}\log_2\left(3\rho^{\max\{\alpha_c,\alpha_{d3}\}} + \frac{\rho^{\alpha_d}}{\rho^{\alpha_c}}\right) + \frac{1}{2}\log_2\left(2\rho^{\alpha_c} + \frac{\rho^{\alpha_d}}{\rho^{\alpha_c}}\right)$$

$$\leq \frac{1}{2}\log_2\left(4\rho^{\max\{\alpha_c,\alpha_{d3},\alpha_d-\alpha_c\}}\right)+\frac{1}{2}\log_2\left(3\rho^{\max\{\alpha_c,\alpha_d-\alpha_c\}}\right)$$
$$=\frac{1}{2}\log_2\left(\rho^{\max\{\alpha_{d3},\alpha_d-\alpha_c\}}\right)+\frac{1}{2}\log_2\left(\rho^{\alpha_d-\alpha_c}\right)+1+\frac{1}{2}\log_2 3. \tag{4.25}$$

Now, dividing this upper bound by $\frac{1}{2}\log_2(\rho)$ and letting $\rho \to \infty$, we obtain the following upper bound for the GDoF

$$d(\boldsymbol{\alpha}) \leq \max\{\alpha_{d3}, \alpha_d - \alpha_c\} + \alpha_d - \alpha_c. \tag{4.26}$$

Similarly, the upper bounds in (4.18)–(4.20) are converted into GDoF upper bounds as follows

$$d(\boldsymbol{\alpha}) \leq \max\{\alpha_d, \alpha_{d3} - \alpha_{c3}\} + \max\{\alpha_{c3}, \alpha_d - \alpha_c\}, \tag{4.27}$$
$$d(\boldsymbol{\alpha}) \leq \max\{\alpha_c, \max\{\alpha_d, \alpha_{d3}\} - \max\{\alpha_c, \alpha_c - \alpha_d + \alpha_{d3}\}\} + \max\{\alpha_{c3}, \alpha_d - \alpha_c\} \quad \text{if} \quad \alpha_{d3} - \alpha_d \leq \alpha_{c3} - 2\alpha_c, \tag{4.28}$$
$$d(\boldsymbol{\alpha}) \leq \max\{\alpha_c, \max\{\alpha_d, \alpha_{d3}\} - \max\{\alpha_{c3}, \alpha_{c3} - \alpha_{d3} + \alpha_d\}\} + \max\{\alpha_{c3}, \alpha_d - \alpha_c\} \quad \text{if} \quad \alpha_d - \alpha_{d3} \leq \alpha_c - 2\alpha_{c3}. \tag{4.29}$$

Now, we need to compare the upper bounds for the GDoF of the PIMAC given in (4.26)–(4.29) with the achievable GDoF in (4.22)–(4.24). The scheme $\mathcal{S} \in$ {TIN, IA, IA-TIN} is an optimal scheme in terms of GDoF when its achievable GDoF coincides with the upper bound. For sake of simplicity in presenting the results, we divide the parameter space of the PIMAC into several regimes based on the optimality of the considered transmission scheme in the respective regime. In our analysis, we exclude the case $\alpha_{d3} - \alpha_{c3} = \alpha_d - \alpha_c$, which is equivalent to $\frac{h_{d3}}{h_{c3}} = \frac{h_d}{h_c}$, since in this case, by defining $\tilde{X}_1 = X_1 + \frac{h_{d3}}{h_d} X_3$, the received signal at Rx1 and Rx2 is given by

$$Y_1 = h_d \tilde{X}_1 + h_c X_2 + Z_1 \tag{4.30}$$
$$Y_2 = h_d X_2 + h_c \tilde{X}_1 + Z_2. \tag{4.31}$$

In this degenerate case, the input-output relationship is similar to that of the IC. Hence, if $\alpha_{d3} - \alpha_{c3} = \alpha_d - \alpha_c$ holds, the PIMAC can be treated similar to the IC as in [5]. Thus, we exclude this case in the analysis to focus our study on the non-degenerate PIMAC.

For a PIMAC with $\alpha_c \leq \frac{\alpha_d}{2}$, we define regimes 1 to 3 (shown in Fig. 4.6) as follows:

- **Regime (1) (Tx3-off)**: $\alpha_{d3} \leq \alpha_d - \alpha_c$ or $\alpha_{d3} - (\alpha_d - 2\alpha_c) \leq \alpha_{c3} \leq \alpha_d - \alpha_c$
  - **Regime (1A)**: $\alpha_{d3} \leq \alpha_d - \alpha_c$ and $\alpha_{c3} \leq \alpha_c$,
  - **Regime (1B)**: $\alpha_{d3} \leq \alpha_d - \alpha_c$ and $\alpha_{c3} > \alpha_c$,
  - **Regime (1C)**: $\alpha_{d3} > \alpha_d - \alpha_c$, $\alpha_{c3} \leq \alpha_d - \alpha_c$ and $\alpha_{d3} - \alpha_{c3} \leq \alpha_d - 2\alpha_c$.

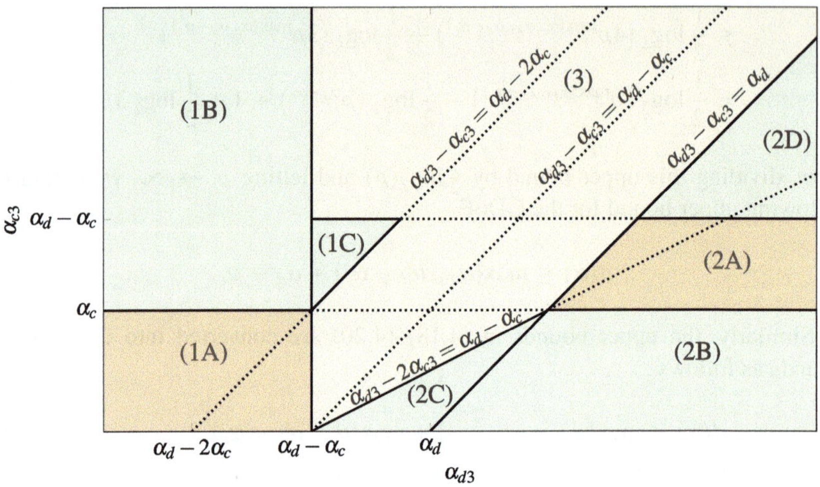

Fig. 4.6 The $(\alpha_{d3}, \alpha_{c3})$-plane of the parameter space of the PIMAC with $\alpha_c \leq \frac{\alpha_d}{2}$ divided into 3 regimes

- **Regime (2) (Tx1-off)**: $\min\{\alpha_{c3}, \alpha_c\} + \alpha_d - \alpha_c \leq \alpha_{d3} - \alpha_{c3}$
  - **Regime (2A)**: $\alpha_{d3} - \alpha_{c3} \geq \alpha_d$ and $\alpha_c \leq \alpha_{c3} \leq \alpha_d - \alpha_c$,
  - **Regime (2B)**: $\alpha_{d3} - \alpha_{c3} \geq \alpha_d$ and $\alpha_{c3} < \alpha_c$,
  - **Regime (2C)**: $\alpha_{d3} - 2\alpha_{c3} \geq \alpha_d - \alpha_c$ and $\alpha_{d3} - \alpha_{c3} < \alpha_d$,
  - **Regime (2D)**: $\alpha_{c3} + \alpha_d \leq \alpha_{d3}$ and $\alpha_{c3} > \alpha_d - \alpha_c$.
- **Regime (3) (All Tx's active)**: All remaining cases excluding $\alpha_{d3} - \alpha_{c3} = \alpha_d - \alpha_c$.

Now, we are ready to present the GDoF-optimal regimes of TIN, IA, or IA-TIN. Consider regime (1) first. Applying the conditions of this regime to the upper bounds in (4.26)–(4.29), we obtain

$$d(\boldsymbol{\alpha}) \leq 2(\alpha_d - \alpha_c). \tag{4.32}$$

By comparing this upper bound with the achievable GDoF in (4.22)–(4.24), we conclude that while IA-TIN is optimal within the whole regime (1) (Tx3-off), TIN is optimal only in sub-regime (1A). Moreover, the scheme IA is always sub-optimal. Note that in regime (1), the optimal strategy is to switch Tx3 off. Roughly speaking, this is due to the fact that in this regime, the capacity of the desired link between Tx3 to Rx1 is lower than the capacity of the desired link from Tx1 to Rx1 even after treating the interference of Tx2 as noise in the latter case. Notice that only, in the cases where the interference link between Tx3 to Rx2 is also weak (sub-regime (1A)), switching Tx2 on achieves the same GDoF as the IA-TIN. This is due to the fact that in this case, the interference caused by Tx3 at Rx2 is weaker that of Tx1.

Now, we consider regime (2) (Tx1-off). In this regime, we obtain the minimum of the upper bounds in (4.27) and (4.29) to write

$$d(\boldsymbol{\alpha}) \leq \max\{\alpha_{d3} - \alpha_{c3}, \alpha_d\} + \max\{\alpha_{c3}, \alpha_d - \alpha_c\}. \tag{4.33}$$

Now, we compare this upper bound with the achievable GDoF in (4.22)–(4.24). With the conditions given in this regime, we conclude that TIN is optimal only in sub-regime (2A). IA-TIN, on the other hand, achieves $\alpha_{d3} - \alpha_{c3} + \alpha_d - \alpha_c$ by switching Tx1 off, which is optimal in sub-regimes (2A), (2B), and (2C). This is due to the fact that in these sub-regimes the desired link from Tx1 to Rx1 is so weak that it is optimal to switch this user off. Now, consider sub-regime (2D). In this sub-regime, the desired link between Tx3 to Rx1 is very strong. Hence, switching Tx3 off is sub-optimal. On the other hand the interference caused by Tx3 is so strong that Rx2 is not able to decode its desired signal reliably. Therefore, it is optimal to switch Tx2 off in order to avoid the unnecessary interference of Tx2 at Rx1. This scheme is IA, which performs optimally in sub-regime (2D).

Now, we consider regime (3) (all Tx's active). In this regime, by using interference alignment in addition to private and common signalling, we can achieve a higher GDoF than using TIN, IA, or IA-TIN. This is partially shown in [14]. An interesting insight from this regime is that schemes like TIN, IA, or IA-TIN can be suboptimal even when the interference is very weak, in the sense that the interference parameter is smaller than the desired channel parameters, i.e.,

$$\alpha_c \leq \max\left\{\frac{\alpha_d}{2}, \frac{\alpha_{d3}}{2}\right\}, \tag{4.34}$$

$$\max\{\alpha_c, \alpha_{c3}\} \leq \frac{\alpha_d}{2}. \tag{4.35}$$

At this point, it is worth to note that the optimality of TIN has been studied in [15] for the $M \times N$ X channel, where M transmitters communicate with N receivers. As PIMAC is a special case of 3×2 X channel where the rate of some messages is equal to zero, a comparison of the statements above with the results in [15] is insightful. By applying the TIN optimality condition in [15, Theorem 3] to the PIMAC, we conclude that IA-TIN is optimal in sub-regimes (1A), part of (1B) ($\max\{n_{d3}, n_{c3}\} \leq n_d - n_c$), (2A), and (2B). Note that these regimes are sub-sets of the IA-TIN optimality regime of the PIMAC presented above. This is partly due to the structure of the channel (the achievable sum-rate of 3×2 X channel is an upper bound for that of the PIMAC) and partly due to the upper bounds (4.28) and (4.29) which have not been considered in [15]. These upper bounds have been extended in [16] for the 3×2 X channel. Doing this, the upper bound on the GDoF gets tighter and hence the identified TIN-optimality regime in [16] is extended to new regimes. These new regimes are equivalent to the sub-regimes (1C) and (2C) in the PIMAC.

4.4 Constant Gap Analysis

In the previous section, the optimality of simple schemes such as TIN, IA, and IA-TIN has been studied in terms of GDoF. As it can be understood from (4.13), GDoF is an approximation for the capacity at high SNR. The question which arises is the optimality of these schemes in terms of sum-rate at finite SNR. To answer this question, we assess the optimality of each scheme by bounding the gap between the achievable sum-rate and the upper bound in regimes where the scheme performs optimally in terms of GDoF.

We know from the discussion in previous section that TIN is optimal in terms of GDoF in sub-regimes (1A) and (2A). In what follows, we show the constant gap optimality of TIN in these sub-regimes.

First, by considering the assumption that the channel is interference limited (see (4.9)), we lower bound the achievable sum-rate of TIN as follows

$$\begin{aligned} R_{\Sigma,\text{TIN}} &\geq \frac{1}{2}\log_2\left(\frac{\rho^{\max\{\alpha_d,\alpha_{d3}\}}}{2\rho^{\alpha_c}}\right)+\frac{1}{2}\log_2\left(\frac{\rho^{\alpha_d}}{3(\rho^{\max\{\alpha_c,\alpha_{c3}\}})}\right)\\ &=\frac{1}{2}\left[\max\{\alpha_d,\alpha_{d3}\}-\alpha_c+(\alpha_d-\max\{\alpha_c,\alpha_{c3}\})^+\right]\log_2(\rho)-\frac{1}{2}\log_2(6). \end{aligned} \tag{4.36}$$

Next, we relax the upper bounds (4.17) and (4.18) by keeping the conditions of sub-regime (1A) and (2A) in mind. First, consider the upper bound in (4.18). By relaxing this upper bound, we write

$$\begin{aligned} C_\Sigma &\leq \frac{1}{2}\log_2\left(1+\rho^{\alpha_d}+\rho^{\alpha_c}+\frac{\rho^{\alpha_{d3}}}{1+\rho^{\alpha_{c3}}}\right)+\frac{1}{2}\log_2\left(1+\rho^{\alpha_{c3}}+\frac{\rho^{\alpha_d}}{1+\rho^{\alpha_c}}\right)\\ &\leq \frac{1}{2}\log_2\left(3\rho^{\alpha_d}+\frac{\rho^{\alpha_{d3}}}{\rho^{\alpha_{c3}}}\right)+\frac{1}{2}\log_2\left(2\rho^{\alpha_{c3}}+\frac{\rho^{\alpha_d}}{\rho^{\alpha_c}}\right)\\ &\leq \frac{1}{2}\log_2\left(4\rho^{\max\{\alpha_d,(\alpha_{d3}-\alpha_{c3})^+\}}\right)+\frac{1}{2}\log_2\left(3\rho^{\max\{\alpha_{c3},\alpha_d-\alpha_c\}}\right)\\ &=\frac{1}{2}\left[\max\{\alpha_d,(\alpha_{d3}-\alpha_{c3})^+\}+\max\{\alpha_{c3},\alpha_d-\alpha_c\}\right]\log_2(\rho)+\frac{1}{2}\log_2(12). \end{aligned} \tag{4.37}$$

Notice that we obtain a tight upper bound on GDoF in sub-regime (2A) by converting (4.18) into a GDoF expression. Hence, by applying the conditions of sub-regime (2A) to the bound in (4.37), we write

$$C_\Sigma \leq \frac{1}{2}\left[(\alpha_{d3}-\alpha_{c3})^+ +\alpha_d-\alpha_c\right]\log_2(\rho)+\frac{1}{2}\log_2(12). \tag{4.38}$$

Now, by comparing the upper bound (4.38) with the lower bound in (4.36), we conclude that the achievable sum-rate of the TIN is within a gap of $\frac{3}{2}+\log_2 3$ bits of the sum-capacity of the PIMAC in regime (2A). Interestingly, this gap in independent of the SNR. Hence, the gap between achievable sum-rate of TIN and the sum-capacity is bounded in for all SNR within the sub-regime (2A).

Consider sub-regime (1A) now. As we know from previous section, the GDoF expression obtained from upper bound in (4.17) is tight in sub-regime (1A). Hence, for studying the constant gap optimality in sub-regime (1A), we relax this upper bound as follows

$$C_\Sigma \leq (\alpha_d - \alpha_c)\log_2(\rho) + \frac{1}{2}\log_2(12). \tag{4.39}$$

By comparing (4.39) with (4.36), we conclude that the gap between the achievable sum-rate of TIN and the sum-capacity is bounded by $\frac{3}{2} + \log_2 3$ bits in sub-regime (1A). As we know, TIN is not GDoF optimal in the remaining sub-regimes. Hence, it cannot achieve the sum-capacity within a constant gap in the remaining sub-regimes.

Now, we consider the IA scheme. This scheme is optimal in sub-regime (2D) in terms of GDoF. Due to the fact that in sub-regime (2D) $\alpha_d \leq \alpha_{d3}$, the achievable sum-rate using IA can be lower bounded by

$$\begin{aligned} R_{\Sigma,\text{IA}} &= \frac{1}{2}\log_2\left(1 + \rho^{\alpha_{d3}}\right) \\ &\geq \frac{\alpha_{d3}}{2}\log_2(\rho). \end{aligned} \tag{4.40}$$

Now, by using the upper-bound for the sum-capacity given in (4.37) and keeping the conditions of sub-regime (2D) in mind and the fact that in this regime $\alpha_{c3} \leq \alpha_{d3}$, we obtain

$$C_\Sigma \leq \frac{\alpha_{d3}}{2}\log_2(\rho) + 1 + \frac{1}{2}\log_2 3. \tag{4.41}$$

By comparing this upper bound for the sum-capacity with the lower bound for the sum-rate in (4.40), we conclude that IA achieves the sum-capacity within a constant gap of $1 + \frac{1}{2}\log_2 3$ bits. Obviously, this scheme cannot achieve the sum-capacity within a constant gap in the remaining regimes.

Finally, we assess the gap between the achievable sum-rate using IA-TIN and the sum-capacity. First, we need to lower bound the achievable sum-rate of IA-TIN in regimes where this scheme performs optimally in terms of GDoF. Doing this, we obtain

$$R_{\Sigma,\text{IA-TIN}} \geq \frac{1}{2}\max\{2(\alpha_d - \alpha_c), (\alpha_{d3} - \alpha_c)^+ + \alpha_d - \alpha_{c3}\}\log(\rho) - 1. \tag{4.42}$$

Now, we need to compare the lower bound in (4.42) with the upper for the sum-capacity. In sub-regimes {(1A),(1B)} and {(2A),(2B)} the sum-capacity is upper bounded by (4.39) and (4.38), respectively. By comparing these upper bounds, with the lower bound in (4.42), we see that the gap between the achievable sum-rate using

IA-TIN and the sum-capacity does exceed $2+\frac{1}{2}\log_2 3$ bits. Now, we need to evaluate the gap between sum-rate using IA-TIN and the sum-capacity for the sub-regimes (1C) and (2C). In sub-regime (1C) and (2C), the tight upper bound on the GDoF is derived from (4.19) and (4.20), respectively. By relaxing these upper bounds, we obtain

$$C_\Sigma < (\alpha_d - \alpha_c)\log_2(\rho) + \frac{5}{2}, \quad \text{in sub-regime (1C)}, \tag{4.43}$$

$$C_\Sigma < \frac{1}{2}(\alpha_d - \alpha_c) + (\alpha_{d3} - \alpha_{c3})\log_2(\rho) + \frac{5}{2}, \quad \text{in sub-regime (2C)}. \tag{4.44}$$

Comparing the upper bounds in (4.43) and (4.44) with the lower bound in (4.42), we see that the gap between the achievable sum-rate of IA-TIN and the sum-capacity is bounded by $\frac{7}{2}$ bits in sub-regimes (1C) and (2C).

The following figures shows a numerical comparison of the achievable rates of the proposed schemes with the sum-capacity upper bound. Figure 4.7 shows the normalized achievable sum-rate $\frac{R_\Sigma}{\frac{1}{2}\log_2(\rho)}$ versus α_{d3} for a PIMAC with $\alpha_d = 1$, $\alpha_c = 0.4$, and $\alpha_{c3} = 0.2$. Note that in this numerical comparison, the parameters α_d, α_c, and α_{c3} are chosen such that by increasing α_{d3}, the PIMAC passes regimes (1A), 3, (2C), and (2D). The parameter ρ is 100 and 30 dB for all dashed and solid curves, respectively. While from dashed curves, the GDoF optimality of schemes can be understood, the solid curves indicate the gap between the sum-capacity and the achievable sum-rate in a moderate SNR regime. As it is shown in this figure, while IA-TIN performs optimally (within a constant gap) for small and large values of α_{d3}, it cannot achieve the upper bound within a constant gap when α_{d3} is in the moderate range. Moreover, the constant gap optimality of TIN can be seen when $\alpha_{d3} < 0.6$. This corresponds to the case where PIMAC performs in sub-regime (1A). In the next section, the sum-rate optimality of TIN will be studied in more detail.

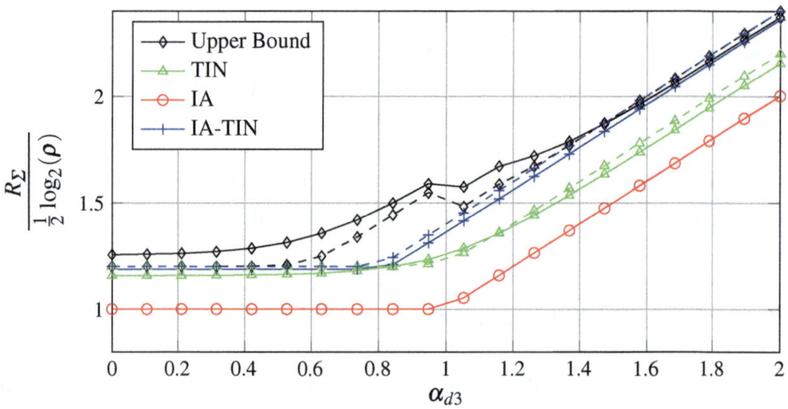

Fig. 4.7 Normalized achievable sum-rate as a function of α_{d3} for a PIMAC with $\alpha_d = 1, \alpha_c = 0.4$, $\alpha_{c3} = 0.2$. The parameter ρ is 30 and 100 dB for the *solid* and *dashed curves*, respectively

4.5 TIN Is Always Sub-optimal

In the previous section, it was shown that there are some cases when TIN achieves the sum-capacity of PIMAC within a constant gap. Nevertheless, we can show that TIN is always sub-optimal. To do this, we propose a scheme which outperforms TIN for all channel values. This scheme is a combination of time division multiple access (TDMA) and IA-TIN in which Tx1 transmits with the power P/τ in a fraction $\tau \in [0, 1]$ of time, and Tx3 transmits in the remaining fraction $1 - \tau$ of the time with the power $P/(1 - \tau)$. In other words, we share the time between two 2-user IC's operating over orthogonal time slots. The required time scheduling for establishing the coordination between Tx1 and Tx3 makes this scheme slightly complexer than TIN and IA-TIN. Using this scheme, the following sum-rate is achievable

$$R_{\Sigma,\text{TDMA-IA-TIN}} = \max_{\substack{\tau\in[0,1]\\ \bar{\tau}=1-\tau}} \frac{\tau}{2}\left[\log_2\left(1+\frac{\rho^{\alpha_d}}{\tau(1+\rho^{\alpha_c})}\right)+\log_2\left(1+\frac{\tau\rho^{\alpha_d}}{\tau+\rho^{\alpha_c}}\right)\right]+ \frac{\bar{\tau}}{2}\left[\log_2\left(1+\frac{\rho^{\alpha_{d3}}}{\bar{\tau}(1+\rho^{\alpha_c})}\right)+\log_2\left(1+\frac{\bar{\tau}\rho^{\alpha_d}}{\bar{\tau}+\rho^{\alpha_{c3}}}\right)\right].$$

In [17], the sum-rate of the TDMA-IA-TIN is compared with TIN. It is shown that TDMA-IA-TIN strictly outperforms TIN except in the case where $\frac{h_d}{h_{d3}} = \frac{h_c}{h_{c3}}$. Note that, this case corresponds to the special case where $\alpha_d - \alpha_c = \alpha_{d3} - \alpha_{c3}$ which is excluded from our analysis. In this section, we do not present the proof of this claim but we show a numerical comparison to highlight this result. In Fig. 4.8, the achievable sum-rates using TIN, TDMA-IA-TIN, and IA-TIN are computed versus α_{d3} for a PIMAC with $\alpha_d = 1$, $\alpha_c = 0.2$, and $\alpha_{c3} = 0.1$ when $\rho = 10\,\text{dB}$. Interestingly, this plot shows that there are some cases where TIN outperforms IA-TIN. However, by using optimal time sharing between two cases of IA-TIN, TIN will be strictly outperformed.

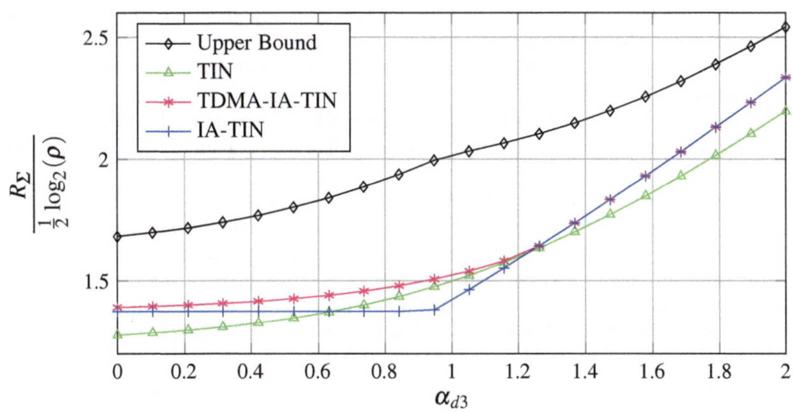

Fig. 4.8 Normalized achievable sum-rate as a function of α_{d3} for a PIMAC with $\alpha_d = 1, \alpha_c = 0.2$, $\alpha_{c3} = 0.1$. The parameter ρ is 10 dB

4.6 Summary

In this chapter, we started with the discussion on whether the simple scheme of treating interference (TIN) as noise has the potential to be optimal in terms of being capacity achieving. We then discussed recent results showing that indeed for the basic two user IC there are regimes in which TIN is optimal. Lately, optimality was also shown for the K user interference channel, which holds as long as a certain condition on the interference strength is satisfied. We then relaxed the condition of exact optimality to approximate optimality. In more details, we asked whether there are regimes in which variants of TIN are constant-gap optimal, i.e., is the additive gap between upper bounds on the capacity and the rate obtained with variants of TIN bounded by a constant? We answered this question in the affirmative for the partially interfering multiple access channel (PIMAC). The analytical results are verified by means of numerical results. In consequence there are certain regimes in interference limited networks, in which the system designer could save computational resources by employing TIN strategies without loss of optimality. As an outlook, it would be interesting to investigate the existence of such regimes for more general (e.g., relays, multi-hop, multi-antenna) networks than discussed in this chapter.

Acknowledgments This work was supported by the German Research Foundation (DFG) under Grant SE 1697/7-1 and SE 1697/10-1.

References

1. Cover T, Thomas J (2006) Elements of information theory, 2nd edn. Wiley, New York
2. Jafar SA, Interference alignment: a new look at signal dimensions in a communication network. Found Trends Commun Inf Theory 7(1):1–136. http://dx.doi.org/10.1561/010000047
3. Carleial AB (1978) Interference channels. IEEE Trans Inf Theory IT-24:60–70
4. Shannon CE (1948) A mathematical theory of communication. Bell Syst Tech J 27:379–423
5. Etkin RH, Tse DNC, Wang H (2008) Gaussian interference channel capacity to within one bit. IEEE Trans Inf Theory 54(12):5534–5562
6. Kramer G (2004) Outer bounds on the capacity of Gaussian interference channel. IEEE Trans Inf Theory 50(3):581–586
7. Annapureddy VS, Veeravalli VV (2009) Gaussian interference networks: sum capacity in the low interference regime and new outer bounds on the capacity region. IEEE Trans Inf Theory 55(9):3032–3050
8. Shang X, Kramer G, Chen B (2009) A new outer bound and the noisy-interference sum-rate capacity for Gaussian interference channels. IEEE Trans Inf Theory 689–699
9. Motahari AS, Khandani AK (2009) Capacity bounds for the Gaussian interference channel. IEEE Trans Inf Theory 55(2):620–643
10. Etkin RH (2009) New sum-rate upper bound for the two-user Gaussian interference channel. In: Proceedings of the IEEE ISIT, Seoul, Jun-Jul 2009, pp. 2582–2586
11. Chaaban A, Sezgin A (2011) An extended etkin-type outer bound on the capacity of the Gaussian interference channel. In: Proceedings of the 16th Asilomar conference on SSC, Pacific Grove 2011

12. Shang X, Kramer G, Chen B (2008) Throughput optimization in multi-user interference channels. In: Proceedings of IEEE military communications conference (MILCOM), San Diego, CA, Nov 2008
13. Huang C, Cadambe VR, Jafar SA (2012) Interference alignment and the generalized degrees of freedom of the *X* channel. IEEE Trans Inf Theory 58(8):5130–5150
14. Gherekhloo S, Di C, Chaaban A, Sezgin A (2014) (Sub-)optimality of treating interference as noise in the cellular uplink with weak interference. IEEE Trans Inf Theory, in revision. arXiv:1401.8265
15. Geng C, Sun C, Jafar SA, On the optimality of treating interference as noise: general message sets. arXiv:1401.2592
16. Gherekhloo S, Chaaban A, Sezgin A (2014) Extended generalized DoF optimality regime of treating interference as noise in the X channel. In: 11th international symposium on wireless communications systems (ISWCS), Aug 2014, pp. 971–975
17. Chaaban A, Sezgin A (2012) Sub-optimality of treating interference as noise in the cellular uplink. In: Proceedings of the 16th international ITG workshop on smart antennas WSA, Dresden, Germany, Mar 2012

Chapter 5
Interference-Aware Analog Computation over the Wireless Channel: Fundamentals and Strategies

Mario Goldenbaum, Sławomir Stańczak and Holger Boche

Abstract State of the art wireless network solutions are mainly concerned with forwarding human generated high-resolution image, audio, and video content. In doing so, interference is usually seen as a limitation to overcome so that the channel access of users is coordinated in time or frequency. In many emerging technologies, however, forwarding raw data messages is not necessary and should be avoided for efficiency improvements. Instead, there is a need for a reliable and efficient *computation of functions* thereof (e.g., computing the maximum flue gas concentration for fire detection, the average frequency drift in a Smart Grid, the minimum humidity in a greenhouse). This chapter deals with this type of problems and provides an overview of some recent results. In particular, it is demonstrated that for reliably computing *real-valued* functions over the wireless channel, harnessing interference rather than avoiding or canceling it can lead to huge performance gains. It is shown which real-valued functions can be essentially computed by following this paradigm and why this question is closely related to the famous *13th Hilbert problem*. Then, the question is addressed of how efficiently such computations can be done in terms of the number of transmissions needed per function value. Subsequently, corresponding encoding and decoding strategies are presented and evaluated in its performance. The chapter concludes with a brief discussion of related and open problems.

M. Goldenbaum (✉) · S. Stańczak
Network Information Theory Group, Technische Universität Berlin,
10587 Berlin, Germany
e-mail: mario.goldenbaum@tu-berlin.de

S. Stańczak
e-mail: slawomir.stanczak@tu-berlin.de

H. Boche
Lehrstuhl für Theoretische Informationstechnik, Technische Universität München,
80333 Munich, Germany
e-mail: boche@tum.de

W. Utschick (ed.), *Communications in Interference Limited Networks*,
Signals and Communication Technology, DOI 10.1007/978-3-319-22440-4_5

5.1 Introduction

Over the last two decades, research and development activities in the realm of communications engineering were mainly driven by the demand for high speed wireless data access anywhere and anytime. Starting in the early 90s with data rates of not more than 10 kbit/s, we recently arrived with the advent of the 4th cellular generation at rates of up to 100 Mbit/s [8]. The lion's share of data to be transmitted is currently governed by human-generated high resolution image, audio, and video content. With the ongoing advances in microelectronics, however, smart wireless technologies increasingly find their way into other areas such as industrial automation [28]. This trend not only dramatically changes the way we interact with our physical environment but also the type and amount of data to be transmitted. Accordingly, the 5th cellular generation is predicted to provide an Internet of Things that interconnects up to 1 trillion products, machines, and devices by 2022 [25]. In this regard, a plenty of challenges have to be overcome [9] before some of the most promising technologies (e.g., Internet of Things, Industry 4.0, Smart Grids, Tactile Internet, Mobile Crowd Sensing, eHealth) can seamlessly be supported.

An integral part of each of these emerging technologies will be a *reliable and efficient computation of functions* that depend on the data available at spatially distributed sources. Functions of interest can be, for instance, the maximum flue gas concentration in a building for fire detection, the average frequency drift in a Smart Grid, the maximum noise level in an urban area by means of Mobile Crowd Sensing, or the optimal resource allocation in a 5G cellular network. Compared with current network solutions, this will result in a *paradigm shift* as the efficient transmission of raw data is no longer a top priority.

This chapter deals with this type of problems and provides an overview of our recent results, which are published in [14–17] as well as in the first author's Ph.D. thesis [13]. We demonstrate that for reliably computing *real-valued* functions over the wireless channel, harnessing interference rather than avoiding or canceling it can lead to huge performance gains. In particular, it is determined which real-valued functions can be essentially computed by following this paradigm and why this question is closely related to the famous *13th Hilbert problem*. Then, we address the question of how efficiently such computations can be done in terms of the number of transmissions needed per function value. Subsequently, we present corresponding encoding and decoding strategies and evaluate their performance. The chapter concludes with a brief discussion of related and open problems.

5.1.1 *Related Work*

In the seminal paper [23], Nazer and Gastpar lay the information-theoretic foundation for a general theory of distributed function computation over unreliable channels. The big difference to standard information theory is that the intended receiver decodes function values immediately from the channel output. In other words, the receiver does not care about individual messages and penalizes itself only when the function is incorrectly decoded. The main results of [23] are concerned with the special case of reliably computing linear functions over multiple-access channels (MACs) having a linear structure. In this regard, Nazer and Gastpar show that in many cases, the performance gains over separation-based computation strategies are proportional to the number of concurrently active transmitters, where a separation-based strategy is a strategy in which the receiver first tries to reliably decode all the individual messages and subsequently computes the sought function value. A remarkable fact is that the gains over separation-based strategies have their origin in a perfect match between the desired function and the *algebraic structure* of the MAC.

In this regard, Karamchandani, Niesen, and Diggavi show in [20] that in case of a structural mismatch between a given noiseless MAC and some function to be computed, a separation-based strategy is almost always rate-optimal.[1] On the other hand, Keller, Karamchandani, and Fragouli provide in [21] conditions under which linear and certain nonlinear functions are efficiently computable by harnessing the structure of a noiseless linear vector-channel with finite input and output alphabets. We can find similar conclusions in [1, 19].

5.1.2 Notation

For some $p \in \mathbb{N}$, $\mathbb{Z}_p = \{0, \ldots, p-1\}$ denotes the integers modulo p and $\bigoplus$ summation modulo p. The n-times Cartesian product of some set $\mathcal{A}$ is written as $\mathcal{A}^n$. Lowercase italic letters are used to denote scalars, bold lowercase letters to denote vectors, and bold uppercase letters to denote matrices, respectively (e.g., x, $\mathbf{x}$, $\mathbf{X}$). Scalar random variables are described by uppercase italic letters and their realizations by lowercase italic letters (e.g., X and x). The expected value and the variance of some random element X are denoted, respectively, by $\mathbb{E}\{X\}$ and $\mathbb{V}\text{ar}\{X\}$, whereas $\mathbb{P}(A)$ designates the probability of some random event A. Finally, $\log_2^+(x) := \max\{\log_2(x), 0\}$.

5.2 Communication Versus Computation

In this section, we first introduce the "wireless multiple-access channel" as a reasonable model for the scenario in which multiple sources communicate with a common destination over a shared radio channel. Subsequently, we provide a formal description of what we mean when referring to the communication or the computation problem over a wireless MAC. Using two simple examples, we demonstrate that the

[1] Note that the results reviewed in this chapter show that this statement does not apply to the wireless multiple-access channel.

problems can be substantially different from each other so that they should also be differently treated.

5.2.1 The Wireless Multiple-Access Channel

The wireless channel is a shared broadcast medium so that a simultaneous access of multiple users to the available frequency spectrum results in interference. In other words, the electromagnetic waves radiated in the same frequency band by a set of distributed transmitters superimpose at some receiver. This motivates the model of the (memoryless) wireless multiple-access channel [14].

Definition 5.2.1 (*WMAC*) Let $\mathcal{X}_i \subset \mathbb{R}$ be the channel input alphabet of terminal i, $i = 1, \ldots, N$, $N \geq 2$, and $\mathcal{Y} \subseteq \mathbb{R}$ some channel output alphabet. Then, the mapping

$$W : \mathcal{X}_1 \times \cdots \times \mathcal{X}_N \to \mathcal{Y},\ Y = W(x_1, \ldots, x_N) = \sum_{i=1}^{N} H_i x_i + Z, \tag{5.1}$$

is referred to as the *wireless multiple-access channel (WMAC)*. Here and hereafter, Z denotes zero-mean white Gaussian noise of variance σ_Z^2 (i.e., $Z \sim \mathcal{N}_{\mathbb{R}}(0, \sigma_Z^2)$) and $H_i \in \mathbb{R}$ is the fading coefficient between the destination and terminal i. For the degenerate case $\sigma_Z^2 = 0$ and $H_1 = \cdots = H_N \equiv 1$, (5.1) simplifies to the *ideal WMAC*

$$y = w(x_1, \ldots, x_N) = \sum_{i=1}^{N} x_i. \tag{5.2}$$

The input sequences of a WMAC are typically constrained with respect to some nonnegative cost function.

Definition 5.2.2 (*Input Cost Constraint*) Let $n \in \mathbb{N}$ be an arbitrary number of channel uses and $\rho_n : \mathbb{R}^n \to \mathbb{R}_+$ some input cost function that is common to all the WMAC terminals. Then, for an arbitrary nonnegative real number P_i, the requirement

$$\rho_n\big(x_i[1], \ldots, x_i[n]\big) \leq P_i,$$

for all sequences $(x_i[1], \ldots, x_i[n]) \in \mathcal{X}_i^n$, is referred to as the *input cost constraint* of terminal i, $i = 1, \ldots, N$.

Example 5.2.1 Input cost constraints primarily used in wireless network design are:

- Average transmit power constraint

$$\rho_n\big(x_i[1], \ldots, x_i[n]\big) = \frac{1}{n}\sum_{j=1}^{n}\big|x_i[j]\big|^2 \leq P_i \qquad i = 1, \ldots, N, \tag{5.3}$$

- Maximum transmit power constraint

$$\rho_n\big(x_i[1], \dots, x_i[n]\big) = \max_{1 \le j \le n} \big|x_i[j]\big|^2 \le P_i \qquad i = 1, \dots, N. \tag{5.4}$$

5.2.2 The Communication Problem

The classical *communication problem* has a long standing history and its first systematic treatment goes back to Shannon and his groundbreaking work "A Mathematical Theory of Communication" [26]. Since then, Shannon's ideas have been extended in various directions (see [10] and references therein). In the following, we give a formal description of the communication problem over a WMAC.

Consider a network consisting of a destination and $N \ge 2$ distributed *analog* sources. More precisely, every sequence of source symbols is drawn from a compact alphabet $\mathcal{S} := [s_{\min}, s_{\max}] \subset \mathbb{R}$, which could be, for instance, an interval of temperatures, pressures, or voltages. In this context, the communication problem is as follows: Given a WMAC with fixed input cost constraints and some integer T, the destination intends to *reliably* and *efficiently* reconstruct any length-T sequences of source symbols $\mathbf{s}_i := (s_i[1], \dots, s_i[T]) \in \mathcal{S}^T$, $i = 1, \dots, N$. To be more precise, let

$$\mathscr{E}_i : \mathcal{S}^T \to \mathcal{X}_i^n, \ \mathbf{s}_i \mapsto \mathbf{x}_i \tag{5.5}$$

be the encoder of source i, $i = 1, \dots, N$, which maps each length-T sequence of source symbols to a length-n sequence of channel inputs $\mathbf{x}_i := \big(x_i[1], \dots, x_i[n]\big)$. Correspondingly, the mapping

$$\mathscr{D} : \mathcal{Y}^n \to \mathcal{S}^{NT}, \ \mathbf{y} \mapsto (\hat{\mathbf{s}}_1, \dots, \hat{\mathbf{s}}_N) \tag{5.6}$$

represents the decoder of the destination, where $\mathbf{y} := \big(Y[1], \dots, Y[n]\big)$ denotes the length-n sequence of WMAC outputs. Let $\varepsilon > 0$ be some given accuracy and

$$d_{\mathcal{S}^N} : \mathcal{S}^N \times \mathcal{S}^N \to \mathbb{R}_+ \tag{5.7}$$

some distortion measure. Then, the communication problem consists in designing encoders (5.5) along with decoder (5.6) such that

$$\mathbb{P}\left(\bigcup_{t=1}^{T} \Big\{ d_{\mathcal{S}^N}\big((\hat{s}_1[t], \dots, \hat{s}_N[t]), (s_1[t], \dots, s_N[t])\big) > \varepsilon \Big\} \right) \le \delta, \tag{5.8}$$

for some arbitrary small $\delta > 0$. If a solution to this problem exists, a communications engineer is generally interested in solutions (i.e., encoder-decoder combinations) that are efficient in the sense that (5.8) can be achieved with as little *channel uses* as possible.

Remark 5.2.1 The communication problem is subject to a fundamental trade-off between the desired accuracy $(d_{\mathcal{S}^N}, \varepsilon, \delta)$, the channel input constraints $(\rho_n, \{P_i\}_{i=1}^N)$, and the numbers T and n.

5.2.3 The Computation Problem

Instead of reconstructing the source sequences $\mathbf{s}_1, \ldots, \mathbf{s}_N$, many emerging technologies need that the destination reliably and efficiently computes some pre-defined *function* thereof [11, 12]. This can be, for instance, the maximum flue gas concentration for fire detection, the average pressure in a steam boiler, or the minimum humidity in a greenhouse. In what follows, we call such functions "desired functions".

Definition 5.2.3 (*Desired Function*) Let $\mathcal{S} \subset \mathbb{R}$ be some source alphabet and $s_i \in \mathcal{S}$ an instantaneous symbol of source i, $i = 1, \ldots, N$. Then, every function

$$f : \mathcal{S}^N \to \mathbb{R}, \ (s_1, \ldots, s_N) \mapsto f(s_1, \ldots, s_N)$$

to be computed at the destination is called *desired function*.

In accordance with the definition, whenever the sources have values $s_1, \ldots, s_N$ to be transmitted, the destination would like to have a sufficiently accurate estimate $\hat{f}(s_1, \ldots, s_N)$ of the corresponding function value $f(s_1, \ldots, s_N)$. The distributed communication problem discussed in Sect. 5.2.2 therefore turns into a distributed *computation problem*, formally specified as follows: Let decoder (5.6) be modified to

$$\mathcal{D} : \mathcal{Y}^n \to \mathbb{R}^T, \ \mathbf{y} \mapsto \left(\hat{f}(s_1[1], \ldots, s_N[1]), \ldots, \hat{f}(s_1[T], \ldots, s_N[T])\right). \quad (5.9)$$

Then, given a WMAC with fixed input cost constraints, predefined accuracy $\varepsilon > 0$, and some distortion measure

$$d_{\mathbb{R}} : \mathbb{R} \times \mathbb{R} \to \mathbb{R}_+, \quad (5.10)$$

the computation problem consists in designing encoders (5.5) along with decoder (5.9) such that

$$\mathbb{P}\left(\bigcup_{t=1}^{T} \left\{ d_{\mathbb{R}}\left(\hat{f}(s_1, \ldots, s_N), f(s_1, \ldots, s_N)\right) > \varepsilon \right\}\right) \leq \delta, \quad (5.11)$$

for $\delta > 0$ arbitrary small. As for the communication problem, efficient solutions to the computation problem (if existent) are of particular interest, which achieve (5.11) with as little channel uses as possible. In doing so, there is a fundamental trade-off between the tuples $(d_{\mathbb{R}}, f, \varepsilon, \delta)$, $(\rho_n, \{P_i\}_{i=1}^N)$, and the numbers T and n.[2]

5.2.4 Two Insightful Examples

Having defined underlying the communication or the computation problem, we demonstrate in the following that the problems substantially differ in general. Therefore, let us take a look on two simple but insightful examples.

Example 5.2.2 Consider the ideal WMAC (5.2) without input cost constraints[3] and let the desired function be the sum of the sources, that is,

$$f(s_1, \dots, s_N) = \sum_{i=1}^{N} s_i. \tag{5.12}$$

Furthermore, let $T = 1$ and the sequence of channel inputs of source i be of the form

$$\mathbf{x}_i = \mathcal{E}_i(s_i) = \big(a_i[1], a_i[2], \dots, a_i[n]\big)s_i \qquad i = 1, \dots, N, \tag{5.13}$$

$(a_i[1], \dots, a_i[n]) \in \mathbb{R}^n$, which due to the memoryless property of the WMAC leads to the channel output sequence

$$\mathbf{y} = \begin{pmatrix} y[1] \\ y[2] \\ \vdots \\ y[n] \end{pmatrix} = \sum_{i=1}^{N} \mathbf{x}_i = \underbrace{\begin{pmatrix} a_1[1] & a_2[1] & \cdots & a_N[1] \\ a_1[2] & a_2[2] & \cdots & a_N[2] \\ \vdots & \vdots & \ddots & \vdots \\ a_1[n] & a_2[n] & \cdots & a_N[n] \end{pmatrix}}_{=:\mathbf{A}} \begin{pmatrix} s_1 \\ s_2 \\ \vdots \\ s_N \end{pmatrix}. \tag{5.14}$$

For computing the desired function-value (5.12), the destination first intends to *uniquely* infer the N unknowns $s_1, \dots, s_N$ from $\mathbf{y}$. It is one of the basic facts of linear algebra that a unique solution to (5.14) exists if and only if $\mathbf{A} \in \mathbb{R}^{n \times N}$ is invertible. Thus, the number of channel uses has to be equal to the number of sources (i.e., $n = N$) and the encoders (5.13) have to be designed such that the vectors $(a_i[1], \dots, a_i[N])^\mathsf{T}$, $i = 1, \dots, N$, span $\mathbb{R}^N$. In this case, the decoding rule is simply

$$\mathcal{D}(\mathbf{y}) = \mathbf{1}^\mathsf{T}\mathbf{A}^{-1}\mathbf{y} = f(s_1, \dots, s_N). \tag{5.15}$$

---

[2]Note that for neither of the two problems we introduced a probability distribution on the source symbols, which is in contrast to the standard information theoretic setting. From the perspective of a communications engineer, however, the problem statements given here are more relevant as any practical coding scheme has to work for any choice of source symbols.

[3]The ideal WMAC is noiseless so that it is not necessary to incorporate input cost constraints (i.e., channel inputs and outputs can be arbitrarily scaled without performance degradation).

In what follows, a strategy as in Example 5.2.2 is called *separation-based computation strategy* as it separates the process of reliably communicating source symbols from the process of computing the desired function thereof. In other words, a separation-based strategy first solves the communication problem (if feasible) and the computation problem afterwards.

Example 5.2.3 In contrast to Example 5.2.2, consider a *single* use of the ideal WMAC (i.e., $n = 1$) with encoders to be chosen as

$$x_i[1] = \mathscr{E}_i(s_i) = s_i \qquad i = 1, \ldots, N. \tag{5.16}$$

Since the corresponding channel output is

$$y[1] = \sum_{i=1}^{N} x_i[1] = \sum_{i=1}^{N} s_i,$$

choosing the decoding rule as

$$\mathscr{D}\big(y[1]\big) = y[1] = f(s_1, \ldots, s_N) \tag{5.17}$$

immediately results in the desired function-value (5.12).

The conclusion of the examples is that separation-based computation strategies can be highly inefficient. In particular, the strategy of Example 5.2.2 is capable of computing $1/N$ function-values per channel use in a distortion-free manner whereas the strategy of Example 5.2.3, referred to as *Computation over MAC (CoMAC)*, allows for the distortion-free computation of one function-value per channel use. Thus, in the considered example, the efficiency is improved by a factor of N (i.e., the number of sources). The reason for this N-fold gain is that the CoMAC strategy *exploits the interference* of the ideal WMAC for computing (5.12) rather than avoiding or canceling it. In other words, the processes of communication and computation are *merged*.

5.3 Analog Computation over the WMAC—Fundamentals

Essential for the fact that harnessing interference leads in Example 5.2.3 to an N-fold increase in efficiency is that the desired function (5.12) *perfectly matches* the algebraic structure of the ideal WMAC (i.e., the WMAC itself computes the desired function). A question that immediately arises in this context is whether there is a chance that harnessing interference can be advantageous over separation-based computation even if the desired function *does not perfectly match* the structure of the WMAC. In order to allow an in-depth treatment of this question, in this section we consider the ideal WMAC model in greater detail. It turns out that these efforts are closely related to the famous *13th Hilbert problem* [18]. Certainly a surprising fact that not only draws attention to a beautiful mathematical theory that might be less known in the communications engineering community but also leads to some surprising insights.

5.3.1 Computation with a Single Channel Use

Karamchandani, Niesen, and Diggavi show in [20] that if there is a *structural mismatch* between a given ideal MAC and some desired function, separation-based computation is almost always optimal. In other words, for almost all mismatched combinations of MACs and desired functions, an attempt of exploiting interference to increase efficiency fails. This result may appear discouraging but at least for the ideal WMAC we are interested in, we provide a kind of remedy by employing encoders and decoders that appropriately establish the desired matches. Towards this end, each source transforms its outcomes prior to transmission by an individual pre-processing function defined as follows:

Definition 5.3.1 (*Pre-Processing Function*) The univariate function

$$\varphi_i : \mathcal{S} \to \mathbb{R}, \; s_i \mapsto \varphi_i(s_i) \tag{5.18}$$

is referred to as the *pre-processing function* of source $i, i = 1, \ldots, N$.

On the other hand, the destination can pass channel output symbols through some post-processing function:

Definition 5.3.2 (*Post-Processing Function*) Let $y \in \mathbb{R}$ be the output of the ideal WMAC at an arbitrary channel use. Then, the univariate function

$$\psi : \mathbb{R} \to \mathbb{R}, \; y \mapsto \psi(y) \tag{5.19}$$

is referred to as the *post-processing function*.

According to these definitions, we simply choose the encoder of source i to be

$$\mathcal{E}_i(s_i) = x_i[1] \equiv \varphi_i(s_i) \qquad i = 1, \ldots, N \tag{5.20}$$

and the decoder to be

$$\mathcal{D}\big(y[1]\big) \equiv \psi\big(y[1]\big), \tag{5.21}$$

respectively. This particular combination allows the destination to compute every desired function for which there exist N pre-processing functions along with a post-processing function such that it can be represented in the form

$$f(s_1, \ldots, s_N) = \psi\left(\sum_{i=1}^{N} \varphi_i(s_i)\right), \tag{5.22}$$

for all $(s_1, \ldots, s_N) \in \mathcal{S}^N$.

Remark 5.3.1 The pre- and post-processing functions transform the ideal WMAC such that the resulting overall channel is *matched* to the desired function (see Fig. 5.1).

It is an interesting coincidence that the space containing functions of the form (5.22) is in multivariate calculus known as the space of *nomographic functions* [4]. They owe their name to *nomographs*. Nomographs are graphical aids that were primarily used in the pre-computer era in order to solve complicated equations manually [7]. A well-known example is the Smith chart, which is often used in microwave engineering. In what follows, we denote the space of nomographic functions with domain $\mathcal{S}^N$ as $\mathsf{N}(\mathcal{S}^N)$.

Example 5.3.1 (Nomographic Functions) Let $s_i \in \mathcal{S}, i = 1, \ldots, N$.

- Arithmetic Mean: $f(s_1, \ldots, s_N) = \frac{1}{N}\sum_{i=1}^{N} s_i$, with $\varphi_i(s) = s, i = 1, \ldots, N$, and $\psi(y) = y/N$.
- Polynomial: $f(s_1, \ldots, s_N) = a_0 + a_1 s_1^{\ell_1} + a_2 s_2^{\ell_2} + \cdots + a_N s_N^{\ell_N}$, with $\varphi_i(s_i) = a_i s_i^{\ell_i}$, $\ell_i \in \{0, \ldots, N\}$ arbitrary, $i = 1, \ldots, N$, and $\psi(y) = y + a_0$.
- Euclidean Norm: $f(s_1, \ldots, s_N) = \sqrt{s_1^2 + \cdots + s_N^2}$, with $\varphi_i(s) = s^2$, $i = 1, \ldots, N$, and $\psi(y) = \sqrt{y}$.
- Number of Active Nodes: $f(s_1, \ldots, s_N) = N$, with $\varphi_i(s) \equiv C$ for some $C \in \mathbb{R}$, $i = 1, \ldots, N$, and $\psi(y) = y/C$.

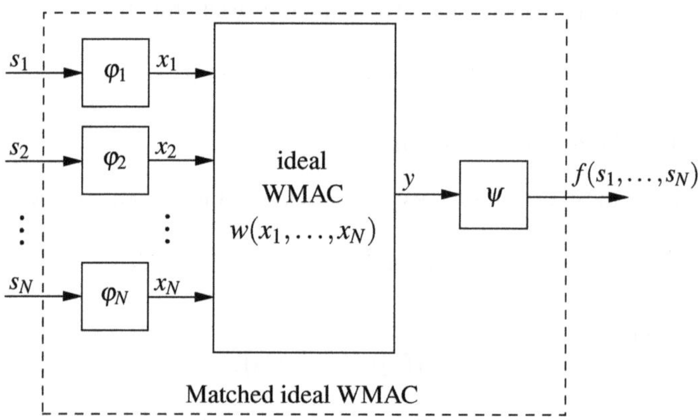

Fig. 5.1 The ideal WMAC matched to the desired function f by appropriately chosen pre- and post-processing functions $\varphi_1, \ldots, \varphi_N, \psi$. The resulting overall channel has source symbols as inputs and desired function-values as output

A closer look at the examples reveals that the pre-processing functions depend on the choice of the desired function so that a desirable property is the following.

Definition 5.3.3 (*Universality* [15]) A fixed set of pre-processing functions $\{\varphi_i\}_{i=1}^N$ is said to be *universal* with respect to some function space $\mathsf{A}(\mathcal{S}^N)$ if and only if *every* $f \in \mathsf{A}(\mathcal{S}^N)$ can be represented in the form of (5.22) by a proper choice of post-processing function ψ.

In this context, we were able to prove the following surprising theorem, which completely determines the set of functions that are computable over an ideal WMAC.

Theorem 5.3.1 ([15]) *Every desired function is universally computable over the ideal WMAC with a single channel use and zero distortion.*

In the meaning of Sect. 5.2.3, the theorem states that essentially *every* computation problem can be solved by using the ideal WMAC only once. Note that the denotation "universally computable" refers to the fact that the pre-processing functions can be chosen to be universal and therefore *independent* of the function to be computed. For the communication between the sources and the destination this implies that any feedback additional to the mandatory one (e.g., for providing channel state information) is not needed. It has to be emphasized, however, that in the proof of Theorem 5.3.1, the construction of universal pre-processing functions depends explicitly on the number of sources N. It is still an open problem if there are constructions that are independent of N. With such construction, the universality property would remain even if sources leave or enter the network, which is of high practical relevance. See [15] for a more detailed discussion.

Remark 5.3.2 A system engineer is always confronted with the fundamental question of which technology he should employ to implement a solution to a given problem (e.g., analog versus digital). With regard to the problem of computing a real-valued function of N inputs, Theorem 5.3.1 suggests that the solution can be optimally implemented[4] by means of a very simple circuitry of analog components: (i) N univariate pre-processing functions, (ii) an adder, (iii) an univariate post-processing function. Of course, the function computation can also be implemented by using standard digital components but this will at best result in an approximation. From a practical perspective, a question that naturally arises is whether the above three analog components can actually be realized.

In Theorem 5.3.1, no restrictions on the choice of pre- and post-processing functions were imposed so that they can be arbitrary. Since continuity can be useful for practical implementations, it is interesting to ask whether the theorem remains valid when pre- and post-processing functions are required to be continuous. In this regard, we denote in contrast to $\mathsf{N}(\mathcal{S}^N)$ the space of nomographic functions with the additional continuity property as $\mathsf{N}^0(\mathcal{S}^N)$. The following theorem, proven by Buck [5], gives a clear answer to this question.

[4]Optimal in the sense that a given desired function cannot be distinguished from its implementation.

Theorem 5.3.2 ([5]) *Nomographic functions with continuous pre- and post-processing functions are nowhere dense.*

In simple terms, the theorem states that for continuous desired functions that vanish on, and only on, some subset of $\mathcal{S}^N$ there do not exist nomographic representations in which pre- and post-processing functions are continuous.

Example 5.3.2 (*Geometric Mean*) Let $\mathcal{S}$ be the unit interval $\mathbb{E} = [0, 1]$ and the desired function be chosen to be the "geometric mean"

$$f : \mathbb{E}^N \to \mathbb{R},\ (s_1, \ldots, s_N) \mapsto f(s_1, \ldots, s_N) = \left(\prod_{i=1}^{N} s_i\right)^{\frac{1}{N}}. \tag{5.23}$$

Then, it can be shown [15] that there do not exist continuous functions $\varphi_1, \ldots, \varphi_N, \psi$ such that f belongs to $\mathsf{N}^0(\mathbb{E}^N)$.

At first glance, Theorem 5.3.2 seems a bit discouraging as requiring pre- and post-processing functions to be continuous significantly reduces the amount of functions computable with a single channel use. A stringent prerequisite of the statement is that the computations are distortion free, which means in terms of Sect. 5.2.3 that $d_{\mathbb{R}}\big(f(s_1, \ldots, s_N), \hat{f}(s_1, \ldots, s_N)\big) \equiv 0$, for some given distortion measure $d_{\mathbb{R}}$. However, if we tolerate some small distortion $\varepsilon > 0$, many desired functions that do not belong to $\mathsf{N}^0(\mathbb{E}^N)$ can accurately be approximated by a nomographic representation. For instance, consider again the "geometric mean" of Example 5.3.2. It can be easily verified that

$$\forall \ell \geq \ell_0\, \forall (s_1, \ldots, s_N) \in \mathbb{E}^N : \left|\left(\prod_{i=1}^{N} s_i\right)^{\frac{1}{N}} - \psi\left(\sum_{i=1}^{N} \varphi_i^{(\ell)}(s_i)\right)\right| \leq \varepsilon, \tag{5.24}$$

with pre-processing functions $\varphi_i^{(\ell)}(s) = \log(s + 1/p)$ and post-processing function $\psi(y) = \exp(y/N)$, respectively. Thus, choosing $d_{\mathbb{R}}$ to be the supremum norm over $\mathbb{E}^N$ and ℓ sufficiently large, the "geometric mean" can be uniformly approximated within $\varepsilon > 0$ by a single use of the ideal WMAC. Further examples that are of high relevance for environmental monitoring applications, are the following.

Example 5.3.3 (*Nomographic Approximations*) Let $\mathcal{S} = \mathbb{E}$, $\varepsilon > 0$ be arbitrary but fixed, distortion measure $d_{\mathbb{R}}$ be the supremum norm $\|\cdot\|_\infty$ over $\mathbb{E}^N$, and $\ell_0 = \ell_0(\varepsilon)$ be chosen such that (5.24) is fulfilled for all $\ell \geq \ell_0$. Furthermore, let $f \approx \hat{f}$ denote that $\hat{f}$ is close to f in $\|\cdot\|_\infty$.

- Maximum Value: $f(s_1, \ldots, s_N) = \max_{1\leq i\leq N}\{s_i\} \approx \psi\big(\sum_{i=1}^{N} \varphi_i(s_i)\big)$, with $\varphi_i(s) = s^{\ell_0}, i = 1, \ldots, N$, and $\psi(y) = y^{\frac{1}{\ell_0}}$.
- Minimum Value: $f(s_1, \ldots, s_N) = \min_{1\leq i\leq N}\{s_i\} \approx \psi\big(\sum_{i=1}^{N} \varphi_i(s_i)\big)$, with $\varphi_i(s) = s^{-\ell_0}, i = 1, \ldots, N$, and $\psi(y) = y^{-\frac{1}{\ell_0}}$.

5.3.2 *Computation with Multiple Channel Uses: Hilbert's 13th Problem*

As explained in the previous subsection, the continuity of pre- and post-processing functions has an enormous impact on the space of computable functions. In particular, if the pre- and post-processing functions are required to be continuous, a single use of the ideal WMAC is not sufficient to compute every continuous desired function. In order to better understand the interrelations, in the following we allow sources to use the channel more than once so that they have more degrees of freedom.

Towards this end, let $n > 1$ and the encoders in (5.20) be modified to

$$\mathscr{E}_i(s_i) = \big(x_i[1], \ldots, x_i[n]\big) = \big(\varphi_{i1}(s_i), \ldots, \varphi_{in}(s_i)\big) \qquad i = 1, \ldots, N, \tag{5.25}$$

with φ_{ij} being some continuous pre-processing functions to be specified later. Then, the corresponding length-n sequence of ideal WMAC outputs yields

$$\mathbf{y} = (y[1], \ldots, y[n]) = \left(\sum_{i=1}^{N} \varphi_{i1}(s_i), \ldots, \sum_{i=1}^{N} \varphi_{in}(s_i)\right). \tag{5.26}$$

If the destination is allowed to apply to each of the outputs an individual continuous post-processing function ψ_j, a very general decoder is of the form

$$\mathscr{D}(\mathbf{y}) = \Theta\Big(\psi_1\big(y[1]\big), \ldots, \psi_n\big(y[n]\big)\Big), \tag{5.27}$$

where the continuous function $\Theta : \mathbb{R}^n \to \mathbb{R}$ has the purpose to reasonably combine the sequence of post-processed ideal WMAC outputs.

In order to characterize which functions can be represented as on the right-hand side of (5.27), one could fix the number of channel uses to some $n > 1$ and then try to construct such functions with certain additional properties. On the other hand, one could assume some fixed, and perhaps well-understood, function space (e.g., the space of continuous functions) and then try to figure out how to choose n (if feasible) such that every element of this space can be represented as in (5.27).

It is somewhat surprising that the latter approach is closely related to the 13th of the famous list of 23 problems posed by David Hilbert in 1900 [18]. The original 13th problem involves the study of solutions of algebraic equations and Hilbert conjectured that a solution of the general equation of degree seven cannot be represented as a superposition of continuous functions of two variables. In his own words [18, p. 462]:

> … Now it is probable that the root of the equation of the seventh degree is a function of its coefficients which does not belong to this class of functions capable of nomographic construction, i.e., that it cannot be constructed by a finite number of insertions of functions of two arguments. In order to prove this, the proof would be necessary that the equation of the seventh degree $f^7 + xf^3 + yf^2 + zf + 1 = 0$ is not solvable with the help of any

continuous functions of only two arguments. I may be allowed to add that I have satisfied myself by a rigorous process that there exist analytical functions of three arguments x, y, z which cannot be obtained by a finite chain of functions of only two arguments.

Using our terminology, if Hilbert's conjecture is true, then it would not be possible to compute every continuous desired function with any *finite* number of channel uses n by harnessing the interference property of the WMAC. Fortunately, the conjecture was disproven by Kolmogorov in his landmark paper [22]. Using this result, in [15] we were able to prove the following theorem, which gives a complete answer to the question of which functions are computable when pre- and post-processing functions are required to be continuous.

Theorem 5.3.3 ([15]) *Let $N \geq 2$ be arbitrary. Then,* every continuous *desired function of N variables is universally computable over an ideal WMAC with $2N+1$ channel uses, continuous pre- and post-processing functions, and zero distortion.*

Remark 5.3.3 From the proof of the theorem it can be concluded that in (5.27), Θ can be chosen to be simply the sum:

$$\Theta\Big(\psi_1\big(y[1]\big), \ldots, \psi_{2N+1}\big(y[2N+1]\big)\Big) = \sum_{j=1}^{2N+1} \psi_j\big(y[j]\big). \tag{5.28}$$

Remark 5.3.4 Along with (5.26), representations in the form of (5.28) are called *Kolmogorov's superpositions*.

As we used the proof technique of Kolmogorov in [22], the statement of Theorem 5.3.3 can be seen as a sufficient condition on the number of channel uses n. Thus, one might expect that $n = 2N + 1$ could be reduced in order to save wireless resources. Unfortunately, Sternfeld has proven in [27] that this is not possible, which leads to the following result.

Theorem 5.3.4 ([13]) *Let the encoders be chosen as in* (5.25) *and the decoder as in* (5.27)*, respectively, with pre- and post-processing functions required to be continuous. Then, in order to universally compute every continuous desired function at zero distortion by harnessing interference, $n \geq 2N+1$ channel uses are necessary.*

5.4 Analog Computation over the WMAC—Strategies

The results of the previous section have shown that for many real-valued functions of practical relevance, harnessing interference rather than avoiding or canceling it can lead to huge performance gains, often in the order of the number of sources. Realizing this in a reliable manner over a channel with noise, however, requires the application of coding techniques, where according to Sect. 5.2.3, a computation code is composed of the following elements [23].

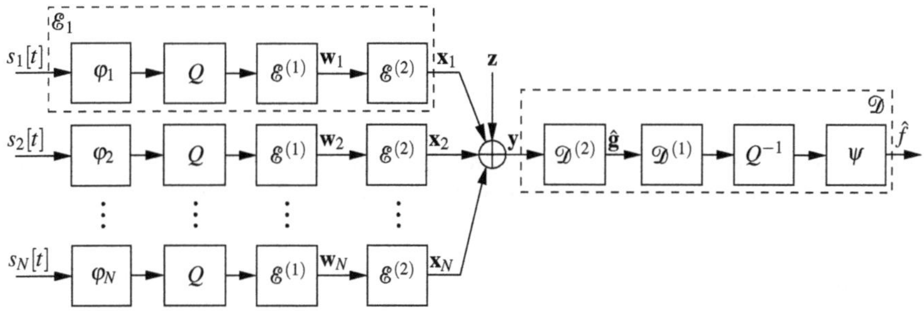

Fig. 5.2 Block diagram of the coding scheme of Sect. 5.4.1 for efficiently computing a nomographic function over the WMAC (5.1), where $H_1 = H_2 = \cdots = H_N = 1$

Definition 5.4.1 (*Computation Code*) Let $T, n \in \mathbb{N}$ be chosen arbitrarily. An (f, T, n) *computation code* consists of:

- A desired function f.
- Encoders (5.5) that map T source symbols $s_i[1], \ldots, s_i[T]$, $i = 1, \ldots, N$, to n channel input symbols each.
- A decoder (5.9) that assigns T estimates $\hat{f}(\mathbf{s}[1]), \ldots, \hat{f}(\mathbf{s}[T])$ of desired function-values to each length-n sequence of channel output symbols.

In this section, we summarize two computation codes recently proposed in [14, 16] and discuss their performance in terms of achievable computation rates. For the sake of simplicity, we assume in the remainder of the chapter that the fading coefficients fulfill $H_1 = H_2 = \ldots H_N = 1$ almost surely.

5.4.1 Strategy I

It is matter of fact that the (non-ideal) WMAC defined in Sect. 5.2.1 is for finite input cost constraints and finite bandwidth also of *finite capacity*. This entails that communicating real values with infinite precision at some nonzero rate is generally not possible. To account for this, in [16] we proposed a coding scheme in which the encoders $\mathscr{E}_1, \ldots, \mathscr{E}_N$ and the decoder $\mathscr{D}$ are decomposed into several parts as depicted in Fig. 5.2. In the following subsection, we provide some details.

5.4.1.1 Encoding and Decoding

Encoding After applying its individual pre-processing function, each source employs a *quantizer* Q in order to form for each t, $t = 1, \ldots, T$, a length-b binary representation $\mathbf{w}_i[t] = Q(\varphi_i(s_i[t]))$, $i = 1, \ldots, N$.[5] Then, the quantizers are followed by the same *source encoder* $\mathscr{E}^{(1)}$, which efficiently combines, for some prime

[5]Note that $b \in \mathbb{N}$ is a crucial design parameter that dictates the quantization error.

p, the binary representations to a length-k message over $\mathbb{Z}_p$, that is,

$$\mathbf{w}_i = \mathscr{E}^{(1)}\big(\mathbf{w}_i[1], \ldots, \mathbf{w}_i[T]\big) \in \mathbb{Z}_p^k \qquad i = 1, \ldots, N.$$

The last component of each encoder consists of the same *channel encoder*, which maps the message on its input to a length-n sequence of channel input symbols: $\mathbf{x}_i = \mathscr{E}^{(2)}(\mathbf{w}_i), i = 1, \ldots, N$.

Decoding On the receiving end, the *channel decoder* $\mathscr{D}^{(2)}$ first provides a reliable estimate $\hat{\mathbf{g}}$ of the modulo p sum of messages $\mathbf{g} := \bigoplus_{i=1}^N \mathbf{w}_i$. The estimate $\hat{\mathbf{g}}$ is then passed through a *source decoder* that decomposes it into the modulo p sums of the binary representations:

$$\mathscr{D}^{(1)}(\mathbf{g}) = \left(\bigoplus_{i=1}^{N} \mathbf{w}_i[1], \ldots, \bigoplus_{i=1}^{N} \mathbf{w}_i[T]\right).$$

Finally, the inverse of quantizer Q along with post-processing function ψ provides reliable estimates of the sequence of function values $\hat{f}(\mathbf{s}[1]), \ldots, \hat{f}(\mathbf{s}[T])$.

Note that each of the encoder- and decoder-components has to respect the linear structure of the WMAC in order to ensure that the modulo p sum of messages can immediately be decoded from the sum over the reals of channel input symbols. For instance, the tuple $(\mathscr{E}^{(2)}, \mathscr{D}^{(2)})$ is chosen as a nested lattice code, which posses in addition to linearity excellent statistical properties [6, 24]. See [16] for more details on the explicit construction of the coding scheme outlined above.

5.4.1.2 Achievable Computation Rate

The performance of a computation code is typically determined in terms of achievable computation rates [23]. Given a particular coding scheme, an achievable computation rate specifies how many function values can be computed per function value within some pre-defined accuracy.

Definition 5.4.2 ([16]) Let f be some fixed desired function, $d_{\mathbb{R}}$ a given distortion measure, and $\varepsilon > 0$ some given computation accuracy. Then, $R^{\mathrm{C}}(f, d_{\mathbb{R}}, \varepsilon)$ is said to be an *achievable computation rate* for f and ε if for every rate

$$R' := \frac{T}{n} < R^{\mathrm{C}}(f, d_{\mathbb{R}}, \varepsilon) \quad \frac{\text{function values}}{\text{channel use}}$$

and every $\delta > 0$ there exists an (f, T, n) computation code such that (5.11) is fulfilled for sufficiently large block length n.

The following result, proven in [16], provides the computation rate achievable with the computation code outlined in the preceding subsection under the average transmit power constraint (5.3).

Theorem 5.4.1 ([16]) *Let f be a nomographic function with continuous pre- and post-processing functions, $\varepsilon > 0$ some given desired accuracy, and $b_0(f, \varepsilon)$ the smallest number of bits with which f can be represented within accuracy ε in the supremum norm $\|\cdot\|_\infty$. Then,*

$$R^C(f, \|\cdot\|_\infty, \varepsilon) = \frac{\frac{1}{2}\log_2^+\left(\frac{P}{\sigma_Z^2}\right)}{b_0(f, \varepsilon) + \log_2(N)} \tag{5.29}$$

is an achievable computation rate for f and ε.

For different nomographic examples, the left part of Fig. 5.3 depicts the computation rate (5.29) achievable over a WMAC with $N = 7$ sources. It can be seen that at a signal-to-noise ratio (SNR) (i.e., P/σ_Z^2) of 15 dB, the "arithmetic mean" can be computed approximately 1.2 times more often (per channel use) than the "geometric mean" and approximately 1.9 times more often than the "Euclidean norm", respectively. The reason is that approximating the "arithmetic mean" within ε requires fewer bits (i.e., a smaller $b_0(f, \varepsilon)$) than for the "geometric mean" or the "Euclidean norm".

Now, in order to illustrate how good Theorem 5.4.1 actually is, the right part of Fig. 5.3 compares (5.29) with the best possible separation-based strategy. Recall that a separation-based strategy first decodes all the individual source symbols from the channel output and then computes the desired function values. Thus, any separation-based strategy is limited to the WMAC capacity region. Comparing the plots reveals that except for small SNR values, many linear and nonlinear functions can be reliably computed at rates that are not achievable with any separation-based strategy. See [16] for a more detailed discussion.

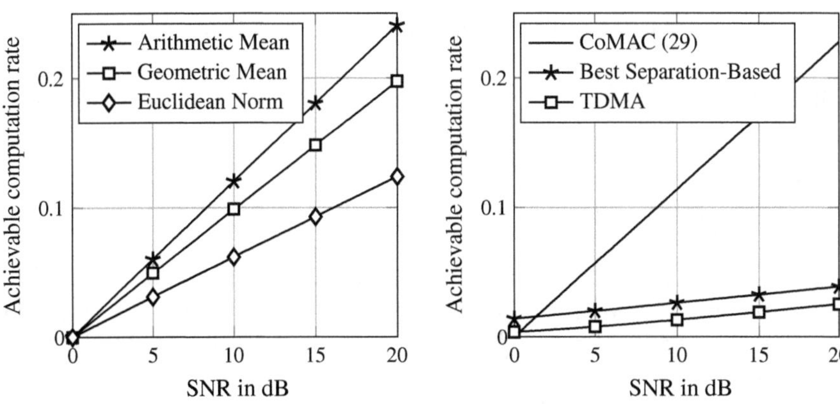

Fig. 5.3 (*left*) Achievable computation rates in a network of $N = 7$ sources, $\varepsilon = 10^{-3}$, and different nomographic functions; (*right*) Achievable computation rates in a network of $N = 12$ sources where the destination wishes to know the "arithmetic mean" within accuracy $\varepsilon = 10^{-3}$. TDMA refers to naive time sharing [10, p. 96]

5.4.2 *Strategy II*

In this section, we outline an alternative coding strategy that was recently proposed in [14]. Unlike the previous one, the strategy is completely *analog* in the sense that real-valued pre-processed source symbols are directly mapped to real-valued channel input symbols without some quantization operation in between. This not only significantly reduces the complexity of the encoding and decoding operations but also the need for precise synchronization between sources. Especially the latter was assumed in previous sections without explicitly mentioning. See Fig. 5.4 for a block diagram.

5.4.2.1 Encoding and Decoding

The basic idea behind the strategy is to let each source encode its real-valued pre-processed symbols into the transmit power of a sequence of random pulses whereas the destination simply determines the receive energy followed by some analog post-processing. In the following, we give a brief summary.

Encoding Once source i, $i = 1, \dots, N$, has applied pre-processing function φ_i to its sequence of source symbols, the outcomes undergo another function g, which has the objective to ensure that each sequence of channel input symbols satisfies the maximum transmit power constraint (5.4). This is why we refer to the time series $P_i[t] := g(\varphi_i(s_i[t]))$, $t = 1, \dots, T$, as the power profile of source i. Now, each power profile modulates an independently generated random length-n sequence $\mathbf{u}_i$, $i = 1, \dots, N$, that is zero mean and of unit norm. As a consequence, the information to be conveyed to the destination is encoded into sequences $\mathbf{u}_1, \dots, \mathbf{u}_N$ as time varying transmit powers. Note that since each $\mathbf{u}_i$ is assumed to have unit norm, we have

$$\max_{1 \le m \le n} |X_i[m]|^2 = \max_{1 \le t \le T} P_i[t] \le P \qquad i = 1, \dots, N$$

as desired.

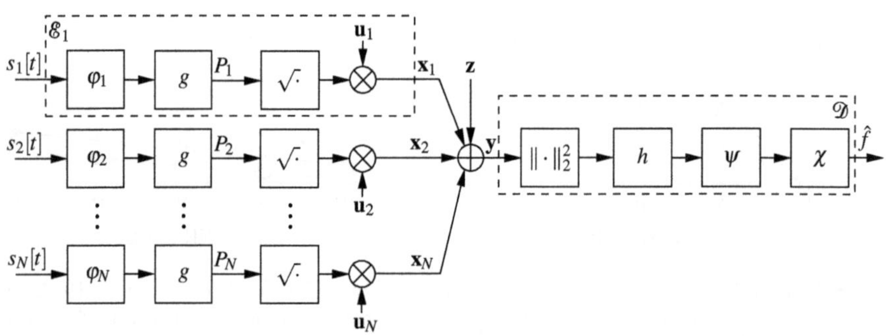

Fig. 5.4 Block diagram of the analog coding scheme of Sect. 5.4.2 for efficiently computing a nomographic function over the WMAC (5.1), where $H_1 = H_2 = \cdots = H_N = 1$

Decoding The destination first determines the receive energy $\|\mathbf{y}\|_2^2$, which results in [14]

$$\tilde{Y}[t] := \frac{n}{T}\sum_{i=1}^{N} g\big(\varphi_k(s_k[t])\big) + \Delta[t] \qquad t = 1,\ldots,T. \tag{5.30}$$

Equation (5.30) can be seen as an effective WMAC with inputs $g\big(\varphi_k(s_k[t])\big)$ and effective noise $\Delta[t]$. Based on this, for each t the decoding rule is of the form (see Fig. 5.4)

$$\hat{f}\big(s_1[t],\ldots,s_K[t]\big) = \chi\big[\psi\big(h(\tilde{Y}[t])\big)\big].$$

Here, function h forms the counterpart to g in the sense that $h\left(\frac{n}{T}\sum_i g\big(\varphi_i(s_i)\big)\right) \equiv \sum_i \varphi_i(s_i)$. On the other hand, the final function block χ is responsible for ensuring certain estimation properties such as unbiasedness and consistency. See [14] for further details.

5.4.2.2 Achievable Computation Rate

In contrast to Theorem 5.4.1, determining for the above-outlined strategy a general achievable computation rate expression is notoriously difficult. Therefore, in the following we provide a result that deals with nomographic functions that have continuous pre-processing functions along with an affine post-processing function. That is,

$$f(s_1,\ldots,s_N) = a_1\sum_{i=1}^{N}\varphi_i(s_i) + a_0, \tag{5.31}$$

with $a_0, a_1 \in \mathbb{R}$ some finite constants.

Theorem 5.4.2 ([17]) *Let f be a nomographic function of the kind* (5.31), $\varepsilon > 0$ *some given desired accuracy, and* $\varphi_{\min} := \min_i \min_{s\in\mathcal{S}} \varphi_i(s)$, $\varphi_{\max} := \max_i \max_{s\in\mathcal{S}} \varphi_i(s)$. *Furthermore, let the sources be wide-sense stationary stochastic processes $S_i[t]$, $i = 1,\ldots,N$. Then,*

$$R^{\mathrm{C}}(f,\varepsilon) = \frac{\varepsilon P^2}{a_1^2(\varphi_{\max}-\varphi_{\min})^2}\,\mathrm{Var}\{\tilde{\Delta}\}^{-1}, \tag{5.32}$$

with

$$\mathrm{Var}\{\tilde{\Delta}\} := 2\sum_{i=1}^{N}\sum_{\substack{j=1\\ j\neq i}}^{N}\mathbb{E}\{g\big(\varphi_i(S_i[1])\big)g\big(\varphi_j(S_j[1])\big)\} + 2\sigma_Z^2\sum_{i=1}^{N}\mathbb{E}\{g\big(\varphi_i(S_i[1])\big)\} + \sigma_Z^4,$$

is an achievable computation rate for f and ε.

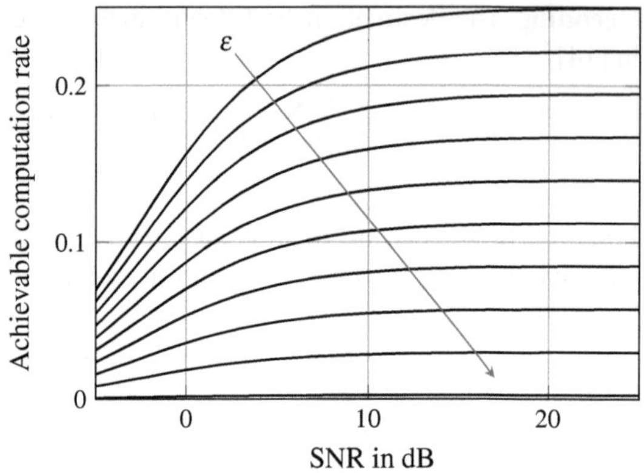

Fig. 5.5 Achievable computation rates in a network of $N = 5$ sources where the destination wishes to know the "arithmetic mean". The targeted computation accuracy, ε, ranges from 10^{-1} (*top line*) to 10^{-3} (*bottom line*)

The theorem states that the computation rates achievable with the coding scheme of Sect. 5.4.2.1 are proportional to the accuracy ε and that it depends through $\varphi_{\max}$ and $\varphi_{\min}$ on the source alphabet $\mathcal{S}$ and the choice of the desired function (i.e., the choice of pre-processing functions). The dependency on the SNR, P/σ_Z^2, and the number of sources N is mainly captured by the term $\mathbb{V}\text{ar}\{\tilde{\Delta}\}$.

Figure 5.5 plots (5.32) for f chosen as the "arithmetic mean", $N = 5$, and a computation accuracy ranging from 10^{-1} to 10^{-3}. Consider, for instance, the destination wishes to compute the arithmetic mean within accuracy $\varepsilon = 10^{-1}$ (i.e., the first plot from above) at an SNR of 20 dB. Then, on average about 4 transmissions are required per function value.

Note that (5.32) depends through $\mathbb{V}\text{ar}\{\tilde{\Delta}\}$ on the number of sources, which is responsible for the fact that the rates in Fig. 5.5 are bounded away from 1 even if the SNR tends to infinity. It can be shown, however, that (5.32) generally does not vanish with $N \to \infty$ as in case of any separation-based strategy.

5.5 Conclusions

In this chapter, we addressed the problem of reliably and efficiently computing real-valued linear and nonlinear functions of spatially distributed analog sources that communicate with the destination over a wireless channel. Instead of avoiding interference that occurs when several of the sources concurrently access the channel, we considered it as a freely accessible *computational resource*. In this context, we showed that every real-valued function of the sources can essentially be computed directly *in the air*. By using an appropriate signal pre- and post-processing strategy, every function of the source symbols can be matched to the structure of the wireless channel and therefore computed with just a single channel use. Compared with standard separation-based strategies, which reconstruct individual source symbols at the destination and compute desired function values afterwards, rate improvements in the order of the number of sources can be achieved.

As for noisy channels, we briefly outlined two recently proposed coding strategies and evaluated its performance in terms of achievable computation rates. It turns out that up to a small logarithmic penalty, the gains in computation rate observed in the noiseless case carry over to noisy channels. However, even though in the noiseless case the achievable rates do not depend on the function to be computed, when adding noise they do.

Note that the analog computation problem reviewed in this chapter can be considered as an instance of a very general point of view. For computing a multivariate function over an unreliable channel, we have seen that a system engineer could follow a standard separation-based approach, which is driven by the classical paradigm of communicating digitized messages. On the other hand, in order to improve efficiency he could also decide for an analog implementation as it was already discussed in Remark 5.3.2. Which implementation/technology (i.e., digital or analog) should be chosen by a practitioner finally depends on the existing experience and preferences. For instance, in the realm of image processing, the processing of images is usually based on a two-dimensional FFT implemented on a desktop computer. However, by simply using the laws of Fourier optic, a two-dimensional Fourier transform (plus filtering) could in the analog domain even be implemented in real time. The fact that the former (digital) approach has been established as a standard relies mainly on the assumption that it can approximate the latter with arbitrary precision.

Recently it was shown that such an approach is not always crowned with success. For instance, in [3] Boche and Pohl have shown that the Hilbert transform, which is a central operation in baseband signal processing for communication systems, cannot be well approximated with classical computing hardware, not even if the computational power would be unlimited. On the other hand, a precise computation of the Hilbert transform is possible by simple analog methods [3]. Thus, the example demonstrates that if an approach can be stably implemented in the analog domain it does not necessarily imply that it can also be approximated with arbitrary precision on a digital computer. For a more detailed discussion the reader is referred to [2].

Acknowledgments The work summarized in this chapter was mainly supported by the German Research Foundation (DFG) under grants STA 864/3-1, STA 864/3-2, and BO 1743/20-1, respectively.

References

1. Appuswamy R, Franceschetti M (2014) Computing linear functions by linear coding over networks. IEEE Trans Inf Theory 60(1):422–431
2. Boche H, Mönich UJ (2014) Signal and system approximation from general measurements. In: New perspectives on approximation and sampling theory—Festschrift in honor of Paul Butzer's 85th birthday. Applied and numerical harmonic analysis, Birkhäuser. Springer, New York

3. Boche H, Pohl V (2008) On the calculation of the Hilbert transform from interpolated data. IEEE Trans Inf Theory 54(5):2358–2366
4. Buck RC (1979) Approximate complexity and functional representation. J Math Anal Appl 70:280–298
5. Buck RC (1982) Nomographic functions are nowhere dense. Proc Am Math Soc 85(2):195–199
6. Erez U, Zamir R (2004) Achieving $\frac{1}{2}\log(1+\mathrm{SNR})$ on the AWGN channel with lattice encoding and decoding. IEEE Trans Inf Theory 50(10):2293–2314
7. Evesham HA (1982) The history and development of nomography. Docent Press, Boston
8. Fettweis G, Alamouti S (2014) 5G: personal mobile internet beyond what cellular did to telephony. IEEE Commun Mag 52(2):140–145
9. Fettweis G, Boche H, Wiegand T et al (2014) The tactile internet. ITU-T technical watch report, International Telecommunication Union (ITU). http://www.itu.int/oth/T2301000023/en
10. Gamal AE, Kim YH (2011) Network information theory. Cambridge University Press, Cambridge
11. Giridhar A, Kumar PR (2005) Computing and communicating functions over sensor networks. IEEE J Sel Areas Commun 23(4):755–764
12. Giridhar A, Kumar PR (2006) Toward a theory of in-network computation in wireless sensor networks. IEEE Commun Mag 44(4):98–107
13. Goldenbaum M (2014) Computation of real-valued functions over the channel in wireless sensor networks. PhD thesis, Technische Universität München, Department of Electrical and Computer Engineering, Munich, Germany
14. Goldenbaum M, Stańczak S (2013) Robust analog function computation via wireless multiple-access channels. IEEE Trans Commun 61(9):3863–3877
15. Goldenbaum M, Boche H, Stańczak S (2013) Harnessing interference for analog function computation in wireless sensor networks. IEEE Trans Signal Process 61(20):4893–4906
16. Goldenbaum M, Boche H, Stańczak S (2015) Nomographic functions: efficient computation in clustered Gaussian sensor networks. IEEE Trans Wirel Commun 14(4):2093–2105
17. Goldenbaum M, Stańczak S, Boche H (2015) On achievable rates for analog computing real-valued functions over the wireless channel. In: Proceedings of the IEEE international conference on communications (ICC), London, UK, 8–12 June 2015
18. Hilbert D (1902) Mathematical problems. Bull Am Math Soc 8(10):437–479
19. Karamchandani N, Keller L, Fragouli C, Franceschetti M (2013) Function computation via subspace coding. Phys Commun 6:134–141
20. Karamchandani N, Niesen U, Diggavi S (2013) Computation over mismatched channels. IEEE J Sel Areas Commun 31(4):666–677
21. Keller L, Karamchandani N, Fragouli C (2010) Function computation over linear channels. In: Proceedings of the IEEE international symposium on network coding (NetCod), Toronto, Canada, 9–11 June 2010
22. Kolmogorov AN (1957) On the representation of continuous functions of several variables by superposition of continuous functions of one variable and addition. Dokl Akad Nauk SSSR (114):953–956
23. Nazer B, Gastpar M (2007) Computation over multiple-access channels. IEEE Trans Inf Theory 53(10):3498–3516
24. Nazer B, Gastpar M (2011) Compute-and-forward: harnessing interference through structured codes. IEEE Trans Inf Theory 57(10):6463–6486
25. Pretz K (2014) Tech topic: smart sensors—making the internet of things soar. The Institute, IEEE, pp 6–7
26. Shannon CE (1948) A mathematical theory of communications. Bell Syst Tech J 27:379–423, 623–656
27. Sternfeld Y (1985) Dimension, superposition of functions and separation of points, in compact metric spaces. Israel J Math 50:13–52
28. Willig A (2008) Recent and emerging topics in wireless industrial communications: a selection. IEEE Trans Ind Inf 4(2):102–124

Chapter 6
Bounds on the Outage Constrained Capacity of the Gaussian Relay Channel

Andreas Gründinger, Lennart Gerdes, Michael Joham and Wolfgang Utschick

Abstract This study focuses on the probabilistically constrained capacity of the Gaussian relay channel. Since the capacity itself remains unknown in general, we study the achievable epsilon-outage rates that are based on direct source-to-destination transmission, the *decode-and-forward* (DF) strategy, and the *compress-and-forward* (CF) strategy as well as two upper bounds that are based on the mutual information expressions of the *cut-set bound* (CSB). The probabilistic outage constraint is necessary due to Rayleigh fading and the absence of *channel state information* (CSI) at the transmitting nodes. We derive closed-form expressions for the outage rates of direct transmission and the DF lower bound. A closed-form expression for the probability of a successful transmission with CF is also provided. Since closed-form CSB probability terms for coherent source and relay transmission are difficult to obtain, we use a genie-aided and a loosened CSB formulation to obtain the upper bounds. Alternatively, we approximate the probability that the CSB exceeds a target rate assuming noncoherent transmission from the source and the relay to the destination. The numerical results verify that the CF lower bound meets the capacity if the relay is at the destination and that the DF scheme achieves the CSB for noncoherent transmission if the relay is close to the source.

A. Gründinger (✉) · L. Gerdes · M. Joham · W. Utschick
Fachgebiet Methoden der Signalverarbeitung, Technische Universität München, 80290 München, Germany
e-mail: gruendinger@tum.de

L. Gerdes
e-mail: gerdes@tum.de

M. Joham
e-mail: joham@tum.de

W. Utschick
e-mail: utschick@tum.de

W. Utschick (ed.), *Communications in Interference Limited Networks*,
Signals and Communication Technology, DOI 10.1007/978-3-319-22440-4_6

6.1 Introduction

In this work, we consider the probabilistically constrained capacity, i.e., the outage capacity, of the single-antenna Gaussian relay channel, which is a simple three-node network where one source wants to convey information to one destination with the help of one relay. More specifically, since the outage capacity of the Gaussian relay channel is unknown, we study upper and lower bounds on the outage capacity that are based on the *cut-set bound* (CSB) and the achievable *decode-and-forward* (DF) and *compress-and-forward* (CF) rates, respectively.

The concept of relaying traces back to van der Meulen [12], who introduced the first information theoretic model for the relay channel. Unfortunately, the capacity of the relay channel remains unknown in general, even if perfect *channel state information* (CSI) is available at all nodes. Nevertheless, substantial advances towards the information theoretic understanding of the relay channel have since been made. The most important work is by Cover and El Gamal [3], who derived a capacity upper bound and achievable rates based on a then new CSB and the two fundamental relay strategies that are now usually referred to as DF and CF, respectively. The DF strategy requires the relay to decode the entire source message, which is then re-encoded and, in cooperation with the source, forwarded to the destination. When the relay uses the CF strategy, on the other hand, it does not decode any part of the source message, but rather it reliably forwards a compressed version of its received signal to the destination.

Instead of assuming perfect CSI at all nodes, this work addresses the case where only channel *distribution* information is available at the transmitting nodes, whereas the receiving nodes are still assumed to be aware of the channels' states. In particular, for this setup and slow Rayleigh fading channels, we employ the CSB and the DF and CF strategies to derive upper and lower bounds on the outage capacity of the Gaussian relay channel, respectively. We remark that while the ergodic capacity, i.e., the maximum achievable average rate, is the most important performance criterion for fast fading channels, the appropriate figure-of-merit for slow fading channels is the outage probability [16, Sect. 5.4].[1]

Approaches aimed at finding the minimum outage probability for a given rate target have already been considered in various works. For example, the minimum outage probability for the Gaussian relay channel with phase fading was bounded using the CSB and the DF strategy in [10]. Moreover, these outage probability bounds were later extended to the half-duplex case in [9], where Monte Carlo simulations were performed to obtain numerical results. Closed-form solutions for the outage probabilities of half-duplex DF relay systems with Rayleigh fading channels connecting the single-antenna source, relay(s), and destination were presented in [2, 18]. Yet other works considered either the low-SNR regime to derive tight bounds on the (outage) capacity [1, 8] or the high-SNR regime to simplify the analysis for

[1]In the considered relay channel, an outage occurs when the channel conditions are so poor that no error-free communication from the source to the destination is possible at the desired rate.

the outage probability and to elaborate on the diversity-multiplexing tradeoff for relaying, cf. [11, 13, 17], for example.

In contrast to these works, we are not interested in finding the minimum outage probability for a given rate target. Rather, we address the inverse problem, i.e., we seek to maximize the achievable rate for a given outage requirement in the Gaussian relay channel. Note that if we could actually solve this problem, we would obtain the outage capacity. Because this is not possible, however, we derive bounds on the outage capacity which, as previously mentioned, are based on the CSB and the DF and CF strategies. The quality of these bounds is analyzed by means of numerical simulations for a simple line network.

6.2 Relay System Model

Let $x_S \in \mathbb{C}$ and $x_R \in \mathbb{C}$ denote the transmit signals of the source and the relay, respectively. In the considered single-antenna Gaussian relay channel, which is illustrated in Fig. 6.1, the received signals of the relay and the destination are given by

$$\begin{aligned} y_R &= h_{SR} x_S + n_R, \\ y_D &= h_{SD} x_S + h_{RD} x_R + n_D, \end{aligned} \tag{6.1}$$

where $h_{SR} \in \mathbb{C}$, $h_{SD} \in \mathbb{C}$, and $h_{RD} \in \mathbb{C}$ represent the complex-valued channel coefficients. Furthermore, $n_R \sim \mathcal{N}_{\mathbb{C}}(0, 1)$ and $n_D \sim \mathcal{N}_{\mathbb{C}}(0, 1)$ denote zero-mean proper (circularly symmetric) complex Gaussian noise components of unit variance, which are assumed to be mutually independent as well as independent of the transmitted signals. In addition, we assume perfectly synchronized transmission and reception between all nodes, and it is implicit in (6.1) that the relay is able to completely cancel its own self-interference.

Without further conditions on the channel inputs, the capacity of the Gaussian relay channel would be infinite. This is because one can then choose infinite subsets of inputs arbitrarily far apart such that they are distinguishable at the outputs with arbitrarily small probability of error, cf. [4, Chap. 9]. We therefore impose the transmit power constraints

$$\mathrm{E}\left[|X_S|^2\right] \le P_S, \qquad \mathrm{E}\left[|X_R|^2\right] \le P_R \tag{6.2}$$

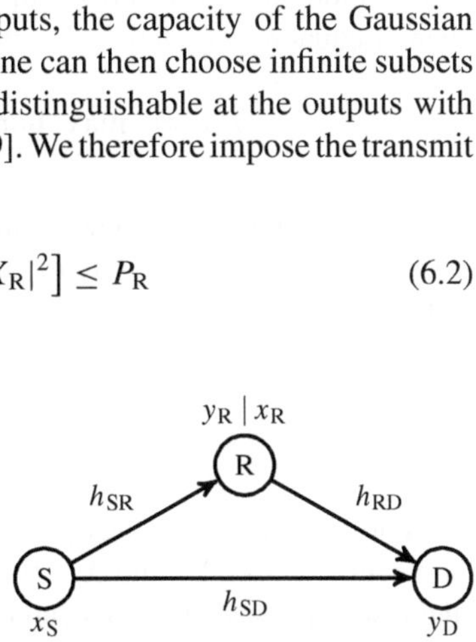

Fig. 6.1 Illustration of the Gaussian relay channel

on the channel inputs, where $P_S > 0$ and $P_R > 0$ denote the power budgets available at the source and the relay, respectively. Additionally, note that without loss of generality, it suffices to consider zero-mean channel inputs since it is clear that the optimal X_S and X_R are always zero-mean. The reason for optimality of zero-mean inputs is that channel inputs with non-zero mean consume more transmit power to convey the same amount of information as translations do not change the differential entropy of continuous random variables [4, Theorem 8.6.3].

Finally, we consider Rayleigh fading channels in this work, which means that the channel coefficients are modeled as independent zero-mean proper complex Gaussian random variables. In particular,

$$h_{SR} \sim \mathcal{N}_{\mathbb{C}}(0, \sigma_{SR}^2), \qquad h_{SD} \sim \mathcal{N}_{\mathbb{C}}(0, \sigma_{SD}^2), \qquad h_{RD} \sim \mathcal{N}_{\mathbb{C}}(0, \sigma_{RD}^2), \tag{6.3}$$

and we assume the channel coefficients to change slowly enough such that the outage probability is the relevant performance criterion.

6.3 Capacity Bounds for Perfect Channel State Information

Before we consider the outage capacity of the Gaussian relay channel, let us briefly revisit the CSB and the achievable DF and CF rates for the Gaussian relay channel with perfect CSI. In their pioneering work, Cover and El Gamal [3] proved that the capacity of the relay channel is upper bounded by

$$C_{CSB} = \max_{p(x_S, x_R)} \min \{ I(X_S; Y_R, Y_D | X_R), I(X_S, X_R; Y_D) \} . \tag{6.4}$$

Furthermore, they also showed that if the relay uses the DF strategy, all rates smaller than or equal to

$$R_{DF} = \max_{p(x_S, x_R)} \min \{ I(X_S; Y_R | X_R), I(X_S, X_R; Y_D) \} \tag{6.5}$$

can be achieved. For the Gaussian relay channel, it follows from the entropy maximizing property of the circularly symmetric complex Gaussian distribution [14, 15] that both C_{CSB} and R_{DF} are attained for $\left[\begin{smallmatrix} x_S \\ x_R \end{smallmatrix}\right] \sim \mathcal{N}_{\mathbb{C}}(\mathbf{0}, \boldsymbol{C})$, where

$$\boldsymbol{C} = \begin{bmatrix} P_S & \beta\sqrt{P_S P_R} \\ \beta^* \sqrt{P_S P_R} & P_R \end{bmatrix} \tag{6.6}$$

is the joint covariance matrix and $\beta = \mathrm{E}\left[X_S X_R^*\right] / \sqrt{P_S P_R} \in \mathbb{C}$ denotes the correlation coefficient of X_S and X_R. If the channel state $\boldsymbol{h} = (h_{SR}, h_{SD}, h_{RD})$ is perfectly and instantaneously known at all three nodes, the CSB and the maximum achievable

DF rate for the Gaussian relay channel with the power constraints given in (6.2) are hence equal to

$$C_{\mathrm{CSB}}(\boldsymbol{h}) = \max_{0\le|\beta|\le 1} \min\left\{C_{\mathrm{CSB}}^{(1)}(\beta,\boldsymbol{h}),\, C_{\mathrm{CSB}}^{(2)}(\beta,\boldsymbol{h})\right\}, \tag{6.7}$$

$$R_{\mathrm{DF}}(\boldsymbol{h}) = \max_{0\le|\beta|\le 1} \min\left\{R_{\mathrm{DF}}^{(1)}(\beta,\boldsymbol{h}),\, R_{\mathrm{DF}}^{(2)}(\beta,\boldsymbol{h})\right\}, \tag{6.8}$$

where the two achievable DF rate and CSB expressions read as

$$C_{\mathrm{CSB}}^{(1)}(\beta,\boldsymbol{h}) = \log_2\big(1 + (1-|\beta|^2)(|h_{\mathrm{SR}}|^2 + |h_{\mathrm{SD}}|^2)P_{\mathrm{S}}\big), \tag{6.9}$$

$$R_{\mathrm{DF}}^{(1)}(\beta,\boldsymbol{h}) = \log_2\big(1 + (1-|\beta|^2)|h_{\mathrm{SR}}|^2 P_{\mathrm{S}}\big), \tag{6.10}$$

and

$$\begin{aligned} C_{\mathrm{CSB}}^{(2)}(\beta,\boldsymbol{h}) &= R_{\mathrm{DF}}^{(2)}(\beta,\boldsymbol{h}) \\ &= \log_2\left(1 + |h_{\mathrm{SD}}|^2 P_{\mathrm{S}} + |h_{\mathrm{RD}}|^2 P_{\mathrm{R}} + 2\,\mathrm{Re}\{\beta h_{\mathrm{SD}} h_{\mathrm{RD}}^*\}\sqrt{P_{\mathrm{S}}P_{\mathrm{R}}}\right). \end{aligned} \tag{6.11}$$

In order to solve the remaining optimization for $C_{\mathrm{CSB}}(\boldsymbol{h})$ and $R_{\mathrm{DF}}(\boldsymbol{h})$ with perfect CSI, the correlation coefficient β must be chosen as $\beta = |\beta|\,\mathrm{e}^{\mathrm{j}\phi_\beta}$ with

$$\mathrm{e}^{\mathrm{j}\phi_\beta} = \frac{h_{\mathrm{SD}}^* h_{\mathrm{RD}}}{|h_{\mathrm{SD}}||h_{\mathrm{RD}}|}, \tag{6.12}$$

which results in the second CSB and DF rate expressions

$$\begin{aligned} C_{\mathrm{CSB}}^{(2)}(\beta,\boldsymbol{h}) &= R_{\mathrm{DF}}^{(2)}(\beta,\boldsymbol{h}) \\ &= \log_2\big(1 + |h_{\mathrm{SD}}|^2 P_{\mathrm{S}} + |h_{\mathrm{RD}}|^2 P_{\mathrm{R}} + 2|\beta||h_{\mathrm{SD}}||h_{\mathrm{RD}}|\sqrt{P_{\mathrm{S}}P_{\mathrm{R}}}\big). \end{aligned} \tag{6.13}$$

If possible, the optimal scalar $|\beta|$ then equalizes $C_{\mathrm{CSB}}^{(1)}(\beta,\boldsymbol{h})$ and $C_{\mathrm{CSB}}^{(2)}(\beta,\boldsymbol{h})$ or $R_{\mathrm{DF}}^{(1)}(\beta,\boldsymbol{h})$ and $R_{\mathrm{DF}}^{(2)}(\beta,\boldsymbol{h})$, respectively, or it is zero otherwise.

The CF strategy was also originally derived in [3]. In particular, Cover and El Gamal proved that all rates satisfying

$$R \le R_{\mathrm{CF}} = \max_{\substack{p(x_{\mathrm{S}})p(x_{\mathrm{R}}) \\ p(\hat{y}_{\mathrm{R}}|y_{\mathrm{R}},x_{\mathrm{R}})}} I(X_{\mathrm{S}};\hat{Y}_{\mathrm{R}}Y_{\mathrm{D}}|X_{\mathrm{R}}) \quad \text{s.t.} \quad I(\hat{Y}_{\mathrm{R}};Y_{\mathrm{R}}|X_{\mathrm{R}}Y_{\mathrm{D}}) \le I(X_{\mathrm{R}};Y_{\mathrm{D}}) \tag{6.14}$$

can be achieved if the relay uses the CF strategy, and El Gamal et al. [6] later showed that R_{CF} can equivalently be expressed as

$$R_{\mathrm{CF}} = \max_{\substack{p(x_{\mathrm{S}})p(x_{\mathrm{R}}) \\ p(\hat{y}_{\mathrm{R}}|y_{\mathrm{R}},x_{\mathrm{R}})}} \min \left\{ I(X_{\mathrm{S}}; \hat{Y}_{\mathrm{R}} Y_{\mathrm{D}} | X_{\mathrm{R}}), I(X_{\mathrm{S}}, X_{\mathrm{R}}; Y_{\mathrm{D}}) - I(\hat{Y}_{\mathrm{R}}; Y_{\mathrm{R}} | X_{\mathrm{R}} Y_{\mathrm{D}}) \right\}. \tag{6.15}$$

For the Gaussian relay channel, the joint distribution of the channel inputs X_{S}, X_{R} and the relay quantization $\hat{Y}_{\mathrm{R}}$ that attains R_{CF} is unknown. However, an achievable CF rate for the Gaussian relay channel can be obtained by choosing $(X_{\mathrm{S}}, X_{\mathrm{R}}, \hat{Y}_{\mathrm{R}})$ to be jointly proper complex Gaussian. More precisely, the channel inputs are chosen as $x_{\mathrm{S}} \sim \mathcal{N}_{\mathbb{C}}(0, P_{\mathrm{S}}), x_{\mathrm{R}} \sim \mathcal{N}_{\mathbb{C}}(0, P_{\mathrm{R}})$ and the relay quantization is determined according to

$$\hat{y}_{\mathrm{R}} = y_{\mathrm{R}} + \hat{n}_{\mathrm{R}} = h_{\mathrm{SR}} x_{\mathrm{S}} + n_{\mathrm{R}} + \hat{n}_{\mathrm{R}}, \tag{6.16}$$

where $\hat{n}_{\mathrm{R}} \sim \mathcal{N}_{\mathbb{C}}(0, \hat{\sigma}^2)$ represents a proper complex Gaussian quantization noise that is independent of the transmitted signals x_{S} and x_{R} and also the additive Gaussian noise components n_{R} and n_{D}. If the channel state $\boldsymbol{h}$ is known to all three nodes and $\xi := 1/(1 + \hat{\sigma}^2) \in [0, 1]$, it can hence be shown that [cf. (6.15)]

$$R_{\mathrm{CF}}(\boldsymbol{h}) = \max_{0 \le \xi \le 1} \min \left\{ R_{\mathrm{CF}}^{(1)}(\xi, \boldsymbol{h}), R_{\mathrm{CF}}^{(2)}(\xi, \boldsymbol{h}) \right\} \tag{6.17}$$

is an achievable CF rate for the Gaussian relay channel, where

$$R_{\mathrm{CF}}^{(1)}(\xi, \boldsymbol{h}) = \log_2\left(1 + (\xi |h_{\mathrm{SR}}|^2 + |h_{\mathrm{SD}}|^2) P_{\mathrm{S}}\right), \tag{6.18}$$

$$R_{\mathrm{CF}}^{(2)}(\xi, \boldsymbol{h}) = \log_2\left(1 + |h_{\mathrm{SD}}|^2 P_{\mathrm{S}} + |h_{\mathrm{RD}}|^2 P_{\mathrm{R}}\right) + \log_2(1 - \xi). \tag{6.19}$$

To conclude this section, we remark that there exist closed-form expressions for $C_{\mathrm{CSB}}(\boldsymbol{h})$, $R_{\mathrm{DF}}(\boldsymbol{h})$, and $R_{\mathrm{CF}}(\boldsymbol{h})$, cf. [6], for example. However, these closed-form expressions are not helpful for deriving bounds on the outage capacity of the Gaussian relay channel, so we do not elaborate on them here.

6.4 Outage Constrained Capacity Bounds

From now on, we take the independent block fading of the channels [see (6.3)] into account and consider the case that only the statistics of the channels are available at the transmitting nodes. The receiving nodes, on the other hand, are assumed to know the channel states because of training, for example. In other words, the source is only aware of σ_{SR}^2 and σ_{SD}^2, the relay has knowledge of σ_{RD}^2 and h_{SR}, and the destination knows h_{SD} and h_{RD}. An outage may occur in this situation if the imposed data transmission for error-free decoding from the source to the destination at a rate $\rho > 0$ remains inaccessible because the actual channels' gains are too low.

For a reliable transmission over the slow block fading channels, we limit the probability for such an outage event to lie below $\varepsilon \in (0, 1)$ while we seek to maximize the targeted rate ρ. The maximum rate is the ε-outage capacity $C^{(\text{out})}$, which can only be calculated in closed form for very simple channel setups, e.g., the Gaussian point-to-point channel, but which is unknown for the Gaussian relay channel at hand. A good introduction into the ε-outage capacity for the point-to-point and the relay channel was provided by Avestimehr and Tse [1], where it was shown that a bursty amplify-and-forward transmit strategy can achieve the ε-outage capacity in the low-SNR and the very low ε regime. Instead of the generally unknown ε-outage capacity, we focus on upper and lower bounds in this work. We obtain these bounds via a maximization of the aforementioned CSB and the DF and CF achievable rates under the ε-outage probability requirement.

6.4.1 *Lower Bounds*

The ε-outage rate, i.e., the achievable error-free data rate under an ε-outage probability constraint, is in general defined as

$$R^{(\text{out})} = \max_{\rho} \left\{ \rho \in \mathbb{R} : \Pr[R(\boldsymbol{h}) \geq \rho] \geq 1 - \varepsilon \right\}, \tag{6.20}$$

where we may equivalently rewrite the probabilistic constraint in terms of the ε-outage probability as

$$p^{(\text{out})} = \Pr[R(\boldsymbol{h}) < \rho] < \varepsilon. \tag{6.21}$$

Note that the probabilistic constraint formulation is necessary due to the unknown channel states at the transmitting source and relay, which decide about their (joint) transmit strategy and the data rate for transmission. Hence, we place any remaining rate maximization operation for the transmit strategy design outside of the probability operator when we replace the rate $R(\boldsymbol{h})$ with either the DF or the CF achievable rate, for example. The direct transmission from the source to the destination shall serve as a benchmark and is discussed next (see also [1]). We remark that the subsequently discussed DF strategy is strict compared to the dynamic alternative in [1] in that the source cannot fall back to the direct source to destination transmission depending on the actual channel realization.

6.4.1.1 Direct Transmission

If we only allow for a point-to-point transmission from the source to the destination, the error-free achievable data rate is equal to

$$R_{\text{SD}}(\boldsymbol{h}) = \log_2(1 + |h_{\text{SD}}|^2 P_{\text{S}}), \tag{6.22}$$

and the maximum ε-outage data rate is then given by

$$R_{\mathrm{SD}}^{(\mathrm{out})} = \max_{\rho} \left\{\rho \in \mathbb{R} : \Pr[R_{\mathrm{SD}}(\boldsymbol{h}) \geq \rho] \geq 1 - \varepsilon\right\}. \tag{6.23}$$

As (6.22) is the rate of an AWGN channel, its outage probability can be obtained from the *cumulative distribution function* (CDF) of the gain $g_{\mathrm{SD}} = |h_{\mathrm{SD}}|^2$ (see [16]), i.e.,

$$p_{\mathrm{SD}}^{(\mathrm{out})}(\rho) = \Pr\left[|h_{\mathrm{SD}}|^2 < \frac{2^\rho - 1}{P_{\mathrm{S}}}\right] = \mathrm{F}_{g_{\mathrm{SD}}}\left(\frac{2^\rho - 1}{P_{\mathrm{S}}}\right). \tag{6.24}$$

The CDF of a standard Rayleigh fading channel with unit variance is $\mathrm{F}(x) = 1 - \mathrm{e}^{-x}$ so that, for the given variance σ_{SD}^2 of h_{SD}, the CDF reads as

$$\mathrm{F}_{g_{\mathrm{SD}}}(x) = \mathrm{F}(x/\sigma_{\mathrm{SD}}^2) = 1 - \mathrm{e}^{-x/\sigma_{\mathrm{SD}}^2}. \tag{6.25}$$

With the inverse CDF $\mathrm{F}_{g_{\mathrm{SD}}}^{-1}(\varepsilon)$, the point-to-point ε-outage rate is [1]

$$R_{\mathrm{SD}}^{(\mathrm{out})} = \log_2\left(1 + \mathrm{F}_{g_{\mathrm{SD}}}^{-1}(\varepsilon) P_{\mathrm{S}}\right). \tag{6.26}$$

With the inverse CDF of a Rayleigh fading channel $\mathrm{F}^{-1}(y) = -\ln(1-y)$, we obtain

$$\mathrm{F}_{g_{\mathrm{SD}}}^{-1}(\varepsilon) = \mathrm{F}^{-1}(\varepsilon)\sigma_{\mathrm{SD}}^2 = -\ln(1-\varepsilon)\sigma_{\mathrm{SD}}^2, \tag{6.27}$$

which implies that we can write the ε-outage rate in (6.26) in closed form as

$$R_{\mathrm{SD}}^{(\mathrm{out})} = \log_2\left(1 + \ln\left(\tfrac{1}{1-\varepsilon}\right) P_{\mathrm{SD}}\right), \tag{6.28}$$

where $P_{\mathrm{SD}} = \sigma_{\mathrm{SD}}^2 P_{\mathrm{S}}$. We see that the inverse CDF essentially serves as a multiplicative degradation for the *signal-to-noise-ratio* (SNR) of the achievable data rate. A detailed discussion is provided by Avestimehr and Tse [1].

6.4.1.2 Decode-and-Forward

If the source and the relay employ the DF strategy for transmission, their achievable ε-outage rate maximization includes the joint optimization over β (cf. [7]), i.e.,

$$R_{\mathrm{DF}}^{(\mathrm{out})} = \max_{\rho, 0 \leq |\beta| \leq 1} \left\{\rho \in \mathbb{R} : p_{\mathrm{DF}}(\rho, \beta) \geq 1 - \varepsilon\right\}. \tag{6.29}$$

The constrained reliable transmission probability in (6.29) reads as

$$p_{\mathrm{DF}}(\rho, \beta) = \Pr\left[\min_{i=1,2}\{R_{\mathrm{DF}}^{(i)}(\beta, \boldsymbol{h})\} \geq \rho\right]. \tag{6.30}$$

It is a function of the target rate ρ and the joint transmit parameter β, and it is separable into the product

$$p_{\mathrm{DF}}(\rho, \beta) = p_{\mathrm{DF}}^{(1)}(\rho, \beta) p_{\mathrm{DF}}^{(2)}(\rho, \beta), \tag{6.31}$$

where the right-hand-side probabilities are given by

$$p_{\mathrm{DF}}^{(i)}(\rho, \beta) = \Pr\left[R_{\mathrm{DF}}^{(i)}(\beta, \boldsymbol{h}) \geq \rho\right], \quad i = 1, 2. \tag{6.32}$$

This is a result of the stochastic independence of the two corresponding data rates, i.e., $R_{\mathrm{DF}}^{(1)}(\beta, \boldsymbol{h})$ only depends on h_{SR} and $R_{\mathrm{DF}}^{(2)}(\beta, \boldsymbol{h})$ only depends on h_{SD} and h_{RD}, and their interrelation over the minimum operator in the joint transmit parameter.

For the considered Rayleigh channels, the first probability in (6.32) is [cf. (6.25)]

$$p_{\mathrm{DF}}^{(1)}(\rho, \beta) = \mathrm{e}^{-\frac{2^{\rho}-1}{(1-|\beta|^2)P_{\mathrm{SR}}}}, \tag{6.33}$$

where we used the short-hand notation $P_{\mathrm{SR}} = P_{\mathrm{S}}\sigma_{\mathrm{SR}}^2$. The second probability in (6.32) may be written in terms of the CDF of a weighted sum of two i.i.d. χ^2-distributed random variables z_1 and z_2, both with degree two, i.e.,

$$\begin{aligned} p_{\mathrm{DF}}^{(2)}(\rho, \beta) &= 1 - \Pr\left[z_1\lambda + z_2(1-\lambda) < \frac{2^{\rho}-1}{P_{\mathrm{RD}} + P_{\mathrm{SD}}}\right] \\ &= 1 - \mathrm{F}_{z_1\lambda + z_2(1-\lambda)}\left(\frac{2^{\rho}-1}{P_{\mathrm{RD}} + P_{\mathrm{SD}}}\right), \end{aligned} \tag{6.34}$$

where we additionally substituted $P_{\mathrm{RD}} = P_{\mathrm{R}}\sigma_{\mathrm{RD}}^2$ and the weight is limited to $\lambda \in [0.5, 1]$. The closed-form expression of (6.34) reads as

$$p_{\mathrm{DF}}^{(2)}(\rho, \beta) = \begin{cases} \frac{1}{2\lambda - 1}\left(\lambda\, \mathrm{e}^{-\frac{2^{\rho}-1}{\lambda}} - (1-\lambda)\, \mathrm{e}^{-\frac{2^{\rho}-1}{1-\lambda}}\right) & \text{if } \lambda > 0.5, \\ \left(1 + \frac{2^{\rho}-1}{\lambda}\right)\mathrm{e}^{-\frac{2^{\rho}-1}{\lambda}} & \text{if } \lambda = 0.5. \end{cases} \tag{6.35}$$

The weight $\lambda \in [0.5, 1]$, for the sum of two χ^2-distributed random variables, equals

$$\lambda = \frac{1}{2} + \sqrt{\frac{1}{4} - \left(1 - |\beta|^2\right)\frac{P_{\mathrm{RD}} P_{\mathrm{SD}}}{(P_{\mathrm{RD}} + P_{\mathrm{SD}})^2}} \tag{6.36}$$

and is the dominant eigenvalue of the positive semidefinite matrix[2]

$$\boldsymbol{C}_{\mathrm{DF}} = \frac{1}{P_{\mathrm{RD}} + P_{\mathrm{SD}}} \begin{bmatrix} P_{\mathrm{SD}} & \beta\sqrt{P_{\mathrm{RD}} P_{\mathrm{SD}}} \\ \beta^* \sqrt{P_{\mathrm{SD}} P_{\mathrm{RD}}} & P_{\mathrm{RD}} \end{bmatrix}. \tag{6.37}$$

Note that λ is increasing in $|\beta|$ and independent of the phase ϕ_β.

We remark that the stochastic independence of the two DF rate expressions and the probability formulations in (6.33) and (6.34) allow us to write the DF ε-outage rate in a similar form to (6.26). In particular,

$$R_{\mathrm{DF}}^{(\mathrm{out})} = \max_{\substack{\varepsilon_1, \varepsilon_2 \in (0,1), \\ 0 \le |\beta| \le 1}} \min_{i=1,2} \left\{ R_{\mathrm{DF},i}^{\mathrm{out}}(\varepsilon_i) \right\} \quad \text{s.t.} \quad (1-\varepsilon_1)(1-\varepsilon_2) = 1 - \varepsilon, \tag{6.38}$$

where the two ε-outage rates inside the minimum operator are given by

$$R_{\mathrm{DF},1}^{\mathrm{out}}(\varepsilon_1) = \log_2 \left(1 + \ln\left(\tfrac{1}{1-\varepsilon_1}\right)(1 - |\beta|^2) P_{\mathrm{SR}}\right), \tag{6.39}$$

$$R_{\mathrm{DF},2}^{\mathrm{out}}(\varepsilon_2) = \log_2 \left(1 + \mathrm{F}^{-1}_{z_1\lambda + z_2(1-\lambda)}(\varepsilon_2)(P_{\mathrm{RD}} + P_{\mathrm{SR}})\right). \tag{6.40}$$

To obtain (6.38), we introduced the two outage probabilities ε_1 and ε_2 in order to bound the two probabilities in (6.32) as $p_{\mathrm{DF}}^{(i)}(\rho, \beta) \ge (1 - \varepsilon_i)$.

Unfortunately, a closed-form expression for the inverse CDF $\mathrm{F}^{-1}_{z_1\lambda + z_2(1-\lambda)}(x)$ is missing. Therefore, we might use a two-dimensional search over ρ and $|\beta|$ to solve the original problem formulation in (6.29). Alternatively, we analyzed the two probabilities of (6.32) in [7]. While it is obvious that $|\beta| = 0$ always maximizes $p_{\mathrm{DF}}^{(1)}(\rho, \beta)$, we observed that $|\beta| = 0$ also maximizes $p_{\mathrm{DF}}^{(2)}(\rho, \beta)$ if ε_2 is sufficiently small. Note that the DF ε-outage rate is obviously lower bounded by noncoherent transmission so that we may also circumvent the search over β and choose $\beta = 0$ in preference of a simplified probability and ε-outage rate computation.

6.4.1.3 Compress-and-Forward

If the destination knows all the channel states perfectly, namely h_{SD}, h_{RD}, and additionally h_{SR},[3] the source and the relay can also employ the CF strategy. The ε-outage rate maximization then reads as

$$R_{\mathrm{CF}}^{(\mathrm{out})} = \max_{\rho, 0 \le \xi \le 1} \left\{ \rho \in \mathbb{R} : p_{\mathrm{CF}}(\rho, \xi) \ge 1 - \varepsilon \right\}. \tag{6.41}$$

[2]Here, (6.37) defines the joint source and relay transmit covariance matrix of an equivalent system with i.i.d. unit variance source- and relay-to-destination channels and a source-to-relay channel with variance $\sigma_{\mathrm{SR}}^2 / \sigma_{\mathrm{SD}}^2$.

[3]The state of h_{SR} may be forwarded from the relay to the destination.

The probability inside of (6.41) is a substitute for the expression

$$p_{\mathrm{CF}}(\rho, \xi) = \Pr\left[\min_{i=1,2}\{R_{\mathrm{CF}}^{(i)}(\xi, \boldsymbol{h})\} \geq \rho\right], \tag{6.42}$$

where the corresponding Gaussian CF rates are given by (6.18) and (6.19), respectively. Note that both CF rates contain $|h_{\mathrm{SD}}|^2$, that is, they are stochastically dependent. Therefore, the CF ε-outage rate does not feature a direct expression as for the direct transmission in (6.26) or for the DF scheme (6.38). However, (6.41) again defines a two-dimensional line search, but now over ρ and ξ. Thus, we may compute the ε-outage CF rate via a golden section search over ξ and a bisection in each golden section step to find the maximum ρ that achieves the probability $1-\varepsilon$ with equality.

To this end, we derive a closed-form expression for the probability in (6.42) by dividing the stochastic event of error-free transmission, i.e.,

$$\min_{i=1,2}\{R_{\mathrm{CF}}^{(i)}(\xi, \boldsymbol{h})\} \geq \rho, \tag{6.43}$$

in three disjoint sub-events depending on the value of the random gain $z_{\mathrm{SD}} = |h_{\mathrm{SD}}|^2/\sigma_{\mathrm{SD}}^2$, which follows a χ^2-distribution with two degrees of freedom:

(i) First, (6.43) is satisfied if

$$\log_2(1 + z_{\mathrm{SD}} P_{\mathrm{SD}}) + \log_2(1-\xi) \geq \rho.$$

(ii) If z_{SD} has a value that achieves

$$\log_2(1 + z_{\mathrm{SD}} P_{\mathrm{SD}}) + \log_2(1-\xi) < \rho \leq \log_2(1 + z_{\mathrm{SD}} P_{\mathrm{SD}}),$$

we additionally need that $z_{\mathrm{RD}} = |h_{\mathrm{RD}}|^2/\sigma_{\mathrm{RD}}^2$ satisfies

$$\log_2(1 + z_{\mathrm{RD}} P_{\mathrm{RD}} + z_{\mathrm{SD}} P_{\mathrm{SD}}) + \log_2(1-\xi) > \rho.$$

(iii) Finally, if the value of z_{SD} results in

$$\log_2(1 + z_{\mathrm{SD}} P_{\mathrm{SD}}) < \rho,$$

we require $z_{\mathrm{SR}} = |h_{\mathrm{SR}}|^2/\sigma_{\mathrm{SR}}^2$ to satisfy

$$\log_2(1 + \xi z_{\mathrm{SR}} P_{\mathrm{SR}} + z_{\mathrm{SD}} P_{\mathrm{SD}}) \geq \rho$$

in addition to the requirement on z_{RD} in (ii).

Summing up the three probabilities $p_{\mathrm{CF}}^{(j)}$, $j = \text{(i), (ii), (iii)}$, that correspond to above case distinction, we obtain the CF probability (6.42) as

$$p_{\mathrm{CF}}(\rho, \xi) = p_{\mathrm{CF}}^{(\mathrm{i})} + p_{\mathrm{CF}}^{(\mathrm{ii})} + p_{\mathrm{CF}}^{(\mathrm{iii})}. \tag{6.44}$$

The closed-form expressions for the three summands in (6.44) are given by

$$p_{\mathrm{CF}}^{(\mathrm{i})} = \mathrm{e}^{-\frac{2^{\rho-\log_2(1-\xi)}-1}{P_{\mathrm{SD}}}}, \tag{6.45}$$

$$p_{\mathrm{CF}}^{(\mathrm{ii})} = \frac{P_{\mathrm{RD}}}{P_{\mathrm{SD}} - P_{\mathrm{RD}}} \left(\mathrm{e}^{-\frac{2^{\rho-\log_2(1-\xi)}-1}{P_{\mathrm{SD}}}} - \mathrm{e}^{-(2^{\rho}-1)(\frac{1}{P_{\mathrm{SD}}} - \frac{1}{P_{\mathrm{RD}}}) - \frac{2^{\rho-\log_2(1-\xi)}-1}{P_{\mathrm{RD}}}} \right), \tag{6.46}$$

$$p_{\mathrm{CF}}^{(\mathrm{iii})} = \frac{\frac{1}{P_{\mathrm{SD}}}}{\frac{1}{P_{\mathrm{SR}}} + \frac{\xi}{P_{\mathrm{RD}}} - \frac{\xi}{P_{\mathrm{SD}}}} \mathrm{e}^{-\frac{2^{\rho-\log_2(1-\xi)}-1}{P_{\mathrm{RD}}}} \left(\mathrm{e}^{-(2^{\rho}-1)(\frac{1}{P_{\mathrm{SD}}} - \frac{1}{P_{\mathrm{RD}}})} - \mathrm{e}^{-\frac{2^{\rho}-1}{\xi P_{\mathrm{RD}}}} \right). \tag{6.47}$$

Note that these probability expressions are only valid for $\xi \in (0, 1)$. If $\xi = 0$, the CF achievable rate results in the achievable rate for direct transmission and we have to use (6.24) for the probability computation. Moreover, the CF rate becomes zero with probability one if $\xi \to 1$ [cf. (6.17)].

6.4.2 *Upper Bounds*

Upper bounds on the ε-outage capacity with only statistical channel information at the transmitting nodes can be based on the CSB in (6.7) as

$$C_{\mathrm{CSB}}^{(\mathrm{out})} = \max_{\rho, |\beta|^2 \leq 1} \left\{ \rho \in \mathbb{R} : p_{\mathrm{CSB}}(\rho, \beta) \geq 1 - \varepsilon \right\}, \tag{6.48}$$

where the CSB probability for achieving the rate target ρ has the expression

$$p_{\mathrm{CSB}}(\rho, \beta) = \Pr\left[\min_{i=1,2} \{ C_{\mathrm{CSB}}^{(i)}(\beta, \boldsymbol{h}) \} \geq \rho \right]. \tag{6.49}$$

Problem (6.48) defines a two-dimensional search similar to the DF ε-outage rate or the CF ε-outage rate. As we showed in [7], the joint transmit parameter β can without loss of optimality be restricted to be real and non-negative. This is because the phases of the zero-mean circularly complex Gaussian channels h_{SD} and h_{RD} are uniformly distributed over $[0, 2\pi)$. Thus, we may apply a golden section search over β and an inner bisection with respect to ρ to find the solution to (6.48).

However, the probability computations for each step of the two-dimensional search would require numerical computations. For example, we might employ a Monte-Carlo method that uses an extensively large number of sample channel realizations to compute the probability in (6.49) with the desired accuracy. This is required because the statistical properties of the CSB are different from that of the DF and the CF rates. The CSB terms in (6.9) and (6.11) depend on $|h_{\mathrm{SD}}|^2$ and $\mathrm{Re}\{\beta h_{\mathrm{SD}} h_{\mathrm{RD}}{}^*\}$.

To circumvent the numerical probability evaluation, we propose two upper bounds for (6.9) and (6.11) to bound the ε-outage value of the CSB from above.

6.4.2.1 Genie-Aided Cut-Set Bound

First, we employ an upper bound approximation for the real part operator in (6.11):

$$\mathrm{Re}\{\beta h_{\mathrm{SD}} h_{\mathrm{RD}}^*\} \leq |\beta||h_{\mathrm{SD}}||h_{\mathrm{RD}}|. \tag{6.50}$$

This bound could be attained if we are able to choose the phase of β as for the perfect CSI case and use the corresponding second perfect CSI rate expression in (6.13) instead of (6.11) for the probability in (6.49). Since (6.13) only depends on the absolute value of the channels, we can write the resulting *genie-aided* (GA) CSB probability as the sum of three probabilities

$$p_{\mathrm{GA}}(\rho, \beta) = p_{\mathrm{CSB}}\left(\rho, |\beta| \frac{h_{\mathrm{SD}}^* h_{\mathrm{RD}}}{|h_{\mathrm{SD}}||h_{\mathrm{RD}}|}\right) = p_{\mathrm{GA}}^{(\mathrm{i})} + p_{\mathrm{GA}}^{(\mathrm{ii})} + p_{\mathrm{GA}}^{(\mathrm{iii})}. \tag{6.51}$$

Each of the probabilities $p_{\mathrm{GA}}^{(j)}$, $j = \text{(i), (ii), (iii)}$ corresponds to a disjoint range for the χ^2-distributed random variable $z_{\mathrm{SD}} \in [0, \infty)$ for satisfying $C_{\mathrm{CSB}}^{(i)}(\beta, \boldsymbol{h}) \geq \rho$, $i = 1, 2$ (see [7] for the details). The representations for these probabilities read as

$$p_{\mathrm{GA}}^{(\mathrm{i})} = \mathrm{e}^{-\frac{2^\rho - 1}{(1-|\beta|^2) P_{\mathrm{SD}}}}, \tag{6.52a}$$

$$p_{\mathrm{GA}}^{(\mathrm{ii})} = \frac{\sigma_{\mathrm{SR}}^2}{\sigma_{\mathrm{SR}}^2 - \sigma_{\mathrm{SD}}^2} \left(\mathrm{e}^{-\frac{2^\rho - 1}{(1-|\beta|^2) P_{\mathrm{SD}}}} - \mathrm{e}^{-\frac{2^\rho - 1}{P_{\mathrm{SD}}} - \frac{(2^\rho - 1)|\beta|^2}{(1-|\beta|^2) P_{\mathrm{SR}}}} \right), \tag{6.52b}$$

$$p_{\mathrm{GA}}^{(\mathrm{iii})} = \int_0^{\frac{2^\rho - 1}{P_{\mathrm{SD}}}} \mathrm{e}^{-z_{\mathrm{SD}} - \frac{2^\rho - 1 - (1-|\beta|^2) P_{\mathrm{SD}} z_{\mathrm{SD}}}{(1-|\beta|^2) P_{\mathrm{SR}}}} \mathrm{e}^{-\frac{1}{P_{\mathrm{RD}}}\left(\sqrt{2^\rho - 1 - (1 - \frac{|\beta|^2}{4}) P_{\mathrm{SD}} z_{\mathrm{SD}}} - \sqrt{\frac{|\beta|^2}{4} P_{\mathrm{SD}} z_{\mathrm{SD}}}\right)^2} \mathrm{d}z_{\mathrm{SD}}. \tag{6.52c}$$

We are unfortunately not aware of a closed-form expression for $p_{\mathrm{GA}}^{(\mathrm{iii})}$ and employed a standard numerical integration for the simulations.

6.4.2.2 Loosened Cut-Set Bound

Alternatively to the genie-aided design, we avoid the difficulties with the cooperation term β via a loosened CSB. For example, we loosen the first CSB term (6.9) as

$$C_{\mathrm{LOS}}^{(1)}(\boldsymbol{h}) = \log_2\left(1 + P_{\mathrm{S}}(|h_{\mathrm{SR}}|^2 + |h_{\mathrm{SD}}|^2)\right), \tag{6.53}$$

which implies that $\beta = 0$. To loosen the second CSB term in (6.11), we follow similar steps as in [5, Proof of Theorem 19.2]. In particular, we apply

$$\det(\mathbf{I} + \boldsymbol{A}\boldsymbol{B}) \leq \det\left(\mathbf{I} + \frac{1}{\mu\,\mathrm{tr}(\boldsymbol{A})}\boldsymbol{A}\right)\det\left(\mathbf{I} + \mu\,\mathrm{tr}(\boldsymbol{A})\boldsymbol{B}\right), \tag{6.54}$$

where $\mu \geq 0$ and $\boldsymbol{A}, \boldsymbol{B} \succeq \mathbf{0}$, to obtain the upper bound

$$C_{\mathrm{CSB}}^{(2)}(\beta, \boldsymbol{h}) \leq \log_2\det\left(\mathbf{I} + \frac{1}{\mu(P_{\mathrm{R}} + P_{\mathrm{S}})}\boldsymbol{C}\right) + \log_2\left(1 + \mu(P_{\mathrm{S}} + P_{\mathrm{R}})(|h_{\mathrm{SD}}|^2 + |h_{\mathrm{RD}}|^2)\right) \tag{6.55}$$

with $\boldsymbol{A} = \boldsymbol{C}$ given in (6.6) and $\boldsymbol{B} = [h_{\mathrm{SD}}, h_{\mathrm{RD}}]^{\mathrm{T}}[h_{\mathrm{SD}}, h_{\mathrm{RD}}]^*$. The right-hand-side in (6.55) is maximized for $\beta = 0$, which is independent of the channel realization. Hence, we obtain the loosened expression

$$C_{\mathrm{LOS}}^{(2)}(\mu, \boldsymbol{h}) = \log_2\left(1 + \frac{P_{\mathrm{R}}}{\mu(P_{\mathrm{R}} + P_{\mathrm{S}})}\right) + \log_2\left(1 + \frac{P_{\mathrm{S}}}{\mu(P_{\mathrm{R}} + P_{\mathrm{S}})}\right) + \log_2\left(1 + \mu(P_{\mathrm{S}} + P_{\mathrm{R}})(|h_{\mathrm{SD}}|^2 + |h_{\mathrm{RD}}|^2)\right). \tag{6.56}$$

If we exploit (6.53) and (6.56) instead of the actual expressions (6.9) and (6.11) for the optimization in (6.48), the resulting ε-outage loosened CSB maximization

$$C_{\mathrm{CSB}}^{(\mathrm{out})} \leq C_{\mathrm{LOS}}^{(\mathrm{out})} = \max_{\rho}\min_{\mu\geq 0}\left\{\rho \in \mathbb{R} : p_{\mathrm{LOS}}(\rho, \mu) \geq 1 - \varepsilon\right\} \tag{6.57}$$

gives an upper bound to the solution for (6.48). Note that (6.57) contains a minimization over $\mu \geq 0$, which we realized with a golden section search. This is to tighten the initially loosened CSB from above.

The probability in (6.57) reads as

$$p_{\mathrm{LOS}}(\rho, \mu) = \Pr\left[\min_{i=1,2}\{C_{\mathrm{LOS}}^{(i)}(\mu, \boldsymbol{h})\} \geq \rho\right] \tag{6.58}$$

and allows for a closed-form expression, but not for a direct formulation of the ε-outage loosened CSB expression because (6.53) and (6.56) both depend on h_{SD}.

Similar to the CF probability calculation, we calculate the probability in (6.58) by dividing the stochastic event

$$\min_{i=1,2}\{C^{(i)}_{\mathrm{LOS}}(\mu, \boldsymbol{h})\} \geq \rho \tag{6.59}$$

into three disjoint sub-events depending on the value of z_{SD}:

(i) The event in (6.59) is satisfied if

$$\log_2(1 + z_{\mathrm{SD}} P_{\mathrm{SD}}) \geq \rho. \tag{6.60}$$

(ii) Next, if z_{SD} is too small for (i), but satisfies

$$\begin{aligned}\log_2(1 + z_{\mathrm{SD}} P_{\mathrm{SD}}) < \rho \leq \log_2\left(1 + \frac{P_{\mathrm{R}}}{\mu(P_{\mathrm{R}} + P_{\mathrm{S}})}\right)\\ + \log_2\left(1 + \frac{P_{\mathrm{S}}}{\mu(P_{\mathrm{R}} + P_{\mathrm{S}})}\right) + \log_2\left(1 + \mu(P_{\mathrm{S}} + P_{\mathrm{R}})\sigma^2_{\mathrm{SD}} z_{\mathrm{SD}}\right)\end{aligned} \tag{6.61}$$

instead, then z_{SR} additionally needs to be large enough such that

$$\log_2(1 + z_{\mathrm{SD}} P_{\mathrm{SD}} + z_{\mathrm{SR}} P_{\mathrm{SR}}) \geq \rho. \tag{6.62}$$

(iii) Finally, if z_{SD} is so small that

$$\begin{aligned}\rho > \log_2\left(1 + \frac{P_{\mathrm{R}}}{\mu(P_{\mathrm{R}} + P_{\mathrm{S}})}\right) + \log_2\left(1 + \frac{P_{\mathrm{S}}}{\mu(P_{\mathrm{R}} + P_{\mathrm{S}})}\right)\\ + \log_2\left(1 + \mu(P_{\mathrm{S}} + P_{\mathrm{R}})\sigma^2_{\mathrm{SD}} z_{\mathrm{SD}}\right),\end{aligned} \tag{6.63}$$

then we need condition (6.62) for z_{SR} and a sufficiently large z_{RD}, i.e.,

$$\begin{aligned}\rho \leq \log_2\left(1 + \frac{P_{\mathrm{R}}}{\mu(P_{\mathrm{R}} + P_{\mathrm{S}})}\right) + \log_2\left(1 + \frac{P_{\mathrm{S}}}{\mu(P_{\mathrm{R}} + P_{\mathrm{S}})}\right)\\ + \log_2\left(1 + \mu(P_{\mathrm{S}} + P_{\mathrm{R}})(\sigma^2_{\mathrm{SD}} z_{\mathrm{SD}} + \sigma^2_{\mathrm{RD}} z_{\mathrm{SR}}\right).\end{aligned} \tag{6.64}$$

We remark that events (ii) and (iii) in this form suppose that

$$\rho > \log_2\left(1 + \frac{P_{\mathrm{R}}}{\mu(P_{\mathrm{R}} + P_{\mathrm{S}})}\right) + \log_2\left(1 + \frac{P_{\mathrm{S}}}{\mu(P_{\mathrm{R}} + P_{\mathrm{S}})}\right). \tag{6.65}$$

If (6.65) is not satisfied by a given pair (ρ, μ), event (iii) and the upper bound in (6.61) can be neglected for the probability computation.

The probability in (6.58) is then obtained by adding the probabilities that correspond to above three disjoint events, i.e.,

$$p_{\mathrm{LOS}}(\rho,\mu) = p_{\mathrm{LOS}}^{(\mathrm{i})} + p_{\mathrm{LOS}}^{(\mathrm{ii})} + p_{\mathrm{LOS}}^{(\mathrm{iii})}. \tag{6.66}$$

The probabilities in (6.66) read as

$$p_{\mathrm{LOS}}^{(\mathrm{i})} = \mathrm{e}^{-\frac{2^\rho-1}{P_{\mathrm{SD}}}}, \tag{6.67}$$

$$p_{\mathrm{LOS}}^{(\mathrm{ii})} = \frac{\sigma_{\mathrm{SR}}^2}{\sigma_{\mathrm{SD}}^2-\sigma_{\mathrm{SR}}^2}\left(\mathrm{e}^{-\frac{2^\rho-1}{P_{\mathrm{SD}}}} - \mathrm{e}^{-\frac{2^\rho-1}{P_{\mathrm{SR}}}-\frac{\zeta_2(\rho,\mu)}{\mu(P_{\mathrm{SD}}+P_{\mathrm{RSD}})}\frac{\sigma_{\mathrm{SR}}^2-\sigma_{\mathrm{SD}}^2}{\sigma_{\mathrm{SR}}^2}}\right), \tag{6.68}$$

$$p_{\mathrm{LOS}}^{(\mathrm{iii})} = \frac{\sigma_{\mathrm{SD}}^{-2}}{\sigma_{\mathrm{SR}}^{-2}+\sigma_{\mathrm{RD}}^{-2}-\sigma_{\mathrm{SD}}^{-2}}\,\mathrm{e}^{-\frac{2^\rho-1}{P_{\mathrm{SR}}}}\left(\mathrm{e}^{-\frac{\zeta_2(\rho,\mu)}{\mu(P_{\mathrm{SD}}+P_{\mathrm{RSD}})}\frac{\sigma_{\mathrm{SR}}^2-\sigma_{\mathrm{SD}}^2}{\sigma_{\mathrm{SR}}^2}} - \mathrm{e}^{-\frac{\zeta_2(\rho,\mu)}{\mu(P_{\mathrm{SRD}}-P_{\mathrm{RD}})}}\right) \tag{6.69}$$

if (6.65) is satisfied and where $P_{\mathrm{RSD}} = P_{\mathrm{R}}\sigma_{\mathrm{SD}}^2$, $P_{\mathrm{SRD}} = P_{\mathrm{S}}\sigma_{\mathrm{RD}}^2$ and

$$\zeta_2(\rho,\mu) = 2^{\rho-\log_2\left(1+\frac{P_{\mathrm{R}}}{\mu(P_{\mathrm{R}}+P_{\mathrm{S}})}\right)-\log_2\left(1+\frac{P_{\mathrm{S}}}{\mu(P_{\mathrm{R}}+P_{\mathrm{S}})}\right)} - 1. \tag{6.70}$$

Otherwise, if (6.65) is violated by the given pair (ρ,μ), then $p_{\mathrm{LOS}}^{(\mathrm{iii})} = 0$ and

$$p_{\mathrm{LOS}}^{(\mathrm{ii})} = \frac{\sigma_{\mathrm{SR}}^2}{\sigma_{\mathrm{SD}}^2-\sigma_{\mathrm{SR}}^2}\left(\mathrm{e}^{-\frac{2^\rho-1}{P_{\mathrm{SD}}}} - \mathrm{e}^{-\frac{2^\rho-1}{P_{\mathrm{SR}}}}\right). \tag{6.71}$$

6.5 Bounds for Noncoherent Transmission

For practical and also various theoretical reasons, noncoherent transmission, i.e., $\beta = 0$, may be imposed for the relay-assisted source-to-destination transmission:

- First, coherent transmission is difficult to establish in practice since it requires synchronized transmission of the source and relay nodes.
- Besides, $\beta = 0$ appears to be optimal for the DF strategy with only statistical information about the channels at the transmitting nodes if ε is sufficiently small [7]. For this small ε, $\beta = 0$ may also maximize

$$\Pr\left[C_{\mathrm{CSB}}^{(2)}(\rho,\beta) \geq \rho\right]. \tag{6.72}$$

- Finally, noncoherent transmission helps only for fifty percent of the channel realizations, namely, when the uniform phase fading h_{SD} and h_{SR} result in

$$\mathrm{Re}\{\beta h_{\mathrm{SD}} h_{\mathrm{RD}}^*\} > 0. \tag{6.73}$$

If we consider only noncoherent transmission, the resulting ε-outage CSB value computation (6.48) becomes a bisection search over ρ, until the corresponding probability (6.49) meets the requirement $1-\varepsilon$. In this simplified scenario, we can write (6.49) in closed-form as

$$p_{\mathrm{CSB}}(\rho,0) = \mathrm{e}^{-\frac{2^\rho-1}{P_{\mathrm{SD}}}} + \frac{\frac{1}{P_{\mathrm{SD}}}}{\frac{1}{P_{\mathrm{SR}}}+\frac{1}{P_{\mathrm{RD}}}-\frac{1}{P_{\mathrm{SD}}}}\left(\mathrm{e}^{-\frac{2^\rho-1}{P_{\mathrm{SD}}}} - \mathrm{e}^{-(2^\rho-1)\left(\frac{1}{P_{\mathrm{SR}}}+\frac{1}{P_{\mathrm{RD}}}\right)}\right). \tag{6.74}$$

6.6 Numerical Results

The simulation setup consists of the three-node line network in Fig. 6.2. The relay's location is between the source and the destination. The source destination distance is normalized to one and the source relay distance is d. The channels are independent zero-mean proper complex Gaussian [see (6.3)], with variances chosen according to a simplified path-loss model, i.e., $\sigma_{\mathrm{SD}}^2 = 1$, $\sigma_{\mathrm{SR}}^2 = d^{-\alpha}$, and $\sigma_{\mathrm{RD}}^2 = (1-d)^{-\alpha}$, with the path-loss coefficient $\alpha = 3$.

In Fig. 6.3, we depict the discussed outage constrained capacity bounds for $\varepsilon = 0.25$ for the line network model. Four figures are shown for equal source and relay transmit power $P_{\mathrm{S}} = P_{\mathrm{R}}$ that we varied from 0 to 30 dB. The ε-outage upper and lower capacity bounds are normalized with respect to the direct transmission ε-outage capacity $R_{\mathrm{SD}}^{(\mathrm{out})}$, whose y-axis value is henceforth 1. From the first view, we see that the proportional gains over the direct source-to-destination transmission decrease with the available transmit power.

The genie-aided CSB $C_{\mathrm{GA}}^{(\mathrm{out})}$ and the loosened CSB $C_{\mathrm{LOS}}^{(\mathrm{out})}$ give a tight ε-outage CSB approximation if d is between 0.75 and 1. Both curves become equal to the noncoherent CSB curve $C_{\mathrm{CSB}}^{(\mathrm{out})}$ for these relay positions, which merely provides a lower bound on the actual ε-outage CSB because we sub-optimally imposed $\beta = 0$ for its computation (see Sect. 6.5). This is due to the fact that the first CSB term in (6.9) limits the data transmission for most of the channel realizations in this case. This term remains the same for the genie-aided bound and is only slightly increased for the loosened CSB.

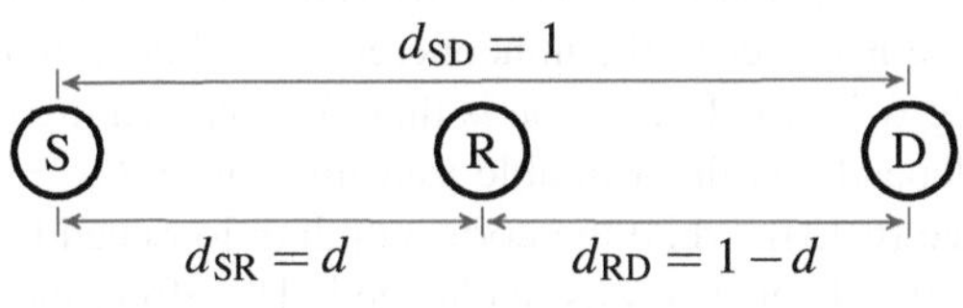

Fig. 6.2 Line network model for a Gaussian relay channel with Rayleigh fading channel coefficients h_{SR}, h_{RD}, and h_{SD} and variances $\sigma_{\mathrm{SR}}^2 = d_{\mathrm{SR}}^{-\alpha}$, $\sigma_{\mathrm{RD}}^2 = d_{\mathrm{RD}}^{-\alpha}$, and $\sigma_{\mathrm{SD}}^2 = d_{\mathrm{SD}}^{-\alpha}$, respectively

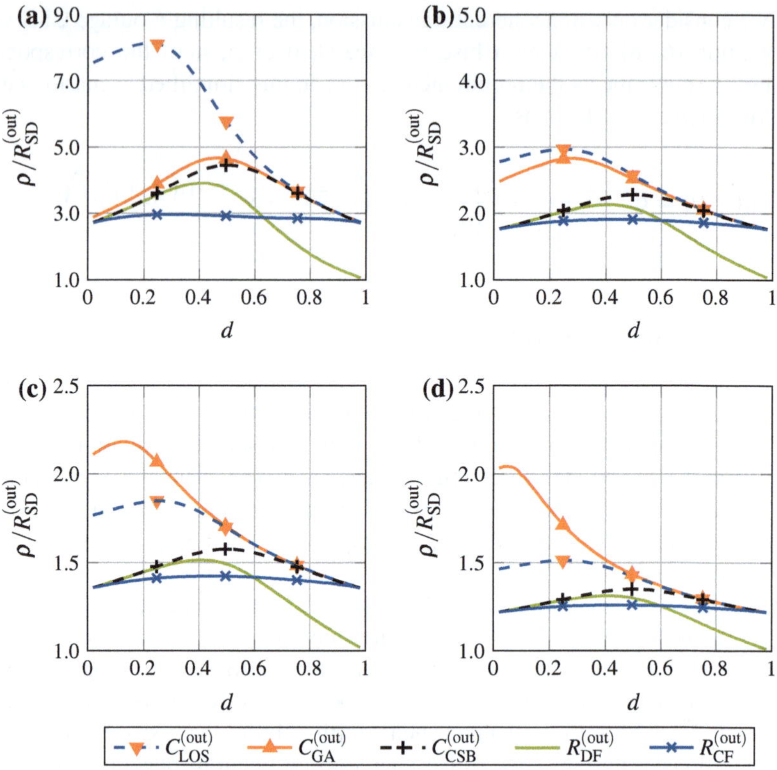

Fig. 6.3 Outage constrained capacity bounds for a line network relay model with Rayleigh fading channels and variances $\sigma_{SD}^2 = 1$, $\sigma_{SR}^2 = d^{-\alpha}$, and $\sigma_{RD}^2 = (1-d)^{-\alpha}$, where $\alpha = 3$. **a** $P_S = P_R = 0\,\text{dB}$. **b** $P_S = P_R = 10\,\text{dB}$. **c** $P_S = P_R = 20\,\text{dB}$. **d** $P_S = P_R = 30\,\text{dB}$

In contrast, for a d between 0 and 0.5, the second CSB term in (6.11) limits the achievable CSB for an overwhelming number of channel realizations. Since we employed upper bound approximations for this expression in the genie-aided and the loosened CSB, the corresponding ε-outage CSB curves are generally loose when the relay is closer to the source than to the destination. Here, the genie-aided approximation gives a tighter upper bound than the loosened CSB for small transmit powers [see Fig. 6.3a, b]. However, if the transmit power is 20 dB or larger, the loosened CSB curve is closer to the noncoherent CSB bound than the genie-aided CSB curve [see Fig. 6.3c, d]. The reason is that the obtained increase in the second CSB expression depends on the available transmit power $P_S = P_R$ for the genie-aided approach [see (6.13)], while the corresponding loosened CSB term in (6.56) resulted in a constant offset if μ was unchanged. Therefore, the genie-aided CSB approximation is close to the noncoherent CSB approximation in the low power regime, e.g., for 0 dB, but leads to a loose ε-outage bound for 30 dB, i.e., where the almost fixed gap to the loosened CSB is small compared to the achieved rate value.

For the imposed noncoherent transmission with $\beta = 0$, the ε-outage CSB curve gives a very tight upper bound to the maximized DF achievable ε-outage rate when the relay is close to the source. We remark that $\beta = 0$ was optimal in the DF ε-outage probability optimization for the given scenario. This is a consequence of the sufficiently small $\varepsilon = 0.25$ as we observed in [7].

The gap between the DF curve and the noncoherent CSB curve increases with d. We see that the performance of the DF strategy falls back to that of direct transmission for the probabilistic optimization if the relay is close to the destination. Since $d \approx 1$ in this case, the source-to-destination channel has the same statistical properties as the source-to-relay channel, which limits the DF transmission for an overwhelming number of channel realizations. Hence, the DF strategy gives a loose lower capacity bound for large d also for the statistically constrained optimization.

For a d that is approximately larger than 0.6, the DF scheme is outperformed by the computed CF ε-outage curves. The CF performance curves promise nearly constant gain in comparison to the direct transmission scheme over $d \in [0, 1]$, even though this gain is proportionally decreasing from about 3.0 to 1.25 times that of the direct transmission, respectively, for increasing transmit power from 0 to 30 dB. The CF curves even meet the noncoherent CSB curves for $d \approx 1$. Hence, our employed CF formulation achieves the ε-outage capacity when we place the relay directly at the destination.

6.7 Conclusion

We studied lower bounds and upper bounds on the probabilistically constrained capacity of a Gaussian relay channel with Rayleigh fading. For the chance constrained DF lower bound, we were able to derive a direct expression for the ε-outage constrained rate. This formulation only requires a line search over the coherent transmission parameter, which we may set to zero without a significant performance loss. We also used an ε-outage CF based rate formulation that allows for a closed-form outage probability calculation. The results indicate that the DF scheme outperforms the CF scheme when the relay is close to the source, but the CF scheme may achieve the ε-outage capacity when the relay is at the destination.

Upper bounds on the ε-outage capacity that led us to this observation are based on the CSB. We used a genie-aided and a loosened approximation for the second CSB mutual information term to obtain a (closed-form) probability computation without the difficulties that come along with the coherent transmission parameter. Both of them give a tight CSB approximation when the relay is closer to the destination than to the source, but they are generally loose if the relay is close to the source. The genie-aided bound is tighter than the loosened bound for low transmit power but looser for high transmit power.

Alternatively, we approximated the ε-outage CSB by imposing noncoherent transmission, which also allows for a closed-form outage probability calculation. The corresponding numerically computed CSB curves meet the DF achievable rates from above when the relay is close to the source. This facilitates also the search for other noncoherent transmission schemes if only statistical channel information is available at the transmitting nodes and the allowed outage probability is small.

Acknowledgments The work of Lennart Gerdes and Andreas Gründinger was supported by the German Research Foundation (DFG) under Grant Jo 724/2-1.

References

1. Avestimehr AS, Tse DNC (2007) Outage capacity of the fading relay channel in the low-SNR regime. IEEE Trans Inf Theory 53(4):1401–1415
2. Beaulieu NC, Hu J (2006) A closed-form expression for the outage probability of decode-and-forward relaying in dissimilar Rayleigh fading channels. IEEE Commun Lett 10(12):813–815
3. Cover TM, El Gamal A (1979) Capacity theorems for the relay channel. IEEE Trans Inf Theory 25(5):572–584
4. Cover TM, Thomas JA (2006) Elements of information theory, 2nd edn. Wiley, Hoboken
5. El Gamal A, Kim YH (2012) Network information theory. Cambridge University Press, New York, NY, USA
6. El Gamal A, Mohseni M, Zahedi S (2006) Bounds on capacity and minimum energy-per-bit for AWGN relay channels. IEEE Trans Inf Theory 52(4):1545–1561
7. Gründinger A, Gerdes L, Joham M, Utschick W (2015) Bounds on the outage constrained capacity of the single-antenna Gaussian relay channel. In: Proceedings of the annual conference on information sciences and systems
8. Gómez-Vilardebó J, Perez-Neira AI (2006) Upper bound on outage capacity of orthogonal relay networks. In: Proceedings of the IEEE international workshop on signal processing advances in wireless communications
9. Høst-Madsen A, Zhang J,(2005) Capacity Bounds and power allocation for wireless relay channels. IEEE Trans Inf Theory 51(6):2020–2040
10. Kramer G, Gastpar M, Gupta P (2005) Cooperative strategies and capacity theorems for relay networks. IEEE Trans Inf Theory 51(9):3037–3063
11. Laneman JN, Wornell GW (2003) Distributed space-time-coded protocols for exploiting cooperative diversity in wireless networks. IEEE Trans Inf Theory 49(10):2415–2425
12. van der Meulen EC (1971) Three-terminal communication channels. Adv Appl Probab 3(1):120–154
13. Nabar RU, Bölcskei H, Kneubühler FW (2004) Fading relay channels: performance limits and space-time signal design. IEEE J Sel Areas Commun 22(6):1099–1109
14. Neeser FD, Massey JL (1993) Proper complex random processes with applications to information theory. IEEE Trans Inf Theory 39(4):1293–1302
15. Telatar E (1999) Capacity of multi-antenna Gaussian channels. Eur Trans Telecomm 10(6):585–595
16. Tse D, Viswanath P (2008) Fundamentals of wireless communications, 4th edn. Camebridge University Press, New York, NY, USA
17. Yüksel M, Erkip E (2007) Multiple-antenna cooperative wireless systems: a diversity-multiplexing tradeoff perspective. IEEE Trans Inf Theory 53(10):3371–3393
18. Zhao Y, Adve R, Lim TJ (2005) Outage probability at arbitrary SNR With cooperative diversity. IEEE Commun Lett 9(8):700–702

Chapter 7
Balancing for Interference-Limited Multi-User Satellite Communications

Andreas Gründinger, Michael Joham and Wolfgang Utschick

Abstract The challenges in satellite communication for serving many (mobile) terminals simultaneously in higher carrier frequencies with reuse one are due to the interference and the various fading sources. Thus, the imposed downlink beamforming for the interference management has to account also for the link reliability. This chapter details the forward-link model of a modern multi-spotbeam satellite with generalized transmit power limitations, that also includes the basic rain-fading and scattering that is visible for the transmission to mobile terminals. In particular, the channel model consists of an additive and a multiplicative random channel error. Two kinds of rate balancing beamformer design strategies are studied when the satellite has statistical information about the fading channels, namely, the epsilon-outage rate balancing optimization and the average rate balancing optimization. Each of these robust approaches is detailed for fading that leads to a rank-one channel covariance or (at worst) a full-rank channel covariance due to scattering. While the epsilon-outage rate for the rank-one channel covariance can straightforwardly be rewritten into a standard rate balancing problem, but with increased noise, this is impossible for full-rank channel covariances. Then, the idea is to split the outage probability and the optimization into two parts for the multiplicative and the additive channel errors. An inner problem takes conservatively the additive channel errors into account, assuming prior information of the multiplicative errors, and the outer optimization adjusts this kind of information for the inner problem. For the average rate balancing optimization, already rank-one channel fading leads to unattractive rate expressions for a direct optimization. However, rate bounds with a standard logarithmic structure and

A. Gründinger (✉) · M. Joham · W. Utschick
Fachgebiet Methoden der Signalverarbeitung, Technische Universität München, 80290 München, Germany
e-mail: gruendinger@tum.de

M. Joham
e-mail: joham@tum.de

W. Utschick
e-mail: utschick@tum.de

W. Utschick (ed.), *Communications in Interference Limited Networks*,
Signals and Communication Technology, DOI 10.1007/978-3-319-22440-4_7

an adapted effective noise power (and small offsets) are almost tight. This changes for a (worst-case) scattering that leads to full-rank channel covariance matrices. Then, a mean square error lower bound for the rate is sequentially optimized instead.

7.1 Aims and Challenges in Satellite Communication

Terabit on-demand data transmission is the aim in in multi-spotbeam *satellite communication* (SatCom) [26, 61, 63]. To pursue this goal, researchers have strengthened investigations to increase the number of spotbeams, apply highly adaptive physical layer strategies, and use higher frequency bands. Only these changes allow the reuse of the same resources, e.g., bandwidth segment and time slots for MF-TDMA, to serve a large number of terminals in adjacent regions of the coverage area without harmful losses due to interference.

With the technologies for higher frequency bands, smaller transmit and receive aperture at the satellite and the (mobile) terminals, the S-band (2–4 GHz) and Ka-band (20–30 GHz) gained importance for mobile and fixed terminal applications, respectively, compared to the L-band (1–2 GHz) and the Ku-band (12–14 GHz) [48]. The frequency bands above 10 GHz have also been investigated for mobile SatCom services with vehicles [5], e.g., trains, trucks, or boats. Thus, mobile and fixed terminals may be served simultaneously in the same bands and design methods need to be developed that can cope with the different demands for these users.

The fading of the channels changes its characteristics with the frequency and depends on the employed antennas for fixed or mobile terminals. For example, the atmospheric fading effects may become dominant over the scattering at obstacles in the environment surrounding one terminal when the frequency lies above 10 GHz and directive reflector antennas are employed (see Sect. 7.2.2). Therefore, physical layer design techniques must be investigated that are adaptive to the different channel fading properties. This work focuses on statistically robust beamformer designs that are able to cope with the inherent fading characteristics.

The large number of spotbeams and the increased frequency reuse to achieve the terabit target are additionally challenging. Interference from neighboring spotbeams becomes a major issue.[1] To keep the system tractable and manage the resources and interference, the physical layer design needs to be adaptive and scalable to some extent. For example, a first attempt is the use of adaptive phased antenna arrays that are able to vary the focus within the serving spotbeam areas. By jointly coordinating multiple spotbeams and switching the settings between different illumination patterns, the interference can be limited [58]. This is a first step to exploit fully adaptive antenna arrays for the illumination in each spotbeam and to apply a joint design for

[1] Interference becomes especially severe if the frequency band reuse factor is one, i.e., the same bandwidth is used in all cells, or the increased number of spotbeams is obtained by decreasing the beamwidth below the usually assigned 3 dB area.

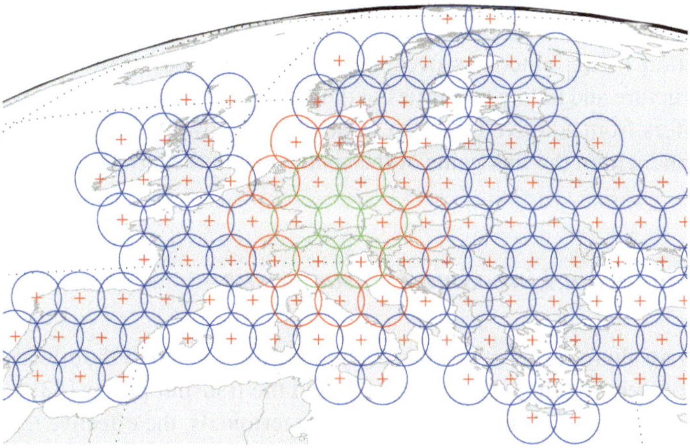

Fig. 7.1 Exemplary beam structure with *center* positions to cover Europe with 121 spotbeams. Clusters of 7 and 19 beams for the simulations, respectively, are highlighted in *green* and *red*

multiple spotbeam areas. Note that the more spotbeams are taken into account for the design and the more flexible the physical layer design should be, the more the computational complexity becomes important. Therefore, mainly the area of seven neighboring spotbeams is addressed for the joint physical layer design.

7.2 Multi-Spotbeam Satellite Communication Channel

In multi-spotbeam systems, a large number of beams with narrow beamwidth is employed to cover Europe [21] (see Fig. 7.1) or the USA [49], for example. Each spotbeam covers the area that is bounded by 3 dB loss of the radiation gain from the used directed antennas [43], e.g., a horn antenna with circular polarization that is focused by a circular reflector [14].[2]

The satellite forward link—transmission from the satellite to the (fixed or mobile) terminals on the earth's surface—has the properties of a vector *broadcast channel* (BC) if K terminals are served simultaneously in the same frequency band (e.g., [18, 21, 24, 70]). The received signal at the k-th terminal (fixed or mobile) reads as

$$y_k = \boldsymbol{h}_k^{\mathrm{H}} \boldsymbol{t}_k s_k + \boldsymbol{h}_k^{\mathrm{H}} \sum_{i \neq k} \boldsymbol{t}_i s_i + n_k$$

[2]Instead of a single antenna, a spotbeam may also be created by an active or semi-active phased-array architecture, so that the spotbeams become (partly) adaptive.

when the satellite linearly precodes the independent zero-mean unit-variance data signals s_k with the beamforming vectors $\boldsymbol{t}_k \in \mathbb{C}^N$, $k = 1, \ldots, k$ and transmits the superposition of the outcomes over the channels $\boldsymbol{h}_k \in \mathbb{C}^N$ to the terminals. The special structure and fading properties of the channels are detailed next. The received signal suffers from additive Gaussian noise $n_k \sim \mathcal{N}_{\mathbb{C}}(0, 1)$, which is normalized to variance one within the link gains of the spotbeams to the terminals.

7.2.1 *Spotbeam Link-Gain Characteristic*

The link-gain along the central path of each spotbeam is given by the effective radiated isotropic power, i.e., the antenna gain G_{tx} and the transmit power [43], the various losses between the satellite's antennas and the terminals, the effective terminal gain G_{rx}, and the thermal noise level at the terminals. Exemplary values for a mobile terminal system that operates in the S-band are given in Table 7.2 (cf. [18]). The large satellite antenna gain therein is required to overcome the *free space loss* (FSL) $g_{\text{FS},k}^{-1}$ between the satellite and terminal k on the earth's surface. It is proportional to the squared of the carrier frequency and the distance, e.g., about 36,000 km for the geostationary satellite orbit.

The gain characteristics offside the central path of a spotbeam depends on the exact antenna and reflector aperture [14]. We use the model of an undeformed (symmetric) tapered-aperture beamstructure (e.g., see [43]) for each spotbeam. Then, the gain off the central path is a function of the angular distance to the beamcenter axis and the one-sided half-power beamwidth $\theta_{3\,\text{dB}}$. For multiple spotbeams, the gain parameter $g_{i,k} = g(\theta_{i,k})$ is a function of the angle $\theta_{i,k}$ between the i-th beamcenter axis and the location of the k-th user (see Fig. 7.2) and is calculated as in [24, 69].

The relative angular position of the users and the geometric structure of the antenna and reflector aperture also specify the phase shifts at the satellite for a specific angle of departure. Let $\phi_{i,k}$ be the phase shift of the i-th spotbeam to the k-th terminal in

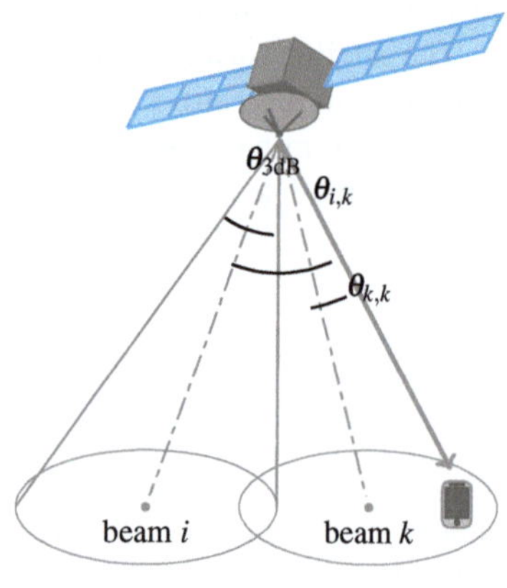

Fig. 7.2 Angular model for a multi-spotbeam satellite scenario with one user

the k-th spotbeam area. These phase shifts are assumed to be small due to the narrow angular distance of the spotbeams within the coverage area. We approximate them using the assumption that the antenna array is located in a plane orthogonal to the pointing direction of the satellite. Thus, we obtain the characteristic

$$b_{i,k} = \sqrt{G_{\text{rx}} G_{\text{tx}}} \sqrt{g_{i,k}(\theta_{i,k})}\, \mathrm{e}^{-\mathrm{j}\phi_{i,k}} \tag{7.1}$$

for the relative angular gain from the satellite's i-th antenna element to the k-th user.

7.2.2 Fading Models

The fading of the SatCom channels stems mainly from the atmospheric attenuation effects along the *line-of-sight* (LoS) path and the effects in the vicinity of the terminal on earth. The strength of these effects depend on the operating frequency and the employed antennas at the users (cf. [4]).

For example, above 10 GHz in the Ku- and Ka-bands and for appropriately placed *(fixed) terminals*[3] with reflector antennas, scattering can be neglected. However, the atmospheric influence and the effects at small water drops [48], also known as *rain-fading*, are increased in these bands. The detailed fading impairments can be simulated via the methods provided in the ITU-R Recommendations P.618 [51], P.840 [54], and P.676 [52], and were summarized by Arapoglou et al. in [3]. The dominant effects are the slow-fading rain attenuation $a_{\text{rain},k}$ and the fast-fading scintillation due to variations of the refractive index in the atmosphere. The attenuation's value in dB has a log-normal distribution [12], i.e., $\ln(a^{(\text{dB})}_{\text{rain},k}) \sim \mathcal{N}(m_{\text{rain},k}, \sigma^2_{\text{rain},k})$, and shall be independent for terminals in different spotbeams. Therewith, the channel to the k-th terminal reads as

$$\boldsymbol{h}_k = \sqrt{\frac{g_{\text{FS},k}}{a_{\text{rain},k}}}\, \mathrm{e}^{\mathrm{j}\psi_k}\, \boldsymbol{b}_k \tag{7.2}$$

where $\psi_k \in [0, 2\pi)$ is the uniformly-distributed random phase from above effects and $\boldsymbol{b}_k = [b_{1,k}, \ldots, b_{N,k}]^{\mathrm{T}}$ comprises the beam gains to terminal k.

For carrier frequencies below 10 GHz, *mobile stations* operate in the L-band and the S-band with only weakly directive small antenna apertures.[4] This results in a changed fading model with weakened rain-fading, but with other effects that are typically observed in terrestrial communications, where the link quality depends on the terminals' vicinity (cf. [4]).[5]

[3]Fixed terminal data services work with the DVB-S2 standard [45], for example.

[4]Mobile data services are according to the DVB-SH standard [40], for example.

[5]A probabilistic model for the environment and fading changes, e.g., via a two- or three-state Markov chain (cf. [42] and ITU-Recommendation R.681 [53]), are beyond the scope of this work.

- *Slow (shadow) fading* along the LoS path due to trees and houses near the mobile terminals, e.g., in a rural environment, and the movement.
- *Fast multi-path fading* for the non-LoS signal path components due to scattering and dispersion, e.g., in urban environments or if the elevation angle is low.

The S-band channel model from [18] (see also [69]) accounts for these fading properties with a (quasi) LoS characteristic based on Rician fading. Accordingly, the channel to the k-th (mobile) terminal can be written as

$$\boldsymbol{h}_k = \sqrt{\frac{g_{\mathrm{FS},k}}{a_{\mathrm{rain},k}}}\,\mathrm{e}^{\mathrm{j}\psi_k}\left(\sqrt{\frac{\kappa}{\kappa+1}}\boldsymbol{b}_k + \sqrt{\frac{1}{\kappa+1}}\boldsymbol{B}_k\boldsymbol{w}_k\right) \tag{7.3}$$

where $\boldsymbol{B}_k = \mathrm{diag}(b_{1,k}, \ldots, b_{N,k})$, $\boldsymbol{w}_k \in \mathbb{C}^N$ is zero-mean Gaussian distributed, and κ is the Rice factor, whose value lies between 5 and 15 dB [43].

The effective beamgain characteristic fluctuates with $\boldsymbol{w}_k \sim \mathcal{N}_{\mathbb{C}}(\boldsymbol{0}, \boldsymbol{C}_{\boldsymbol{w}_k})$ in the neighborhood of $\boldsymbol{b}_k$. If the scattered rays in the vicinity of the mobile terminal vary the signal intensity at the user, but originate in the same beamgain characteristic, the correlation matrix has rank one, e.g., $\boldsymbol{C}_{\boldsymbol{w}_k} = \frac{1}{N}\boldsymbol{1}_{N\times N}$ where $\boldsymbol{1}_{N\times N}$ denotes an $N \times N$ dimensional all-ones matrix. We exploited this assumption in [32, 34] to reduce the computational complexity of the robust physical layer designs. The other extreme with no correlations for the entries in $\boldsymbol{w}_k$, e.g., $\boldsymbol{C}_{\boldsymbol{w}_k} = \mathbf{I}_N$, might result from a rich scattering environment with also far distant scattering obstacles. We considered this worst-cast model in [28, 35]. A compromise between between these extremes is obtained via an exponential model, where the correlation decays with the cells' distance, i.e., the strongest correlation for each beam is to the 6 surrounding spotbeam elements [41].

7.2.3 Generalized Error Model

In the forward-link of SatCom—from the satellite to the (mobile) terminals—channel estimation at the terminals is possible, e.g., via a pilot based training. However, the long delay for the link quality information feedback [6] results in only imperfect CSI at the satellite, as the channels in (7.2) and (7.3) are subject to fading. Rainfall may lead to imperfect CSI also for fixed terminal systems. If the satellite knows at least the statistics of the fading channel, e.g., it has access to an estimate $\bar{\boldsymbol{h}}_k$ via feedback and knows that this estimate may contain an additive error $\boldsymbol{e}_k \sim \mathcal{N}_{\mathbb{C}}(\boldsymbol{0}, \boldsymbol{C}_k)$ and a multiplicative uncertainty ξ_k, the model reads as

$$\boldsymbol{h}_k = (1 + \xi_k)(\bar{\boldsymbol{h}}_k + \boldsymbol{e}_k), \quad k = 1, \ldots, k. \tag{7.4}$$

The inverse multiplicative factor $(1+\xi_k)^{-1}$ models the attenuation and $\boldsymbol{e}_k$ the Gaussian part for mobile terminals [cf. (7.3)], while $\boldsymbol{e}_k = \boldsymbol{0}$ for fixed terminals [cf. (7.2)].

7.3 Balancing for Satellite Communications

Since television broadcast and on-demand (data and video) services are the future of SatCom [25], the performance metric of interest is the spectral efficiency. The spectral efficiency for the k-th terminal is

$$r_k = \log_2(1 + \mathrm{SINR}_k), \tag{7.5}$$

which inherently corresponds to Gaussian signals, i.e., $s_k \sim \mathcal{N}_{\mathbb{C}}(0, 1)$, and the *signal-to-interference-and-noise-ratio* (SINR) is given by

$$\mathrm{SINR}_k = \frac{|\boldsymbol{h}_k^{\mathrm{H}} \boldsymbol{t}_k|^2}{1 + \sum_{i \neq k} |\boldsymbol{h}_k^{\mathrm{H}} \boldsymbol{t}_i|^2}. \tag{7.6}$$

The achievable data rate in (7.5) is also closely connected to other metrics, e.g., the *minimum mean square error* (MMSE) was exploited for SatCom in [19, 23, 28].

Besides the throughput maximization and rate matching optimization to evaluate how close the system capacity and certain demands can be met [69], respectively, rate balancing is a very important physical layer design goal in SatCom. The aim is to maximize a ratio of the rates to their *quality of service* (QoS) target requirements $\rho_k, k = 1, \ldots, K$. A mathematical max-min formulation of balancing reads as

$$\max \; \min_k \frac{r_k(\boldsymbol{t})}{\rho_k} \quad \text{s.t.} \quad \boldsymbol{t} \in \mathcal{P}.$$

The transmit powers at the satellite are bounded from above by the constraints on $\boldsymbol{t} = \mathrm{vec}([\boldsymbol{t}_1, \ldots, \boldsymbol{t}_K])$ in $\mathcal{P}$, which limits the achievable data rates to the terminals. Also note that the balancing problem formulation is closely connected to QoS power minimization (e.g., [32, 55, 66]). The solution of the balancing optimization problem can be obtained via series of power minimizations.

7.3.1 *Power Limitations*

For a complex transmission system with high power *radio frequency* (RF) amplifiers as it can be found at the satellite, various power constraints need to be taken into account. A typical categorization of these constraints is into the following groups:

- per spotbeam or per antenna powers;
- per data stream (served terminal) power;
- per antenna array power;
- a total (sum) transmit power.

A general conic formulation for the power constraint set $\mathcal{P}$, that includes all above types as special cases, reads as

$$\mathcal{P} = \left\{ \boldsymbol{t} = \text{vec}([\boldsymbol{t}_1, \ldots, \boldsymbol{t}_K]) \in \mathbb{C}^{KN} \,\middle|\, \sum_{i=1}^{K} \boldsymbol{t}_i^{\mathrm{H}} \boldsymbol{A}_{i,\ell} \boldsymbol{t}_i \leq P_\ell,\ \ell = 1, \ldots, L \right\} \quad (7.7)$$

where $\boldsymbol{A}_{i,\ell} \in \mathbb{C}^{N \times N}$, $\boldsymbol{A}_{i,\ell} \succeq \mathbf{0}$ has the square root representation $\boldsymbol{A}_{i,\ell} = \boldsymbol{A}_{i,\ell}^{\mathrm{H}/2} \boldsymbol{A}_{i,\ell}^{1/2}$ and rank $\left\{ \sum_{\ell=1}^{L} \boldsymbol{A}_{i,\ell} \right\} = N$. For either of the above power constraint cases, the matrices $\boldsymbol{A}_{i,\ell}$ are diagonal with either a one or a zero as diagonal elements [28].

Even though above special cases are the most important ones for the physical layer design, the generalized form in (7.7) also allows to model correlations for the transmit powers in neighboring spotbeams and amplifier supply chains. Then, the matrices $\boldsymbol{A}_{i,\ell}$ have also non-zero entries on their off-diagonals. For example, the per antenna array power constraint matrices may have the form of a block-diagonal matrix $\boldsymbol{A}_{i,\ell} = \text{bdiag}\{\mathbf{0}, \boldsymbol{A}'_{i,\ell}, \mathbf{0}\}$ with a general positive definite block $\boldsymbol{A}'_{i,\ell} \succ \mathbf{0}$ if there are correlations in the power supply for the transmit array.

Furthermore, note that non-linearities in the RF amplifiers may result in a non-linear function for the input to output powers [69]. An example are concave functions for the output power depending on the input power for modeling saturation effects [43]. Linear approximations may be applied for these power constraints.

7.3.2 *Balancing with Statistical Channel Knowledge*

The balancing problem in (7.3) cannot directly be used for the physical layer design. For the satellite, the channel states $\boldsymbol{h}_k$ are only available for clear sky conditions and fixed terminals. Otherwise, model (7.4) holds, i.e., the channels are (7.2) and (7.3) for the fixed and mobile terminals, respectively, which are subject to fading. Therefore, the achievable rate in (7.5) is a random quantity for the satellite.

If the channels change sufficiently slow and are constant during a transmission block, the outage probability $\Pr(r_k/\rho_k \leq \tau)$—the probability that the channel is too poor for an error-free transmission at the target rate $\tau\rho_k$—is the key measure of interest. This probability has to be limited from above by ε_k to avoid plenty of outages when maximizing τ and ensure reliable transmission. The resulting problem formulation is an outage (chance) constrained balancing optimization and requires a careful treatment of the probabilistic constraints.

For very fast fading and long codes—each transmission experiences multiple channel realizations—the ergodic mutual information $R_k = \mathrm{E}[r_k]$ replaces the actual rate (7.5) for the beamformer design. The expectation is with respect to the random channels. The ergodic balancing optimization can also be seen as a relaxed form of the conservative optimization with outage constraints.

7.4 Chance-Constrained Rate Balancing

The probabilistically constrained rate balancing problem reads as (cf. [29, 30, 35])

$$\max\ \tau \quad \text{s.t.} \quad \boldsymbol{t} \in \mathcal{P}, \quad \Pr(r_k(\boldsymbol{t}) \geq \tau\rho_k) \geq 1 - \varepsilon_k, \quad k = 1, \ldots, K.$$

The difficulty for solving (7.4) depends on the form of the channel model. If the additive error in (7.7) is either zero or has a rank-one covariance matrix that is a scalar multiple of $\bar{\boldsymbol{h}}_k\bar{\boldsymbol{h}}_k^{\mathrm{H}}$, then the probabilistic constraint can be reformulated as a standard rate constraint, but with increased noise power (e.g., [30]).

In contrast, if the additive errors feature a covariance matrix with a rank larger than one, the optimal solution is unknown. The known approaches to the problem aim at a conservative (convex) reformulation of the chance constraints [35]. Alternatively, the problem may be divided into a two-step approach, where the beamformers are found via an approximation in the first step and a power allocation in the second step ensures that the probabilistic rate constraints are satisfied [29].

7.4.1 Channels with Rank-One Covariance Matrix

For fixed terminals and mobile terminals with a rank-one channel covariance matrix, the generalized channel model in (7.4) reduces to

$$\boldsymbol{h}_k = \zeta_k^{-1/2}\, \mathrm{e}^{\mathrm{j}\phi_k}\, \bar{\boldsymbol{h}}_k. \tag{7.8}$$

The additive error in (7.4) is zero for fixed terminals [cf. (7.2)], $\zeta_k = |1-\xi_k|^{-2}$ models the attenuation, and ϕ_k the phase. For mobile terminals with a covariance matrix $\boldsymbol{C}_k$ that is a scaled version of $\bar{\boldsymbol{h}}_k\bar{\boldsymbol{h}}_k^{\mathrm{H}}$, e.g., if $\boldsymbol{C}_{\boldsymbol{w}_k} = \frac{1}{N}\mathbf{1}_{N\times N}$ for the S-band channel model in (7.3), the additive error in (7.4) is $\boldsymbol{e}_k = e_k\bar{\boldsymbol{h}}_k$. Then, $\zeta_k^{-1} = |1 + e_k|^2|1 + \xi_k|^2$ is a χ^2-distributed random variable.

Substituting (7.8) into the probabilistic requirements of (7.4), they read as

$$\Pr\left(\bar{\boldsymbol{h}}_k^{\mathrm{H}}\boldsymbol{B}_k\bar{\boldsymbol{h}}_k \geq \zeta_k\right) \geq 1 - \varepsilon_k \tag{7.9}$$

where $\boldsymbol{B}_k = \frac{1}{2^{\tau\rho_k}-1}\boldsymbol{t}_k\boldsymbol{t}_k^{\mathrm{H}} - \sum_{i\neq k}\boldsymbol{t}_i\boldsymbol{t}_i^{\mathrm{H}}$ is indefinite with at most one positive eigenvalue. In (7.9), the inequality sign within the probability operator separates the optimization variables $\boldsymbol{t}$ from the scalar random variable ζ_k, which forms the effective random noise power. Therefore, the left-hand-side probability in (7.9) can be calculated via the *cumulative distribution function* (CDF) of the effective noise. In other words, the quadratic form $\bar{\boldsymbol{h}}_k^{\mathrm{H}}\boldsymbol{B}_k\bar{\boldsymbol{h}}_k$ has to lie above the $(1 - \varepsilon_k)$-quantile of the CDF for the effective noise variance. If the inverse CDF for ζ_k can be computed, i.e.,

$q_k = \mathrm{F}_{\xi_k}^{-1}(1-\varepsilon_k)$, the probabilistic outage constraint transforms to the deterministic requirement $\bar{\boldsymbol{h}}_k^{\mathrm{H}} \boldsymbol{B}_k \bar{\boldsymbol{h}}_k \geq q_k$ (cf. [30]), which can be rewritten as the rate constraint

$$\tilde{r}_k(\boldsymbol{t}) = \log_2\left(1 + \frac{|\bar{\boldsymbol{h}}_k \boldsymbol{t}_k|^2}{q_k + \sum_{i \neq k} |\bar{\boldsymbol{h}}_k \boldsymbol{t}_i|^2}\right) \geq \tau \rho_k. \tag{7.10}$$

Despite the increased noise variance q_k, the rate $\tilde{r}_k$ has obviously a similar representation as r_k in (7.5). Hence, the same balancing optimization methods as for perfect CSI can be applied for the deterministic counterpart

$$\max\ \tau \quad \text{s.t.} \quad \boldsymbol{t} \in \mathcal{P}, \quad \tilde{r}_k(\boldsymbol{t}) \geq \tau \rho_k, \quad k = 1, \ldots, K \tag{7.11}$$

to the chance constrained formulation in (7.4). For example, one might employ uplink-downlink duality [56] and Yates' fixed point framework for standard interference functions [67] to obtain the optimal balanced rates if only a sum-power constraint limits $\boldsymbol{t} \in \mathcal{P}$ [36]. Using Lagrangian uplink-downlink duality [68], the solution strategy can also be extended to the generalized power constraints (e.g., see [22, 28]). Alternatively, one may employ a bisection over τ, with a search for a feasibile $\boldsymbol{t} \in \mathcal{P}$ in each bisection step [62].

7.4.2 Channels with Full-Rank Covariance Matrix

If the mobiles' channel error covariance matrices $\boldsymbol{C}_k$ in (7.4) have a rank larger than one, e.g., when the covariance matrix of $\boldsymbol{w}_k$ in the S-band channel model (7.3) is $\boldsymbol{C}_{\boldsymbol{w}_k} = \mathbf{I}_N$, there is no deterministic reformulation for the constraints in (7.4) as for rank-one covariance matrices. Only the multiplicative error can be separated from the optimization variables in the stochastic requirement of (7.4), that transforms to

$$\Pr\left(b_k(\boldsymbol{t}, \boldsymbol{e}_k) \geq \frac{1}{|1+\xi_k|^2}\right) \geq 1 - \varepsilon_k. \tag{7.12}$$

The left hand side of the stochastic (inner) inequality is an indefinite quadratic form in the Gaussian channel errors and reads as

$$b_k(\boldsymbol{t}, \boldsymbol{e}_k) = (\bar{\boldsymbol{h}}_k + \boldsymbol{e}_k)^{\mathrm{H}} \boldsymbol{B}_k (\bar{\boldsymbol{h}}_k + \boldsymbol{e}_k). \tag{7.13}$$

Therefore, the probability computation for the left-hand-side in (7.12) requires a nested numerical integration for the additive Gaussian errors [2, 38] and the effective random noise variance $\sigma_k^2 = 1/|1+\xi_k|^2$ (cf. [35]).

Note that the separation of the additive errors $\boldsymbol{e}_k$ from the multiplicative errors ξ_k inside the left-hand-side probability of (7.12) allows the following interpretations:

1. If we were aware of $\boldsymbol{e}_k$, the probability would be the CDF of σ_k^2 at $b_k(\boldsymbol{t}, \boldsymbol{e}_k)$.
2. Vice versa, if we were aware of σ_k^2, the randomness would only be due to $\boldsymbol{e}_k$.

The interpretations motivate a conservative approximation approach. With Interpretation 2, for example, a restriction of the noise power is introduced as

$$\sigma_k^2 \leq a_k, \qquad \Pr(\sigma_k^2 \leq a_k) = \mathrm{F}_{\sigma_k^2}(a_k) = \alpha_k. \tag{7.14}$$

With the inverse CDF $a_k(\alpha_k) = \mathrm{F}_{\sigma_k^2}^{-1}(\alpha_k), \alpha_k \in [1-\varepsilon_k, 1]$, that is analytically available for a log-normal attenuation, the left-hand-side of (7.12) is approximated by

$$\Pr\big(b_k(\boldsymbol{t}, \boldsymbol{e}_k) \geq \sigma_k^2\big) > \Pr\big(b_k(\boldsymbol{t}, \boldsymbol{e}_k) \geq a_k(\alpha_k) \big| a_k(\alpha_k) \geq \sigma_k^2\big)\alpha_k. \tag{7.15}$$

This approximation is conservative, but it provides the degree of freedom to choose $\alpha_k \in [1-\varepsilon_k, 1]$. Therefore, an optimization over $\boldsymbol{\alpha} = [\alpha_1, \ldots, \alpha_K]^{\mathrm{T}}$ is added when using (7.15) as constraints for the optimization problem (7.4) instead of (7.12), i.e.,

$$\begin{aligned} \max \tau \quad \text{s.t.} \quad & \Pr\big(b_k(\boldsymbol{t}, \boldsymbol{e}_k) \geq a_k \big| a_k \geq \sigma_k^2\big) \geq \frac{1-\varepsilon_k}{\alpha_k}, \quad k = 1, \ldots, K \\ & \mathbf{1} - \boldsymbol{\varepsilon} \leq \boldsymbol{\alpha} \leq \mathbf{1}, \quad \boldsymbol{t} \in \mathcal{P} \end{aligned} \tag{7.16}$$

where $\boldsymbol{\varepsilon} = [\varepsilon_1, \ldots, \varepsilon_K]^{\mathrm{T}}$, $\mathbf{1}$ is the all ones vector, and the vector inequalities for $\boldsymbol{\alpha}$ are elementwise. The joint optimization in (7.16) is divided into an inner optimization w.r.t. the beamformers in $\boldsymbol{t}$ for fixed probabilities in $\boldsymbol{\alpha}$ and an outer optimization w.r.t. the probability vector $\boldsymbol{\alpha}$. This represents the maximization of a lower bound on the outage constrained rates.

7.4.2.1 Inner Beamformer Design

A direct inner beamformer design with fixed $\boldsymbol{\alpha}$ is still non-convex in $\boldsymbol{t}$ due to the probabilistic constraints, whose computation involves a numerical integration [38]. However, the available approximations in [11, 47, 65] allow to replace the chance constraints by deterministic constraints and apply a *semidefinite relaxation* (SDR) approach. For example, $\boldsymbol{e}_k$ may be restricted to lie in a sphere with probability $(1-\varepsilon_k)/\alpha_k$. Then, the obtained quadratic uncertainty optimization [8] can be solved via a semidefinite relaxation formulation [35] (see also [65, Table 1]).

A tighter approximation of (7.16) than above error restiriction was found by Wang et al. [64] with the Bernstein's type inequality from [7]. The inequality bounds the probability that the quadratic form $Q = \boldsymbol{z}^{\mathrm{H}}\boldsymbol{A}\boldsymbol{z} + 2\,\mathrm{Re}(\boldsymbol{b}^{\mathrm{T}}\boldsymbol{z}) + c$, with $\boldsymbol{z} \sim \mathcal{N}(\mathbf{0}, \mathbf{I})$ and Hermitian $\boldsymbol{A}$, deviates from its mean $\bar{Q} = \mathrm{tr}(\boldsymbol{A}) + c$ from below [65, Lemma 1][6]:

$$\Pr\left(Q \geq \bar{Q} - \sqrt{2y}\big\|\big[\mathrm{vec}(\boldsymbol{A})^{\mathrm{T}}, \sqrt{2}\boldsymbol{b}^{\mathrm{T}}\big]\big\|_2 - y\lambda\right) \geq 1 - \mathrm{e}^{-y} \tag{7.17}$$

[6]Here, already the complementary event is defined.

where $\lambda = \min\{\lambda \in \mathbb{R}_+ : \lambda \mathbf{I} + \boldsymbol{A} \succeq \mathbf{0}\}$ is the maximum eigenvalue of $-\boldsymbol{A}$ if it is positive and zero otherwise. For the quadratic Gaussian form $b_k(\boldsymbol{t}, \boldsymbol{e}_k)$ in (7.13), the parameters are $\boldsymbol{A} = \boldsymbol{C}_k^{\mathrm{H}/2} \boldsymbol{B}_k \boldsymbol{C}_k^{1/2}$, $\boldsymbol{b} = \boldsymbol{C}_k^{\mathrm{H}/2} \boldsymbol{B}_k \bar{\boldsymbol{h}}_k$, and $c = \bar{\boldsymbol{h}}_k^{\mathrm{H}} \boldsymbol{B}_k \bar{\boldsymbol{h}}_k$.

A deteriministic approximiation of the probabilistic constraint in (7.16) is obtained with (7.17) by setting $y = -\ln(1-\beta_k)$ and $\beta_k = \frac{1-\varepsilon_k}{\alpha_k}$ and requiring the lower bound approximation of Q to lie above σ_k^2. The deterministic approximate of the original chance constraint thus reads as

$$\begin{aligned} \operatorname{tr}\left(\boldsymbol{C}_k^{\mathrm{H}/2} \boldsymbol{B}_k \boldsymbol{C}_k^{1/2}\right) + \bar{\boldsymbol{h}}_k^{\mathrm{H}} \boldsymbol{B}_k \bar{\boldsymbol{h}}_k - \sqrt{2 \ln\left(\tfrac{1}{1-\beta_k}\right)} x_k + \ln\left(1-\beta_k\right) \lambda_k &\geq \sigma_k^2, \\ \left\| \operatorname{vec}\left(\boldsymbol{C}_k^{\frac{\mathrm{H}}{2}} \boldsymbol{B}_k \left[\boldsymbol{C}_k^{\frac{1}{2}}, \sqrt{2} \bar{\boldsymbol{h}}_k\right]\right) \right\|_2 &\leq x_k, \\ \lambda_k \mathbf{I} + \boldsymbol{C}_k^{\frac{\mathrm{H}}{2}} \boldsymbol{B}_k \boldsymbol{C}_k^{\frac{1}{2}} &\succeq \mathbf{0}. \end{aligned} \tag{7.18}$$

Even though these constraints are deterministic and convex in $\boldsymbol{B}_k$, they are still nonconvex in $\boldsymbol{t}$ [see below (7.9)]. Wang et. al [65] use the semidefinite substitutes $\boldsymbol{W}_i = \boldsymbol{t}_i \boldsymbol{t}_i^{\mathrm{H}}$ for a relaxation. Therewith, $\boldsymbol{B}_k = \frac{1}{2^{\tau \rho_k}-1} \boldsymbol{W}_k - \sum_{i \neq k} \boldsymbol{W}_i$ and the generalized power constraint set $\mathcal{P}$ in (7.7) is rewritten in terms of covariance matrices as

$$\mathcal{W} = \left\{ \boldsymbol{W}_k \in \mathbb{C}^{N \times N},\ k = 1, \ldots, K \,\middle|\, \boldsymbol{W}_k \succeq \mathbf{0}_{N \times N},\ \sum_{k=1}^{K} \operatorname{tr}(\boldsymbol{W}_k \boldsymbol{A}_\ell) \leq P_\ell,\ \ell = 1, \ldots, L \right\}.$$

The resulting transmit covariance optimization problem reads as (cf. [65, Table 1])

$$\max \tau \quad \text{s.t.} \quad y_k \geq 0,\ x_k \geq 0,\ (8.18),\ k = 1, \ldots, K,\ (\boldsymbol{W}_1, \ldots, \boldsymbol{W}_K) \in \mathcal{W}. \tag{7.19}$$

Since the constraints are non-convex in τ, but convex in $\boldsymbol{W}_k$, λ_k, and x_k, $k = 1, \ldots, K$, a bisection over τ suffices to find a solution. A given τ' lies below the optimum if a feasible tuple of covariance matrices $\boldsymbol{W}_k$ is found, e.g., with the disciplined convex programming toolbox CVX [27]. Otherwise, τ' lies above the optimum of (7.19).

7.4.2.2 Outer Attenuation Restriction

The remaining outer optimization over $\boldsymbol{\alpha}$ reads as

$$\max\ \tau(\boldsymbol{\alpha}) \quad \text{s.t.} \quad \mathbf{1} - \boldsymbol{\varepsilon} \leq \boldsymbol{\alpha} \leq \mathbf{1} \tag{7.20}$$

where $\tau(\boldsymbol{\alpha})$ denotes the value of the inner design in (7.19). This optimization is non-convex. An equal choice of the probabilities, i.e., $\boldsymbol{\alpha} = \alpha_0 \mathbf{1}$ with optimized α_0, may provide a good lower bound for the optimum of (7.20) if the required outage probabilities, the rate targets, and the channels' fading parameters are almost the same for all users [35]. As an alternative, we suggest a sequential update of the α_k, $k = 1, \ldots, K$ to maximize $\tau(\boldsymbol{\alpha})$, assuming the other variables to be fixed. Remarkably, this iterative search only provides better rates if the statistics of the multiplicative errors lead to a high attenuation for some of the channels and a low one for the others (see [35]).

After convergence of the outer optimization, the beamformers need to be reconstructed from the covariance matrices $\boldsymbol{W}_k$. If these are rank-one, $\boldsymbol{W}_k = \boldsymbol{t}_k \boldsymbol{t}_k^{\mathrm{H}}$.[7] Otherwise, a Gaussian randomization shows a good match when non-rank-one transmit covariances were obtained (cf. [20]). The potential mismatch in the power constraints and the achievable outage constrained rate may be overcome by a subsequent power allocation [35]. If an accurate numerical calculation of the probabilities in (7.12) is possible and the beamformers are only limited by their sum power, a subsequent power allocation may even further increase the rates until all requirements are met exactly [29].

7.5 Average Rate Balancing

With average rate requirements, the balancing problem reads as (cf. [28, 32–34])

$$\max \ \min_k \frac{R_k(\boldsymbol{t})}{\rho_k} \quad \text{s.t.} \quad \boldsymbol{t} \in \mathcal{P}.$$

The ergodic rates are nonconvex in the beamformers and a direct convex reformulation is so far unknown. However, various tight lower and upper bounds on the ergodic rates can be used to obtain a conservative or an opportunistic beamformer design, respectively. While the proposed bounds have a similar structure as the actual rate expressions in (7.5) for fixed and mobile terminals with the rank-one channel model (cf. Sect. 7.4.1), an MMSE approximation of R_k is exploited if the channels follow the general model in (7.4), with no restrictions for the rank of the covariance matrix.

7.5.1 *Channels with Rank-One Covariance Matrix*

Inserting the rank-one channel covariance model (7.8), the average rate reads as

$$R_k(\boldsymbol{t}) = \mathrm{E}\left[\log_2\left(1 + \frac{|\bar{\boldsymbol{h}}_k^{\mathrm{H}} \boldsymbol{t}_k|^2}{\zeta_k + \sum_{i \neq k} |\bar{\boldsymbol{h}}_k^{\mathrm{H}} \boldsymbol{t}_i|^2}\right)\right] \tag{7.21}$$

---

[7]Necessary requirements for rank-one solutions consider a sum transmit power constraint (e.g., see [16, 60]). For other power constraints, rank-one solutions are likely if the ε_k are small (cf. [20]).

where only the effective noise variance ζ_k is random [cf. (7.8)]. For (7.21), tight upper and lower bounds are obtained via Jensen's inequality (e.g., [13, p. 77]). Their structure is similar to that of a standard rate expression, which allows us to solve the balancing problem (7.5) with the standard optimization frameworks of the literature. Besides these bounds, a sequential approximation approach is introduced that provides a locally optimal solution upon convergence, but increases the computational complexity for the optimization.

7.5.1.1 Upper and Lower Rate Bounds

The considered lower and upper bounds of R_k in (7.21) have the structure (cf. [32])

$$R_k^{(\mathrm{B})}(\boldsymbol{t}) = \log_2\left(1 + \frac{|\bar{\boldsymbol{h}}_k^{\mathrm{H}}\boldsymbol{t}_k|^2}{\nu_k^{(\mathrm{B})} + \sum_{i\neq k}|\bar{\boldsymbol{h}}_k^{\mathrm{H}}\boldsymbol{t}_i|^2}\right) + \mu_k^{(\mathrm{B})} \tag{7.22}$$

where the parameters $\nu_k^{(\mathrm{B})}$ and $\mu_k^{(\mathrm{B})}$, respectively, represent the effective noise variance and the offset. Examples for these parameters are shown in Table 7.1 for various upper and lower bounds. The first lower bound (LB1) and the first upper bound (UB1) are obtained with Jensen's inequality and the convex and concave functions $\ln(1+a/(b+z))$ and $\ln(1+a/(b+1/z))$, respectively, in z for $a, b, z \geq 0$.

Note that either of the expected values $\mathrm{E}[\zeta_k]$ or $\mathrm{E}[\zeta_k^{-1}]$ may not exist. For example, $\mathrm{E}[\zeta_k]$ is infinity if $z_k = \zeta_k^{-1}$ is standard exponentially distributed with the *probability density function* (PDF) e^{-z_k}. Similarly, $\mathrm{E}[\zeta_k^{-1}]$ would be infinity if the PDF of ζ_k was $\mathrm{e}^{-\zeta_k}$. Then, either the loosened lower bound LB2 or the loosened upper bound UB2, respectively, with the parameters of Table 7.1 can be used. To derive LB2, we exploited that $\ln(1 + a\,\mathrm{E}[z]) - \mathrm{E}[\ln(1 + az)]$ is monotonically increasing with a, 0 if $a = 0$, and approaches $\mathrm{E}[\ln(\mathrm{E}[z]/z)]$ for $a \to \infty$ if $z \geq 0$. Thus,

$$\ln(1+a\,\mathrm{E}[z]) - \mathrm{E}[\ln(1+az)] - \big(\ln(1+b\,\mathrm{E}[z]) - \mathrm{E}[\ln(1+bz)]\big) \leq \mathrm{E}[\ln(z/\mathrm{E}[z])]$$

Table 7.1 Effective noise variance $\nu_k^{(\mathrm{B})}$ and offset $\mu_k^{(\mathrm{B})}$ of the ergodic downlink beamforming rate with rank-one channel covariances for various upper and lower bounds

Bound (B)	Noise $\nu_k^{(\mathrm{B})}$	Offset $\mu_k^{(\mathrm{B})}$
LB1	$\mathrm{E}[\zeta_k]$	0
UB1	$1/\mathrm{E}[\zeta_k^{-1}]$	0
LB2	$1/\mathrm{E}[\zeta_k^{-1}]$	$\mathrm{E}\left[\log_2\left(\zeta_k^{-1}/\mathrm{E}[\zeta_k^{-1}]\right)\right]$
UB2	$\mathrm{E}[\zeta_k]$	$\mathrm{E}\left[\log_2\left(\mathrm{E}[\zeta_k]/\zeta_k\right)\right]$
ALB	$2^{-\mu_k^{(\mathrm{LB2})}}$	$-d_k/\ln(2) + \mu_k^{(\mathrm{LB2})}$
AUB	$2^{-\mu_k^{(\mathrm{LB2})}}$	$d_k/\ln(2) + \mu_k^{(\mathrm{LB2})}$

d_k denotes the maximum distance value $d_k = \max_{a\geq 0} \mathrm{E}[\log_2(1+a\zeta_k^{-1})] - \mathrm{E}[\log_2(1+2^{-\mu_k^{(\mathrm{LB2})}}a\zeta_k^{-1})]$

holds for all $a \geq b \geq 0$, which leads us to LB2 via some algebraic reformulation and replacing a and b by the signal plus interference power and the interference power, respectively. Similarly, we derived UB2 with the knowledge that $\mathrm{E}[\ln(1 + a/z)] - \ln(1 + a/\mathrm{E}[z])$ is monotonically increasing in a, 0 if $a = 0$, and approaches $\mathrm{E}[\ln(\mathrm{E}[z]/z)]$ for $a \to \infty$ if $z \geq 0$. For an exponentially distributed z with PDF e^{-z}, the offset is $\mathrm{E}[\ln(1/z)] \approx \gamma \approx 0.577216$, where γ denotes the Euler-Mascheroni constant (cf. [32]). If z is non-central χ^2-distributed instead, the expected value $\mathrm{E}[\ln(z)]$ can be calculated with [46] (see also [34]).

In [34], we derived a further pair of approximate lower and upper bounds (ALB and AUB in Table 7.1). These bounds are motivated by the approximation

$$\mathrm{E}[\ln(1 + az)] \approx \ln\left(1 + \exp\left(-\mathrm{E}\left[\ln\left(\tfrac{\mathrm{E}[z]}{z}\right)\right]\right) a\,\mathrm{E}[z]\right),$$

which holds with equality for either $a = 0$ or $a \to \infty$. In particular, the right hand side appears to be a lower bound since $-\mathrm{E}[ln(\mathrm{E}[z]/z)] \leq 0$ for $z \geq 0$. Thus, the functions slope in a is smaller for the left hand side than for the right hand side. Denoting the maximum distance of these functions as d, which we computed numerically for the simulations, the (approximate) bounds derivation is straightforward.

Replacing the ergodic rates in (7.5) by either of the bounds corresponding to (7.22), the resulting balancing problem can equivalently be written as

$$\max \quad \tau \quad \text{s.t.} \quad \boldsymbol{t} \in \mathcal{P}, \quad R_k^{(\mathrm{B})}(\boldsymbol{t}) \geq \tau \rho_k, \quad k = 1, \ldots, K \tag{7.23}$$

where the slack variable τ denotes the minimum common balancing factor. This problem can be solved in the same way as the perfect CSI balancing problem (7.3). For example with a bisection over τ, where in each bisection step it is checked whether a feasible beamforming set $\boldsymbol{t}$ is found, i.e., one that satisfies the power constraints and a convex reformulation of the rate constraints (e.g., similar to [62]).

Alternatively, an SINR uplink-downlink duality is used to transform the downlink problem into the dual uplink domain. Therewith, the precoder design is replaced by an uplink receive filter computation and a power allocation for the dual transmitters. If a single sum power constraint forms $\mathcal{P}$, this procedure is detailed in [32]. For the generalized power constraints in this work (see Sect. 7.3.1), the uplink-downlink duality of Yu and Lan [68] is appropriate, which results in an additional outer worst-case noise search for the dual rate balancing problem [22].

7.5.1.2 Sequential Approximation Strategy

Above mentioned balancing optimization techniques can be applied iteratively for a sequential local search. For example, a weighted MSE approximation strategy (e.g., see [17, 50]) can be used for a locally tight approximation of the average rate in (7.21). Alternatively, we proposed a sequential approximation strategy in [30]. To this end, we equivalently rewrite the average balancing problem in (7.5) as

$$\max\ \tau \quad \text{s.t.} \quad \boldsymbol{t} \in \mathcal{P}, \quad R_k(\boldsymbol{t}) \geq \tau\rho_k, \quad k = 1, \ldots, K \tag{7.24}$$

and remark that the ergodic rates are given by

$$R_k = h_k\left(|\bar{\boldsymbol{h}}_k^{\mathrm{H}} \boldsymbol{t}_k|^2 + I_k\right) - h_k\left(I_k\right), \tag{7.25}$$

where $I_k = \sum_{i \neq k} |\bar{\boldsymbol{h}}_k^{\mathrm{H}} \boldsymbol{t}_i|^2$ and the function $h_k : \mathbb{R}_+ \rightarrow \mathbb{R}_+$ is defined as

$$h_k(x) = \mathrm{E}\left[\log_2\left(1 + \zeta_k^{-1} x\right)\right].$$

For example, if $\zeta_k^{-1} = |w_k|^2$ with $w_k \sim \mathcal{N}_{\mathbb{C}}(0, 1)$, then $h_k(x) \triangleq \mathrm{e}^{\frac{1}{x}}\, \mathrm{E}_1\left(\frac{1}{x}\right)$, where $\mathrm{E}_1(x) = \int_x^{\infty} \frac{\mathrm{e}^{-t}}{t}\, \mathrm{d}t$ denotes the exponential integral function [1]. Unfortunately, the sublogarithmic functions $h_k(x)$ prevents us from a direct reformulation of (7.24) into an SINR constraint. We only obtained the SINR like term

$$\frac{|\bar{\boldsymbol{h}}_k^{\mathrm{H}} \boldsymbol{t}_k|^2}{g_k(\tau\rho_k, I_k)} \geq 1, \tag{7.26}$$

where the continuous function $g_k(c, x) = h_k^{-1}\left(c + h_k(x)\right) - x$ is non-linear increasing in x. We observed from the simulations in [30, 34], that $g_k(c, x)$ is concave in x and close to linear if ζ_k^{-1} is (non-)central χ^2-distributed with degree two.

To obtain a tractable linear denominator in (7.26), we apply the Taylor expansion

$$g_k(c, x) \approx g_k(c, x') + \left.\frac{\partial g_k(c, x)}{\partial x}\right|_{x=x'} (x - x'), \tag{7.27}$$

with the partial derivative

$$\frac{\partial g_k(c, x)}{\partial x} = \frac{\left(h_k^{-1}(c + h_k(x))\right)^2}{h_k^{-1}(c + h_k(x)) - c - h_k(x)} \frac{x - h_k(x)}{x^2} - 1.$$

Then, the ergodic rate balancing optimization (7.24) can be approximated via

$$\max\ \tau \quad \text{s.t.} \quad \boldsymbol{t} \in \mathcal{P}, \quad \frac{|\bar{\boldsymbol{h}}_k^{\mathrm{H}} \boldsymbol{t}_k|^2}{\frac{\beta_k(\tau)}{\alpha_k(\tau)} + I_k} \geq \alpha_k(\tau), \quad k = 1, \ldots, K, \tag{7.28}$$

where $\alpha_k = \frac{\partial g_k(\tau\rho_k, I_k)}{\partial I_k}\Big|_{I_k=I_k'}$ and $\beta_k = g_k(\tau\rho_k, I_k') - \alpha_k I_k'$ depend both on the ergodic rate target ρ_k and the approximate interference value $I_k' = \sum_{i\neq k} |\bar{\boldsymbol{h}}_k \boldsymbol{t}_i'|^2$. This SINR constraint balancing problem can in turn be solved via a bisection over τ and a feasibility test for $\boldsymbol{t}$ in each bisection step similar to [62].[8]

Due to the linearization of $g_k(\tau\rho_k, I_k)$, the approximation is accurate if the obtained I_k from (7.28) is close to the approximation point I_k'. This motivates the sequential approximation procedure, which solves (7.28) in each iteration with an approximation for $\boldsymbol{t}'$ that is the solution of the previous iteration. If the obtained sequence $\tau^{(i)}$ converges, a stationary point of the actual ergodic rate balancing problem in (7.5) is found. Note that convergence is ensured if (7.27) results in an inner restriction of the constraint set [44, 57], i.e., if $g_k(\tau\rho_k, I_k)$ is concave in I_k. A rigorous proof for concavity of $g_k(\tau\rho_k, I_k)$ in I_k is still missing, even though the plots in [30, 34] indicate this for χ^2-distributed ζ_k^{-1}.

7.5.2 Channels with Full-Rank Covaraince Matrix

The average rate bounds and the convex restriction of the rank-one channel model in Sects. 7.5.1.1 and 7.5.1.2, respectively, are unfortunately invalid if a full-rank channel covariance is assumed for the model in (7.4), e.g., when $\boldsymbol{C}_{\boldsymbol{w}_k} = \frac{1}{N}\mathbf{I}_N$ for (7.3). Furthermore, a direct average rate optimization is difficult since already the rate computation requires a numerical integration. Therefore, a *mean square error* (MSE) balancing optimization is applied in [28] instead to maximize a lower bound of the average rates. The MSE of the k-th terminal with equalizer $f_k \in \mathbb{C}$ reads as

$$\mathrm{MSE}_k = 1 - 2\,\mathrm{Re}(f_k^* \boldsymbol{h}_k^{\mathrm{H}} \boldsymbol{t}_k) + |f_k|^2 \Big(\sum_{i=1}^{K} |\boldsymbol{h}_k^{\mathrm{H}} \boldsymbol{t}_k|^2 + 1 \Big), \tag{7.29}$$

which average lower bounds the average rate as

$$R_k = -\,\mathrm{E}\big[\log_2\big(\mathrm{MMSE}_k\big)\big] \geq -\log_2\big(\mathrm{E}[\mathrm{MSE}_k]\big). \tag{7.30}$$

The inequality is due to Jensen's inequality for convex functions and $\mathrm{MMSE}_k = \min_{f_k} \mathrm{MSE}_k$, which is obtained by the perfect CSI MMSE receive filters

$$f_{k,\mathrm{MMSE}} = \frac{\boldsymbol{h}_k^{\mathrm{H}} \boldsymbol{t}_k}{\sum_{i=1}^{K} |\boldsymbol{h}_k^{\mathrm{H}} \boldsymbol{t}_i|^2 + 1}. \tag{7.31}$$

[8] The power minimization reformulations in [66] or the iterative SINR constrained power minimizations via uplink-downlink duality in [22, 55, 68] may be employed for the feasibility test.

Balancing the average MSEs with the filters in (7.31), i.e.,

$$\mathrm{E}[\mathrm{MMSE}_k] = 1 - \mathrm{E}\left[|\boldsymbol{h}_k^{\mathrm{H}}\boldsymbol{t}_k|^2 \Big/ \Big(\sum_{i=1}^{K} |\boldsymbol{h}_k^{\mathrm{H}}\boldsymbol{t}_i|^2 + 1\Big)\right], \tag{7.32}$$

remains difficult for imperfect transmitter CSI. The expectation in (7.32) is over a ratio of correlated random parameters. Even though closed form expressions may be found for the average MMSEs, balancing them is still a non-convex problem. Therefore, an upper bound optimization was proposed in [28], where the beamformers are optimized under the assumption that the terminals have the same imperfect channel knowledge as the transmitter. This upper bound minimization can be employed for the beamformer update of an *alternating convex search* (ACS) to solve the aimed MMSE minimization with perfect CSI at the receiving terminals (cf. [31]).

7.5.2.1 Mean Square Error Minimization

If the terminals have the same imperfect CSI as the satellite, the average MSE balancing with equal targets reads as

$$\min \max_k \widehat{\mathrm{MSE}}_k \quad \text{s.t.} \quad \boldsymbol{t} \in \mathcal{P}. \tag{7.33}$$

The average MSEs $\mathrm{E}[\mathrm{MSE}_k]$ are given by

$$\widehat{\mathrm{MSE}}_k = 1 - 2\,\mathrm{Re}(f_k^* \hat{\boldsymbol{h}}_k^{\mathrm{H}} \boldsymbol{t}_k) + |f_k|^2 \Big(\sum_{i=1}^{K} \boldsymbol{t}_i^{\mathrm{H}} \boldsymbol{R}_k \boldsymbol{t}_i + \sigma_k^2\Big). \tag{7.34}$$

Here, $\hat{\boldsymbol{h}}_k = \mathrm{E}[\boldsymbol{h}_k]$ and $\boldsymbol{R}_k = \mathrm{E}[\boldsymbol{h}_k \boldsymbol{h}_k^{\mathrm{H}}]$ are the first and (full-rank) second order moments of the random channels $\boldsymbol{h}_k$ and $\sigma_k^2 = 1$.

The solution for (7.33) can be found with a convex solver [27]. Inserting the MMSE filters for imperfect receiver CSI, the resulting MMSE problem formulation is a quasiconvex program as is shown in [28]. Alternatively, a dual uplink MSE balancing formulation and an iterative solution approach for the dual problem were proposed in [28]. The dual average MSE problem reads as

$$\max_{\boldsymbol{\mu} \geq \boldsymbol{0}} \min_{\boldsymbol{\lambda} \geq \boldsymbol{0}, \boldsymbol{u}} \max_i \widehat{\mathrm{MSE}}_i^{(\mathrm{UL})} \quad \text{s.t.} \quad \sum_{i=1}^{K} \sigma_k^2 \lambda_i \leq 1, \quad \sum_{\ell=1}^{L} \mu_\ell P_\ell \leq 1 \tag{7.35}$$

where the average uplink MSE that corresponds to terminal i is given by

$$\widehat{\mathrm{MSE}}_i^{(\mathrm{UL})} = 1 - \sqrt{\lambda_i} 2\,\mathrm{Re}\{\hat{\boldsymbol{h}}_i^{\mathrm{H}} \boldsymbol{u}_i\} + \boldsymbol{u}_i^{\mathrm{H}} \Big(\sum_{k=1}^{K} \lambda_k \boldsymbol{R}_k + \sum_{\ell=1}^{L} \mu_\ell \boldsymbol{A}_{i,\ell} \Big) \boldsymbol{u}_i. \tag{7.36}$$

The uplink power allocation vector $\boldsymbol{\lambda} = [\lambda_1, \ldots, \lambda_K]^{\mathrm{T}} \geq \mathbf{0}$ comprises the dual variables associated with the downlink MSEs. The vector $\boldsymbol{\mu} = [\mu_1, \ldots, \mu_L]^{\mathrm{T}} \geq \mathbf{0}$, that defines the worst-case noise covariance matrix $\sum_{\ell=1}^{L} \mu_\ell \boldsymbol{A}_{i,\ell}$ in (7.36), contains the dual variables for the generalized transmit power constraints.

The iterative solution for the uplink MSE balancing problem consists of two nested loops (see [28]). The inner loop controls the power allocation $\boldsymbol{\lambda}$ and the MMSE equalizers in the uplink via a globally convergent standard interference function update that is normalized by a target MSE value to satisfy the power constraint in (7.35). The outer loop uses a subgradient projection method similar to [15] to control the worst-case uplink noise covariance via the Lagrangian multipliers $\boldsymbol{\mu}$.

The downlink beamformers and receive filters follow from the uplink MMSE equalizers $\boldsymbol{u}_k$ and powers λ_k of (7.35) by the linear transformation (cf. [28, 37])

$$\boldsymbol{t}_i = \sqrt{\beta_i} \boldsymbol{u}_i, \quad f_i = \sqrt{\lambda_i} / \sqrt{\beta_i}, \quad i = 1, \ldots, K \tag{7.37}$$

and $\boldsymbol{\beta} = [\beta_1, \ldots, \beta_K]^{\mathrm{T}}$ is the solution to a linear system of equations that results from requiring equality between the MSEs of (7.34) and (7.36) and inserting (7.37).

7.5.2.2 Average Mean Square Error Minimization

The solution for the upper bound minimization in (7.33) can be employed for solving the initial average MSE minimization in (7.32) with an ACS. The alternation is between the terminal equalizer update in the downlink and the satellites beamformer design [37, 59], assuming the other variables are fixed. In other words, the following two steps are performed in each iteration (cf. [59]):

1. First, the equalizers in $\boldsymbol{f} = [f_1, \ldots, f_K]^{\mathrm{T}} \in \mathbb{C}^K$ are found in the downlink for fixed $\boldsymbol{t}$ as in (7.31).
2. Second, the beamformers in $\boldsymbol{t}$ are optimized as equalizers in the dual uplink system for given downlink equalizers in $\boldsymbol{f}$.

Note that the employed uplink-downlink duality in Step 2 of the ACS reduces the complexity for implementation and computations. Otherwise, the precoder update requires a convex solver for including the generalized power constraints.

For imperfect transmitter CSI, Step 1 remains the same. However, the update of $\boldsymbol{t}$ in Step 2 has to take into account that the equalizer $f_{k,\mathrm{MMSE}}$ in (7.31) is a random variable as it is a function of the channel $\boldsymbol{h}_k$, $k = 1, \ldots, K$ [39].[9] In the proposed

[9]The work by Bogale and Vandendorpe [9, 10] is for imperfect transmitter and terminal CSI.

ACS for a sum MSE minimization of [31], the total equalizers $\tilde{f}_k = v_k f_{k,\text{MMSE}}$, $k = 1, \ldots, K$ are virtually split for the beamformer update. The first part v_k is controlled by the transmitter, thus it is independent of the channel's state, and the other part $f_{k,\text{MMSE}}$ is the (fixed) MMSE equalizer that depends on $\boldsymbol{h}_k$. Therewith, Step 2 can be performed with the upper bound minimization using the following substitutes:

$$\hat{\boldsymbol{h}}_k = \mathrm{E}[\boldsymbol{h}_k f_{k,\text{MMSE}}], \tag{7.38}$$

$$\boldsymbol{R}_k = \mathrm{E}[\boldsymbol{h}_k |f_{k,\text{MMSE}}|^2 \boldsymbol{h}_k^{\mathrm{H}}], \tag{7.39}$$

$$\sigma_k^2 = \mathrm{E}[|f_{k,\text{MMSE}}|^2]. \tag{7.40}$$

Note that the expectations to compute $\hat{\boldsymbol{h}}_k$, $\boldsymbol{R}_k$, and σ_k^2 require numerical evaluations for the considered imperfect CSI model, e.g., via a Monte-Carlo approach.

7.6 Satellite Simulation Setup and Numerical Results

The forward-link SatCom setup under consideration corresponds to a geostationary satellite with S-band for mobile terminals according to the multi-spot beam satellite communication channel detailed in Sect. 7.2. The parameters for this satellite communication scenario are provided in Table 7.2. Similar to the previous sections, the results are divided into a part for the outage constrained balancing optimizations (see Sect. 7.6.1) and the average performance measures (see Sect. 7.6.2).

Table 7.2 Link budget parameters for a mobile terminal SatCom system (cf. [18])

Parameter	S-band
Satellite configuration	GEO
Beamwidth $\theta_{3\text{dB}}$ (in degree)	0.4
Frequency reuse	1
Max satellite/user antenna gain	52 dBi/3 dBi
Approximate FSL	190 dB
Base receive noise power	−133 dBW
Rician fading factor κ	8, 10 dB
Log-normal fading $m_{\text{rain},k}/\sigma_{\text{rain},k}^2$ [18]	−2.62/1.63, 3.26 dB
SNR P	$0, \ldots, 30$ dB

7.6.1 Results for Chance-Constrained Rate Balancing

For the outage based optimizations, a SatCom system with $N = K = 3$ antennas and users is employed, where the communication takes place in the S-band (see Table 7.2). The small number of antennas and mobile terminals is to keep the computationally most complex method tractable, i.e., the outage constrained optimization with iterative prior optimization. If not stated otherwise, the target rates are fixed to $\rho_k = 1$ and the outage probability is limited to $\varepsilon_k = 0.1$ for all terminals. Moreover, per-antenna constraints with $P_\ell = P/3$ are employed to model the per-feed power constraints at the satellite. For this setup, 10 random positions of the terminals within the beams are created and, therewith, 10 realizations of the beamgain matrices $\boldsymbol{G}_k$ and FSLs $g_{\text{FSL},k}$ according to the model in Sect. 7.2.1. For the Rician fading in SatCom, $\kappa = 10, 15\,\text{dB}$ are considered as two typical values and $\sigma_{\text{rain},k} = 1.63, 3.26\,\text{dB}$ serve as weak and strong values, respectively, for the log-normally distributed rain-fading (e.g., see [18]).

7.6.1.1 Equal Fading Conditions

Figure 7.3 shows the averaged balanced achievable rate ρ_0 versus the transmit power P, where the averaging is for the 10 channel realizations. The four plots in the figures are for only rain fading (blue line), i.e., no additive error, the conservative inner approximation in Sect. 7.4.2 with Bernstein's type inequality for the inner problem and either equal or iterative prior probability optimization for the outer problem (orange line and dashed black line, respectively), and the results for using a sphere bounding for the additive errors instead of Bernstein's type inequality (cf. [35]).

As expected, the smaller κ and the larger $\sigma^2_{\text{rain},k}$ are, the stronger is the influence of the additive and the multiplicative channel error, respectively, and the smaller are the reliably achievable balanced rates. If no Rician fading is present and the channels have essentially a rank-one covariance matrix, the obtained rates lie strictly above the results with Rician fading. Thus, the scattering that leads to a non-rank one covariance matrix for the SatCom channel tremendously reduces the achievable epsilon-outage rates. The inner Bernstein's inequality approximation outperforms the sphere bounded additive channel error approximation (cf. [65]). Only for strong rain and Rician fading parameters, i.e., $\sigma^2_{\text{rain},k} = 3.26\,\text{dB}$ and $\kappa = 10\,\text{dB}$, there is a slight loss from the Bernstein's type inequality to the bounded channel error.

Figure 7.4 sketches the rate maximizing $\alpha_0 \in [1 - \min_k(\varepsilon_k)]$ (equal prior probabilities) versus P in dB for the inner Bernstein's inequality approximation, which lies between 0.9 and 1.0 due to the fixed $\varepsilon_k = 0.1$. The prior probability α_0 increases with the transmit power. With weak rain fading $\sigma^2_{\text{rain},k} = 1.63\,\text{dB}$ and Rician fading $\kappa = 15\,\text{dB}$, the increase is from about 94 to 98 % in the considered interval for P. The value is only slightly lower if the more conservative sphere bounding method is employed instead. If the Rician factor increases, therewith, the influence of the additive multi-path channel errors, the prior α_0 is larger and only increases from 96.5

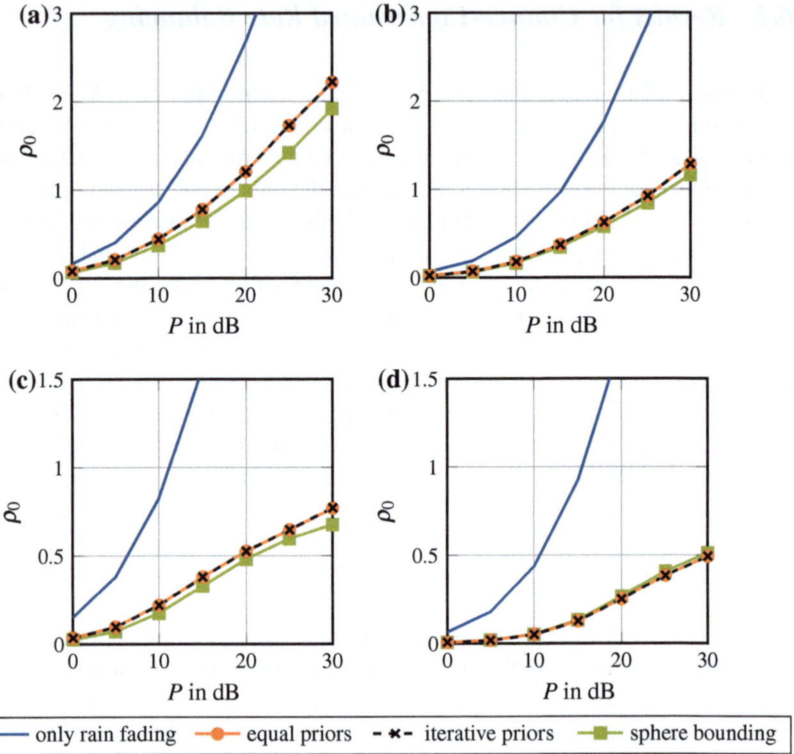

Fig. 7.3 Maximum ρ_0 versus P for a $N = K = 3$ antenna and terminal system with equal targets $\rho_k = 1$, equal outage requirements $\varepsilon_k = 0.1$, and per-antenna power constraints $P_\ell = P/3$. **a** $\kappa = 15\,\text{dB}$, $\sigma^2_{\text{rain},k} = 1.63\,\text{dB}$. **b** $\kappa = 15\,\text{dB}$, $\sigma^2_{\text{rain},k} = 3.26\,\text{dB}$. **c** $\kappa = 10\,\text{dB}$, $\sigma^2_{\text{rain},k} = 1.63\,\text{dB}$. **d** $\kappa = 10\,\text{dB}$, $\sigma^2_{\text{rain},k} = 3.26\,\text{dB}$

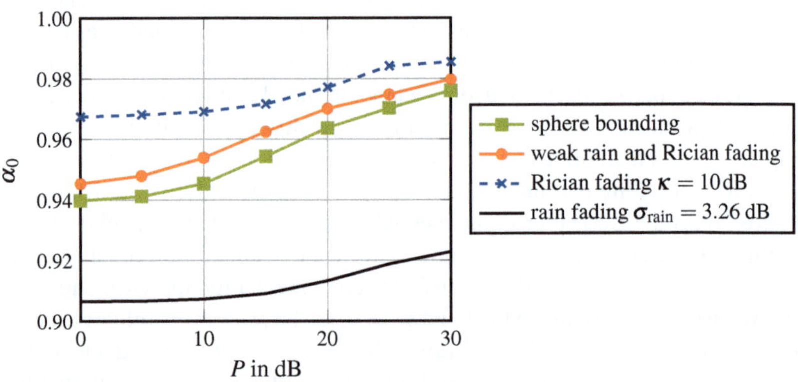

Fig. 7.4 Maximizing α_0 versus P for a $N = K = 3$ system with equal targets $\rho_k = 1$, requirements $\varepsilon_k = 0.1$, and priors $\alpha_k = \alpha_0$

to 98.5 %. When the rain fading is strong $\sigma^2_{\text{rain},k} = 3.26$ dB instead, the optimal prior α_0 is smaller than for strong rain fading and only increases from 90.7 to 92.2 %.

Note that the larger the prior probability α_0 becomes with increasing transmit power P, the smaller the posterior probabilities $\frac{1-\varepsilon_k}{\alpha_0}$ are. In other words, the outage probability shifts from the rate limiting influence of the additive channel error to the rate limiting influence of the multiplicative error, i.e., the effective noise is enhanced.

7.6.1.2 Effects for Distinct Fading Conditions

Comparing the optimized ρ_0 and α_k's of the equal and the iterative prior optimization, e.g., as in [35], one recognizes that the influence of the rain fading and changed outage requirements may lead to completely different results. For example, the achievable balanced rates for the iterative optimization of the α_k's lie one bit above the balanced rates for equal α_k's for $P = 30$ dB if only terminal 2 experiences strong rain fading with $\sigma_{\text{rain},2} = 3.26$ dB and has the loosened requirement $\varepsilon_2 = 0.20$, while receivers 1 and 3 are subject to low rain fading with the requirements $\varepsilon_1 = \varepsilon_3 = 0.10$ (see Fig. 7.5). The balanced rates benefit from an iterative prior optimization since the obtained difference between the optimized α_2 and α_1, α_3 is more than 10 %, while the equally optimization α_0 is restricted to lie above 90 %. Therefore, an adaptive prior optimization needs to be implemented if the statistics of the multiplicative channel errors strongly differ among the served receivers while the equal prior choice performs sufficiently well for channels with similar fading properties.

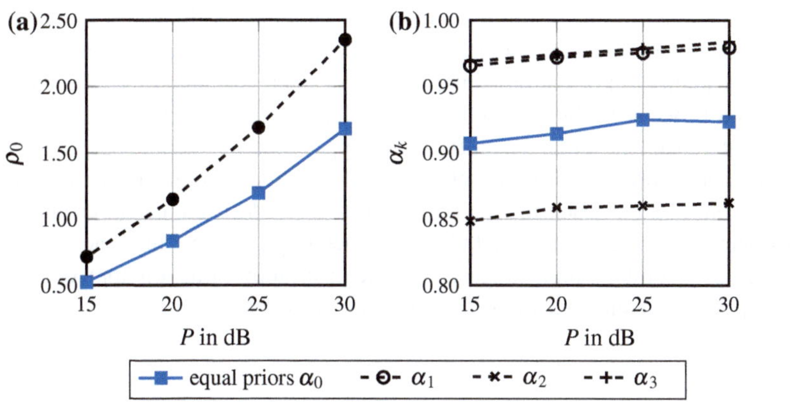

Fig. 7.5 Comparison of the iterative and the equal prior optimizations for a $N = K = 3$ system with equal $\rho_k = 1$, but increased rain fading and outage margin for terminal 2, i.e., $\sigma_{\text{rain},2} = 3.26$ dB and $\varepsilon_2 = 0.20$ while $\sigma_{\text{rain},1} = 1.63$ dB and $\varepsilon_1 = \varepsilon_3 = 0.10$. **a** Achievable rate. **b** Optimal priors

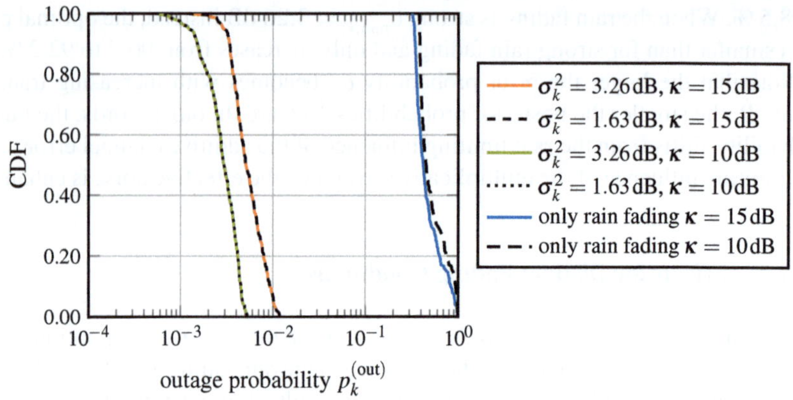

Fig. 7.6 Actual outage probabilities for a $N = K = 3$ system with $\rho_k = 1$, $\varepsilon_k = 0.1$, $\sigma_{\text{rain},k} = 1.63$ dB, and equal priors

7.6.1.3 Achieved Outage Probabilities

When calculating the actually achieved outage probabilities $p_k^{(\text{out})} = \Pr(r_k \leq \rho_0 \rho_k)$ via a numerical integration for the computed beamformers as is detailed in [35], we observe that the inner semidefinite approximation based on Bernstein's type inequality is loose. In contrast, if only rain fading is taken into account and the Rician fading is neglected, the result is over-optimistic in the sense that the required outage limits are missed. This is also seen in Fig. 7.6, where the empirical CDF of the probabilities $p_k^{(\text{out})}$ is plotted for the beamformers that we computed with the proposed conservative approach in Sect. 7.4.2, with an equal outer prior restriction and inner Bernstein's type inequality.[10] For the plots, the fading parameters κ and $\sigma_{\text{rain},k}^2$ are varied from medium to weak Rician fading and from weak to strong rain fading, respectively. The differences show that the inner convex approximation method becomes less tight for increased Rician fading. In other words, the inner optimizations are statistically more conservative when the likelihood of a large additive channel error increases. In contrast, changes in the rain fading parameter $\sigma_{\text{rain},k}^2$ do not influence the obtained outage probability, which suggests that the prior approximation approach takes the multiplicative channel errors sufficiently into account.

7.6.2 *Results for Average Rate Balancing*

The average balancing optimizations are made for the rank-one and the full-rank channel covariance matrix models in Sect. 7.2.2, with the S-band model in Table 7.2.

[10]The outer equal prior restriction and the inner Bernstein's type inequality approach are used, which is less conservative than directly restricting the additive channel errors to lie in a sphere [65].

While simulations for the full-rank channel covariance matrix scenario are made only for a cluster of 7 neighboring spotbeam antennas, the rank-one channel covariance matrix scenario allows also a low-complex computation of the 19 spotbeam and the 121 spotbeam scenarios depicted in Fig. 7.1. Moreover, per-antenna constraints are imposed for SatCom if not otherwise stated.

7.6.2.1 Results for Rank-One Channel Covariance Matrices

For the simulations to the rank-one channel covariance results, the target rates are $\rho_k = 2$ for a subset of $\lfloor K/2 \rfloor$ terminals and $\rho_k = 1$ for the the remaining $\lceil K/2 \rceil$ ones. The channels are for the S-band SatCom system with parameters in Table 7.2 and for three setups with either, $N = K = 7$, $N = K = 19$, or $N = K = 121$ spotbeams and terminals. For simplicity, we imposed only a sum power constraint for balancing the ergodic rates according to (7.5).

The average balancing level τ is computed for some of the bounds in Sect. 7.5.1.1 and the sequential approximation approach in Sect. 7.5.1.2 for 100 different channel realizations. Each of these channels corresponds to a randomly drawn placement of the terminal within the spotbeams and independently drawn rain attenuation coefficients for each terminal. The satellite is assumed to know the slow fading rain attenuation and the statistic of the Rician fading with factor κ.

Figure 7.7 depicts the resulting τ over the sum transmit power P for the ergodic rate upper and lower bounds UB1, LB2, and ALB of Sect. 7.5.1.1, *zero-forcing* (ZF) beamforming with power adaption, and the sequential approximation strategy (SEQ) in Sect. 7.5.1.2 for four setups. The first two graphics (Fig. 7.7a, b) are for $K = N = 7$, Fig. 7.7c is for a scenario with $K = N = 19$ antennas and terminals, and Fig. 7.7d is for the complete coverage of Europe with $K = N = 121$ (cf. Fig. 7.1). Note that the sequential approximation is computationally to complex for $K = N = 121$.

The obtained curve for the sequential approximation strategy lies between the upper and lower bounds as expected. Moreover, the ALB curves outperform the LB2 curves. This is due to the tighter approximation of the average rate for this bound. The ZF strategy obviously falls behind the obtained performance of the lower bounds. The reason is that a strict ZF leads to a remarkably high drop of the useful signal power. Only for a very large transmit power (over 50 dB for $K = N = 7$), it reaches the lower bound curves. The loss due to the strict interference avoidance with ZF increases with the dimension of the scenario and interference management gain in importance.

When the Rician fading is decreased by increasing κ from 5 to 10 dB, the distance between the UB1 curve and the LB curves becomes very small. In contrast, the distance between the bounds remains almost constant, when increasing the system dimension to either $N = K = 19$ or $N = K = 121$. This indicates that these bounds are also useful for very large spot-beam scenarios if the channel covariance matrices are rank-one. To reach a similar rate target range, the focus of the plots is shifted by 10 dB from the $N = K = 7$ case in Fig. 7.7a to the $N = K = 19$ case in Fig. 7.7c, and by another 10 dB compared to the $N = K = 121$ scenario in Fig. 7.7d. This

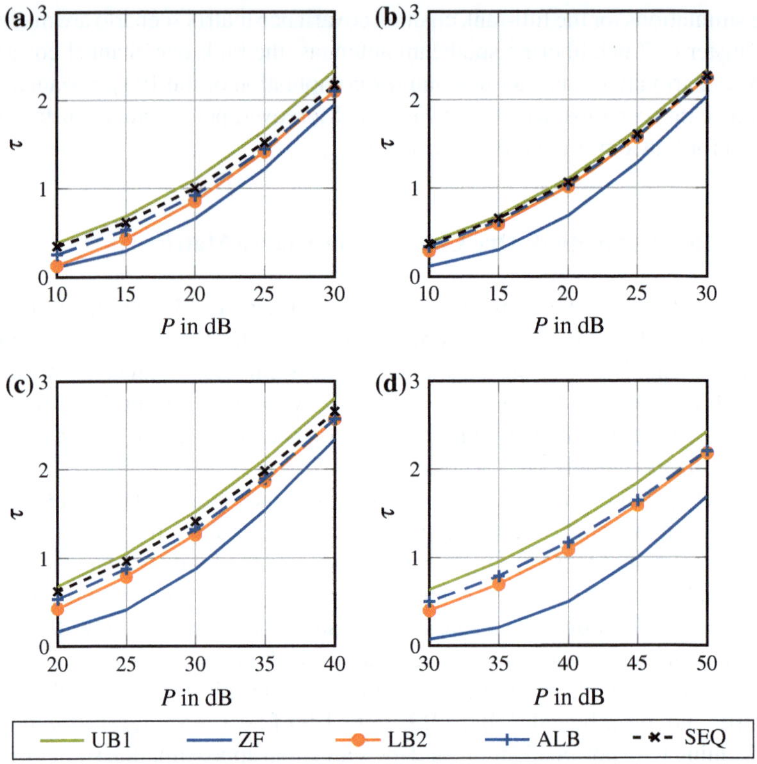

Fig. 7.7 Optimized τ over the sum power P in [dB] for the average balancing optimization with a rank-one channel covariance matrix in an S-band SatCom downlink with known rain fading attenuation but unknown Rician fading with factor κ. **a** $N = K = 7, \kappa = 5$ dB. **b** $N = K = 7, \kappa = 10$ dB. **c** $N = K = 19, \kappa = 5$ dB. **d** $N = K = 121, \kappa = 5$ dB

observation highlights the increasing influence of the interference with the system dimensions for the downlink of a multi-spotbeam SatCom scenario.

7.6.2.2 MSE Results for Full-Rank Channel Covariance Matrices

The balanced average MSEs for the S-band SatCom system setups in Table 7.2 are computed for a cluster of 7 neighboring spotbeams (see Fig. 7.1). Per-antenna power constraints are imposed for the simulations, i.e., one antenna forms one spotbeam ($L = N$), and the (mobile) terminals are randomly placed within the 3 dB area of the SatCom beams, i.e., one terminal per spotbeam ($K = N$). In this way, 100 different channel characteristics are created for the average MSE balancing computations. Additionally, one terminal realization for an 121 spotbeam system is drawn, which represents a typical terminal placement in the coverage of Europe (see Fig. 7.1).

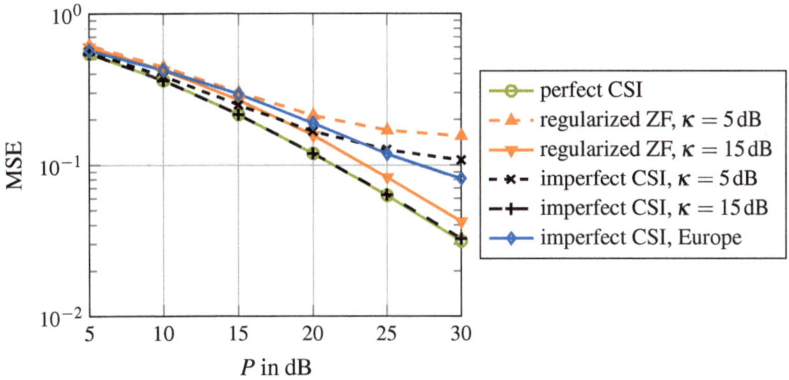

Fig. 7.8 Average MSE balancing results versus P for a S-band spotbeam SatCom model with either 7 neigboring spotbeams or the 121 spotbeams that cover Europe

In Fig. 7.8, the mean of the balanced (average) MSEs for the 100 characteristic realizations are depicted versus P, where $P_\ell = P/L$ for $\ell, k = 1, \ldots, K$. The perfect CSI curve for the SatCom system (solid green) provides a lower bound for the achievable average MSE curves with imperfect CSI. While all imperfect CSI curves are close to each other and to the perfect CSI curve for small P, the MSE curves saturate for a small Rician fading factor κ and large P. The saturation level is about 0.1 for $\kappa = 5\,\text{dB}$. The higher κ, the lower the saturation level until the imperfect CSI curve meets the perfect CSI curve for $\kappa = 15\,\text{dB}$ or larger. Then, the multi-path scattering may be neglected for the considered SatCom channel model.

The optimized average MSE curves are strictly below the curves with regularized *zero-forcing* (ZF) beamformers [18]. This is a consequence of neglecting the SatCom beamgain characteristic in the regularization matrix. The beamgain characteristic deforms the channel mean and error covariance alike [see (7.4)]. For the 7 spotbeam system with $\kappa = 15\,\text{dB}$ and $P = 20\,\text{dB}$, this scheme has an 26 % higher MSE per user than the proposed optimal MSE balancing scheme due to the suboptimal beamformer design.

In addition to the 7 cell MSE curves, Fig. 7.8 shows also the balanced MSE for one exemplary channel realization in the 121 spotbeam SatCom scenario that covers Europe for $\kappa = 15\,\text{dB}$. This curve decreases apparently slower than the 7 spotbeam curves. The reason is that users in boundary cells experience less interference (on average) than users within cells in the inner region (with 6 neighbor cells) and the percentage of inner to boundary cells is increased for the 121 cell scenario.

7.6.2.3 Limits for Per-Antenna Power Constraints

To show the effects of the various power constraints, a sum MSE minimization is assumed where N is strictly larger than K. For simplicity, $K = 2$ is fixed and only a row of spotbeams is assumed, which leads to a simple exponential power profile

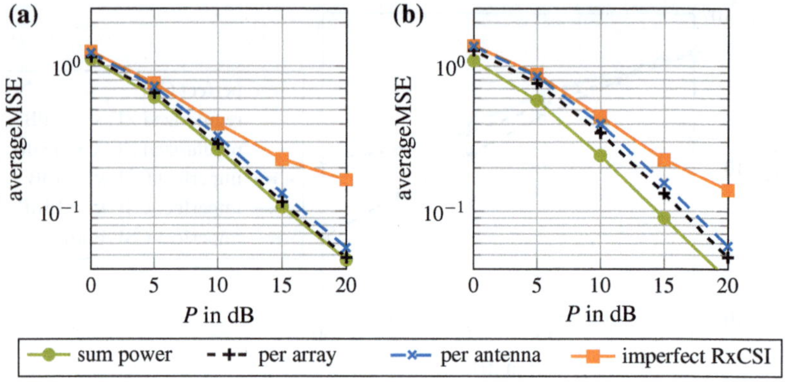

Fig. 7.9 Average sum MSE versus P for a *row* of N spotbeam antennas with imperfect CSI, per-antenna, per-array, and a sum power constraint at the satellite, $K = 2$ terminals with perfect CSI (or imperfect) CSI within these spotbeams, log-normal rain fading, and Rician fading with $\kappa = 10$ dB. **a** $N = 4$ Satellite antennas. **b** $N = 8$ Satellite antennas

when the users are centrally placed in their beams [41]. This power profile forms the channels' gain characteristic, which is then corrupted by the log-normal rain fading attenuation and the Rician fading. Hence, the transmitter on average prefers to serve the k-th user with the k-th antenna element, while the rain fading determines the terminals' link quality.

Figure 7.9 depicts the minimized (average) sum MSE curves over the sum-power for $N = 4$ antennas on the left hand side and for $N = 8$ antennas on the right hand side. The four lines correspond to a sum power limitation P (solid green line), per-array power constraints for antennas 1, 2 and 3, 4 with $P_\ell = P/2$ (dashed line), and per-antenna constraints with $P_\ell = P/N$ (long-dashed line). The data for the lines to the per-array and per-antenna constraint case are created with an alternating optimization according to Sect. 7.5.2.2 (see also [31]).

In this case, the per-antenna constraints are stricter than the per-array constraints and the sum power constraint, which results in an increased average sum MSE for more power constraints. In other words, the sum power limitation gives a lower bound for the achievable average sum MSE with per-array or per-antenna constraints, which require more complex beamformer designs. An upper bound is obtained by assuming that the terminals have the same imperfect CSI as the satellite (solid orange line), that is subject to per-antenna power constraints. The difference between the sum power and per-antenna power constraint lines further increases when N becomes larger (for constant K), e.g., $N = 8$ as in the right-hand side plot of Fig. 7.9.

Even though the sum MSE results may imply only a small loss for the realistic per-antenna or per-array constraints compared to the less complex sum power constraint, using the sum power constraint for the beamformer computation dramatically increases the dynamic range of the per-antenna or per-array gains. Figure 7.10 depicts the empirical CDF curves of the largest per-antenna power when minimizing the average sum MSE with a sum power constraint or per-array constraints. The

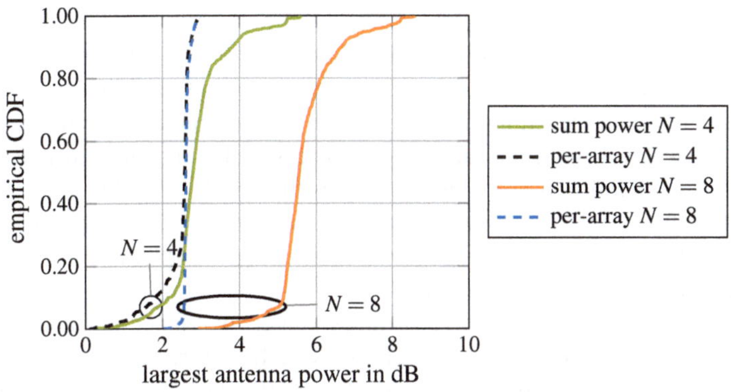

Fig. 7.10 Empirical CDF of the largest per-antenna power in dB (normalized to P/N) for the sum MSE optimal beamformers with a sum power constraint or per-antenna power constraints and $\kappa = 10\,\text{dB}$

antenna power is normalized such that an associated per-antenna power $P_\ell = P/N$ represents 0 dB.

For $N = 4$ and a sum power constraint, a double of the per-antenna bound P_ℓ (i.e., 3 dB) is surpassed in more than 20 % of the channel realizations. This 20 % bound increases to 6 dB for $N = 8$ transmit antennas because the exponential channel characteristic leads preferably to an antenna power profile with a large power for only two of the eight antennas, those with the highest channel gains. Moreover, since the channels are subject to shadow fading, their gains are likely to differ, which results in an unbalanced power distribution for these two main antennas. This is the reason for the increased sum MSE gap between the sum power and the per-antenna curves for $N = 8$ in Fig. 7.9.

7.7 Conclusion

This work focused on a suitable channel model for the satellite downlink with mobile terminals and multi-user beamformer designs, that accounts for rain-fading as well as multi-path scattering in terms of multiplicative and additive channel errors, respectively. The increased importance of the rain-fading stems from the aimed carrier frequencies. The beamformer design was based on a max-min rate optimization to overcome potential limitations due to interference with a frequency reuse of one. To remain reliable against the statistical channel fading, the epsilon-outage rate balancing optimization and the average rate balancing optimization are addressed in particular as two statistical robust formulations.

For sufficiently correlated channels with a rank-one channel covariance, the epsilon-outage rate balancing problem was directly solved and tight rate bounds are found for the ergodic rate optimization with a standard logarithmic form.

In contrast, the epsilon-outage constrained balancing optimization with full-rank channel covariances is difficult to solve directly. The presented approach uses a conservative inner beamformer design to be robust against the additive channel errors and takes care about the multiplicative channel errors, which result in an increased effective noise level, with an outer search. The simulations showed that an equal effective noise adaption is sufficient as long as the rain fading characteristic differs only slightly for jointly processed cells. Note that there is still room for improving the complexity of the epsilon-outage constrained design in order to cope with a larger number of jointly processed multi-spotbeams.

The ergodic rate balancing with full-rank channel covariance was approached via an ergodic mean square error minimization, which features a beamformer design based on uplink-downlink duality and fixed point iterations also for per-feed power constraints. Numerical results for the mean square error balancing in satellite communication suggests that imperfectly known multi-path fading may result in limited performance, but it becomes almost negligible for a sufficiently strong line-of-sight component due to the pointing characteristic of the spotbeams.

Acknowledgments The work of Andreas Gründinger was supported by the German Research Foundation (DFG) under Grants Jo 724/1-1 and Jo 724/1-2.

References

1. Abramowitz M, Stegun IA (1964) Handbook of mathematical functions, 1st edn. Dover Publications Inc
2. Al-Naffouri T, Hassibi B (2009) On The distribution of indefinite quadratic forms in Gaussian random variables. In: Proceedings of the IEEE international symposium on information theory, pp 1744–1748
3. Arapoglou PD, Burzigotti P, Bertinelli M, Bolea Alamanac A, De Gaudenzi R (2011a) To MIMO or not to MIMO in mobile satellite broadcasting systems. IEEE Trans Wirel Commun 10(9):2807–2811
4. Arapoglou PD, Liolis K, Bertinelli M, Panagopoulos A, Cottis P, De Gaudenzi R (2011b) MIMO over satellite: a review. IEEE Commun Surv Tutorials 13(1):27–51
5. Arapoglou PD, Liolis K, Panagopoulos A (2012) Railway satellite channel at Ku band and above: composite dynamic modeling for the design of fade mitigation techniques. Int J Satell Commun Netw 30(1):1–17
6. Arnau J, Devillers B, Mosquera C, Perez-Neira A (2012) Performance study of multiuser interference mitigation schemes for hybrid broadband multibeam satellite architectures. EURASIP J Wirel Commun Netw 1(132):1–19
7. Bechar I (2009) A Bernstein-type inequality for stochastic processes of quadratic forms of Gaussian variables. (Preprint version online available at ArXiv e-prints) 1:1–6
8. Ben-Tal A, El Ghaoui L, Nemirovski A (2009) Robust optimization. Princeton University Press, Princeton, Princeton series in applied mathematics
9. Bogale T, Vandendorpe L (2011) Sum MSE optimization for downlink multiuser MIMO systems with per antenna power constraint: downlink-uplink duality approach. In: Proceedings of the IEEE 22nd international symposium on personal indoor and mobile radio communications, pp 2035–2039

10. Bogale T, Vandendorpe L (2012) Robust sum MSE optimization for downlink multiuser MIMO systems with arbitrary power constraint: generalized duality approach. IEEE Trans Signal Process 60(4):1862–1875
11. Botros Shenouda M, Lampe L (2009) Quasi-convex designs of max-min linear BC precoding with outage QoS constraints. In: Proceedings of the 6th international symposium on wireless communication systems, pp 156–160
12. Boulanger X, Feral L, Castanet L, Jeannin N, Carrie G, Lacoste F (2013) A rain attenuation time-series synthesizer based on a dirac and lognormal distribution. IEEE Trans Antennas Propag 61(3):1396–1406
13. Boyd S, Vandenberghe L (2004) Convex Optimization, 1st edn. Cambridge University Press, New York, NY, USA
14. Braun T (2012) Antenna, chap 3. Wiley, IEEE Press, pp 33–77
15. Cai D, Quek T, Tan CW, Low S (2011) Max-min weighted SINR in coordinated multicell MIMO downlink. In: Proceedings of the 9th international symposium on modeling and optimization in mobile, Ad Hoc, and wireless networks, pp 286–293
16. Chang TH, Ma WK, Chi CY (2011) Worst-case robust multiuser transmit beamforming using semidefinite relaxation: duality and implications. In: Proceedings of the 45th Asilomar conference on signals, systems, and computers, pp 1579–1583
17. Christensen S, Agarwal R, Carvalho E, Cioffi J (2008) Weighted sum-rate maximization using weighted MMSE for MIMO-BC beamforming design. IEEE Trans Wirel Commun 7(12):4792–4799
18. Christopoulos D, Chatzinotas S, Zheng G, Grotz J, Ottersten B (2012) Linear and nonlinear techniques for multibeam joint processing in satellite communications. EURASIP J Wirel Commun Network 1(162):1–13
19. Christopoulos D, Arnau J, Chatzinotas S, Mosquera C, Ottersten B (2013) MMSE performance analysis of generalized multibeam satellite channels. IEEE Commun Lett 17(7):1332–1335
20. Christopoulos D, Chatzinotas S, Ottersten B (2014) Weighted fair multicast multigroup beamforming under per-antenna power constraints. IEEE Trans Signal Process 62(19):5132–5142
21. Cottatellucci L, Debbah M, Casini E, Rinaldo R, Mueller R, Neri M, Gallinaro G (2006) Interference mitigation techniques for broadband satellite system. In: Proceedings of 24th AIAA international communications satellite systems conference
22. Dartmann G, Gong X, Afzal W, Ascheid G (2013) On the duality of the max-min beamforming problem with per-antenna and per-antenna-array power constraints. IEEE Trans Veh Technol 62(2):606–619
23. Devillers B, Perez-Neira A, Mosquera C (2011) Joint linear precoding and beamforming for the forward link of multi-beam broadband satellite systems. In: IEEE global telecommunications conference
24. Diaz M, Courville N, Mosquera C, Liva G, Corazza G (2007) Non-linear interference mitigation for broadband multimedia satellite systems. In: Proceedings of the international workshop on satellite and space communications, pp 61–65
25. Duflos A, Evans B, Thompson P, Tomatis A, Amos S, Laurent A, Noeldeke C, Verlest G (2012) Approaching the terabit/s satellite: a system study. Executive summary 1, revision 1, ESA contract no: 4000103563
26. Gayrard JD (2009) Terabit satellite: myth or reality? In: Proceedings of the 1st international conference on advances in satellite and space communications
27. Grant M, Boyd S (2009) CVX: Matlab software for disciplined convex programming
28. Gründinger A, Barthelme A, Joham M, Utschick W (2014a) Mean square error beamforming in SatCom: uplink-downlink duality with per-feed constraints. In: Proceedings of the 6th international symposium on wireless communication systems, pp 600–605
29. Gründinger A, Bethenod R, Joham M, Riemensberger M, Utschick W (2013a) Optimal power allocation for the chance-constrained vector broadcast channel and rank-one channel approximation. In: Proceedings of the 12th IEEE international workshop on signal processing advances in wireless communications, pp 31–35

30. Gründinger A, Butabayeva A, Joham M, Utschick W (2012) Chance constrained and ergodic robust QoS power minimization in the satellite downlink. In: Proceedings of the 46th Asilomar conference on signals, systems, and computers, pp 1147–1151
31. Gründinger A, Joham M, González-Coma JP, Castedo L, Utschick W (2014b) Average sum MSE minimization in the multi-user downlink with multiple power constraints. In: Proceedings of the 48th Asilomar conference on signals, systems, and computers
32. Gründinger A, Joham M, Utschick W (2012) Bounds on optimal power minimization and rate balancing in the satellite downlink. In: Proceedings of the IEEE international conference on communications, pp 3600–3605
33. Gründinger A, Joham M, Utschick W (2012) Feasibility test and globally optimal beamformer design in the satellite downlink based on instantaneous and ergodic rates. In: Proceedings of the 16th international ITG workshop on smart antennas, pp 217–224
34. Gründinger A, Leiner D, Joham M, Hellings C, Utschick W (2013 b) Ergodic robust rate balancing for rank-one vector broadcast channels via sequential approximations. In: Proceedings of the 38th international conference on acoustics, speech, and signal processing, pp 4744–4748
35. Gründinger A, Pickart J, Joham M, Utschick W (2015) A probabilistic downlink beamforming approach with multiplicative and additive channel errors. In: Proceedings of the 49th annual conference on information sciences and systems
36. Hellings C, Joham M, Utschick W (2011) Gradient-based rate balancing for MIMO broadcast channels with linear precoding. In: Proceedings of the 15th international ITG workshop on smart antennas
37. Hunger R, Joham M, Utschick W (2009) On the MSE-duality of the broadcast channel and the multiple access channel. IEEE Trans Signal Process 57(2):698–713
38. Imhof JP (1961) Computing the distribution of quadratic forms in normal variables. Biometrika 48(3/4):419–426
39. Joham M, Vonbun M, Utschick W (2010) MIMO BC/MAC MSE duality with imperfect transmitter and perfect receiver CSI. In: Proceedings of the 11th IEEE international workshop on signal processing advances in wireless communications
40. Kelley P (2009) Overview of the DVB-SH specifications. Int J Satell Commun Network 27(4–5):198–214
41. Letzepis N, Grant A (2008) Capacity of the multiple spot beam satellite channel with Rician fading. IEEE Trans Inf Theory 54(11):5210–5222
42. Lutz E (1996) A markov model for correlated land mobile satellite channels. Int J Satell Commun 14(4):333–339
43. Maral G, Bousquet M (2009) Satellite communications systems, 5th edn. Wiley, Chichester, England
44. Marks B, Wright G (1978) Technical Note–A general inner approximation algorithm for nonconvex mathematical programs. Oper Res 26(4):681–683
45. Morello A, Mignone V (2006) DVB-S2: the second generation standard for satellite broad-band services. Proc IEEE 94(1):210–227
46. Moser S (2007) Some expectations of a non-central chi-square distribution with an even number of degrees of freedom. In: IEEE region 10 conference
47. Mutapcic A, Kim SJ, Boyd S (2007) A tractable method for robust downlink beamforming in wireless communications. In: Proceedings of the 41st Asilomar conference on signals, systems, and computers, pp 1224–1228
48. Panagopoulos A, Arapoglou PD, Cottis P (2004) Satellite communications at KU, KA, and V bands: propagation impairments and mitigation techniques. IEEE Commun Surv Tutorials 6(3):2–14
49. Rao S (2003) Parametric design and analysis of multiple-beam reflector antennas for satellite communications. IEEE Antennas Propag Mag 45(4):26–34
50. Razaviyayn M, Hong M, Luo ZQ (2011) Linear transceiver design for a MIMO interfering broadcast channel achieving max-min fairness. In: Proceedings of the 45th Asilomar conference on signals, systems, and computers, pp 1309–1313

51. Recommendation ITU-R, P618–11, (2013) Propagation data and prediction methods required for the design of earth-space telecommunication systems. Switzerland, Geneva
52. Recommendation ITU-R, P676–10, (2013) Attenuation by atmospheric gases. Switzerland, Geneva
53. Recommendation ITU-R, P681–7, (2009) Propagation data required for the design of earth-space land mobile telecommunication systems. Switzerland, Geneva
54. Recommendation ITU-R, P840–3, (1999) Attenuation due to clouds and fog. Switzerland, Geneva
55. Schubert M, Boche H (2004) Solution of the multiuser downlink beamforming problem with individual SINR sonstraints. IEEE Trans Veh Technol 53(1):18–28
56. Schubert M, Boche H (2005) Iterative multiuser uplink and downlink beamforming under SINR constraints. IEEE Trans Veh Technol 53(7):2324–2334
57. Scutari G, Facchinei F, Song P, Palomar D, Pang JS (2013) Decomposition by partial linearization in multiuser systems. In: Proceedings of the 38th international conference on acoustics, speech, and signal processing, pp 4424–4428
58. Sharma S, Chatzinotas S, Ottersten B (2014) Cognitive beamhopping for spectral coexistence of multibeam satellites. Int J Satell Commun Network 1–23
59. Shi S, Schubert M, Boche H (2007) Downlink MMSE transceiver optimization for multiuser MIMO systems: duality and sum-MSE minimization. IEEE Trans Signal Process 55(11):5436–5446
60. Song E, Shi Q, Sanjabi M, Sun R, Luo ZQ (2011) Robust SINR-constrained MISO downlink beamforming: when is semidefinite programming relaxation tight? In: Proceedings of the 36th international conference on acoustics, speech, and signal processing, pp 3096–3099
61. Thompson P, Evans B, Castenet L, Bousquet M, Mathiopoulos T (2011) Concepts and technologies for a terabit/s satellite. In: Proceedings of the 3rd international conference on advances in satellite and space communications, IARIA XPS Press, pp 12–19
62. Tölli A, Codreanu M, Juntti M (2008) Linear multiuser MIMO transceiver design with quality of service and per-antenna power constraints. IEEE Trans Signal Process 56(7):3049–3055
63. Vidal O, Verelst G, Lacan J, Alberty E, Radzik J, Bousquet M (2012) Next generation high throughput satellite system. In: Proceedings of the 1st AESS European conference on satellite telecommunications
64. Wang KY, Chang TH, Ma WK, So AC, Chi CY (2011) Probabilistic SINR constrained robust transmit beamforming: a Bernstein-type inequality based conservative approach. In: Proceedings of the 36th international conference on acoustics, speech, and signal processing, pp 3080–3083
65. Wang KY, So AC, Chang TH, Ma WK, Chi CY (2014) Outage constrained robust transmit optimization for multiuser MISO downlinks: tractable approximations by conic optimization. IEEE Trans Signal Process 62(21):5690–5705
66. Wiesel A, Eldar Y, Shamai S (2006) Linear precoding via conic optimization for fixed MIMO receivers. IEEE Trans Signal Process 54(1):161–176
67. Yates RD (1995) A framework for uplink power control in cellular radio systems. IEEE J Sel Areas Commun 13(7):1341–1347
68. Yu W, Lan T (2007) Transmitter optimization for the multi-antenna downlink with per-antenna power constraints. IEEE Trans Signal Process 55(6):2646–2660
69. Zheng G, Chatzinotas S, Ottersten B (2012) Generic optimization of linear precoding in multibeam satellite systems. IEEE Trans Wirel Commun 11(6):2308–2320
70. Zorba N, Realp M, Perez-Neira A (2008) An improved partial CSIT random beamforming for multibeam satellite systems. In: Proceedings of the 10th international workshop on signal processing for space communications

Chapter 8
Distributed Channel Selection for Underlay Device-to-Device Communications: A Game-Theoretical Learning Framework

Setareh Maghsudi and Sławomir Stańczak

Abstract Device-to-device (D2D) communications underlaying a cellular infrastructure is one of the key technology enablers for future wireless networks. The basic idea consists in enabling suitably-selected nearby device pairs to reuse the cellular spectrum for direct data transfer, while ensuring that there is no detrimental impact on cellular transmissions via base stations. Despite its great potential, D2D communications poses some challenges to system designers, which are exacerbated by the lack of channel state information for direct D2D links. Therefore there is a strong need for distributed D2D resource allocation solutions that beneficially exploit some side information made available through the cellular network assistance. Such D2D solutions must be also capable of dealing with inherent uncertainty and competition in wireless networks. We model a distributed D2D wireless network as a multi-agent system, in which a set of self-interested agents with bounded rationality share limited spectrum resources, by taking actions according to some decision making strategy, targeting maximum utility. By incorporating a learning model into a game-theoretical formulation, the agents' actions evolve over time as a function of past outcomes and observed side information. Therefore the distributed channel selection problem can be rephrased as a sequential online decision making problem, to be solved by strategies whose outcomes are some sort of equilibria. In this chapter we study some decision making strategies that can be applied to solve a distributed channel selection problem.

S. Maghsudi (✉) · S. Stańczak
Chair for Communications and Information Theory, Technical University of Berlin, Berlin, Germany
e-mail: setareh.maghsudi@tu-berlin.de

S. Stańczak
Fraunhofer Institute for Telecommunications Heinrich Hertz Institute, Berlin, Germany
e-mail: slawomir.stanczak@hhi.fraunhofer.de

W. Utschick (ed.), *Communications in Interference Limited Networks*,
Signals and Communication Technology, DOI 10.1007/978-3-319-22440-4_8

8.1 Introduction

Due to the ever-increasing need for mobile services, we expect a massive growth in demand for wireless access in the years to come. In order to keep pace with this growth, future mobile networks will have to accommodate new communications and networking paradigms, such as device-to-device (D2D) communications as an underlay for cellular wireless networks. The basic idea of such underlay D2D communications is to replace a part of traditional end-to-end connections via access points, base stations (BS) or relays, by direct short-distance communications links between suitably-selected nearby wireless devices that reuse the cellular spectrum. The major potential advantages of this approach stem from the proximity-, hop-, and reuse gains which can be translated to a better network performance expressed in terms of coverage, capacity, energy efficiency and quality-of-service (QoS) [3].

In order to make the vision of underlay D2D communications a reality, researchers and system designers need to address some fundamental research challenges. Some of the central challenges pertain to the difficulty in obtaining channel state information (CSI). Indeed, while reliable CSI for cellular users[1] is usually available at the serving BS, acquiring such information for D2D links at the network side is expected to be a highly challenging task. This is mainly because pilot-based measurements for D2D channels are costly in terms of communications resources and control overhead, and therefore highly undesired. A consequence is that the allocation of resources to D2D links (users) has to be performed autonomously at the level of D2D devices. While doing so, it must be ensured that there is no detrimental impact on the prioritized cellular transmissions. Consequently, it is evident that efficient and robust D2D communications design needs to deal with the following issues:

- uncertainty and the lack of knowledge, as well as the existence of different types of randomness in a network, and
- competition between users that attempt to access strictly limited wireless resources.

The above discussion shows that an underlay D2D communications network can be seen as a multi-agent system, in which each device pair (agent), independently from other agents, takes actions to access available shared radio resources (such as frequency channels). Each joint action profile corresponds to some utility vector so that for any given agent, each action is associated with a utility (or cost) that is a function of (possibly random) network characteristics, as well as other agents' actions. Due to the lack of information, however, the actions' reward generating functions are not known to agents a priori. In the absence of prior information and a centralized network controller, we model device pairs as selfish learning agents that play a strategic game repeatedly.

To be more precise, consider a set $\mathcal{K}$ of agents that are indexed from 1 to K, where every agent has access to a set $\mathcal{M}$ of actions indexed from 1 to M.

[1]In this chapter, any wireless device that operates in the traditional cellular mode is referred to as a cellular user. Accordingly, any *pair* of wireless devices that communicate directly are counted as a D2D user.

Then a typical game-theoretical learning model can be formally defined as $\mathfrak{L}_{\mathfrak{G}} := \{\mathcal{K}, \mathcal{M}, \{f_m^{(k)}\}_{k\in\mathcal{K},m\in\mathcal{M}}, \{\mathbf{x}^{(k)}\}_{k\in\mathcal{K}}\}$, where

- $f_m^{(k)}$ is the mean reward process of action $m \in \mathcal{M}$ for player $k \in \mathcal{K}$, and
- $\mathbf{x}^{(k)}$ is the observations vector of player $k \in \mathcal{K}$.

Since there is no prior information, every agent interacts with a random environment and other agents in order to solve an optimization problem that arises in an unknown reactive model. Common examples of such a problem are the long-term accumulated reward maximization and the average regret minimization. Both approaches give rise to a distributed sequential decision making problem in a multi-agent system, which is a clear instance of *exploration-exploitation dilemma* [12]. The dilemma boils down to finding a balance between selecting actions that yield immediate reward (exploitation or control) and selecting actions that might provide benefit in the future; for instance, gathering information (exploration or learning) does not provide immediate reward but the gathered additional information can be utilized to improve future rewards. A highly relevant instance of the exploration-exploitation dilemma is the trade-off between data transmission and pilot-based channel estimation in wireless networks. In this case, each player is a wireless user that gathers channel state information by using a pilot signal for estimating its time-varying channel states. This information can be exploited to enhance future data transmissions. However, since the channel estimation requires communication resources that cannot be used for the data transmission, utility maximization requires finding an optimal trade-off between the data transmission that exploits the wireless communication environment and the channel estimation that explores this environment to have a better knowledge about available resources. For the learning agents that strive for asymptotic optimality, the underlying resource allocation problem can therefore be stated as a problem of developing decision making strategies, according to which the players' actions evolve sequentially, by observing the past outcomes of the game and (probably) some side information. In doing so, equilibrium arises as an asymptotic outcome of repeated interactions in a random environment among learning agents with bounded rationality that aim at achieving long run optimality in a pre-defined sense. This chapter reviews some recently-developed convergent online decision making strategies that can be utilized to solve the channel selection problem.

8.2 Problem Statement

In this section, we first describe the system model and then formulate the channel selection problem as a game with incomplete information.

8.2.1 System Model

We consider a distributed D2D communications system as an underlay to a single-cell wireless network with a set $\mathcal{M}$ of M licensed orthogonal channels (frequency bands). The D2D system consists of a set $\mathcal{K}$ of K suitably-selected single-antenna transmitter-receiver pairs. Each pair is referred to as a D2D user, and is denoted either by $k \in \mathcal{K}$ or by the pair (k, k').

Assumption 8.2.1 (*No collision among D2D and cellular users*) We eliminate adverse effects of D2D transmissions on cellular users by assuming that a channel is available to D2D users only if it is not occupied by any cellular user.

Moreover, since the D2D data is not forwarded via the BS, which implies that a conventional pilot signal broadcasted by the BS cannot be used for channel estimation, it is reasonable to make the following assumption.

Assumption 8.2.2 (*Channel and network knowledge*) The BS has no knowledge of the D2D channel states/qualities, while the D2D users have neither channel (quality and availability) nor network knowledge. In contrast, the BS has the knowledge of channel availability since it observes the transmission channels of all D2D and cellular users.

D2D users do not exchange any information but they might have some side information.

Assumption 8.2.3 (*Side information*) There *might* exist a channel through which the BS broadcasts some informative signals referred to as side information. Based on the physical characteristics of the radio propagation medium, the broadcast signal, if any at all, is assumed to be heard by all D2D users.

Throughout the chapter, $|h_{uv,t,x}|^2 \in \mathbb{R}^+$ is used to denote the *average* gain of some channel x between u and v at time t, including path loss and fast fading effects. The variance of zero-mean additive white Gaussian noise (AWGN) is denoted by N_0. We assume that all users transmit at some fixed average power P. Furthermore, suppose that each user, say user k, selects a channel $i^{(k)} \in \mathcal{M}$ to sense. The joint selection profile of all D2D users therefore is denoted by $\mathbf{i} = \left(i^{(1)}, \ldots, i^{(k)}, \ldots, i^{(K)}\right)$, which can be also written as $\mathbf{i} = \left(i^{(k)}, \mathbf{i}^{(-k)}\right)$, with $\mathbf{i}^{(-k)}$ being the joint selection profile of all D2D users except for user k.

Assumption 8.2.4 (*Perfect sensing*) We assume that sensing is perfect and therefore, for any channel i, we define θ_i to be the expected value of a Bernoulli random variable that indicates whether channel i is occupied by some cellular user or not.

If the selected channel is not occupied by any cellular user, then the D2D user transmits. We define the bounded mean utility (reward) function as[2]

$$f_t^{(k)}(\mathbf{i}) = \log\left(\frac{P|h_{kk',t,i^{(k)}}|^2}{\sum_{q\in\mathcal{Q}^{(k)}} P|h_{qk',t,i^{(k)}}|^2 + N_0}\right)\theta_{i^{(k)}}, \tag{8.1}$$

where $\mathcal{Q}^{(k)}$ denotes the set of D2D users that share a channel with user k. For simplicity, we assume $f_t^{(k)}(\mathbf{i}) \in [0, 1]$.

8.2.2 Game-Theoretical Problem Formulation

It follows from (8.1) that the utility of each D2D user is in general a function of the joint action profile of all users. Moreover, each user is able to observe only its own utility (e.g., transmission rate) achieved by transmitting through the selected channel, and not those of other channels. To account for this, we model the network as a multi-agent system. Under this model, the resource allocation problem can be formulated as a game with incomplete information, where each user is a player that competes for higher utility by selecting the optimal action, here being the transmission channel. Each player observes only its own reward and the rewards of other actions, i.e., those not played by the player, are not revealed. Such a setting is in general referred to as the *bandit setting*.

Let $g_t^{(k)}(\mathbf{i}_t)$ denote the achieved reward of player k at time t, as a function of joint action profile at time t, i.e., $\mathbf{i}_t$. We consider a game with noisy rewards so that for all $\mathbf{i}$ and every time t, we have $g_t^{(k)}(\mathbf{i}) = f_t^{(k)}(\mathbf{i}) + \mathrm{c}^{(k)}(\mathbf{i})$, with $\mathrm{c}^{(k)}(\mathbf{i})$ being some zero-mean random variable with bounded variance, independent of all other random variables. Ideally, the objective of each D2D user is to maximize its own accumulated reward, which is formally expressed as

$$\underset{\left\{i_t^{(k)}\right\}_{t=1}^{n}}{\text{maximize}} \sum_{t=1}^{n} g_t^{(k)}(\mathbf{i}_t), \tag{8.2}$$

where $i_t^{(k)} \in \mathcal{M}$ is the selected channel of D2D user k at time t. Due to the lack of information, however, the objective function in (8.2) is not available. For this reason, we argue for a less ambitious goal, which is known as *regret minimization*. First, we assume that at each time t, any player k selects a probability distribution (mixed strategy) over all frequency channels (actions), $\mathbf{p}_t^{(k)} = \left(p_{1,t}^{(k)}, \ldots, p_{i,t}^{(k)}, \ldots, p_{M,t}^{(k)}\right)$, according to which a channel is chosen. We also define the external regret as follows.

[2]Throughout this chapter logarithms are natural unless otherwise is stated.

Definition 8.2.1 (*External regret* [2]) The external cumulative regret of player $k \in \mathcal{K}$ up to time n is defined to be

$$\begin{aligned} R_{\text{Ext}}^{(k)} := R_{\text{Ext}}^{(k)}(n) &= \max_{i \in \mathcal{M}} \sum_{t=1}^{n} g_t^{(k)}\left(i, \mathbf{i}_t^{(-k)}\right) - \sum_{t=1}^{n} \bar{g}_t^{(k)}\left(\mathbf{p}_t^{(k)}, \mathbf{i}_t^{(-k)}\right) \\ &= \max_{i \in \mathcal{M}} \sum_{t=1}^{n} \sum_{j \in \mathcal{M}} p_{j,t}^{(k)}\left(g_t^{(k)}\left(i, \mathbf{i}_t^{(-k)}\right) - g_t^{(k)}\left(j, \mathbf{i}_t^{(-k)}\right)\right), \end{aligned} \tag{8.3}$$

where $\bar{g}_t^{(k)}(\cdot)$ denotes the *expected* reward at round t by using mixed strategy $\mathbf{p}_t^{(k)}$, defined as $\bar{g}_t^{(k)}(\cdot) = \sum_{j \in \mathcal{M}} g_t^{(k)}(j, \cdot) p_{j,t}^{(k)}$.

D2D user (player) k then attempts to achieve a per-round vanishing external regret in the sense that

$$\lim_{n \to \infty} \frac{1}{n}\left(\max_{i \in \mathcal{M}} \sum_{t=1}^{n} g_t^{(k)}\left(i, \mathbf{i}_t^{(-k)}\right) - \sum_{t=1}^{n} \bar{g}_t^{(k)}\left(\mathbf{p}_t^{(k)}, \mathbf{i}_t^{(-k)}\right)\right) = 0. \tag{8.4}$$

The formulation in (8.4) gives rise to a sequential decision making problem in a multi-player competitive environment in the absence of channel and network knowledge, as described in Sect. 8.1. In addition, at the network level, there is a need for a global goal to achieve some steady state, i.e., equilibrium.

In the remainder of this chapter, we present some distributed algorithmic solutions to the formulated channel selection problem that satisfy a twofold objective: (i) the regret of each user vanishes asymptotically according to (8.4), and (ii) the actions of all players converge to equilibrium in some sense. Since the corresponding strategy for resource allocation depends in general on various network's characteristics such as type of randomness and information availability, we consider two specific types of networks, namely: (i) adversarial network without side information and (ii) stochastic network with side information.

8.3 Adversarial Network

Assume that $f_t^{(k)}(\mathbf{i})$, given by (8.1), does not follow any specific density function, which means that the series of instantaneous rewards achieved by playing actions cannot be attributed to a specific distribution. In this case, the network is referred to as an adversarial network, as the players' rewards change adversarially and not according to a specific rule. Also, in order to maintain the most general formulation, it is assumed that *no* side information is available to the players. Under these assumptions, we describe two selection (decision making) strategies that not only satisfy (8.4) for all players, but also guarantee that the game converges to equilibrium in some sense.

8.3.1 Preliminaries

Before we proceed with the design of the selection strategies, we state some important definitions and results.

By Definition 8.2.1, external regret compares the expected reward of the current mixed strategy with that of the best fixed action; however, it fails to compare actions in pairs. For this reason, *internal regret* is introduced, which is closely related to the concept of equilibrium in games.

Definition 8.3.1 (*Internal regret* [2]) The internal cumulative regret of player k up to time n is defined as

$$\begin{aligned} R^{(k)}_{\text{Int}} := R^{(k)}_{\text{Int}}(n) &= \max_{i,j\in\mathcal{M}} R^{(k)}_{(i\to j),n} \\ &= \max_{i,j\in\mathcal{M}} \sum_{t=1}^{n} p^{(k)}_{i,t}\left(g^{(k)}_t\left(j,\mathbf{i}^{(-k)}_t\right) - g^{(k)}_t\left(i,\mathbf{i}^{(-k)}_t\right)\right). \end{aligned} \quad (8.5)$$

In the right-hand side of (8.5), $r^{(k)}_{(i\to j),t} = p^{(k)}_{i,t}\left(g^{(k)}_t(j,\cdot) - g^{(k)}_t(i,\cdot)\right)$ denotes the expected regret caused by selecting action i instead of action j. A comparison of (8.3) and (8.5) shows that the external regret is bounded above by the internal one since we have [13]

$$R^{(k)}_{\text{Ext}} = \max_{i\in\mathcal{M}} \sum_{j\in\mathcal{M}} R^{(k)}_{(i\to j),n} \le M \max_{i,j\in\mathcal{M}} R^{(k)}_{(i\to j),n} = M R^{(k)}_{\text{Int}}. \quad (8.6)$$

Remark 8.3.1 Throughout this chapter, *vanishing (zero-average) external and internal regret* means that $\lim_{n\to\infty}\frac{1}{n}R_{\text{Ext}} = 0$ and $\lim_{n\to\infty}\frac{1}{n}R_{\text{Int}} = 0$ or, in other words, $R_{\text{Ext}} \in o(n)$ and $R_{\text{Int}} \in o(n)$, respectively. By (8.6), $R_{\text{Int}} \in o(n)$ implies $R_{\text{Ext}} \in o(n)$. We call any strategy with $R_{\text{Int}} \in o(n)$ as "no-regret strategy".

As aforementioned, internal regret is closely related to the concept of equilibria in games, described formally by the following theorem.

Theorem 8.3.1 ([2]) *Consider a K-player game, where players are provided with an action set $\mathcal{M}$. Denote the internal regret of player k by $R^{(k)}_{Int}$, and the set of correlated equilibria by $\mathcal{C}$. At time n, define the empirical joint distribution of the game as*

$$\hat{\pi}_n(\mathbf{j}) = \frac{1}{n}\sum_{t=1}^{n} \mathbf{1}_{\{\mathbf{i}_t=\mathbf{j}\}}, \quad \mathbf{j} = \left(j^{(1)},\ldots,j^{(K)}\right) \in \bigotimes_{k=1}^{K}\mathcal{M}, \quad (8.7)$$

where $\mathbf{1}_{\{x\}}$ is the indicator function that returns one if x holds and zero otherwise. If all players $k \in \mathcal{K}$ play according to some strategy so that

$$\lim_{n\to\infty}\frac{1}{n}R^{(k)}_{Int} = 0, \quad (8.8)$$

then the distance $\inf_{\pi \in \mathcal{C}} \sum_{\mathbf{j}} \left|\hat{\pi}_n(\mathbf{j}) - \pi(\mathbf{j})\right|$ *between the empirical joint distribution of plays and the set of correlated equilibria converges to* 0 *almost surely.*

8.3.1.1 From Vanishing External Regret to Vanishing Internal Regret

Reference [13] proposed an approach for converting any selection strategy with vanishing external regret to another approach with vanishing internal regret. We describe this approach briefly in what follows. The player index (k) is omitted for convenience.

Consider some selection strategy κ that at each time t selects one of the M actions according to some probability distribution $\mathbf{p}_t$. Let $\mathbf{p}_1$ be the uniform distribution. In order to calculate $\mathbf{p}_t$ for $t > 1$, κ constructs a meta-strategy κ' with $M(M-1)$ virtual actions $(i \to j)$, $(i, j \in \mathcal{M}, i \neq j)$. Assume that κ' uses some mixed strategy $\mathbf{w}_t$ over $M(M-1)$ virtual actions, where the probability of the virtual action $(i \to j)$, i.e., $w_{(i\to j),t}$, depends on its past performance in some way.[3] Given $\mathbf{w}_t$ and $\mathbf{p}_{t-1} = \left(p_{1,t-1}, \ldots, p_{i,t-1}, \ldots, p_{j,t-1}, \ldots, p_{M,t-1}\right)$, κ defines $\mathbf{p}_{(i\to j),t-1}$, which has 0 and $p_{j,t-1} + p_{i,t-1}$ at the place of $p_{i,t-1}$ and $p_{j,t-1}$, respectively, and all other elements remain unchanged; that is, $\mathbf{p}_{(i\to j),t-1} = \left(p_{1,t-1}, \ldots, 0, \ldots, p_{j,t-1} + p_{i,t-1}, \ldots, p_{M,t-1}\right)$. Then $\mathbf{p}_t = \sum_{(i,j):i\neq j} \mathbf{p}_{(i\to j),t-1} w_{(i\to j),t}$. As a result, κ has the property that its internal regret is upper-bounded by the external one of κ'. So if κ' exhibits a vanishing external regret, then the internal regret of κ vanishes as well. In the next Section, we use this property in order to develop no-regret selection strategies.

8.3.2 No-Regret Bandit Exponential-Based Weighted Average Strategy

The selection strategy presented in this section is an adapted version of the general *exponential-based weighted average* strategy. The core concept is to play each action with some probability that is directly related to its exponentially-weighted past performance [7]. In other words, if playing an action has resulted in large regret in the past, its future selection probability is small, and *vice versa*. However, as described in Sect. 8.2, in the bandit setting considered here, players only observe the reward of the played action, and not those of others. Therefore the reward of each action i is estimated as [2][4]

$$\tilde{g}_t^{(k)}(i) = \begin{cases} \frac{g_t^{(k)}(i)}{p_{i,t}^{(k)}} & i = i_t^{(k)} \\ 0 & \text{o.w.} \end{cases}, \tag{8.9}$$

[3]The gains of virtual actions cannot be calculated explicitly. Later we see that the gain achieved by a virtual action $(i \to j)$ is calculated based on the gain achieved by playing true actions i and j.

[4]It is clear that such estimation is not required in the *full information* setting, where after each trial, the rewards of all actions, even those not played by the player, are revealed.

which is an unbiased estimate of the true reward of action i; that is, $\mathrm{E}\{\tilde{g}^{(k)}(i)\} = g^{(k)}(i)$, where the expectation, denoted by $\mathrm{E}\{\cdot\}$, is with respect to the distribution $\mathbf{p}_t$ of the random variable $i_t^{(k)}$. Estimated rewards are afterwards used in order to calculate regret values. For example, the regret of *not* playing action j instead of action i yields

$$\tilde{R}^{(k)}_{(i\to j),t-1} = \sum_{s=1}^{t-1} \tilde{r}^{(k)}_{(i\to j),s} = \sum_{s=1}^{t-1} p^{(k)}_{i,s}\left(\tilde{g}^{(k)}_s(j) - \tilde{g}^{(k)}_s(i)\right). \tag{8.10}$$

It is known that despite exhibiting a vanishing external regret, exponential-based weighted average strategy lead in general to a large internal regret [2]; consequently, it does not guarantee a convergence to equilibrium. However, the bandit version of this algorithm [1] can be improved to exhibit small internal regret, using the approach of Sect. 8.3.1.1. The strategy is called no-regret bandit exponentially-weighted average strategy (NR-BEWAS) [9], and is described in Algorithm 1.

Algorithm 1 NR-BEWAS [9]

1: Define γ_t and η_t as given in Proposition 8.3.1.
2: Define $\Phi(\mathbf{u}) = \frac{1}{\eta_t}\log\left(\sum_{i\in\mathcal{M}}\exp(\eta_t u_i)\right)$, where $\mathbf{u} = (u_1, ..., u_M) \in \mathbb{R}^M$.
3: Let $\mathbf{p}_1^{(k)} = \left(\frac{1}{M}, ..., \frac{1}{M}\right)$, i.e., uniform distribution.
4: Select an action using $\mathbf{p}_1^{(k)}$.
5: Play and observe the reward.
6: **for** $t = 2, ..., n$ **do**
7: Let $\mathbf{p}_{t-1}^{(k)} = \left(p^{(k)}_{1,t-1}, ..., p^{(k)}_{i,t-1}, ..., p^{(k)}_{j,t-1}, ..., p^{(k)}_{M,t-1}\right)$ be the mixed strategy at time $t-1$. Construct $\mathbf{p}^{(k)}_{(i\to j),t-1}$ as follows: replace $p^{(k)}_{i,t-1}$ in $\mathbf{p}^{(k)}_{t-1}$ by zero, and instead increase $p^{(k)}_{j,t-1}$ to $p^{(k)}_{j,t-1} + p^{(k)}_{i,t-1}$. Other elements remain unchanged. We obtain $\mathbf{p}^{(k)}_{(i\to j),t-1} = \left(p^{(k)}_{1,t-1}, ..., 0, ..., p^{(k)}_{j,t-1} + p^{(k)}_{i,t-1}, ..., p^{(k)}_{M,t-1}\right)$.
8: Define [2]

$$w^{(k)}_{(i\to j),t} = \frac{\exp\left(\eta_t \tilde{R}^{(k)}_{(i\to j),t-1}\right)}{\sum_{(m\to l):m\neq l}\exp\left(\eta_t \tilde{R}^{(k)}_{(m\to l),t-1}\right)}, \tag{8.11}$$

where $\tilde{R}^{(k)}_{(i\to j),t-1}$ is calculated by using (8.9) and (8.10).
9: Given $w^{(k)}_{(i\to j),t}$, solve the following fixed point equation to find $\mathbf{p}_t^{(k)}$:

$$\mathbf{p}_t^{(k)} = \sum_{(i\to j):i\neq j} \mathbf{p}^{(k)}_{(i\to j),t} w^{(k)}_{(i\to j),t}. \tag{8.12}$$

10: Final probability distribution yields

$$\mathbf{p}_t^{(k)} = (1-\gamma_t)\mathbf{p}_t^{(k)} + \frac{\gamma_t}{M}. \tag{8.13}$$

11: Using the final $\mathbf{p}_t^{(k)}$, given by (8.13), select an action.
12: Play and observe the reward.
13: **end for**

Algorithm 1 needs two parameters γ_t and η_t. The following proposition states how to select these parameters so as to guarantee a vanishing internal regret.

Proposition 8.3.1 ([9]) *If* $\eta_t = \frac{\gamma_t^3}{M^2}$ *and* $\gamma_t = t^{-\frac{1}{3}}$, *then Algorithm 1 exhibits a vanishing internal regret so that* $R_{Int}^{(k)} \in o(n)$.

Corollary 8.3.1 *Consider a K-player game where players have access to a set* $\mathcal{M}$ *of actions. Let* $\mathcal{C}$ *denote the set of correlated equilibria and* $\mathbf{i} = \left(i^{(1)}, ..., i^{(K)}\right) \in \bigotimes_{k=1}^{K} \mathcal{M}$. *Let the empirical joint frequencies of play be defined by* (8.7). *If all players play according to the NR-BEWAS, then the distance* $\inf_{\pi \in \mathcal{C}} \sum_{\mathbf{i}} \left|\hat{\pi}_n(\mathbf{i}) - \pi(\mathbf{i})\right|$ *between the empirical joint distribution of plays and the set of correlated equilibria converges to* 0 *almost surely as* $n \to \infty$.

8.3.3 No-Regret Bandit Follow the Perturbed Leader Strategy

As in the case of exponential-based weighted average strategy presented in the previous section, the strategy *follow the perturbed leader* is an approach to solve online decision making problems. In the basic version of this approach, called *follow the leader* [5], the action with the best past performance is selected at each time. This rule is however deterministic so that it does not achieve vanishing regret against non-oblivious opponents. Therefore, in *follow the perturbed leader*, the player adds a random perturbation to the vector of accumulated regrets, and the action with the minimum perturbed regret in the past is selected [2]. Reference [8] developed a bandit version of this algorithm with vanishing external regret. Similar to NR-BEWAS, the algorithm in [8] can be modified to ensure a vanishing internal regret, and with it, a convergence to equilibrium. The approach is called no-regret bandit follow the perturbed leader strategy (NR-BFPLS), and is summarized in Algorithm 2.

Algorithm 2 needs the knowledge of the probability assigned to each action by the follow the perturbed leader strategy at each time, which is calculated as described below in the following. From (8.14), the selection probability of virtual action $(i \to j) \in \{1, \ldots, M(M-1)\}$ is the probability that $\tilde{R}_{(i \to j),t-1}$ plus perturbation $q_{(i \to j),t}$ is larger than those of other actions, i.e.,

Algorithm 2 NR-BFPLS [9]

1: Define $\varepsilon_t = \varepsilon_n = \frac{\sqrt{\log(n)}}{3\sqrt{Mn}}$, and $\gamma_t = \min(1, M\varepsilon_t)$. Note that unlike NR-BEWAS, here we know the game horizon, n, in advance.
2: Let $\mathbf{p}_1^{(k)} = \left(\frac{1}{M}, \ldots, \frac{1}{M}\right)$, i.e., uniform distribution.
3: Select an action using $\mathbf{p}_1^{(k)}$.
4: Play and observe the reward.
5: **for** $t = 2, \ldots, n$ **do**
6: Let $\mathbf{p}_{t-1}^{(k)} = \left(p_{1,t-1}^{(k)}, \ldots, p_{i,t-1}^{(k)}, \ldots, p_{j,t-1}^{(k)}, \ldots, p_{M,t-1}^{(k)}\right)$ be the mixed strategy at time $t-1$. Construct $\mathbf{p}_{(i\to j),t-1}^{(k)}$ as follows: replace $p_{i,t-1}^{(k)}$ in $\mathbf{p}_{t-1}^{(k)}$ by zero, and instead increase $p_{j,t-1}^{(k)}$ to $p_{j,t-1}^{(k)} + p_{i,t-1}^{(k)}$. Other elements remain unchanged. We obtain $\mathbf{p}_{(i\to j),t-1}^{(k)} = \left(p_{1,t-1}^{(k)}, \ldots, 0, \ldots, p_{j,t-1}^{(k)} + p_{i,t-1}^{(k)}, \ldots, p_{M,t-1}^{(k)}\right)$.
7: Calculate $\tilde{R}_{(i\to j),t-1}^{(k)}$ using (8.9) and (8.10).
8: Define $\sigma_{(i\to j),t-1} = \left(\sum_{s=1}^{t-1} \frac{1}{w_{(i\to j),s}^{(k)}}\right)^{\frac{1}{2}}$, which is the upper-bound of conditional variances of random variables $\tilde{R}_{(i\to j),t-1}^{(k)}$ [8].
9: Let $\tilde{R}_{(i\to j),t-1}^{(k)} = \tilde{R}_{(i\to j),t-1}^{(k)} - \sqrt{1+\sqrt{2/M}}\, \sigma_{(i\to j),t-1} \sqrt{\log(t)}$ [8].
10: Randomly select a perturbation vector $\mathbf{q}_t$ with $M(M-1)$ elements from two-sided exponential distribution with width ε_t.
11: Consider a selection rule which selects the action $(i \to j)$ given by

$$\operatorname{argmax} \left\{\tilde{R}_{(i\to j),t-1}^{(k)} + q_{(i\to j),t}\right\}, \quad (i \to j) \in \{1, \ldots, M(M-1)\} \tag{8.14}$$

Note that in our setting $\tilde{R}_{(i\to j)}$ denotes the estimated regret of *not* playing action $(i \to j)$; hence we find the action with largest $\tilde{R}$.
12: By using (8.17), calculate the probability $w_{(i\to j),t}^{(k)}$ assigned to each pair $(i \to j)$.
13: Given $w_{(i\to j),t}^{(k)}$, solve the following fixed point equation to find $\mathbf{p}_t^{(k)}$.

$$\mathbf{p}_t^{(k)} = \sum_{(i\to j):i\neq j} \mathbf{p}_{(i\to j),t}^{(k)} w_{(i\to j),t}^{(k)}. \tag{8.15}$$

14: Final probability distribution yields

$$\mathbf{p}_t^{(k)} = (1-\gamma_t)\mathbf{p}_t^{(k)} + \frac{\gamma_t}{M}. \tag{8.16}$$

15: Using the final $\mathbf{p}_t^{(k)}$, given by (8.16), select an action.
16: Play and observe the reward.
17: **end for**

$$
\begin{aligned}
&\Pr[i_t = (i \to j)] = w^{(k)}_{(i\to j),t-1} \\
&= \Pr[\tilde{R}^{(k)}_{(i\to j),t-1} + q_{(i\to j),t} \geq \tilde{R}^{(k)}_{(i'\to j'),t-1} + q_{(i'\to j'),t}\ \forall (i \to j) \neq (i' \to j')] \\
&= \int_{-\infty}^{\infty} \Pr[\tilde{R}^{(k)}_{(i\to j),t-1} + q_{(i\to j),t} = m \wedge \tilde{R}^{(k)}_{(i'\to j'),t-1} + q_{(i'\to j'),t} \\
&\leq m\ \forall (i \to j) \neq (i' \to j')]dm \\
&= \int_{-\infty}^{\infty} \Pr[\tilde{R}^{(k)}_{(i\to j),t-1} + q_{(i\to j),t} = m] \prod_{(i'\to j')\neq(i\to j)} \Pr[\tilde{R}^{(k)}_{(i'\to j'),t-1} + q_{(i'\to j'),t} \\
&\leq m]dm. \qquad (8.17)
\end{aligned}
$$

Since $\mathbf{q}_t$ is distributed according to a two-sided exponential distribution with width ε_n, the terms under integral can be calculated easily (see [6], for example).

Now we are in a position to state some properties of NR-BFPLS (Algorithm 2).

Proposition 8.3.2 ([9]) *Let* $\varepsilon_t = \varepsilon = \frac{\sqrt{\log(n)}}{3\sqrt{Mn}}$ *and* $\gamma_t = \gamma = \min(1, M\varepsilon_t)$. *Then Algorithm 2 yields vanishing internal regret, i.e.,* $R^{(k)}_{Int} \in o(n)$.

Corollary 8.3.2 ([9]) *Consider a K-player game where players are provided with a set* $\mathcal{M}$ *of actions. Let* C *denote the set of correlated equilibria and* $\mathbf{i} = \left(i^{(1)}, \ldots, i^{(K)}\right) \in \bigotimes_{k=1}^{K} \mathcal{M}$. *Define the empirical joint frequencies of play as* (8.7). *If all players play according to the NR-BFPLS, then the distance* $\inf_{\pi\in C} \sum_{\mathbf{i}} \left|\hat{\pi}_n(\mathbf{i}) - \pi(\mathbf{i})\right|$ *between the empirical joint distribution of plays and the set of correlated equilibria converges to* 0 *almost surely as* $n \to \infty$.

8.4 Stochastic Network

In this part, we assume that the mean reward functions defined by (8.1) is time-invariant so that $f^{(k)}_t(\mathbf{i}) = f^{(k)}(\mathbf{i})$. Moreover, it is assumed that the instantaneous rewards of any given action to any player can be attributed to some (unknown) density distribution. In this case, the network is referred to as stochastic. In addition, in this section, we assume that some side information is available to players. More precisely, the players are able to observe the actions of each other. This setting is called *full monitoring*, which stands in contrast to the *no monitoring* setting considered in the previous section. In an underlay D2D network, such side information becomes available to D2D users through *network assistance*, as described in the following. Recall that by Assumptions 8.2.2 and 8.2.3, the BS is able to observe the selected channel of all users and there might be a control channel through it the BS broadcasts some information. In particular, in this section, we assume that *after* each round of transmissions, the BS shares its knowledge (about joint action profile, i.e., selected channels) with all D2D users, by broadcasting this information through the control channel. Since the broadcast signal is heard by all D2D users, at each time, every user is aware of the past joint action profiles. For such a scenario, we aim at developing an efficient decision making strategy that converges to equilibrium in some sense.

8.4.1 Preliminaries

We make a small change in notation and use $f_i^{(k)}\big(\mathbf{i}^{(-k)}\big) := f^{(k)}\big(i, \mathbf{i}^{(-k)}\big)$ to denote the time-invariant mean reward process of some action $i \in \mathcal{M}$ to some player $k \in \mathcal{K}$. Moreover, at trial t, let $i_t^{*(k)} := \arg\max_{i\in\mathcal{M}} f_i^{(k)}\big(\mathbf{i}_t^{(-k)}\big)$, which results in $g_t^{*(k)}\big(\mathbf{i}_t^{(-k)}\big) := g_{i_t^{*(k)},t}^{(k)}\big(\mathbf{i}_t^{(-k)}\big)$. Additionally, we assume that $i_t^{*(k)}$ yields $f^{*(k)}\big(\mathbf{i}_t^{(-k)}\big) := \max_{i\in\mathcal{M}} f_i^{(k)}\big(\mathbf{i}_t^{(-k)}\big)$. In [14], it is shown that (8.4) is equivalent to

$$\lim_{n\to\infty} \frac{1}{n} \sum_{t=1}^{n} \Big(f_{i_t^{(k)}}^{(k)}\big(\mathbf{i}_t^{(-k)}\big) - f^{*(k)}\big(\mathbf{i}_t^{(-k)}\big)\Big) = 0, \tag{8.18}$$

provided that $g_{i,t}^{(k)}\big(\mathbf{i}^{(-k)}\big)$ is bounded above and away from zero. Hence, if a selection strategy satisfies (8.18), then it also satisfies (8.4). Here we assume that $f_i^{(k)}\big(\mathbf{i}^{(-k)}\big)$ and $f^{*(k)}\big(\mathbf{i}^{(-k)}\big)$ obey the following assumption [14].

Assumption 8.4.1 $\forall\, k \in \mathcal{K}, i \in \mathcal{M}$, and $\mathbf{i}^{(-k)} \in \bigotimes_{k=1}^{K-1} \mathcal{M}$,

(a) $f_i^{(k)}\big(\mathbf{i}^{(-k)}\big) \in [0, A]$ for some $A > 0$,
(b) $B = \sup_i \sup_{\mathbf{i}^{(-k)}} \big(f^{*(k)}\big(\mathbf{i}^{(-k)}\big) - f_i^{(k)}\big(\mathbf{i}^{(-k)}\big)\big) < \infty$,
(c) $\mathrm{E}\big\{f^{*(k)}\big(\mathbf{i}_1^{(-k)}\big)\big\} > 0$.

8.4.2 Calibrated Forecaster

In this section, we briefly describe some basic elements of calibrated forecasting. We also clarify how calibrated forecasting is related to the concept of equilibria in games, which is later used for developing a convergent selection strategy.

Following [11], consider a random experiment with a finite set of outcomes $\mathcal{D}$ of cardinality D, and let δ_{d_t} be the Dirac probability distribution on some outcome d at time t. The set of probability distributions over $\mathcal{D}$ is denoted by $\mathcal{P} \subseteq \mathbb{R}^D$. Equip $\mathcal{P}$ with some norm $\|\cdot\|$. At time t, the forecaster outputs a probability distribution $\mathbf{p}_t$ over the set of outcomes.

Definition 8.4.1 (*Calibrated forecaster* [11]) A forecaster is said to be calibrated if $\forall\, \varepsilon > 0$ and $\mathbf{p} \in \mathcal{P}$, almost surely,

$$\lim_{n\to\infty} \left\| \frac{1}{n} \sum_{t=1}^{n} \mathbf{1}_{\{\|\mathbf{p}_t - \mathbf{p}\| \le \varepsilon\}} \big(\mathbf{p}_t - \delta_{d_t}\big) \right\| = 0. \tag{8.19}$$

A relaxed notion of calibration is ε-calibration. Given $\varepsilon > 0$, an ε-calibrated forecaster considers some finite covering of $\mathcal{P}$ by N_ε balls of radius ε. Denoting the centers of these balls by $\mathbf{p}^{(1)}, \ldots, \mathbf{p}^{(N_\varepsilon)}$, the forecaster selects only forecasts $\mathbf{p}_t \in \{\mathbf{p}^{(1)}, \ldots, \mathbf{p}^{(N_\varepsilon)}\}$. Using this, ε-calibration is defined as follows.

Definition 8.4.2 (ε-*Calibrated forecaster* [11]) Define Q_t to be the index in $\{1, \ldots, N_\varepsilon\}$, such that $\mathbf{p}_t = \mathbf{p}^{(Q_t)}$. A forecaster is said to be ε-calibrated if almost surely,

$$\limsup_{n\to\infty} \sum_{q=1}^{N_\varepsilon} \left\| \frac{1}{n} \sum_{t=1}^{n} \mathbf{1}_{\{Q_t=q\}} \left(\mathbf{p}^{(q)} - \delta_{d_t}\right) \right\| \leq \varepsilon. \tag{8.20}$$

The following result describes the relation between calibrated forecasting and correlated equilibria in games.

Theorem 8.4.1 ([4]) *Consider a game with K players provided with an action set $\mathcal{M}$. Let $\mathcal{C}$ stand for the set of correlated equilibria, and define the joint empirical frequencies of play as* (8.7). *For a player k, let $\mathcal{D} = \mathcal{I}^-$ be the set of joint action profiles of opponents. Assume that each player plays by best responding to a calibrated forecast of the opponents joint action profile in a sequence of plays; that is, for each player k we have*

$$i_t^{(k)} = \arg\max_{i\in\mathcal{M}} \sum_{d=1}^{D} p_{d,t}^{(k)} f_i^{(k)}(d), \tag{8.21}$$

where $\mathbf{p}_t^{(k)} = \left(p_{1,t}^{(k)}, \ldots, p_{D,t}^{(k)}\right)$ stands for the output of the forecaster, which is a probability distribution over $D = M^{K-1}$ possible joint action profiles of opponents. Accordingly, each d represents a realization of the joint action profile of opponents of player k, i.e., $\mathbf{i}^{(-k)}$. Then the distance $\inf_{\pi\in\mathcal{C}} \sum_{\mathbf{i}} \left|\hat{\pi}_n(\mathbf{i}) - \pi(\mathbf{i})\right|$ between the empirical joint distribution of plays and the set of correlated equilibria converges to 0 *almost surely as $n \to \infty$.*

8.4.3 Construction of a Calibrated Forecaster

For constructing a calibrated forecaster, in the first step, an ε-calibrated forecaster is constructed for some $\varepsilon > 0$. Then, the time is divided into periods of increasing length, and the procedure of ε-calibration is repeated as a subroutine over the periods, where ε decreases gradually to zero (that is, N_ε-grid becomes finer at each period). In Algorithm 3, we briefly review this procedure. See [11] for details and the proof of calibration.

Theorem 8.4.2 ([11]) *The forecasting procedure (Algorithm 3) is calibrated. That is, it satisfies* (8.19).

Algorithm 3 A Calibrated Forecaster [11]

Define $T_j = 2^j$ where T_j is the length of period $j = 1, 2, ...$

For any period j, define a two-player game, where the first player is the forecaster with the action set $\mathcal{Z} = \{1, ..., N_{\varepsilon_j}\}$ and the second player is the nature with action set $\mathcal{D}$. With respect to our model, the first player is some agent k, while the second player is the set of all other $K-1$ agents; hence $\mathcal{D} = \mathcal{I}^-$ and $D = M^{K-1}$. Accordingly, any outcome d is the realization of a joint action profile of $K-1$ players, that is $\mathbf{i}^{(-k)}$.

For each period j, let $\varepsilon_j = 2^{-j/(D+1)}$.

Define the vector-valued regret of the first player as $\mathbf{u}\{q, d\} = \left(\mathbf{0}, ..., \mathbf{0}, \mathbf{p}^{(q)} - \delta_d, \mathbf{0}, ..., \mathbf{0}\right)$ for each $q \in \{1, ..., N_{\varepsilon_j}\}, d \in \mathcal{D}$.

Define the target set $\mathcal{F}$ as follows:

- Write (DN_{ε_j})-dimensional vectors of $\mathbb{R}^{DN_{\varepsilon_j}}$ as N_{ε_j}-dimensional vectors with components in $\mathbb{R}^D$, i.e., $\mathbf{x} = \left(\mathbf{x}_1, ..., \mathbf{x}_{N_{\varepsilon_j}}\right)$, where $\mathbf{x}_l \in \mathbb{R}^D$ for all $l \in \{1, ..., N_{\varepsilon_j}\}$.
- $\mathcal{F}$ is a subset of the ε_j-ball around $(\mathbf{0}, ..., \mathbf{0})$ for the calibration norm $\|\cdot\|$, which is a closed convex set.

Define the sequence of the vector-valued regrets up to time T $(1 \le T \le T_j)$ as

$$\mathbf{u}_T = \frac{1}{T}\sum_{t=1}^{T}\mathbf{u}\{Q_t, d_t\} = \frac{1}{T}\left(\sum_{t=1}^{T}\mathbf{1}_{\{Q_t=1\}}\left(\mathbf{p}^{(1)} - \delta_t\right), ..., \sum_{t=1}^{T}\mathbf{1}_{\left\{Q_t=N_{\varepsilon_j}\right\}}\left(\mathbf{p}^{(N_{\varepsilon_j})} - \delta_t\right)\right). \tag{8.22}$$

Now, (8.20) (condition of ε_j-calibration) can be restated as the convergence of $\mathbf{u}_{T_j}$ to the set $\mathcal{F}$ almost surely. In the following, $\mathbf{u}^{(j)} := \mathbf{u}_{T_j}$ denotes the final regret of period j.

repeat
  for $t = 1 \to T_j$ **do**
    if $(j = 1 \wedge t = 1)$ **then**
      Select an action Q_t from $\mathcal{Z}$ according to the uniform distribution over the action set, i.e., let $\psi_1 = \left(\frac{1}{N_{\varepsilon_j}}, ..., \frac{1}{N_{\varepsilon_j}}\right)$. Note that ψ_t is the mixed strategy at time t, while $\psi^{(j)} := \psi_{T_j}$ denotes the final mixed strategy of period j.
    else if $(j > 1 \wedge t = 1)$ **then**
      Select an action Q_t from $\mathcal{Z}$ according to a probability distribution in a small neighborhood of $\psi^{(j-1)}$ (localization of search).
    else
      Select an action Q_t from $\mathcal{Z}$ according to a probability distribution $\psi_t = \left(\psi_{t,1}, ..., \psi_{t,N_{\varepsilon_j}}\right)$ on $\{1, ..., N_{\varepsilon_j}\}$ such that $\forall d \in \mathcal{D}$,

$$(\mathbf{u}_{t-1} - \Pi_{\mathcal{F}}(\mathbf{u}_{t-1})) \cdot (\mathbf{u}\{\psi_t, d\} - \Pi_{\mathcal{F}}(\mathbf{u}_{t-1})) \le 0, \tag{8.23}$$

      where $\Pi_{\mathcal{F}}$ denotes the projection in l_2-norm onto $\mathcal{F}$ and $\cdot$ denotes the inner product in $\mathbb{R}^{DN_{\varepsilon_j}}$. See [11] for details.
    end if
  end for
  Calculate the final regret of the current period, $\mathbf{u}^{(j)} = \mathbf{u}_{T_j}$. Also, let $\psi^{(j)} = \psi_{T_j}$.
  if $\mathbf{u}^{(j)} > \varepsilon_j$, **then**
  • Let $j = 1$ and $t = 1$.
  else
  • Let $j = j + 1$ and $t = 1$.
  end if
until convergence (j is large enough so that $\varepsilon_j \approx 0$)

8.4.4 Calibrated Bandit Strategy

By the definition of utility function in Sect. 8.2, the performance of each D2D user depends on two factors: (1) channel quality and availability, and (2) the joint channel selection profile of all D2D users. Given no initial information, in a stochastic setting, the impact of these factors on the average reward are learned over time, and the average reward function is estimated by means of a regression process. Here we make the following assumption [10].

Assumption 8.4.2 The regression process is strongly consistent in L_∞ norm for each $f_i^{(k)}(\mathbf{i}^{(-k)})$; that is, $\| \hat{f}_{i,t}^{(k)}(\mathbf{i}^{(-k)}) - f_i^{(k)}(\mathbf{i}^{(-k)}) \|_\infty \to 0$, for all $i \in \mathcal{M}$, $k \in \mathcal{K}$ and $\mathbf{i}^{(-k)} \in \bigotimes_{k=1}^{K-1} \mathcal{M}$, almost surely as $t \to \infty$, where $\hat{f}_{i,t}^{(k)}(\mathbf{i}^{(-k)})$ denotes the regression estimate of $f_i^{(k)}(\mathbf{i}^{(-k)})$ at the t-th trial.

In what follows, we describe a selection strategy to solve the resource allocation game. The game horizon is first divided into periods $j = 1, 2, \ldots$ of increasing length T_j'. We also define a sequence Z_j for $j = 1, 2, \ldots$, so that T_j' and Z_j satisfy the following assumption [10].

Assumption 8.4.3 We assume that $\{T_j'\}_{j=1,2,\ldots}$ and $\{Z_j\}_{j=1,2,\ldots}$ are selected so that

(a) $\{\lceil T_j' Z_j \rceil\}_{j=1,2,\ldots}$ is an increasing sequence of integers,

(b) $\lim_{J\to\infty} \sum_{j=1}^{J} \lceil T_j' Z_j \rceil \to \infty$,

(c) $\lim_{J\to\infty} \frac{\sum_{j=1}^{J} \lceil T_j' Z_j \rceil}{\sum_{j=1}^{J} T_j'} = 0.$

At each period j, we randomly select $\lceil T_j' Z_j \rceil$ trials for exploration, and the rest of trials are used for exploitation. In an exploitation trial, every player k receives a probability distribution over all joint action profiles of its opponents, i.e., the output of its forecaster (See Sect. 8.4.2). By using this information as well as the estimated mean reward functions, the player selects the action with the highest estimated mean reward; that is, it acts with the best response to the predicted joint action profile of its opponents. In an exploration trial, with probability $0 < \gamma \ll 1$, the best response to the predicted joint action profile is played, whereas with probability $1-\gamma$, an action is selected uniformly at random. At each round, after receiving the reward, the player's estimation of the mean reward process of the selected action is updated. Moreover, opponents' actions are observed (as aforementioned, by hearing the broadcast message). This observation is used by a calibrated forecaster, described in Sect. 8.4.2. The procedure is called calibrated bandit strategy (CBS) and is summarized in Algorithm 4.

Algorithm 4 CBS [10]

```
Define an increasing sequence of integers, T'_j = 2^j for j = 1, 2, .... Each member T'_j of this
  sequence is the length of period j, i.e. the number of trials included in it.
Define a decreasing sequence of numbers, Z_j = j/2^j for j = 1, 2, ....
Set the period j = 1 and select the exploration parameter 0 < γ ≪ 1.
repeat
  Select ⌈T'_j Z_j⌉ exploration trials belonging to [1 + Σ_j T'_{j-1}, Σ_j T'_j] uniformly at random.
  for t = s + Σ_j T'_{j-1}, 1 ≤ s < T'_j, do
    if t is an exploring trial, then
      with probability 1 − γ, select an action equally at random;
      with probability γ,
      1. receive the output of the forecaster (Sect. 8.4.2),
      2. using this information, select the action with the highest estimated mean reward.
    else
      Receive the input from the forecaster.
      Using this information, select the action with the highest estimated mean reward.
    end if
    Play the selected action and observe the reward.
    Observe the actions of other players and inform the forecaster (forecaster's input).
    Improve the estimation of the mean reward function of the played action.
  end for
  j = j + 1.
until convergence (j is sufficiently large)
```

Theorem 8.4.3 ([10]) *Under Assumptions* 8.4.1, 8.4.2 *and* 8.4.3, *Algorithm 4 (CBS) satisfies* (8.18) *and thus* (8.4).

Theorem 8.4.4 ([10]) *Consider a K-player game in which the players have access to a set $\mathcal{M}$ of actions. Let $\mathcal{C}$ denote the set of correlated equilibria and let* $\mathbf{i} = \left(i^{(1)}, \ldots, i^{(K)}\right) \in \bigotimes_{k=1}^{K} \mathcal{M}$. *Suppose that the empirical joint frequencies of play is given by* (8.7). *If all players play according to the CBS, then the distance between the empirical joint distribution of plays and the set of correlated equilibria, i.e.,* $\inf_{\pi \in \mathcal{C}} \sum_{\mathbf{i}} \left|\hat{\pi}_n(\mathbf{i}) - \pi(\mathbf{i})\right|$, *converges to* 0 *almost surely as* $n \to \infty$.

Acknowledgments The work was supported by the German Research Foundation (DFG) under grant STA 864/3-3.

References

1. Cesa-Bianchi N, Lugosi G (2003) Potential-based algorithms in on-line prediction and game theory. J Mach Learn 51(3):239–261
2. Cesa-Bianchi N, Lugosi G (2006) Prediction, learning, and games. Cambridge University Press, Cambridge
3. Doppler K, Rinne M, Wijting C, Ribeiro CB, Hugl K (2009) Device-to-device communication as an underlay to LTE-advanced networks. IEEE Commun Mag 47(12):42–49

4. Foster DP, Vohra R (1997) Calibrated learning and correlated equilibrium. Games Econ Behav 21:40–55
5. Hannan J (1957) Approximation to Bayes risk in repeated play. Contrib Theory Games 3(39):97–139
6. Hutter M, Poland J (2005) Adaptive online prediction by following the perturbed leader. J Mach Learn Res 6:639–660
7. Hart S, Mas-colell A (2001) A general class of adaptive strategies. Econ Theory 98:26–54
8. Kujala J, Elomaa T (2007) Following the perturbed leader to gamble at multi-armed bandits. Algorithmic Learn Theory 4754:166–180
9. Maghsudi S, Stanczak S (2014) Joint channel selection and power control in infrastructureless wireless networks: a multi-player multi-armed bandit framework. IEEE Trans Veh Technol. doi:10.1109/TVT.2014.2369425
10. Maghsudi S, Stanczak S (2014) Channel selection for network-assisted D2D communication via no-regret bandit learning with calibrated forecasting. IEEE Trans Wirel Commun. doi:10.1109/TWC.2014.2365803
11. Mannor S, Stoltz G (2010) A geometric proof of calibration. Math Oper Res 35(4):721–727
12. Robbins H (1952) Some aspects of the sequential design of experiments. Bull Am Math Soc 58(5):527–535
13. Stoltz G, Lugosi G (2005) Internal regret in on-line portfolio selection. J Mach Learn 59(1):125–159
14. Yang Y, Zhu D (2002) Randomized allocation with nonparametric estimation for a multi-armed bandit problem with covariates. Ann Stat 30(4):100–121

Chapter 9
Cyclic Interference Alignment Via Polynomial Rings for Multi-user Communication Channels

Henning Maier, Johannes Schmitz and Rudolf Mathar

Abstract This chapter deals with the basic concept of *interference alignment* (IA) as introduced by Cadambe et al., particularly focusing on *IA by propagation delay* (IAP). This specific type of IA addresses the problem of interference in multi-user communication systems with long propagation delays as observed in satellite or underwater communication systems for instance. So far, IAP was conceptually described by the *linear deterministic channel model* (LDCM) by Avestimehr et al. The impact of channel gains is described by linear shifted and superimposed signal vectors between transmitters and receivers. However, a closed-form information-theoretic capacity analysis of such channels is challenging for asymmetric channel gains with more than two transmitter-receiver pairs communicating simultaneously. Our main contribution in this chapter is a novel channel model extending the LDCM to cyclic shifts—the *cyclic polynomial channel model* (CPCM). It is inspired by the algebraically convenient description of cyclic codes. We investigate the 3-user interference channel, the 2-user X-channel and the $2 \times 2 \times 2$ relay-interference channel in terms of the CPCM with arbitrary cyclic shifts. We derive upper bounds and formulate optimal *interference alignment* and *interference neutralization* schemes to show their achievability.

9.1 Introduction

Interference alignment (IA), as introduced by Cadambe and Jafar in [3], is an innovative strategy recently introduced to counteract the detrimental effects of multi-user interference in the realm of information theory and wireless communications. In the

H. Maier (✉) · J. Schmitz · R. Mathar
Chair of Theoretical Information Technology, Kopernikusstraße 16,
52074 Aachen, Germany
e-mail: maier@ti.rwth-aachen.de

J. Schmitz
e-mail: schmitz@ti.rwth-aachen.de

R. Mathar
e-mail: mathar@ti.rwth-aachen.de

W. Utschick (ed.), *Communications in Interference Limited Networks*,
Signals and Communication Technology, DOI 10.1007/978-3-319-22440-4_9

present work we investigate the special case of *IA by propagation delay* (IAP). This particular class of IA focuses on wireless channels with long propagation delays as observed in satellite or underwater communication systems for instance. The conceptual description of IAP was first established by some initial examples in [3] and by the *linear deterministic channel model* (LDCM) as introduced by Avestimehr et al. in [1]. It has turned out that the closed-form information-theoretic capacity analysis of such linear deterministic interference channels is quite challenging for asymmetric channel gains, especially when considering more than two dedicated user-pairs.

To elaborate this problem in the context of the LDCM, we first consider a 3-user interference channel with normalized dedicated links, symmetrically parameterized interference links, and a single exceptionally parameterized interference link. A greedy IA strategy for max-min fair rate-allocations from a graph-based perspective is applied. The achieved sum-rate shows structured symmetries with a regularly oscillating behavior towards moderate interference power. Furthermore, this result reveals certain singularities if all channel gains are fully symmetric. However, a closed-form description of the sum-rate is difficult to formulate due to the exponentially increasing number of parameters. Approaching multi-user capacity remains elusive from this perspective, even for the 3-user case.

The aforementioned problem of considering arbitrary channel gains is tackled by introducing an alternative algebraic channel description with polynomial rings. This approach is mainly inspired by cyclic codes. We term it the *cyclic polynomial channel model* (CPCM). To elaborate the proposed model, we consider the 2-user X-channel and provide a simple and representative example of cyclic IA. A polynomial ring is used to describe how multiple interfering signals are sent over a channel with arbitrary cyclic shifts. In cyclic IA signals are allocated such that interference is aligned to a common interference space for non-dedicated signals at each receiver. Dedicated signals however, are segregated and thus decodable. We derive a set of separability conditions such that cyclic IA applies. These conditions classify types of interference patterns that allow for linear decodability at the receivers. Moreover, the singular behavior of the related LDCM is revealed by the separability conditions. This novel approach provides an approximate capacity analysis in terms of the *Degrees-of-Freedom* (DoF), which is an accurate metric for capacity of Gaussian networks at high SNR. We formulate upper bounds on the DoF and optimal closed-form schemes proving their achievability. Besides a characterization of the DoF, a *complementary reciprocal symmetry* property of IA is observed. It describes how pair-wise alignment patterns at different receivers are related. We also briefly investigate cyclic IA for the 3-user interference channel.

Actually, the proposed CPCM is not confined to the application of IA only. It also permits a generalization to multi-user and multi-relay networks with arbitrary channel gains. We discuss *cyclic interference neutralization* (IN) for a $2 \times 2 \times 2$ relay-interference channel and for a related two-way relaying setup.

In summary, multi-user interference channels for IAP with arbitrary delays are investigated. As our main result, we established the CPCM with convenient algebraic properties for the computation of information-theoretic channel capacity.

9.1.1 Multi-user Interference Channels

In wireless communications, the impact of co-channel interference has been an elementary bottleneck in the query for high data rates. The characterization of the capacity for the K-user Gaussian interference channel is known as a central problem and has been open over several decades. In order to yet approach this problem at least from a conceptual perspective, the authors of [3] used a simplified channel model based on discrete and long propagation delays for a basic illustration of IA. They showed that, even for the K-user interference channel, it is possible to achieve $\frac{K}{2}$ DoF [3], given all interfering signals exactly overlap at the undesired receivers. The dedicated signals remain distinct and hence decodable. This approach contrasts all the conventional multiple-access schemes which can only achieve 1 DoF [3].

Although the given example in [3] is a very restrictive scenario, it has stimulated the idea of IA to such an extend that IAP was mathematically translated to IA schemes on the highly relevant Gaussian channel model. An immense bulk of research dedicated to IA is mainly concerned with determining the DoF of an asymptotic variant of IA on the highly relevant Gaussian channel model. However, this asymptotic representation still includes several drawbacks for practical systems. The aforementioned instructive example[1] in [3] only considers a particular channel with an IAP scheme over a specific configuration of discrete propagation delays. These examples do not elaborate how the proposed IA schemes apply to channels with arbitrary discrete delays.

9.2 Linear Deterministic Channel Model

In this work, our main goal is to explore and discuss potential models for the elementary scheme of IAP. To approximate the wireless multi-user capacity, we first briefly discuss the LDCM as introduced by Avestimehr et al. [1]. This model will also serve as a well-established reference to the CPCM which will be introduced afterwards. The underlying concept of the LDCM is to eclipse the impact of noise and to highlight the influence of wireless *broadcast*, *interference* and *signal scale* in the shared medium. This model is also useful to investigate the quantity of the *generalized degrees of freedom* (GDoF) as introduced in [6]. The GDoF refine the DoF w.r.t. the impact of different channel gains per user with higher accuracy. It provides a first insight into the behavior of interfering communications in wireless multiple-user channels by allowing a more traceable analysis of structured coding strategies [2] in contrast to probabilistic coding methods.

To a certain degree, IAP can be described by the LDCM as well. An initial connection of the linear deterministic K-user interference channel with the LDCM is given in [5]. A more generalized result for the GDoF of a symmetric K-user interference channel is addressed in [14]. Furthermore an elaborate linear coding strategy using

[1] There is also a related example in [4] for a 2-user X-channel.

interference alignment is proposed for the symmetric 2-user X-channel in [11]. However, these and plenty of other works in the literature restrict themselves to highly symmetric channel gains. This is due to the severe complexity of the model when the channel gains are asymmetric. We intend to relax these symmetry constraints in the course of this work.

9.2.1 3-User Interference Channel

For a first insight, we begin with the 3-user interference channel with only a slight asymmetry of the channel gains. There are 3 transmitter-receiver pairs involved in the communication system. A pair of transmitters and receivers is denoted by Tx_i and Rx_i, respectively, where the indices i are in the set $\mathcal{K}_3 = \{1, 2, 3\}$. The input signal vector of Tx_i is denoted by $\mathbf{x}_i \in \mathbb{F}_2^q$, with the vector length $q \in \mathbb{N}$. The vector $\mathbf{y}_j \in \mathbb{F}_2^q$ is the received output signal at Rx_j. In order to model *signal scale* in a wireless scenario, the channel is represented by a $q \times q$ linear downshift matrix $\mathbf{S}$. A non-negative exponent $q - n_{ji}$ determines the channel gain n_{ji} between Tx_i and Rx_j. The exponent is assumed to be constant throughout the whole duration of the communication. Hence we omit the time-index to keep our notation simple. The l-th element of a signal vector, e. g., $\mathbf{x}_m(l)$, $l \in \mathbb{N}$, is called the l-th *signal level*. The n_{ji} most significant levels from Tx_i are exactly mapped onto the n_{ji} least significant levels of Rx_j. All other levels received below the least significant level are considered as noise and truncated. The received signal of the fully-connected linear deterministic 3-user interference channel is a superposition of linearly shifted signal vectors:

$$\mathbf{y}_j = \sum\nolimits_{m=1}^{3} \mathbf{S}^{q-n_{mj}} \mathbf{x}_m. \tag{9.1}$$

Summation and multiplication is performed in the finite field $\mathbb{F}_2$.

Such a 3-user interference channel is termed as *fully-connected* indicating that all receivers will receive signals from each transmitter. The influence of wireless *broadcast* from each transmitter and of *interference* at each receiver is inherent in the model. These properties are due to the individually shifted superposition of multiple transmitted signals as observed at each receiver.

As introduced in [1], an equivalent graph representation establishes a clear intuition of the LDCM applied to the 3-user interference channel. The *undirected graph* is completely defined by a set of *vertices* (circles) and the set of *edges* (lines). Transmitters Tx_i and receivers Rx_j are represented as *supernodes* (boxes). The given linear transformation $\mathbf{S}^{q-n_{mj}}$ corresponds to edges connecting certain levels between each pair of transmitters and receivers. An according example of the channel is depicted in Fig. 9.1 for $q = 4$ levels and the channel gains $n_{11} = n_{22} = n_{33} = 4$, $n_{21} = n_{31} = n_{12} = n_{13} = n_{23} = 3$, and $n_{32} = 1$.

In the following part, an *active* level indicates that this level is used to transmit/decode symbols at the related transmitter/receiver. An *inactive* level corresponds

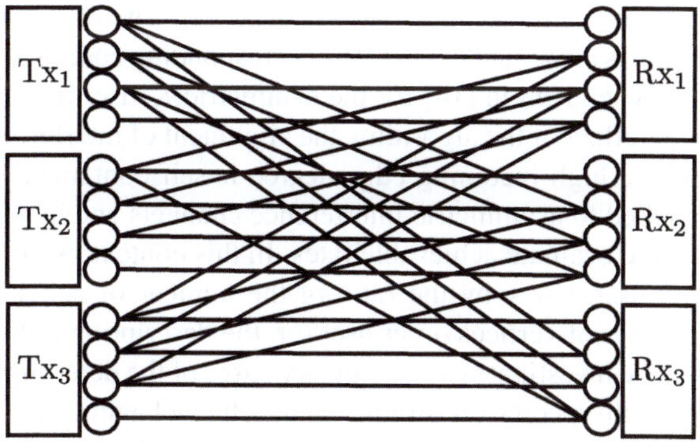

Fig. 9.1 An example of the graph representation depicting the linear deterministic 3-user interference channel

to conducting no transmission/to discard incoming received signals. From a graph-theoretical viewpoint, the allocation of active and inactive levels translates to a *colouring* of vertices by two different colours. In this case, active levels are red and inactive levels are blue. In order to provide interference-free signals at each vertex carrying a dedicated signal, each pair of vertices that is connected by an edge in a desired link is always jointly selected as active or jointly selected as inactive. Further vertices adjacent to such a pair of active vertices must be selected as inactive to avoid interference. Since the dedicated transmitters and receivers always appear in pairs, we will remove this redundancy in the graph in the following step. Dedicated links from each Tx_i to Rx_i with the same index i are merged using the *edge-contraction operation* of graph theory. The resulting merged and coloured graph of the above example is depicted in Fig. 9.2.

Inspired by [9], the resulting communication problem is narrowed down to a graph-theoretic problem: The *maximum independent set problem*. In other words, the target function is dedicated to maximize the number of active vertices subject to not being connected to each other, only to inactive vertices. This sort of problems is computationally extensive in general. We allocate the active vertices via a greedy algorithm, which is known to be quite efficient with minimal effort, especially for symmetric cases. In our proposed algorithm, vertices with the minimal number of incoming edges are iteratively assigned as active, as long as there is no

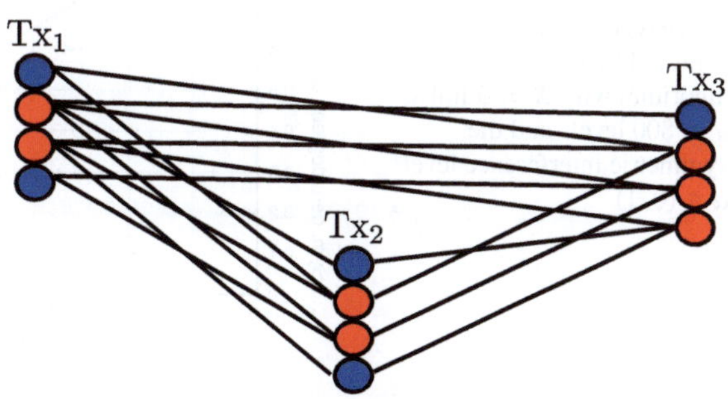

Fig. 9.2 This is an example of the merged graph of the linear deterministic 3-user interference channel. In this case, the rates $d_1 = d_2 = \frac{1}{2}$ and $d_3 = \frac{1}{4}$ are achieved after colouring

active neighbouring vertex assigned yet. The detailed scheme of our greedy algorithm is presented in [19]. The corresponding coding scheme yields congruent results to those given in [14] for the symmetric K-user interference channel. As a positive side-effect of the algorithm, the alignment of interfering signals at the inactive levels with a high incoming edge-degree is self-contained. The greedy algorithm is also applied to asymmetric interference channels and hence a simple tool to perform IA and to evaluate achievable rates. In this context, we measure the achieved GDoF d_k per user-pair k by the ratio of the number of active vertices divided by the total number of vertices q per user (cf. the description in Fig. 9.2).

We investigate two exemplary cases of the deterministic 3-user interference channel with specific constraints on the channel gains. The first case describes symmetric channel gains for each user pair. The *dashed black lines* in Fig. 9.3 represent the achieved GDoF for the reference coding scheme, as derived in [14, Theorem 3.1]. These GDoF are described by piecewise defined linear functions, depending on the interference level $\alpha = \frac{n_{mj}}{q}$.

The symmetric rate achieved by our greedy algorithm is indicated by *solid red lines*. For $\alpha \in (0, \frac{2}{3}]$, this coding scheme is exactly congruent to the reference scheme. In the remaining part with $\frac{2}{3} < \alpha < 1$, we observe a consistent oscillation of the achieved sum-rate. Apart from local maxima, the coding schemes differ within $\alpha \in (\frac{2}{3}, 1)$ due to omitting *interference cancellation* used in the reference scheme. Here, we only treat the (aligned) interference at inactive levels as noise. At the point $\alpha = 1$, there is a singularity, where the achieved GDoF are only $\frac{1}{3}$ [14].

The achieved GDoF d_k per link M_k, for a symmetric channel with interference level α for non-negative integers $n \in \mathbb{N}_0$, as obtained by the proposed greedy algorithm, are:

$$d(\alpha) = \begin{cases} -(n+1)\alpha + n + 1 & , \alpha \in [\kappa_1(n), \kappa_2(n)] \\ (n+1)\alpha - n & , \alpha \in [\kappa_2(n), \kappa_3(n)] \end{cases}, \tag{9.2}$$

and the interval bounds $\kappa_1(n), \kappa_2(n), \kappa_3(n)$ with $n \in \mathbb{N}_0$ are:

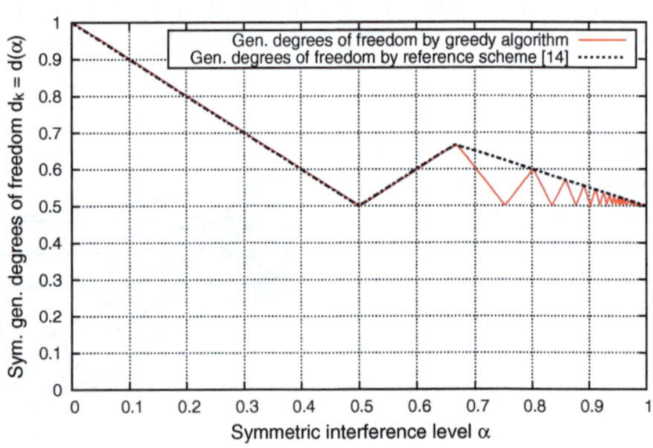

Fig. 9.3 Generalized degrees of freedom d_k per link M_k for the symmetric linear deterministic interference channel as achieved by the greedy algorithm with $K = 3$ links, $q = 800$ levels and the symmetric interference level $\alpha \in (0, 1)$

$$\kappa_1(n) := \frac{2n}{2n+1}, \ \kappa_2(n) := \frac{2n+1}{2n+2}, \ \kappa_3(n) := \frac{2n+2}{2n+3}. \tag{9.3}$$

In the first part of (9.2), each segment with a negative slope is characterized for the intervals $\alpha \in [\kappa_1(n), \kappa_2(n)]$. In the second part, the segments with a positive slope for the intervals $\alpha \in [\kappa_2(n), \kappa_3(n)]$ are described. Note $\kappa_3(n) = \kappa_1(n+1)$ holds. A closely related oscillating behaviour has also been reported by the authors of [29].

In the next example we consider a slightly asymmetric 3-user interference channel. Instead of only a single parameter for the symmetric interference level α, we permit exactly one interference link that is parameterized separately with an interference level β. We use the *achieved sum-rate of the generalized degrees of freedom*, normalized to the number of direct links, for evaluation, i.e., $\overline{d}(\alpha, \beta) := \frac{1}{K} \sum_{k \in \mathcal{K}} d_k$. The application of the greedy algorithm reveals that $\overline{d}(\alpha, \beta)$ behaves as illustrated in Fig. 9.4.

For the special case $\beta = \alpha$, these results evidently reduce to those of the symmetric channel given in (9.2). For all other values of β and α, the GDoF achieved by the greedy algorithm still display well-structured symmetries of the considered channel. Clear local maxima occur at dedicated turning points similar to those in the symmetric case. We identify particular regions that imply the weak interference regime (left/bottom-left area) and moderate interference regime (right area). Especially in the region of moderate interference strengths, the aforementioned oscillation of the achieved rates occurs again. However a closed form description of these achieved rates is analytically already hard to track for multiple users and asymmetric channel gains. Furthermore, the additional application of interference cancellation does not directly carry over from the symmetric case.

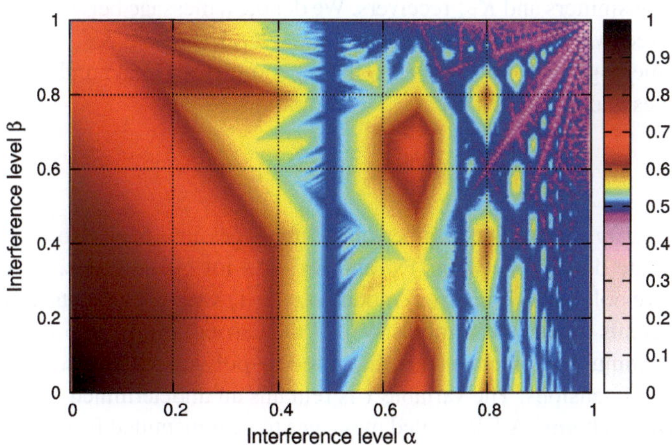

Fig. 9.4 Normalized achieved sum-rate $\overline{d}(\alpha, \beta)$ for the symmetric interference level $\alpha \in (0, 1)$ and a single asymmetric interference level $\beta \in (0, 1)$ with $i \neq j$ and $K = 3$ direct links, each link containing $q = 200$ levels

Nonetheless, there is an interesting aspect to this model when high values of q and moderate interference strengths are considered. Then, an increasing subset of levels at each user provides a repetitive pattern. So far, a graph-theoretic analysis of this phenomenon seems to be quite cumbersome for closed-form sum-rate expressions. However this observation motivates another approach wherein we may focus on cyclically shifted signals with an infinite amount of signal levels instead.

9.3 Cyclic Polynomial Channel Model

The main drawback of the previous LDCM describing IAP concerns the extensive involvement of numerous different parameters when considering non-symmetric channels gains. The truncation of the linear shifted vectors for a limited number of levels is one of the main reasons for this problem. It is even more important to note that the truncation of the signal levels below the noise threshold is not fully accurate for the description of IAP.

In the following part, we assume that the entire duration of the communication is infinite, and hence longer than the longest propagation delays. Thus, we modify the linear deterministic channel model. The previous linear shifts are now replaced by corresponding *cyclic shifts*. There are several reasons for modelling a channel with cyclic shifts: The description of cyclic shifts using polynomial rings is a well-known mathematical tool for cyclic codes in coding theory [28]. Such an approach captures the periodical repetitions with a simple algebraic representation. Several of the observed structures gained from the LDCM remain still intact. In some special cases, the two models and the related optimal communication schemes even coincide.

Our proposed model provides a conceptual description of a multi-user channel with K_{Tx} transmitters and K_{Rx} receivers. We denote a message between a transmitter Tx_i and a receiver Rx_j for $i \in \mathcal{K}_{\text{Tx}} := \{1, ..., K_{\text{Tx}}\}$ and $j \in \mathcal{K}_{\text{Rx}} := \{1, ..., K_{\text{Rx}}\}$ by $\mathbf{w}_{ji}$. A message vector $\mathbf{w}_{ji}$ from Tx_i to Rx_j comprises a total number of $m_{ji} \in \mathbb{N}_0$ submessages and is denoted by:

$$\mathbf{w}_{ji} = (W_{ji}^{[0]}, ..., W_{ji}^{[m_{ji}-1]}), \ W_{ji}^{[l]} \in \mathbb{F}, \tag{9.4}$$

for $l = 1, ..., m_{ji} - 1$. The number of submessages m_{ji} may also be interpreted as a fixed rate demand between Tx_i and Rx_j. As for cyclic codes, we consider a commutative polynomial ring $\mathbb{F}(x)$ modulo $(x^n - 1)$ with coefficients over a field $\mathbb{F}$. The coefficients denote the symbols $W_{ji}^{[l]}$ of the carried information in the message $\mathbf{w}_{ji}$. The communication channel is described by polynomials in x of degree $n - 1$ with $n \in \mathbb{N}$ dimensions. The variable x is remains an undetermined value.

Encoding Scheme: All dedicated messages to be transmitted from Tx_i to Rx_j are encoded into the coefficients of a polynomial $v_i(x)$ by the linear encoding function:

$$e_i : (\mathbf{w}_{1i}, ..., \mathbf{w}_{K_{\text{Rx}}i}) \mapsto v_i(x).$$

The dedicated submessages are allocated to the coefficients of $v_i(x)$ with a linear coding scheme. A single position within a period of n dimensions is addressed by an offset $x^0, x^1, ..., x^{n-1}$, from 0 (zero-offset) to $n-1$ (maximal offset).

Channel Transfer Matrix: An individual cyclic shift of the signal from Tx_i to Rx_j is denoted by the multiplication of the transmitted polynomial with a monomial $d_{ji} \in \mathcal{D}$, with $\mathcal{D} := \{x^k | k \in \mathbb{N}\} \cup \{0\}$. A zero-valued entry $d_{ji} = 0$ means that there is no direct link between Tx_i to Rx_j. The channel transfer matrix $\mathbf{D}$ is assumed to be static and fully known to all users during the whole duration of the communication, and it is defined by the $K_{\text{Rx}} \times K_{\text{Tx}}$ matrix:

$$\mathbf{D} = (d_{ji})_{1 \leq j \leq K_{\text{Rx}}, 1 \leq i \leq K_{\text{Tx}}}. \tag{9.5}$$

The (cyclic polynomial) transfer function for K_{Tx} transmitters and K_{Rx} receivers is compactly described by a matrix multiplication of the vector of input polynomials $\mathbf{v} = (v_1(x), ..., v_{K_{\text{Tx}}}(x))$ with the channel matrix $\mathbf{D}$, such that the vector of received polynomials $\mathbf{r}$ yields:

$$\mathbf{r}^{\mathsf{T}} \equiv \mathbf{D}\mathbf{v}^{\mathsf{T}} \bmod (x^n - 1), \tag{9.6}$$

with $\mathbf{v} = (v_1(x), ..., v_{K_{\text{Tx}}})$ and $\mathbf{r} = (r_1(x), ..., r_{K_{\text{Rx}}}(x))$. The modulo $x^n - 1$ operation is applied element-wise. The received signal at Rx_j is a superposition of transmitted polynomials:

$$r_j(x) = \sum\nolimits_{i \in \mathcal{K}_{\text{Tx}}} d_{ji} v_i(x) \bmod (x^n - 1). \tag{9.7}$$

Decoding Scheme: The received polynomial $r_j(x)$ are linearly decoded to obtain an estimate of the dedicated messages $\widehat{\mathbf{w}}_{ji}$. The decoding function f_j for $r_j(x)$ is:

$$f_j : r_j(x) \mapsto (\mathbf{w}_{j1}, \mathbf{w}_{j2}, ..., \mathbf{w}_{jK_{\text{Tx}}}). \tag{9.8}$$

The coefficients carrying the dedicated symbols are mapped to the dedicated messages. The dedicated messages are correctly decoded if the messages can be linearly resolved at the dedicated receiver. Otherwise, there exists no error-free decoding scheme.

We would like to remark that it is also possible to illustrate the CPCM by a graph with vertices (offsets) and edges (cyclic shifts), as similarly shown for the LDCM in Fig. 9.1. The truncation of the linear shifts LDCM would be replaced by the cyclic shift operation, correspondingly. But for the benefit of readability, we will use tables instead to illustrate some elementary examples. Two of the properties demanded in the LDCM are already fulfilled: wireless *broadcast* and multi-user *interference*. Although the effect of *signal scale* is not covered as such, the CPCM provides a comparable feature of *signal phase* instead.

For a particular application of this CPCM to a propagation delay based channel with only a single line-of-sight component and no further echoes in the delay profile, as in [3, 4, 9], the n dimensions can be interpreted as a cyclically repeated period of n discrete time-slots of normalized length. The discrete propagation delay of δ_{ji} between Rx_j and Tx_i is a multiple of a normalized time-slot and related to the cyclic shift by $d_{ji} = x^{\delta_{ji}}$. Note that in the decoding process, these cyclically repeated time-slots are unrolled over time. Furthermore, the decoders must also take the transient settling time at the beginning and ending of the transmission into account. Even a limited measurement inaccuracy of the propagation delays due to fluctuations in the synchronization is permitted. The induced *leakage interference* from such an imprecision can be mitigated via proper guard intervals for instance. For sufficiently large-scaled delays and small-scaled time-slots, the impact if leakage interference on the achievable rate is considered as negligible.

9.3.1 Upper Bounds on the Degrees of Freedom

In order to evaluate the performance of a communication scheme on the given CPCM, we consider an analogous DoF-measure[2] as defined in [3, 4]. Assuming i.i.d. zero mean unit variance complex Gaussian noise at the receivers and an average power constraint P per message within each dimension, a single interference-free link between Tx_i and Rx_j has a capacity of $\log(P)$ bits per dimension at a high *signal-to-noise-ratio* (SNR). The *degrees of freedom* (DoF) are defined as the pre-log of the achieved approximate sum-capacity $C_\Sigma(P)$ in the high SNR-regime:

$$\mathrm{DoF} = \lim_{P\to\infty} \frac{C_\Sigma(P)}{\log(P)},$$
$$C_\Sigma(P) = \mathrm{DoF}\cdot\log(P) + o(P).$$

In the presented CPCM, we define the achieved DoF by the total number of submessages M received in interference-free dimensions after linear decoding within the full period of n dimensions:

$$\mathrm{DoF} = \lim_{P\to\infty} \frac{\frac{M}{n}\log(P)}{\log(P)} = \frac{M}{n}. \tag{9.9}$$

The *messaging matrix* $\mathbf{M}$ describes the number of dedicated submessages from Tx_i to Rx_j per period of n dimensions:

$$\mathbf{M} = (m_{ij})_{1\le j\le K_{\mathrm{Rx}},1\le i\le K_{\mathrm{Tx}}},\; m_{ij} \in \mathbb{N}_0. \tag{9.10}$$

[2]The DoF are also known as multiplexing gain as in [10, 12].

The total number of messages in the system is computed by:

$$M = \sum_{j=1}^{K_{\mathrm{Rx}}} \sum_{i=1}^{K_{\mathrm{Tx}}} m_{ji}. \tag{9.11}$$

From the number of dedicated and interfering submessages at each receiver, upper bounds on the DoF can be formulated. The upper bounds on the DoF are closely related to the bounds provided in [4, Thm. 1] and basically adapt to the CPCM by:

$$\mathrm{DoF} \leq \frac{\sum_{j=1}^{K_{\mathrm{Rx}}} \sum_{i=1}^{K_{\mathrm{Tx}}} m_{ji}}{\max\limits_{j \in \mathcal{K}_{\mathrm{Rx}}, i \in \mathcal{K}_{\mathrm{Tx}}} \left(\sum_{k=1}^{K_{\mathrm{Tx}}} m_{jk} + \sum_{l=1}^{K_{\mathrm{Rx}}} m_{li} - m_{ji} \right)}. \tag{9.12}$$

Note that the term in the denominator provides a lower bound on the minimal number of dimensions necessary to accomodate the total number of all dedicated signals. The full proof of these upper bounds is omitted here and given in [24]. The main task to pursue is to design optimal encoding and decoding schemes for a particular setup of users and rate demands within the CPCM such that these upper bounds are achieved for almost all arbitrary cyclic shifts in $\mathbf{D}$. However, it is permitted to exclude subsets of infeasible cases if it is necessary.

9.3.2 2-User X-Channel

The 2-user X-channel is very useful to illustrate the basic behaviour of the given channel model and also to describe the elementary concept of cyclic IA. The X-channel has been considered in [16] and in [13] for the initial development of the IA principle for time-varying MIMO X-channels, as well. In terms of the LDCM, the X-channel was studied in [11]. A 2×2 X-channel has a number of 4 independent messages $\mathbf{w}_{11}$, $\mathbf{w}_{21}$, $\mathbf{w}_{12}$ and $\mathbf{w}_{22}$. The transmitter and receiver indices are in the sets $\mathcal{K}_{\mathrm{Tx}} = \mathcal{K}_{\mathrm{Rx}} = \mathcal{K} = \{1, 2\}$. The X-channel setup is depicted in Fig. 9.5.

The task is to convey and decode 4 dedicated messages $\mathbf{w}_{11}$, $\mathbf{w}_{12}$, $\mathbf{w}_{21}$ and $\mathbf{w}_{22}$ interference-free within n dimensions. The proposed communication scheme is optimal if n is minimal and decoding of all messages is still feasible at each receiver.

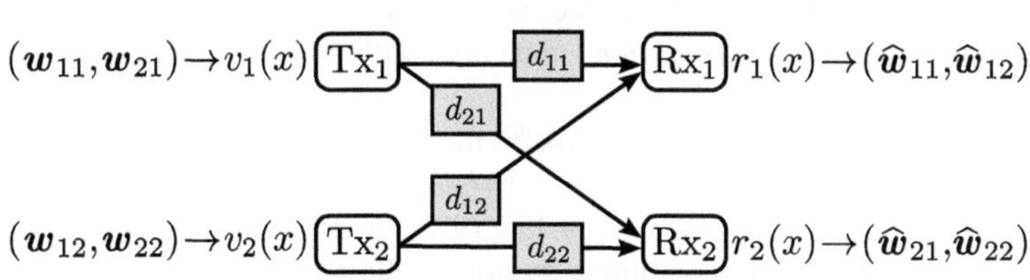

Fig. 9.5 The polynomial X-channel with messages $\mathbf{w}_{11}$, $\mathbf{w}_{12}$, $\mathbf{w}_{21}$ and $\mathbf{w}_{22}$ between Tx_1, Tx_2, Rx_1 and Rx_2 and the cyclic shifts d_{11}, d_{12}, d_{21} and d_{22} of the transfer matrix $\mathbf{D}$

9.3.2.1 Perfect Cyclic Interference Alignment

For an introductory discussion of cyclic IA, we begin with a simplified messaging matrix $\mathbf{M} = \mathbf{1}_{2\times 2}$ with only a single submessage per user-pair. To simplify our notation, and since $m_{ji} = 1$ for all $j, i \in \mathcal{K}$, the superscripts of W_{ji} and of p_{ji} are omitted for this special case. We call two interfering submessages to be *aligned*, if they are received within the same dimension at an undesired receiver. In order to align the submessages W_{22} and W_{21} of $v_1(x)$ and $v_2(x)$, they must overlap in the same dimension at Rx_1, but they must remain distinct in different dimensions at Rx_2. To align the submessages W_{11} and W_{12} of $v_1(x)$ and $v_2(x)$, they must overlap at Rx_2, but remain distinct at Rx_1, accordingly. The following polynomials with the offset parameters $p_{11}, p_{12}, p_{21}, p_{22} \in \mathbb{N}_0$ are used for transmission:

$$v_1(x) = W_{11}x^{p_{11}} + W_{21}x^{p_{21}}, \tag{9.13}$$

$$v_2(x) = W_{12}x^{p_{12}} + W_{22}x^{p_{22}}. \tag{9.14}$$

Then by (9.6), the received polynomials at Rx_1 and Rx_2 yield:

$$r_1(x) \equiv d_{11}W_{11}x^{p_{11}} + d_{12}W_{12}x^{p_{12}} + d_{11}W_{21}x^{p_{21}} + d_{12}W_{22}x^{p_{22}} \bmod (x^n - 1), \tag{9.15}$$

$$r_2(x) \equiv d_{21}W_{21}x^{p_{21}} + d_{22}W_{22}x^{p_{22}} + d_{21}W_{11}x^{p_{11}} + d_{22}W_{12}x^{p_{12}} \bmod (x^n - 1). \tag{9.16}$$

Messages dedicated for Rx_j must be linearly decodable. Thus, we demand that the *multiple-access interference conditions* hold with the indices $j \in \mathcal{K}$, and $i \neq l \in \mathcal{K}$:

$$d_{ji}x^{p_{ji}} \not\equiv d_{jl}x^{p_{jl}} \bmod (x^n - 1). \tag{9.17}$$

The two messages to be transmitted from Tx_i dedicated for different receivers Rx_j and Rx_k must also be linearly decodable. Accordingly, we demand that the *intra-user interference conditions* with $j \neq k \in \mathcal{K}$, and $i \in \mathcal{K}$, hold:

$$x^{p_{ji}} \not\equiv x^{p_{ki}} \bmod (x^n - 1). \tag{9.18}$$

Moreover, the interfering messages from other transmitters to other receivers must also be received separately from the dedicated messages at each Rx_j. More precisely, the *inter-user interference conditions* must hold with $j \neq k \in \mathcal{K}, i \neq l \in \mathcal{K}$:

$$d_{ji}x^{p_{ji}} \not\equiv d_{jl}x^{p_{kl}} \bmod (x^n - 1). \tag{9.19}$$

We call these negated congruences (9.17), (9.18) and (9.19) the *separability conditions*. These separability conditions do not preclude to align the intra- and inter-user interference to a single dimension at each Rx_j. This particular alignment is expressed by:

$$d_{ji}x^{p_{ki}} \equiv d_{jl}x^{p_{kl}} \mod(x^n - 1), \tag{9.20}$$

with $j \neq k \in \mathcal{K}$ and $i \neq l \in \mathcal{K}$. Such an alignment is also called *perfect* because the two different interference signals perfectly overlap in exactly one dimension per receiver. We further remark that (9.20) substituted into (9.19) yields (9.18):

$$\begin{aligned} d_{ji}x^{p_{ji}} &\not\equiv d_{jl}x^{p_{kl}} \mod(x^n - 1) \\ \Rightarrow d_{ji}x^{p_{ji}} &\not\equiv d_{ji}x^{p_{ki}} \mod(x^n - 1) \\ \Rightarrow x^{p_{ji}} &\not\equiv x^{p_{ki}} \mod(x^n - 1), \end{aligned}$$

with the indices $j \neq k \in \mathcal{K}$ and $i \neq l \in \mathcal{K}$. Hence, we can neglect the condition (9.19) if both (9.18) and (9.20) hold.

An example of such an alignment is shown in the Table 9.1. Note that this theorem also covers the example of the 2-user X-channel considered in [4, Fig. 1] for $d_{11} = d_{21} = x^1$, $d_{12} = x^0$, $d_{22} = x^2$ and $p_{21} = p_{22} = 0$, $p_{11} = p_{22} = 1$.

Theorem 9.3.1 *A perfect cyclic IA scheme for the cyclic polynomial X-channel satisfying the separability conditions exists, if and only if* $\det(\mathbf{D}) \not\equiv 0 \mod(x^n - 1)$ *and $n = 3$ hold. Then, cyclic IA achieves the upper bound of $\frac{4}{3}$ DoF.*

Proof (a) Necessity of $\det(\mathbf{D}) \not\equiv 0 \mod(x^n - 1), n \in \mathbb{N}$:
Assuming $\det(\mathbf{D}) \equiv 0 \mod(x^n - 1)$ yields:

$$\begin{aligned} \det(\mathbf{D}) &\equiv 0 \mod(x^n - 1) \\ \Rightarrow d_{11}d_{22} - d_{21}d_{12} &\equiv 0 \mod(x^n - 1) \\ \Rightarrow d_{11}d_{22} &\equiv d_{21}d_{12} \mod(x^n - 1) \\ \Rightarrow d_{ji}d_{kl} &\equiv d_{ki}d_{jl} \mod(x^n - 1), \end{aligned} \tag{9.21}$$

with the indices $j \neq k \in \mathcal{K}, i \neq l \in \mathcal{K}$. Including (9.21) into condition (9.20) yields:

$$\begin{aligned} d_{ji}x^{p_{ki}} &\equiv d_{jl}x^{p_{kl}} \mod(x^n - 1) \\ \Rightarrow d_{ki}x^{p_{ki}} &\equiv d_{kl}x^{p_{kl}} \mod(x^n - 1). \end{aligned}$$

Table 9.1 Cyclic IA is applied on an X-channel with $n = 3$ dimensions

	x^0	x^1	x^2
$v_1(x)$	W_{11}	0	W_{21}
$v_2(x)$	0	W_{12}	W_{22}
$r_1(x)$	$\mathbf{W_{21}+W_{22}}$	W_{11}	W_{12}
$r_2(x)$	$\mathbf{W_{11}+W_{12}}$	W_{22}	W_{21}

The following cyclic 'right-shifts' $d_{11} = d_{12} = x^1$, $d_{21} = x^3$ and $d_{22} = x^2$ are assumed. The parameters are $p_{11} = 0$, $p_{12} = 1$, $p_{21} = p_{22} = 2$ and $n = 3$ and $\frac{4}{3}$ DoF are achieved. Aligned interference is highlighted in bold

Relabeling the indices $j \leftrightarrow k$ provides:

$$\Rightarrow d_{ji}x^{p_{ji}} \equiv d_{jl}x^{p_{jl}} \mod(x^n - 1),$$

and contradicts (9.17) for any $n \in \mathbb{N}$.

(b) Necessity of $n > 2$ dimensions:
Assume $\det(\mathbf{D}) \not\equiv 0 \mod(x^n - 1)$ holds. We consider the right-hand sides of (9.17) and (9.18). The previous congruence (9.18) is expanded by d_{ji}.

$$(9.17) : d_{ji}x^{p_{ji}} \not\equiv d_{jl}x^{p_{jl}} \mod(x^n - 1),$$
$$(9.18) : d_{ji}x^{p_{ji}} \not\equiv d_{ji}x^{p_{ki}} \mod(x^n - 1).$$

These right-hand side terms must also be pair-wise distinct, since we can relabel the indices $i \leftrightarrow l$ in (9.22) to obtain (9.19):

$$d_{jl}x^{p_{jl}} \not\equiv d_{ji}x^{p_{ki}} \mod(x^n - 1)$$
$$\Leftrightarrow d_{ji}x^{p_{ji}} \not\equiv d_{jl}x^{p_{kl}} \mod(x^n - 1). \tag{9.22}$$

Thus, there is no solution to satisfy all three conditions on $d_{ji}x^{p_{ji}}$ with only $n = 1$ or $n = 2$ dimensions.

(c) Sufficiency of $n = 3$ *and* $\det(\boldsymbol{D}) \not\equiv 0 \mod(x^n - 1)$:
From the perfect IA condition (9.20), the following holds:

$$x^{p_{12}} \equiv d_{22}^{-1}d_{21}x^{p_{11}} \mod(x^3 - 1), \tag{9.23}$$
$$x^{p_{21}} \equiv d_{11}^{-1}d_{12}x^{p_{22}} \mod(x^3 - 1). \tag{9.24}$$

Furthermore, the condition (9.18) must hold:

$$x^{p_{11}} \not\equiv x^{p_{21}} \mod(x^3 - 1), \tag{9.25}$$
$$x^{p_{12}} \not\equiv x^{p_{22}} \mod(x^3 - 1). \tag{9.26}$$

The insertion of (9.23) and (9.24) into condition (9.25) yields:

$$x^{p_{22}} \not\equiv d_{22}d_{11}d_{21}^{-1}d_{12}^{-1}x^{p_{12}} \mod(x^3 - 1). \tag{9.27}$$

Due to (9.21), the following holds:

$$d_{12}^{-1}d_{21}^{-1}d_{22}d_{11} \not\equiv 1 \mod(x^3 - 1).$$

W.l.o.g., we can fix p_{11} and compute p_{12} using (9.23). We can determine a solution for p_{22} from (9.26) and (9.27) only if $n > 2$. For $n = 3$ the solution of p_{22} is unique.

The remaining parameter p_{21} is derived using (9.26). The validity of condition (9.17) is yet to check. Inserting (9.23) and (9.24) into (9.17) for all cases provides:

$$x^{p_{11}} \not\equiv d_{12}d_{21}d_{22}^{-1}d_{11}^{-1}x^{p_{11}} \mod(x^3 - 1),$$
$$x^{p_{21}} \not\equiv d_{22}d_{11}d_{12}^{-1}d_{21}^{-1}x^{p_{21}} \mod(x^3 - 1).$$

Both conditions are satisfied by prerequisite, since (9.21) holds. Altogether, there is a solution for cyclic IA on the X-channel with $n = 3$ dimensions and $M = 4$ messages satisfying the separability conditions and achieving the upper bound of $\frac{4}{3}$ DoF. □

Corollary 9.3.1 *If* $\det(\boldsymbol{D}) \equiv 0 \mod(x^n - 1)$ *holds, only* 1 *DoF can be achieved on the cyclic polynomial* 2×2*-user* X*-channel.*

In Theorem 9.3.1 (a), it was shown that interference signals can not be aligned if $\det(\mathbf{D}) \equiv 0 \mod(x^n - 1)$ holds. Each message must be received distinctly within its own dedicated offset, i. e., a MA scheme demands $n \geq 4$. Using a multiple-access scheme, only 1 DoF are achievable. Note that, if $\det(\mathbf{D}) \equiv 0 \mod(x^n - 1)$ holds, the cyclic polynomial X-channel does not provide *relativity of IA*. In other words, the influence of each subchannel on the receivers does not provide a sufficiently diverse view of the transmitted signals, so that the observed signals behave identically and the benefits of cyclic IA are precluded. An according exceptional case (the singularity in the GDoF) is also observed in the symmetric 2-user X-channel in terms of the LDCM in [11] at full symmetry.

9.3.2.2 Complementary Reciprocal Symmetry of Cyclic IA

Interestingly, aligning two messages from different transmitters at one receiver provides an inherent symmetry property in the given X-channel, if the condition $\det(\mathbf{D}) \not\equiv 0 \mod(x^n - 1)$ holds. To elaborate this, we consider two parameters $p_1, p_2 \in \mathbb{N}$ for messages $\tilde{W}_1, \tilde{W}_2$ from Tx_1 and Tx_2, respectively. Let p_1 and p_2 be chosen such that these messages are aligned at Rx_2 at a fixed offset $\lambda_2 \in \mathbb{N}$:

$$d_{21}x^{p_1} \equiv d_{22}x^{p_2} \equiv x^{\lambda_2} \mod(x^n - 1). \tag{9.28}$$

This alignment results in the following received signals:

$$\tilde{r}_1(x) \equiv x^{\lambda_2}(d_{11}d_{21}^{-1}\tilde{W}_1 + d_{12}d_{22}^{-1}\tilde{W}_2) \mod(x^n - 1), \tag{9.29}$$
$$\tilde{r}_2(x) \equiv x^{\lambda_2}(\tilde{W}_1 + \tilde{W}_2) \mod(x^n - 1). \tag{9.30}$$

Conversely, we may choose p_1, p_2 such that two other messages, say, $\mathring{W}_1, \mathring{W}_2$ align at Rx_1 at a fixed offset $\lambda_1 \in \mathbb{N}$ instead:

$$d_{12}x^{p_2} \equiv d_{11}x^{p_1} \equiv x^{\lambda_1} \mod(x^n - 1). \tag{9.31}$$

The received signals yield:

$$\mathring{r}_1(x) \equiv x^{\lambda_1}(\mathring{W}_1 + \mathring{W}_2) \mod(x^n - 1), \tag{9.32}$$

$$\mathring{r}_2(x) \equiv x^{\lambda_1}(d_{11}^{-1}d_{21}\mathring{W}_1 + d_{12}^{-1}d_{22}\mathring{W}_2) \mod(x^n - 1). \tag{9.33}$$

By comparing (9.29) with (9.33) and (9.30) with (9.32), we observe that the offsets of the polynomials $\tilde{r}_i(x)$ and $\mathring{r}_j(x)$ can be mutually converted by the following transformation:

$$\mathring{r}_j(x) \equiv x^{\lambda_i+\lambda_j}\, \tilde{r}_i(x^{-1})\big|_{W_1\to\mathring{W}_1, W_2\to\mathring{W}_2} \mod(x^n - 1), \tag{9.34}$$

$$\tilde{r}_j(x) \equiv x^{-\lambda_i-\lambda_j}\, \mathring{r}_i(x^{-1})\big|_{\mathring{W}_1\to W_1, \mathring{W}_2\to W_2} \mod(x^n - 1), \tag{9.35}$$

for $i \neq j \in \mathcal{K}$. The coefficients with the messages are substituted correspondingly.

We call these two alignments *complimentary reciprocal symmetric*,[3] since the dedicated signals and the aligned interference signals are swapped at the receivers and the offsets are reciprocally inverted along λ_i and shifted to λ_j (or vice versa). In other words, the complimentary reciprocal symmetry basically states that aligning signals from both transmitters at receiver one provides a particular signal pattern at receiver two, while, vice versa, aligning two signals at receiver two provides the reciprocal signal pattern observed before at receiver one. This property can already be observed for the elementary X-channel treated in Theorem 9.3.1. When progressing to the 2-user X-channel with general message lengths as given in [17], this reciprocal symmetry property turns out to be a key ingredient for achieving the upper bounds.

9.3.3 3-User Interference Channel

We briefly return to the fully-connected 3-user interference channel as previously discussed in Sect. 9.2.1, but now in terms of the CPCM. We assume that there is a number of $K = 3$ independent messages W_i dedicated to be conveyed pair-wise from transmitter Tx_i to receiver Rx_i with indices $i \in \mathcal{K} := \{1, 2, 3\}$. The 3×3 channel matrix between the transmitters Tx_i and the receivers Rx_j is defined as $\mathbf{D} = (d_{ji})_{1\le i\le 3, 1\le j\le 3}$. The cyclic shifts are $d_{ji} \in \mathcal{D}$ as in the case of the 2-user X-channel. The polynomial $v_i(x)$ contains the message W_i for the dedicated Rx_i with the parameters $p_i \in \mathbb{N}_0$ and $i \in \mathcal{K}$:

$$v_i(x) = W_i x^{p_i}. \tag{9.36}$$

[3]This property should not be confused with the related concept of the *reciprocity of alignment* in [7], where IA on the reciprocal channel with swapped transmitter-receiver pairs is considered instead.

The received polynomials at Rx_j with indices $j \in \mathcal{K}$ are:

$$r_j(x) = \sum_{i=1}^{3} d_{ji} W_i x^{p_i} \bmod(x^n - 1).$$

The task of cyclic IA on the given 3-user interference channel is to convey and decode these 3 dedicated messages W_1, W_2, W_3 interference-free within a period of n dimensions. The proposed scheme is optimal in the sense of cyclic IA if the number of dimensions n is minimal and still feasible. In contrast to the X-channel, neither the *multiple-access interference conditions* nor the *intra-user interference conditions* are required for the 3-user interference channel, since there is only one message W_i per user-pair i. Thus, only the following *inter-user interference conditions* are to be considered for $j \neq i \in \mathcal{K}$:

$$d_{jj} x^{p_j} \not\equiv d_{ji} x^{p_i} \bmod(x^n - 1). \tag{9.37}$$

We count 6 of these separability conditions in total for the 3 users, i.e., a single receiver perceives 2 inter-user interference signals from undesired transmitters. In the given case of 3 user-pairs, *perfect IA* means to align all 2 interfering signals received at Rx_j into a single dimension for all $j \in \mathcal{K}$:

$$d_{ji} x^{p_i} \equiv d_{jk} x^{p_k} \bmod(x^n - 1), \tag{9.38}$$

with pair-wise distinct $i, j, k \in \mathcal{K}$. Cyclic IA on the 3-user interference channel can achieve the upper bound of $\frac{3}{2}$ DoF if all conditions of the following Theorem 9.3.2 hold.

Theorem 9.3.2 *A perfect cyclic IA scheme for the 3-user interference channel with $\boldsymbol{M} = \boldsymbol{I}_{3\times 3}$ exists, if the three condition types:*

- $d_{jj} d_{ki} \not\equiv d_{kj} d_{ji} \bmod(x^n - 1)$,
- $d_{ji} d_{kj} d_{ik} \equiv d_{ij} d_{jk} d_{ki} \bmod(x^n - 1)$,
- *and* $n = 2$,

hold with pair-wise distinct $i, j, k \in \mathcal{K}$. Cyclic IA achieves the upper bound of $\frac{3}{2}$ DoF.

A full proof of this theorem is given in [24] for the more general K-user interference channel with $\mathbf{M} = \mathbf{I}_{K\times K}$ and omitted here.

Furthermore, we would like to point to our work on the 3-user X-channel in [22]. That channel is physically equivalent to the 3-user interference channel. Therein, 9 messages are to be conveyed. Each user has 3 dedicated messages, one to each receiver. We show that it is infeasible to achieve the upper bound of $\frac{9}{5}$ DoF, since there is no valid channel matrix left that satisfies all separability conditions and still allows a perfect cyclic IA solution. In order to enable feasibility we incorporate minimal backhaul networks on the transmitter and/or the receiver side. An alternative approach would be to consider an asymptotic and non-perfect alignment scheme with $n \to \infty$ dimensions. However this step has not been pursued yet.

9.3.4 *Cyclic Interference Neutralization on the* $2 \times 2 \times 2$ *Relay-Interference Channel*

Besides the previous application of cyclic IA, the CPCM also provides an opportunity to investigate the novel concept of *interference neutralization* (IN). In this light, we consider a system of two transmitters, two parallel intermediate relays and two destinations, i. e., a $2 \times 2 \times 2$ *relay-interference channel* as depicted in Fig. 9.6.

On the basis of the LDCM discussed above, the idea of IN has been developed in [27] and also applied in terms of asymptotic IA in [8]. IN is a cooperative signalling scheme for both the sources and relays such that the interfering signals at undesired destinations are literally 'erased over the air'. The effective communication from a source to a dedicated destination is received entirely interference-free. In other words, IN extends the concept of IA and achieves the min-cut upper bounds on the approximate capacity in channels with multiple relays.

In this section, we investigate the $2 \times 2 \times 2$ relay-interference channel in terms of the proposed CPCM as depicted in Fig. 9.6. Our main goal is to convey the maximal number of interference-free messages from each source to each dedicated destination. Instead of decoding all messages at the relays separately, only functions of superimposed messages are decoded. Then, these superimposed messages are forwarded from the relays to the destinations using a proper relaying function. Let the dedicated messages from Tx_i to Rx_i be represented by a vector $\mathbf{w}_i$ of n submessages $W_i^{[k]}$, for $i \in \mathcal{K} = \{1, 2\}$. The transmission is composed of two hops, the *first hop* from the sources to the relays, and the *second hop* from the relays to the destinations. There is no direct link between the sources and the destinations. The channel matrix of the first hop is denoted by $\mathbf{D}$ and the channel matrix of the second hop is $\mathbf{E}$. Let $d_{ij} = x^{\delta_{ij}}$ and $e_{ij} = x^{\eta_{ij}}$, with $\delta_{ij}, \eta_{ij} \in \mathbb{N}$. The received vectors are:

$$\mathbf{r}^{\mathsf{T}} = \mathbf{D}\mathbf{u}^{\mathsf{T}} \operatorname{mod}(x^n - 1), \tag{9.39}$$

$$\mathbf{t}^{\mathsf{T}} = \mathbf{E}\mathbf{r}^{\mathsf{T}} \operatorname{mod}(x^n - 1). \tag{9.40}$$

First hop: Each of the n submessages $W_i^{[k]}$ from source Tx_i is allocated to the corresponding dimension at offset x^k. The transmitted polynomial from Tx_i yields:

$$u_i(x) = \sum\nolimits_{k=0}^{n-1} W_i^{[k]} x^k. \tag{9.41}$$

The two relays R_j receive a superposition of two submessages per dimension k:

$$r_j^{[k]} = W_1^{[k-\delta_{j1}]} + W_2^{[k-\delta_{j2}]}. \tag{9.42}$$

Let the superscript indices denoted in squared brackets, i. e., $[\cdot]$, be reduced modulo n for notational convenience.

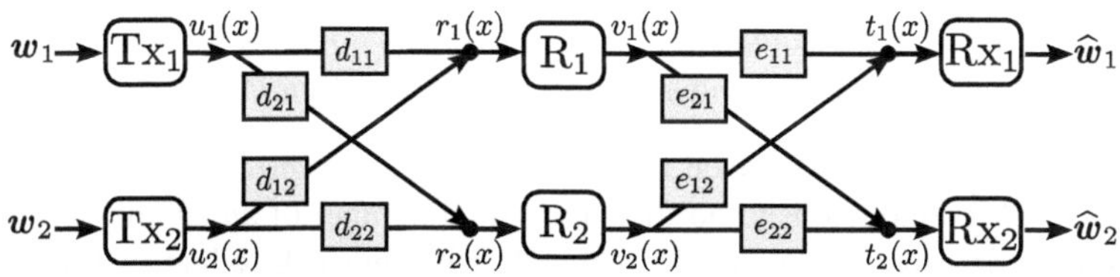

Fig. 9.6 The $2 \times 2 \times 2$ relay-interference channel

Second hop: The two relays forward their previously received polynomials by $v_1(x) = x^{\gamma_1} r_1(x) \bmod (x^n - 1)$ and $v_2(x) = -x^{\gamma_2} r_2(x) \bmod (x^n - 1)$, using the shifting parameters $\gamma_1, \gamma_2 \in \{0, \ldots, n-1\}$, respectively. The destinations Rx_1 and Rx_2 receive four submessages per dimension:

$$t_1^{[k]} = W_1^{[k-\delta_{11}-\gamma_1-\eta_{11}]} + W_2^{[k-\delta_{12}-\gamma_1-\eta_{11}]} - W_1^{[k-\delta_{21}-\gamma_2-\eta_{12}]} - W_2^{[k-\delta_{22}-\gamma_2-\eta_{12}]}, \tag{9.43}$$

$$t_2^{[k]} = W_1^{[k-\delta_{11}-\gamma_1-\eta_{21}]} + W_2^{[k-\delta_{12}-\gamma_1-\eta_{21}]} - W_1^{[k-\delta_{21}-\gamma_2-\eta_{22}]} - W_2^{[k-\delta_{22}-\gamma_2-\eta_{22}]}. \tag{9.44}$$

At both destinations, the dedicated submessages are clearly superimposed by interference. The idea of IN is to combine two identical inter-user interference signals with complementary signs within the same dimension k at the receivers, such that their sum is zero. Thus, to suppress the inter-user interference at both destinations, these two *interference-neutralization conditions* must hold:

$$\delta_{12} + \gamma_1 + \eta_{11} \equiv \delta_{22} + \gamma_2 + \eta_{12} \bmod n, \tag{9.45}$$

$$\delta_{11} + \gamma_1 + \eta_{21} \equiv \delta_{21} + \gamma_2 + \eta_{22} \bmod n. \tag{9.46}$$

In other words, inter-user interference is properly aligned at the relays and the destinations, so that it is *neutralized* over two hops.

On the other hand, we must also ensure that the desired signals remain intact and are not neutralized. Thus, we must impose the *no-signal-neutralization conditions*:

$$\delta_{11} + \gamma_1 + \eta_{11} \not\equiv \delta_{21} + \gamma_2 + \eta_{12} \bmod n, \tag{9.47}$$

$$\delta_{12} + \gamma_1 + \eta_{21} \not\equiv \delta_{22} + \gamma_2 + \eta_{22} \bmod n. \tag{9.48}$$

If these separability conditions are satisfied, the superposition of submessages in (9.43) and (9.44) is reduced to:

$$t_1^{[k]} = W_1^{[k-\delta_{11}-\gamma_1-\eta_{11}]} - W_1^{[k-\delta_{21}-\gamma_2-\eta_{12}]}, \tag{9.49}$$

$$t_2^{[k]} = W_2^{[k-\delta_{12}-\gamma_1-\eta_{21}]} - W_2^{[k-\delta_{22}-\gamma_2-\eta_{22}]}. \tag{9.50}$$

However, this approach leads to the following problem.

Lemma 9.3.1 *Linear decoding at destination* T_j *can not resolve n desired submessages* $W_j^{[k]}$ *from the received polynomial* $t_j(x)$ *for the given IN scheme.*

The corresponding proof is given in [21]. Basically, it is shown that the conditions (9.45) to (9.48) are obviously too strict for a total number of $2n$ submessages.

However, in order to yet enable cyclic IN with linear decoding, we relax these conditions by proposing an *asymptotic IN scheme* for a total number of $2n-1$ instead of $2n$ submessages:

First Hop: Let Tx_1 transmit n submessages as in (9.41) but let Tx_2 transmit only $n-1$ submessages, discarding submessage $W_2^{[\tau]}$ for a parameter $\tau \in \{0,\ldots,n-1\}$:

$$u_1(x) = \sum_{k=0}^{n-1} W_1^{[k]} x^k, \tag{9.51}$$

$$u_2(x) = \sum_{k=0,k\neq\tau}^{n-1} W_2^{[k]} x^k. \tag{9.52}$$

Now, the $k = 0,\ldots,n-1$ received offsets at both relays R_j are:

$$r_j^{[k]} = W_1^{[k-\delta_{j1}]} + W_2^{[k-\delta_{j2}]}, \tag{9.53}$$

$$r_j^{[k]} = W_1^{[k-\delta_{j1}]}, \text{ if } k \equiv \tau + \delta_{j2}. \tag{9.54}$$

Second hop: Relay R_1 forwards all n dimensions and R_2 forwards only $n-1$ of the n received dimensions. In particular, relay R_2 discards forwarding the dimension received at $k_2 \equiv \tau + \delta_{22} \bmod n$. One γ_1, γ_2 is arbitrarily chosen and the other is computed by (9.45). The transmitted polynomials are:

$$v_1(x) = x^{\gamma_1} r_1(x) \bmod (x^n - 1), \tag{9.55}$$

$$v_2(x) = -x^{\gamma_2} \sum_{k=0,k\neq k_2}^{n-1} r_2^{[k]} x^k \bmod (x^n - 1). \tag{9.56}$$

The received signals at $\mathrm{Rx}_1, \mathrm{Rx}_2$ correspond to (9.43), (9.44). The discarded messages for $\sigma_{ji} = \tau + \delta_{i2} + \gamma_i + \eta_{ji} \bmod n$ yield:

$$t_1^{[\sigma_{11}]} = W_1^{[\sigma_{11}-\delta_{11}-\gamma_1-\eta_{11}]} - W_1^{[\sigma_{11}-\delta_{21}-\gamma_2-\eta_{12}]} - W_2^{[\sigma_{11}-\delta_{22}-\gamma_2-\eta_{12}]}, \tag{9.57}$$

$$t_1^{[\sigma_{12}]} = W_1^{[\sigma_{12}-\delta_{11}-\gamma_1-\eta_{11}]} + W_2^{[\sigma_{12}-\delta_{12}-\gamma_1-\eta_{11}]}, \tag{9.58}$$

$$t_2^{[\sigma_{21}]} = W_1^{[\sigma_{21}-\delta_{11}-\gamma_1-\eta_{21}]} - W_1^{[\sigma_{21}-\delta_{21}-\gamma_2-\eta_{22}]} - W_2^{[\sigma_{21}-\delta_{22}-\gamma_2-\eta_{22}]}, \tag{9.59}$$

$$t_2^{[\sigma_{22}]} = W_1^{[\sigma_{22}-\delta_{11}-\gamma_1-\eta_{21}]} + W_2^{[\sigma_{22}-\delta_{12}-\gamma_1-\eta_{21}]}. \tag{9.60}$$

Theorem 9.3.3 *The asymptotic interference neutralization scheme achieves* $\frac{2n-1}{n}$ *DoF on the cyclic polynomial channel, if the interference-neutralization conditions* (9.45), (9.46) *and no-signal-neutralization conditions* (9.47), (9.48) *hold.*

Due to neglecting only a single element in each hop at Tx_2 and Rx_2, the conditions are sufficiently relaxed. The receivers are already enabled to decode each dedicated signal, now. The full proof of this theorem is given in [21]. Note that our proposed scheme generalizes the scheme of [8] for arbitrary cyclic shifts in the CPCM. Furthermore, we introduce particular separability conditions for this scenario in accordance to our previous methodology applied on the 2-user X-channel.

For certain multi-user multi-way communication scenarios, such as two-way relaying, it became apparent that it is useful to combine IA with network coding [15]. An approach of aligning bidirectional pairs of dedicated signals at a relay is termed *signal alignment*. We apply a corresponding combination of cyclic IN and signal alignment for a $2 \times 2 \times 2$ two-way relay channel in [20]. In that case, the separability conditions must hold for 4 messages, i.e., 2 bidirectional messages per user-pair. This demand further restrains the number of valid channel matrices. Using a related asymptotic scheme, the cut-set upper bounds of 4 DoF can be achieved.

When considering these relay-interference channels from the perspective of propagation delays, it is important to maintain the causality constraint at the relays. Thus, the second hop may not proceed before the longest propagation delay of the first hop is traversed. For a sufficiently long duration of the communication, the capacity is nonetheless still achieved asymptotically. However, an immense latency must be taken into account.

9.4 Further Related Works

While our primary focus in this chapter is on the application of IAP, we would like to point to further works considering the CPCM for a variety of networks. For the CPCM of a ring of concatenated two-way relay channels, we observe a User-Relay Duality in [23]. Moreover, inspired by the well-known Δ-Y transformation in electrical circuits, we investigate a corresponding Δ-Y relationship on 3-way and 3-user Y-channels on the CPCM in [17] and the capacity region of the reciprocal linear deterministic 3-way channel in [18]. In this regard, we would like to acknowledge the excellent contributions of Anas Chaaban and Aydin Sezgin at Ruhr-University Bochum, Germany.

If propagation delay is proportional to distance, the concept of IAP is also closely related to placing users in the Euclidean space in order to artificially generate the discrete propagation delays. This approach has been pursued in the works [25, 26].

9.5 Conclusions

In this work we have investigated the concept of interference alignment by propagation delay. To obtain a motivational insight into this concept for the 3-user interference channels, we have discussed it in terms of the *linear deterministic channel model* (LDCM). We have proposed an algorithmic scheme based on interference alignment and we identified several profound challenges that originated from the given channel description.

Although the proposed scheme was neither shown to be optimal, nor fully presented in a closed form, we have observed that the achieved rates yield a consistent oscillating behaviour at moderate interference gains, even for asymmetric channel gains. But the achieved results were actually congruent to results in the literature which are shown to be optimal for the fully symmetric case. This peculiar oscillating property has been traced back to cyclically repeated structures within the LDCM at each user. Using this new insight, we have formulated a modified system model that is based on the LDCM, but with cyclic shifts instead of linear shifts - the *cyclic polynomial channel model* (CPCM). We have utilized the algebraically convenient description of polynomials inspired by cyclic codes to describe our model. In our investigations of the 2-user X-channel and the 3-user interference channel, we were able to obtain closed-form solutions for optimal communication schemes with reasonable effort. These results inherently include asymmetric channel gains. Furthermore, the optimal solution is again based on (cyclic) *interference alignment* and also on (cyclic) *interference neutralization*.

There are several reasons why the proposed CPCM is valuable. First of all, this approach may serve as an intermediate step to derive optimal communication schemes for the LDCM for asymmetric channel gains. Furthermore, key methods like interference alignment and interference neutralization have been included to highlight the flexibility of the proposed model. We have also observed that our results are closely connected to the *Degrees of Freedom*-results of the asymptotic IA schemes as, e.g., in [3, 4]. Our approach has enabled us to focus on detecting some quite general phenomena in multi-user networks that appear hidden in the LDCM for instance. Some particular examples of such phenomena are the *separability conditions*, the *relativity of alignment property* and the *reciprocal symmetry property*.

Altogether, we have established the CPCM for the theoretical investigation interference alignment by propagation delay and we have provided results on the Degrees of Freedom for the 2-user X-channel, the 3-user interference channel and we apply interference neutralization the $2 \times 2 \times 2$ relay-interference channel.

Acknowledgments The work of H. Maier, J. Schmitz and R. Mathar was supported by the German Research Foundation (DFG) under Grants PACIA Ma 1184/15.

References

1. Avestimehr A, Diggavi S, Tse D (2007) A deterministic approach to wireless relay networks. In: Proceedings of 45th allerton conference on communication, control, and computing, Monticello, IL, USA, pp 721–728, Sept 2007
2. Bresler G, Parekh A, Tse D (2010) The approximate capacity of the many-to-one and one-to-many Gaussian interference channels. IEEE Trans Inf Theory 56(9):4566–4592

3. Cadambe V, Jafar S (2008) Interference alignment and degrees of freedom of the K-user interference channel. IEEE Trans Inf Theory 54(8):3425–3441
4. Cadambe V, Jafar S (2009) Interference alignment and the degrees of freedom of wireless X networks. IEEE Trans Inf Theory 55(9):3893–3908
5. Cadambe V, Jafar S, Shamai (Shitz) S, (2009) Interference alignment on the deterministic channel and application to fully connected Gaussian interference networks. IEEE Trans Inf Theory 55(1):269–274
6. Etkin R, Tse D, Wang H (2008) Gaussian interference channel capacity to within one bit. IEEE Trans Inf Theory 54(12):5534–5562
7. Gomadam K, Cadambe V, Jafar S (2008) Approaching the capacity of wireless networks through distributed interference alignment. In: Global communications conference, New Orleans, LO, USA, Nov 2008
8. Gou T, Jafar S, Jeon S, Chung S (2012) Aligned interference neutralization and the degrees of freedom of the $2 \times 2 \times 2$ interference channel. IEEE Trans Inf Theory 58(7):4381–4395
9. Grokop L, Tse D, Yates R (2011) Interference alignment for line-of-sight channels. IEEE Trans Inf Theory 57(9):5820–5839
10. Host-Madsen A, Nosratinia A (2005) The multiplexing gain of wireless networks. In: Proceedings of IEEE international symposium on information theory, Adelaide, SA, Australia, pp 2065–2069, Sept 2005
11. Huang C, Jafar S, Vishwanath S (2012) Interference alignment and the generalized degrees of freedom of the X channel. IEEE Trans Inf Theory 58(8):5130–5150
12. Jafar S, Fakhereddin M (2007) Degrees of freedom for the MIMO interference channel. IEEE Trans Inf Theory 53(7):2637–2642
13. Jafar S, Shamai (Shitz) S (2008) Degrees of freedom region of the MIMO X channel. IEEE Trans Inf Theory 54(1):151–170
14. Jafar S, Vishwanath S (2010) Generalized degrees of freedom of the symmetric Gaussian K user interference channel. IEEE Trans Inf Theory 56(7):3297–3303
15. Lee N, Lim J, Chun J (2010) Degrees of freedom of the MIMO Y channel: signal space alignment for network coding. IEEE Trans Inf Theory 56(7):3332–3342
16. Maddah-Ali M, Motahari A, Khandani A (2006) Signaling over MIMO multi-base systems: Combination of multi-access and broadcast schemes. In: Proceedings of IEEE international symposium on information theory, Seattle, WA, USA, pp 2104–2108, July 2006
17. Maier H (2015) Cyclic interference alignment in multi-user communication networks. Dissertation, RWTH Aachen University
18. Maier H, Chaaban A, Mathar R, Sezgin A (2014) Capacity region of the reciprocal deterministic 3-way channel via Δ transformation. In: Proceedings of 52th allerton conference on communication, control and computing (Allerton), Monticello, IL, USA, pp 167–174, Oct 2014
19. Maier H, Mathar R (2011) Max-min greedy interference alignment on linear deterministic K-user interference channels. In: Proceedings of IEEE international conference on communications, Kyoto, Japan, June 2011
20. Maier H, Mathar R (2013) Cyclic interference neutralization on the $2 \times 2 \times 2$ full-duplex two-way relay-interference channel. In: Proceedings of IEEE information theory workshop, Seville, Spain, pp 644–648, Sept 2013
21. Maier H, Mathar R (2013) Cyclic interference neutralization on the full-duplex relay-interference channel. In: Proceedings of IEEE international symposium on information theory, Istanbul, Turkey, pp 2409–2413, July 2013
22. Maier H, Mathar R (2014) Cyclic interference alignment and cancellation in 3-networks with minimal backhaul. In: Proceedings of IEEE information theory workshop, Hobart, TAS, Australia, pp 87–91, Nov 2014
23. Maier H, Mathar R (2014) User-relay duality in cyclic interference alignment for cascaded two-way relay networks. In: Proceedings of IEEE international symposium on information theory and its applications (ISITA), Melbourne, VIC, Australia, pp 471–475, Oct 2014

24. Maier H, Schmitz J, Mathar R (2012) Cyclic interference alignment by propagation delay. In: Proceedings of 50th allerton conference on communication, Control, and Computing, Monticello, IL, USA, pp 1761–1768, Oct 2012
25. Mathar R, Böcherer G (2009) On spatial patterns of transmitter-receiver pairs that allow for interference alignment by delay. In: Proceedings of 3rd international conference on signal processing and communication systems (ICSPCS), Omaha, USA, Sept 2009
26. Mathar R, Zivkovic M (2009) How to position *n* dimensions such that each can use half of the channel with zero interference from the others. In: Proceedings of global communications conference (GLOBECOM), Honolulu, HI, USA, Dec 2009
27. Mohajer S, Diggavi S, Fragouli C, Tse D (2008) Transmission techniques for relay-interference networks. In: Proceedings of 46th allerton conference on communication, control, and computing, Monticello, IL, USA, pp 467–474, Sept 2008
28. Peterson W, Brown D (1961) Cyclic codes for error detection. Proc Inst Radio Eng 49(1):228–235
29. Zhao Y, Tan C, Avestimehr A, Diggavi S, Pottie G (2012) On the maximum achievable sum-rate with successive decoding in interference channels. IEEE Trans Inf Theory 58(6):3798–3820

Chapter 10
Multicast in Networks of Broadcast Channels—Part I: Submodular Models and Optimization

Maximilian Riemensberger and Wolfgang Utschick

Abstract We present and discuss three multicast rate region models, which are suitable for representing inner bounds, approximations, and outer bounds of the multicast capacity region of wireless networks. These models are the submodular cut model, the hypergraph model, which is an extension of the graph model to include the wireless broadcast advantage, and the polymatroid broadcast model, which is kind of a local cut model. We establish that the hypergraph model is strictly generalized by the polymatroid broadcast model, which is in turn strictly generalized by the submodular cut model. The polymatroid broadcast model also admits a multicast max-flow min-cut theorem that is similar to the hypergraph multicast max-flow min-cut theorem. Furthermore, the polymatroid broadcast model allows for a generic dual decomposition approach to concave network utility maximization problems, which exploits the polymatroid structure to efficiently deal with the large number of (local) cuts in the network. The combination of these properties with the observation that the hypergraph model is not capable of representing some important multicast rate regions, e.g., the multicast capacity region of networks of deterministic broadcast channels, see also Part II of this chapter, makes the polymatroid broadcast model a suitable choice for representing multicast rate regions for networks of broadcast channels and for modeling the wireless broadcast advantage.

10.1 Introduction

Network coding [1] introduced a new communication paradigm where each node codes over its incoming data and forwards coded data instead of just storing and forwarding its plain incoming data as in routed networks. The main result establishes a max-flow min-cut theorem for multicast communication in graphical networks where the single-source multicast capacity coincides with the minimum cut separating the

M. Riemensberger (✉) · W. Utschick
Technische Universität München, 80290 München, Germany
e-mail: riemensberger@tum.de

W. Utschick
e-mail: utschick@tum.de

W. Utschick (ed.), *Communications in Interference Limited Networks*,
Signals and Communication Technology, DOI 10.1007/978-3-319-22440-4_10

source from any terminal in the graphical network. The theorem establishes also a multicast version of the traditional max-flow min-cut theorem in capacitated graphs.

Graphical networks and capacitated graphs are well suited to model wired communication networks, in particular, networks of point-to-point channels, which can be represented as capacitated graphs with the arc capacities given by to the underlying point-to-point channel capacities [37]. However, both are quite unsuitable for wireless communication networks. The reason is that the *wireless broadcast advantage*, which models the effect that transmissions from any node are potentially received by multiple other nodes due to the shared and omnidirectional nature of the wireless medium, cannot adequately be represented by any graph model.

As a remedy to this situation, the graph model has been augmented with virtual nodes and arcs to model transmission of the same information to multiple nodes [43]. This is equivalent to the subsequently introduced hypergraph model for wireless networks [23]. Each hyperarc has a single tail node (transmitter) and multiple head nodes (receivers) and models the transmission of common information from the transmitter to all receivers. The hypergraph model can represent Shannon rate physical layer models, e.g., [5, 18, 39], and packet erasure broadcast network models [23, 24]. However, it has also some severe shortcomings. First, the number of hyperarcs is generally exponential in the number of nodes. This can only be alleviated by local models, where each node has a constant number of neighbors and only hyperarcs to a nodes' neighbors can carry any information, or by exploiting additional mathematical structure in the hyperarc rates [39]. Second, it cannot represent all types of wireless broadcasts. For example, the multicast capacity region of networks of deterministic broadcast channels [30] and the cut-set outer bound of networks of Gaussian broadcast channels cannot be represented by the hypergraph model, cf. [31, 32].

The polymatroid broadcast model was proposed in [32] as a generalization to the hypergraph model. Instead of dividing each transmitters' broadcast into multiple hyperarcs each of which carries some amount of information to a dedicated set of receivers (transmitter-centric model), the polymatroid broadcast model accounts for the total received information for each subset of receivers (receiver-centric model). This amount is quantified by the broadcast rate function, which is required to be a polymatroid rank function [7]. The polymatroid broadcast model shares the property of the graph and hypergraph model that its corresponding cut rate function, which quantifies the total information that can cross each cut in the network, is a submodular function. Furthermore, it strictly generalizes the hypergraph model. In particular, it is capable of representing the multicast capacity region of networks of deterministic broadcast channels [30] and the cut-set outer bound of networks of Gaussian broadcast channels with multiple antennas [32]. Finally, the polymatroid structure of the model can be exploited to derive a multicast max-flow min-cut theorem for this model, and it allows to efficiently deal with the exponential number of receiver subsets of each transmitter. In Sect. 10.2, we give a short overview over the hypergraph model and various variants of the polymatroid broadcast model, most of which were proposed in [31, 32].

Section 10.3 exemplary shows how the polymatroid structure of the polymatroid broadcast model can be exploited to devise efficient strategies for network utility maximization problems. The key idea is to decompose the problem along the flow conservation law into a utility part and a physical layer part using dual decomposition techniques. Then, we exploit Edmonds' greedy algorithm for linear maximization on polymatroid polyhedra [7] to explicitly characterize the physical layer dual function part in terms of the broadcast rates. This allows for an efficient computation of the physical layer dual function despite the exponential number of receiver subsets of each transmitter. Similar techniques apply to other models that feature some suitable submodular structure, cf. [31, 32, 35].

In Part II of this work, which is also included in this book, the polymatroid broadcast models are applied to bounds on the multicast capacity regions of networks of broadcast channels. In particular, the cut-set outer bound and a variant of the noisy network coding inner bound are expressed in terms of the polymatroid broadcast model, cf. [31]. The focus in Part II is on networks of deterministic broadcast channels and networks of erasure broadcast channels. For the former class of networks, the multicast capacity region was established in [30]. It coincides with the cut-set outer bound and admits a simple representation via a polymatroid broadcast rate region, see also [31]. For the latter class of networks, the multicast capacity region is only known if side information on the erasure locations is available to all terminals [6]. This region also coincides with the cut-set outer bound and admits a representation by a singleton set in the polymatroid broadcast model [31]. However, this region need not be achievable without side information and serves as an outer bound in this case. An achievable multicast rate region that can be formulated in the polymatroid broadcast model [31] follows from the noisy network coding scheme [21].

Notation Let N be a finite set and 2^N its power set. Inclusion, strict inclusion, set difference, set intersection, and set union are denoted by $\subset$, $\subsetneq$, $-$, $\cap$, and $\cup$, respectively. The complement of $A \subset N$ in N is denoted by $A^c = N - A$. $\mathscr{R}^N$ denotes the vector space of tuples of reals indexed by elements of N and equivalently the vector space of functions from N to $\mathbb{R}$. $\mathscr{F}^N$ denotes the vector space of tuples indexed by ordered pairs of elements of N and equivalently the vector space of functions from $N \times N$ to $\mathbb{R}$. $\mathscr{S}^N$ denotes the vector space of tuples indexed by subsets of N of and equivalently the vector space of real-valued set functions on N, i.e., the vector space of functions from 2^N to $\mathbb{R}$. $\mathscr{N}^N$ denotes the subspace of $\mathscr{S}^N$ of normalized set functions, i.e., $v(\emptyset) = 0$ for all $v \in \mathscr{N}^N$. $\mathscr{Z}^N$ denotes the vector space of set function vectors on N, i.e., the vector space of N tuples of normalized set functions, e.g., $f = (f_a : a \in N)$ with $f_a \in \mathscr{N}^N$ for all $a \in N$. $\mathscr{R}_+^N$, $\mathscr{F}_+^N$, $\mathscr{S}_+^N$, $\mathscr{N}_+^N$, $\mathscr{Z}_+^N$ denote the respective elementwise and pointwise nonnegative cones.

10.2 Rate Models for Multicast in Networks of Broadcast Channels

We begin this exposition with an abstract definition of a multicast network and a multicast rate region. Let N and T be two nonempty finite sets satisfying $T \subset N$. We refer to (N, T) as a multicast network with *node set* N and *terminal set* T. Some or all nodes $a \in N$ want to send independent messages to the terminals such that all terminals can recover all messages from all nodes. For any particular network and channel model that defines how the nodes can communicate with each other, the corresponding *multicast capacity region* $\mathcal{C} \subset \mathscr{R}_+^N$ represents the set of achievable rate vectors. A rate vector $r \in \mathcal{C}$ collects simultaneously achievable communication rates r_a, $a \in N$, at which all nodes $a \in N$ can simultaneously send messages such that all terminals $t \in T$ can recover all messages. This means that the multicast capacity region abstractly quantifies the maximum multicast information flow from all nodes in N to the terminals in T.

Since the multicast capacity region is unknown except for special classes of networks, one usually has to deal with multicast rate regions that represent inner bounds on the multicast capacity region, i.e., sets of achievable rate vectors using some particular communication strategy, approximations of the multicast capacity region, and outer bounds on the multicast capacity region, i.e., set of rate vectors that contain all achievable rate vectors. In the remainder of this section, we present different multicast rate region formulations based on generic high-level structures that can represent various types of inner bounds, approximations, and outer bounds across large classes of channel and network models. Some of these models are particularly suited to networks of broadcast channels.

10.2.1 Submodular Cut Model

The cut-set structure for rate regions originates in the min-cut upper bound on the maximum commodity and information flow in graphs and the equivalence of the max-flow and the min-cut in graphs [10, 11]. The cut-set structure is also a fundamental property of the cut-set outer bound for arbitrary communication demands in networks [8], see also [4, 9]. Therefore, we first introduce a multicast rate region model based on this structure.

A *cut rate region* $\mathcal{U} \subset \mathscr{N}^N$ on a set of nodes N is a set of normalized set functions $u\colon 2^N \to \mathbb{R}$, $u(\emptyset) = 0$, which are called *cut rate functions*, and generates a *multicast rate region* $\mathcal{R}(\mathcal{U})$ as[1]

$$\mathcal{R}(\mathcal{U}) = \bigcup_{u \in \mathcal{U}} \left(\bigcap_{t \in T} \left\{ r \in \mathscr{R}_+^N : \sum_{a \in A} r_a \le u(A) \quad \forall A \subset \{t\}^c \right\} \right). \tag{10.1}$$

[1]Note that we do not explicitly indicate the node set N and the terminal set T to which $\mathcal{R}(\mathcal{U})$ corresponds since they are always denoted by N and T, and thus there is no source of confusion.

This definition inherently incorporates both the multi-source and the multicast nature of the communication demand. Any rate vector $r \in \mathscr{R}_+^N$ represents simultaneous transmission of messages at rates $r_a \geq 0$ from all nodes $a \in N$ to all terminal nodes $t \in T$. All nodes may assist each other in achieving this communication rate. The multicast nature of the communication can directly be observed in (10.1) since the multicast rate region is defined as the intersection of the rate regions corresponding to all terminals $t \in T$. This is due to the requirement for multicast communication that all terminals must recover all messages. Finally, the union over all cut rate functions $u \in \mathcal{U}$ models availability of multiple communication strategies, e.g., medium access strategies, physical layer strategies, channel input distributions, auxiliary random variable distributions, etc. For details on how the cut rate regions modeling different inner bounds, approximations, and outer bounds are parameterized, we refer the reader to Part II and [31, 32].

The definition of cut rate regions contains no particular structural restrictions. Therefore, even computing the multicast rate region generated by a single cut rate function $u \in \mathscr{N}^N$ generally requires of the order of $2^{|N|}$ set function evaluations. Therefore, we focus on a smaller class of cut rate functions, namely, submodular cut rate functions. A cut rate region $\mathcal{U} \subset \mathscr{N}^N$ is submodular if any cut rate function $u \in \mathcal{U}$ is a submodular set function, i.e.,

$$u(A) + u(B) \geq u(A \cap B) + u(A \cup B) \quad \forall A, B \subset N. \tag{10.2}$$

The set of all submodular set functions is a polyhedral convex cone and is denoted by $\mathscr{K}^N \subset \mathscr{N}^N$. Submodular set functions play an important role since they can be dealt with much easier than general set functions. In the context of discrete optimization, submodularity is often referred to as the *discrete analogous of convexity* [26]. This interpretation stems from the convexity of the Lovász extension [22] and the equivalence of the submodular function minimization problem on 2^N to the convex Lovász extension minimization problem on the unit hypercube in $\mathscr{R}_+^N$. In particular, the minimum cut separating a node $s \in N$ from $t \in N$ with respect to a submodular cut rate function $u \in \mathscr{N}^N$, i.e., the minimum of $u(A)$ over all sets $A \subset N$ such that $s \in A$ and $t \notin A$, can be found in polynomial-time, cf. [13, 19, 25]. A submodular function minimization based optimization framework for point-to-point communication in networks was proposed in [28], and a submodular dual decomposition approach for weighted sum rate maximization in multi-source multicast communication scenarios in was proposed in [31, 32, 35].

10.2.2 Hypergraph Model

Wired networks are usually modeled as directed graphs with each arc $(a, b) \in N \times N$ representing a perfect but capacitated communication link from its tail node a to its head node b. Thereby the flow capacity of each arc represents the maximum information flow that can be communicated over the arc. The graph model can be adapted to wireless networks by allowing that the flow capacities on all arcs depend on each other and by considering a whole region of vectors of flow capacities to represent the shared nature of the wireless medium and the available resources. However, the graph model is not capable of properly modeling the wireless broadcast advantage since the situation where the same information is received by multiple nodes simultaneously cannot be represented. This can be partially alleviated by generalizing the graph model to a hypergraph model, where the broadcast of each node a is represented by a set of capacitated directed hyperarcs $(a, B) \in N \times 2^N$. Information that is conveyed from a over the hyperarc (a, B) is considered to be available to all nodes in B but to none of the nodes in B^c. Therefore, the hypergraph model can represent the potential dependencies in the received information across all receivers of any node $a \in N$, while being closely related to the graph model. The hypergraph model has been quite frequently used in the literature, e.g., [3, 5, 18, 23, 24, 27, 29, 33, 38–40, 43–46].

In order to model multicast rate regions in the hypergraph model, we define a *hyperarc rate function* $g_a : \mathscr{N}^N \to \mathbb{R}$ originating at node a as a set function that satisfies the following three properties:

$$g_a(\emptyset) = 0, \qquad \textit{(normalized)}$$
$$g_a(B) \geq 0 \quad \forall B \subset N, \qquad \textit{(nonnegative)}$$
$$g_a(B \cup \{a\}) = 0 \quad \forall B \subset N. \qquad \textit{(loop-free)}$$

A *hyperarc rate vector* $g = (g_a : a \in N) \in \mathscr{Z}^N$ assigns each node $a \in N$ a hyperarc rate function g_a. The set of all hyperarc rate vectors is denoted as $\mathscr{H}_+^N \subset \mathscr{Z}^N$ and is a convex cone. Each hyperarc rate vector $g \in \mathscr{H}_+^N$ represents a simultaneous allocation of flow capacities to all hyperarcs $(a, B) \in N \times 2^N$ with respect to some underlying channel and network model. A *hyperarc rate region* $\mathcal{G} \subset \mathscr{H}_+^N$ collects multiple such hyperarc rate vectors and can be used to model either an inner bound, an outer bound, or an approximation of the multicast capacity region of the underlying channel and network model, cf. Part II.

The graph model is directly related to the cut model as the value of a cut A through a directed graph is defined as the sum of the capacities of all those arcs that originate in A and end in A^c. This amounts to the maximum amount of information that can be transferred from the source side A of the cut to the terminal side A^c of the cut. Building on this interpretation, for any hyperarc rate vector g, which represents the simultaneous flow capacities between all nodes in $a \in N$ and all subgroups of nodes $B \subset N$, the value of a cut $A \subset N$ through a hypergraph is determined as the sum of the hyperarc rate function values $g_a(B)$ for all hyperarcs (a, B) such that $a \in A$ and $B \cap A^c \neq \emptyset$. That is, it is the sum of the flow capacities of all hyperarcs (a, B) whose tail node is on the source side A of the cut and whose head node set B contains at least one node on the terminal side A^c of the cut, so that the information from $a \in A$ is received by at least one node $b \in B \cap A^c$.

Formally, we define the hyperarc-to-cut transformation $\Omega\colon \mathscr{L}^N \to \mathscr{N}^N, g \mapsto u = \Omega g$ such that

$$u(A) = \sum_{a\in A} \sum_{B\cap A^c \neq \emptyset} g_a(B) \quad \forall A \subset N. \tag{10.3}$$

Each hyperarc rate region $\mathcal{G} \subset \mathscr{H}_+^N$ generates a cut rate region as its image under the map Ω, i.e., $\mathcal{U} = \Omega(\mathcal{G})$ with

$$\mathcal{U} = \bigcup_{g\in\mathcal{G}} \left\{ u \in \mathscr{N}^N : u(A) = \sum_{a\in A} \sum_{B\cap A^c \neq \emptyset} g_a(B) \quad \forall A \subset N \right\}. \tag{10.4}$$

As an immediate consequence, any hyperarc rate region $\mathcal{G}$ generates also a multicast rate region $\mathcal{R}(\mathcal{G})$ through the cut rate region $\Omega(\mathcal{G})$ by defining $\mathcal{R}(\mathcal{G}) = \mathcal{R}(\Omega(\mathcal{G}))$. The properties of the hyperarc-to-cut transformation are discussed in more detail in Sect. 10.2.4 and in [31].

In analogy to the graph model, the multicast rate region $\mathcal{R}(\mathcal{G})$ admits also a direct formulation using a flow in the hypergraph, see [31] for a detailed discussion. The equivalence of both formulations is essentially a multicast max-flow min-cut theorem. This result has directly and indirectly been established in [17, 23, 24, 43]. The proof of this theorem is omitted in this work.

Theorem 10.2.1 *Let $\mathcal{G} \subset \mathscr{H}_+^N$ be a hyperarc rate region. The multicast rate region $\mathcal{R}(\mathcal{G}) = \mathcal{R}(\Omega(\mathcal{G}))$ defined by* (10.1) *and* (10.4) *is given by*

$$\begin{aligned}\mathcal{R}(\mathcal{G}) = \bigcup_{g\in\mathcal{G}} \Bigg(\bigcap_{t\in T} \Bigg\{ & r \in \mathscr{R}_+^N : \;\; x^t \in (\mathscr{R}^N \otimes \mathscr{S}^N \otimes \mathscr{R}^N)_+, \\ & \sum_{b\in C} x^t_{aCb} \le g_a(C), \sum_{b\in C^c} x^t_{aCb} = 0 \quad \forall a \in N, C \subset N, \\ & \qquad \sum_{b\in N}\sum_{C\subset N} x^t_{aCb} - \sum_{b\in N}\sum_{C\subset N} x^t_{bCa} = r_a \quad \forall a \in \{t\}^c, \Bigg\}\Bigg).\end{aligned} \tag{10.5}$$

The nonnegative vector $x^t \in (\mathscr{R}^N \otimes \mathscr{S}^N \otimes \mathscr{R}^N)_+$ for all $t \in T$ represents a nonnegative information flow through the hypergraph.[2] In particular, x^t_{aCb} is the information flow directed to terminal $t \in T$ from node $a \in N$ to node $b \in N$ over the hyperarc (a, C) such that $b \in B$. The second line in (10.5), referred to as the *hyperarc capacity constraint*, states that the total flow $\sum_{b\in C} x^t_{aCb}$ directed to each terminal $t \in T$ over each hyperarc (a, C) cannot exceed the flow capacity of the hyperarc given by $g_a(C)$. Furthermore, it ensures that $x^t_{aCb} = 0$ for all $b \notin C$, i.e., any node $b \notin C$

[2] $\mathscr{R}^N \otimes \mathscr{S}^N \otimes \mathscr{R}^N$ denotes the tensor product of $\mathscr{R}^N$, $\mathscr{S}^N$, and $\mathscr{R}^N$, i.e., real-valued functions on $N \times 2^N \times N$ or, equivalently, vectors indexed by a node $a \in N$, a subset of nodes $C \subset N$, and another node $b \in N$. $(\mathscr{R}^N \otimes \mathscr{S}^N \otimes \mathscr{R}^N)_+$ denotes the elementwise nonnegative cone of $\mathscr{R}^N \otimes \mathscr{S}^N \otimes \mathscr{R}^N$.

cannot receive information from a through the hyperarc (a, C). Additionally, the loop-free condition in the definition of a hyperarc rate function guarantees $x^t_{aCa} = 0$ for all $a \in N$ and $C \subset N$. Finally, the third line in (10.5), referred to as the *flow conservation law*, states that for each terminal $t \in T$ and at each node $a \in \{t\}^c$, the difference between the total outgoing flow directed to $t \in T$ and the total incoming flow directed to t must be equal to the rate r_a at which data are injected into the network by node a. This constraint is the direct analogue of the flow conservation law in the graph model applied to hypergraph flows $x^t \in (\mathscr{R}^N \otimes \mathscr{S}^N \otimes \mathscr{R}^N)_+$ for all $t \in T$.

In [38] a simplified flow formulation for the multicast rate region $\mathcal{R}(\mathcal{G})$ was introduced. It significantly reduces the number of flow variables by summing the flows that leave a and enter b over all hyperarcs $C \subset N$. The equivalence of the simplified formulation to the original formulation follows from the inherent polymatroid structure of the flow capacity constraint and the stability of polymatroids under Minkowski sums.[3] The following theorem establishes this simplified flow formulation of $\mathcal{R}(\mathcal{G})$. For the formal proof of the equivalence of both flow formulations, we refer the reader to [38].

Theorem 10.2.2 *Let $\mathcal{G} \subset \mathscr{H}^N_+$ be a hyperarc rate region. The multicast rate region $\mathcal{R}(\mathcal{G}) = \mathcal{R}(\Omega(\mathcal{G}))$ defined by* (10.1) *and* (10.4) *is given by*

$$\mathcal{R}(\mathcal{G}) = \bigcup_{g \in \mathcal{G}} \Bigg(\bigcap_{t \in T} \Bigg\{ r \in \mathscr{R}^N_+ : x^t \in \mathscr{F}^N_+,\ \sum_{b \in B} x^t_{ab} \leq \sum_{C \cap B \neq \emptyset} g_a(C) \quad \forall a \in N, B \subset N,$$

$$\sum_{b \in N} x^t_{ab} - \sum_{b \in N} x^t_{ba} = r_a \quad \forall a \in \{t\}^c \Bigg\} \Bigg). \tag{10.6}$$

A notable feature of this formulation is the new flow capacity constraint (10.6) (last part of first line) with respect to the new flow vector $x^t \in \mathscr{F}^N_+$. In particular, the right hand side of this constraint appears identically in the definition of the cut rate region generated by a hyperarc rate region (10.4). Furthermore, as a function of $B \subset N$, the right hand side is a polymatroid rank function, i.e., a nonnegative increasing submodular set function [31, 32, 38]. Therefore, the question arises whether the hyperarc sum structure of the right hand side is necessary or whether there is any more general model that fully includes the hyperarc model. The subsequent section introduces such a model, which was proposed in [31, 32].

10.2.3 Polymatroid Broadcast Model

The polymatroid broadcast model [31, 32] leaves the arc or hyperarc structure behind and instead entirely focuses on the amount of received information of each group of receivers from any transmitter. In this sense, it is kind of a local cut model, cf. [36],

[3]Details on the polymatroid structure of the hypergraph model are deferred to Sect. 10.2.4.

which accounts for each cut A through the network, how much information each transmitter $a \in A$ can get through to the terminal side A^c of the cut A. However, the polymatroid broadcast model additionally requires that the contribution of each node to the cut $A \subset N$ is a polymatroid rank function [7] as a function of the terminal side A^c of the cut A, which is for example the case for a local cut model that is derived from the hypergraph model. The polymatroid structure turns out to be crucial to obtain a max-flow min-cut result similar to those for the graph model and the hypergraph model. The variants of the polymatroid broadcast model have been successfully exploited for various unicast and multicast problems, e.g., [27, 34, 41, 42] for unicast communication in lossy wireless packet networks without explicitly mentioning the polymatroid structure.

We define a *broadcast rate function* $f_a\colon \mathscr{N}^N \to \mathbb{R}$ originating at node a as a set function that satisfies the following four properties:

$$
\begin{aligned}
&f_a(\emptyset) = 0, && \textit{(normalized)}\\
&f_a(B) \le f_a(C) \quad \forall B \subset C \subset N, && \textit{(monotone)}\\
&f_a(B) + f_a(C) \ge f_a(B \cup C) + f_a(B \cap C) \quad \forall B, C \subset N. && \textit{(submodular)}\\
&f_a(B) = f_a(B \cup \{a\}) \quad \forall B \subset N. && \textit{(loop-free)}
\end{aligned}
$$

A *broadcast rate vector* $f = (f_a : a \in N) \in \mathscr{X}^N$ assigns each node $a \in N$ a broadcast rate function f_a. The set of all broadcast rate vectors is denoted as $\mathscr{B}_+^N \subset \mathscr{X}^N$ and is a convex cone. Each broadcast rate vector $f \in \mathscr{B}_+^N$ represents a simultaneous allocation of local cut value contributions $f_a(A^c)$ of all nodes $a \in N$ to all cuts $A \subset N$ with respect to some underlying channel and network model. A *broadcast rate region* $\mathcal{F} \subset \mathscr{B}_+^N$ collects multiple broadcast rate vectors and can be used to model either an inner bound, an outer bound, or an approximation of the multicast capacity region of the underlying channel and network model, cf. Part II.

The relation of the polymatroid broadcast model to the cut model is directly established by its interpretation as local cut model. We define the broadcast-to-cut transformation $\Lambda\colon \mathscr{X}^N \to \mathscr{N}^N$, $f \mapsto u = \Lambda f$ such that

$$
u(A) = \sum_{a \in A} f_a(A^c) \quad \forall A \subset N. \tag{10.7}
$$

Each broadcast rate region $\mathcal{F} \subset \mathscr{B}_+^N$ generates a cut rate region as its image under the map Λ, i.e., $\mathcal{U} = \Lambda(\mathcal{F})$ with

$$
\mathcal{U} = \bigcup_{f \in \mathcal{F}} \left\{ u \in \mathscr{N}^N : u(A) = \sum_{a \in A} f_a(A^c) \quad \forall A \subset N \right\}. \tag{10.8}
$$

Consequently, any broadcast rate region $\mathcal{F}$ generates also a multicast rate region $\mathcal{R}(\mathcal{F})$ through the cut rate region $\mathcal{U} = \Lambda(\mathcal{F})$ by defining $\mathcal{R}(\mathcal{F}) = \mathcal{R}(\Lambda(\mathcal{F}))$.

Comparing the two cut rate regions (10.4) and (10.7) reveals that maps generating these regions are tightly connected through a third map converting hyperarc rate vectors to broadcast rate vectors. We define the hyperarc-to-broadcast transformation $\Gamma\colon \mathscr{L}^N \to \mathscr{L}^N, g \mapsto f = \Gamma g$ such that

$$f_a(B) = \sum_{C \cap B \neq \emptyset} g_a(C) \quad \forall a \in N, B \subset N. \tag{10.9}$$

Note that the three transformations Γ, Λ, and Ω are well-defined in the sense that $\Omega = \Lambda\Gamma$, i.e., the cut rate function generated by applying Ω to $g \in \mathscr{L}^N$ is identical to the one generated by applying the first Γ to g and then Λ to the resulting $f = \Gamma g$. Therefore, each hyperarc rate region $\mathcal{G} \subset \mathscr{H}_+^N$ generates not only cut rate region $\mathcal{U} = \Omega(\mathcal{G})$ but also broadcast rate region $\mathcal{F} = \Omega(\mathcal{G})$, which is given by

$$\mathcal{F} = \bigcup_{g \in \mathcal{G}} \left\{ f \in \mathscr{L}^N : f_a(B) = \sum_{C \cap B \neq \emptyset} g_a(C) \quad \forall a \in N, B \subset N \right\} \tag{10.10}$$

In Sect. 10.2.4, we show that $\mathcal{F}$ indeed satisfies the requirements for being a broadcast rate region according to the aforementioned definition provided $\mathcal{G}$ is a hyperarc rate region according to the definition in Sect. 10.2.2. The properties of the transformations are analyzed in greater detail in Sect. 10.2.4.

The multicast rate region $\mathcal{R}(\mathcal{F})$ also admits a direct formulation using a flow model that is similar to the simplified flow model of the hypergraph model (10.6). The equivalence of the flow formulation to the cut rate region formulation based on $\mathcal{U} = \Lambda(\mathcal{F})$ is essentially a multicast max-flow min-cut theorem. It is solely a consequence of the polymatroid structure of the broadcast rate functions f_a. It can be proved by introducing a virtual super-source node and then applying the general polymatroid max-flow min-cut theorem for polymatroid flow networks [14, 15, 20]. The proof is omitted in this work, see [31, 32].

Theorem 10.2.3 *Let $\mathcal{F} \subset \mathscr{B}_+^N$ be a broadcast rate region. The multicast rate region $\mathcal{R}(\mathcal{F}) = \mathcal{R}(\Lambda(\mathcal{F}))$ defined by* (10.1) *and* (10.4) *is given by*

$$\mathcal{R}(\mathcal{F}) = \bigcup_{f \in \mathcal{F}} \left(\bigcap_{t \in T} \left\{ r \in \mathscr{R}_+^N : x^t \in \mathscr{F}_+^N, \ \sum_{b \in B} x_{ab}^t \leq f_a(B) \quad \forall a \in N, B \subset N, \right.\right.$$
$$\left.\left. \sum_{b \in N} x_{ab}^t - \sum_{b \in N} x_{ba}^t = r_a \quad \forall a \in \{t\}^{\mathrm{c}} \right\} \right). \tag{10.11}$$

The vector $x^t = (x_{ab}^t : a, b \in N)$ represents the virtual information flow vector from all sources except t to the terminal t. The flow is closely connected to the rate at which each node injects coded information. This injection rate is occasionally referred to as the network coding subgraph [17, 23] if the multicast rate region $\mathcal{R}(\mathcal{F})$ models an inner bound on the multicast capacity region. Similar to the simplified max-flow formation for hypergraphs, the formulation (10.11) consists of a *flow capacity constraint* (last part of first line) and a *flow conservation law* (second line). The latter is identical to the flow conservation law in the simplified max-flow formulation in the hypergraph model (10.6). The flow capacity constraint upper bounds the total information flow from any node $a \in N$ to any group of nodes $B \subset N$ by the local cut value $f_a(B)$, which represents the total amount of information that can be extracted by the group of nodes B from a transmission of a. The constraint is structurally similar to the flow capacity constraint in (10.6) in the sense that the left hand sides are identical and the right hand sides are connected by the hyperarc-to-broadcast transformation Γ. Finally, we remark that for each $a \in N$, the flow capacity constraint for all $B \subset N$ defines precisely a polymatroid polyhedron on the ground set N, i.e.,

$$P(f_a) = \left\{ x_a \in \mathscr{R}_+^N : \sum_{b \in B} x_{ab} \leq f_a(B) \ \forall B \subset N \right\}. \tag{10.12}$$

10.2.4 Model Inclusion

Three different models for multicast rate regions have been introduced so far. These are tightly connected to each other through the transformations Ω, Λ, and Γ. In this section, we analyze the connections among these models by establishing some fundamental properties of these transformations.

Consider the hyperarc-to-broadcast transformation $\Gamma \colon \mathscr{Z}^N \to \mathscr{Z}^N$ defined on the vector space of set function vectors. The transformation Γ is linear and invertible with its inverse $\Gamma^{-1} \colon \mathscr{Z}^N \to \mathscr{Z}^N$, $f \mapsto g = \Gamma^{-1} f$ given by

$$g_a(C) = \sum_{B \subset C} (-1)^{|B|+1} f_a(B \cup C^c) \quad \forall a \in N, C \subset N. \tag{10.13}$$

The proof that this expression inverts Γ can be found in [31, 32]. The invertibility of Γ on $\mathscr{Z}^N$ implies that we can go back and forth between the broadcast rate function domain and the hyperarc rate function domain. However, it does not establish whether each hyperarc rate vector $g \in \mathscr{H}_+^N$ is transformed into a valid broadcast rate function $f = \Gamma g \in \mathscr{B}_+^N$ and vice versa. This means that we are particularly interested in the properties of the restrictions of the maps Γ and Γ^{-1} to the cones $\mathscr{H}_+^N$ and $\mathscr{B}_+^N$, respectively.

Theorem 10.2.4 *The polymatroid broadcast model strictly generalizes the hypergraph model if* $|N| \geq 4$*, i.e.,* $\Gamma(\mathscr{H}_+^N) \subsetneq \mathscr{B}_+^N$*. For* $|N| = 3$*, both are equivalent, i.e.,* $\Gamma(\mathscr{H}_+^N) = \mathscr{B}_+^N$.

This theorem, whose proof can be found in [31, 32], establishes that the polymatroid broadcast model is a strict generalization of the hypergraph model. This means that there are broadcast rate vectors $f \in \mathscr{B}_+^N$ that cannot be represented by hyperarc rate vectors, i.e., $\Gamma^{-1} f \notin \mathscr{H}_+^N$. However, any hyperarc rate vector $g \in \mathscr{H}_+^N$

corresponds to a broadcast rate vector $\Gamma g \in \mathscr{B}_+^N$. In combination with the maps Ω and Λ, which satisfy $\Omega = \Lambda\Gamma$, we can conclude that any multicast rate region that can be represented by a hyperarc rate region $\mathcal{G}$ can equivalently be represented by the broadcast rate region $\Gamma(\mathcal{G})$, but not vice versa. Part II shows an example for a broadcast rate function taken from a deterministic broadcast channel model that cannot be represented by any hyperarc rate function.

One particularly useful property of the cut rate function introduced by a graph is that it is submodular, see for example [12]. The following theorem establishes the same result for the hypergraph model and for the polymatroid broadcast model. The proof can be found in [31].

Theorem 10.2.5 *The hypergraph model and the polymatroid broadcast model are strictly generalized by the submodular cut model if* $|N| \geq 3$, *i.e.,* $\Omega(\mathscr{H}_+^N) \subsetneq \mathscr{K}^N$ *and* $\Lambda(\mathscr{B}_+^N) \subsetneq \mathscr{K}^N$.

The inclusion theorems Theorems 10.2.4 and 10.2.5 imply a strict hierarchy of the models, which in order of ascending generality is the graph model,[4] the hypergraph model, the polymatroid broadcast model, the submodular cut model, and the general cut model. This means that any multicast rate region $\mathcal{R}(\mathcal{G})$ for some $\mathcal{G} \subset \mathscr{H}_+^N$ admits a representation by a broadcast rate region $\mathcal{F} = \Gamma(\mathcal{G}) \subset \mathscr{B}_+^N$, and any multicast rate region $\mathcal{R}(\mathcal{F})$ admits a representation by a submodular cut rate region $\mathcal{U} = \Lambda(\mathcal{F}) \subset \mathscr{K}^N$. Since Theorems 10.2.4 and 10.2.5 state strict inclusions, the opposite direction is not true in general. That is, one can easily create counter examples of submodular cut rate functions that cannot be represented by broadcast rate vectors and broadcast rate vectors that cannot be represented by hyperarc rate vectors. Such counter examples can be found in [31, 32] and in Part II.

10.2.5 Penalized Polymatroid Broadcast Model

We consider an extension of the polymatroid broadcast model that takes into account that some portion of the multicast rate may be reserved to communicate side information to the terminals to facilitate decoding of the source messages and is thus not available for data transmission. Since this rate portion directly reduces the available multicast rate, we refer to it as *penalty rate*.

A tuple $(f, h) \in \mathscr{Z}^N \times \mathscr{R}^N$ is called a *penalized broadcast rate vector* if $f \in \mathscr{B}_+^N$ is a broadcast rate vector and the penalty rate vector $h \in \mathscr{R}_+^N$ is a nonnegative vector. A set $\mathcal{E} \subset \mathscr{B}_+^N \times \mathscr{R}_+^N$ is referred to as *penalized broadcast rate region*. The penalized broadcast-to-cut transformation is defined as $\Sigma\colon \mathscr{Z}^N \times \mathscr{R}^N \to \mathscr{N}^N$, $(f, h) \mapsto u = \Sigma(f, h)$ such that

$$u(A) = \sum_{a \in A} f_a(A^c) - \sum_{a \in A} h_a \quad \forall A \subset N. \tag{10.14}$$

[4] The graph model can easily be represented as a special case of the hypergraph model by setting the hyperarc rates $g_a(B) = 0$ for all $a \in N$ and $B \subset N$ such that $|B| \geq 2$.

We remark that plugging this cut rate function expression into the multicast rate constraint in (10.1) yields the constraint

$$\sum_{a\in A}(r_a + h_a) \leq \sum_{a\in A} f_a(A^c) \tag{10.15}$$

for any $t \in T$ and all $A \subset \{t\}^c$. The penalty interpretation of the rate vector h is evident from this expression since it reduces the multicast rate vector r that would otherwise be supported by f without the penalty rate vector h. For example, the penalty rate vector plays an important role when modeling noisy network coding schemes where it quantifies the amount of rate that is needed to communicate the output quantization to from each node to the terminals.

The multicast rate region $\mathcal{R}(\mathcal{E})$ generated by a penalized broadcast rate region $\mathcal{E}$ is defined via the cut model as $\mathcal{R}(\mathcal{E}) = \mathcal{R}(\Sigma(\mathcal{E}))$. Furthermore, we also obtain a flow formulation of this region given by

$$\mathcal{R}(\mathcal{E}) = \bigcup_{(f,h)\in\mathcal{E}} \Bigg(\bigcap_{t\in T}\Bigg\{ r \in \mathscr{R}_+^N : x^t \in \mathscr{F}_+^N,\ \sum_{b\in B} x_{ab}^t \leq f_a(B) \quad \forall a \in N, B \subset N,$$
$$\sum_{b\in N} x_{ab}^t - \sum_{b\in N} x_{ba}^t - h_a = r_a \quad \forall a \in \{t\}^c \Bigg\}\Bigg). \tag{10.16}$$

The equivalence of this formulation to the cut formulation can be shown by a straightforward extension of the proof of Theorem 10.2.3 since the penalty rate vector h just reduces the multicast rate vector r in the cut and the flow formulations. We refer to [31] for further details. Finally, we remark that the hypergraph model can be augmented by a penalty rate vector in an analogous manner. Nevertheless, the resulting penalized hypergraph model is less general than the penalized polymatroid broadcast model in the same way as this is the case for the two models without penalty rate vectors.

10.2.6 Extension to Per-Terminal Models

Considering the cut formulation of the multicast rate region $\mathcal{R}(\mathcal{U})$ given in (10.1), we observe that the multicast rate region depends on the terminal set T only through the restriction of the evaluation of the cut rate function u to the sets $A \subset \{t\}^c$. A natural extension is to consider a specific cut rate function u^t for any potential terminal node $t \in N$. Therefore, we can directly extend the formulation to vectors of cut rate functions $u = (u^t : t \in N) \subset \prod_{t\in N} \mathscr{N}^N$. For any *per-terminal cut rate region* $\mathcal{U}^N \subset \prod_{t\in N} \mathscr{N}^N$, we define the corresponding multicast rate region as

$$\mathcal{R}(\mathcal{U}^N) = \bigcup_{u\in\mathcal{U}^N} \Bigg(\bigcap_{t\in T}\Bigg\{ r \in \mathscr{R}_+^N : \sum_{a\in A} r_a \leq u^t(A) \quad \forall A \subset \{t\}^c \Bigg\}\Bigg). \tag{10.17}$$

We observe in (10.17) that the values $u^t(A)$ for all $A \not\subset \{t\}^c$ are irrelevant since they do not bound any rates. Consequently, we define the submodularity of $\mathcal{U}^N$ only based on the relevant cuts. That is, a per-terminal cut rate region is submodular if any $u \in \mathcal{U}^N$ satisfies

$$u^t(A) + u^t(B) \geq u^t(A \cap B) + u^t(A \cup B) \quad \forall t \in N, A \subset \{t\}^c. \tag{10.18}$$

Similarly to the per-terminal cut model, we can also introduce a per-terminal penalized polymatroid broadcast model with vectors of broadcast rate functions $f = (f^t : t \in N) \in \prod_{t\in N} \mathscr{B}_+^N$ and penalty rate vectors $h = (h^t : t \in N) \in \prod_{t\in N} \mathscr{R}_+^N$ consisting of one specific broadcast rate function vector f^t and penalty rate vector h^t for each potential terminal $t \in N$. A set $\mathcal{E}^N \subset \prod_{t\in N} \mathscr{B}_+^N \times \prod_{t\in N} \mathscr{R}_+^N$ is referred to as *per-terminal penalized broadcast rate region*. It generates a cut rate region though the per-terminal broadcast-to-cut transformation $\Sigma^N : \prod_{t\in N} \mathscr{Z}^N \times \prod_{t\in N} \mathscr{R}_+^N \to \prod_{t\in N} \mathscr{N}^N, (f, h) \mapsto u = \Sigma^N(f, h)$ such that

$$u^t(A) = \sum_{a\in A} f_a^t(A^c) - \sum_{a\in A} h_a^t \quad \forall t \in N, A \subset N. \tag{10.19}$$

The map Σ^N is the straightforward extension of Σ to the per-terminal setting.

The multicast rate region $\mathcal{R}(\mathcal{E}^N)$ generated by a per-terminal broadcast rate region $\mathcal{E}^N$ directly follows from $\mathcal{R}(\mathcal{E}^N) = \mathcal{R}(\Sigma^N(\mathcal{E}^N))$. Finally, since the flows in (10.11) are separate for each terminal, the flow formulation for the per-terminal penalized polymatroid broadcast model is readily defined as

$$\mathcal{R}(\mathcal{E}^N) = \bigcup_{(f,h)\in\mathcal{E}^N} \Bigg(\bigcap_{t\in T} \Bigg\{ r \in \mathscr{R}_+^N : x^t \in \mathscr{F}_+^N, \ \sum_{b\in B} x_{ab}^t \leq f_a^t(B) \quad \forall a \in N, B \subset N, \\ \sum_{b\in N} x_{ab}^t - \sum_{b\in N} x_{ba}^t - h_a^t = r_a \quad \forall a \in \{t\}^c \Bigg\} \Bigg). \tag{10.20}$$

The equivalence of this formulation to the cut formulation can be shown by a straightforward extension of the proof of Theorem 10.2.3 to the per-terminal setting. We remark that the per-terminal penalized broadcast-to-cut transformation and the observation that $u^t(A)$ for all $A \not\subset \{t\}^c$ are irrelevant implies that h_t^t for all $t \in N$, $f_t^t(B)$ for all $t \in N$ and $B \subset N$, and $f_a^t(B)$ for all $t \in N$, $a \in \{t\}^c$, and $B \subset \{t\}^c$ are irrelevant to the definition of the multicast rate region $\mathcal{R}(\mathcal{E}^N)$.

Finally, from the cut model for the multicast rate region (10.17) we can directly conclude that if $r \in \mathcal{R}(\mathcal{U}^N)$, then $\{r' \in \mathscr{R}_+^N : r' \leq r\} \subset \mathcal{R}(\mathcal{U}^N)$. That is, the multicast rate region $\mathcal{R}(\mathcal{U}^N)$ is a downward comprehensive set. Using this observation and the equivalence of the cut and the flow formations of $\mathcal{R}(\mathcal{E}^N)$, we find another equivalent flow formulation

$$\mathcal{R}(\mathcal{E}^N) = \bigcup_{(f,h)\in\mathcal{E}^N} \left(\bigcap_{t\in T} \left\{ r \in \mathscr{R}_+^N : x^t \in \mathscr{F}_+^N, \ \sum_{b\in B} x_{ab}^t \le f_a^t(B) \ \ \forall a \in N, B \subset N, \right.\right.$$
$$\left.\left. \sum_{b\in N} x_{ab}^t - \sum_{b\in N} x_{ba}^t - h_a^t \ge r_a \ \ \forall a \in \{t\}^{\mathrm{c}} \right\} \right). \tag{10.21}$$

The only difference to (10.20) is the equality flow conservation law has been replaced by an inequality flow conservation law. It states that for each terminal $t \in T$, the multicast rate for each node $a \subset \{t\}^{\mathrm{c}}$ must be no greater than the total outgoing flow minus the total incoming flow and the penalty rate. Therefore, it explicitly accounts for $\{r' \in \mathscr{R}_+^N : r' \le r\} \subset \mathcal{R}(\mathcal{E}^N)$ whereas the original flow formulation does this only implicitly.

10.3 Dual Decomposition for Network Utility Maximization

The mathematical structure of the proposed submodular cut and polymatroid broadcast models can be exploited to solve certain network utility maximization problems. For example, Parvaresh and Etkin [28] proposed a submodular function minimization based approach to rate maximization and power minimization for certain point-to-point capacity approximations in Gaussian relay networks. Nevertheless, the submodularity can be exploited in a more general context, especially with multiple sources, using a dual decomposition approach. In [31, 32, 35], dual decomposition approaches based on the submodular cut and (penalized) polymatroid broadcast models were proposed. In this section, we focus on a dual decomposition approach based on the per-terminal penalized polymatroid broadcast model.

We consider the *network utility maximization problem*

$$\max \ U(r) \quad \text{s.t.} \quad r \in \mathcal{R}(\mathcal{E}^N) \tag{10.22}$$

with *utility function* $U : \mathscr{R}_+^N \to \mathbb{R}$ and per-terminal penalized broadcast rate region $\mathcal{E}^N \subset \prod_{t\in N} \mathscr{B}_+^N \times \prod_{t\in N} \mathscr{R}_+^N$. We assume $0 \in \mathcal{R}(\mathcal{E}^N)$, U is concave, $\mathcal{E}^N$ is closed and convex, and $\mathcal{E}^N$ has upper bounded broadcast rates, i.e., there exists $\bar{f} \in \prod_{t\in N} \mathscr{B}_+^N$ such that $f_a^t(B) \le \bar{f}_a^t(B)$ for all $t \in N, a \in N$ and $B \subset N$ holds for any $(f, h) \in \mathcal{E}^N$. Using the flow formulation of $\mathcal{R}(\mathcal{E}^N)$ in (10.21), the network utility problem (10.22) can explicitly be formulated as

$$\begin{aligned} &\max \ U(r) \\ &\text{s.t.} \ \sum_{b\in N} x_{ab}^t - \sum_{b\in N} x_{ba}^t - h_a^t \ge r_a \ \ \forall a \in \{t\}^{\mathrm{c}}, \\ &\qquad \sum_{b\in B} x_{ab}^t \le f_a^t(B) \ \ \forall a \in N, B \subset N, \\ &\qquad (f, h) \in \mathcal{E}^N, \ r \in \mathscr{R}_+^N, \ x^t \in \mathscr{F}_+^N \ \ \forall t \in T. \end{aligned} \tag{10.23}$$

Note that we have chosen the inequality-based flow formulation over the equality-based flow formulation since this slightly simplifies the analysis in the following sections.

10.3.1 Polymatroid Broadcast Dual Decomposition Approach

We decompose this optimization problem with respect to the flow conservation constraint, i.e., the first constraint in (10.23). Therefore, we introduce nonnegative dual variables μ_a^t for each $a \in N$ and $t \in T$ such that $\mu_t^t = 0$ for all $t \in T$ since there is no constraint for $a = t$. The vectors $\mu^t = (\mu_a^t : a \in N) \in \mathscr{R}_+^N$ and $\mu = (\mu^t : t \in T) \in \prod_{t \in T} \mathscr{R}_+^N$ denote the vectors of all dual variables for terminal $t \in T$ and the vector of all dual variables. The corresponding Lagrangian function is given by

$$\Phi(r, x, f, h, \mu) = U(r) - \sum_{a \in N} \sum_{t \in T} \mu_a^t r_a + \sum_{a \in N} \sum_{t \in T} \left(\sum_{b \in N} (\mu_a^t - \mu_b^t) x_{ab}^t - \mu_a^t h_a^t \right). \tag{10.24}$$

The dual function $\Theta(\mu)$ is defined as the supremum of Φ over the primal variables (r, x, f, h) taking into account the remaining constraints in (10.23), i.e., the constraints in the last two lines in (10.23). Inspecting the Lagrangian function and the remaining constraints reveals that the dual function Θ decomposes into two parts Θ_{U} and Θ_{P} such that $\Theta(\mu) = \Theta_{\mathrm{U}}(\mu) + \Theta_{\mathrm{P}}(\mu)$. The utility part Θ_{U} is obtained from the supremum over r and is given by

$$\Theta_{\mathrm{U}}(\mu) = \sup_{r \in \mathscr{R}_+^N} \left\{ U(r) - \sum_{a \in N} \sum_{t \in T} \mu_a^t r_a \right\}. \tag{10.25}$$

The physical layer part Θ_{P} is obtained from the supremum over (x, f, h) and is given by

$$\begin{aligned} \Theta_{\mathrm{P}}(\mu) = \sup_{(f,h) \in \mathcal{E}^N} \Bigg\{ \sum_{a \in N} \sum_{t \in T} \Bigg(& -\mu_a^t h_a^t \\ & + \sup_{x_a^t \in \mathscr{R}_+^N} \left\{ \sum_{b \in N} (\mu_a^t - \mu_b^t) x_{ab}^t : \sum_{b \in B} x_{ab}^t \le f_a^t(B) \; \forall B \subset N \right\} \Bigg) \Bigg\}, \end{aligned} \tag{10.26}$$

where $x_a^t = (x_{ab}^t : b \in N) \in \mathscr{R}_+^N$ denotes the vector of flows directed to $t \in T$ leaving node $a \in N$.

Finally, the dual problem to the primal network utility maximization problem (10.23) is given by

$$\min\ \Theta_{\mathrm{U}}(\mu) + \Theta_{\mathrm{P}}(\mu) \quad \text{s.t.} \quad \mu^t \in \mathscr{R}_+^N,\ \mu_t^t = 0\ \ \forall t \in T. \tag{10.27}$$

Note that the dual problem is convex and has only a few simple constraints, i.e., constraints that involve only individual elements of the dual variable vector μ. The weak duality theorem [2] states that the dual problem defines an upper bound on the optimal value of the primal problem. Provided the per-terminal penalized broadcast rate region $\mathcal{E}^N$ is convex, the primal and dual problem are feasible, and a few technical conditions hold, in particular, the conditions stated at the beginning of Sect. 10.3, the strong duality theorem [2] ensures that the optimal value of both problems coincides. In this case, we can solve the network utility maximization problem (10.22) or (10.23) via its dual problem (10.27).

Assuming a value and subgradient evaluation oracle for both dual function parts, the dual problem can be readily solved by standard approaches to nondifferentiable convex optimization, e.g., subgradient, cutting plane, and bundle methods, see [16], since the dual function parts are convex functions. Evaluating $\Theta_{\mathrm{U}}(\mu)$ is a standard convex optimization problem on $\mathscr{R}_+^N$ and can be easily solved for many typical concave utility functions. On the other hand, evaluating Θ_{P} is more involved since the inner maximization problems with respect to x_a^t for all $a \in N$ and $t \in T$ have of the order of $2^{|N|}$ constraints. However, the polymatroid structure of the broadcast rate vectors f leads to a quasi closed-form solution to the inner problem in (10.26), since the inner problem is a linear program on a polymatroid polyhedron. The solution is stated in the following theorem, see [31, 32], which is an immediate corollary to Edmonds' greedy algorithm [7].

Theorem 10.3.1 *Let $\mu^t \in \mathscr{R}_+^N$ and $\mu_t^t = 0$ for all $t \in T$. Define the enumerations e_m^t, $m = 1, \ldots, |N|$, of the node set N for all $t \in T$ such that $e_1^t = t$ and*

$$\mu_{e_1^t}^t \le \cdots \le \mu_{e_{|N|}^t}^t, \tag{10.28}$$

define $E_m^t = \{e_1^t, \ldots, e_m^t\}$ with $E_0^t = \emptyset$, and define o_a^t, $a \in N$, such that $o_a^t = m$ if and only if $e_m^t = a$ for all $a \in N$. The optimal flows $x^\star$ in (10.26) *are given by*

$$x_{ae_m^t}^{\star,t} = \begin{cases} f_a(E_m^t) - f_a(E_{m-1}^t) & \text{if } 1 \le m \le o_a^t - 1, \\ 0 & \text{if } o_a^t \le m \le |N| \end{cases} \tag{10.29}$$

for all $t \in T$ and $a \in N$. The physical layer dual function (10.26) *satisfies*

$$\Theta_{\mathrm{P}}(\mu) = \sup_{(f,h)\in\mathcal{E}^N} \sum_{a\in N}\sum_{t\in T}\left(-\mu_a^t h_a^t + \sum_{m=1}^{o_a^t-1}\left(\mu_{e_{m+1}^t}^t - \mu_{e_m^t}^t\right) f_a^t(E_m^t)\right). \tag{10.30}$$

The optimal flows directly follow from the greedy algorithm in [7] and the physical layer dual function expression (10.30) from plugging the flow solution into (10.26) and some simple algebraic manipulations. Note that the remaining optimization in (10.30) over the penalized broadcast rate vectors $(f, h) \in \mathcal{E}^N$ has a linear objective function, whose coefficients for $f_a^t(B)$ are nonnegative for all $a \in N$, $t \in T$, and $B \subset N$ and whose coefficients for h_a^t are nonpositive for all $a \in N$ and $t \in T$. Therefore, if the supremum in (10.30) is attained, it is attained at a Pareto efficient extreme point of $\mathcal{E}^N$ where Pareto efficiency is defined elementwise increasing for f and elementwise decreasing for h. Whether and how this optimization can efficiently be solved depends entirely on the structure and parameterization of the per-terminal penalized broadcast rate region $\mathcal{E}^N$.

The dual variable vector μ^t for each $t \in T$ admits an interpretation as a node potential, which for example represents a virtual information queue length. The potential of node $a \in N$ with respect to terminal $t \in T$ is $\mu_a^t \geq 0$, and $\mu_t^t = 0$ is the reference potential of terminal t. The enumeration e_m^t is chosen such that the nodes are enumerated in order of ascending potential, i.e., $e_1^t = t$ is the node with the lowest potential and $e_{|N|}^t$ is the node with the highest potential. At each nodes $a \in N$, the optimal outgoing flow vector $x_a^{t,\star}$ is constructed as follows: Flows are only assigned to nodes with lower index than a itself. The node with the lowest potential is assigned as much flow as the broadcast rate function f_a^t permits, i.e.,

$$x_{ae_1^t}^{t,\star} = f_a^t(\{e_1^t\}). \tag{10.31}$$

Note that the monotonicity property of broadcast rate functions ensures that this assignment is feasible with respect to all constraints of the inner problem in (10.26). The node with the second lowest potential e_2^t is assigned as much flow as the broadcast rate function f_a^t permits without decreasing the flow to e_1^t, i.e.,

$$x_{ae_2^t}^{t,\star} = f_a^t(\{e_1^t, e_2^t\}) - f_a^t(\{e_1^t\}). \tag{10.32}$$

The feasibility of this assignment is again ensured by the monotonicity property in conjunction with the submodularity property of the broadcast rate function f_a^t. This procedure continues to the $(o_a^t - 1)$-th lowest potential node. We conclude from this interpretation that the optimal solution routes as much (virtual) information flow as possible from each node $a \in N$ to the nodes with the lowest potential with respect to the terminal $t \in T$, to which all information has eventually to be conveyed, and which it the lowest potential with respect to itself.

10.3.2 Special Cases of the Physical Layer Dual Function Part

In this section, we specialize the physical layer dual function part the physical layer to two simpler models: the polymatroid broadcast model in Sect. 10.2.3 and the hypergraph model in Sect. 10.2.2. The polymatroid broadcast model can be directly represented in the per-terminal penalized polymatroid broadcast model by mapping each broadcast rate vector $f \in \mathscr{B}_+^N$ onto a per-terminal penalized broadcast rate vector $(\hat{f}, 0) \in \prod_{t \in N} \mathscr{B}_+^N \times \prod_{t \in N} \mathscr{R}_+^N$ with $\hat{f}^t = f$ for all $t \in N$. Therefore, the physical layer dual function part corresponding to a broadcast rate region $\mathcal{F} \subset \mathscr{B}_+^N$ is given by

$$\Theta_{\mathrm{P}}(\mu) = \sup_{f \in \mathcal{F}} \sum_{a \in N} \sum_{t \in T} \sum_{m=1}^{o_a^t - 1} \left(\mu^t_{e^t_{m+1}} - \mu^t_{e^t_m} \right) f_a(E^t_m). \tag{10.33}$$

This dual function was proposed in [32].

Since each hyperarc rate vector $g \in \mathscr{H}_+^N$ maps onto a broadcast rate vector $f = \Gamma g \in \mathscr{B}_+^N$, the physical layer dual function part corresponding to a hyperarc rate region $\mathcal{G} \subset \mathscr{H}_+^N$ is obtained from (10.33) via arithmetic manipulations as

$$\Theta_{\mathrm{P}}(\mu) = \sup_{g \in \mathcal{G}} \sum_{a \in N} \sum_{B \subset N} \left(\sum_{t \in T} \left[\mu^t_a - \min_{b \in B} \mu^t_b \right]_+ \right) g_a(B), \tag{10.34}$$

where $[\,\cdot\,]_+ = \max\{0, \,\cdot\,\}$. We observe that a hyperarc (a, B) is assigned a nonzero weight if B contains at least one node with strictly lower potential than a with respect to at least one terminal $t \in T$. Furthermore, note that in this model of the order of $N2^{|N|-1}$ weights need to be computed, i.e., one for each hyperarc (a, B), $a \in N$, $B \subset \{a\}^{\mathrm{c}}$ with potentially nonzero hyperarc rate function.

The weight formula inside the parenthesis has been derived in the literature directly from the hyperarc model [5, 18]. Nevertheless, it turns out to be just one special case of the proposed max-flow dual decomposition approach for penalized broadcast rate functions for a very specific broadcast model, which is generally used in conjunction with Shannon rate broadcast models and hypergraph-based decomposition inner bounds, see for example [3, 5, 18, 23, 27, 29, 33, 38–40, 44–46]. Part II, which is also included in this book, applies the models discussed in this part to bounds on the multicast capacity region of discrete memoryless networks of broadcast channels with particular focus on deterministic and erasure broadcast channels.

10.4 Conclusions

In this part, we developed and discussed various models for multicast rate regions in wireless networks. The central feature of these models is the submodularity of their associated cut rate functions. The penalized polymatroid broadcast model is of particular interest since it is strictly more general than the graph and hypergraph model and features a flow formulation via a multicast max-flow min-cut theorem, which is analogous to the (hyper-)graph max-flow min-cut result. Furthermore, the polymatroid structure allows for an efficient dual decomposition approach to network utility maximization. A similar approach has also been proposed for the submodular cut model in [31, 35].

Acknowledgments This work was supported by the German Research Foundation (DFG) under grants UT 36/10-1 and UT 36/10-2.

References

1. Ahlswede R, Cai N, Li SYR, Yeung RW (2000) Network information flow. IEEE Trans Inf Theory 46(4):1204–1216
2. Bazaraa M, Sherali H, Shetty C (2006) Nonlinear programming: theory and algorithms, 3rd edn. Wiley, New York
3. Chen L, Ho T, Low S, Chiang M, Doyle J (2007) Optimization based rate control for multicast with network coding. In: IEEE INFOCOM, pp 1163–1171
4. Cover TM, Thomas JA (2006) Elements of information theory, 2nd edn. Wiley, New York
5. Cui T, Chen L, Ho T (2010) On distributed scheduling in wireless networks exploiting broadcast and network coding. IEEE Trans Commun 58(4):1223–1234
6. Dana A, Gowaikar R, Palanki R, Hassibi B, Effros M (2006) Capacity of wireless erasure networks. IEEE Trans Inf Theory 52(3):789–804
7. Edmonds J (1970) Submodular functions, matroids, and certain polyhedra. In: Guy R, Hanani H, Sauer N, Schönheim J (eds) Combinatorial structures and their applications. Gordon and Breach, New York, pp 69–87
8. El Gamal A (1981) On information flow in relay networks. In: IEEE national telecommunications conference. New Orleans, LA, USA, vol 2, pp D4.1.1–D4.1.4
9. El Gamal A, Kim YH (2011) Network information theory. Cambridge University Press
10. Elias P, Feinstein A, Shannon C (1956) A note on the maximum flow through a network. IRE Trans Inf Theory 2(4):117–119
11. Ford LR Jr, Fulkerson DR (1962) Flows in networks. Princeton University Press
12. Fujishige S (2005) Submodular functions and optimization, annals of discrete mathematics, vol 58, 2nd edn. Elsevier
13. Grötschel M, Lovász L, Schrijver A (1981) The ellipsoid method and its consequences in combinatorial optimization. Combinatorica 1(2):169–197
14. Hassin R (1978) On network flows. Ph.d. thesis. Yale University
15. Hassin R (1982) Minimum cost flow with set-constraints. Networks 12(1):1–21
16. Hiriart-Urruty JB, Lemaréchal C (1993) Convex analysis and minimization algorithms II: advanced theory and bundle methods, grundlehren der mathematischen wissenschaften, vol 306. Springer, Berlin
17. Ho T, Lun D (2008) Network coding: an introduction. Cambridge University Press
18. Ho T, Viswanathan H (2009) Dynamic algorithms for multicast with intra-session network coding. IEEE Trans Inf Theory 55(2):797–815
19. Iwata S (2008) Submodular function minimization. Math Program 112:45–64
20. Lawler EL, Martel CU (1982) Computing maximal "Polymatroidal" network flows. Math Oper Res 7(3):334–347
21. Lim S, Kim YH, El Gamal A, Chung SY (2011) Noisy network coding. IEEE Trans Inf Theory 57(5):3132–3152
22. Lovász L (1983) Submodular functions and convexity. In: Korte B, Grötschel M (eds) Bachem A. The state of the art. Springer, Berlin, Mathematical programming, pp 235–257
23. Lun D, Ratnakar N, Médard M, Kötter R, Karger D, Ho T, Ahmed E, Zhao F (2006) Minimum-cost multicast over coded packet networks. IEEE Trans Inf Theory 52(6):2608–2623

24. Lun D, Médard M, Kötter R, Effros M (2008) On coding for reliable communication over packet networks. Phys Commun 1(1):3–20
25. McCormick ST (2006) Handbook on discrete optimization. Chapter submodular function minimization. Elsevier, pp 321–391
26. Murota K (2003) Discrete convex analysis. Society for industrial and applied mathematics (SIAM), philadelphia, PA, SIAM monographs on discrete mathematics and applications
27. Neely MJ, Urgaonkar R (2009) Optimal backpressure routing for wireless networks with multi-receiver diversity. Ad Hoc Netw 7(5):862–881
28. Parvaresh F, Etkin R (2014) Efficient capacity computation and power optimization for relay networks. IEEE Trans Inf Theory 60(3):1782–1792
29. Rajawat K, Gatsis N, Giannakis G (2011) Cross-layer designs in coded wireless fading networks with multicast. IEEE/ACM Trans Netw 19(5):1276–1289
30. Ratnakar N, Kramer G (2006) The multicast capacity of deterministic relay networks with no interference. IEEE Trans Inf Theory 52(6):2425–2432
31. Riemensberger M (2015) Submodular rate region models for multicast communication in wireless networks. Dissertation (submitted), Technische Universität München
32. Riemensberger M, Utschick W (2014) A polymatroid flow model for network coded multicast in wireless networks. IEEE Trans Inf Theory 60(1):443–460
33. Riemensberger M, Dotzler A, Utschick W (2009) Factorization for advanced physical layer techniques in network-coded wireless communication networks. In: Wireless network coding (WiNC), pp 1–6
34. Riemensberger M, Wiese T, Utschick W (2013) Network coded wireless multicast with advanced receiver capabilities. In: International ITG conference on systems, communications and coding (SCC), pp 1–6
35. Riemensberger M, Gerdes L, Utschick W (2014) Submodular structure and optimal quantization in gaussian multiple access relay networks. In: IEEE workshop on signal processing advances in wireless communications (SPAWC), pp 319–323
36. Sagduyu YE, Ephremides A (2007) On joint MAC and network coding in wireless Ad Hoc networks. IEEE Trans Inf Theory 53(10):3697–3713
37. Song L, Yeung RW, Cai N (2006) A separation theorem for single-source network coding. IEEE Trans Inf Theory 52(5):1861–1871
38. Traskov D, Heindlmaier M, Médard M, Kötter R (2012) Scheduling for network-coded multicast. IEEE/ACM Trans Netw 20(5):1479–1488
39. Wan L, Luo J (2012) On the complexity of wireless multicast optimization. IEEE Wirel Commun Lett 1(6):593–596
40. Wan L, Luo J, Ephremides A (2008) An iterative framework for optimizing multicast throughput in wireless networks. In: IEEE international symposium on information theory (ISIT), pp 196–200
41. Wiese T (2011) Scheduling with interference in coded wireless packet networks. Tech. Rep. TUM-MSV-TR-12-07, Associate institute for signal processing, Technische Universität München
42. Wiese T, Riemensberger M, Utschick W (2014) Scheduling for network-coded multicast with interference (unpublished as of Dec 2014)
43. Wu Y, Chou PA, Zhang Q, Jain K, Zhu W, Kung SY (2005) Network planning in wireless Ad Hoc networks: a cross-layer approach. IEEE J Sel areas Commun 23(1):136–150
44. Yuan J, Li Z, Yu W, Li B (2006) A cross-layer optimization framework for multihop multicast in wireless mesh networks. IEEE J Sel Areas Commun 24(11):2092–2103
45. Zhao F, Médard M, Lun D, Ozdaglar A (2009) Minimum cost subgraph algorithms for static and dynamic multicasts with network coding. In: Tarokh V (ed) New directions in wireless communications research. Springer, pp 317–349
46. Zhao F, Médard M, Ozdaglar A, Lun D (2014) Convergence study of decentralized min-cost subgraph algorithms for multicast in coded networks. IEEE Trans Inf Theory 60(1):410–421

Chapter 11
Multicast in Networks of Broadcast Channels—Part II: Representation of Bounds on the Multicast Capacity Region

Maximilian Riemensberger and Wolfgang Utschick

Abstract We apply the polymatroid broadcast model to inner and outer bounds on the multicast capacity region of networks of broadcast channels, which are a suitable network model to represent the wireless broadcast advantage in wireless networks. The considered model allows for a channel state such that the state sequence is noncausally known to all nodes to model for example medium access mechanisms. We establish that the cut-set outer bound and the noisy network coding inner bound with independent output quantization admit a formulation in the polymatroid broadcast model proposed in Part I for networks of broadcast channels with independent noise, i.e., the channel outputs are independent across all nodes given the channel inputs and the channel state. This applies in particular to networks of deterministic broadcast channels and networks of erasure broadcast channels with independent erasures given the channel state. The polymatroid broadcast structure inherent to these two bounds enables us to characterize the corresponding multicast rate regions by means of the weighted sum multicast rate maximization problem, which can be solved using the dual decomposition approach in Part I, which is based on the polymatroid broadcast model. For networks of erasure broadcast channels, we propose a simple erasure quantization strategy for noisy network coding based on this dual decomposition approach and submodular maximization. This approach achieves a sum rate performance that is close to the cut set outer bound, which is demonstrated in a bidirectional communication example.

11.1 Introduction

Network coding in wireless packet networks has shown significant gains in terms of throughout and reliability [11, 17]. Networks of erasure broadcast channels, potentially with channel state to represent medium access, are well suited to model lossy

M. Riemensberger (✉) · W. Utschick
Technische Universität München, 80290 München, Germany
e-mail: riemensberger@tum.de

W. Utschick
e-mail: utschick@tum.de

W. Utschick (ed.), *Communications in Interference Limited Networks*,
Signals and Communication Technology, DOI 10.1007/978-3-319-22440-4_11

wireless packet networks [5, 18]. The single-source multicast capacity of networks of erasure broadcast channels with independent erasures has been established in [5] under the assumption of additional side information on the erasure locations at the terminals. A similar capacity result has been obtained for coded wireless packet networks in [18] using random linear network coding [12] based on a nonstandard packet scaling argument.

Another important class of networks are networks of deterministic broadcast channels [1], occasionally also referred to as Aref networks. The single-source multicast capacity of such networks was established in [20]. Networks of deterministic broadcast channels are a special class of deterministic networks, which include linear deterministic networks [2]. The latter have attracted attention due to their theoretical accessibility and their relation to general Gaussian networks [2].

Both classes of networks can be jointly modeled as networks of discrete memoryless broadcast channels with noncausally known state sequence. The multicast capacity regions of such networks are in general unknown except for the aforementioned special cases. Nevertheless, they can be studied through outer and inner bounds, in particular the cut-set outer bound [6, 14] and the noisy network coding inner bound [15, 25]. Both bounds admit a formulation in the proposed polymatroid broadcast models in Part I, also included in this book. We remark that similar results apply also to networks of Gaussian broadcast channels, which are omitted in this work, see [21].

In Part I, we introduced among other models the per-terminal cut model and the per-terminal penalized polymatroid broadcast model to represent arbitrary multicast rate regions. We very briefly review some multicast rate region formulations and refer to Part I for the details. The per-terminal cut model represents a multicast rate region through a per-terminal cut rate region $\mathcal{U}^N \subset \prod_{t \in N} \mathscr{N}^N$ as[1]

$$\mathcal{R}(\mathcal{U}^N) = \bigcup_{u \in \mathcal{U}^N} \Big(\bigcap_{t \in T} \Big\{ r \in \mathscr{R}_+^N : \sum_{a \in A} r_a \le u^t(A) \quad \forall A \subset \{t\}^{\mathrm{c}} \Big\} \Big), \tag{11.1}$$

where $u^t(A)$ represents the cut rate of cut A for a communication to terminal $t \in T$. The per-terminal cut model includes the standard cut model represented by cut rate regions $\mathcal{U} \subset \mathscr{N}^N$ as the special case where $u^t = u^s$ for all $t, s \in N$.

The most general polymatroid broadcast model for a network with node set N in Part I represents multicast rate regions through a per-terminal penalized broadcast rate region $\mathcal{E}^N \subset \prod_{t \in N} \mathscr{B}_+^N \times \prod_{t \in N} \mathscr{R}_+^N$ consisting of pairs (f, h) where $f_a^t(B)$ denotes

[1] $\mathscr{R}^N$ and $\mathscr{F}^N$ denote the vector spaces of real-valued functions (tuples) on N and $N \times N$, respectively, and $\mathscr{R}_+^N$ and $\mathscr{F}_+^N$ denote the corresponding cones of pointwise (elementwise) nonnegative functions (tuples). $\mathscr{N}^N$ denotes the vector space of normalized real-valued set functions on N, i.e., functions $u\colon 2^N \to \mathbb{R}$ such that $u(\emptyset) = 0$. $\mathscr{Z}^N$ denotes the vector space of normalized real-valued set function vectors, i.e., $f = (f_a : a \in N)$ with $f_a \in \mathscr{N}^N$ for all $a \in N$, and $\mathscr{Z}_+^N$ denotes the corresponding cone of elementwise and pointwise nonnegative vectors in $\mathscr{Z}^N$. $\mathscr{B}_+^N \subset \mathscr{Z}_+^N$ denotes the cone of set function vectors f whose set function elements f_a are broadcast rate functions according to the definition in Part I, i.e., essentially polymatroid rank functions on N.

the broadcast rate from $a \in N$ to a subset of nodes $B \subset N$ for a communication to any potential terminal $t \in N$ and h_a^t denotes the penalty rate that node $a \in N$ has to invest to communicate side information for a communication to $t \in N$. The region $\mathcal{E}^N$ directly generates a per-terminal cut rate region through the penalized broadcast-to-cut transformation Σ^N defined by

$$u^t(A) = \sum_{a \in A} f_a^t(A^c) - \sum_{a \in A} h_a^t \tag{11.2}$$

for all $t \in N$ and $A \subset N$. Consequently, $\mathcal{E}^N$ generates a multicast rate region $\mathcal{R}(\mathcal{E}^N) = \mathcal{R}(\Sigma^N(\mathcal{E}^N))$. The details on the requirements that f^t and h^t and their interpretation are given in Part I. The per-terminal penalized broadcast model contains the standard polymatroid broadcast model [22], which is represented by broadcast rate regions $\mathcal{F} \subset \mathscr{B}_+^N$ as the special case where $f^t = f^s$ and $h^t = 0$ for all $t, s \in N$. The broadcast-to-cut transformation without penalties is denoted by Λ and is defined by

$$u(A) = \sum_{a \in A} f_a(A^c) \tag{11.3}$$

for all $A \subset N$.

Due to the polymatroid structure, the multicast rate region $\mathcal{R}(\mathcal{E}^N) \subset \mathscr{R}_+^N$ can be formulated using a flow model as

$$\mathcal{R}(\mathcal{E}^N) = \bigcup_{(f,h) \in \mathcal{E}^N} \left(\bigcap_{t \in T} \left\{ r \in \mathscr{R}_+^N : x^t \in \mathscr{F}_+^N, \ \sum_{b \in B} x_{ab}^t \leq f_a^t(B) \ \ \forall a \in N, B \subset N, \right. \right.$$
$$\left. \left. \sum_{b \in N} x_{ab}^t - \sum_{b \in N} x_{ba}^t - h_a^t \geq r_a \ \ \forall a \in \{t\}^c \right\} \right). \tag{11.4}$$

The auxiliary variables x^t for all $t \in T$ represent virtual information flows between all nodes $a, b \in N$. The second line requires conservation of the flows in the sense that each node cannot inject data at a rate r_a that is higher than the total outgoing flow minus the total incoming flow and the penalty rate h_a^t that is required to communicate side information to the terminals. The last part of the first line models that the virtual information flow to each subgroup of nodes may not exceed the supported broadcast rate $f^t(B)$. The multicast character of the communication stems from the intersection over all terminals $t \in T$ since all data must be communicated to all terminals. The outer union over the per-terminal penalized broadcast rate region $\mathcal{E}^N$ models potential choices of physical layer parameters, e.g., medium access parameters, channel input distributions, quantization distributions for noisy network coding, etc. We remark that the flow formulation $\mathcal{R}(\mathcal{F})$ for a broadcast rate region $\mathcal{F} \subset \mathscr{B}_+^N$ according to the standard polymatroid broadcast model directly follows from setting $f^t = f^s$ for all $t, s \in N$ and $h^t = 0$ for all $t \in N$. Furthermore, we point out that the polymatroid broadcast structure of $\mathcal{R}(\mathcal{E}^N)$ (or $\mathcal{R}(\mathcal{F})$) leads directly to a dual decomposition approach for concave network utility maximization problems [21, 22], which is discussed in detail in Part I. The focus in this chapter will be on the weighted sum multicast rate maximization problem.

In Sect. 11.2, networks of discrete memoryless broadcast channels with non-causally known state sequence are introduced. We establish that the cut-set outer bound and a variant of the noisy network coding inner bound admit a formulation in the polymatroid broadcast model and the per-terminal penalized polymatroid broadcast model, respectively, provided that the network is an independent noise network, i.e., all channel outputs are independent across all nodes given the channel inputs and the channel state. Additionally, we establish the dual formulation of the weighted sum multicast rate maximization problem on those outer and inner bounds. Sect. 11.3 particularizes these results to networks of deterministic broadcast channels. In this case the cut-set outer bound is achievable [20] so that all results for the cut-set outer bound apply also to the multicast capacity region.

Networks of erasure broadcast channels are studied in Sect. 11.4. The cut-set outer bound turns out to be representable by a singleton broadcast rate region since independent inputs are optimal regardless of the erasure channel parameters. The polymatroid broadcast formulation for the noisy network coding inner bound is particularized to erasure broadcast channels with focus on independent erasure quantization and independent inputs. This gives relatively simple closed-from expressions for the penalized broadcast rate function vectors in terms of the erasure probabilities before and after quantization. Both inner and outer bound formulations are again used to define dual functions for the weighted sum multicast rate maximization problem. For the cut-set outer bound, the dual function reduces to a maximization over the state only, whereas for noisy network coding with independent erasure quantization, the erasure probabilities after quantization can additionally be optimized for each state. However, this problem is in general nonconvex. Therefore, we briefly introduce a simple suboptimal erasure quantization optimization scheme based on a local search algorithm for submodular maximization problems, which was proposed in [21]. Finally, this chapter concludes with a simple numerical example for a network of erasure broadcast channels with twelve nodes and a singleton state space.

11.2 Networks of Discrete Memoryless Broadcast Channels

We consider a node set N, finite channel input alphabets $\mathcal{X}_a$ for all $a \in N$, channel input random variables X_a for all $a \in N$, finite channel output alphabets $\mathcal{Y}_{ab}$ for all $a, b \in N$, channel output random variables Y_{ab} for all $a, b \in N$, and a channel state random variable S chosen from a finite state space $\mathcal{S}$. We denote $\mathcal{X}_A = \prod_{a \in A} \mathcal{X}_a$ and $X_A = (X_a : a \in A)$ for all $A \subset N$ and $\mathcal{Y}_{AB} = \prod_{a \in A} \prod_{b \in B} \mathcal{Y}_{ab}$ and $\mathcal{Y}_{AB} = (Y_{ab} : a \in A, b \in B)$ for all $A, B \subset N$. The network is memoryless, i.e., the channel outputs at any time τ depend only on the current channel inputs and channel state at the same time, but are independent from past and future inputs and states.

Furthermore, the conditional distribution of the outputs given the inputs and the state is constant for all times. Therefore, we refrain from explicitly denoting a time index.

The channel inputs and outputs are connected to each other by the conditional probability distribution represented by its probability mass function $p(Y_{NN}|X_N S)$ that factors across all nodes as

$$p(Y_{NN}|X_N S) = \prod_{a \in N} p(Y_{aN}|X_a S). \tag{11.5}$$

This means that the vector of output random variables Y_{aN} represents the outputs of the broadcast channel rooted at node $a \in N$ and is independent of all other inputs $X_{\{a\}^c}$ and outputs $Y_{\{a\}^c N}$ given the input X_a and the state S. Furthermore, we require that Y_{aa} is a constant so that the broadcast channels are loop free. We refer to such a network as *independent noise* network if the conditional distribution factors as

$$p(Y_{NN}|X_N S) = \prod_{a \in N} \prod_{b \in N} p(Y_{ab}|X_a S), \tag{11.6}$$

i.e., the outputs are independent across all broadcast channels and nodes given the respective input and the state.

The channel states at all times are assumed to be noncausally available to all nodes and its distribution can be freely chosen by the network to facilitate communication. This represents a very simple model for some interference and medium access model where the state S represents the channels resulting, for example, from a prearranged medium access schedule. It is quite frequently, but usually implicitly, used for medium access optimization in wireless packet networks to model for example packet collision effects and half-duplex constraints, e.g., see [17, 24]. Finally, we remark that $|S| = 1$ corresponds to a setting without any channel states.

For any particular network of discrete memoryless broadcast channels, the *multicast capacity region* $\mathcal{C} \subset \mathscr{R}_+^N$ represents the closure of the set of achievable rate vectors. A rate vector $r \in \mathscr{R}_+^N$ is *achievable* if each node $a \in N$ can send one out of 2^{nr_a} messages within n time units (channel usages) such that all terminals can recover all messages with vanishing probability of error as $n \to \infty$. This is, the multicast capacity region quantifies the maximum set of information rate vectors for reliable multicast communication from all nodes in N to the terminals in T.

11.2.1 Cut-Set Outer Bound

The general outer bound on the multicast capacity region is the cut-set outer bound, which was originally derived in [6], see also [4, 7]. In [21], it was shown that the cut-set outer bound admits a formulation as multicast rate region generated by a broadcast rate region as follows:

Theorem 11.2.1 *For any discrete memoryless network of broadcast channels with nodes N and channel $p(Y_{NN}|X_N S)$, if a rate vector $r \in \mathscr{R}_+^N$ of independent data rates is achievable for the multicast to the terminal set $T \subset N$, i.e., if $r \in \mathcal{C}$, then it satisfies $r \in \mathcal{R}(\Lambda(\mathcal{F}_o))$ where the multicast rate region $\mathcal{R}(\Lambda(\mathcal{F}_o))$* (11.1) *is generated through the broadcast-to-cut transformation Λ by the region of set function vectors $\mathcal{F}_o \subset \mathscr{L}_+^N$ given by*

$$\mathcal{F}_o = \bigcup_{p(S)\prod_{a\in N} p(X_a|S)} \left\{ f \in \mathscr{L}_+^N : f_a(B) = I(X_a; Y_{aB}|S) \ \forall a \in N, B \subset N \right\}. \tag{11.7}$$

The proof of the general cut-set outer bound can be found in [4, 7] and the application to networks of broadcast channels in [21]. The original cut-set outer bound requires the rate vector r to satisfy

$$\sum_{a\in A} r_a \le I(X_A; Y_{NA^c}|X_{A^c}S) \quad \forall t \in T, A \subset \{t\}^c \tag{11.8}$$

for some joint distribution $p(X_N S)$, where $I(X_A; Y_{NA^c}|X_{A^c}S)$ denotes the conditional mutual information of X_A and Y_{NA^c} given X_{A^c} and S. However, exploiting the decomposition of the channel conditional distribution $p(Y_{NN}|X_N S)$ in (11.5), we observe that the right hand side in (11.8) is maximized by independent inputs, i.e., $p(X_N S) = p(S)\prod_{a\in N} p(X_a|S)$, for which it decomposes as

$$I(X_A; Y_{NA^c}|X_{A^c}S) = \sum_{a\in A} I(X_a; Y_{aA^c}|S) \tag{11.9}$$

for any $A \subset N$. The right hand side in (11.9) has the same form as a cut rate function that is generated from a set function vector by the broadcast-to-cut rate transformation Λ. The resulting set function vector f with $f_a(B) = I(X_a; Y_{aB}|S)$ for all $a \in N$ and $B \subset N$ is clearly nonnegative and increasing in $B \subset N$ due to the nonnegativity of the conditional mutual information and the chain rule, see [4, Ch. 2]. However, the submodularity of $f_a(B)$ cannot be established in general. However, the following conditions are helpful to establish the submodularity of $f_a(B)$ for special classes of networks, see also [21]:

Proposition 11.2.1 *The set $\mathcal{F}_o$ is a broadcast rate region, i.e., $\mathcal{F}_o \subset \mathscr{B}_+^N$, if and only if $-H(X_a|Y_{aB}S)$ is a submodular set function in B. The set $\mathcal{F}_o$ is a broadcast rate region if the network of broadcast channels is an independent noise network, i.e., if the conditional distribution $p(Y_{NN}|X_N S)$ factors across all nodes as in* (11.6).

Note that if $\mathcal{F}_o$ is a broadcast rate region, then the flow formulation (11.4) (without penalties) is applicable to the multicast rate region $\mathcal{R}(\mathcal{F}_o)$. Otherwise, only the cut formulation (11.1) represents the cut-set outer bound on the multicast capacity region. Furthermore, we remark that the necessary and sufficient condition that the conditional entropy $-H(X_a|Y_{aB}S)$ is submodular as a set function in B follows directly from the definition of the conditional mutual information in terms of conditional entropies as

$$I(X_a; Y_{aB}|S) = H(X_a|S) - H(X_a|Y_{aB}S). \tag{11.10}$$

The second condition, which is sufficient but not necessary, is established in [21] based on the polymatroid structure of the joint entropy function $H(Y_{aB}|S)$ as a set function in B [9] and the decomposition

$$I(X_a; Y_{aB}|S) = H(Y_{aB}|S) - H(Y_{aB}|X_aS) = H(Y_{aB}|S) - \sum_{b\in B} H(Y_{ab}|X_aS), \tag{11.11}$$

where the second equality is due to the factorization of the conditional distribution $p(Y_{NN}|X_NS)$ in (11.6). Furthermore, note that even for independent noise networks, the cut-set outer bound need not be representable by the hypergraph model. An exemplary counter example based on a simple network of deterministic broadcast channels is given in Sect. 11.3, see also [21, 22].

11.2.2 Noisy Network Coding Inner Bound

The noisy network coding inner bound was proposed in [15, 25] as an extension and generalization of the compress-and-forward scheme [3] to arbitrary networks. The key concept of this coding scheme is that each node independently encodes a quantized version of its received signals together with its own messages and each terminal of the multicast jointly decodes all messages and some quantization indices from its received signals. Assuming optimal encoders and decoders, the performance of the scheme is governed by the choice of the input and quantization distributions. These are required to factor across all nodes since the noisy network coding scheme does not consider transmitter cooperation as opposed to other relaying schemes such as (partial) decode-and-forward, see for example [3, 7, 16].

We consider a network of discrete memoryless broadcast channels as introduced in Sect. 11.2. We denote by $\hat{Y}_{NN}$ the quantized channel output at all nodes and by Q a timesharing random variable to allow for convex combinations of multiple input and quantization distributions. The noisy network coding scheme applied to networks of broadcast channels requires the following structure of the joint distribution

$$p(\hat{Y}_{NN}Y_{NN}X_NQS) = p(S)p(Q|S)p(Y_{NN}|X_NS)\prod_{a\in N} p(X_a|QS)\prod_{b\in N} p(\hat{Y}_{Nb}|X_bY_{Nb}QS). \tag{11.12}$$

Since we focus on the polymatroid broadcast model in this work, we additionally require that at each node, all received signals are quantized independently and independently of its own transmit signal, i.e.,

$$p(\hat{Y}_{Nb}|X_b Y_{Nb} QS) = \prod_{a\in N} p(\hat{Y}_{ab}|Y_{ab} QS) \quad \forall b \in N. \tag{11.13}$$

Furthermore, we restrict our analysis to independent noise networks (11.6), and therefore, the joint distribution satisfies

$$p(\hat{Y}_{NN} Y_{NN} X_N QS) = p(S)p(Q|S) \prod_{a\in N} p(X_a|QS) \prod_{b\in N} p(Y_{ab}|X_a S)p(\hat{Y}_{ab}|Y_{ab} QS). \tag{11.14}$$

Applying this quantization scheme yields the following noisy network coding inner bound:

Theorem 11.2.2 *For any independent noise discrete memoryless network of broadcast channels with node set N and channel $p(Y_{NN}|X_N S)$, any rate vector $r \in \mathscr{R}_+^N$ is achievable for a multicast to the terminal set $T \subset N$, i.e., $r \in \mathcal{C}$, if it satisfies $r \in \mathcal{R}(\Sigma^N(\mathcal{E}_{\text{nnc}}^N))$ where the multicast rate region $\mathcal{R}(\Sigma^N(\mathcal{E}_{\text{nnc}}^N))$ is generated through the per-terminal penalized broadcast-to-cut transformation Σ^N by the region $\mathcal{E}_{\text{nnc}}^N \subset \prod_{t\in N} \mathscr{Z}_+^N \times \prod_{t\in N} \mathscr{R}_+^N$ given by*

$$\mathcal{E}_{\text{nnc}}^N = \bigcup_{\substack{p(S)p(Q|S)\prod_{a\in N} p(X_a|QS) \\ \times \prod_{a\in N}\prod_{b\in N} p(\hat{Y}_{ab}|Y_{ab}QS)}} \Bigg\{ (f,h) \in \prod_{t\in N} \mathscr{Z}_+^N \times \prod_{t\in N} \mathscr{R}_+^N :$$

$$f_a^t(B) = I(X_a; \hat{Y}_{aB} Y_{at}|QS),\ h_a^t = \sum_{b\in N} I(Y_{ba}; \hat{Y}_{ba}|X_b QS) \quad \forall t \in N, a \in N, B \subset N \Bigg\}. \tag{11.15}$$

This theorem straightforwardly extends the penalized broadcast formulation for noisy network coding from [21]. The proof is omitted. The achievability directly follows from the achievability of the noisy network coding inner bound for general networks [15]. The general bound for networks of broadcast channels states that $r \in \mathscr{R}_+^N$ is achievable if it satisfies

$$\sum_{a\in A} r_a \le I(X_A; \hat{Y}_{NA^c} Y_{Nt}|X_{A^c} QS) - I(Y_{NA}; \hat{Y}_{NA}|\hat{Y}_{NA^c} Y_{Nt} X_N QS) \tag{11.16}$$

for all $t \in T$ and $A \subset \{t\}^c$ and some distribution with the required structure (11.14). The code construction and decoding procedure are in detail explained in [13, 15, 25]. The intuition behind this rate bound is that for any cut $A \subset \{t\}^c$, the first term quantifies the achievable communication rate over the cut from all nodes A to the terminal t, where the terminal has access to its own received signals Y_{Nt} and all quantized received signals $\hat{Y}_{NA^c}$ on its side of the cut. The second term quantifies the amount of cut rate that needs to be allocated to communicate the quantized received signals $\hat{Y}_{NA}$ from the source side of the cut to the terminal and is thus not available for source information communication.

The structure of the channel, quantization, and input distributions allows for a decomposition of the right hand side in (11.16). That is, for any $t \in T$ and $A \subset \{t\}^c$, the first term factors as

$$I(X_A; \hat{Y}_{NA^c} Y_{Nt} | X_{A^c} QS) = \sum_{a \in A} I(X_a; \hat{Y}_{aA^c} Y_{at} | QS) \tag{11.17}$$

and the second term as

$$I(Y_{NA}; \hat{Y}_{NA} | \hat{Y}_{NA^c} Y_{Nt} X_N QS) = \sum_{a \in A} \sum_{b \in N} I(Y_{ba}; \hat{Y}_{ba} | X_b QS). \tag{11.18}$$

Both factorizations are obtained via lengthy chains of Shannon information inequalities where various Markov chains, which are due to the distribution structure, need to be exploited. The derivation follows the lines of the penalized broadcast model for noisy network coding in [21] and is omitted in this work. Comparing the factorizations to the structure of the per-terminal penalized broadcast model yields the rate region (11.15).

However, Theorem 11.2.2 does not establish that $\mathcal{E}^N_{\text{nnc}}$ is indeed a per-terminal broadcast rate region, in particular, that the set functions $f^t_a(B) = I(X_a; \hat{Y}_{aB} Y_{at} | QS)$ are submodular. This is the statement of the following proposition, whose proof follows again the lines of the corresponding result in [21] and is also omitted in this work.

Proposition 11.2.2 *The set $\mathcal{E}^N_{\text{nnc}}$ is a per-terminal penalized broadcast rate region, i.e., $\mathcal{E}^N_{\text{nnc}} \subset \prod_{t \in N} \mathscr{B}^N_+ \times \prod_{t \in N} \mathscr{R}^N_+$, for any independent noise network of broadcast channels.*

This result is again tightly connected to the polymatroid structure of the entropy function. In particular, the broadcast rate function mutual information expression can be decomposed as

$$\begin{aligned} I(X_a; \hat{Y}_{aB} Y_{at} | QS) &= H(\hat{Y}_{aB} Y_{at} | QS) - H(\hat{Y}_{aB} Y_{at} | X_a QS) \\ &= H(\hat{Y}_{aB} Y_{at} | QS) - \sum_{b \in B} H(\hat{Y}_{ab} Y_{at} | X_a QS). \end{aligned} \tag{11.19}$$

The second inequality follows from the independent noise assumption and the considered factorization of the quantization distribution. The first term in this expression is submodular in B [9] and the second term is modular in B. Consequently, f^t_a is a submodular set function and thus a polymatroid rank function since it is also non-negative and increasing in B.

An important special case of a noisy network coding quantization scheme is *perfect quantization*, which means that $\hat{Y}_{ab} = Y_{ab}$ is chosen for all $a, b \in N$.

This yields per-terminal penalized broadcast rate function vectors (f, h) of the form $f_a^t(B) = I(X_a; Y_{at} Y_{aB} | QS)$ and $h_a^t = \sum_{b \in N} H(Y_{ba} | X_b QS)$ for all $t \in T$, $a \in N$, and $B \subset N$. That is, essentially the same broadcast rate vectors as in the cut-set outer bound can be achieved, except for the additional Y_{at} term, which turns out to be irrelevant for the perfect quantization case. However, the supported multicast rate at each node is reduced by the total entropy of all receive signals given the corresponding inputs.

11.2.3 Weighted Sum Multicast Rate Maximization

We consider the network utility maximization problem as presented in Part I on the multicast rate regions representing the aforementioned inner and outer bounds with a weighted sum multicast rate utility. For a convex per-terminal penalized broadcast rate region $\mathcal{E}^N$ and a weighted sum multicast rate utility function $U(r) = \sum_{a \in N} w_a r_a$ with weight vector $w \in \mathscr{R}_+^N$ such that $w_t = 0$ if $T = \{t\}$ for some $t \in N$, the problem is given by

$$\max \sum_{a \in N} w_a r_a \quad \text{s.t.} \quad r \in \mathcal{R}(\mathcal{E}^N). \tag{11.20}$$

Applying the polymatroid broadcast dual decomposition approach in Part I yields the corresponding dual problem

$$\min \ \Theta_{\text{P}}(\mu) \quad \text{s.t.} \quad \mu^t \in \mathscr{R}_+^N,\ \mu_t^t = 0\ \forall t \in T,\ \sum_{t \in T} \mu_a^t \geq w_a\ \forall a \in N. \tag{11.21}$$

The physical layer dual function part Θ_{P} for all $\mu^t \in \mathscr{R}_+^N$ and $\mu_t^t = 0$ for all $t \in T$ follows from the polymatroid structure of $\mathcal{E}^N$, see Part I, as

$$\Theta_{\text{P}}(\mu) = \sup_{(f,h) \in \mathcal{E}^N} \sum_{a \in N} \sum_{t \in T} \left(-\mu_a^t h_a^t + \sum_{m=1}^{o_a^t - 1} \left(\mu_{e_{m+1}^t}^t - \mu_{e_m^t}^t \right) f_a^t(E_m^t) \right), \tag{11.22}$$

where for each $t \in T$, the enumeration e_m^t, $m = 1, \ldots, |N|$, is chosen such that e_m^t sorts μ^t in ascending order and $e_1^t = t$, and the sets E_m^t are defined as $E_m^t = \{e_1^t, \ldots, e_m^t\}$.

11.2.3.1 Cut-Set Outer Bound

Plugging the broadcast rate region $\mathcal{F}_{\text{o}}$, which represents the cut-set outer bound, into the definition of the physical layer dual function part Θ_{P} (11.22) yields

$$\begin{aligned}\Theta_P(\mu) &= \sup_{p(S)\prod_{a\in N} p(X_a|S)} \sum_{a\in N}\sum_{t\in T}\sum_{m=1}^{o_a^t-1}\left(\mu^t_{e^t_{m+1}} - \mu^t_{e^t_m}\right) I(X_a; Y_{aE^t_m}|S)\\ &= \max_{\sigma\in\mathcal{S}} \sum_{a\in N} \sup_{p(X_a|S=\sigma)} \sum_{t\in T}\sum_{m=1}^{o_a^t-1}\left(\mu^t_{e^t_{m+1}} - \mu^t_{e^t_m}\right) I(X_a; Y_{aE^t_m}|S=\sigma). \qquad (11.23)\end{aligned}$$

The second expression follows from the first since the objective decomposes across all nodes $a \in N$ and each summand depends only on one input distribution factor. Furthermore, since the objective is linear in the distribution $p(S)$, the optimum with respect to $p(S)$ is achieved at an extreme point of the set of distributions, and therefore the optimization is equivalent to the maximization of over all states. Note that the input distribution optimization for each state is a convex optimization problem, since the mutual information expressions are concave in the input distribution, see [4].

11.2.3.2 Noisy Network Coding Inner Bound

Plugging the per-terminal penalized broadcast rate region $\mathcal{E}^N_{\text{nnc}}$, which represents the noisy network coding inner bound with independent quantization across all broadcast channels at each node for an independent noise network, into the physical layer dual function Θ_P (11.22) yields

$$\begin{aligned}\Theta_P(\mu) = \sup_{\substack{p(S)p(Q|S)\prod_{a\in N} p(X_a|QS)\\ \times\prod_{a\in N}\prod_{b\in N} p(\hat{Y}_{ab}|Y_{ab}QS)}} \sum_{a\in N}\sum_{t\in T}\Bigg(&-\mu^t_a \sum_{b\in N} I(Y_{ba}; \hat{Y}_{ba}|X_b QS)\\ &+\sum_{m=1}^{o_a^t-1}\left(\mu^t_{e^t_{m+1}} - \mu^t_{e^t_m}\right) I(X_a; \hat{Y}_{aE^t_m} Y_{at}|QS)\Bigg).\end{aligned} \qquad (11.24)$$

This expression can be significantly simplified by performing the following three steps: We convert the optimization over the state distribution $p(S)$ into a maximization over the state $\sigma \in \mathcal{S}$, which is equivalent since all terms are linear in $p(S)$. The timesharing random variable can be altogether dropped since it is equivalent to a convex hull operation over the mutual information terms in the objective with respect to all input and quantization distributions. Finally, we can exchange the summations with respect to a and b is the first term and thus move the remaining optimization with respect to the input and quantization distribution inside the first sum over all nodes. Combining all three simplifications yields

$$\Theta_{\mathrm{P}}(\mu) = \max_{\sigma \in \mathcal{S}} \sum_{a \in N} \sup_{p(X_a|S)} \sup_{\prod_{b \in N} p(\hat{Y}_{ab}|Y_{ab}S)} \sum_{t \in T} \Bigg(-\sum_{b \in N} \mu_b^t I(Y_{ab}; \hat{Y}_{ab}|X_a S) + \sum_{m=1}^{o_a^t - 1} \Big(\mu_{e_{m+1}^t}^t - \mu_{e_m^t}^t \Big) I(X_a; \hat{Y}_{aE_m^t} Y_{at}|S) \Bigg). \tag{11.25}$$

In order to evaluate $\Theta_{\mathrm{P}}(\mu)$, we need to solve $|N|$ joint input and quantization distribution optimization problems for each state, which is in general difficult since the mutual information expressions are usually nonconcave. This is in contrast to the evaluation of the physical layer dual function for the cut-set outer bound, which requires only the solution of convex optimization problems.

11.3 Networks of Deterministic Broadcast Channels

A network of broadcast channels as defined in Sect. 11.2 is called a *network of deterministic broadcast channels* if the channel outputs are deterministic functions of the channel inputs and the state. That is, there exist maps $G_{ab} \colon \mathcal{X}_a \times \mathcal{S} \to \mathcal{Y}_{ab}$ such that

$$Y_{ab} = G_{ab}(X_a, S) \tag{11.26}$$

for all $a, b \in N$. An immediate consequence is that the entropy of the channel outputs given the inputs and the state is zero, i.e., $H(Y_{NN}|X_N S) = 0$. Therefore, networks of deterministic broadcast channels satisfy the independent noise property (11.6) regardless of the functions G_{ab}. Such networks have been introduced by Aref [1] and are occasionally referred to as Aref networks. The multicast capacity region of networks of deterministic broadcast channels was established in [20].

For networks of deterministic broadcast channels, the region $\mathcal{F}_{\mathrm{o}}$, which characterizes the cut-set outer bound, is a broadcast rate region due to Proposition 11.2.1 since such networks are trivially independent noise networks. Furthermore, the broadcast rate region for networks of deterministic broadcast rate regions specializes to

$$\mathcal{F}_{\mathrm{o}} = \bigcup_{p(S) \prod_{a \in N} p(X_a|S)} \big\{ f \in \mathscr{B}_+^N : f_a(B) = H(Y_{aB}|S) \ \forall a \in N, B \subset N \big\}. \tag{11.27}$$

since $I(X_a; Y_{aB}|S) = H(Y_{aB}|S) - H(Y_{aB}|X_a S)$ and the second term is zero due to the deterministic nature of the channel. The polymatroid structure of $f_a(B) = H(Y_{aB}|S)$ in $B \subset N$ directly follows from the polymatroid structure of the joint entropy function with respect to sets of random variables [9]. Finally, it was shown in [20] that the cut-set outer bound $\mathcal{R}(\mathcal{F}_{\mathrm{o}})$ is achievable in networks of deterministic broadcast channels. The same achievability result also follows from [15, 25], see Sect. 11.2.2, by using noisy network coding with perfect output quantization.

We remark that multicast capacity regions of networks of deterministic broadcast channels cannot be represented by the hypergraph model in general. The following simple counter example has been presented in [21, 22]: Consider $N = \{1, 2, 3, 4\}$, $|\mathcal{S}| = 1$, $X_a = [X_a^1, X_a^2] \in \{0, 1\}^2$ and $Y_{ab} \in \{0, 1\}$ for all $a, b \in N$, and

$$Y_{11} = 0, \qquad Y_{21} = X_2^2, \qquad Y_{31} = X_3^1 \oplus X_3^2, \qquad Y_{41} = X_4^1, \tag{11.28}$$

$$Y_{12} = X_1^1, \qquad Y_{22} = 0, \qquad Y_{32} = X_3^2, \qquad Y_{42} = X_4^1 \oplus X_4^2, \tag{11.29}$$

$$Y_{13} = X_1^1 \oplus X_1^2, \qquad Y_{23} = X_2^2, \qquad Y_{33} = 0, \qquad Y_{43} = X_4^2, \tag{11.30}$$

$$Y_{14} = X_1^2, \qquad Y_{24} = X_2^1 \oplus X_2^2, \qquad Y_{34} = X_3^1, \qquad Y_{44} = 0. \tag{11.31}$$

The broadcast rate vector $f \in \mathscr{B}_+^N \subset \mathscr{X}^N$ according to (11.27) and the candidate hyperarc rate vector $g \in \mathscr{X}^N$ that corresponds to f according to the hyperarc-to-broadcast transformation $\Gamma\colon \mathscr{X}^N \to \mathscr{X}^N$ defined in Part I are given by

$$f_a(B) = \begin{cases} 0 & \text{if } |B - \{a\}| = 0, \\ 1 & \text{if } |B - \{a\}| = 1, \\ 2 & \text{if } |B - \{a\}| \geq 2, \end{cases} \qquad g_a(C) = \begin{cases} 0 & \text{if } a \in C \text{ or } |C| \leq 1, \\ 1 & \text{if } a \notin C \text{ and } |C| = 2, \\ -1 & \text{if } a \notin C \text{ and } |C| = 3 \end{cases} \tag{11.32}$$

for all $a \in N$ and $B, C \subset N$. Note that g is the unique set function vector in $\mathscr{X}^N$ such that $f = \Gamma g$ since Γ is invertible on $\mathscr{X}^N$. Furthermore, f is clearly a valid broadcast rate vector according to the definition in Part I since $f_a(B) = H(Y_{aB}|S)$ for all $a \in N$ and $B \subset N$ and the joint entropy function is a polymatroid rank function [9]. However, the candidate hyperarc vector g violates the nonnegativity condition for hyperarc rate vectors (Part I), i.e., $g_a(C) \geq 0$ for all $a \in N$ and $C \subset N$, which is due to the hyperarc rate vector representing the generalization of an arc flow capacity to hyperarcs. Therefore, we conclude that there exists no valid hyperarc rate vector that can generate the broadcast rate vector f.

Finally, the weighted sum multicast rate maximization problem on the multicast capacity region of networks of deterministic broadcast channels according to Sect. 11.2.3 is characterized by the physical layer dual function part

$$\Theta_{\mathrm{P}}(\mu) = \max_{\sigma \in \mathcal{S}} \sum_{a \in N} \sup_{p(X_a|S=\sigma)} \sum_{t \in T} \sum_{m=1}^{o_a^t - 1} \left(\mu^t_{e^t_{m+1}} - \mu^t_{e^t_m}\right) H(Y_{aE^t_m}|S = \sigma). \tag{11.33}$$

The only difference to the general case is that the mutual information expression in (11.23) has been replaced by a entropy expression for networks of deterministic broadcast channels. As for the general cut-set outer bound in Sect. 11.2.3.1, evaluating $\Theta_{\mathrm{P}}(\mu)$ amounts to solving one convex optimization problem for each state $\sigma \in \mathcal{S}$.

11.4 Networks of Erasure Broadcast Channels

A network of broadcast channels as defined in Sect. 11.2 is called a *network of erasure broadcast channels* if for all $a, b \in N$, the channel outputs Y_{ab} satisfy $Y_{ab} \in \{X_a, \mathrm{e}\}$ with "e" denoting the *erasure* symbol. The channel output alphabets are given by $\mathcal{Y}_{ab} = \mathcal{X}_{ab} \cup \{\mathrm{e}\}$. Furthermore, the probability that an erasure occurs is required to be symmetric with respect to the channel input symbols, i.e., whether $Y_{ab} = X_a$ or $Y_{ab} = \mathrm{e}$ given the state S is independent of X_a. The channel conditional distribution of the outputs given the inputs and the state can therefore be fully determined from the probability expressions $\Pr[Y_{ab} \neq \mathrm{e}\ \forall b \in B,\ Y_{ab} = \mathrm{e}\ \forall b \in B^{\mathrm{c}} | S = \sigma]$ for all $a \in N$, $B \subset N$, and $\sigma \in \mathcal{S}$. Finally, if the erasures are independent across all nodes for each broadcast channel given the state, i.e., if the network is an independent noise network, then there exist $\varepsilon_{ab}^{\sigma} \in [0, 1]$ for all $a, b \in N$ and $\sigma \in \mathcal{S}$ such that

$$\Pr[Y_{ab} \neq \mathrm{e}\ \forall b \in B,\ Y_{ab} = \mathrm{e}\ \forall b \in B^{\mathrm{c}} | S = \sigma] = \prod_{b \in B} (1 - \varepsilon_{ab}^{\sigma}) \prod_{b \in B^{\mathrm{c}}} \varepsilon_{ab}^{\sigma}. \quad (11.34)$$

The multicast capacity region of networks of erasure broadcast channels is unknown. Outer bounds that are based on side-information at the terminals or are based on packet scaling arguments were studied in [5, 11, 17, 18]. An inner bound follows from the general noisy network coding scheme [15, 25].

11.4.1 Cut-Set Outer Bound

For networks of erasure broadcast channels, the cut-set outer bound $\mathcal{R}(\mathcal{F}_{\mathrm{o}})$ is characterized by the region of set function vectors $\mathcal{F}_{\mathrm{o}}$ and each set function vector $f \in \mathcal{F}_{\mathrm{o}}$ satisfies

$$\begin{aligned} f_a(B) &= H(X_a|S) - H(X_a|Y_{aB}S) \\ &= \sum_{\sigma \in \mathcal{S}} \Pr[S = \sigma] H(X_a|S = \sigma)\big(1 - \Pr[Y_{ab} = \mathrm{e}\ \forall b \in B | S = \sigma]\big) \\ &\leq \log_2(|\mathcal{X}_a|) \sum_{\sigma \in \mathcal{S}} \Pr[S = \sigma]\big(1 - \Pr[Y_{ab} = \mathrm{e}\ \forall b \in B | S = \sigma]\big) \quad (11.35) \end{aligned}$$

for all $a \in N$ and $B \subset N$ with equality if and only if the inputs given the state are uniformly distributed. Furthermore, we remark that $(1 - \Pr[Y_{ab} = \mathrm{e}\ \forall b \in B | S = \sigma])$ is a submodular set function in B. Since both the left and the right hand side of the inequality in (11.35) are nonnegative combinations of such expressions, both sides of the inequality define broadcast rate functions. Therefore, $\mathcal{F}_{\mathrm{o}}$ is a broadcast rate region.

Furthermore, it also follows from (11.35) that $\mathcal{F}_o$ can be replaced by a broadcast rate region $\mathcal{F}_{eo}$ such that $\mathcal{R}(\mathcal{F}_{eo}) = \mathcal{R}(\mathcal{F}_o)$ with $\mathcal{F}_{eo}$ given by

$$\mathcal{F}_{eo} = \text{conv} \bigcup_{\sigma \in \mathcal{S}} \Big\{ f \in \mathscr{B}_+^N : \\ f_a(B) = \log_2(|\mathcal{X}_a|)\big(1 - \Pr[Y_{ab} = \mathrm{e}\ \forall b \in B | S = \sigma]\big)\ \forall a \in N, B \subset N \Big\}, \tag{11.36}$$

and for independent noise networks by

$$\mathcal{F}_{eo} = \text{conv} \bigcup_{\sigma \in \mathcal{S}} \Big\{ f \in \mathscr{B}_+^N : f_a(B) = \log_2(|\mathcal{X}_a|)\Big(1 - \prod_{b \in B} \varepsilon_{ab}^{\sigma}\Big)\ \forall a \in N, B \subset N \Big\}. \tag{11.37}$$

Note that $\mathcal{F}_{eo}$ is simply the convex hull of $|\mathcal{S}|$ points in $\mathscr{B}_+^N$, where each point corresponds to a different channel state $\sigma \in \mathcal{S}$. Finally, we remark that the cut-set outer bound for networks of erasure broadcast channels can also be represented in the hyperarc model. The hyperarc rate functions are nonnegative combinations of probability expressions of the form $\Pr[Y_{ab} \neq \mathrm{e}\ \forall b \in B,\ Y_{ab} = \mathrm{e}\ \forall b \in B^c | S = \sigma]$.

11.4.2 Noisy Network Coding with Erasure Quantization

For networks of erasure broadcast channels with independent erasures across all nodes, i.e., independent noise networks, we consider a specific quantization scheme, where the quantization distribution $p(\hat{Y}_{ab}|Y_{ab}QS)$ has the structure of a symmetric erasure channel. This means that, if $Y_{ab} = \xi \in \mathcal{X}_a$, then either $\hat{Y}_{ab} = \xi$ or $\hat{Y}_{ab} = \mathrm{e}$ (erasure) with the probability of erasure being independent of ξ, and if $Y_{ab} = \mathrm{e}$, then $Y_{ab} = \mathrm{e}$ with probability one. As a result of this erasure quantization scheme the virtual channel between the inputs X_a and quantized outputs $\hat{Y}_{aN}$ is again an erasure broadcast channel with independent erasures across all nodes. From this structure, we immediately see that uniform inputs are optimal again for all states. Furthermore, ignoring the timesharing variable Q, which will subsequently be replaced by a convex hull operation, we may define the effective erasure probabilities after quantization $\hat{\varepsilon}_{ab}^{\sigma} = \Pr[\hat{Y}_{ab} = \mathrm{e}|S = \sigma] \in [\varepsilon_{ab}^{\sigma}, 1]$ for all $a, b \in N$ and $\sigma \in \mathcal{S}$ such that

$$\Pr[\hat{Y}_{ab} \neq \mathrm{e}\ \forall b \in B,\ \hat{Y}_{ab} = \mathrm{e}\ \forall b \in B^c | S = \sigma] = \prod_{b \in B}(1 - \hat{\varepsilon}_{ab}^{\sigma}) \prod_{b \in B^c} \hat{\varepsilon}_{ab}^{\sigma}. \tag{11.38}$$

We can explicitly characterize a per-terminal penalized broadcast rate region $\mathcal{E}_{\text{ennc}}^{N} \subset \mathcal{E}_{\text{nnc}}^{N}$ for noisy network coding with erasure quantization based on $\varepsilon^{\sigma} = (\varepsilon_{ab}^{\sigma} : a, b \in N)$ and $\hat{\varepsilon}^{\sigma} = (\hat{\varepsilon}_{ab}^{\sigma} : a, b \in N)$ as

$$\mathcal{E}_{\text{ennc}}^{N} = \text{conv} \bigcup_{\sigma \in \mathcal{S}} \bigcup_{\hat{\varepsilon}^{\sigma} \in [\varepsilon^{\sigma}, 1]} \Bigg\{ (f, h) \in \prod_{t \in N} \mathscr{L}_{+}^{N} \times \prod_{t \in N} \mathscr{R}_{+}^{N} :$$
$$f_{a}^{t}(B) = \log(|\mathcal{X}_{a}|) \left(1 - \varepsilon_{at}^{\sigma} \prod_{b \in B - \{t\}} \hat{\varepsilon}_{ab}^{\sigma} \right) \quad \forall t \in N, a \in N, B \subset N,$$
$$h_{a}^{t} = \sum_{b \in N} \mathrm{H}_{2}(\hat{\varepsilon}_{ba}^{\sigma}) - (1 - \varepsilon_{ba}^{\sigma}) \mathrm{H}_{2}\left(\frac{\hat{\varepsilon}_{ba}^{\sigma} - \varepsilon_{ba}^{\sigma}}{1 - \varepsilon_{ba}^{\sigma}} \right) \quad \forall t \in N, a \in N \Bigg\}. \tag{11.39}$$

The convex hull and the union over all states represent the possibility to choose an arbitrary state distribution, in particular, an arbitrary sequence of states over time which is noncausally known to all nodes. The union over the erasure probability vectors after quantization models the potential choices of erasure quantization distributions in each state. Note that the timesharing random variable is not necessary since it is already included in the convex hull operation. The broadcast rate function expression in (11.39) is very similar to the one in the broadcast rate regions $\mathcal{F}_{\text{eo}}$ (11.37) that represents the cut-set outer bound. Essentially, the erasure probabilities of the channel are replaced by the equivalent erasure probabilities after quantization. Furthermore, the broadcast rate functions are clearly decreasing in the erasure probabilities after quantization.

The penalty rate expression in (11.39) quantifies the rate necessary to communicate some side information about the erasure locations to the terminals. Choosing $\hat{\varepsilon}_{ba}^{\sigma} = \varepsilon_{ba}^{\sigma}$ means that all erasure locations in Y_{ba} are communicated to the terminal, which yields a penalty rate contribution of node b to h_{a}^{t} of $\mathrm{H}_{2}(\varepsilon_{ba}^{\sigma})$. On the other hand, choosing $\hat{\varepsilon}_{ba}^{\sigma} = 1$ means that no erasure locations in Y_{ba} are communicated to the terminal at all, i.e., $\hat{Y}_{ba} = \mathrm{e}$ regardless of Y_{ba}. Thus, b contributes zero penalty rate to h_{a}^{t}. Between these extreme cases, h_{a}^{t} is decreasing in $\hat{\varepsilon}_{ba}^{\sigma}$. Therefore, there is clearly a fundamental trade-off between increasing the broadcast rate functions, which contribute positively to the cut rate, and increasing the penalty rates, which contribute negatively to the cut rate.

One central aspect in (11.39) is that the broadcast rate functions increase linearly with the logarithm of the alphabet size of the channel input, which represents the length of an input symbol (or packet) in bits. However, the penalty rates are independent of size of the input alphabet. This means that for large input alphabets, the quantization overhead may be negligible. Additionally, if the logarithm of the input alphabet is sufficient high, then perfect erasure quantization performs relatively close to the cut-set outer bound.

11.4.3 Weighted Sum Multicast Rate Maximization

In this section, we apply the dual decomposition approach to weighted sum multicast rate maximization to the aforementioned outer and inner bounds for networks of erasure broadcast channels with independent erasures across all nodes given the state.

11.4.3.1 Cut-Set Outer Bound

Using the broadcast rate region $\mathcal{F}_{\text{eo}}$ (11.37), which generates the cut-set outer bound for independent erasures, in the expression for the physical layer dual function (11.23) yields

$$\Theta_{\mathrm{P}}(\mu) = \max_{\sigma \in \mathcal{S}} \sum_{a \in N} \sum_{t \in T} \sum_{m=1}^{o_a^t - 1} \left(\mu^t_{e^t_{m+1}} - \mu^t_{e^t_m} \right) \log_2(|\mathcal{X}_a|) \left(1 - \prod_{b \in E^t_m} \varepsilon^\sigma_{ab} \right). \tag{11.40}$$

Note that no input optimization needs to be performed since uniform input are optimal for any state. Therefore, the evaluation of the objective for each state requires just a few arithmetic operations. The complexity of evaluating the physical layer dual function part is entirely determined by the size of the state space $\mathcal{S}$.

11.4.3.2 Noisy Network Coding

Plugging the per-terminal broadcast rate region $\mathcal{E}^N_{\text{ennc}}$ into the physical layer dual function part expression (11.25), dropping the optimization over the input distribution since $\mathcal{E}^N_{\text{ennc}}$ considers only uniform inputs, and replacing the optimization with respect to the quantization distribution by an optimization with respect to the erasure probabilities after quantization yields

$$\begin{aligned}\Theta_{\mathrm{P}}(\mu) = \max_{\sigma \in \mathcal{S}} \sum_{a \in N} \max_{\hat{\varepsilon}^\sigma_a \in [\varepsilon^\sigma_a, 1]} \sum_{t \in T} \Bigg[&- \sum_{b \in N} \mu^t_b \left(\mathrm{H}_2(\hat{\varepsilon}^\sigma_{ab}) - (1 - \varepsilon^\sigma_{ab}) \mathrm{H}_2\left(\frac{\hat{\varepsilon}^\sigma_{ab} - \varepsilon^\sigma_{ab}}{1 - \varepsilon^\sigma_{ab}} \right) \right) \\ &+ \sum_{m=1}^{o_a^t - 1} \left(\mu^t_{e^t_{m+1}} - \mu^t_{e^t_m} \right) \log_2(|\mathcal{X}_a|) \left(1 - \varepsilon^\sigma_{at} \prod_{b \in E^t_m - \{t\}} \hat{\varepsilon}^\sigma_{ab} \right) \Bigg]\end{aligned} \tag{11.41}$$

Note that in analogy to the quantization distribution in (11.25), the optimization over the probability vector $\hat{\varepsilon}^\sigma = (\hat{\varepsilon}^\sigma_{ab} : a, b \in N)$ decomposes across all nodes $a \in N$ into $|N|$ optimizations with respect to the vectors $\hat{\varepsilon}^\sigma_a = (\hat{\varepsilon}^\sigma_{ab} : b \in N)$. We remark that despite the decomposition, the optimization problem over the erasure probabilities remains a difficult one since the second term in (11.41) is nonconcave in the erasure probabilities.

11.4.4 Suboptimal Erasure Quantization

This section briefly introduces a very simple suboptimal erasure quantization approach based on a approximate local search method suitable for submodular function maximization [8]. This approach serves as a proof of concept that a very primitive erasure quantization scheme achieves a performance close to the cut-set outer bound even for small to moderate input alphabet sizes. Furthermore, it illustrates how submodularity properties can further be exploited beyond the dual decomposition approach in Part I.

We restrict the optimization with respect to $\hat{\varepsilon}^{\sigma}_{ab}$ for all $a, b \in N$ to the two discrete options $\{\varepsilon^{\sigma}_{ab}, 1\}$ instead of the interval $[\varepsilon^{\sigma}_{ab}, 1]$. Note that this corresponds to restricting the union over the hypercube $[\varepsilon^{\sigma}, 1]$ in (11.39) to the union over its vertices. To this end we consider the maximization of the function

$$\Psi^{\sigma}_{a}(B;\mu) = \sum_{t\in T}\sum_{m=1}^{o^t_a-1}\left(\mu^t_{e^t_{m+1}} - \mu^t_{e^t_m}\right)\log_2(|\mathcal{X}_a|)\left(1 - \varepsilon^{\sigma}_{at}\prod_{\substack{b\in B\\ b\in E^t_m-\{t\}}}\hat{\varepsilon}^{\sigma}_{ab}\right) - \sum_{t\in T}\sum_{b\in B}\mu^t_b \mathrm{H}_2(\hat{\varepsilon}^{\sigma}_{ab}) \tag{11.42}$$

with respect to $B \subset N$. This yields a lower bound on the physical layer dual function part given by

$$\Theta_{\mathrm{P}}(\mu) \geq \max_{\sigma\in\mathcal{S}}\sum_{a\in N}\max_{B\subset N}\Psi^{\sigma}_{a}(B;\mu). \tag{11.43}$$

We remark that $\Psi^{\sigma}_{a}(B;\mu)$ is a normalized submodular set function in $B \subset N$ given $\mu \in \prod_{t\in T}\mathscr{R}^N_+$ since the first term a nonnegatively weighted sum of normalized, increasing, submodular functions and the second term is a modular set function. Therefore, the inner maximization problem is an instance of a submodular set function maximization problem.

Submodular set function maximization is an NP-hard problem and has been studied in the literature, see [19]. Our specific problem is an instance of an unconstrained submodular function maximization problem with not necessarily increasing objective function. A local search algorithm for this type of problem was proposed in [8]. It is guaranteed to achieve the optimal value to within a factor of $\frac{1}{3}$. Since this is a worst-case, the local search algorithm may perform much better in practice for the considered problem. The local search algorithm applied to this problem performs the following steps, see [21]:

Step 1. Let $i = 0$, $\delta > 0$, $n = |N|$, and $B^0 = \emptyset$.
Step 2. If $\max_{b \in N - B^i} \Psi_a^\sigma(B^i \cup \{b\}; \mu) > (1 + \frac{\delta}{n})\Psi_a^\sigma(B^i; \mu)$ with $b^\star$ denoting a maximizer, then set $B^{i+1} = B^i \cup \{b^\star\}$ and goto Step 5, otherwise goto Step 3.
Step 3. If $\max_{b \in B^i} \Psi_a^\sigma(B^i - \{b\}; \mu) > (1 + \frac{\delta}{n})\Psi_a^\sigma(B^i; \mu)$ with $b^\star$ denoting a maximizer, then set $B^{i+1} = B^i - \{b^\star\}$ and goto Step 5, otherwise goto Step 4.
Step 4. If $\Psi_a^\sigma(N - B^i; \mu) > (1 + \frac{\delta}{n})\Psi_a^\sigma(B^i; \mu)$, then set $B^{i+1} = N - B^i$ and goto Step 5, otherwise stop and return $\hat{B} = B^i$.
Step 5. Increment i by one and goto Step 2.

The algorithm is discussed in more detail in [21]. The proof that this algorithm achieves an optimality ratio of $(\frac{1}{3} - \delta)$ is due to Feige et al. [8], see also [21]. Note that they showed that the number of set function evaluations of this algorithm is polynomially upper bounded in $|N|$ and $\frac{1}{\delta}$.

11.4.5 Numerical Example

In this section, we consider the bidirectional communication scenario depicted in Fig. 11.1 with $S = T = \{1, 12\}$ and present sum multicast rate results, i.e., $U(r) = \sum_{a \in S} r_a$, for for different sizes of the input alphabet. The erasure probabilities are computed using a Rayleigh fading outage model with distance based channel strength as

$$\varepsilon_{ab} = 1 - \exp\big(-0.1 \cdot \operatorname{dist}(a, b)^3\big). \tag{11.44}$$

We consider that the network has one state, i.e., $|S| = 1$. All nodes use the same input alphabet $\mathcal{X}_a = \mathcal{X}$ for all $a \in N$ and we denote by $L = \log_2(|\mathcal{X}|)$ the number of bits necessary to represent any symbol in $\mathcal{X}$, i.e., the packet length in a coded wireless packet network. Finally, we consider that the erasure probabilities are independent of the packet length.[2]

Figure 11.2 shows the sum rate curves for the considered bidirectional communication demand vs. the packet length L. All sum rate curves are obtained via the proposed dual decomposition approach where the dual problem has been solved with a standard cutting plane method, see e.g. [10]. The cut-set outer bound (CSB) sum rate curve is an upper bound on the achievable sum rate. It is obtained by solving the dual problem (11.21) with the physical layer dual function part (11.40). The perfect quantization (PQ) sum rate curve is achievable using noisy network coding with perfect quantization. It is obtained by solving the dual problem (11.21) with the physical layer dual function part (11.41) but with the maximization over the erasure probabilities after quantization replaced by the fixed choice $\hat{\varepsilon}^\sigma = \varepsilon^\sigma$. Finally, the exhaustive quantization search (EQS) and the local quantization search (LQS) sum

[2]This assumption is most certainly not satisfied in any particular real-world communication system [23]. However, it serves as a good approximation for wireless packet networks with sufficiently large packet lengths.

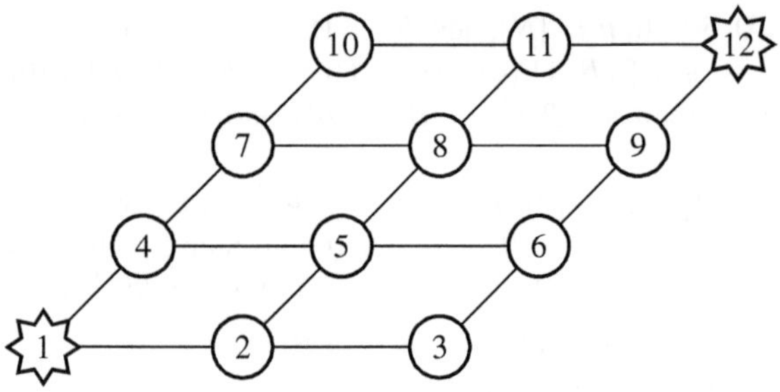

Fig. 11.1 Parallelogram network topology with 12 nodes $N = \{1, \ldots, 12\}$ for a bidirectional communication scenario. The source and terminal nodes $S = T = \{1, 12\}$ (*star shapes*) are the southwest and northeast positions of the regular parallelogram grid, which is indicated by the thin *lines*. The dedicated relay nodes $N - T = \{2, \ldots, 11\}$ (*circle shapes*) are place at the remaining grid positions. The distance between nodes 1 and 2 is given by $\operatorname{dist}(1, 2) = 2$, the distance between nodes 1 and 4 is given by $\operatorname{dist}(1, 2) = \sqrt{2}$, and the angle between the lines from 1 to 2 and 1 to 4 is $\frac{\pi}{4}$

Fig. 11.2 Sum rate versus packet length L in bits for bidirectional communication between nodes 1 and 12 in the parallelogram network, see Fig. 11.1

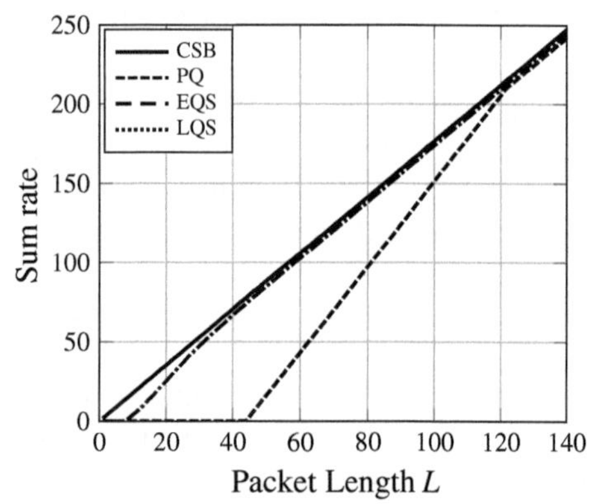

rate curves are also achievable by using noisy network coding. The EQS curve is obtained by solving the dual problem (11.21) with the physical layer dual function part given by the right hand side of (11.43) and exhaustively searching for the optimal $B \subset N$. For the LQS curve the exhaustive search is replaced by the proposed local search algorithm.

The CSB sum rate curve increases linearly with the packet length L. This is an immediate consequence of the physical layer dual function (11.40) being linear in the packet length $L = \log_2(|\mathcal{X}|)$ and $\mathcal{X} = \mathcal{X}_a$ for all $a \in N$. The PQ sum rate curve cannot support any nonzero sum rate for small packet lengths up to 40 bits. This is due to the excessive erasure quantization at all nodes, i.e., $\hat{\varepsilon}^{\sigma}_{ab} = \varepsilon^{\sigma}_{ab}$ for all $a, b \in N$. In this regime the rate required to communicated the quantized outputs to the terminals is larger than the rate supported by the network with known erasure locations. As the packet length increases the price of quantization becomes negligible compared to the gain by knowing the erasure locations, and therefore, the PQ curve approaches the upper bound curve CSB to a certain extent. This behavior fundamentally is due to the linear scaling of the broadcast rate vectors f in L and the independence of the penalty rate vector h of L. The EQS and LQS scheme trades-off between cost and gain of erasure quantization by perfectly quantizing only a few outputs for small packet lengths L and increasing the set of quantized outputs as the L increases. We observe that the EQS and LQS sum rate curves are indistinguishable, which means that there is no loss in performance for the suboptimal local search. Furthermore, both curves are very close to the upper bound curve CSB over the whole range of packet lengths with the only notable difference being observable for very small packet lengths. This shows that the CSB sum rate curve can serve as a simple and reasonable approximation of the achievable sum rate for all reasonable packet lengths in wireless packet networks.

11.5 Conclusions

In this part, we applied the polymatroid broadcast models from Part I to networks of discrete memoryless broadcast channels. For independent noise networks, the cut-set outer bound and the noisy network coding inner bound with independent output quantization admit a formulation within the (per-terminal penalized) polymatroid broadcast model. This applies in particular also to the multicast capacity region of deterministic networks, which coincides with the cut-set outer bound [20], and the cut-set outer bound and the noisy network coding inner bound with independent erasure quantization on the multicast capacity region of networks of erasure broadcast channels with independent erasures. We showed that the hypergraph model, which is also presented in Part I, is not capable of representing these regions in some cases. Furthermore, the polymatroid structure leads directly to a weighted sum multicast rate maximization characterization of these bounds by exploiting the dual decomposition approach from Part I. This characterization is exemplary used to provide a suboptimal, yet efficient optimization approach to find good erasure quantization parameters for noisy network coding. The aforementioned results also apply to networks of Gaussian broadcast channels, see [21, 22]. Finally, we remark that submodular cut models, which generalize the polymatroid broadcast models, but have not been the focus of this work, are capable of modeling bounds on the capacity regions for even more general networks, e.g., independent noise discrete memoryless networks and independent noise Gaussian networks, see [21].

Acknowledgments This work was supported by the German Research Foundation (DFG) under grants UT36/10-1 and UT36/10-2.

References

1. Aref M (1980) Information flow in relay networks. Ph.D. thesis. Stanford University, Stanford
2. Avestimehr A, Diggavi S, Tse D (2011) Wireless network information flow: a deterministic approach. IEEE Trans Inf Theory 57(4):1872–1905
3. Cover T, El Gamal A (1979) Capacity theorems for the relay channel. IEEE Trans Inf Theory 25(5):572–584
4. Cover TM, Thomas JA (2006) Elements of information theory, 2nd edn. Wiley, New York
5. Dana A, Gowaikar R, Palanki R, Hassibi B, Effros M (2006) Capacity of wireless erasure networks. IEEE Trans Inf Theory 52(3):789–804
6. El Gamal A (1981) On information flow in relay networks. In: IEEE national telecommunications conference. New Orleans, LA, USA, vol 2, pp D4.1.1–D4.1.4
7. El Gamal A, Kim YH (2011) Network information theory. Cambridge University Press
8. Feige U, Mirrokni V, Vondrák J (2011) Maximizing non-monotone submodular functions. SIAM J Comput 40(4):1133–1153
9. Fujishige S (1978) Polymatroidal dependence structure of a set of random variables. Inf Control 39(1):55–72
10. Hiriart-Urruty JB, Lemaréchal C (1993) Convex analysis and minimization algorithms ii: advanced theory and bundle methods, grundlehren der mathematischen wissenschaften, vol 306. Springer, Berlin
11. Ho T, Lun D (2008) Network coding: an introduction. Cambridge University Press
12. Ho T, Médard M, Kötter R, Karger DR, Effros M, Shi J, Leong B (2006) A random linear network coding approach to multicast. IEEE Trans Inf Theory 52(10):4413–4430
13. Hou J, Kramer G (2013) Short message noisy network coding with a decode-forward option. IEEE Trans Inf Theory. arXiv:1304.1692 [cs.IT]
14. Khojastepour M, Sabharwal A, Aazhang B (2003) Bounds on achievable rates for general multi-terminal networks with practical constraints. In: Zhao F, Guibas L (eds) Information processing in sensor networks, vol 2634., Lecture notes in computer scienceSpringer, Berlin, pp 146–161
15. Lim S, Kim YH, El Gamal A, Chung SY (2011) Noisy network coding. IEEE Trans Inf Theory 57(5):3132–3152
16. Lim S, Kim K, Kim YH (2014) Distributed decode-forward for multicast. In: IEEE international symposium on information theory (ISIT), pp 636–640
17. Lun D, Ratnakar N, Médard M, Kötter R, Karger D, Ho T, Ahmed E, Zhao F (2006) Minimum-cost multicast over coded packet networks. IEEE Trans Inf Theory 52(6):2608–2623
18. Lun D, Médard M, Kötter R, Effros M (2008) On coding for reliable communication over packet networks. Phys Commun 1(1):3–20
19. Nemhauser G, Wolsey L, Fisher M (1978) An analysis of approximations for maximizing submodular set functions–I. Math Program 14(1):265–294
20. Ratnakar N, Kramer G (2006) The multicast capacity of deterministic relay networks with no interference. IEEE Trans Inf Theory 52(6):2425–2432
21. Riemensberger M (2015) submodular rate region models for multicast communication in wireless networks. Dissertation (submitted). Technische Universität München
22. Riemensberger M, Utschick W (2014) A polymatroid flow model for network coded multicast in wireless networks. IEEE Trans Inf Theory 60(1):443–460
23. Shrader B, Ephremides A (2007) On packet lengths and overhead for random linear coding over the erasure channel. In: International wireless communications and mobile computing conference (IWCMC), ACM, New York, NY, USA, IWCMC'07, pp 314–318
24. Traskov D, Heindlmaier M, Médard M, Kötter R (2012) Scheduling for network-coded multicast. IEEE/ACM Trans Netw 20(5):1479–1488
25. Yassaee M, Aref M (2011) Slepian-wolf coding over cooperative relay networks. IEEE Trans Inf Theory 57(6):3462–3482

Chapter 12
Arbitrarily Varying Channels—A Model for Robust Communication in the Presence of Unknown Interference

Rafael F. Schaefer, Holger Boche and H. Vincent Poor

Abstract The emerging paradigm shift in wireless networks from an exclusive allocation of resources such as frequency bands to a shared use of resources motivates the development of new techniques to deal with interference as a ubiquitous phenomenon. Since such interference is man-made and conceptually different from entropic noise, new concepts are needed which capture these effects. The concept of *arbitrarily varying channels (AVCs)* has been shown to be suitable to model the impact of such unknown interference. Within this framework, capacity can be zero if unassisted encoders and decoders are used although the corresponding mutual information terms are non-zero in general. This necessitates the use of more sophisticated coordination schemes in which the choice of encoders and decoders is additionally coordinated based on coordination resources such as common randomness or correlated sources. In this paper, a brief introduction to the concept of AVCs is given and recent results specifically for broadcast channels with receiver side information and multiple access channels with conferencing encoders are discussed. A complete characterization of the corresponding capacity regions is given and the impact of coordination resources is discussed. In particular, the unassisted capacity regions can no longer be solely characterized by entropic quantities.

R.F. Schaefer (✉) · H.V. Poor
Department of Electrical Engineering, Princeton University, Princeton, NJ 08544, USA
e-mail: rafaelfs@princeton.edu

H.V. Poor
e-mail: poor@princeton.edu

H. Boche
Lehrstuhl für Theoretische Informationstechnik, Technische Universität München, 80333 München, Germany
e-mail: boche@tum.de

W. Utschick (ed.), *Communications in Interference Limited Networks*,
Signals and Communication Technology, DOI 10.1007/978-3-319-22440-4_12

12.1 Introduction

Ongoing research activities in wireless networks envision a paradigm shift from an exclusive allocation of certain resources to shared usage. While many current wireless communication systems such as cellular systems usually operate on exclusive frequency bands, there is a trend for future systems such as sensor or ad-hoc networks to operate on the same frequency bands in a shared but uncoordinated and self-organizing manner. With this development comes the issue of interference which has been identified as one of the major impairments of such future wireless networks. Since there is no way to coordinate such induced interference between coexisting networks, there is a need for new concepts especially pertaining to frequency use.

In a wireless network consisting of several transmitters and receivers, each receiver receives its intended signal but is also confronted with interfering signals from other transmitters. If these transmitter-receiver pairs or links are uncoordinated among themselves and if there is no a priori knowledge about transmit strategies such as coding or modulation schemes, there is little or no knowledge about the induced interference. Thus, users in the network experience channels that may vary from channel use to channel use in an arbitrary and unknown manner. Accordingly, it is no longer valid to treat this interference as noise in the ordinary sense and new concepts are needed to capture these effects as motivated in detail in Sect. 12.2. The concept of *arbitrarily varying channels (AVCs)* [3, 8, 18] has been shown to be a suitable and robust model for such communication scenarios.

The framework of AVCs for robust communication in the presence of unknown interference is developed in Sect. 12.3. Interestingly, for AVCs the performance of a communication system highly depends on how the encoder and decoder are coordinated within one transmitter-receiver link. The traditional approach of using prespecified encoders and decoders can result in zero capacity [3], which can explicitly be characterized by the concept of symmetrizability. In [18] it has been shown that zero capacity happens for symmetrizable AVCs (for which the corresponding mutual information quantity is strictly greater than zero in general). The reason is that for such AVCs the interfering sequences can look like a valid codeword making it impossible for the decoder to decide on the correct one. This issue can be resolved by using more sophisticated approaches in which the choices of encoder and decoder are coordinated based on certain coordination resources such as *common randomness* [8] or *correlated sources* [4]. The corresponding assisted capacity is then non-zero also for symmetrizable channels and completely characterized by entropic quantities.

In this chapter, we develop the motivation for and basic framework of AVCs in Sects. 12.2 and 12.3. Then we consider its applications to certain multi-user scenarios to demonstrate that corresponding results, ideas, and insights can be obtained for more complex scenarios as well. In particular, the *arbitrarily varying broadcast channel (AVBC) with receiver side information* [29, 30, 37, 39–41] is studied in Sect. 12.4 and the *arbitrarily varying multiple access channel (AVMAC) with conferencing encoders* [12, 32, 33] in Sect. 12.5. The capacity regions and the impact of coordination resources are investigated. To the best of our knowledge, these two scenarios are the only multi-user systems for which such a complete characterization is known. In these scenarios, the influence of the unknown interference induced by other coexisting networks is studied. Such coexisting transmitters are benign in general in the sense that they only try to establish reliable communication to their respective receivers. Usually, they do not have the intention to harm or jam other transmissions outside their own network.

In recent years, the framework of AVCs has also been used to model and to analyze secure communication in the presence of adversarial attacks. In particular the framework of AVCs has been applied to the so-called *arbitrarily varying wiretap channel (AVWC)*. Together with a conclusion, this is briefly discussed as an outlook for the future in Sect. 12.6.

12.2 Communication in the Presence of Unknown Interference

In this section we motivate the concept of arbitrarily varying channels as a suitable model for communication in the presence of unknown interference. To highlight the most important points, we consider the simplest communication scenario in which man-made interference appears: the classical interference channel with two transmitters and two receivers. Although each receiver receives signals from both transmitters, each is only interested in what its own transmitter sends. The resulting communication problem is shown in Fig. 12.1.

Throughout the paper we assume finite input and output alphabets denoted by $\mathcal{X}_i$ and $\mathcal{Y}_i$ for transmitter-receiver pair i, $i = 1, 2$. Then for input and output sequences $x_i^n \in \mathcal{X}_i^n$ and $y_i^n \in \mathcal{Y}_i^n$ of length n, the discrete memoryless channel to receiver i is completely characterized by a stochastic matrix as

$$W_i^n(y_i^n|x_1^n, x_2^n) := \prod_{k=1}^{n} W_i(y_{i,k}|x_{1,k}, x_{2,k}). \tag{12.1}$$

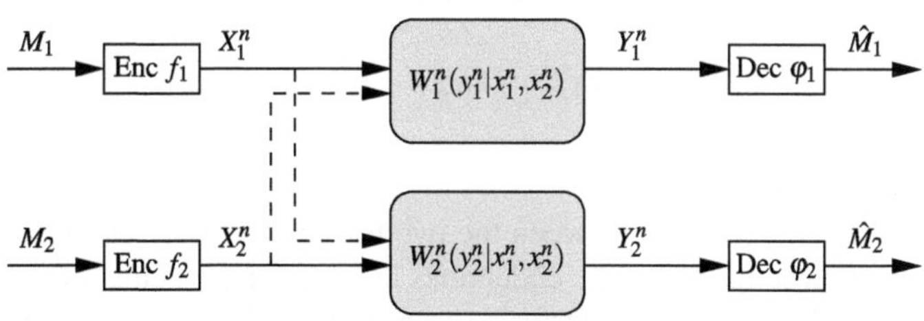

Fig. 12.1 Interference channel with two transmitters and two receivers. The channel output of each receiver depends on the channel input of its intended transmitter, but also on the other interfering transmitter

The effect of additive noise is captured in this model by considering stochastic matrices instead of deterministic ones. A closer look at (12.1) reveals that the communication to receiver 1 is basically governed by a multiple access channel, as the received signal y_1^n depends both on the codeword x_1^n of the intended message m_1 and on the interfering codeword x_2^n of the message m_2 (with a similar observation for receiver 2).

Remark 12.2.1 Note that treating the interference as noise would lead to modified stochastic matrices $\widetilde{W}_i^n(y_i^n|x_i^n)$, $i = 1, 2$, where the received signal y_i^n depends only on the intended codeword x_i^n, while the interfering codeword is captured by the modified stochastic matrix.

The traditional approach for the communication of transmitter-receiver pair i is then to use an $(n, M_{i,n})$-code $\mathcal{C}_i$ which consists of an encoder $E_i : \mathcal{M}_i \to \mathcal{X}_i^n$ with a set of messages $\mathcal{M}_i := \{1, \dots, M_{i,n}\}$ and a decoder $\varphi_i : \mathcal{Y}_i^n \to \mathcal{M}_i$, $i = 1, 2$.

Now, when $x_1^n(m_1)$ and $x_2^n(m_2)$ have been sent according to fixed codebooks $\mathcal{C}_1$ and $\mathcal{C}_2$ and y_1^n and y_2^n have been received, the decoder at receiver i is in error if $\varphi_i(y_i^n) \neq m_i$. With this, the probability of decoding error at receiver 1 when messages m_1 and m_2 have been sent is

$$e_1(m_1, x_2^n(m_2)) := \sum_{y_1^n : \varphi(y_1^n) \neq m_1} W_1^n\big(y_1^n | x_1^n(m_1), x_2^n(m_2)\big). \tag{12.2}$$

Averaged over all messages of transmitter-receiver pair 1, this becomes

$$\bar{e}_1(x_2^n(m_2)) := \frac{1}{|\mathcal{M}_1|} \sum_{m_1 \in \mathcal{M}_1} e_1(m_1, x_2^n(m_2)) \tag{12.3}$$

with similar expressions $e_2(m_2, x_1^n(m_1))$ and $\bar{e}_2(x_1^n(m_1))$ for receiver 2.

The crucial observation is that the probability of error in (12.2) and (12.3) depends on the codebooks that both transmitter-receiver pairs use and, in particular, on the specific message and codeword the interfering transmitter sends. Thus, to ensure reliable communication we have to control the average probabilities of error for *all possible interfering codewords*, i.e., we need

$$\max_{x_2^n(m_2)} \bar{e}_1(x_2^n(m_2)) \leq \lambda_{1,n} \quad \text{and} \quad \max_{x_1^n(m_1)} \bar{e}_2(x_1^n(m_1)) \leq \lambda_{2,n}$$

with $\lambda_{1,n}, \lambda_{2,n} \to 0$ as $n \to \infty$.

If there is no coordination between the two transmitter-receiver pairs, then there is no a priori knowledge about the codebooks used and in particular about the codewords that are chosen by the interfering transmitter. Although we do not assume malicious behavior on the part of the interfering transmitter (indeed, we assume that it is also only interested in establishing reliable communication by itself), the receiver must be prepared for the worst: arbitrary and unknown interfering sequences. This corresponds to the concept of arbitrarily varying channels [3, 8, 18] and reliable communication is possible only if there is a universal strategy that works for all possible codebooks and interfering codewords simultaneously.

12.3 Arbitrarily Varying Channels

Although the interference is motivated by other coexisting transmitters as discussed in the previous section, we concentrate on the single-user case in the following, i.e., we consider a single transmitter-receiver pair only as visualized in Fig. 12.2. Other transmitters act then as jammers for this link and are captured by the state sequence.

12.3.1 System Model

We model the appearance of arbitrary interfering sequences with the help of a finite state set $\mathcal{S}$. For a fixed state sequence $s^n \in \mathcal{S}^n$ of length n, the discrete memoryless channel to the receiver is given by $W^n_{s^n}(y^n|x^n) = W^n(y^n|x^n, s^n) := \prod_{k=1}^n W(y_k|x_k, s_k)$.

Definition 12.3.1 The discrete memoryless *arbitrarily varying channel* $\mathfrak{W}$ is the family of channels for all $s^n \in \mathcal{S}^n$, i.e.,

$$\mathfrak{W} := \left\{ W^n_{s^n} : s^n \in \mathcal{S}^n \right\}.$$

For characterizing the capacity of such a channel, we need the definition of an *averaged channel*. For any probability distribution $q \in \mathcal{P}(\mathcal{S})$ the averaged channel is given by

$$\overline{W}_q(y|x) = \sum_{s \in \mathcal{S}} W(y|x, s)q(s) \tag{12.4}$$

for all $x \in \mathcal{X}$ and $y \in \mathcal{Y}$. We also need the concept of symmetrizability which is defined next.

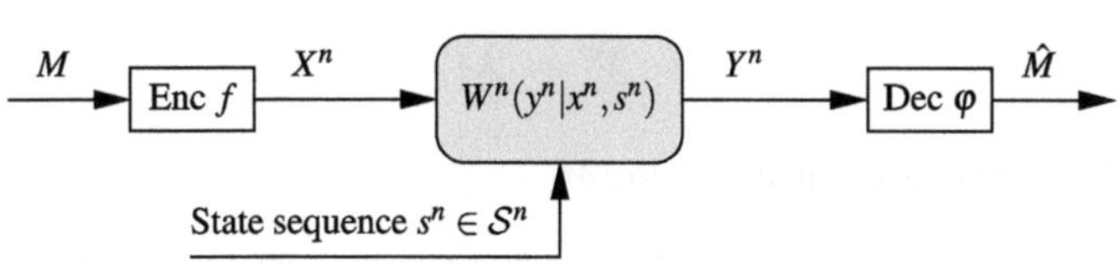

Fig. 12.2 Arbitrarily varying channel. The actual channel is governed by an unknown state sequence $s^n \in \mathcal{S}^n$ of length n, which may vary in an arbitrary and unknown manner from channel use to channel use

Definition 12.3.2 An AVC is called *symmetrizable* if there exists a channel (stochastic matrix) $\sigma : \mathcal{X} \to \mathcal{P}(\mathcal{S})$ such that

$$\sum_{s\in\mathcal{S}} W(y|x,s)\sigma(s|x') = \sum_{s\in\mathcal{S}} W(y|x',s)\sigma(s|x) \tag{12.5}$$

holds for all $x, x' \in \mathcal{X}$ and $y \in \mathcal{Y}$.

Now, if we write the left hand side of (12.5) as $\widetilde{W}(y|x,x') = \sum_{s\in\mathcal{S}} W(y|x,s)\sigma(s|x')$, we see that symmetrizability basically means that the following is true: The resulting channel $\widetilde{W}$ is symmetric in both inputs x and x' so that $\widetilde{W}(y|x,x') = \widetilde{W}(y|x',x)$. Roughly speaking, a symmetrizable AVC can "simulate" a valid channel input which makes it impossible for the decoder to decide on the correct codeword send by the transmitter.

12.3.2 *Code Concepts*

Communication over AVCs is much more involved than over ordinary discrete memoryless channels. In particular, it makes a substantial difference whether unassisted (deterministic) or more sophisticated code concepts utilizing coordination resources are used. It turns out that the unassisted capacity of an AVC can be zero, while assisted codes allow for communication at non-zero rates by utilizing available coordination resources such as *common randomness (CR)* or *correlated sources (CS)* [3, 4, 8, 18].

12.3.2.1 Unassisted Codes

Unassisted codes refer to the traditional approach of pre-specified encoder and decoder which are fixed prior to transmission as shown in Fig. 12.2.

Definition 12.3.3 An *unassisted* (n, M_n)*-code* $\mathcal{C}$ consists of an encoder at the transmitter

$$f : \mathcal{M} \to \mathcal{X}^n \tag{12.6}$$

with a set of messages $\mathcal{M} := \{1, \dots, M_n\}$ and a decoder at the receiver

$$\varphi : \mathcal{Y}^n \to \mathcal{M}. \tag{12.7}$$

The rate of this code is defined as $\frac{1}{n}\log M_n$.

The term "unassisted" refers to the fact that the encoder (12.6) and the decoder (12.7) are chosen and fixed prior to the transmission of the message. This means that they must be universal such that they work for all $s^n \in \mathcal{S}^n$ simultaneously.

Now, the reliability requirement has to take all possible state sequences $s^n \in \mathcal{S}^n$ of length n into account. Accordingly, we define the average probability of error as

$$\bar{e} = \max_{s^n \in \mathcal{S}^n} \frac{1}{|\mathcal{M}|} \sum_{m \in \mathcal{M}} \sum_{y^n : \varphi(y^n) \neq m} W^n(y^n | x^n(m), s^n). \tag{12.8}$$

Then the *unassisted capacity* $C(\mathfrak{W})$ of the AVC $\mathfrak{W}$ is given by the supremum of all achievable rates for which an unassisted code of Definition 12.3.3 can be found that ensures reliability (12.8), i.e., $\bar{e} \to 0$ as $n \to \infty$.

Remark 12.3.1 Note that we only consider the average probability of error (12.8) and not the maximum probability of error. While for the classical discrete memoryless channel both error criteria lead to the same capacity result, the situation changes dramatically for AVCs. Here, the capacities under average and maximum probability of error differ and the latter is still unknown. Moreover, it has been shown that the AVC under the maximum error includes the famous zero-error problem of Shannon as a special case [1], which makes it an extremely challenging problem.

Unfortunately, it has been shown that unassisted approaches with pre-determined encoder and decoder work only for certain channel configurations. In particular, if the AVC is symmetrizable, cf. Definition 12.3.2, reliable communication is not possible with unassisted codes and its corresponding capacity is zero [18]. In such situations, more sophisticated code concepts based on coordination resources are needed.

12.3.2.2 CR-Assisted Codes

A powerful coordination resource is given by CR which can be realized for example by common synchronization procedures or by a common satellite signal. CR is modeled by a random variable Γ which takes values in a finite set $\mathcal{G}_n$ according to a distribution $P_\Gamma \in \mathcal{P}(\mathcal{G}_n)$. This enables the transmitter and the receiver to coordinate their choice of the encoder (12.6) and the decoder (12.7) according to the actual realization $\gamma \in \mathcal{G}_n$. This scenario is depicted in Fig. 12.3.

Definition 12.3.4 A *CR-assisted* (n, M_n, Γ)*-code* $\mathcal{C}_{\mathrm{CR}}$ is given by a family of unassisted codes

$$\{\mathcal{C}(\gamma) : \gamma \in \mathcal{G}_n\}$$

together with a random variable Γ taking values in $\mathcal{G}_n$ with $|\mathcal{G}_n| < \infty$ according to $P_\Gamma \in \mathcal{P}(\mathcal{G}_n)$. The rate of this code is defined as $\frac{1}{n} \log M_n$.

A natural way of extending the reliability requirement from unassisted to CR-assisted codes is done by averaging over all possible CR realizations $\gamma \in \mathcal{G}_n$, i.e.,

$$\bar{e}_{\mathrm{CR}} = \max_{s^n \in \mathcal{S}^n} \frac{1}{|\mathcal{M}|} \sum_{m \in \mathcal{M}} \sum_{\gamma \in \mathcal{G}_n} \sum_{y^n : \varphi_\gamma(y^n) \neq m} W^n(y^n | x^n_\gamma(m), s^n) P_\Gamma(\gamma). \tag{12.9}$$

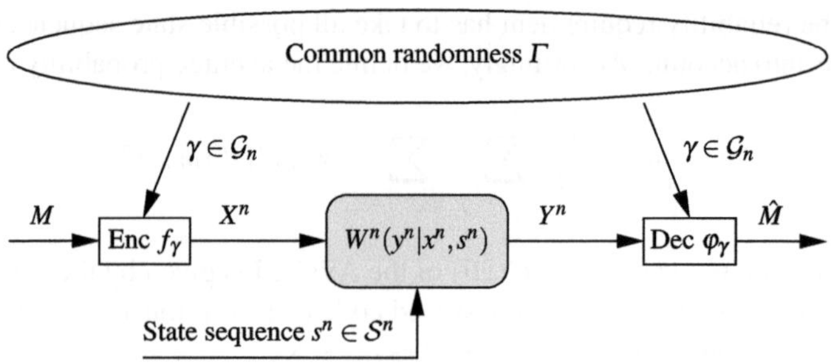

Fig. 12.3 Arbitrarily varying channel with CR available to transmitter and receiver. The choice of encoder and decoder can now be coordinated accordingly to the realization $\gamma \in \mathcal{G}_n$ of the CR Γ distributed according to $P_\Gamma \in \mathcal{P}(\mathcal{G}_n)$

Then the *CR-assisted capacity* $C_{\text{CR}}(\mathfrak{W})$ of the AVC $\mathfrak{W}$ is given by the supremum of all achievable rates for which a CR-assisted code of Definition 12.3.4 can be found that ensures reliability (12.9), i.e., $\bar{e}_{\text{CR}} \to 0$ as $n \to \infty$.

Remark 12.3.2 One may wonder how much CR is needed to actually achieve the CR-assisted capacity (as $\mathcal{G}_n$ is allowed to depend on the block length n). Luckily, in [3] it has been shown that for transmitting at rate R with exponentially many messages, i.e., $M_n = 2^{nR}$, it suffices to have CR of size $|\mathcal{G}_n| \approx n^2$. Thus, the required CR scales only polynomially with the block length, which is negligible compared to the exponential scaling of the message size.

12.3.2.3 CS-Assisted Codes

A weaker form of coordination resources is given by correlated sources $(U^n, V^n)_{n=1}^{\infty}$ with $I(U; V) > 0$, where the transmitter observes U^n and the receiver V^n. Then, the encoder (12.6) at the transmitter depends only on $u^n \in \mathcal{U}^n$ and the decoder (12.7) at the receiver only on $v^n \in \mathcal{V}^n$.

Remark 12.3.3 Note that CS are in fact a weaker resource than CR, since in general it is impossible to extract common randomness from correlated sources. In addition, this is not robust in the sense that the set of all probability distributions allowing CR extraction is closed, nowhere dense, and has zero Lebesgue measure, cf. [9].

Definition 12.3.5 A *CS-assisted* $(n, M_n, (U, V))$*-code* $\mathcal{C}_{\text{CS}}$ is given by a family of unassisted codes

$$\left\{\mathcal{C}(u^n, v^n) : u^n \in \mathcal{U}^n, v^n \in \mathcal{V}^n\right\}$$

such that the encoder f_{u^n} depends on $u^n \in \mathcal{U}^n$ only and the decoder φ_{v^n} on $v^n \in \mathcal{V}^n$ only. The realizations $(u^n, v^n) \in \mathcal{U}^n \times \mathcal{V}^n$ are chosen according to $P_{U^nV^n} = \prod_{i=1}^n P_{UV} \in \mathcal{P}(\mathcal{U}^n \times \mathcal{V}^n)$.

Then the reliability requirement becomes

$$\bar{e}_{CS} = \max_{s^n \in \mathcal{S}^n} \frac{1}{|\mathcal{M}|} \sum_{m \in \mathcal{M}} \sum_{(u^n, v^n) \in \mathcal{U}^n \times \mathcal{V}^n} \sum_{y^n : \varphi_{v^n}(y^n) \neq m} W^n(y^n | x^n_{u^n}(m), s^n) P_{U^n V^n}(u^n, v^n) \tag{12.10}$$

and the *CS-assisted capacity* $C_{CS}(\mathfrak{W})$ of the AVC $\mathfrak{W}$ is defined accordingly.

12.3.3 Capacity Results

The AVC has been well studied and the capacity has been established for all code concepts introduced above. The unassisted capacity is given in the following theorem.

Theorem 12.3.1 *The unassisted capacity $C(\mathfrak{W})$ of the AVC $\mathfrak{W}$ is*

$$C(\mathfrak{W}) = \begin{cases} \max_{P_X \in \mathcal{P}(\mathcal{X})} \inf_{q \in \mathcal{P}(\mathcal{S})} I(X; \overline{Y}_q) & \text{if } \mathfrak{W} \text{ is non-symmetrizable} \\ 0 & \text{if } \mathfrak{W} \text{ is symmetrizable} \end{cases}$$

where $\overline{Y}_q$ denotes the random variable associated with the output of the averaged channel $\overline{W}_q$, $q \in \mathcal{P}(\mathcal{S})$, cf. (12.4).

The capacity has been characterized by Ahlswede [3] and Csiszár and Narayan [18] and a complete proof can also be found in [17]. In the following we want to give some insights into why the unassisted capacity $C(\mathfrak{W})$ of an AVC $\mathfrak{W}$ displays such a dichotomous behavior. To answer the question of zero capacity for symmetrizable AVCs, we consider the following. Let $x^n(m) \in \mathcal{X}^n$, $m = 1, \dots, M_n$ with $M_n \geq 2$ be arbitrary codewords. Now, for a symmetrizable AVC we can think of interfering sequences that look like valid codewords. For this purpose, we set $s^n(m) = x^n(m)$, $m = 1, \dots, M_n$. For the probability of error, we obtain for each pair of codewords $(k, l) \in \mathcal{M} \times \mathcal{M}$ with $k \neq l$ the following:

$$\begin{aligned}
&\mathbb{E}\big[e(k, S^n(l))\big] + \mathbb{E}\big[e(l, S^n(k))\big] \\
&\quad = \mathbb{E}\big[W^n\big((\varphi^{-1}(k))^c | x^n(k), S^n(l)\big)\big] + \mathbb{E}\big[W^n\big((\varphi^{-1}(l))^c | x^n(l), S^n(k)\big)\big] \\
&\quad = \mathbb{E}\big[W^n\big((\varphi^{-1}(k))^c | x^n(k), S^n(l)\big)\big] + \mathbb{E}\big[W^n\big((\varphi^{-1}(l))^c | x^n(k), S^n(l)\big)\big] \\
&\quad \geq \mathbb{E}\big[W^n\big((\varphi^{-1}(k))^c | x^n(k), S^n(l)\big)\big] + \mathbb{E}\big[W^n\big(\varphi^{-1}(k) | x^n(k), S^n(l)\big)\big] \\
&\quad = \mathbb{E}\big[W^n\big((\varphi^{-1}(k))^c \cup \varphi^{-1}(k) | x^n(k), S^n(l)\big)\big] \\
&\quad = 1
\end{aligned}$$

where the first equality follows from the definition of probability of error, cf. (12.2), and the second equality from the fact that the AVC $\mathfrak{W}$ is symmetrizable, cf. Definition 12.3.2. Averaging over all codewords this leads to

$$\begin{aligned}\frac{1}{|\mathcal{M}|}\sum_{l=1}^{M_n}\mathbb{E}\big[\bar{e}(S^n(l))\big] &= \frac{1}{(M_n)^2}\sum_{k=1}^{M_n}\sum_{l=1}^{M_n}\mathbb{E}\big[e(k,S^n(l))\big]\\ &\geq \frac{1}{(M_n)^2}\cdot\frac{M_n(M_n-1)}{2}\\ &= \frac{M_n-1}{2M_n}\\ &\geq \frac{1}{4}\end{aligned}$$

which implies that $\mathbb{E}[\bar{e}(S^n(l))] \geq \frac{1}{4}$ holds for at least one $l \in \mathcal{M}$. But if the average probability of error is bounded from below by a positive constant, reliable communication is not possible so that $C(\mathfrak{W}) = 0$ in this case.

Intuitively, what happens is the following: If the AVC is symmetrizable, the interfering sequence can look like another valid codeword. But then the receiver observes a superposition of two valid codewords which makes it impossible for the decoder to decide on the correct one.

The main observation is that the traditional approach of unassisted codes fails in the case of symmetrizable channels. This is where the coordination resources enter the picture. The CR-assisted capacity is given in the following theorem.

Theorem 12.3.2 *The CR-assisted capacity $C_{CR}(\mathfrak{W})$ of the AVC $\mathfrak{W}$ is*

$$C_{CR}(\mathfrak{W}) = \max_{P_X\in\mathcal{P}(\mathcal{X})}\inf_{q\in\mathcal{P}(\mathcal{S})} I(X;\overline{Y}_q).$$

This result goes back to Blackwell et al. [8] and shows that CR-assisted approaches achieve the same rate as unassisted approaches, but also in the case of symmetrizable AVCs. Subsequently, the requirement of having CR at the transmitter and receiver was weakened by Ahlswede and Cai who showed in [4] that CS suffices as a coordination resource to overcome symmetrizable channels. Moreover, the capacity does not change and is given in the following theorem.

Theorem 12.3.3 *The CS-assisted capacity $C_{CS}(\mathfrak{W})$ of the AVC $\mathfrak{W}$ is*

$$C_{CS}(\mathfrak{W}) = C_{CR}(\mathfrak{W}) = \max_{P_X\in\mathcal{P}(\mathcal{X})}\inf_{q\in\mathcal{P}(\mathcal{S})} I(X;\overline{Y}_q).$$

Now, Theorems 12.3.1–12.3.3 give a complete picture of the capacity behavior under different coding strategies and how they relate to each other. In particular, the unassisted capacity can now be expressed in terms of its assisted versions.

Corollary 12.3.1 *The unassisted capacity $C(\mathfrak{W})$ of the AVC $\mathfrak{W}$ is*

$$C(\mathfrak{W}) = \begin{cases} C_{CS}(\mathfrak{W}) = C_{CR}(\mathfrak{W}) & \text{if } \mathfrak{W} \text{ is non-symmetrizable}\\ 0 & \text{if } \mathfrak{W} \text{ is symmetrizable,}\end{cases}$$

i.e., $C(\mathfrak{W})$ displays a dichotomous behavior and is either zero or equals its assisted counterparts $C_{CS}(\mathfrak{W})$ and $C_{CR}(\mathfrak{W})$.

12.3.4 Stability, Continuity, and Super-Activation

The framework of AVCs can also be interpreted from an adversarial point of view by considering the state sequence $s^n \in \mathcal{S}^n$ to be chosen by an adversary who wants to maliciously influence or jam the transmission. Then a desirable property of the capacity is the following: it should depend in a continuous way on the uncertainty set. Small changes in the uncertainty set (strategy space of the adversary) should result in small variations of the capacity only. Since otherwise, if small changes would lead to dramatic losses in capacity, the approach at hand will most likely not be used. Such a continuous behavior of the capacity is particularly relevant in the context of secure communication as discussed later in Sect. 12.6.

For the following discussion we need a concept to measure the distance between two channels. We define the distance between two channels W_1 and W_2 based on the total variation distance as

$$d(W_1, W_2) = \max_{x \in \mathcal{X}} \sum_{y \in \mathcal{Y}} \big| W_1(y|x) - W_2(y|x) \big|. \tag{12.11}$$

Now, this generalizes to a distance between two AVCs as follows. Assume we have two finite AVCs $\mathfrak{W}_1$ and $\mathfrak{W}_2$ with two state sets $\mathcal{S}_1$ and $\mathcal{S}_2$ respectively. Then we can define two uncertainty sets $\mathcal{W}_1 = \{W_{s_1} : s_1 \in \mathcal{S}_1\}$ and $\mathcal{W}_2 = \{W_{s_2} : s_2 \in \mathcal{S}_2\}$. We define distances between these two sets of channels as

$$d_1(\mathcal{W}_1, \mathcal{W}_2) = \max_{s_2 \in \mathcal{S}_2} \min_{s_1 \in \mathcal{S}_1} d(W_{s_1}, W_{s_2})$$
$$d_2(\mathcal{W}_1, \mathcal{W}_2) = \max_{s_1 \in \mathcal{S}_1} \min_{s_2 \in \mathcal{S}_2} d(W_{s_1}, W_{s_2})$$

so that the distance between the two AVCs $\mathfrak{W}_1$ and $\mathfrak{W}_2$ is

$$D(\mathfrak{W}_1, \mathfrak{W}_2) = \max \big\{ d_1(\mathcal{W}_1, \mathcal{W}_2), d_2(\mathcal{W}_1, \mathcal{W}_2) \big\}.$$

Roughly speaking, the distance $D(\mathfrak{W}_1, \mathfrak{W}_2)$ between two AVCs $\mathfrak{W}_1$ and $\mathfrak{W}_2$ is given by the largest distance in (12.11) for all possible channel realizations in the corresponding uncertainty sets.

The following result shows the stability of positivity of the unassisted capacity $C(\mathfrak{W})$ of an AVC $\mathfrak{W}$: Whenever $C(\mathfrak{W}) > 0$, it remains positive in a certain neighborhood. This is a direct consequence of [26] where the corresponding result is shown in the context of secure communication over AVWCs.

Theorem 12.3.4 *If the unassisted capacity $C(\mathfrak{W})$ of the finite AVC $\mathfrak{W}$ satisfies $C(\mathfrak{W}) > 0$, then there is an $\varepsilon > 0$ such that for all AVCs $\mathfrak{W}_\varepsilon$ satisfying $D(\mathfrak{W}, \mathfrak{W}_\varepsilon) \leq \varepsilon$, we have $C(\mathfrak{W}_\varepsilon) > 0$.*

This is a desirable property of the unassisted capacity as it means that small variations in the adversary's strategy result in only small variations in the unassisted capacity. But we want to further explore the question of continuity of the capacity. For this purpose, we define

$$F(\mathfrak{W}) = \min_{\sigma:\mathcal{X}\to\mathcal{P}(\mathcal{S})} \left(\max_{x\neq x'} \sum_{y\in\mathcal{Y}} \left| \sum_{s\in\mathcal{S}} W(y|x,s)\sigma(s|x') - \sum_{s\in\mathcal{S}} W(y|x',s)\sigma(s|x) \right| \right)$$

which is a continuous function of the AVC $\mathfrak{W}$. One can easily check that the AVC $\mathfrak{W}$ is symmetrizable if and only if $F(\mathfrak{W}) = 0$, cf. also Definition 12.3.2. With this definition we get the following result.

Theorem 12.3.5 *The AVC $\mathfrak{W}$ is a point of discontinuity of $C(\mathfrak{W})$ if and only if the following holds:*

1. $C_{CR}(\mathfrak{W}) > 0$
2. $F(\mathfrak{W}) = 0$ *and for every* $\varepsilon > 0$ *there is a finite* $\mathfrak{W}_\varepsilon$ *with* $D(\mathfrak{W}, \mathfrak{W}_\varepsilon) \leq \varepsilon$ *and* $F(\mathfrak{W}_\varepsilon) > 0$.

Note that both the CR-assisted capacity $C_{CR}(\mathfrak{W})$ and the function $F(\mathfrak{W})$ are continuous. Thus, interestingly, the *discontinuity* of $C(\mathfrak{W})$ can be completely characterized by two *continuous* functions.

To give an example of an AVC whose capacity has a point of discontinuity, we construct a suitable AVC with the help of a similar example in the context of AVWCs [13, 15] (which goes back to an example first given in [8] and later discussed in [3]). To do so, we consider a communication scenario with alphabets of sizes $|\mathcal{X}| = 2$, $|\mathcal{Y}| = 3$, and $|\mathcal{S}| = 2$. For the first AVC $\mathfrak{W}_1$ we construct the uncertainty set

$$\mathcal{W}_1 = \{W_1, W_2\}$$

where

$$W_1 := \begin{pmatrix} 1 & 0 & 0 \\ 0 & 0 & 1 \end{pmatrix} \quad \text{and} \quad W_2 := \begin{pmatrix} 0 & 0 & 1 \\ 0 & 1 & 0 \end{pmatrix}.$$

We know from [3] that $\mathcal{W}_1$ defines a symmetrizable AVC $\mathfrak{W}_1$ so that its unassisted capacity is $C(\mathfrak{W}_1) = 0$. For the second AVC $\mathfrak{W}_2$ we set its uncertainty set to $\mathcal{W}_2 = \{\overline{W}\}$ with

$$\overline{W} := \begin{pmatrix} 1 & 0 & 0 \\ 0 & 1 & 0 \end{pmatrix}.$$

Now we define $\mathfrak{W}(\lambda)$ as a convex combination of $\mathfrak{W}_1$ and $\mathfrak{W}_2$ by

$$\mathcal{W}(\lambda) = \{W_1(\lambda), W_2(\lambda)\} \qquad \text{for } 0 \le \lambda \le 1$$

with

$$W_1(\lambda) = (1-\lambda)W_1 + \lambda\overline{W}$$
$$W_2(\lambda) = (1-\lambda)W_2 + \lambda\overline{W}.$$

Then, the unassisted capacity $C(\mathfrak{W}(\lambda))$ of the AVC $\mathfrak{W}(\lambda)$ is discontinuous in $\lambda = 0$. This can be shown similarly as in [13, Theorem 5]:

Proposition 12.3.1 *The following hold for the previous example:*

1. *The CR-assisted capacity $C_{CR}(\mathfrak{W}(\lambda))$ is continuous in λ for all $\lambda \in [0, 1]$ and it holds that*
$$\min_{\lambda\in[0,1]} C_{CR}(\mathfrak{W}(\lambda)) > 0.$$
2. *The unassisted capacity $C(\mathfrak{W}(\lambda))$ is continuous in λ for all $\lambda \in (0, 1]$. It holds that $C(\mathfrak{W}(0)) = 0$ and further that*
$$\lim_{\lambda\searrow 0} C(\mathfrak{W}(\lambda)) > 0,$$
i.e., $\lambda = 0$ is a discontinuity point of $C(\cdot)$.

This example showed that the unassisted capacity of an AVC can have a discontinuity point. In more detail, symmetrizable AVCs can appear as a discontinuity point within sets of non-symmetrizable AVCs. However, with the next example we want to further show that whole sets of symmetrizable AVCs are possible as well.

For this purpose, we first construct a suitable AVC $\mathfrak{W}^*$ such that a whole set of AVCs $\mathfrak{W}$ around this channel $\mathfrak{W}^*$ always satisfy $C_{CR}(\mathfrak{W}) > 0$ and $C(\mathfrak{W}) = C(\mathfrak{W}^*) = 0$. For $\mathfrak{W}^*$, we define the uncertainty set $\mathcal{W}^* = \{W_1^*, W_2^*\}$ with

$$W_1^* = \begin{pmatrix} \frac{1}{2} & \frac{1}{2} & 0 \\ \frac{1}{4} & 0 & \frac{3}{4} \end{pmatrix} \quad \text{and} \quad W_2^* = \begin{pmatrix} 0 & 0 & 1 \\ 0 & 1 & 0 \end{pmatrix}.$$

Further, we define the channel $\sigma^* : \mathcal{X} \to \mathcal{P}(\mathcal{S})$ as

$$\sigma^* = \begin{pmatrix} \frac{4}{5} & \frac{1}{5} \\ \frac{2}{5} & \frac{3}{5} \end{pmatrix}.$$

Then it is easy to show that

$$\sum_{s\in\mathcal{S}} W_s^*(y|1)\sigma^*(s|2) = \sum_{s\in\mathcal{S}} W_s^*(y|2)\sigma^*(s|1)$$

holds for all $y \in \mathcal{Y}$, which means that the AVC $\mathfrak{W}^*$ is symmetrizable so that $C(\mathfrak{W}^*) = 0$, cf. Theorem 12.3.1. Now, the following result says that there exists a whole set around this AVC which has zero capacity as well. This can be shown similarly as in [13, Theorem 7]):

Proposition 12.3.2 *There exists an $\varepsilon_0 > 0$ such that for all finite AVCs $\mathfrak{W}$ with $D(\mathfrak{W}^*, \mathfrak{W}) < \varepsilon_0$ it always holds that*

$$C_{CR}(\mathfrak{W}) > 0 \quad \textit{and} \quad C(\mathfrak{W}) = 0.$$

This example demonstrates the usefulness of coordination resources as they help to overcome the issue of symmetrizable channels: the CR-assisted capacity is non-zero while the unassisted capacity is zero. Moreover, with coordination resources the capacity becomes a function that depends continuously on the uncertainty set.

Finally, we want to address the question of super-activation. This question is closely related to resource allocation in wireless networks, which is an important issue as it determines the overall performance. Taking an orthogonal frequency division multiplexing (OFDM) system as an example, the overall capacity is given by the sum of the capacities of all orthogonal sub-channels. To this end, for a system consisting of two orthogonal ordinary discrete memoryless channels, where both are "useless" in the sense of having zero capacity, the capacity of the whole system is zero as well. This reflects the world view of classical additivity of resources in the sense that "$0 + 0 = 0$."

Recently, it was shown in [10] that the classical additivity of basic resources does not hold anymore in the context of secure communication over AVCs. It was demonstrated that there are AVWCs which are individually useless but allow for transmission if they are used together, i.e., "$0+0 > 0$." The corresponding AVWCs for which super-activation is possible were completely characterized in [26]. In addition, the discontinuity behavior of the secrecy capacity is characterized in [26] (indeed, the previous Theorems 12.3.4 and 12.3.5 are direct consequences of [26]).

The observations in [10, 26] regarding the super-activation for secure communication over AVCs lead to the question of whether super-activation can also happen for non-secure communication over AVCs. However, in [31] it has been shown that whenever $C(\mathfrak{W}_1) = 0$ and $C(\mathfrak{W}_2) = 0$ for two AVCs $\mathfrak{W}_1$ and $\mathfrak{W}_2$, we have $C(\mathfrak{W}_1 \otimes \mathfrak{W}_2) = 0$, where $\mathfrak{W}_1 \otimes \mathfrak{W}_2$ denotes the joint use of the two orthogonal AVCs. As a consequence, we see that the question of whether a system can be super-activated or not is not related to the question of discontinuity. Hence, super-activation is a unique feature of the secrecy capacity.

12.4 Broadcast Channels with Receiver Side Information

In the previous section we developed the framework of AVCs and motivated why this is a suitable model for communication under unknown interference. Now, we want to make use of this framework by applying it to *bidirectional relaying* (or two-way

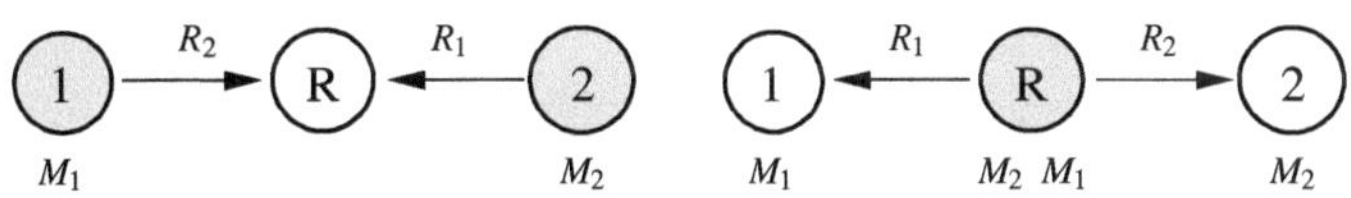

Fig. 12.4 Bidirectional relaying in a three-node network. Nodes 1 and 2 exchange their messages M_1 and M_2 at rates R_2 and R_1 via the relay node using a decode-and-forward protocol

relaying) [21, 23, 27, 28, 38]. This concept refers to a three-node network, in which two nodes want to exchange information with the help of a relay node. A decode-and-forward protocol divides the entire information exchange into two phases as shown in Fig. 12.4: a initial multiple access phase and a succeeding broadcast phase.

Since the relay is required to decode both messages nodes 1 and 2 have sent, the first phase corresponds to the classical multiple access channel (MAC) which is well understood [2, 24]. The corresponding arbitrarily varying MAC has been subsequently studied in [5, 6, 19, 20].

The succeeding broadcast phase is known under different names: the *bidirectional broadcast channel (BBC)* or *broadcast channel (BC) with receiver side information*. Both names have in common that they underline the fact that both receiving nodes have their own messages from the previous phase as side information available for decoding. The capacity region of this broadcast channel has been found in [21, 22, 27] for discrete memoryless channels and in [38] for multiple-input multiple-output (MIMO) Gaussian channels. The corresponding arbitrarily varying BC with receiver side information [29, 30, 37, 39–41] is studied in the following.

12.4.1 System Model

Similar to the single-user AVC discussed in Sect. 12.3 we model the unknown interference with the help of a finite state set $\mathcal{S}$. For a fixed state sequence $s^n \in \mathcal{S}^n$ of length n, the discrete memoryless BC is given by $W^n_{s^n}(y^n_1, y^n_2|x^n) = W^n(y^n_1, y^n_2|x^n, s^n) := \prod_{k=1}^n W(y_{1,k}, y_{2,k}|x_k, s_k)$. Since we do not allow any cooperation between the receiving nodes, it suffices to consider marginal channels $W^n_{i,s^n}(y^n_i|x^n) = W^n_i(y^n_i|x^n, s^n) := \prod_{k=1}^n W(y_{i,k}|x_k, s_k)$, $i = 1, 2$ only. The communication problem at hand is visualized in Fig. 12.5.

Definition 12.4.1 The discrete memoryless *arbitrarily varying broadcast channel* $\mathfrak{W}$ is the family of channels for all $s^n \in \mathcal{S}^n$ as

$$\mathfrak{W} := \left\{W^n_{s^n} : s^n \in \mathcal{S}^n\right\}.$$

Similar to the single-user AVC, we define for any probability distribution $q \in \mathcal{P}(\mathcal{S})$ the *averaged BC* as $\overline{W}_q(y_1, y_2|x) = \sum_{s\in\mathcal{S}} W(y_1, y_2|x, s)q(s)$ for all $x \in \mathcal{X}$, $y_1 \in \mathcal{Y}_1$, and $y_2 \in \mathcal{Y}_2$, cf. (12.4). The averaged marginal channels $\overline{W}_{1,q}(y_1|x)$

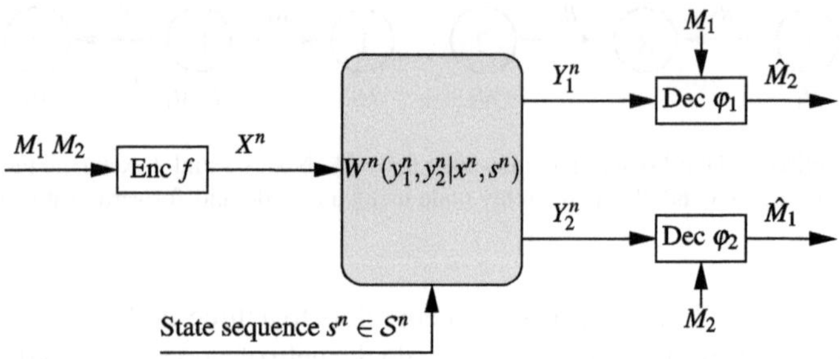

Fig. 12.5 Arbitrarily varying broadcast channel with receiver side information. The transmitter encodes both messages M_1 and M_2 into one codeword X^n based on the network coding idea. The receiving nodes decode the intended message having the other message as side information available

and $\overline{W}_{2,q}(y_2|x)$ follow then accordingly. We also need an appropriate version of symmetrizability, cf. Definition 12.3.2, which is given next.

Definition 12.4.2 An AVBC is called $\mathcal{Y}_i$*-symmetrizable* if there exists a channel (stochastic matrix) $\sigma_i : \mathcal{X} \to \mathcal{P}(\mathcal{S})$ such that

$$\sum_{s\in\mathcal{S}} W_i(y_i|x,s)\sigma_i(s|x') = \sum_{s\in\mathcal{S}} W_i(y_i|x',s)\sigma_i(s|x)$$

holds for all $x, x' \in \mathcal{X}$ and $y_i \in \mathcal{Y}_i$, $i = 1, 2$.

The concept of unassisted codes of Definition 12.3.3 extends to the AVBC with receiver side information in the following way.

Definition 12.4.3 An *unassisted* $(n, M_{1,n}, M_{2,n})$*-code* $\mathcal{C}$ for the AVBC with receiver side information consists of an encoder at the relay node

$$f : \mathcal{M}_1 \times \mathcal{M}_2 \to \mathcal{X}^n$$

with message sets $\mathcal{M}_i := \{1, \dots, M_{i,n}\}$, $i = 1, 2$, which are also known at corresponding receiving nodes, and decoders

$$\varphi_1 : \mathcal{Y}_1^n \times \mathcal{M}_1 \to \mathcal{M}_2$$
$$\varphi_2 : \mathcal{Y}_2^n \times \mathcal{M}_2 \to \mathcal{M}_1.$$

The main difference from unassisted codes for the single-user case in Definition 12.3.3 are in the decoders which now have one message as side information available to decode the other intended message.

Then the CR-assisted and CS-assisted versions of this concept follow Definitions 12.3.4 and 12.3.5. The average probabilities of error for these codes are defined analogously to (12.8)–(12.10) and the *unassisted capacity region* $\mathcal{R}(\mathfrak{W})$, *CR-assisted capacity region* $\mathcal{R}_{\text{CR}}(\mathfrak{W})$, and *CS-assisted capacity region* $\mathcal{R}_{\text{CS}}(\mathfrak{W})$ follow accordingly.

12.4.2 Capacity Results

The AVBC with receiver side information has been studied for unassisted, CR-assisted, and CS-assisted codes. To state the corresponding capacity regions we need the following rate region:

$$\overline{\mathcal{R}}(\mathfrak{W}) := \bigcup_{P_X \in \mathcal{P}(\mathcal{X})} \left\{ \begin{array}{l} R_1 \leq \inf_{q \in \mathcal{P}(\mathcal{S})} I(X; \overline{Y}_{1,q}) \\ R_2 \leq \inf_{q \in \mathcal{P}(\mathcal{S})} I(X; \overline{Y}_{2,q}) \end{array} \right\}$$

where $\overline{Y}_{i,q}$ denotes the random variable associated with the output of the averaged marginal channel $\overline{W}_{i,q}$, $q \in \mathcal{P}(\mathcal{S})$.

The unassisted capacity region of the AVBC with receiver side information has been established in [39, 40] and it displays a similar behavior to the unassisted capacity of the AVC in Theorem 12.3.1: it either equals a certain rate region or else has an empty interior. Again, this dichotomous behavior can be completely characterized in terms of symmetrizability.

Theorem 12.4.1 *If the AVBC* $\mathfrak{W}$ *with receiver side information is non-*$\mathcal{Y}_1$*-symmetrizable and non-*$\mathcal{Y}_2$*-symmetrizable, then the unassisted capacity region* $\mathcal{R}$ *is*

$$\mathcal{R}(\mathfrak{W}) = \overline{\mathcal{R}}(\mathfrak{W}).$$

We have $\text{int}(\mathcal{R}(\mathfrak{W})) = \emptyset$ *if and only if the AVBC* $\mathfrak{W}$ *with receiver side information is* $\mathcal{Y}_1$*-symmetrizable or* $\mathcal{Y}_2$*-symmetrizable.*

Similarly, for the AVBC with receiver side information coordination resources such as CR or CS help to overcome the problem of symmetrizable channels and the corresponding capacities have been established in [29, 39]. Similar to the single-user AVC in Theorems 12.3.2 and 12.3.3, CR and CS both lead to the same capacity region, i.e., $\mathcal{R}_{\text{CR}}(\mathfrak{W}) = \mathcal{R}_{\text{CS}}(\mathfrak{W})$.

Theorem 12.4.2 *The CR-assisted capacity region* $\mathcal{R}_{CR}(\mathfrak{W})$ *and the CS-assisted capacity region* $\mathcal{R}_{CS}(\mathfrak{W})$ *of the AVBC* $\mathfrak{W}$ *with receiver side information are*

$$\mathcal{R}_{CR}(\mathfrak{W}) = \mathcal{R}_{CS}(\mathfrak{W}) = \overline{\mathcal{R}}(\mathfrak{W}).$$

Theorems 12.4.1 and 12.4.2 give a complete picture of the capacity regions of the AVBC with receiver side information and show the relation between them: The unassisted capacity region $\mathcal{R}(\mathfrak{W})$ of the AVBC $\mathfrak{W}$ with receiver side information either equals its CR-assisted and CS-assisted capacity regions $\mathcal{R}_{\text{CR}}(\mathfrak{W})$ and $\mathcal{R}_{\text{CS}}(\mathfrak{W})$ or else has an empty interior. This establishes a dichotomous behavior for the AVBC with receiver side information as well.

The AVBC with receiver side information has been studied under other situations as well. So far the input sequences (codewords) $x^n \in \mathcal{X}^n$ and state sequences $s^n \in \mathcal{S}^n$ are not constrained in any way and the question of whether the unassisted capacity region has an empty interior or equals its CR-assisted and CS-assisted versions solely depends on the symmetrizability of the channel, cf. Definition 12.4.2. Now, if one considers constraints (such average power constraints) on the input and state sequences, the situation changes completely and the unassisted capacity region can be non-empty although the channel is symmetrizable according to Definition 12.4.2. The reason for this is that the state sequences that symmetrize the channel must now satisfy a power constraint as well. Obviously this must not always be the case and the whole analysis becomes much more involved. A detailed discussion is given in [37] including a complete characterization of the capacity region of the AVBC with receiver side information under input and state constraints.

Theorems 12.4.1 and 12.4.2 show that coordination resources such as CR or CS suffice to establish reliable communication even for symmetrizable channels. Unfortunately, if there are no coordination resources available, communication breaks down completely and the capacity region is empty. Another approach to overcome this issue in the absence of coordination resources is the concept of list decoding. Here, the decoder is allowed to decode the received signal into a list of potential messages. In [30] the corresponding list capacity region of the AVBC with receiver side information is established and it is shown that list decoding helps to reliably communicate over symmetrizable channels. In particular, one finds always list decoders at the receivers with a finite list size which resolve the symmetrizability of the channel.

To the best of our knowledge, the questions of stability, continuity, and superactivation (similarly to Sect. 12.3.4) have not been addressed so far for the AVBC with receiver side information. These are an interesting and open problems for future work.

12.5 Multiple Access Channels with Conferencing Encoders

In this section we discuss another scenario to which the framework of AVCs can successfully be applied: the *MAC with conferencing encoders* [16, 34, 36]. This refers to a MAC in which the transmitters are allowed to exchange information (of any kind) over noiseless cooperation links prior to the transmission. Such a communication scenario applies for example to cellular systems in which neighboring base stations cooperate via high-speed backbones; so-called *coordinated multipoint (CoMP) transmission*. The corresponding arbitrarily varying MAC with conferencing encoders [12, 32, 33] is studied in the following.

12.5.1 System Model

Similar to the single-user AVC in Sect. 12.3 and to the AVBC with receiver side information in Sect. 12.4 we model the unknown interference with the help of a finite state set $\mathcal{S}$. For a fixed state sequence $s^n \in \mathcal{S}^n$ of length n, the discrete memoryless MAC is given by $W^n_{s^n}(y^n|x_1^n, x_2^n) = W^n(y^n|x_1^n, x_2^n, s^n) := \prod_{k=1}^n W(y_k|x_{1,k}, x_{2,k}, s_k)$. The communication problem at hand is visualized in Fig. 12.6.

Definition 12.5.1 The discrete memoryless *arbitrarily varying multiple access channel* $\mathfrak{W}$ is the family of channels for all $s^n \in \mathcal{S}^n$ given by

$$\mathfrak{W} := \{W^n_{s^n} : s^n \in \mathcal{S}^n\}.$$

Similar to the single-user AVC and AVBC with receiver side information, we define for any probability distribution $q \in \mathcal{P}(\mathcal{S})$ the *averaged MAC* as $\overline{W}_q(y|x_1, x_2) = \sum_{s\in\mathcal{S}} W(y|x_1, x_2, s)q(s)$ for all $x_1 \in \mathcal{X}_1$, $x_2 \in \mathcal{X}_2$, and $y \in \mathcal{Y}$, cf. (12.4). We also need an appropriate version of symmetrizability, cf. Definition 12.3.2, which is given next.

Definition 12.5.2 An AVMAC is called $(\mathcal{X}_1, \mathcal{X}_2)$*-symmetrizable* if there exists a channel (stochastic matrix) $\sigma : \mathcal{X}_1 \times \mathcal{X}_2 \to \mathcal{P}(\mathcal{S})$ such that

$$\sum_{s\in\mathcal{S}} W(y|x_1, x_2, s)\sigma(s|x_1', x_2') = \sum_{s\in\mathcal{S}} W(y|x_1', x_2', s)\sigma(s|x_1, x_2)$$

holds for all $x_1, x_1' \in \mathcal{X}_1$, $x_2, x_2' \in \mathcal{X}_2$, and $y \in \mathcal{Y}$.

Remark 12.5.1 The classical AVMAC without conferencing encoders has been analyzed in [5, 19]. There it has been shown that one needs not only $(\mathcal{X}_1, \mathcal{X}_2)$-symmetrizability as in Definition 12.5.2 but also so-called $\mathcal{X}$-symmetrizability and $\mathcal{Y}$-symmetrizability conditions that operate on one input only. While these “marginal” conditions are necessary to characterize the unassisted capacity region of

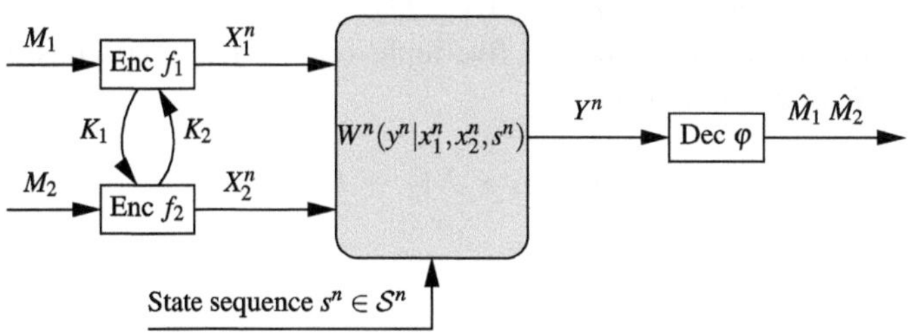

Fig. 12.6 Arbitrarily varying multiple access channel with conferencing encoders. Prior to transmission, both transmitters exchange information K_1 and K_2 via the cooperation links. Then each transmitter encodes its message M_1 (and M_2) into a codeword X_1^n (and X_2^n) with the help of the exchanged information K_2 (and K_1) it received during the conference prior to the transmission

the AVMAC without conferencing encoders, only $(\mathcal{X}_1, \mathcal{X}_2)$-symmetrizability will be needed for the case with conferencing encoders.

In the classical AVMAC scenario, neither transmitter has any knowledge about the message the other one will transmit. In our problem at hand, both transmitters can cooperate by exchanging limited information using Willems conferencing [36]. Such information can be (part of) the messages to be transmitted, but is not necessarily restricted to them.

The term *Willems conference* refers to an iterative protocol that allows both transmitters to exchange information via noiseless cooperation links. In each time slot, a transmitter sends some information to the other one based on its own knowledge but also on what it received in the previous time slots. We model this information exchange via a pair of functions

$$(c_1, c_2) : \mathcal{M}_1 \times \mathcal{M}_2 \to \mathcal{K}_1 \times \mathcal{K}_2 \tag{12.12}$$

where $\mathcal{K}_1$ and $\mathcal{K}_2$ refer to the information the transmitters obtained after the conference.

Remark 12.5.2 If the Willems conference is unrestricted in the sense that an arbitrary amount of information can be exchanged, both messages can be made available at both transmitters turning the AVMAC into a single-user AVC.

For practical reasons we consider only limited exchange capabilities. This means the following: Assume that a Willems conference pair (c_1, c_2) as given in (12.12) is used for a block length n. Then the conference (c_1, c_2) has conferencing capacities $C_1, C_2 > 0$ if

$$\frac{1}{n} \log |\mathcal{K}_i| \le C_i, \quad i = 1, 2$$

and we call this an (n, C_1, C_2)-*Willems conference.*

The concept of unassisted codes of Definition 12.3.3 extends to the AVMAC with conferencing encoders in the following way.

Definition 12.5.3 An *unassisted* $(n, M_{1,n}, M_{2,n}, C_1, C_2)$-*code* $\mathcal{C}$ for the AVMAC $\mathfrak{W}$ with conferencing encoders is a five-tuple $(c_1, c_2, f_1, f_2, \varphi)$ consisting of an (n, C_1, C_2)-Willems conference

$$(c_1, c_2) : \mathcal{M}_1 \times \mathcal{M}_2 \to \mathcal{K}_1 \times \mathcal{K}_2,$$

encoders at transmitters 1 and 2

$$f_1 : \mathcal{M}_1 \times \mathcal{K}_2 \to \mathcal{X}_1^n$$
$$f_2 : \mathcal{M}_2 \times \mathcal{K}_1 \to \mathcal{X}_2^n,$$

and a decoder at the receiver

$$\varphi : \mathcal{Y}^n \to \mathcal{M}_1 \times \mathcal{M}_2.$$

Then the CR-assisted and CS-assisted versions of this concept follow Definitions 12.3.4 and 12.3.5. The average probabilities of error for these codes are defined analogously to (12.8)–(12.10). For given conferencing capacities $C_1, C_2 > 0$, the unassisted capacity region $\mathcal{R}(\mathfrak{W}, C_1, C_2)$, CR-assisted capacity region $\mathcal{R}_{\mathrm{CR}}(\mathfrak{W}, C_1, C_2)$, and CS-assisted capacity region $\mathcal{R}_{\mathrm{CS}}(\mathfrak{W}, C_1, C_2)$ follow accordingly.

12.5.2 Capacity Results

The AVMAC with conferencing encoders has been studied for unassisted, CR-assisted, and CS-assisted codes. To state the corresponding capacity regions, we need some further definitions.

Let $\mathcal{U}$ be a finite (auxiliary) set and $p \in \mathcal{P}(\mathcal{U} \times \mathcal{X}_1 \times \mathcal{X}_2)$ a probability distribution of the form $p(u, x_1, x_2) = P_U(u)P_{X_1|U}(x_1|u)P_{X_2|U}(x_2|u)$. We define the region $\overline{\mathcal{R}}(\mathfrak{W}, p, q, C_1, C_2)$ as

$$\overline{\mathcal{R}}(\mathfrak{W}, p, q, C_1, C_2) := \left\{ \begin{array}{r} R_1 \le I(X_1; \overline{Y}_q|X_2, U) + C_1 \\ R_2 \le I(X_2; \overline{Y}_q|X_1, U) + C_2 \\ R_1 + R_2 \le \min\{I(X_1, X_2; \overline{Y}_q), I(X_1, X_2; \overline{Y}_q|U) + C_1 + C_2\} \end{array} \right\}$$

where $\overline{Y}_q$ is the random variable associated with the output of the averaged channel $\overline{W}_q$, $q \in \mathcal{P}(\mathcal{S})$. Now, we set

$$\overline{\mathcal{R}}(\mathfrak{W}, C_1, C_2) := \bigcup_{p \in \mathcal{P}(\mathcal{U} \times \mathcal{X}_1 \times \mathcal{X}_2)} \bigcap_{q \in \mathcal{P}(\mathcal{S})} \overline{\mathcal{R}}(\mathfrak{W}, p, q, C_1, C_2)$$

which allows us to state the unassisted capacity region $\mathcal{R}(\mathfrak{W}, C_1, C_2)$ of the AVMAC $\mathfrak{W}$ with conferencing encoders [32].

Theorem 12.5.1 *The unassisted capacity region $\mathcal{R}(\mathfrak{W}, C_1, C_2)$ of the AVMAC $\mathfrak{W}$ with conferencing capacities $C_1, C_2 > 0$ is*

$$\mathcal{R}(\mathfrak{W}, C_1, C_2) = \overline{\mathcal{R}}(\mathfrak{W}, C_1, C_2)$$

if and only if the AVMAC $\mathfrak{W}$ is non-$(\mathcal{X}_1, \mathcal{X}_2)$-symmetrizable. If the AVMAC $\mathfrak{W}$ is $(\mathcal{X}_1, \mathcal{X}_2)$-symmetrizable, then $\mathcal{R}(\mathfrak{W}, C_1, C_2) = \{(0, 0)\}$.

Remark 12.5.3 From the proof of Theorem 12.5.1 it follows that a one-shot non-iterative Willems conference is sufficient to achieve capacity, cf. [32, 36].

Similar to the previously considered scenarios, the unassisted capacity region of the AVMAC with conferencing encoders displays a dichotomous behavior: it either equals a certain rate region or else is zero. This behavior is completely characterized in terms of symmetrizability.

The following result shows then that coordination resources such as CR or CS allow reliable communication even in the case of symmetrizable channels. The corresponding CR-assisted and CS-assisted capacity regions have been established in [32, 33] respectively and, again, both coordination resources yield the same capacity region, i.e., $\mathcal{R}_{\text{CR}}(\mathfrak{W}, C_1, C_2) = \mathcal{R}_{\text{CS}}(\mathfrak{W}, C_1, C_2)$.

Theorem 12.5.2 *The CR-assisted capacity region $\mathcal{R}_{CR}(\mathfrak{W}, C_1, C_2)$ and the CS-assisted capacity region $\mathcal{R}_{CS}(\mathfrak{W}, C_1, C_2)$ of the AVMAC $\mathfrak{W}$ with conferencing capacities $C_1, C_2 > 0$ are*

$$\mathcal{R}_{CR}(\mathfrak{W}, C_1, C_2) = \mathcal{R}_{CS}(\mathfrak{W}, C_1, C_2) = \overline{\mathcal{R}}(\mathfrak{W}, C_1, C_2).$$

Theorems 12.5.1 and 12.5.2 give a complete picture of the capacity region of the AVMAC with conferencing encoders. The unassisted capacity region $\mathcal{R}(\mathfrak{W}, C_1, C_2)$ for conferencing capacities $C_1, C_2 > 0$ either equals its CR-assisted and CS-assisted capacity regions $\mathcal{R}_{\text{CR}}(\mathfrak{W}, C_1, C_2)$ and $\mathcal{R}_{\text{CS}}(\mathfrak{W}, C_1, C_2)$ or else is zero. This establishes a dichotomous behavior for the AVMAC with conferencing encoders as well.

The AVMAC with conferencing encoders has been studied under list decoding as well. Similarly to the AVBC with receiver side information in Sect. 12.4, allowing the receiver to decode the received signal into a list of potential messages help to overcome the problem of symmetrizable channels if there are no coordination resources available. In [12] the corresponding list capacity region of the AVMAC with conferencing encoders is established and it is shown that list decoding helps to reliably communicate over symmetrizable channels.

To the best of our knowledge, the questions of stability, continuity, and super-activation (similarly to Sect. 12.3.4) have not been addressed so far for the AVMAC with conferencing encoders. These are an interesting and open problems for future work.

12.6 Conclusion and Outlook

The framework of AVCs has been developed as a suitable and robust model for communication in the presence of unknown interference. This is particularly relevant for future wireless networks which will act in an uncoordinated and self-organizing way sharing wireless resources with other coexisting systems. The main issue of such an approach is interference which will be a ubiquitous phenomenon and one of the limiting factors.

Unknown varying interference has a substantial impact on communication over such uncoordinated wireless systems. In particular, if unassisted strategies without additional coordination are used, the unknown interference can lead to situations in which no communication is possible at all. This is based on the assumption of treating interference as noise as is usually done in such unassisted approaches. However, for so-called symmetrizable channels the unknown interference terms can act as valid channel inputs making it impossible to reliably decode on the right codeword. Consequently, unassisted strategies are only reasonable in scenarios in which the unknown interference can be made small enough. The good news is that communication can be stabilized by additional coordination of encoders and decoders based on certain coordination resources such as CR and CS. Then, reliable communication at positive rates is possible even in the case of symmetrizable channels.

In contrast to ordinary channels, new phenomena appear for AVCs. Most noteworthy, the unassisted capacity of an AVC can no longer be solely characterized by entropic quantities. In general, the unassisted capacity of an AVC is now discontinuous in the uncertainty set and coordination resources help to overcome the problems of symmetrizable channels. The corresponding CR-assisted capacity is then continuous.

The framework of AVCs has been well developed for the point-to-point channel and it can be extended to more complex multi-user scenarios as well. This has been demonstrated for the AVBC with receiver side information and the AVMAC with conferencing encoders.

Another interesting direction in which to extend these ideas is for secure communication; and particularly for the so-called wiretap channel. Here, a transmitter wants to transmit confidential information to a legitimate receiver in the presence of a non-legitimate eavesdropper. The presence of unknown varying interference, which can even originate from active adversaries, leads then to the arbitrarily varying wiretap channel [7, 10, 11, 13–15, 25, 26, 35]. There are similar effects such as a dichotomous behavior of the unassisted secrecy capacity: it either equals its CR-assisted secrecy capacity or else is zero. But interestingly, there are other new phenomena as well which have not been observed for non-secure communication. Most noteworthy is the phenomenon of super-activation for which two orthogonal, useless AVWCs (in the sense of zero unassisted secrecy capacity) can be used together to super-activate the whole system allowing for secure communication at non-zero rates.

Acknowledgments This work of R.F. Schaefer was supported by the German Research Foundation (DFG) under Grants BO 1734/25-1 and WY 151/2-1. This work of H. Boche was supported by the German Research Foundation (DFG) under Grant BO 1734/25-1. This work of H.V. Poor was supported by the U.S. National Science Foundation under Grant CMMI-1435778.

References

1. Ahlswede R (1970) A note on the existence of the weak capacity for channels with arbitrarily varying channel probability functions and its relation to Shannon's zero error capacity. Ann Math Stat 41(3):1027–1033
2. Ahlswede R (1971) Multi-way communication channels. In: Proceedings international symposium information theory, Tsahkadsor, Armenian, pp 23–52

3. Ahlswede R (1978) Elimination of correlation in random codes for arbitrarily varying channels. Z Wahrscheinlichkeitstheorie verw Gebiete 44:159–175
4. Ahlswede R, Cai N (1997) Correlated sources help transmission over an arbitrarily varying channel. IEEE Trans Inf Theory 43(4):1254–1255
5. Ahlswede R, Cai N (1999) Arbitrarily varying multiple-access channels part I—Ericson's symmetrizability is adequate, Gubner's conjecture is true. IEEE Trans Inf Theory 45(2):742–749
6. Ahlswede R, Cai N (1999) Arbitrarily varying multiple-access channels part II—correlated senders' side information, correlated messages, and ambiguous transmission. IEEE Trans Inf Theory 45(2):749–756
7. Bjelaković I, Boche H, Sommerfeld J (2013) Information theory, combinatorics, and search theory, Springer, chap. Capacity results for arbitrarily varying wiretap channels, pp 123–144
8. Blackwell D, Breiman L, Thomasian AJ (1960) The capacities of certain channel classes under random coding. Ann Math Stat 31(3):558–567
9. Boche H, Nötzel J (2013) Arbitrarily small amounts of correlation for arbitrarily varying quantum channels. J Math Phys 54(11):112202
10. Boche H, Schaefer RF (2013) Capacity results and super-activation for wiretap channels with active wiretappers. IEEE Trans Inf Forensics Security 8(9):1482–1496
11. Boche H, Schaefer RF (2014) Arbitrarily varying wiretap channels with finite coordination resources. In: Proceedings of IEEE international conference communications workshops, Sydney, Australia, pp 746–751
12. Boche H, Schaefer RF (2014) List decoding for arbitrarily varying multiple access channels with conferencing encoders. In: Proceedings of IEEE international conference on communications, Sydney, Australia, pp 1934–1940
13. Boche H, Schaefer RF, Poor HV (2015) On the continuity of the secrecy capacity of compound and arbitrarily varying wiretap channels. IEEE Trans Inf Forensics Security 10(12):2531–2545
14. Boche H, Schaefer RF, Poor HV (2014) On arbitrarily varying wiretap channels for different classes of secrecy measures. In: Proceedings IEEE international symposium information theory, Honolulu, HI, USA, pp 2376–2380
15. Boche H, Schaefer RF, Poor HV (2015) On the continuity of the secrecy capacity of wiretap channels under channel uncertainty. In: Proceedings of IEEE international conference on communications, London, UK, pp 4163–4168
16. Bross SI, Lapidoth A, Wigger MA (2008) The Gaussian MAC with conferencing encoders. In: Proceedings of the IEEE international symposium on information theory, Toronto, Canada, pp 2702–2706
17. Csiszár I, Körner J (2011) Information theory: coding theorems for discrete memoryless systems, 2nd edn. Cambridge University Press, Cambridge
18. Csiszár I, Narayan P (1988) The capacity of the arbitrarily varying channel revisited: positivity. Constraints. IEEE Trans Inf Theory 34(2):181–193
19. Gubner JA (1990) On the deterministic-code capacity of the multiple-access arbitrarily varying channel. IEEE Trans Inf Theory 36(2):262–275
20. Jahn JH (1981) Coding of arbitrarily varying multiuser channels. IEEE Trans Inf Theory 27(2):212–226
21. Kim SJ, Mitran P, Tarokh V (2008) Performance bounds for bidirectional coded cooperation protocols. IEEE Trans Inf Theory 54(11):5235–5241
22. Kramer G, Shamai (Shitz) S (2007) Capacity for classes of broadcast channels with receiver side information. In: Proceedings of the IEEE information theory workshop, Tahoe City, CA, USA, pp 313–318
23. Larsson P, Johansson N, Sunell KE (2005) Coded bi-directional relaying. In: Proceedings of the 5th Scandinavian workshop on ad hoc networks, Stockholm, Sweden, pp 851–855
24. Liao H (1972) Multiple access channels. PhD thesis, University of Hawaii, Honolulu, USA
25. MolavianJazi E, Bloch M, Laneman JN (2009) Arbitrary jamming can preclude secure communication. In: Proceedings of the 47th annual Allerton conference communication control, computing, Monticello, IL, USA, pp 1069–1075

26. Nötzel J, Wiese M, Boche H (2015) The arbitrarily varying wiretap channel—secret randomness, stability and super-activation. In: Proceedings IEEE international symposium information theory, Hong Kong, pp 2151–2155
27. Oechtering TJ, Schnurr C, Bjelaković I, Boche H (2008) Broadcast capacity region of two-phase bidirectional relaying. IEEE Trans Inf Theory 54(1):454–458
28. Rankov B, Wittneben A (2007) Spectral efficient protocols for half-duplex fading relay channels. IEEE J Sel Areas Commun 25(2):379–389
29. Schaefer RF, Boche H (2014) How much coordination is needed for robust broadcasting over arbitrarily varying bidirectional broadcast channels. In: Proceedings of the IEEE international conference communications, Sydney, Australia, pp 1872–1877
30. Schaefer RF, Boche H (2014) List decoding for arbitrarily varying broadcast channels with receiver side information. IEEE Trans Inf Theory 60(8):4472–4487
31. Schaefer RF, Boche H, Poor HV (2015) Super-activation as a unique feature of secure communication in malicious environments, submitted for publication
32. Wiese M, Boche H (2013) The arbitrarily varying multiple-access channel with conferencing encoders. IEEE Trans Inf Theory 59(3):1405–1416
33. Wiese M, Boche H (2014) On the weakest resource for coordination in arbitrarily varying multiple access channels with conferencing encoders. Probl Inf Transmission 50(1):15–26
34. Wiese M, Boche H, Bjelaković I, Jungnickel V (2011) The compound multiple access channel with partially cooperating encoders. IEEE Trans Inf Theory 57(5):3045–3066
35. Wiese M, Nötzel J, Boche H (2015) The arbitrarily varying wiretap channel—deterministic and correlated random coding capacities under the strong secrecy criterion. In: Proceedings of the IEEE international symposium information theory, Hong Kong, pp 2146–2150
36. Willems FMJ (1983) The discrete memoryless multiple access channel with partially cooperating encoders. IEEE Trans Inf Theory 29(3):441–445
37. Wyrembelski RF, Boche H (2012) Impact of interference in coexisting wireless networks with applications to arbitrarily varying bidirectional broadcast channels. Entropy 14(8):1357–1398
38. Wyrembelski RF, Oechtering TJ, Bjelaković I, Schnurr C, Boche H (2008) Capacity of Gaussian MIMO bidirectional broadcast channels. In: Proceedings of the IEEE international symposium information theory, Toronto, Canada, pp 584–588
39. Wyrembelski RF, Bjelaković I, Boche H (2009) Coding strategies for bidirectional relaying for arbitrarily varying channels. In: Proceedings of the IEEE global communications conference, Honolulu, HI, USA, pp 1–6
40. Wyrembelski RF, Bjelaković I, Boche H (2009) On the capacity of bidirectional relaying with unknown varying channels. In: Proceedings of the IEEE workshop computational advances in multi-sensor adaptive processing, Aruba, Dutch Antilles, pp 269–272
41. Wyrembelski RF, Bjelaković I, Boche H (2010) Bidirectional relaying in wireless networks—impact of degree of coordination. In: Proceedings of the IEEE international conference on acoustics, speech, signal processing, Dallas, TX, USA, pp 3234–3237

26. Nötzel J, Wiese M, Boche H (2015) The arbitrarily varying wiretap channel—secret randomness, stability and super-activation. In: Proceedings IEEE international symposium information theory, Hong Kong, pp 2151–2155
27. Oechtering TJ, Schnurr C, Bjelakovic I, Boche H (2008) Broadcast capacity region of two-phase bidirectional relaying. IEEE Trans Inf Theory 54(1):454–458
28. Rankov B, Wittneben A (2007) Spectral efficient protocols for half-duplex fading relay channels. IEEE J Sel Areas Commun 25(2):379–389
[illegible]

Chapter 13
Resource Allocation and Pricing in Non-cooperative Interference Networks with Malicious Users

Fei Shen, Anil Kumar Chorppath, Eduard Jorswieck and Holger Boche

Abstract Strategic users in a wireless network cannot be assumed to follow the network algorithms blindly. Moreover, some of these users aim to use their knowledge about network algorithms to maliciously gain more resources and also to create interference to other users. We consider a general model of Multiple Access Channel (MAC) without successive interference cancellation (SIC) under Quality of Service (QoS) requirement of each user where malicious behavior exists. We model the heterogeneous behavior of users, which ranges from altruistic to selfish and then to malicious, within the analytical framework of game theory. To ensure the QoS requirements with efficient resource allocation, the noncooperative game in normal form is formulated and the Nash Equilibrium (NE) power allocation is derived in closed form. The effects of malicious behavior in network resource allocation mechanisms such as auctions and pricing schemes are studied. We consider firstly the problem of net utility maximization and then individual user QoS requirement satisfaction. We show that the well-known Vicrey-Clarke-Groves (VCG) mechanism loses its efficiency property in the presence of malicious users, which motivates the need to quantify the effect of malicious behavior. Then, the *Price of Malice* of the VCG mechanism and of some other network mechanisms are derived. Differentiated pricing as a method to counter adversarial behaviors is discussed. Next, we consider power allocation in wireless networks subject to QoS requirements of the users.

F. Shen (✉)
Large Systems and Networks Group (LANEAS), CentraleSupélec,
Université Paris-Saclay, 3 rue Joliot-Curie, 91192 Gif-sur-Yvette, France
e-mail: fei.shen@centralesupelec.fr

E. Jorswieck
Communications Theory, Communications Laboratory,
Technical University of Dresden, 01062 Dresden, Germany
e-mail: Eduard.Jorswieck@tu-dresden.de

A.K. Chorppath · H. Boche
Institute of Theoretical Information Technology, Technical University of Munich,
Theresienstrasse 90, 80333 Munich, Germany
e-mail: anil.chorppath@tum.de

H. Boche
e-mail: boche@tum.de

W. Utschick (ed.), *Communications in Interference Limited Networks*,
Signals and Communication Technology, DOI 10.1007/978-3-319-22440-4_13

Given the designed individual prices, the best response (BR) power converges to the unique NE power allocation rapidly, where the QoS requirement of each transmitter is satisfied. The impact of the malicious behavior on other users with QoS requirements is analyzed and the punishment prices are designed. We show that in the proposed noncooperative power allocation game, the user misbehavior is predicted, detected and prevented. As a result, all rate requirements in the capacity region of the general MAC are achieved at the NE point. Next we consider a scenario, in which a mechanism designer and legitimate users gather probabilistic information about the presence of malicious users by observing the network over a long time period and modify their actions accordingly. We analyze Bayesian mechanisms, both pricing schemes and auctions, and obtain the Bayesian Nash Equilibrium (BNE) points. The BNE points provide conditions indicating when the uncertainty about their nature (type) is better for regular users. Finally, we extend Bayesian pricing mechanisms to wireless networks subject to QoS requirements of the users.

13.1 Introduction

There has been a shifting focus in wireless network research for the past few years on decision making and incentives in the presence of malicious users [2, 6]. The malicious users create inefficiencies in allocation of network resources or even disrupt the networks. Wireless devices are becoming highly capable and at the same time more prone to security threats from increasingly sophisticated attacks.

The classical Vicrey-Clarke-Groves (VCG) mechanism [45] is an efficient and strategy-proof (truth revealing) mechanism in the presence of selfish players. In [16], an efficient, truth revealing and almost budget balanced (sum of payments zero) VCG mechanism is obtained for divisible resource allocation. We first show the influence of the adversarial behavior of some users on the efficiency of VCG mechanism for allocation of divisible resources, which we use to motivate the need for quantifying the effect of adversarial behavior on network mechanisms.

In networked systems with selfish users, a loss in overall social welfare was identified and referred to as *Price of Anarchy* [24, 31]. In [28], with the presence of malicious users this concept was extended and *Price of Byzantine Anarchy* and *Price of Malice* were first introduced and bounds were obtained on these metrics, which are parametrized by the number of malicious users for a virus inoculation game in social networks. A modified definition was proposed in [4] for congestion games based on the delay experienced at Nash equilibrium point with and without the presence of a malicious player. Both of these works observed a *Windfall of Malice*, where malicious behavior actually improves the social welfare of non-oblivious selfish users due to the better cooperation resulting from the 'fear factor' or effects similar to Braess's paradox [4]. To quantify the effects of adversarial behavior, we analyze the robustness of some known network mechanisms with respect to the adversarial behavior of (some of) their participants. The metric called Price of Malice proposed in [4, 28] needs to be modified for network resource allocation games. In the cases analyzed in the chapter, the malicious players are assumed to take the maximum possible resource share without being detected and try to disrupt others.

When there are malicious users who may deviate from equilibrium point by showing extreme selfishness, i.e. they behave 'irrationally' if modeled using this class of utility functions. We retain the rationality assumption by associating them with different utility functions. We first develop an auction mechanism for rate control in networks without interference coupling in user utility functions and then extend it for interference coupled problems. We consider the interference management by uplink power control in the CDMA wireless networks. In the presence of these adversarial agents, the mechanisms employed will have Nash equilibrium different from the efficient point and this deviation is captured in the metric *Price of Malice*. In this chapter, the Price of Malice is quantified for some specific network mechanisms and these mechanisms are modified to punish the adversarial users.

To counter the adversarial behavior, Micali and Valiant [27], developed a modified VCG mechanism, taking into account collusive, irrational, and adversarial user behavior for combinatorial auctions. In the proposed mechanism [27], the price charged to an agent is increased from VCG price by a scaled factor of the maximum social welfare of other agents. We employ a differentiated pricing scheme in which both aggressively selfish and malicious players with disproportional usage of resources are made to pay higher prices than regular selfish players. Clearly, when the malicious users do not abide by the rules and vandalize the system, a stronger response such as blocking the users suspected of malicious behavior after detection is required.

In [32], Bayesian jamming games are considered and the NE points for different jamming scenarios are obtained. Unlike the work in [32], we propose a mechanism design framework for pricing mechanisms and auctions [3] in the presence of malicious users and modify the mechanisms to counter malicious behavior. Our model captures the fact that, in addition to the resource allocation the malicious users affect the regular users through the prices charged to the users. We also use a different modeling of the malicious behavior in this chapter than in [32], which is proposed in our earlier work in [19], where the malicious user is interested in his own utility and has a degree of maliciousness, i.e. the malicious user aims to harm other regular users. In the scenario considered in this chapter, malicious users hide among the crowd of regular users accepting the prices and allocations from the designer, but use resources for their purpose and harm others.

We model malicious user in a wireless network, where compromised devices act like regular (selfish) users accepting the mechanism rules, which are the prices and allocation determined by the network designer. This way, they avoid immediate detection and continue having access to resources such as transmission power and spectrum. The legitimate users and the designer know only the probability with which a mobile device could be a malicious user and study the effect of this scenario on the wireless network (mechanism). The designer and the regular users obtain the probabilistic information about the number of malicious users in the network by observing the network over a long period of time.

In realistic situations, the users are uncertain about the nature of other users, i.e. whether others are regular users or malicious users [18]. In this chapter, we study the conditions under which uncertainty in the network is beneficial for regular users. The boundary conditions are based on wireless system parameters. A malicious user does not want to harm other malicious users by unnecessarily spending more energy and paying more price for the extra power. Therefore, by creating the uncertainty about their nature, the regular users confuse the malicious users. The uncertainty created in the network is a way for the regular users to counter the malicious users and have better utility for themselves. By observing the network over a long period time, the designer forms probability distributions on the nature of users and then design prices based on them.

We analyze Bayesian mechanisms [21] which have a designer (network) who designs allocation and prices based on information expressed as a probabilistic distribution over the type of the users. The utility a user derives is a function of the Signal-to-Interference plus Noise Ratio (SINR) which is a Quality of Service (QoS) metric in wireless network. The impact of the malicious behavior in interference limited wireless network is quantified within a Bayesian framework and malicious behavior resistant mechanisms are designed. The designer knows the probability of malicious user's existence and counters them by updating the prices using the probabilistic information [18]. The additional pricing by the designer and the hiding strategy of the regular users in the pricing mechanism counter malicious activities.

In this chapter, we also develop the distributed power allocation with individual pricing for the general MAC system without successive interference cancellation (SIC). The pricing is given such that the BR power converges to achieve the QoS requirement of each user and the malicious behavior of the users is prevented. We then propose Bayesian mechanisms for net utility maximization with QoS requirement for the users [20].

For different resource allocation algorithms in wireless networks involving games, the users need to know the channel gain of other users for finding their best response. Sometimes, the designer also need to know the channel gains of all the users to find the prices and allocations. For the efficient resource allocation to the users, the designer need to know their utility functions which are usually infinite dimensional. In distributed mechanism design, the designer finds allocation and pricing rules based on the one dimensional scalar signals from the users. Alternatively, the designer can assume a surrogate utility function for the users and make them report a scalar parameter of the utility function [23]. Learning schemes can be used to obtain utility functions from the signals of the users for designing prices and allocation. The paper [15] utilizes Gaussian process regression learning techniques to infer general user utilities to maximize the social welfare by a designer in a mechanism design setting. In pricing mechanisms, the price taking players are charged with the appropriate value of Lagrange multiplier which corresponds to the marginal utility functions of the users. In this chapter, we apply the regression learning technique to pricing mechanisms, to learn the utility function of regular and malicious users and then the malicious users are detected by observing the anomalies in utility functions [20].

The results of detections provide a better estimate of the statistics of malicious users to implement the Bayesian pricing .

First, in Sect. 13.2, we give the network mechanism model, utility model and Bayesian mechanism model with malicious users as well as the general MAC model under user QoS requirements. Next in Sect. 13.3, the mechanisms including VCG mechanism with malicious users are analysed. Initially, A centralized auction mechanism with only selfish users is proposed and its efficiency is proven. Then the values of *PoM* for auction mechanisms with malicious users are obtained with and without interference coupled utility functions. Next in Sect. 13.5, we relax the assumption that the users and the designer know the nature of users and we design Bayesian mechanisms. Then pricing mechanisms with malicious users are extended to a case where the Shannon rate based QoS requirement of each user is guaranteed by the wireless system in Sect. 13.6 for the general MAC. The power allocation is controlled by the carefully designed individual prices and the user potential misbehavior is predicted, detected and prevented by the punishment pricing mechanism. The chapter ends with concluding remarks and summary in Sect. 13.7.

13.2 Model

We denote vectors with block letters. The following Table 13.1 provides the notations and variables used in the chapter.

At the center of the mechanism design [3] model is the *designer D* who influences *N users*, denoted by the set *A*, who engage in a **strategic (noncooperative) game** [5] with each other. Define a *N*-player *strategic game* which results due to the interaction of users, $\mathcal{G}(\mathcal{A}, x \in \mathcal{X}, U)$, where each user or player $i \in \mathcal{A}$ has a respective scalar **decision variable** x_i such that

$$\mathbf{x} = [x_1, \ldots, x_N] \in \mathcal{X} \subset \Re^N_+,$$

and $\mathcal{X}$ is the decision space of all players. The decision variable x_i may represent, depending on the specific problem formulation, *i*th player's flow rate, power level, investment, or bidding in an auction.

Assumption 13.2.1 We assumes that the strategy space $\mathcal{X}$ has scalar decision variables, is compact, convex and has a nonempty interior.

Due to the inherent coupling between the players, the decisions of players directly affect each others performance as well as the aggregate allocation of limited resources. For example, the players may share fixed divisible resource X_{max}, such that $\sum_i x_i \leq X_{max}$. In the context of power control, x_i denotes the transmit power of user *i*.

Table 13.1 Table of Notations

Parameter	Description
$N; N^m$	Total number of users; Total number of malicious users
x_i	Transmit power of user i in the pricing game
X	The decision space of all users
$\underline{u}_i$	QoS requirement of user i
C_i	Payment of user i
γ_i	SINR of user i
P_i	Price per unit power of user i
β_i	Price per unit SINR of user i
U_i	Utility of user i
J_i	Cost of user i
B	Energy cost per unit transmitted power
h_i	Channel gain of user i
ψ^s	Probability that the other user is malicious for a regular user
ψ^m	Probability that the other user is regular for a malicious user
$\mu^s(N, N^m)$	Joint pmf of N and N^m as observed by regular user
$\mu^m(N, N^m)$	Joint pmf of N and N^m as observed by malicious user
$\mu^d(N, N^m)$	Joint pmf of N and N^m as observed by the designer
ψ_i^d	Probability that user i is malicious as observed by the designer
u_i	QoS requirement of user i
θ_i	Private type of user behavior

We consider a CDMA system, for which the received SINR of a user is given by

$$\gamma_i(\mathbf{x}) = \frac{h_i x_i}{I_i(\mathbf{x}_{-i})} = \frac{h_i x_i}{\frac{1}{L}\sum_{j \neq i} h_j x_j + \sigma^2}, \tag{13.1}$$

where h_i is the channel gain, L is the bandwidth and σ^2 denotes noise power.

The **preferences** of the users are captured by utility functions

$$U_i(\gamma_i(\mathbf{x})) : \mathcal{X} \to \Re, \quad \forall i \in \mathcal{A}.$$

Assumption 13.2.2 The utility function of the ith user, $U_i(\mathbf{x})$, is jointly continuous in all its arguments and twice continuously differentiable, non-decreasing and strictly concave in x_i.

13.2.1 Interference Modeling

Yates [46] proposed standard interference functions using an axiomatic approach. A different class of functions known as general interference functions were proposed in [9] and defined as follows.

Definition 13.2.1 (*General interference functions*) These are interference functions, $I : \Re_+^{K+1} \rightarrow \Re_+$, which satisfy following properties,
A1 conditional positivity: $I(x) > 0$ if $x > 0$
A2 scale invariance: $I(\alpha x) = \alpha I(x), \forall \alpha \in \Re_+$
A3 monotonicity: $I(x) \geq I(\tilde{x})$ if $x \geq \tilde{x}$
A4 strict monotonicity: $I(x) > I(\tilde{x})$ if $x \geq \tilde{x}, x_{N+1} \geq \tilde{x}_{N+1}$.

In [11], both the framework in [46] and the framework of general interference functions were compared and it was proved every standard interference function is a special case of framework of general interference functions. It means that any problem involving standard interference functions can be reformulated in terms of the framework $A1, A2, A3$. Therefore, the structural results obtained for general interference functions in [7, 8] can be applied also for standard interference functions.

The characterization of all convex resource allocation problems with cooperative users and their optimal solution are given in [12]. A Nash bargaining solution for users with log convex interference functions, in which proportionally fair optimizer depends on the interference coupling among the users, is proposed in [10].

In [33], the interference modeling is used for jointly designing the beamformers and transmission powers according to individual SINR requirements. In [35] the transceiver design for a multiuser network utilizes the interference modeling. A comprehensive theoretical framework for SIR balancing, with and without noise is derived in [34]. Reference [13] shows that under certain natural assumptions, the exponential transformation is the unique transformation (up to a positive constant) for convexification of resource allocation problems for linear interference functions. It also characterizes the largest class of utility functions and the largest class of interference functions, which allow a convex and concave formulation of a problem for interference coupled wireless systems.

The results in interference modeling are proven to be true assuming users as cooperative. We focus on non-cooperative users with the class of *log-convex* interference functions [7] which are a subset of general interference functions. They satisfy $A1 - A3$ and additionally $I(e^x)$ is log-convex on $\Re^{N+1}$. In [8], it was proven that every convex interference function is a log-convex interference function, however the converse is not true.

13.2.2 Mechanism Model with Malicious Users

The modified utility function [19] to model malicious users is obtained by a convex combination of user utilities

$$U_i^m(\gamma_i(\mathbf{x})) = U_i(\gamma_i(\mathbf{x})) + \theta_i \sum_{k \in S} U_k(\gamma_k(\mathbf{x})), \tag{13.2}$$

where θ_i is the parameter between –1 and 0. The variable θ captures the range of behavior of a user from malicious to selfish. The first term on the right-hand side is the self utility and the second term captures the malicious goal of the user. For a malicious user, the variable $\theta < 0$ and is called *degree of maliciousness*. We assume that malicious users do not gain anything by harming each other. The designer imposes a payment $C_i(\mathbf{x})$ on the actions of players, which is formulated by adding it as a cost term to utility.

Assumption 13.2.3 The payment function of the ith user, $C_i(\mathbf{x})$, is jointly continuous in all its arguments and twice continuously differentiable, non-decreasing and convex in x_i.

The player i has the cost function

$$J_i(\mathbf{x}) = C_i(\mathbf{x}) - U_i(\mathbf{x}), \quad \forall i \in \mathcal{A} \tag{13.3}$$

and solves the individual optimization problem

$$\min_{x_i} J_i(\mathbf{x}) \quad \forall i \in \mathcal{A}. \tag{13.4}$$

Assumption 13.2.4 The cost function of the ith user, $J_i(\mathbf{x})$, is twice continuously differentiable in all its arguments and strictly convex in x_i.

Definition 13.2.2 (Mechanism) A mechanism $\mathcal{M}$ is a tuple $(\mathcal{X}_1, \mathcal{X}_2, \ldots, \mathcal{X}_N, f(.))$, where function f specify an outcome for every strategy vector $\mathbf{x} \in \mathcal{X} \subset \Re^N$, of the players.

The function f is implemented through allocation and pricing rules.

We differentiate between two kinds of mechanisms, auctions and pricing, which differ in the assumption on the nature of the users and the interaction rules.

Definition 13.2.3 (*Auction Mechanism*) In auction mechanisms, the designer D imposes on a price anticipating user $i \in \mathcal{A}$ a (possibly user-specific)

- resource allocation rule, $Q_i(\mathbf{x})$,
- resource pricing, $C_i(\mathbf{x})$,

based on user bids $\mathbf{x}$.

In auctions, $f(\mathbf{x}) = (Q_1(\mathbf{x}), Q_2(\mathbf{x}), \ldots, Q_N(\mathbf{x}), C_1(\mathbf{x}), C_2(\mathbf{x}), \ldots, C_N(\mathbf{x}))$.

Definition 13.2.4 (*Pricing Mechanism*) In pricing mechanisms, the price taking users decide on their allocation as a best response to the (user-specific) price P_i induced by the designer and there is no explicit allocation rule dictated.

In the pricing mechanism case, the cost function is

$$J_i^m(\mathbf{x}) = P_i(\mathbf{x})x_i - U_i^m(\mathbf{x})\,\forall i. \tag{13.5}$$

The **designer objective**, e.g. maximization of aggregate user utilities or social welfare, can be formulated using an objective function

$$V(\mathbf{x}, U_i(\mathbf{x}), C_i(\mathbf{x})) : \mathcal{X} \to \Re,$$

where $C_i(\mathbf{x})$ and $U_i(\mathbf{x})$, $i = 1, \ldots, N$ are user-specific pricing terms and player utilities, respectively. Thus, the objective function V characterizes the desirability of an outcome $\mathbf{x}$ from the designers perspective. In some cases when the designer objective is to satisfy certain minimum performance constraints such as players achieving certain quality-of-service levels, the objective can be characterized by a region (a subset of the game domain $\mathcal{X}$). For the net utility maximization, the designer objective is

$$V(\mathbf{x}) = \sum_{i \in \mathcal{A}} U_i(\mathbf{x}). \tag{13.6}$$

We redefine the metric $PoM(\mathcal{M})$ of mechanism $\mathcal{M}$ in order to make it suitable for resource sharing mechanisms. In [4], for congestion games with malicious flow concentrated on one malicious player, Price of Malice was defined, based on the delay experienced at Nash equilibrium point with and without the malicious player. We now redefine *PoM* for network games and mechanisms with discrete set of players similar to the definition given in [4].

Definition 13.2.5 (*Price of Malice (PoM)*) The metric Price of Malice (PoM) of a mechanism $\mathcal{M}$ is defined as:

$$PoM(\mathcal{M}) := \frac{\sum_{j \in \mathcal{S}} U_j(\tilde{\mathbf{x}}) - \sum_{j \in \mathcal{S}} U_j(\mathbf{x}^*)}{\sum_{j \in \mathcal{S}} U_j(\tilde{\mathbf{x}})},$$

where $\tilde{\mathbf{x}}$ is the Nash equilibrium when none of the users are malicious and $\mathbf{x}^*$ is Nash equilibrium in the presence of malicious users.

13.2.3 Bayesian Mechanism Model

Now we give the model of the mechanism with an arbitrary number of regular and malicious users. Let $\mu^m(N, N^m)$ and $\mu^s(N, N^m)$ be the joint probability mass function (pmf) of N and N^m as observed by malicious and regular user respectively. The users do not know the nature of the users around them and evaluate their costs based on the pmfs. In addition to the price, the users have battery energy cost for transmission in the uplink of a wireless link. Let a user spends energy B for transmission per unit of transmit power. The cost function of the regular user will be,

$$J_i^s(x^s, x^m) = \sum_{N^m=0}^{N} \mu^s(N, N^m)(\beta_i \gamma_i^s(N, N^m) + B\frac{x_i^s}{h_i} - U(\gamma_i^s(N, N^m)). \tag{13.7}$$

For the symmetric case, when the channel gains of all the regular users and malicious users are equal to h^s and h^m respectively, the SINR of regular users become

$$\gamma^s(N, N^m) = \frac{h^s x^s}{\frac{1}{L}((N - N^m - 1)h^s x^s + N^m h^m x^m) + \sigma^2}$$

where x^s and x^m are the symmetric power strategies for selfish and malicious users respectively.

The utility function of malicious user i is

$$U_i^m(\mathbf{x}) = \sum_{N^m=0}^{N} \mu^m(N, N^m)\left(U(\gamma_i^m) + \theta_i \gamma_i^m\right).$$

For the symmetric case,

$$\gamma^m(N, N^m) = \frac{h^m x^m}{\frac{1}{L}((N - N^m)h^s x^s + (N^m - 1)h^m x^m) + \sigma^2}$$

is the SINR of malicious user.

Then the cost function of the malicious user for the symmetric case with $\theta_i = \theta^m, \ \forall\, i$ is

$$\begin{aligned} J^m(\mathbf{x}) = &\sum_{N^m=0}^{N} \mu^m(N, N^m)(\alpha(\beta\gamma^m(N, N^m) + B\frac{x^m}{h^m}) \\ &- U(\gamma^m(N, N^m)) + \theta^m \gamma^m(N, N_m)), \end{aligned} \tag{13.8}$$

where α_i is the parameter which indicates how much the malicious user i is sensitive towards the payment. Some malicious users would like to disrupt the network even by taking the chance of getting detected and for them $\alpha = 0$. A malicious user does not gain anything by creating interference to other malicious users. For this malicious users need to spend more energy and pay price for the extra power. Therefore, when two malicious users encounter each other in the game they have only payment and energy cost.

Next, as a special case we propose the model for the mechanism with two users and obtain the utility functions and cost functions. Let the probability belief of a regular user that another user is malicious be ψ^s and the probability belief of a malicious user that another user is regular be ψ^m. The total cost of regular user i including the price and energy cost will be,

$$\begin{aligned} J_i(\mathbf{x}) = & \frac{Bx_i^s}{h_i} + \psi^s(C_i(x_i^s, x_j^m) - U_i\left(\gamma_i(x_i^s, x_j^m)\right)) \\ & + (1-\psi^s)(C_i(x_i^s, x_j^s) - U_i\left(\gamma_i(x_i^s, x_j^s)\right)). \end{aligned} \tag{13.9}$$

The case when $\psi^s = 0$ will be the one with complete information. The cost function of malicious user i,

$$\begin{aligned} J_i^m(\mathbf{x}) = & \frac{\alpha_i Bx_i^m}{h_i} + \psi^m(\alpha_i C_i(x_i^m, x_j^s) - U_i\left(\gamma_i(x_i^m, x_j^s)\right) \\ & - \theta_i U_j\left(\gamma_j(x_i^m, x_j^s)\right) + (1-\psi^m)\alpha_i C_i(x_i^m, x_j^m). \end{aligned} \tag{13.10}$$

We consider SINR pricing in this chapter,

$$C_i(\mathbf{x}) = \beta_i \gamma_i(\mathbf{x}), \ \forall\, i.$$

For simplifying the analysis, let us take a modified utility function without energy cost. The maliciousness term is the utility function of user i is replaced in [41] with the SINR of user i. The reason is that the malicious user affects all the other users with his SINR. The total cost function regular user i including the price will be,

$$\begin{aligned} J_i(\mathbf{x}) = B & \frac{x_i^s}{h_i} + \psi^s(\beta_i\gamma_i(x_i^s, x_j^m) - U_i\left(\gamma_i(x_i^s, x_j^m)\right)) \\ & + (1-\psi^s)(\beta_i\gamma_i(x_i^s, x_j^s) - U_i\left(\gamma_i(x_i^s, x_j^s)\right)). \end{aligned} \tag{13.11}$$

and for malicious user

$$\begin{aligned} J_i^m(\mathbf{x}) = \alpha_i B & \frac{x_i^m}{h_i} + \psi^m(\alpha_i\beta_i\gamma_i(x_i^m, x_j^s) - U_i(x_i^m, x_j^s) \\ & - \theta_i\gamma_i(x_i^m, x_j^s)) + (1-\psi^m)\alpha_i\beta_i\gamma_i(x_i^m, x_j^m). \end{aligned}$$

The Nash Equilibrium (NE) of the game is the solution point where no player gains anything by changing their own strategies unilaterally.

Definition 13.2.6 (*Nash Equilibrium*) The strategy profile $\mathbf{x}^* = [x_1^*, \ldots, x_N^*]$ is in Nash Equilibrium if the cost of each player is minimized at the equilibrium given the best strategies of other players.

$$J_i(x_i^*, \mathbf{x}_{-i}^*) \le J_i(x_i, \mathbf{x}_{-i}^*), \forall i \in \mathcal{A}, x_i \in \mathcal{X}_i.$$

The NE is at the same time the intersection point of player's best responses obtained by solving (13.4) individually, i.e.,

$$x_i^* := \arg\min_{x_i} J_i(x_i, \mathbf{x}_{-i}^*), \forall i. \tag{13.12}$$

Definition 13.2.7 (*Dominant Strategy Equilibrium*) The strategy profile $\mathbf{x}^D = [x_1^D, \ldots, x_N^D]$ is in dominant strategy equilibrium if the cost of each player is minimized at the equilibrium irrespective of the strategies of other players.

$$J_i(x_i^D, \mathbf{x}_{-i}) \leq J_i(x_i, \mathbf{x}_{-i}), \forall i \in \mathcal{A}, x_i \in \mathcal{X}_i, \mathbf{x}_{-i} \in \mathcal{X}_{-i}.$$

13.2.4 MAC Model with QoS Requirements

The pricing mechanism can be adopted in the distributed power allocation where each user submits a QoS requirement. We specify our analysis for the general MAC system without SIC [41]. The case for MAC system with SIC is further discussed in [42].

We study the general MAC with N transmitters and one receiver as the BS. All the transmitters and the BS are equipped with single antenna. In the rest of the chapter, we will not differentiate users and transmitters. Each user $i \in N$ has a Shannon rate based QoS requirement $\underline{u}_i$ to be guaranteed by the MAC system. The linear receiver without SIC is considered, therefore each user suffers from the interference of all the other users. We assume the system guarantees the rate requirement of each user by providing the individual prices β_i. The non-cooperative game is discussed in the system, where each user minimizes its own cost J_i as a function of the price β_i and the transmit power x_i. We introduce the prices $\boldsymbol{\beta}$ such that all feasible rate requirement of each user in the capacity region of MAC can be achieved at the NE point of the non-cooperative game with minimum power allocation. The strategy set of each user is their power allocation with single power constraint $x_i < x_i^{max}$.

The received signal at the BS is given by $y = \sum_{k=1}^{N} \alpha_k z_k + n$, where z_k is the transmit signal of user k, n is the additive white Gaussian noise with zero-mean and variance of σ_n^2. The channel gain of user k is denoted by $h_k = |\alpha_k|^2$.

We assume the quasi-static block flat-fading channel gains $\mathbf{h} = h_1, \ldots, h_N$ are independent of each other and remain constant for a sufficiently long period of time.

The Shannon rates are considered as the QoS criterion. For the MAC system without SIC described above, the achievable rate of each user is

$$U_i(x_i, x_{-i}) = \log\left(1 + \frac{h_i x_i}{1 + \sum_{k \neq i} h_k x_k}\right),$$

where $x_{-i} = [x_1, \ldots, x_{i-1}, x_{i+1}, \ldots, x_N]$ denotes the power allocation of all the other users except user i and the noise power is normalized. Here the noise power σ^2 is normalized to be one while the SINR in Eq. (13.1) is with σ^2 and code gain L.

In the game theoretic point of view, the players involved in the noncooperative game have incentives to hide their private types in order to overtake the other players for higher individual utilities. These private types, denoted as θ_i, are certain private information hidden in the individual user utility functions, e.g., channel state information (CSI) or utility preferences. In the proposed noncooperative power allocation game, the malicious users minimize their own costs $J_i(x_i, x_{-i}, \theta_i)$ as a function of θ_i while pretending to other users as $J_i(x_i, x_{-i})$. The resulting power allocation $x_i^{NE}(\theta_i)$ and the achievable rate $U_i(\theta_i)$ are predicted. The user misbehavior can be detected by comparing the rate requirement and the achievable rate of users at the BS. A predefined punishment price β_i^m is announced as a common rule of the game to all the players before the game starts. Being threatened by the potential punishment price, no user in the proposed distributed power allocation game has incentives to behave maliciously. The proposed noncooperative power allocation game results in the unique NE point where all rate requirements of users in the capacity region of the MAC are guaranteed.

13.3 Price of Malice in Mechanisms

We quantify in this section, the resilience of some network mechanisms to malicious behavior. For this purpose, we use the definition of *PoM* in the Definition 13.2.5. Now, we proceed to estimate the value of Price of Malice parameter for different network mechanisms. First, we start with the direct VCG mechanism which was shown to be nonresistant to malicious behavior in the previous section.

13.3.1 Price of Malice in VCG Mechanism

We consider the case where user k is malicious and users have logarithmic utility function,

$$U_i(\mathbf{x}) = \alpha_i \log(Q_i(x)), \forall i. \tag{13.13}$$

We obtain analytical expression for *PoM* in VCG Mechanism which can be generalized to other cases.

Proposition 13.3.1 *For the additive resource sharing with user utility functions given in Eq.* (13.13)*, the Price of Malicious of VCG mechanism PoM*(*VCG*) *is*

$$PoM(VCG) = \frac{\sum_{j\neq k} \alpha_j \log\left(\frac{\alpha_k + (1+\theta_k)\sum_{j\neq k}\alpha_j}{(1+\theta_k)\sum_m \alpha_m}\right)}{\sum_{j\neq k} \alpha_j \log\left(\frac{\alpha_j X_{max}}{\sum_m \alpha_m}\right)}$$

For the case where users are symmetric $\alpha_i = \alpha$, $\forall i$, and only one user is malicious or all the malicious user coordinate to form one entity, this simplifies to

$$PoM(VCG) = \frac{\log\left(\frac{N-1+\frac{1}{1+\theta_k}}{N}\right)}{\log\left(\frac{X_{max}}{N}\right)}.$$

Proof $PoM(VCG)$ is derived directly by substituting the optimal allocations using the utility functions in (13.13) to the Definition 13.2.5 given above.

We could observe that when the maliciousness of users increase, i.e., as θ decreases from 0 to -1, we can see that the Price of Malice increases. We could observe that the $PoM(VCG)$ can be bounded for different possible values of θ_k and is unbounded when θ_k reaches –1.

13.3.2 Price of Malice in Indirect Auction Mechanisms

Here we present indirect auction mechanisms [25] for two network coupling schemes, rate control in wired networks and power allocation in interference coupled wireless networks, and quantify the Price of Malice for both cases. In the indirect mechanisms, instead of reporting their utility function to the designer, the players take a best response to the actions of other players and to the allocation and pricing rules set by the designer. Therefore, the allocation and pricing rules are not a function of utility functions unlike direct VCG mechanism, but rather fixed functions of the player strategies. We consider indirect auction mechanisms with scalar bid here since they have only one dimensional communication requirement which is suitable for network resource allocation. The malicious behavior considered in this section is that the malicious players take maximum possible share of the resources according to their θ value. This way the malicious players aim to disrupt other players by denying their fair share of resources.

13.3.2.1 Auctions for Rate Control in Networks

We consider the rate sharing problem with users having separable utility function of their allocation and quantify the effect of the adversarial behavior on it. Let users with utilities $U_i(Q_i)$ share a fixed bandwidth Q_{max} such that $\sum_{i=1}^{N} Q_i(\mathbf{x}) \leq Q_{max}$, where $x_i \in (0, x_m)$. The vector $\mathbf{x}$ in this case denotes player flow rates and $\mathbf{Q}$ the capacity allocated to them [1, 44]. Consider the utility function given in (13.2) and the cost of ith user is then given by,

$$J_i^m(\mathbf{x}, \theta_i) = C_i(\mathbf{x}) - U_i(Q_i(\mathbf{x})) - \theta_i \sum_{j \in S} U_j(Q_j(\mathbf{x})), \forall i. \tag{13.14}$$

We consider the efficient proportional allocation auction mechanism $\mathcal{M}_a$ introduced in [26] which is an indirect mechanism where the users submit a scalar bid. The proportional allocation which is defined based on the bid vector of players $\mathbf{x}$ is

$$Q_i(\mathbf{x}) := \frac{x_i}{\sum_j x_j + \omega} Q_{max}, \tag{13.15}$$

where ω can be seen as the reserve bid [22] and it removes the singularity of the function. For $\omega = 0$, we could see that the resource is completely utilized, i.e., $\sum_i Q_i = Q_{max}$.

We next briefly show how the pricing rule/function is designed with the use of a generator function, as in [26]. In Sect. 13.4 of this chapter, we detail the procedure with taking into consideration malicious behavior. Let us define $t = \sum_j x_j + \omega$ as a measure of demand for the resource and which allows us to characterize player optimal responses with respect to a parameter which is identical for all players at equilibrium. The generator function is $g(.)$ is a function of t to R^+ and plays the role of Lagrange multiplier to generate the optimal pricing function. The total payment of ith user has several choices, depending on the choice of generator function.

For $g(t) = t^2$, the payment function is derived in [26] as

$$C_i(\mathbf{x}) = x_i \sum_{j \neq i} x_j + \omega, \tag{13.16}$$

which is convex payment function in x_i and is sufficient to guarantee a unique Nash equilibrium. We found a mistake in [26], when using $g(t) = t$. The payment function for this case

$$C_i(\mathbf{x}) = \log\left(1 + \frac{x_i}{\sum_{j \neq i} x_j}\right) \sum_{j \neq i} x_j,$$

is concave in x_i and contradicts with the convexity result in Proposition 1 and 2 in [26]. Therefore, we do not use $g(t) = t$.

Being oblivious to the presence of malicious users, the designer employ the same allocation rule and payment to ith user as the one obtained above for mechanism $\mathcal{M}_a$ assuming all the users are just selfish. First, we characterize the modified Nash equilibrium if some of these users are malicious or altruistic. Let us check for the special case of logarithmic utilities and the mechanism $\mathcal{M}_a$.

Proposition 13.3.2 *The mechanism $\mathcal{M}_a$ defined by* (13.15) *and* (13.16) *with users having logarithmic utilities admits several Nash equilibria and one NE point is given as*

$$x_i^* = \frac{\alpha_i}{t(1 + \theta_i) Q_{max}}, \tag{13.17}$$

where $-1 < \theta_i \leq \alpha_i(\frac{1}{x_m} + \frac{t}{x_m^2})$.

Proof For the allocation in (13.15) the strategy space constrained by the set $0 \leq x_i \leq x_m \; \forall i$ satisfy the Assumption 13.2.1. Then by a standard theorem of game theory (Theorem 4.4, p. 176, in [5]), the network game admits a NE.

For the payment given in (13.16), allocation in (13.15) and logarithmic utility function, the cost functions satisfies Assumption 13.2.4 for $\theta_i < 0$. For altruistic case, i.e. $\theta_i > 0$, the cost functions satisfies Assumption 13.2.4 only for

$$\theta_i \leq \alpha_i\left(\frac{1}{x_m} + \frac{t}{x_m^2}\right).$$

This is obtained by checking for $\frac{d^2 J_i^m}{dx_i^2} \geq 0$. We consider in the game only altruistic users satisfying this condition, in order to obtain an equilibrium. Since the cost function satisfies Assumption 13.2.4 for all the users with $-1 < \theta_i \leq \alpha_i(\frac{1}{x_m} + \frac{t}{x_m^2})$, the best response points obtained from first order conditions gives a Nash equilibrium.

The best response of user becomes

$$\frac{\partial J_i^m}{\partial x_i} = 0 \Longrightarrow x_i^* = \frac{\alpha_i}{t(1+\theta_i)X_{max}},$$

by using the fact that selfish users will have the Nash equilibrium point $x_i^* = \frac{\alpha_i}{tQ_{max}}$, from the incentive compatibility property of the mechanism $\mathcal{M}_a$.

Remark 13.3.1 In [30], the conditions for existence of a unique NE for an $N-$person game is given. In addition to Assumption 13.2.4, the cost functions should satisfy diagonal strict concavity of the weighted nonnegative sum of the cost functions as given in Theorem 2 of [30]. The cost function does not necessarily satisfy this condition in our case. Therefore, the NE is not unique.

We can observe that malicious users having $-1 < \theta < 0$, will have the Nash equilibrium point as $x_i^* > \frac{\alpha_i}{tQ_{max}}$. Therefore, the malicious users bid higher than the selfish users and obtain a disproportionate higher share of resource.

The allocation for the regular selfish users, i.e., users with $\theta_i = 0$ in the presence of malicious users can be written as

$$Q_i' = \frac{\frac{\alpha_i}{tQ_{max}} Q_{max}}{\sum_{j \in \mathcal{S}} \frac{\alpha_j}{tQ_{max}} + \sum_{k \in \mathcal{B}} \frac{\alpha_k}{t(1+\theta_k)Q_{max}} + \omega}. \tag{13.18}$$

Let

$$r_i = \frac{Q_i}{Q_i'} = \frac{\sum_{j \in \mathcal{S}} \alpha_j + \sum_{k \in \mathcal{B}} \frac{\alpha_k}{(1+\theta_k)} + \omega}{\sum_j \alpha_j + \omega} \tag{13.19}$$

be the ratio of allocation of selfish users before and after the presence of malicious users. Now we obtain the value of PoM of the mechanism $\mathcal{M}_a$ at the NE point given in Proposition 13.3.2.

Proposition 13.3.3 *For the additive resource sharing case, the Price of Malicious $PoM(\mathcal{M}_a)$ is*

$$PoM(\mathcal{M}_a) = \frac{\sum_{j\in\mathcal{S}} \alpha_j \log(r_j)}{\sum_{j\in\mathcal{S}} \alpha_j \log\left(\frac{\alpha_j Q_{max}}{\sum_i \alpha_i + \omega}\right)}.$$

For the case where users are symmetric $\alpha_i = \alpha,\ \forall i$ and only one user is malicious or all the malicious user coordinate to form one entity, this simplifies to

$$PoM(\mathcal{M}_a) = \frac{\log\left(\frac{\alpha(N-1+\frac{1}{1+\theta_k})+\omega}{N\alpha+\omega}\right)}{\log\left(\frac{\alpha Q_{max}}{N\alpha+\omega}\right)}.$$

Proof The results follow directly by using the allocation given in Eq. (13.15) and the value of r_i in Eq. (13.19) to the definition of PoM in Definition 13.2.5.

Remark 13.3.2 We could see that $PoM(\mathcal{M}_a)$ is equal to $PoM(VCG)$ when $\omega = 0$ for the special case of the utility function considered. It is because the proportional allocation coincides with the VCG allocation for this case. But we get very different $PoM(\mathcal{M}_a)$ and $PoM(VCG)$ in the case of the other utility functions, for example $U_i(Q_i) = \alpha_i \log(1 + Q_i)$.

We also present another auction-based mechanism, $\mathcal{M}'_a$, for the case when the bid is equal to the payment. The approximately efficient mechanism, $\mathcal{M}'_a$, can be defined based on the bid of player i as,

$$x_i := P_i(\mathbf{x})Q_i(\mathbf{x}), \tag{13.20}$$

the pricing function

$$P_i := \frac{\sum_{j\neq i} x_j + \omega}{Q_{max}}, \tag{13.21}$$

and the resource allocation rule

$$Q_i := \frac{x_i}{\sum_{j\neq i} x_j + \omega} Q_{max}. \tag{13.22}$$

for a scalar $\omega > 0$ sufficiently large such that $\sum_i Q_i \leq Q_{max}$.

Consider now the mechanism $\mathcal{M}'_a$ for the logarithmic case. The cost function in this case is,

$$J_i^m(\mathbf{x}, \theta_i) = x_i - \alpha_i \log\left(\frac{x_i}{I_i}\right) - \theta_i \sum_{j\in\mathcal{S}} \alpha_j \log\left(\frac{x_j}{I_j}\right),$$

where $I_i = \sum_{j \neq i} x_j + \omega$. The best responses of each user will lead to a set of equations,

$$\frac{\alpha_i}{x_i} = 1 - \theta_i \sum_{j \in \mathcal{S}} \frac{\alpha_j}{x_i + \sum_{k \notin i,j} x_k}, \forall i.$$

We can see that for the selfish users, $x_i = \alpha_i$. Therefore, for the case in which there is a single malicious user, the following polynomial of Nth degree is solved by the malicious user i,

$$\frac{\alpha_i}{x_i} = 1 - \theta_i \sum_{j \in \mathcal{S}} \frac{\alpha_j}{x_i + \sum_{k \notin i,j} x_k}.$$

A Nash equilibrium point could be obtained from the intersection of all these points since the cost function satisfies Assumption 13.2.4 for $\theta \leq 0$. But it is not possible to have analytical result for the NE in this case. As above in the case of $\mathcal{M}_a$, the PoM can be calculated in this case also but numerically. Therefore, the variation of values of $PoM(\mathcal{M}_a')$ for different values of θ is given in the simulation section.

Remark 13.3.3 From the Propositions 13.3.3 and 13.3.1, we could see that the Price of Malice of a mechanism can be obtained knowing system parameters and user preferences and can be bounded above and below (if possible) depending on the range and distribution of these values for the specific setting.

13.3.3 Auctions for Interference Coupled Systems

Consider an auction mechanism in the context of a wireless network and uplink power control setting [16, 22] where due to the interference coupling the resource sharing is inherently competitive. Let the user utilities be taken as $U_i(\gamma_i(\mathbf{Q}))$ and the individual power levels, Q, satisfy $\sum_{i=1}^{N} Q_i \leq Q_{max}$, where the SINR received by the base station is given in Eq. (13.1). We propose an auction-based mechanism $\mathcal{M}_b$, defined by Eqs. (13.16) and (13.15) with $\omega = 0$ for interference coupled systems.

Proposition 13.3.4 *The auction-based mechanism, $\mathcal{M}_b$, is an ε-efficient mechanism for system having users with interference coupled utility functions $U_i(\gamma_i(\mathbf{Q}))$ if*

$$\frac{|U_i''|}{U_i}(\gamma_i + L) > 2. \tag{13.23}$$

Proof We decouple the user utilities by rewriting γ_i as

$$\gamma_i(Q_i) = \frac{Q_i(\mathbf{x})}{Q_{max} - Q_i(\mathbf{x}) + \sigma}, \tag{13.24}$$

using the full utilization property of the mechanism $\mathcal{M}_a$ when $\omega = 0$.

In [22], it is observed that in systems with sufficiently high SINR satisfying the Assumption 2 given in Eq. (13.23)

$$U_i(\gamma_i(Q_i)) = U_i(\frac{Q_i}{Q_{max} - Q_i + \sigma})$$

is concave in Q_i. It can be also seen that $U_i(\gamma_i(Q_i))$ is monotonically increasing and twice differentiable in Q_i. Therefore, the sufficient condition for the existence of an ε-efficient unique NE is satisfied along with allocation given in (13.16) and pricing given in (13.15).

For the allocation given in (13.15), the SINR at NE point $\mathbf{x}^*$ is

$$\gamma_i(\mathbf{x}^*) = \frac{x_i^* Q_{max}}{\sum_j x_j^*(Q_{max} + \sigma) - x_i^* Q_{max}}. \tag{13.25}$$

In the presence of malicious and altruistic users, let the SINR obtained by the regular users be $\gamma_i(\mathbf{x}')$ where $\mathbf{x}'$ is the new NE point. Now we give the value of $PoM(\mathcal{M}_b)$ in the following proposition for the interference coupled wireless system.

Proposition 13.3.5 *The PoM of the mechanism $\mathcal{M}_b$ for the interference coupled wireless system is given as*

$$PoM(\mathcal{M}_b) = \frac{\sum_{j \in \mathcal{S}} \alpha_j \log(\frac{\gamma_j(\mathbf{x}^*)}{\gamma_j(\mathbf{x}')})}{\sum_{j \in \mathbf{S}} \alpha_j \log(\gamma_j(\mathbf{x}^*))}.$$

where $\mathbf{x}'$ and $\mathbf{x}^*$ are the NE points with and without the presence of malicious users.

13.3.4 Price of Malice in Pricing Mechanisms

In pricing mechanisms the users choose their allocation as their strategy or action. Pricing mechanisms do not have explicit allocation rule. Their actions reveal only some information about their utility function. The pricing mechanisms are more appropriate for modeling distributed systems where we cannot expect a central authority to allocate resource to the users.

A counterpart of the Price of Malice metric in Definition 13.2.5 for pricing mechanisms, which differ from auctions by their lack of an explicit resource allocation scheme, can be obtained by replacing $Q(\mathbf{x})$ and $Q(\mathbf{x}')$ with the action vector without malicious users $\mathbf{x}$ and with malicious users $\mathbf{x}'$, respectively.

In the case of additive resource sharing, the users with utilities $U_i(x_i) = \alpha_i \log x_i$ share the fixed resource $\sum_{i=1}^{N} x_i = X_{max}$, and $x_i \in (0, x_m)$. Consider an efficient mechanism $\mathcal{M}_c$, for which the efficient allocation is

$$x_i = \frac{\alpha_i}{\lambda}, \tag{13.26}$$

where λ is the Lagrange multiplier. In the case of all selfish users $\lambda = \sum_i \alpha_i / X_{max}$ and it will be set as price to the users.

We can observe that in the case of pricing, the utility function of the malicious user is given by,

$$J_i^m(\mathbf{x}, \theta_i) = P_i x_i - U_i(x_i) - \theta_i \sum_{j \in S} U_j(x_j). \tag{13.27}$$

We can see that the additional third term does not have direct dependence on x_i and does not play a role in malicious user cost minimization. But, that term is indirectly a function of x_i due to the additive coupling in the global objective. The effect of the additive coupling in the global objective is brought by Lagrange multiplier which acts as the price in the user objective.

Let each malicious user take a share x_m which should be less than x_{max}, the maximum share they can use without detection, according to their utility function, in order to disrupt others. Let λ' be the Lagrange multiplier in this case which will be a different point than $\lambda = \sum_i \alpha_i / X_{max}$. The remaining resource $(X_{max} - \sum_B x_m)$ will be shared among good users, under the efficient mechanism $\mathcal{M}_c$, i.e., $x_i = \frac{\alpha_i}{\lambda'}$.

Proposition 13.3.6 *In the additive sharing case $PoM(\mathcal{M}_c)$ is,*

$$PoM(\mathcal{M}_c) = \frac{\sum_{j \in S} \alpha_j \log\left(\frac{X_{max}\lambda'}{\sum_i \alpha_i}\right)}{\sum_{j \in S} \alpha_j \log\left(\frac{\alpha_j X_{max}}{\sum_i \alpha_i}\right)}.$$

For symmetric case, where $\alpha_i = \alpha\ \forall i$, it becomes

$$PoM(\mathcal{M}_c) = \frac{\log\left(\frac{X_{max}\lambda'}{N\alpha}\right)}{\log\left(\frac{X_{max}}{N}\right)}.$$

13.4 Response Mechanisms to Malicious Users

The robustness analyses of mechanisms and quantification of PoM in the Sect. 13.3 only measure the effect of malicious users but does not provide a way to encounter them. In [29] it was shown that the Second Price auction can be made robust to interdependent preferences corresponding to altruistic or malicious behavior by changing just the pricing to that of a First Price auction, augmented by bonus payments. Similar to this approach, in this section we consider a possible response schemes to adversarial behavior, based on a softer punishment scheme using differentiated pricing.

13.4.1 Differentiated Pricing

We propose a softer response scheme than blocking towards malicious users after explicit detection based on any well known (threshold) detection scheme. There are numerous methods of detection already available as given in PART IV of [2]. The response mechanism is implemented by the designer by deploying a differentiated pricing. A necessary assumption we make in this subsection is that malicious users stay within the system and do not have any means to evade the pricing mechanisms imposed by the designer. This assumption is relaxed in the next section. The designer knows the value of θ of malicious user. In practical problems, this is not realistic and the designer needs to make the decision on payment function entirely based on user bids. Therefore, we assume that after detecting the malicious user using a threshold detection scheme based on the bids, the designer punishes the malicious users with a price function assuming $\theta = -1$, i.e., extreme maliciousness. Alternatively, once can couple this parameter with the confidence of the detection scheme used, i.e. low θ values for high probability of malicious behavior and vice versa.

13.4.2 Differentiate Pricing for Proportional Allocation Mechanism

We consider the proportional allocation network mechanism $\mathcal{M}_a$ proposed for network rate sharing in Sect. 13.3.2 and modify it with a new payment function. We propose now a differentiated payment function to counter the malicious behavior of users and propose a new mechanism using this payment function. We propose mechanism $\mathcal{M}_m$ in the following proposition which is efficient in the presence of malicious users, i.e., $PoM(\mathcal{M}_m) = 1$.

Proposition 13.4.1 *The mechanism $\mathcal{M}_m$ defined by the allocation in* (13.15) *with $\omega = 0$ and the payment*

$$C_i(x) = x_i \sum_{j \neq i} x_j - \theta_i (N-1) tC \log\left(1 + \frac{x_i}{\sum_{j \neq i} x_j}\right), \qquad (13.28)$$

is efficient in the presence of malicious users and makes the malicious user take the strategy $x_i^* = \frac{\alpha_i}{tC}$ for network rate sharing with users having logarithmic utility functions.

The proof is given in the Sect. 8 of [19].

Remark 13.4.1 If the designer punishes the users who are detected as malicious with a payment in which $\theta_i = -1$, without knowing the exact θ value in a more realistic situation, the pricing function becomes

$$C_i(x) = x_i \sum_{j \neq i} x_j + \log\left(1 + \frac{x_i}{\sum_{j \neq i} x_j}\right)(N-1). \tag{13.29}$$

For this cost function to be convex, in order to take the best response, from the second order conditions we get

$$N \leq \frac{\sum_{j \neq i} \alpha_j}{C^2} + 1.$$

Note that in this differentiated pricing scheme, the malicious users who will try to bid something higher than its private value will have to pay an additional amount proportional to their bid as in (13.29). Even if the cost function is not convex, it does not affect the equilibrium. This is because anticipating the additional payment, the malicious user will bid taking the best response according to the cost with payment given by Eq. (13.16), which is convex.

13.5 Bayesian Mechanisms for Wireless Network Security

We analyze Bayesian mechanisms [21] which have a designer (network) who designs allocation and prices based on information expressed as a probabilistic distribution over the type of the users. The utility a user derives is a function of the SINR which is a QoS metric in the wireless network. The impact of the malicious behavior in interference limited wireless network is quantified within a Bayesian framework and malicious behavior resistant mechanisms are designed. We analyze an incomplete information case, where the malicious behavior is countered without explicit detection of malicious users or learning their nature. The designer knows the probability of malicious user's existence and counters them by updating the prices using the probabilistic information. The additional pricing by the designer and the hiding strategy of the regular users in the pricing mechanism counter malicious activities. The pricing is given such that the BR power converges to achieve the QoS requirement of each user and the malicious behavior of the users is prevented. We first analyze the complete information case for the comparison.

13.5.1 Pricing Mechanisms with Complete Information

With complete information and for the single carrier case, the optimal prices are proposed in [17]. For the two-users case, the price for User 1 turns out to be,

$$P_1 = \frac{1}{1-\gamma_1\gamma_2}\left(B\gamma_1\gamma_2 + \frac{\gamma_2(\lambda_1+\mu)}{h_1} + \frac{(\lambda_2+\mu)}{h_2}\right) \tag{13.30}$$

and similarly for User 2.

Let us consider the case of linear SINR utility function $U_i(\gamma_i) = \gamma_i(\mathbf{x}),\ \forall i$ and obtain the NE power allocation for this case.

13.5.1.1 Case 1: All Users Are Regular

From the cost function given in Eq. (13.5), we obtain the KKT conditions for BR of user i as, $P_i + \frac{B}{h_i} - \frac{1}{\frac{1}{L}\sum_{j\neq i} x_j + \sigma^2} = 0$ and $x_i \geq 0, \forall i$. The powers of all the users at the NE is obtained as follows, by centrally solving this system of equations.

$$x_i = \frac{1}{N-1}\left[\sum_{j\neq i}\frac{L}{P_j + \frac{B}{h_j}} - (N-2)\left(\frac{L}{P_i + \frac{B}{h_i}}\right) - L\sigma^2\right]^+, \forall i. \tag{13.31}$$

13.5.1.2 Case 2: N^m Users Are Malicious

The cost functions of all the N^m malicious users will be,

$$J_k^m(\mathbf{x}) = \alpha_k(P_k x_k^m + B\frac{x_k^m}{h_k}) - \gamma_k(\mathbf{x}) - \theta_k \sum_{j\neq k}\gamma_j(\mathbf{x}), \forall k. \tag{13.32}$$

The NE power can be only calculated numerically for the general case. Next we obtain the NE points analytically for the two-users case.

Proposition 13.5.1 *For the case, where $N = 2$ and $N^m = 1$, the NE power of User 2 who is malicious is*

$$x_2^m = \left[\frac{L}{P_1 + \frac{B}{h_1}} - L\sigma^2\right]^+, \tag{13.33}$$

and NE power of User 1 who is regular is

$$\theta_2(x_1^s)^2 + x_1^s\left(\theta_2 L\sigma^2 - \alpha_2 L^2\frac{\left(P_2 + \frac{B}{h_2}\right)}{\left(P_1 + \frac{B}{h_1}\right)^2}\right) - (1+\sigma^2)L^3\frac{\left(P_2 + \frac{B}{h_2}\right)}{\left(P_1 + \frac{B}{h_1}\right)^2} = 0. \tag{13.34}$$

Proof The cost function of the malicious user is given by equation in (13.32), but for $N = 2$ and $N^m = 1$. The cost function of the regular user is same as (13.73). The powers of both the users at the NE is obtained by centrally solving the system of equations from the KKT conditions of BR.

When the malicious user does not care about his self utility, the NE power of User 1 who is regular is

$$x_1^s = \frac{\alpha_2 L \left(P_2 + \frac{B}{h_2}\right)}{|\theta_2| \left(P_1 + \frac{B}{h_1}\right)^2}. \tag{13.35}$$

We could observe that when the User 2 is highly malicious, i.e. $|\theta_2|$ is high, the regular User 1 has less power. Apparently, both malicious and regular users takes less power when faced with higher price per unit power. Next, we obtain the *PoM* using the NE points obtained above.

Proposition 13.5.2 *For the two-users case with linear utilities for the case where the malicious user is interested in the self utility, PoM is given by,*

$$PoM(\mathcal{M}) := 1 - \left(\frac{\alpha_2 \left(P_2 + \frac{B}{h_2}\right)^2 (1 - \sigma^2 \left(P_1 + \frac{B}{h_1}\right))}{|\theta_2| \left(P_1 + \frac{B}{h_1}\right)^2 (1 - \sigma^2 \left(P_2 + \frac{B}{h_2}\right))} \right).$$

Proof According to the Definition 13.2.5, for the two-users case,

$$PoM(\mathcal{M}) := \frac{\gamma_1(x_1, x_2) - \gamma_1(x_1^s, x_2^m)}{\gamma_1(x_1, x_2)},$$

where x_1 and x_2 are given by (13.31), x_1^s by (13.35) and x_2^m by (13.33). After the substitutions, we obtain the above result.

Remark 13.5.1 We could observe that *PoM* increases when the user is highly malicious. For the symmetric case with $h_1 = h_2$ and $P_1 = P_2$, the windfall of malice occurs when $\alpha_2 > |\theta_2|$, i.e., when the malicious user cares more about the price and energy cost compared to the maliciousness effect. For $|\theta_2| = 1$, i.e., when the malicious user is extreme malicious, the windfall of malice never happens.

In the next section, we consider the Bayesian case with distributed pricing mechanisms.

13.5.2 BNE with an Arbitrary Number of Malicious Users

We consider the Bayesian case where the users and designer have probabilistic information about others natures. We assume that if a user is regular, it receives the price P^s and if it is malicious P^m. We consider symmetric assumption where each user believes that other nodes of same type choose the same strategy.

For an arbitrary number of malicious users with symmetry assumption, the cost function of a user, if it is regular, is given in Eq. (13.7) and if malicious, in Eq. (13.8). The BR of each regular user is,

$$P^s + \frac{B}{h^s} - \sum_{N^m=0}^{N} \mu^s(N, N^m) \frac{LN^m x^m + L\sigma^2}{\left((N - N^m - 1)x^s + N^m x^m + L\sigma^2\right)^2} = 0.$$

The BR of each malicious user is,

$$\alpha(P^m + \frac{B}{h^m}) - \sum_{N^m=0}^{N} \mu^m(N, N^m) L(\frac{(N - N^m)x^s + L\sigma^2}{\left((N - N^m)x^s + (N^m - 1)x^m + L\sigma^2\right)^2}$$
$$- \theta^m (N - N^m) \frac{Lx^s}{\left((N - N^m - 1)x^s + N^m x^m + L\sigma^2\right)^2}) = 0.$$

The solution of the above two equations subject to $x^s \geq 0,\ x^m \geq 0$ gives the BNE with an arbitrary number of malicious users.

13.5.3 BNE for Two-Users Case

The following proposition gives the BNE power strategies of the regular user and the malicious user with linear utilities.

Theorem 13.5.1 *For the two-users symmetric case, in the Bayesian pricing mechanism $\mathcal{M}_{\text{L}}$, the symmetric power strategy of the regular user at the BNE point for linear SINR utility functions is given by,*

$$x^s = \left[x^{s'}\right]^+, \tag{13.36}$$

and of the malicious user is the solution of

$$\alpha(P^m + Bh^m) - \frac{L^2 |\theta^m| \psi^m x^{s'}}{\left(x^m + L\sigma^2\right)^2} - \frac{\psi^m L}{x^{s'} + L\sigma^2} = 0, \tag{13.37}$$

where

$$x^{s'} = \sqrt{\frac{L^2\sigma^2(1-\psi^s)}{P^s + \frac{B}{h^s} - \frac{\psi^s L}{x^m + L\sigma^2}}} - L\sigma^2. \tag{13.38}$$

Proof For the two-users case, the cost function of a user, if it is regular, is given by Eq. (13.9) and if malicious, by Eq. (13.10). The BR from Eq. (13.9) gives the result in (13.38). From the KKT conditions of the BR using the other cost functions, we could obtain the BNE point.

Without self utility for malicious user, from (13.37),

$$x^{s'} = \frac{(x^m + L\sigma^2)^2\alpha(P^m + \frac{B}{h^m})}{L^2|\theta^m|}. \tag{13.39}$$

We next compare the SINR obtained for the complete information case with x_1^s from Eq. (13.35) and x_2^m from Eq. (13.33) and the SINR for the Bayesian information case with x^s from Eq. (13.36) and x^m from Eq. (13.37). By this we obtain the boundary conditions under which the Bayesian case is better for the regular user.

Proposition 13.5.3 *For the Bayesian pricing mechanism $\mathcal{M}_{\mathsf{L}}$ with 2 users, one of the user is malicious and does not care about self utility and both the users have linear utilities, PoM is given by,*

$$PoM(\mathcal{M}_{\mathsf{L}}) := 1 - \left(\frac{\alpha\psi^s \left(P^m + \frac{B}{h^m}\right)^2 \left(1 - \sigma^2 \left(P^s + \frac{B}{h^s}\right)\right)}{|\theta^m|\left(\left(P^s + \frac{B}{h^s}\right) - \frac{(1-\psi^s)L^2\sigma^2}{(x^s+L\sigma^2)^2}\right)\left(P^s + \frac{B}{h^s}\right)\left(1 - \sigma^2\left(P^m + \frac{B}{h^m}\right)\right)} \right). \tag{13.40}$$

Proof For this case,

$$PoM(\mathcal{M}) := \frac{\gamma_1(x_1, x_2) - \gamma_1(x^s, x^m)}{\gamma_1(x_1, x_2)},$$

where x_1 and x_2 are given by (13.31), x^s by (13.36) and x^m by (13.37). After the substitution we obtain the result in Eq. (13.40).

Remark 13.5.2 We could observe that for this case also *PoM* increases when the user is highly malicious. *PoM* can be reduced by higher price for malicious user compared to the regular user, i.e., $P^m > P^s$. Windfall of malice happens when the second term on the right is greater than 1, i.e., $PoM < 0$. We could observe that when value of θ^m is close to 0 and ψ^s is close to 1, windfall of malice is possible. This is because the user is not malicious enough to have an effect and the user has higher expectation that the other user is regular. At the same time, the regular user is aware that the other user can be malicious. The windfall of malice is also possible when the malicious user is charged very high compared to the regular user by the network and the user is highly price aware, i.e., with high α.

Next, we consider a special case where we look from the perspective of a regular user who tries to benefit from the uncertainty in the network. For this we analyze the case with User 1 who is inherently regular but it is of unknown nature to malicious User 2.

13.5.4 Two-Users Case: User 1 of Unknown Nature

We consider a malicious User 2 who faces User 1 of unknown nature and User 2 has only probabilistic information ψ^m about that user.

Proposition 13.5.4 *The power strategies of and the User* 1 *of unknown type at the Bayesian NE point for a game between two users of linear utility functions are given by,*

$$x_2^m = \left[\frac{L}{P_1 + \frac{B}{h_1}} - L\sigma^2\right]^+, \tag{13.41}$$

x_1^s given by solution of

$$\psi^m\theta_2(x_1^s)^2 + \left(\psi^m\theta_2 L\sigma^2 - \alpha_2 L^2 \frac{\left(P_2 + \frac{B}{h_2}\right)}{\left(P_1 + \frac{B}{h_1}\right)^2}\right) x_1^s - (1+\sigma^2)\psi^m L^3 \frac{\left(P_2 + \frac{B}{h_2}\right)}{\left(P_1 + \frac{B}{h_1}\right)^2} = 0 \tag{13.42}$$

and $x_1^m = 0$, where x_1^s and x_1^m are the powers of User 1 *when it is regular and malicious user respectively, x_2^m be the received power at the base station for malicious User* 2 *and the prices P_1 and P_2 follow from* (13.30).

Proof The cost function of User 1 if it is regular is, $J_1^s = P^s x_1^s + B\frac{x_1^s}{h_1} - \gamma_2(x_1^s, x_2^m)$ and the cost function of User 1 if it is malicious is,$J_1^m = P^m x_1^m + B\frac{x_1^m}{h_1}$. User 2 has the utility function given in Eq. (13.10). The NE points can be obtained from the BRs.

Remark 13.5.3 When the SINR of User 1 calculated from BNE in Eqs. (13.41) and (13.42), is greater than the SINR it obtains from BNE in Eqs. (13.33) and (13.34), the regular user benefits from the uncertainty it creates to the malicious user.

13.5.5 Centralized Bayesian Auctions with Malicious Users

Now we consider auction mechanisms in which the designer (base station) makes centralized decisions on the power level and price for all users. This is the case in the practical wireless networks and standards now, like OFDMA with centrally controlled resource coordination in 3GPP-LTE system [39]. We analyze, how the uncertainty about the type of users, affect the user strategies in the auction mechanisms proposed in Sect. 13.3.3.

Consider the two-users symmetric case with both the users of uncertain type and log utility for all the users. The cost function of the regular User i anticipating that it will receive an SINR given in (13.25) is

$$J_i = \psi^s \left(B \frac{x_i^s}{h_i(x_i^s + x_j^m)} X_t + x_i^s x_j^m - \log \left(\frac{x_i^s X_t L}{x_j^m (X_t + \sigma^2 L) + x_i^s \sigma^2 L)} \right) \right) \tag{13.43}$$
$$+ (1 - \psi^s) \left(B \frac{x_i^s}{h_i(x_i + x_j^s)} X_t + x_i^s x_j^s - \log \left(\frac{x_i^s X_t L}{x_j^s (X_t + \sigma^2 L) + x_i^s \sigma^2 L} \right) \right)$$

where x_j^s and x_j^m are the strategies of User j when it is regular and malicious respectively. User i minimizes J_i subject to $x_i^s \geq 0$. The cost function of User i if it is malicious is,

$$J_i^m = \psi^m \left(x_j^s x_i^m + B \frac{x_i^m}{h_i(x_j^s + x_i^m)} - \log \left(\frac{x_i^m X_t L}{x_j^s (X_t + \sigma^2 L) + x_i^m \sigma^2 L} \right) \right. \tag{13.44}$$
$$\left. - \theta_i \log \left(\frac{x_j^s X_t L}{x_i^m (X_t + \sigma^2 L) + x_j^s \sigma^2 L} \right) \right) + (1 - \psi^m) \left(B \frac{x_i^m}{h_i(x_j^m + x_j^s)} X_t + x_i^m x_j^m \right).$$

The strategies of the users can be obtained by solving the system of equations obtained from the best responses.

13.5.6 Auction Mechanism Resistant to Malicious Users

The designer observing that there are possibilities of users acting as malicious ones, modifies the allocation and pricing rule according to the probability distribution.

Proposition 13.5.5 *For a wireless network with users having logarithmic utilities, if a User j is expected to be malicious by designer with probability ψ^d, the mechanism with the allocation of that user should be modified to,*

$$Q_j = (1 - \psi_j^d) \frac{x_j}{\sum_k x_k + \omega} X_t, \tag{13.45}$$

and the pricing to

$$C_j = x_j \sum_{k \neq j} x_k + \omega - \psi_j^d (N-1) \theta_j t X_t \log \left(1 + \frac{x_j}{\sum_{k \neq j} x_k} \right). \tag{13.46}$$

so that the malicious users are forced to report like regular users in the network and $t = \sum_k x_k$.

The pricing in Eq. (13.46) is a Bayesian version of the differentiated pricing given in Sect. 13.4.

13.6 Pricing Mechanisms for the Security of Wireless Network with QoS Requirements

In this section, we develop the noncooperative game with individual pricing for the general MAC system without SIC. Each user allocates its own power by optimizing the individual utility function with clever price adaptation. We show that by the proposed prices, BR power allocation of each user converges rapidly. The Shannon rate-based QoS requirement of each user is achieved at the unique NE point. We analyse different behavior types of the users, especially the malicious behavior and the resulting NE power allocation and achievable rates of all the users. The strategy-proof mechanism is designed with the punishment prices when the types of the the malicious users are detected in order to guarantee the security of the wireless networks.

13.6.1 System Operation with Selfish Users

Firstly, the distributed power allocation based on the noncooperative game for the general MAC without SIC is discussed, where no malicious behavior exists in the system. Besides the rate requirement, each user in the game minimizes its own cost function by choosing the optimal power allocation. The individual price playing as the tradeoff between maximizing the achievable rate and minimizing the transmit power is introduced into the user utility function to control the distributed power allocation. The individual prices are carefully designed to ensure the existence, uniqueness and stability of the NE transmit power and as a result to guarantee all feasible rate requirement in the capacity region of MAC at the NE point.

The noncooperative game of the MAC system can be formulated as an economic model, where the consumers are the users. The trading good is the power. And the producer provides the individual prices β_i based on the local information of each user. Since each user has a rate requirement $\underline{u}_i$ to be guaranteed and the interferences are coupled among all the users, the demand in power of each user is dependent on other users. In order to better illustrate the properties of the model, we introduce the normalized distributed pricing term $P_i(x_{-i})$ as a function of the individual price β_i and the demand of all the other users $I(x_{-i})$, i.e.,

$$P_i(x_{-i}) = \frac{\beta_i}{I(x_{-i})}. \tag{13.47}$$

The cost function of each user is based on the achievable rate $U_i(x_i, x_{-i})$ and the normalized pricing term as follows.

$$J_i(x_i, x_{-i}) = P_i(x_{-i})x_i - U_i(x_i, x_{-i}). \tag{13.48}$$

In the multiuser case, the interference obviously influences the quality of the good (resource) that user i buys. In order to express the quality loss due to interference, $I(x_{-i})$ is a function of interference. The higher interference, the lower the pricing term, and thus the more power allocation to guarantee the rate requirement. Therefore, the pricing term $P_i(x_{-i})$ is normalized by the noise plus interference caused by all the other users.

$$J_i(x_i, x_{-i}) = \frac{\beta_i}{1 + \sum_{k \neq i} h_k x_k} x_i - U_i(x_i, x_{-i}) \tag{13.49}$$

$$= \frac{\beta_i}{1 + \sum_{k \neq i} h_k x_k} x_i - \log\left(1 + \frac{h_i x_i}{1 + \sum_{k \neq i} h_k x_k}\right). \tag{13.50}$$

The normalized pricing term denotes the quality of the good. If the interference from other users is high, then the price of the power for user i should be lower in order to achieve the rate requirement. β_i can be considered as the SINR pricing.

The game in normal form $\mathcal{G}$ is described by the N players. Their strategy space is $[0, x_i^{max}]$ and their utility function is $J_i(x_i, x_{-i})$. The price controls the trade-off between maximizing the achievable rate and saving as much power as possible. The game best response dynamic (BRD) can be expressed as the N coupled problems

$$\begin{aligned} \min_{x_i} \quad & J_i(x_i, x_{-i}) = \frac{\beta_i}{1 + \sum_{k \neq i} h_k x_k} x_i - U_i(x_i, x_{-i}) \\ \text{subject to} \quad & 0 < x_i < x_i^{max}, \ \forall i = 1, \ldots, N. \end{aligned} \tag{13.51}$$

A basic result from game-theory is that each fixed point of the BRD is an NE of the game, although in general, convergence of the BRD is not guaranteed, nor is the existence of the fixed point.

13.6.1.1 Best Response Power Allocation

If the prices $\beta_i = 0$, transmitting with full power x_i^{max} is the BR of each user. Due to the pricing term for positive $\beta_i > 0$, we can conclude the BR strategy as follows.

Proposition 13.6.1 *For all $i = 1, \ldots, N$, define $\underline{x}_i$ as*

$$\underline{x}_i(x_{-i}) = \left(\frac{1}{\beta_i} - \frac{1}{h_i}\right)(1 + \sum_{k \neq i} h_k x_k). \tag{13.52}$$

The i-th user's best-response is given by $x_i^{BR} = \max(0, \min(\underline{x}_i, x_i^{max}))$. *Moreover, the noncooperative game* $\mathcal{G}$ *always admits an NE* $\{x_i^{BR}\}_{i=1}^N$.

Proof Solve the first derivative of $J_i(x_i, x_{-i})$ to be zero with respect to x_i.

$$\frac{\partial J_i(x_i, x_{-i})}{\partial x_i} = \frac{\beta_i}{1 + \sum_{k \neq i} h_k x_k} - \frac{h_i}{1 + \sum_{k \neq i} h_k x_k + h_i x_i} = 0. \tag{13.53}$$

The positive result $\underline{x}_i$ is achieved in (13.52) if $\beta_i < h_i$. Otherwise it is set to zero to avoid negative power.

The second derivative of $J_i(x_i, x_{-i})$ with respect to x_i is

$$\frac{\partial^2 J_i(x_i, x_{-i})}{\partial x_i^2} = \frac{h_i^2}{(1 + \sum_{k \neq i} h_k x_k + h_i x_i)^2} > 0. \tag{13.54}$$

Therefore, the unique global minimum exists.

By observing that the strategy set of each user is a compact and convex set, $J_i(x_i, x_{-i})$ is a continuous function with respect to the powers of all users, and concave with respect to x_i, which implies the existence of at least one NE.

13.6.1.2 Nash Equilibrium Power Allocation

The noncooperative game $\mathcal{G}$ always admits at least one NE power allocation $\{x_i^{BR}\}_{i=1}^N$. In this part, we figure out the NE point and show that it is unique.

Proposition 13.6.2 *The Nash equilibrium power allocation of each user i in the noncooperative game* $\mathcal{G}$ *in the general MAC system is* $x_i^{NE} = \max(0, \min(\underline{x}_i^{NE}, x_i^{max}))$. *Given the individual prices* β_i,

$$\underline{x}_i^{NE} = \frac{h_i - \beta_i}{h_i^2} \cdot \frac{1}{\sum_{j=1}^N \frac{\beta_j}{h_j} - N + 1}. \tag{13.55}$$

The noncooperative game $\mathcal{G}$ *always admits this unique NE point.*

Proof Due to the space limit, please refer to [42].

13.6.1.3 Pricing for QoS Requirements

As shown in [37], the centralized power allocation to achieve the rate requirement $\underline{u}_i$ of each user is

$$x_i^U = \frac{B_K}{h_i} \cdot \frac{2^{\underline{u}_i} - 1}{2^{\underline{u}_i}}, \tag{13.56}$$

where $B_K = \frac{1}{\sum_{j=1}^{N} \frac{1}{2^{\underline{u}_j}} - N + 1}$ is a constant for given $\underline{u}_j, j = 1, \ldots, N$.

x_i^U is the minimum power allocation to achieve all the rate requirements in the capacity region in the general MAC system without SIC [38]. This energy efficient power allocation x_i^U can be adopted in various wireless systems [36, 40, 43]. In order to determine the individual prices, $\underline{x}_i^{NE}$ should be equal to x_i^U. Therefore, we solve the universal individual prices for the distributed MAC as follows.

Lemma 13.6.1 *In the N-user non-cooperative game $\mathcal{G}$ of the general MAC system, the rate requirement $\underline{u}_i$ of each user i is achieved with the NE power allocation x_i^{NE} if the individual price is*

$$\beta_i = \frac{h_i}{2^{\underline{u}_i}}. \tag{13.57}$$

Proof Solve the equation $\underline{x}_i^{NE} = \frac{h_i - \beta_i}{h_i^2} \cdot \frac{1}{\sum_{j=1}^{N} \frac{\beta_j}{h_j} - N + 1} = x_i^U = \frac{B_K}{h_i} \cdot \frac{2^{\underline{u}_i} - 1}{2^{\underline{u}_i}}$ for β_i.

In order to ensure the positive power allocation and therefore to guarantee the rate requirement of each user, the following conditions regarding the number of users in the wireless system, the individual prices and the channel states should be fulfilled.

Corollary 13.6.1 *In the general N-user MAC system without SIC, the rate requirement of each user i is achieved by the BRD power allocation if and only if*

$$N - 1 < \sum_{i=1}^{N} \frac{\beta_i}{h_i} < N. \tag{13.58}$$

Proof The proof is obtained by guaranteeing $\underline{x}_i^{NE} > 0$ in (13.55).

Remark 13.6.1 The region in (13.58) is equivalent to the feasible utility region in Corollary 1 in [37], if the individual prices β_i are given in (13.57). This price region is the universal pricing region that it guarantees all rate requirements in the capacity region of the general MAC system without SIC.

13.6.2 Malicious Behavior

Let the normalized noise plus interference to user i caused by all the other users be $I_i(x_{-i}) = 1 + \sum_{k \neq i} h_k x_k$. The utility function of each user with type θ_i is defined as

$$J_i(x_i, x_{-i}, \theta_i) = \frac{\beta_i}{I_i(x_{-i})}x_i - U_i(x_i, x_{-i}) - \theta_i \sum_{j\neq i} U_j(x_j, x_{-j}) \tag{13.59}$$

$$= \frac{\beta_i}{I_i(x_{-i})}x_i - U_i(x_i, x_{-i}) - \theta_i\Big(\sum_{j\neq i} \log(1 + \sum_{l=1}^{N} h_l x_l) - \log(1 + \sum_{l\neq j} h_l x_l)\Big), \tag{13.60}$$

where the first term in (13.59) is the pricing term, the second term is its own achievable rate and the third term is its influence on the other users. For malicious users, they get benefit from harming all the other users. Since $\theta_i < 0$, the best response of the malicious user is different from the selfish users. The malicious behavior and its influence on the resulting NE power allocation is interesting and necessary for the mechanism design. In (13.60), the sum rate term $\log(1 + \sum_{l=1}^{N} h_l x_l)$ is not the target of the malicious user because this will also harm its own rate. Therefore, the utility of the malicious user is focused on the other terms. The utility function is modified to

$$J_i(x_i, x_{-i}, \theta_i) = \frac{\beta_i}{I_i(x_{-i})}x_i - U_i(x_i, x_{-i}) + \theta_i \left(\sum_{j\neq i} \log(1 + h_i x_i + \sum_{l\neq j,i} h_l x_l)\right)$$

$$= \frac{\beta_i x_i}{I_i(x_{-i})} - U_i(x_i, x_{-i}) + \theta_i\Big(\sum_{j\neq i} \log\Big((1 + \sum_{l\neq j,i} h_l x_l)\Big(1 + \frac{h_i x_i}{1 + \sum_{l\neq j,i} h_l x_l}\Big)\Big)\Big). \tag{13.61}$$

Since $1 + \sum_{l\neq j,i} h_l x_l$ is independent of the power of the malicious user x_i, it is skipped further on in order to simplify the calculation. The third term in (13.61) becomes $\theta_i \Big(\sum_{j\neq i} \log(1 + \frac{h_i x_i}{1+\sum_{l\neq j,i} h_l x_l})\Big)$. Instead of the complicated $\sum_{j\neq i}$ log term, we linearize the maximization problem as follows. The final cost function of each user with the private types θ_i is

$$J_i(x_i, x_{-i}, \theta_i) = \frac{\beta_i}{I_i(x_{-i})}x_i - U_i(x_i, x_{-i}) + \frac{\theta_i h_i x_i}{I_i(x_{-i})}, \tag{13.62}$$

where the third term reflects the interference to all the other users after linearization.

13.6.2.1 Best Response Power Allocation with Malicious User

Follow a similar procedure as in Sect. 13.6.1, we obtain the distributed BR power allocation of each user with private types θ_i and NE point with malicious users in the system.

Lemma 13.6.2 *For all $i = 1, \ldots, N$ with type θ_i from (13.62), define $\underline{x}_i(\theta_i)$ as*

$$\underline{x}_i(\theta_i) = \left(\frac{1}{\tilde{\beta}_i(\theta_i)} - \frac{1}{h_i}\right)(1 + \sum_{k \neq i} h_k x_k). \tag{13.63}$$

Here $\tilde{\beta}_i(\theta_i) = \beta_i + \theta_i h_i$ is the individual price with type θ_i. The i-th user's best-response with type θ_i is given by $x_i^{BR}(\theta_i) = \max(0, \min(\underline{x}_i(\theta_i), x_i^{max}))$.

Proof Solve $\underline{x}_i(\theta_i)$ for the first derivative $\frac{J_i(x_i, x_{-i}, \theta_i)}{x_i} = 0$ from (13.62).

We observe that the BR power allocation of the malicious user is higher than if it was a regular user, i.e., $\underline{x}_j^m > \underline{x}_j^S$. Since all the selfish users only optimize their utility function in (13.51), their BR power allocation x_i^S remains as in (13.56), and it is exactly $\underline{x}_i(\theta_i)$ in (13.63) when $\theta_i = 0$.

13.6.2.2 Nash Equilibrium Power Allocation with Malicious User

Next, we analyse the NE power allocation of the noncooperative game $\mathcal{G}$ with private type θ_i. From (13.55), we can conclude the following result.

Proposition 13.6.3 *The Nash equilibrium power allocation of each user i in the noncooperative game $\mathcal{G}$ in the general MAC system with private type θ_i is $x_i^{NE}(\theta_i) = \max(0, \min(\underline{x}_i^{NE}(\theta_i), x_i^{max}))$, where given the individual prices $\tilde{\beta}_i(\theta_i)$ with type θ_i,*

$$\underline{x}_i^{NE}(\theta_i) = \frac{h_i - \tilde{\beta}_i(\theta_i)}{h_i^2} \cdot \frac{1}{\sum_{j=1}^{N} \frac{\tilde{\beta}_j(\theta_j)}{h_j} - N + 1}. \tag{13.64}$$

The noncooperative game always admits this unique NE point.

Proof The proof follows the same steps as in the Appendix [42] by replacing the individual price β_i with $\tilde{\beta}_i(\theta_i)$.

In the next Sect. 13.6.3, we propose a way of detecting the malicious users observing the anomalies in the utility function. For this purpose, the designer learns the utility functions of all the users from their BR strategies.

13.6.3 Malicious User Detection

In order to make mechanism to counter the user misbehavior, the system designer needs to predict and detect the private behavior types of heterogeneous users.

13.6.3.1 Detection by Observing Achievable Rate

In this section, we show that the potential user malicious behavior of misrepresenting the user utility function can be detected by comparing the achievable rate and the pre-submitted QoS requirement at the BS. The misbehavior prediction and detection make the design of the punishment price possible.

In order to understand the influence of the malicious behaviour on the resulting NE power and the rate of both the selfish and malicious users comprehensively, we have the following Proposition.

Proposition 13.6.4 *With the individual price $\beta_i = \frac{h_i}{2^{\underline{u}_i}}$, the Nash equilibrium power allocation $x_i^{NE}(\theta_i)$ of each user i in the noncooperative game $\mathcal{G}$ in the general MAC system with private type θ_i is higher than or equal to x_i^U in* (13.56)*, where*

$$\underline{x}_i^{NE}(\theta_i, \theta_{-i}) = \frac{1-\theta_i-2^{-\underline{u}_i}}{h_i}\frac{1}{\sum_{j=1}^{N}(2^{-\underline{u}_i}+\theta_j)-N+1}. \tag{13.65}$$

The resulting rate $U_i(\theta_i)$ is

- $U_i(\theta_i) = \underline{u}_i$, *for selfish users with $\theta_i = 0$*
- $U_i(\theta_i) > \underline{u}_i$, *for malicious users with $\theta_i < 0$.*

Proof Insert $\tilde{\beta}_i(\theta_i) = \beta_i + \theta_i h_i$ with $\beta_i = \frac{h_i}{2^{\underline{u}_i}}$ into (13.64), then (13.65) is proved. It can be observed that the second term in (13.65) is a constant for all the users given the type θ_j and it is larger if there exists at least one user with $\theta_i < 0$. If all the users are selfish, $\underline{x}_i^{NE}(\theta_i, \theta_{-i}) = x_i^U$, which is the minimum power allocation in order to achieve the rate requirement $\underline{u}_i$ of each user i. The power $\underline{x}_i^{NE}(\theta_i, \theta_{-i})$ of malicious users is greater than that of selfish users because the first term is greater if $\theta_i < 0$.

Finally, we calculate the achievable rate of each user with $x_i^{NE}(\theta_i)$. The rate requirement $\underline{u}_i$ can be achieved for the selfish users with $\theta_i = 0$. Since the power allocation of malicious users is larger than that of selfish users, their actual rate is greater than their rate requirements.

The number of the malicious users M and the total users N, the private type θ_i and the rate requirement $\underline{u}_i$ are mutually restricted to ensure the positive NE power allocation $\underline{x}_i^{NE}(\theta_i)$.

Lemma 13.6.3 *In the general MAC system with N users, the rate requirement $\underline{u}_i$ of each user i can be achieved if and only if the following condition is fulfilled,*

$$0 \geq \sum_{j=1}^{N} \theta_j > N - 1 - \sum_{j=1}^{N} 2^{-\underline{u}_j}. \tag{13.66}$$

Proof Since $2^{-\underline{u}_j} < 1$ and $0 \geq \theta_i \geq -1$, the first term in (13.65) is positive. In order to achieve the rate requirement $\underline{u}_i$, positive power allocation must be ensured, therefore the second term in (13.65) should be positive as well. Lemma 13.6.3 is proved.

Remark 13.6.2 Note that if the user types θ_i do not fulfill Lemma 13.6.3, then the NE power allocation with malicious users $\underline{x}_i^{NE}(\theta_i, \theta_{-i})$ of each user i is negative no matter it is selfish or malicious. Thereby, the utility requirements $\underline{\mathbf{u}}$ are not feasible. Then the rates of all users cannot be guaranteed and the misbehaviour is immediately detected by the receiver.

13.6.3.2 Detection by Learning Utilities

In this section, regression techniques are used to learn the user private marginal utilities by the designer. The anomalies in the utility curves are used to obtain the identities of malicious users with a possible error and further for implementation of the differentiated pricing mechanism. The utility function is not assumed to be Shannon rate here and it can be any concave function. The users just give their utility requirements $\underline{u}_i$, $\forall\, i$ and take the best responses.

The regular user optimization problem will be to find the power level which minimizes his individual cost, i.e.,

$$\min_{x_i} \; \beta_i \gamma_i - U_i(x),$$

Consequently, the general condition for player best response obtained from first order derivative is

$$\beta_i \frac{d\gamma_i(x)}{dx_i} - \frac{dU_i(x)}{d\gamma_i} \frac{d\gamma_i(x)}{dx_i} = 0, \; \forall\, i \in \mathcal{A}. \tag{13.67}$$

Let us denote $U_i' = \frac{dU_i}{d\gamma_i}$, $\forall\, i$. Thus,

$$\beta_i = U_i'(\gamma_i^{BR}), \; \forall\, i. \tag{13.68}$$

First, the designer gives sample values of prices β to all the users. Then the designer observes the NE x^{NE} and calculates the SINR at the NE, γ_i^{NE} of all the users. With different values of β, the designer can plot the curve of U_i' against γ_i. For the malicious user,

$$\beta_i = U_i'(\gamma_i^{BR}) - \theta_i \sum_{k \in S} \frac{dU_k(\gamma_k)}{d\gamma_i}, \; \forall\, i. \tag{13.69}$$

The designer will obtain a completely different type of curve U_i' for the malicious users. The designer uses this anomaly in the curve for the detection of malicious users and punishes them with higher price.

From the U_i' curve, the utility function $U_i(\gamma_i)$ can be obtained by integrating. The designer objective is

$$U_i(x) \geq \underline{u}_i, \ \forall\, i.$$

Once the users give their utility requirements $\underline{u}_i$, $\forall\, i$, the designer can find the corresponding γ_i's which satisfy the designer objective with equality. The designer uses the curve obtained above using regression learning for this purpose. The price which moves the NE to the equality point can be also calculated.

13.6.4 Pricing for QoS Requirements with Complete Information

In this section, we design the strategy-proof prices in order to counter the malicious behavior analysed in Sect. 13.6.2. The punishment price is designed after predicting the potential user misbehavior. It is announced as a common rule of the proposed noncooperative game before the game starts. By anticipating the resulting utility after being applied the punishment price, no user in the distributed power allocation game has incentives to behave maliciously.

Denote β_i^m as the trigger price applied on the malicious user i whenever it is detected by the system. In order to counter the malicious behavior, the price given to the malicious users β_i^m should be designed such that the BR power allocation of the malicious users are made smaller than the selfish users. β_i^m can be considered as the punishment price. Therefore, we obtain the following Proposition.

Proposition 13.6.5 *In the N-user non-cooperative game $\mathcal{G}$ of general MAC system without SIC, no user will have incentive to behave maliciously if the punishment price β_i^m is given by*

$$\begin{aligned}\beta_i^m &\geq \beta_i - \theta_i h_{max}\\ &\geq \beta_i - \theta_i h_i,\end{aligned} \tag{13.70}$$

where h_{max} denotes the maximum possible CSI.

Proof With the individual price β_i, $\underline{x}_j^m(\beta_i) > \underline{x}_j(\beta_i)$. Therefore, the punishment price β_i^m should be introduced such that the BR power allocation of malicious users $\underline{x}_j^m(\beta_i^m)$ is smaller than the BR power allocation of the selfish users $\underline{x}_j(\beta_i)$, i.e.,

$$\left(\frac{1}{\beta_i^m + \theta_i h_i} - \frac{1}{h_i}\right) I_i(x_{-i}) \leq \left(\frac{1}{\beta_i} - \frac{1}{h_i}\right) I_i(x_{-i}). \tag{13.71}$$

Since $I_i(x_{-i}) > 0$ and $h_i > 0$, (13.71) becomes

$$\frac{1}{\beta_i^m + \theta_i h_i} \leq \frac{1}{\beta_i}$$
$$\beta_i^m \geq \beta_i - \theta_i h_i.$$

For simplicity, the BS can adopt h_{max} as a defaulted value in the punishment price since $h_{max} \geq h_i$ is always true.

Whenever the malicious behavior is detected, the punishment price is provided to the malicious user. By minimizing its own cost function $J_i(x_i, x_{-i}, \theta_i)$ in (13.62), no user will have incentive to harm the other users. This punishment measure is announced as a common rule before the game really starts. Thus β_i^m acts as a virtual threat to prevent the user misbehavior in the MAC system.

13.6.5 Bayesian Pricing with QoS Requirements

We assume that to implement Bayesian differentiated pricing, the designer observes each user in the network and attach a probability that he is malicious [14]. Let ψ_i^d be the probability that user i is malicious and θ_i^d be the estimate of degree of maliciousness of user i by the designer. Since it is not realistic to estimate the exact value of the degree of maliciousness θ_i by the designer, we assume that he gives the maximum punishment, i.e., with $\theta_i^d = -1$. Each user's Bayesian price according to the probabilities are;

$$\beta_i^m = \frac{h_i}{2^{\frac{u_i}{}}} - \psi_i^d \theta_i^d h_i. \tag{13.72}$$

We consider also that there may be an error in the estimation of probability by the designer. With the Bayesian pricing, for the two-users case, the cost of the regular user becomes

$$J_i = B\frac{x_i^s}{h_i} + \psi^s((\beta_i^s - \psi_i^d \theta_i^d h_i)\gamma_i^{sm}) - U_i\left(\gamma_i^{sm}\right) + (1-\psi^s)((\beta_i^s - \psi_i^d \theta_i^d h_i)\gamma_i(x_i^s, x_j^s) - U_i\left(\gamma_i(x_i^s, x_j^s)\right)), \tag{13.73}$$

and for malicious user

$$J_i^m = \alpha_i B\frac{x_i^m}{h_i} + \psi^m(\alpha_i(\frac{h_i}{2^{\frac{u_i}{}}} - \psi_i^d \theta_i^d h_i)\gamma_i(x_i^m, x_j^s) - U_i(x_i^m, x_j^s) - \theta_i \gamma_i(x_i^m, x_j^s)) + (1-\psi^m)\alpha_i(\frac{h_i}{2^{\frac{u_i}{}}} - \psi_i^d \theta_i^d h_i)\gamma_i(x_i^m, x_j^m), \tag{13.74}$$

where $\gamma_i^{sm} = \gamma_i(x_i^s, x_j^m)$. The BNE can be obtained from these cost functions.

13.7 Summary and Conclusion

Interference limited wireless networks with selfish and malicious users have been considered in this chapter. A new modeling of malicious user utility function has been proposed and a metric PoM for quantifying the effect of malicious users has been defined. Some well known mechanisms like VCG and Kelly mechanisms are shown to be vulnerable to malicious users attack. The PoM is quantified for some auction and pricing mechanisms. Next, we have relaxed the assumption that the users and the designer know the nature of users and designed Bayesian mechanisms. The conditions under which the uncertainty about the nature of the users is beneficial for the regular users and designer have been obtained by comparing to the complete information case. Finally, the optimal prices with malicious users for the complete information case where each user submits an SINR-based QoS requirement are obtained. Then these prices have been modified with Bayesian information. As a future work it is interesting to see, the results in [12, 13] for cooperative users with centralized allocation change, when the users are noncooperative and they take actions in a distributed fashion.

Acknowledgments The work summarized in this chapter was mainly funded by the German Research Foundation (DFG) within the priority program 1397 "Communications in Interference Limited Networks (COIN)" under grants DFG BO 1734/24-3 and JO 801/5-3.

References

1. Alpcan T, Başar T (2005) A utility-based congestion control scheme for internet-style networks with delay. IEEE Trans Networking 13(6):1261–1274
2. Alpcan T, Basar T (2010) Network security: a decision and game theoretic approach. Cambridge University Press, Cambridge
3. Alpcan T, Boche H, Honig M, Poor HV (2013) Mechanisms and games for dynamic spectrum allocation. Cambridge University Press
4. Babaioff M, Kleinberg R, Papadimitriou CH (2007) Congestion games with malicious players. In: Proceedings of the 8th ACM conference on electronic commerce. San Diego, California, pp 103–112
5. Başar T, Olsder GJ (1999) Dynamic noncooperative game theory, 2nd edn. SIAM, Philadelphia
6. Bhattacharya S, Khanafer A, Başar T (2011) Power allocation in team jamming games in wireless ad hoc networks. In: Proceedings of the 5th international ICST conference on performance evaluation methodologies and tools, ICST (Institute for Computer Sciences, Social-Informatics and Telecommunications Engineering), ICST, Brussels, Belgium, Belgium, VALUETOOLS '11, pp 515–524
7. Boche H, Schubert M (2008) A calculus for log-convex interference functions. IEEE Trans Inf Theory 54(12):5469–5490
8. Boche H, Schubert M (2008) Concave and convex interference functions-general characterizations and applications. IEEE Trans Signal Process 56(10):4951–4965

9. Boche H, Schubert M (2008) The structure of general interference functions and applications. IEEE Trans Inf Theory 54(11):4980–4990
10. Boche H, Schubert M (2009) Nash bargaining and proportional fairness for wireless systems. IEEE/ACM Trans Networking 17(5):1453–1466
11. Boche H, Schubert M (2010) A unifying approach to interference modeling for wireless networks. IEEE Trans Signal Process 58(6):3282–3297
12. Boche H, Wiczanowski M, Stanczak S (2004) Characterization of optimal resource allocation in cellular networks. In: IEEE 5th workshop on Signal processing advances in wireless communications, pp 454–458
13. Boche H, Naik S, Alpcan T (2011) Characterization of convex and concave resource allocation problems in interference coupled wireless systems. IEEE Trans Signal Process 59(5):2382–2394
14. Boche H, Naik S, Jorswieck EA (2013) Detecting misbehavior in distributed wireless interference networks. Wireless Netw 19(5):799–810
15. Chorppath AK, Alpcan T (2011) Learning user preferences in mechanism design. In: Proceedings of 50th IEEE conference on decision and control and european control conference, Orlando, Florida
16. Chorppath AK, Bhashyam S, Sundaresan R (2011) A convex optimization framework for almost budget balanced allocation of a divisible good. IEEE Trans Autom Sci Eng 8(3):520–531
17. Chorppath AK, T Alpcan, Boche H (2011) Pricing mechanisms for multi-carrier wireless systems. In: Proceedings of IEEE international dynamic spectrum access networks (DySPAN) symposium, Aachen, Germany
18. Chorppath AK, Alpcan T, Boche H (2014) Bayesian mechanisms for wireless network security. In: IEEE international conference on communications (ICC), Sydney, Australia
19. Chorppath AK, Alpcan T, Boche H (2015) Adversarial behavior in network games. Dyn Games Appl 1(1):26–64
20. Chorppath AK, Shen F, Alpcan T, Jorswieck E, Boche H (2015) Bayesian mechanisms and learning for wireless networks security with qos requirements. In: IEEE international conference on communications (ICC), London, UK
21. Garg D, Narahari Y, Gujar S (2008) Foundations of mechanism design: a tutorial part 1—key concepts and classical results. Sadhana 33(3):83–130
22. Huang J, Berry R, Honig M (2006) Auction-based spectrum sharing. ACM Mobile Netw Appl J 24(5):405–418
23. Johari R, Mannor S, Tsitsiklis J (2005) Efficiency loss in a network resource allocation game: the case of elastic supply. IEEE Trans Autom Control 50(11):1712–1724
24. Koutsoupias E, Papadimitriou C (1999) Worst-case equilibria. In: Proceedings of the 16th annual symposium on theoretical aspects of computer science, Lecture Notes in computer science, pp 404–413
25. Krishna V (2010) Auction theory, 2nd edn. Academic Press
26. Maheswaran RT, Basar T (2004) Social welfare of selfish agents: motivating efficiency for divisible resources. In: 43rd IEEE conference on decision and control (CDC), Paradise Island, Bahamas, pp 1550–1555
27. Micali S, Valiant P (2008) Revenue in truly combinatorial auctions and adversarial mechanism design. Technical report, MIT-Computer Science and Artificial Intelligence Laboratory, http://dspace.mit.edu/handle/1721.1/41872
28. Moscibroda T, Schmid S, Wattenhofer R (2006) When selfish meets evil: byzantine players in a virus inoculation game. In: Proceedings of the twenty-fifth annual ACM symposium on principles of distributed computing, Denver, Colorado
29. Netzer N (2012) An externality-robust auction. Working Paper
30. Rosen JB (1965) Existence and uniqueness of equilibrium points for concave n-person games. Econometrica 33(3):520–534
31. Roughgarden T (2002) The price of anarchy is independent of the network topology. In: Proceedings of the 34th annual ACM symposium on the theory of computing

32. Sagduyu Y, Berry R, Ephremides A (2009) MAC games for distributed wireless network security with incomplete information of selfish and malicious user types. In: International conference on game theory for networks. GameNets '09. , pp 130–139
33. Schubert M, Boche H (2004) Solution of the multiuser downlink beamforming problem with individual sinr constraints. IEEE Trans Veh Technol 53(1):18–28
34. Schubert M, Boche H (2006) QoS-based resource allocation and transceiver optimization. Now Publishers Inc
35. Schubert M, Boche H (2007) A generic approach to qos-based transceiver optimization. IEEE Trans Commun 55(8):1557–1566
36. Shen F, Jorswieck E (2012) User-centric Compensation framework with universal pricing for hybrid femtocell networks. In: IEEE international conference on wireless communications and signal processing-WCSP, pp 1–6
37. Shen F, Jorswieck EA (2012) Universal cheat-proof pricing for multiple access channels without SIC under QoS requirements. In: International conference on communications-ICC, IEEE, pp 3895–3899
38. Shen F, Jorswieck E (2014) Universal non-linear cheat-proof pricing framework for wireless multiple access channels. IEEE Trans Wireless Commun 13(3):1436–1448
39. Shen Z, Papasakellariou A, Montojo J, Gerstenberger D, Xu F (2012) Overview of 3g pp lte-advanced carrier aggregation for 4g wireless communications. IEEE Commun Mag 50(2):122–130
40. Shen F, Zhang M, Jorswieck E (2013) User-centric energy aware compensation framework for hybrid macro-femtocell networks. In: 2013 IEEE global communications conference (GLOBECOM), pp 3089–3094
41. Shen F, Jorswieck E, Chorppath AK, Boche H (2014) Pricing for distributed resource allocation in MAC without SIC under QoS requirements with malicious users. In: WNC, modeling and optimization in mobile, Ad Hoc, and wireless networks (WiOpt'14), Hammamet, Tunisia
42. Shen F, Jorswieck E, Chorppath AK, Boche H (2014) Pricing for distributed resource allocation in MAC under QoS requirements with malicious users. IEEE Trans Wireless Commun (submitted)
43. Shen F, Li D, Lin PH, Jorswieck E (2015) Auction based spectrum sharing for hybrid access in macro-femtocell networks under QoS requirements. In: IEEE international conference on communications-ICC (June)
44. Srikant R (2004) The Mathematics of internet congestion control. Systems and Control: foundations and applications, Birkhauser, Boston, MA
45. Vickrey W (1961) Counterspeculation, auctions and competitive sealed tenders. J Finance 16(1):8–37
46. Yates R (1995) A framework for uplink power control in cellular radio systems. IEEE J Sel Areas Commun 13(7):1341–1347

Chapter 14
Interference Alignment Aided by Non-Regenerative Relays

Rakash SivaSiva Ganesan, Hussein Al-Shatri, Xiang Li, Anja Klein and Tobias Weber

Abstract Traditionally, relays are used for range extension in radio communication networks, i.e., for improving the performance in noise limited scenarios. In contrast to this, we focus on interference limited scenarios, where relays can be used to manipulate the effective channels between the transmitters and the receivers for interference reduction following the ideas of interference alignment. The investigated scenario consists of several pairs of nodes which want to communicate with each other. Especially bidirectional communication is considered. Furthermore, there are several linear, non regenerative relays aiding the communication. Both one-way relaying and two-way relaying protocols are investigated. Going beyond determining the degrees of freedom, questions concerning the required number of relays and relay antennas, the signal processing techniques for the relays, and the precoding at the nodes for achieving interference alignment are answered. Also, numerical algorithms helping the improvement of the performances for not purely interference limited scenarios, but with significant noise are developed.

R. SivaSiva Ganesan (✉) · H. Al-Shatri · A. Klein
Institut für Nachrichtentechnik Fachgebiet Kommunikationstechnik,
Technische Universität Darmstadt, Merckstraße 25, 64283 Darmstadt, Germany
e-mail: r.ganesan@nt.tu-darmstadt.de

H. Al-Shatri
e-mail: h.shatri@nt.tu-darmstadt.de

A. Klein
e-mail: a.klein@nt.tu-darmstadt.de

X. Li · T. Weber
Institut für Nachrichtentechnik Fakultät für Informatik und Elektrotechnik,
Universität Rostock, Richard-Wagner-Straße 31, 18119 Rostock, Germany
e-mail: xiang.li@uni-rostock.de

T. Weber
e-mail: tobias.weber@uni-rostock.de

W. Utschick (ed.), *Communications in Interference Limited Networks*,
Signals and Communication Technology, DOI 10.1007/978-3-319-22440-4_14

14.1 Introduction and Considered Scenario

Interference between communication links is the major performance limiting factor in wireless communication networks. Interference alignment (IA) has been developed as an efficient technique to handle interferences, especially for high signal to noise ratios (SNRs) [9, 20, 25]. Suppose several node pairs want to communicate through a shared medium. It is well known that in order to achieve good performances especially at high SNRs, IA aiming at maximizing the degrees of freedom (DoF) shall be used in such a scenario [9, 26]. The basic idea of IA is that each of the receivers sacrifices half of its signal space for the interferences and uses the other half of the signal space for data transmission. IA can be performed in the dimensions of time [9, 27], frequency [33], space [18, 29, 34] or signal level [8, 26].

Unfortunately, state of the art IA schemes require a large number of antennas at the nodes or lots of transmission phases, i.e., time extensions, in order to achieve high numbers of DoF. To avoid these drawbacks, we introduced non-regenerative relays to aid IA [2, 12, 18, 28]. The considered scenario for relay-aided IA is shown in Fig. 14.1. It consists of $2K$ half-duplex nodes communicating pairwise. Each node has N antennas. Node $k \in \{1, \ldots, K\}$ and its communication partner, node $K + k$, want to exchange information with each other bidirectionally. Q amplify-and-forward half-duplex relays with R antennas each are employed to manipulate the linear end-to-end channel between the nodes in such a way that IA becomes feasible with small numbers of antennas at the nodes and with only few transmission phases required by the relaying protocol, i.e., without additional time extensions.

In the following, both one-way and two-way relaying [7, 30] protocols are employed. Concerning one-way relaying, two transmission phases are required for each direction of transmission. In the first phase, the source nodes transmit to both the relays and the destination nodes and in the second phase, both the source nodes

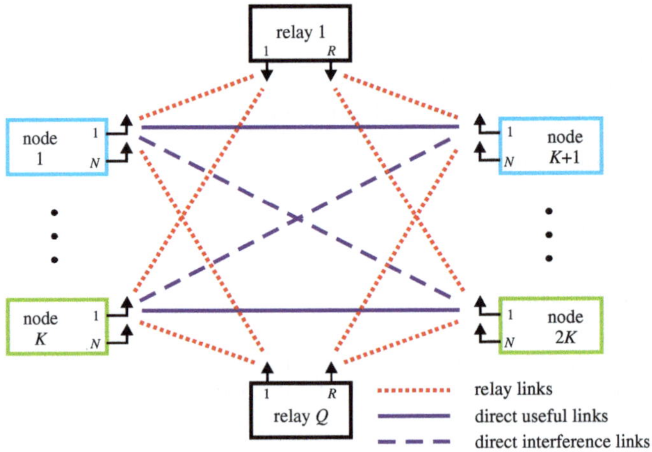

Fig. 14.1 Scenario for relay aided IA

and the relays transmit to the destination nodes. This means that not only the relay links between the nodes and the relays but also the direct links between the nodes are exploited. In the third and fourth phases, the direction of transmission is reversed. In contrast to this, two-way relaying requires only two transmission phases in total for a bidirectional transmission. In the first phase, all the nodes transmit to the relays, and in the second phase, the relays transmit to the nodes. The savings concerning the total number of transmission phases required for two-way relaying come at the price of not being able to use the direct links. We have developed efficient transmission schemes for the considered scenario in the interference limited case based on IA. In particular, precoders for the sources and linear signal processing schemes for the relays have been found.

In the following sections, the obtained results of our investigations will be reported for eight scenarios depending on the relaying protocol, the number of relays, the number of relay antennas, and the type of precoders. In Sects. 14.2 and 14.3, multiple single-antenna relays are considered. In Sects. 14.4 and 14.5, a single relay equipped with multiple antennas is considered. In Sects. 14.6 and 14.7, more general cases including multiple multi-antenna relays are considered. However, the precoders at the transmitters are fixed. In contrast to this, the cases where the precoders can be optimized are considered in Sects. 14.8 and 14.9. A summary concludes our work in Sect. 14.10.

14.2 One-Way Relaying with Multiple Single-Antenna Relays

The scenario shown in Fig. 14.1 with single antennas at the nodes ($N = 1$) and at the relays ($R = 1$) is considered in this section. The one-way relaying protocol is applied. Resulting from the two transmission phases for one direction of transmission, two dimensional signal spaces at every source node and at every destination node can be exploited for IA. Basically, IA is achieved when the interferences are aligned in a one dimensional subspace of the receive signal space at every destination node and the useful signal is not aligned with the interferences. To achieve this, the system variables, namely the relay scaling factors $g^{(q)}$, the transmit filters $(v_1^{(k)}, v_2^{(k)})^{\mathrm{T}}$ at the source nodes, and the receive filters $(u_1^{(l)}, u_2^{(l)})^{\mathrm{T}}$ at the destination nodes are adapted to the channels. Let $h_{\mathrm{DS}}^{(l,k)}$, $h_{\mathrm{DR}}^{(l,q)}$ and $h_{\mathrm{RS}}^{(q,k)}$ denote the channel coefficients between the k-th source node and the l-th destination node, between the q-th relay and the l-th destination node, and between the k-th source node and the q-th relay, respectively. These channel coefficients can be thought of as being the transfer function values of a considered single subcarrier of a multi-carrier system. The effective interference link between a source node and a non-corresponding destination node including the filters at the nodes has to be nullified. Based on this, a system of tri-affine equations can be formulated as

$$\left(u_1^{(l)*}, u_2^{(l)*}\right)\begin{pmatrix} h_{\mathrm{DS}}^{(l,k)} & 0 \\ \sum_{q=1}^{Q} h_{\mathrm{DR}}^{(l,q)} g^{(q)} h_{\mathrm{RS}}^{(q,k)} & h_{\mathrm{DS}}^{(l,k)} \end{pmatrix}\begin{pmatrix} v_1^{(k)} \\ v_2^{(k)} \end{pmatrix} = 0,\ k = 1, \ldots, K,\ l \neq k + K. \tag{14.1}$$

We have proposed two approaches to linearize the system of equations of (14.1) in [2]:

- IA with fixed filters: When the filters at the source nodes and at the destination nodes are fixed, the resulting system of equations is linear in the relay scaling factors.
- IA with partially adapted filters: When only the parts $v_1^{(k)}$ and $u_2^{(l)}$ of the filters are fixed, the resulting system of equations is linear in the other parts $v_2^{(k)}$ and $u_1^{(l)}$ of the filters and the relay scaling factors. This approach follows from the fact that amplifications of the filters at the source and destination nodes are irrelevant to the structure of the signal spaces. Therefore, each filter provides only a single free variable.

We were able to obtain simple closed-form IA solutions with low computational complexity using these linear approaches, which is in general not possible for state of the art IA techniques without relays [18].

For both approaches above, we investigated the feasibility conditions for IA. For IA using fixed filters, as the direct useful links are not considered when aligning the interferences, the useful signals are almost surely not aligned with the interferences. For IA with partially adapted filters, some solutions of (14.1) result in that the useful signals are aligned with the interferences at the destination nodes. We refer to these solutions as invalid solutions. We have shown that the minimum required numbers of relays guaranteeing the existence of valid solutions for IA with fixed filters and for IA with partially adapted filters are $K(K-1)$ and K^2-3K+2, respectively. Furthermore, if the required number of relays is available, a randomly selected solution of (14.1) is almost surely valid. For a detailed discussion, see our paper [22].

For the case of IA with fixed filters, we studied the power allocation problem. Due to the existence of direct links, the relay scaling factors are determined by the IA solution we choose. Thanks to the linearity of the system, one can consider the effective channel including the relays and the filters at the source and destination nodes as an interference-free channel. Therefore, the power allocation problem aiming at maximizing the sum rate with a total power constraint is convex. We used the water-filling algorithm to optimize the power allocation [10].

Based on the approach of IA with partially adapted filters, we solved the sum mean squared error (MSE) minimization problem in a closed form. The estimated data symbols at the destination nodes are linear functions of the system variables, i.e., of the adapted parts of the filters and the relay scaling factors. Consequently, we can formulate the sum MSE as a convex quadratic function of the system variables. By exploiting this, we solved the sum MSE minimization problem with two kinds of constraints. Firstly, we considered the system of linear equations of (14.1) with partially adapted filters as a constraint. Accordingly, we found an IA solution with a minimized sum MSE. Secondly, we considered a constraint on the total transmit power of both the source nodes and the relays. In this case, we reformulated the sum MSE minimization problem as a quadratically constrained quadratic minimization problem. Because of the convexity of both the sum MSE and the total power constraint, we are able to find the optimum sum MSE solution. By simulations, we showed for the latter case that the performance of the sum MSE minimization solution asymptotically converges to the performance of the IA solution at high SNRs if the number of relays is sufficiently large. We published these results in [1, 5].

Furthermore, we considered the sum rate maximization problem for IA with both fixed and partially adapted filters in [1, 6]. The sum rate is not a multi-concave function of the system variables. To make the sum rate maximization problem solvable, we proposed to reformulate it as a multi-convex optimization problem by adding two additional sets of weighting factors. The new formulation proposed by us is quite general and it is applicable to a wide range of networks with linear channels and linear filters. We have shown that this approach outperforms the state of the art techniques which minimize the sum MSE instead of maximizing the sum rate [15, 19, 24, 31, 32].

14.3 Two-Way Relaying with Multiple Single-Antenna Relays

In this section, the scenario consisting of single-antenna nodes and relays considered in Sect. 14.2 is adopted for bidirectional communications using a modified two-way relaying protocol [7, 30]. As each node is assumed to have a single antenna, we introduced a three-phase transmission scheme to perform IA. In the first phase, all nodes transmit simultaneously to the relays. To ensure that the receive signal space at each destination node is two dimensional, the relays transmit twice towards the nodes in the second and third phase. Furthermore, we performed subcarrier permutation for the channels in the second and the third phase so that the channel coefficients would be independent. Based on this transmission scheme, we fixed the receive filters $(u_1^{(l)}, u_2^{(l)})^{\mathrm{T}}$ at the nodes and optimized the transmit filters $(g_1^{(q)}, g_2^{(q)})^{\mathrm{T}}$ at the relays for achieving IA. Assuming that the nodes can perfectly cancel the self-interferences, the interference-nulling conditions for all $2K$ nodes can be written as a homogenous system of linear equations

$$\left(u_1^{(l)*}, u_2^{(l)*}\right)\begin{pmatrix}\sum_q h_{\mathrm{NR},1}^{(l,q)} g_1^{(q)} h_{\mathrm{RN}}^{(q,k)} \\ \sum_q h_{\mathrm{NR},2}^{(l,q)} g_2^{(q)} h_{\mathrm{RN}}^{(q,k)}\end{pmatrix} = 0,\ l \neq k \text{ and } l \neq \begin{cases} k+K \text{ if } k \leq K, \\ k-K \text{ if } k > K, \end{cases} \tag{14.2}$$

where $h_{\mathrm{RN}}^{(q,k)}$ is the channel coefficient from the nodes towards the relays and $h_{\mathrm{NR},1}^{(l,q)}$ and $h_{\mathrm{NR},2}^{(l,q)}$ are the channel coefficients from the relays towards the nodes in the second and in the third phase, respectively. The system of equations of (14.2) is linear in the transmit filters at the relays and it has non-trivial solutions if the number of relays is at least $2K(K-1)+1$. Furthermore, if additional relays are available, we are also able to adapt the filters at the relays such that at some of the nodes, the received useful signal is orthogonal to the interference subspace. In this way, the SNR as well as the sum rate can be improved. We have shown that at least $2K^2 - 1.5K$ relays are required to achieve this at all $2K$ nodes. The scheme we proposed above is able to achieve bidirectional transmission with one less transmission phase as compared with the scheme we proposed in Sect. 14.2. However, this comes at the price of a nearly doubled number of relays.

Extended from the scenario discussed above, we also considered the case where each node has two antennas. Then two dimensional transmit/receive signal spaces can be exploited even without using the third transmission phase. However, using two transmission phases in total results in each relay only providing a single scaling factor as the relays only transmit in the second phase. If the transmit and receive filters are fixed, the required number of relays for achieving IA is nearly doubled as compared to the scenario discussed above, i.e., at least $4K(K-1)+1$ relays are required.

To compare the performances of the two-way relaying scheme considered in the paragraph above and the one-way relaying scheme considered in Sect. 14.2, a scenario consisting of K node pairs equipped with two antennas each and Q single-antenna relays is considered. Fixed filters at the nodes are assumed and the number of relays is chosen as $4K(K-1)+1$, so that there are infinitely many IA solutions for both schemes [2]. Equal power allocation among the source nodes is assumed for both schemes. In the one-way relaying scheme, the solution with the minimum total transmit power at the relays is selected [4]. To ensure that the total transmit powers of both the one-way and the two-way relaying schemes are equal, the filters at the relays for the two-way relaying scheme are scaled accordingly. In four transmission phases, each node can successfully transmit two data symbols to its communication partner either simultaneously in the one-way relaying scheme or sequentially in the two-way relaying scheme. By simulations, we have shown that the two-way relaying scheme outperforms the one-way relaying scheme in terms of the sum rates at all SNRs. The reason is that in the one-way relaying scheme, additional transmit power is required by the relays to compensate for the interferences propagating through the direct links.

14.4 Two-Way Relaying with a Single Multi-Antenna Relay

In this section, a single relay with R antennas assists in the communication between K node pairs each having N antennas, i.e., the scenario of Fig. 14.1 with only a single relay is considered. Each of the $2K$ nodes wants to transmit d data streams to its communication partner. Two-way relaying is considered for the bidirectional communication between the node pairs. A two-phase transmission scheme is exploited. In the first phase called multiple access (MAC) phase, all the nodes transmit to the relay. In the second phase called broadcast (BC) phase, the relay forwards its received signals to the nodes. It is assumed that the relays do not have enough antennas to separate the $2Kd$ data streams of the $2K$ nodes. We have shown that $R \geq Kd$ is a necessary condition to perform IA since the two-way relay channel is a multiple keyhole channel.

For IA based on two-way relaying, exploiting the fact that the nodes can perform self-interference cancelation, we have developed the new concepts of signal alignment and channel alignment [12, 14]. First we considered the special case $R = Kd$ [12]. In the MAC phase, the relay receives the superposition of $2Kd$ data streams. These $2Kd$ data streams cannot be spatially separated in a Kd-dimensional space. However, the nodes can perform self-interference cancelation and, hence, all the nodes shall transmit their signals to the relay in such a way that at the relay, each node's signal subspace aligns with the signal subspace of its communication partner. This is called signal alignment [12]. After signal alignment, we have only K effective data streams at the relay. In the BC phase, each node designs its receive filter such that the effective channel, consisting of the channel between the relay and the node and the receive filter, spans the same subspace as the effective channel of its communication partner. This is called channel alignment [12]. After signal and channel alignment, there are only Kd effective data streams and Kd effective channels. The relay with $R = Kd$ antennas can perform transceive zero forcing (ZF) [12]. In the case $R = Kd$, we have shown that signal and channel alignment are necessary steps to achieve IA. Therefore, without loss of optimality we were able to decouple the trilinear IA problem into three linear problems namely, signal alignment, channel alignment, and transceive ZF [12]. We have shown that in order to perform signal and channel alignment, the nodes need at least $2N \geq Kd + d$ antennas [12].

Secondly, we addressed the general case $R \geq Kd$. In contrast to the case $R = Kd$ above, for the case $R \geq Kd$ the relay has $R - Kd$ additional antennas. Our objective was to identify how the additional antennas at the relay can be utilized. For this, we have proposed a novel IA algorithm called pair-aware interference alignment (PAIA) and we have published it in [14]. In the PAIA algorithm, in the MAC phase, we choose a Kd-dimensional subspace in the R-dimensional relay space such that signal alignment is feasible. Each node aligns its signal at the relay with that of its communication partner in this chosen Kd-dimensional subspace. In the complementary $(R - Kd)$-dimensional subspace, signal alignment does not need to be feasible. This we called partial signal alignment [14]. The Kd-dimensional subspace is chosen such that either partial signal alignment can be performed with a smaller number of antennas at the nodes than required in the case $R = Kd$ or such that the SNR is maximized at the receiver. Similar to partial signal alignment, in the BC phase a Kd-dimensional subspace is chosen such that channel alignment is feasible in this chosen Kd-dimensional subspace. In the complementary $(R-Kd)$-dimensional subspace, channel alignment does not need to be feasible. This we called partial channel alignment [14]. Partial signal alignment followed by partial channel alignment and transceive ZF are performed to achieve IA.

We derived the properness condition for PAIA as $2N + R \geq 2Kd + d$ [14]. By simulations we have shown that whenever the properness condition is satisfied, typically IA is feasible [14]. From the properness condition, one can directly see that when adding more antennas to the relay, the number of DoF $Kd = (2N + R - d)/2$ in the system is increased. Furthermore, for a fixed number of DoF, one can add two more antennas to the relay and reduce the number of antennas at each of the $2K$

nodes by one and the system will still be proper. Hence, additional relay antennas help to reduce the number of antennas required at each of the $2K$ nodes.

We showed that partial signal alignment and partial channel alignment are dual problems [14] and, hence, in the following we consider only partial signal alignment. We have proposed an iterative algorithm to achieve partial signal alignment [14]. In this iterative algorithm, the partial signal alignment conditions are reformulated into the problem of finding a $(R - Kd)$-dimensional subspace which has at least a d-dimensional intersection with each of K given $2N$-dimensional subspaces. We proposed an approach where instead of finding a single $(R - Kd)$-dimensional subspace as described above, $n = (R - Kd)/d$ subspaces of dimension d each are found such that each of these n subspaces has a d-dimensional intersection with a disjoint group of K/n subspaces. In this case, we have found a closed form solution [14]. In addition, when the properness condition is satisfied with strict inequality, multiple IA solutions are possible, and we formulated an optimization problem to find an IA solution that maximizes the SNR [17]. This optimization problem is non-convex and we have proposed a gradient based algorithm to find a local maximum [17].

In the simulation results, we take the sum rate in bits per channel use as a performance measure. Figure 14.2 shows the sum rate as a function of P_{node}/σ^2, where P_{node} is the average transmit power per node and σ^2 is the variance of the white Gaussian noise. The relay retransmit power is chosen to be $K P_{\text{node}}$. I.i.d. Rayleigh fading channels with unit average gains are assumed. Two scenarios with $K = 7$ and $K = 6$, respectively, are considered. In both scenarios, each node has $N = 3$ antennas and the relay has $R = 9$ antennas. For $K = 7$, a PAIA solution can be found using the iterative algorithm. However, serving one node pair less results in fewer constraints which makes it possible to obtain a closed form solution for PAIA. In general, IA based algorithms aim at maximizing the DoF and do not care about the power of the useful signal. Therefore, at high SNRs, PAIA achieves a higher sum rate than the conventional reference algorithm following the idea of pair-wise ZF being

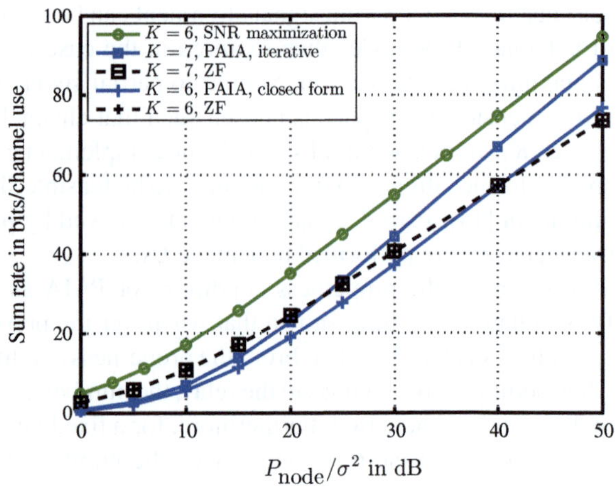

Fig. 14.2 Sum rate performances of the SNR maximization, the PAIA, and the reference ZF algorithms versus P_{node}/σ^2 for $N = 3$, $d = 1$, $Q = 1$, and $R = 9$

described in [21]. At low SNRs, the reference ZF algorithm has better performance. Furthermore, it can be seen that by using the SNR maximization algorithm, higher sum rates are achieved also at low and medium SNRs.

For the special case when the nodes have a single antenna each while the relay still has multiple antennas, we have proposed a projection based algorithm to perform IA along multiple subcarriers [11]. We have shown that even in the case where the relays do not have enough antennas to spatially separate the data streams and the nodes do not have multiple antennas to perform IA along spatial dimensions, a multiple antenna relay can assist the nodes to achieve IA in frequency dimensions. Hence, both the spatial and frequency dimensions at the relays are utilized to assist IA in the frequency dimensions at the nodes [11].

14.5 One-Way Relaying with a Single Multi-Antenna Relay

In this section, we consider the same scenario as in Sect. 14.4 with a single relay with multiple antennas, but from a one-way relaying perspective which utilizes the direct links between the nodes. Depending on the number N of antennas available at the source nodes and the destination nodes, we addressed two cases. In the first case, $N = d$ holds. In this case, $KN/2$ DoF are achieved with only the two channel uses necessary for one-way relaying [3]. In other words, each of the K node pairs achieves $N/2$ DoF per channel use. We have obtained closed form solutions to find the transmit, relay, and receive filters [3].

In the second case, $N > d$ holds. For this case, our algorithm being described in [3] can still be applied, but it does not utilize all the available variables in the system. We developed two new algorithms to perform IA in this case. The first algorithm iteratively minimizes the leakage interference at all the receivers. Each iteration involves three steps. In the first step, the transmit filters and the relay filters are fixed and the receive filters are designed to minimize the leakage interference. In the second step, the relay and the receive filters are fixed and the transmit filters are designed to minimize the leakage interference. Similarly in the third step, the relay filters are designed by fixing the transmit and the receive filters. In this algorithm, the leakage interference is reduced during each iteration and the leakage interference is lower bounded by zero. Hence, the algorithm converges. However, convergence to the global minimum cannot be guaranteed due to the non-convexity of the problem. The second algorithm, in addition to reducing the interference in the system, also maximizes the sum rate in the system. This algorithm is similar to the iterative leakage interference minimization algorithm described above except that after each iteration, the filters are updated to improve the sum rate. This is done as follows: The filters along with the corresponding sum rate objective function form a manifold. The filters designed to minimize the leakage interference are updated by looking in the neighborhood of the manifold and moving towards the maximum gradient along the manifold. By doing this, the sum rate performance of the system is improved compared to the leakage minimization algorithm.

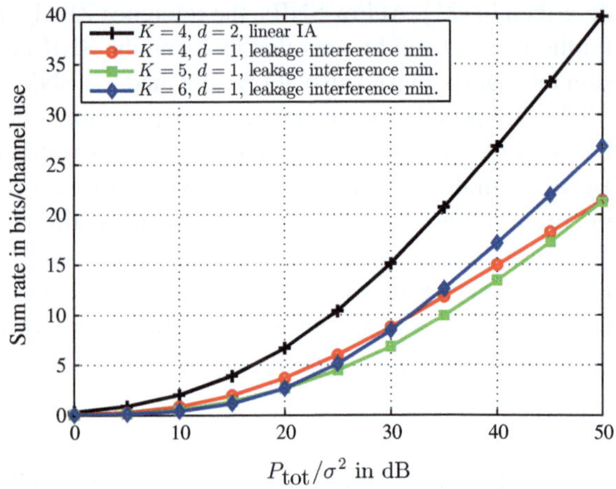

Fig. 14.3 Sum rate performances of the linear IA and the leakage interference minimization algorithms versus P_{tot}/σ^2 for $N = 2$, $Q = 1$, and $R = 5$

Figure 14.3 shows the sum rate in bits per channel use as a function of P_{tot}/σ^2, where P_{tot} is the total transmit power of the network, including the relay retransmit power. Different from two-way relaying, the relay retransmit power for one-way relaying depends on the channel realization and the transmit powers of the source nodes due to the existence of the direct links. Therefore, instead of considering P_{node} as for two-way relaying, P_{tot} is always considered for one-way relaying in the following sections. It can be observed that more DoF can be achieved by the linear IA algorithm as compared to the leakage interference minimization algorithm. This is because the latter may converge to an invalid solution in practice and therefore requires more relay antennas to achieve the same DoF as that achieved by the linear IA algorithm.

We have investigated the feasibility conditions for the considered scenario. For the case where $N = d$ holds, the linearization method proposed in [3] is optimal in DoF sense. Therefore, finding the feasibility conditions is equivalent to determining the dimension of the solution space of the linearized IA conditions [3]. For the case where $N > d$ holds, the number of variables and the number of equations involved in the IA conditions shall be compared to obtain the feasibility conditions.

14.6 One-Way Relaying with Multiple Multi-Antenna Relays, Fixed Precoders

In contrast to Sect. 14.2, the Q relays are now equipped with R antennas each. We have shown that the numbers of relays and relay antennas have to satisfy $QR^2 \geq K(K-1)$. If the equality holds, a unique IA solution can be obtained. If the inequality is strictly satisfied, then there are infinitely many solutions for IA and we proposed to select the one with the minimum total relay transmit power [1, 4] in order to reduce the total transmit power consumed by the system.

To improve the achieved sum rate, especially at low and medium SNRs, the sum MSE minimization algorithm [5] and the sum rate maximization algorithm [6] which we proposed in Sect. 14.2 can be extended and applied in this section. Moreover, we propose to use the least squares method to obtain an approximate solution called imperfect IA [1, 4]. Basically, imperfect IA allows a small amount of interference signals not to be aligned in the interference subspace. By simulations, we have shown that imperfect IA results in less noise forwarded by the relays to the destination nodes and, therefore, higher sum rates can be achieved at low and medium SNRs as compared to perfect IA. However, the remaining interferences cause a saturation of the sum rate at high SNRs. In contrast to this, perfect IA achieves a significantly higher sum rate at high SNRs. Unlike for perfect IA, the inequality $QR^2 \geq K(K-1)$ is not required to be satisfied for imperfect IA. Details on the analysis and the performance of imperfect IA can be found in our paper [4].

We also considered an extended scenario consisting of multi-antenna source and destination nodes. The multiple antennas at the nodes can be exploited for transmitting several data symbols simultaneously, which increases the number of DoF. To achieve IA, the interferences have to be aligned in one half of the signal space at each destination node whereas the other half of the signal space is occupied by the useful signals. In our approach, the relay processing matrices are fully adapted while only parts of the filters at the source nodes and at the destination nodes are adapted to the channel [3]. Based on this, we formulated a system of linear equations which includes both the useful links and the interference links. Among the infinitely many solutions of the system of linear equations, we selected the one with the minimum relay transmit power in a closed form. For more details, see our publication [1, 3].

Figure 14.4 shows the sum rate in bits per channel use as a function of P_{tot}/σ^2. The performances achieved by imperfect and perfect IA are compared. Imperfect IA not only requires less relays and relay antennas, it also achieves a higher sum rate at low SNRs as compared to perfect IA. Moreover, with minimized relay power, the performance achieved by perfect IA can be further improved.

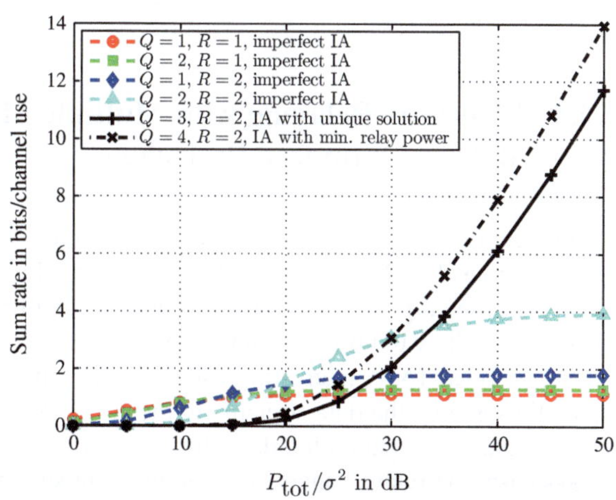

Fig. 14.4 Sum rate performances of imperfect IA, IA with unique solution, and IA with minimized relay power versus P_{tot}/σ^2 for $K=4$ and $N=d=1$

14.7 Two-Way Relaying with Multiple Multi-Antenna Relays, Fixed Precoders

In contrast to Sect. 14.3, now multi-antenna relays are considered for the two-way relaying protocol. It is assumed that the nodes have only receive channel state information (CSI). Each of the relays has R antennas and hence, only sees a set of R linear combinations of all the Kd signals. The number R of linear combinations is less than $2Kd$. However, the relays jointly design their filters to perform IA at the receivers. The required numbers of relays and relay antennas for achieving IA are derived. Since multi-antenna relays are considered, the required number of relays is significantly smaller as compared to the scenario of Sect. 14.3. We have proposed three algorithms to achieve IA [13, 16].

The first proposed algorithm finds a closed form solution to achieve IA. In this case, we have reformulated the problem of IA into a linear problem by treating the self-interference as unknown interference and by sacrificing some of the variables available for performing IA [13]. The properness condition is also derived [13]. In case that the properness condition is satisfied with strict inequality, we have multiple IA solutions. We have generalized the optimization problem to find an IA solution that maximizes the SNR from the single relay case in Sect. 14.4 to the multiple relay case [13]. This optimization problem is non-convex and the gradient based algorithm from Sect. 14.4 has been extended to find a local maximum.

The second and the third proposed algorithms are iterative algorithms, one based on IA and another one based on minimizing the MSE at the receiver [16]. We have derived the properness condition for the IA algorithm. Independent of the system parameters, the minimum mean squared error (MMSE) algorithm always minimizes the MSE and, hence, the number of DoF cannot be derived directly. However, from simulations we observed that the iterative MMSE algorithm achieves effectively the same number of DoF as the iterative IA algorithm [16]. In addition, the iterative MMSE algorithm has a better sum rate performance than the iterative IA algorithm especially in the low to medium SNR regime [16].

14.8 Two-Way Relaying with Multiple Multi-Antenna Relays, Optimized Precoders

In this section, we generalized the scenario from Sect. 14.4 to multiple relays. Therefore, we consider Q relays supporting the bidirectional communication between K node pairs, see Fig. 14.1. Our objective was to identify how to perform IA in such a scenario and how to utilize the multiple relays.

Relay aided IA is interesting especially for the case where the relays cannot spatially separate the data streams. In the single relay case of Sect. 14.4, the relay cannot spatially separate the data streams if $R < 2Kd - d$ holds. In the multiple relays case, the relays cannot separate the data streams if $RQ < 2Kd - d$ holds.

However, also for the case $RQ \geq 2Kd - d$, there are scenarios where all the relays together cannot spatially separate the data streams. Hence, not only $RQ < 2Kd - d$ but also $RQ \geq 2Kd - d$ has been considered in this section. Similar to Sect. 14.4, the two-way relay channel is a multiple keyhole channel and hence, $RQ \geq Kd$ antennas are necessary at the relays.

First, we considered the case $RQ = Kd$. In this case, we have shown that signal alignment and channel alignment are necessary in the MAC and BC phases. We proposed an iterative algorithm to achieve IA. We designed the transmit and the receive filters to perform signal and channel alignment, respectively. As the relays do not share the signals received by them, the relay filter matrix is a block diagonal matrix. Using this block diagonal relay filter matrix, the Kd effective signals as well as the Kd effective channels cannot be spatially separated. Fortunately, the block diagonal relay matrix can be chosen to be of full rank Kd, i.e., there is a set of receive signatures and a corresponding set of transmit signatures which the relays can spatially separate and hence, perform transceive ZF. The nodes jointly choose the signal alignment and channel alignment directions such that the relays can perform transceive ZF. This we called cooperative ZF. Cooperative ZF is a bilinear problem. We proposed an iterative algorithm to find the transmit, relay and receive filters in [13]. The properness condition is also derived.

Furthermore, to improve the performance at low and medium SNRs, we proposed an iterative MMSE algorithm [15]. In this algorithm, the transmit, relay, and receive filters are optimized iteratively to minimize the MSE. The joint optimization of the transmit, receive, and relay filters subject to power constraints is a non-convex problem. We have shown that if we fix two of the three filters, then the MMSE optimization problem is convex with respect to the third filter [15] and, hence, can be solved using convex optimization tools.

Figure 14.5 shows the sum rate in bits per channel use as a function of P_{node}/σ^2. In this scenario, both the cooperative ZF and the iterative IA algorithms can be applied since $QR = Kd$ holds. The reference orthogonalize and forward algorithm has been

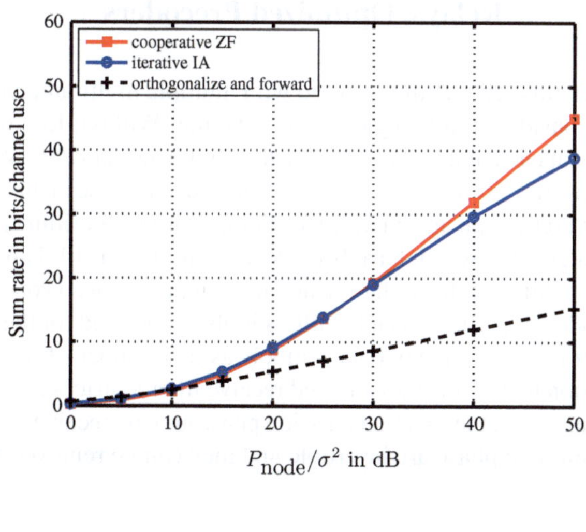

Fig. 14.5 Sum rate performance of the cooperative ZF, the iterative IA, and the reference orthogonalize and forward algorithms versus P_{node}/σ^2 for $K = 4$, $N = 3$, $d = 1$, $Q = 2$, and $R = 2$

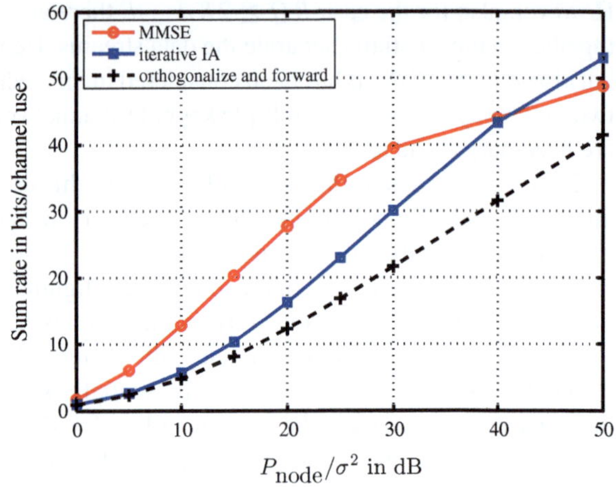

Fig. 14.6 Sum rate performance of the MMSE, the iterative IA, and the reference orthogonalize and forward algorithms versus P_{node}/σ^2 for $K = 5$, $N = 3$, $d = 1$, $Q = 2$, and $R = 4$

proposed in [30]. At low SNR, the noise is the dominating factor and hence, both the cooperative ZF and the iterative IA algorithms have similar performance. At high SNR, the slope of the sum rate curve corresponding to iterative IA reduces. This is due to the residual sum of squared errors at high SNR. Hence, cooperative ZF has better performance than iterative IA.

In the simulation results shown in Fig. 14.6, $QR > Kd$ holds and hence cooperative ZF cannot be applied. It can be observed that iterative IA performs better than the reference orthogonalize and forward algorithm. Furthermore, the proposed MMSE algorithm has superior performance at low and medium SNR as compared to iterative IA.

14.9 One-Way Relaying with Multiple Multi-Antenna Relays, Optimized Precoders

In this section, the scenario with multiple multi-antenna relays shown in Fig. 14.1 is considered utilizing one-way relaying. Without loss of generality, we also assume that each node is equipped with N antennas and $d \leq N$ data symbols are transmitted by every source node. For the multiple relay case in this section, we have extended the iterative leakage interference minimization algorithm and the sum rate maximization algorithm which have been developed in Sect. 14.5 and applied them here.

Adapted from the linearization approach we proposed for Sect. 14.5 [1, 3], the idea of partially adapted filters is also employed for the case where each source node transmits as many data symbols as the number of its antennas. Under the assumption that all the transmit and receive filter matrices are of full rank, the transmit filter matrix in the first transmission phase and the receive filter matrix in the second transmission phase are invertible and they can be removed from the interference-nulling conditions by reformulation. Therefore, the tri-affine interference-nulling conditions can be linearized. Consequently, we are able to describe the IA solution set based on the solution space of the linearized interference-nulling conditions. We have also investigated and determined the dimension of the IA solution space. As compared to the numerical algorithms, the linearization approach provides closed-form solutions which require significantly less computational effort. Furthermore, the closed-form solutions always nullify the interferences perfectly whereas the numerical algorithms may converge to a local extreme, which results in remaining interferences at the destination nodes.

In addition to nullifying the interferences only, we have developed two more schemes which also achieve other objectives. The first scheme minimizes the energy consumption in the system. Here, out of many possible IA solutions, the one that minimizes the transmit power at the relay is chosen [1, 3]. A closed form solution has been derived. The second scheme improves the sum rate at low and medium SNRs. Here, out of many possible IA solutions, the one that minimizes the MSE is chosen [1, 3].

In Fig. 14.7, the sum rates achieved by IA, the sum MSE minimization algorithm, and the sum rate maximization algorithm which is extended from the one introduced in Sect. 14.2 are shown as a function of $P_{\mathrm{tot}}/\sigma^2$. Rayleigh fading channels with realistic path loss model are assumed [1]. The path loss exponent is chosen to be 2. The average gain of the intra-cell effective channels is normalized to 1. All the above three algorithms achieve remarkable performance at moderate to high SNRs. However, the sum MSE minimization algorithm and the sum rate maximization algorithm are not able to achieve the same DoF as that achieved by IA due to the fact that the optimization problems are non-convex and the global optimum solution is not guaranteed.

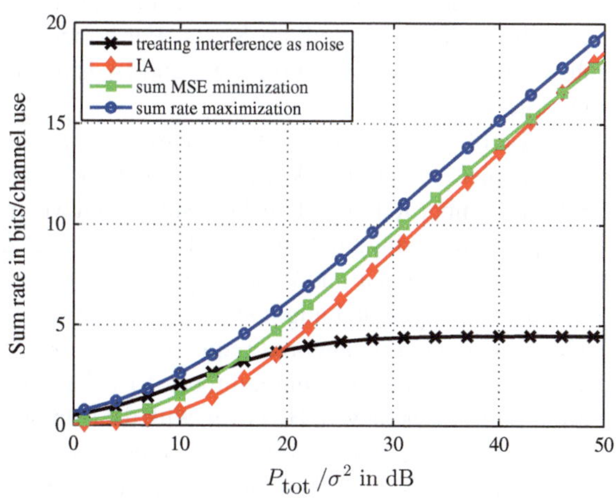

Fig. 14.7 Sum rate performance of the sum rate maximization, the sum MSE minimization, the IA, and the reference algorithm treating interference as noise versus $P_{\mathrm{tot}}/\sigma^2$ for $K = 3$, $N = d = 1$, $Q = 3$, and $R = 2$

14.10 Summary

We have considered bi-directional communication between multiple node pairs. Multiple amplify and forward half duplex relays with multiple antennas aid in the communication. The relays have been used to manipulate the effective channels between the transmitters and the receivers to aid in the IA process. Both one-way relaying and two-way relaying protocols have been investigated. In both relaying protocols, the numbers of antennas required at the relays and at the nodes have been derived. Also, a number of algorithms to design the transmit filters, the relay filters, and the receive filters have been developed.

In case of one-way relaying, we started with a single antenna at the nodes. This made it possible to linearize the problem by considering a single variable per node which defines the transmit signature of the corresponding node. Furthermore, we generalized the linearization method to the case where each node has N antennas and transmits $N = d$ data streams. The method we developed is optimum in the DoF sense, because each node has N antennas and transmits N data streams in two time slots, which is half of the DoF it could achieve in the absence of interferers. We have derived the number of antennas required at the relays to make IA feasible. Also, we have obtained a closed form solution to achieve IA. If the relays have more antennas than required to perform IA, several IA solutions exist. The linear IA algorithm we obtained gives all the IA solutions. Among these IA solutions, we can choose the ones which optimize a given objective function. We investigated different objective functions, like the minimization of transmit power at the relay, the minimization of the sum mean squared error, and the maximization of the sum rate. The first two lead to convex problems and we have proposed closed form solutions. However, sum rate maximization is a non-convex problem. We proposed a new method to reformulate this into a multi-concave problem. Considering the aforementioned objective functions aids in improving the performance of the system especially at low and medium SNRs. Furthermore, when the nodes do not have transmit CSI, we fixed the precoders and IA is achieved through the design of the relay filters only.

In case of two-way relaying, we started with a single relay. Since the two-way relay channel is a multiple keyhole channel, the number R of antennas at the relay limits the number Kd of DoF in the system. First we investigated $R = Kd$. IA is a tri-linear problem. Using the fact that self-interference can be cancelled, we developed the concepts of signal and channel alignment and decoupled IA into three linear problems, namely signal alignment, channel alignment and transceive ZF. We have derived the number of antennas required at the nodes to achieve IA. We have shown that with $R = Kd$, the same number of DoF as in a MIMO interference channel without a relay is achieved. However, with the help of the relay a closed form solution to the IA problem is obtained for any number of node pairs in the system. In addition, when the relay has more than the minimum required number of antennas, i.e., $R > Kd$ holds, we have utilized the additional antennas either to increase the DoF in the system or to reduce the required number of antennas at the nodes. Furthermore, for the case when there are multiple IA solutions, we formulated an optimization problem to find an IA solution that maximizes the SNR. This optimization problem is non-convex and we have proposed a gradient based algorithm to find a local maximum. The properness condition derived and the algorithms developed for the single relay case have been generalized for multiple relays. In addition, to improve the performance at low and medium SNRs, we have proposed an MMSE based scheme. Minimization of the MSE is a non-convex problem. However, by fixing one of the three kinds of variables, namely, the transmit filters, the relay filters or the receive filters, the problem becomes convex. We have proposed an algorithm to iteratively minimize the MSE. All the above mentioned methods require global CSI. Furthermore, scenarios where the nodes have no CSI or limited CSI have been investigated. IA algorithms and the corresponding properness conditions have been derived for the case of no or limited CSI.

Acknowledgments The work of "Interference Alignment Aided by Non-Regenerative Relays" was supported by the German Research Foundation (DFG) under Grants WE2825/11-1 and KL907/5-1.

References

1. Al-Shatri H (2014) Interference reduction in multiuser relay networks. Ph.D. thesis, University of Rostock
2. Al-Shatri H, Weber T (2011) Interference alignment aided by non-regenerative relays for multiuser wireless networks. In: Proceedings of the IEEE international symposium on wireless communication systems, Aachen, pp 271–275
3. Al-Shatri H, Ganesan R, Klein A, Weber T (2012) Interference alignment using a MIMO relay and partially-adapted transmit/receive filters. In: Proceedings of the IEEE wireless communications and networking conference, Paris, pp 459–464
4. Al-Shatri H, Ganesan R, Klein A, Weber T (2012) Perfect versus imperfect interference alignment using multiple MIMO relays. In: Proceedings of international symposium on wireless communication systems, Paris, pp 676–680
5. Al-Shatri H, Li X, Ganesan R, Klein A, Weber T (2013) Closed-form solutions for minimizing sum MSE in multiuser relay networks. In: Proceedings of the IEEE 77th vehicular technology conference, Dresden, pp 1–5
6. Al-Shatri H, Li X, Ganesan R, Klein A, Weber T (2013) Multi-convex optimization for sum rate maximization in multiuser relay networks. In: Proceedings of the IEEE 24th international symposium on personal indoor and mobile radio communications, London, pp 1327–1331
7. Berger S, Kuhn M, Wittneben A, Unger T, Klein A (2009) Recent advances in amplify-and-forward two-hop relaying. IEEE Commun Mag 47(7):50–56
8. Bresler G, Parekh A, Tse D (2010) The approximate capacity of the many-to-one and one-to-many Gaussian interference channels. IEEE Trans Inf Theory 56(9):4566–4592
9. Cadambe V, Jafar S (2008) Interference alignment and degrees of freedom of the K user interference channel. IEEE Trans Inf Theory 54(8):3425–3441
10. Cover T, Thomas J (2006) Elements of information theory. 2nd edn. Wiley
11. Ganesan R, Klein A (2011) Projection based space-frequency interference alignment in a multi-carrier multi-user two-way relay network. In: Proceedings of the IEEE international symposium on wireless communication systems, Aachen, pp 266–270
12. Ganesan R, Weber T, Klein A (2011) Interference alignment in multi-user two way relay networks. In: Proceedings of the IEEE 73rd vehicular technology conference, Budapest, pp 1–5

13. Ganesan R, Al-Shatri H, Weber T, Klein A (2012) Cooperative zero forcing in multi-pair multi-relay networks. In: Proceedings of the IEEE international symposium on personal, indoor and mobile radio communications, Sydney, pp 1740–1745
14. Ganesan R, Al-Shatri H, Kuehne A, Weber T, Klein A (2013) Pair-aware interference alignment in multi-user two-way relay networks. IEEE Trans Wireless Commun 12(8):3662–3671
15. Ganesan R, Al-Shatri H, Weber T, Klein A (2013) Iterative MMSE filter design for multi-pair two-way multi-relay networks. In: Proceedcings of the IEEE international conference on communications, Budapest, pp 5929–5933
16. Ganesan R, Al-Shatri H, Weber T, Klein A (2013) Relay-aided interference alignment for bidirectional communications in multi-pair multi-relay networks. In: Proceedings of the IEEE 14th workshop on signal processing advances in wireless communications, Darmstadt, pp 66–70
17. Ganesan R, Le HQ, Al-Shatri H, Weber T, Klein A (2013) Closed form solution and useful signal power maximization for interference alignment in multi-pair two-way relay networks. In: Proceedings of the IEEE wireless communications and networking conference, Shanghai, pp 3414–3419
18. Gomadam K, Cadambe V, Jafar S (2011) A distributed numerical approach to interference alignment and applications to wireless interference networks. IEEE Trans Inf Theory 57(6):3309–3322
19. Gorski J, Pfeuffer F, Klamroth K (2007) Biconvex sets and optimization with biconvex functions: a survey and extensions. Math Methods Oper Res 66(3):373–407
20. Jafar S, Shamai S (2008) Degrees of freedom region of the MIMO X channel. IEEE Trans Inf Theory 54(1):151–170
21. Leow CY, Ding Z, Leung K, Goeckel D (2011) On the study of analogue network coding for multi-pair, bidirectional relay channels. IEEE Trans Wireless Commun 10(2):670–681
22. Li X, Al-Shatri H, Ganesan R, Klein A, Weber T (2013) Feasibility conditions for relay-aided interference alignment in partially connected networks. In: Proceedings of the IEEE 77th vehicular technology conference, Dresden, pp 1–5
23. Li X, Al-Shatri H, Ganesan R, Klein A, Weber T (2013) Interference alignment aided by locally connected relays. In: Proceedings of the 10th international symposium on wireless communication systems, Illmenau, pp 1–5
24. Ma S, Xing C, Fan Y, Wu Y, Ng T, Poor H (2010) Iterative transceiver design for MIMO AF relay networks with multiple sources. In: Proceedings of the military communications conference, Princeton, pp 369–374
25. Maddah-Ali M, Motahari A, Khandani A (2006) Signaling over MIMO multi-base systems: Combination of multi-access and broadcast schemes. In: Proceedings of the IEEE international symposium on information theory, Seattle, pp 2104–2108
26. Motahari A, Oveis-Gharan S, Maddah-Ali MA, Khandani A (2014) Real interference alignment: Exploiting the potential of single antenna systems. IEEE Trans Inf Theory 60(8):4799–4810
27. Nazer B, Gastpar M, Jafar S, Vishwanath S (2012) Ergodic interference alignment. IEEE Trans Inf Theory 58(10):6355–6371
28. Nourani B, Motahari A, Khandani A (2010) Relay-aided interference alignment for the quasi-static interference channel. In: Proceedings of the IEEE international symposium on information theory, Austin, pp 405–409
29. Peters S, Heath R (2009) Interference alignment via alternating minimization. In: Proceedings of the IEEE international conference on acoustics, speech and signal processing, Taipei, pp 2445–2448
30. Rankov B, Wittneben A (2007) Spectral efficient protocols for half-duplex fading relay channels. IEEE J Sel Areas Commun 25(2):379–389
31. Schmidt D, Shi C, Berry R, Honig M, Utschick W (2009) Minimum mean squared error interference alignment. In: Proceedings of the asilomar conference on signals, systems and computers, Pacific Grove, pp 1106–1110

32. Shen H, Li B, Tao M, Wang X (2010) MSE-based transceiver designs for the MIMO interference channel. IEEE Trans Wireless Commun 9(11):3480–3489
33. Shen M, Host-Madsen A, Vidal J (2008) An improved interference alignment scheme for frequency selective channels. In: Proceedings of the IEEE international symposium on information theory, Toronto, pp 559–563
34. Tresch R, Guillaud M, Riegler E (2009) On the achievability of interference alignment in the K-user constant MIMO interference channel. In: Proceedings of the IEEE/SP 15th workshop on statistical signal processing, Cardiff, pp 277–280

Chapter 15
Stochastic Geometry for Analysis of Coordination and Cooperation in Interference–Limited Ad Hoc Networks

Andrey Skrebtsov, Guido H. Bruck and Peter Jung

Abstract Wireless ad hoc networks with a high density of network nodes have become one of the hottest research topics in communications engineering. Due to their flexibility and relatively low infrastructure and front end expenses, wireless ad hoc networks promise a wide application spectrum. In this contribution, the authors illustrate the simulation framework used for the performance investigation of wireless ad hoc networks. With respect to the stochastic geometry, two major spatial patterns have been considered, namely the spatially homogeneous node distribution and the spatially clustered node distribution.

15.1 Introduction

Wireless ad hoc networks with a high density of network nodes have become one of the hottest research topics in communications engineering. Due to their flexibility and relatively low infrastructure and front end expenses, wireless ad hoc networks promise a wide application spectrum.

To understand their performance, the interference limited case must be considered. The instantaneous interference situation in a wireless ad hoc network depends on the current spatial distribution of the network nodes. The actual spatial distribution can be regarded as the sample of a stochastic geometrical process. It is hence a viable approach to take on the stochastic geometry tools for the generation of spatial distributions. This approach will allow both a theoretical as well as a simulation based performance analysis. The randomness of the spatial distribution of the network

A. Skrebtsov (✉) · G.H. Bruck · P. Jung
University of Duisburg–Essen, Essen, Germany
e-mail: info@kommunikationstechnik.org

G.H. Bruck
e-mail: guido.bruck@kommunikationstechnik.org

P. Jung
e-mail: peter.jung@kommunikationstechnik.org

W. Utschick (ed.), *Communications in Interference Limited Networks*,
Signals and Communication Technology, DOI 10.1007/978-3-319-22440-4_15

nodes can be decoupled from the interference analysis within a wireless ad hoc network, which eases the performance analysis [2, 6].

In this report, the authors illustrate the simulation results obtained for the performance of wireless ad hoc networks. With respect to the stochastic geometry, two major spatial patterns have been considered, namely the spatially homogeneous node distribution and the spatially clustered node distribution. The obtained simulation results for various spatial node densities are shown for both distribution types. In addition, the simulation results are verified by comparing them to available theoretical bounds. The influence of the communications channel is also considered by the model and corresponding results are illustrated in this communication.

Uncoordinated ad hoc wireless networking was studied based on the mathematical model developed in the first part of the project. Smart relaying as a cooperative data transfer method is considered in the last part of this communication.

The remainder of this report is organized as follows: In Sect. 15.2, the authors present the important aspects of the mathematical modelling for spatial network nodes distributions based on either a *Poisson Point Process* (PPP) or a *Cluster Point Process* (CPP). Section 15.3 presents selected results of this study. Section 15.4 describes the location driven smart relaying.

15.2 Stochastic Spatial Distributions

15.2.1 Overview

The spatial node distribution in ad hoc networks is mathematically modelled using the stochastic geometry. The main properties of the considered stochastic point processes are introduced in the following subsections.

15.2.2 Point Processes

The stochastic spatial distribution of the nodes in a wireless ad hoc network can be based on a random point process within the two–dimensional plane [2, 7]. In general, the distribution of the network nodes can be modelled by a PPP. In the case, when the wireless ad hoc network contains two kinds of network nodes, firstly, the dominating ones (*"immigrants"*) which are playing the role of cell centers, and, secondly, the randomly distributed subordinate nodes (*"offsprings"*) within the cells, this network structure can be modelled by a CPP. The basic properties of the both PPP and CPP are described below.

15.2.3 Poisson Point Process (PPP)

A detailed description of a PPP is given in [2, 6]. The PPP is a random process which describes the spatial distribution of the points within a two–dimensional Hilbert space [2, 6]. Although the PPP is generally valid in an infinite plane, software simulations within a plane of limited size, i.e. within a compact (bounded and limited) set within a Hilbert space $B \in \mathbb{R}^N$, N being the dimension of the space, are possible, also. In this study, the two–dimensional space is considered, i.e. N is equal to 2. The points enclosed within the particular compact set are representing a given PPP *realization*. The number of these points is limited due to the properties of the compact set and the definition of the PPP [2].

In a Hilbert space, the coordinates of each point, i.e. of each network node in our case, can be uniquely assigned by the PPP *realization* [2, 6]. The PPP is said to be *homogeneous* if the collection of the i-th coordinate among all point realized by the PPP is randomly distributed and the probability density function is uniform in every i-th dimension [2, 6]. In the remainder of this report, a homogeneous PPP is considered in order to allow a compact analysis.

From a practical point of view, a PPP can be represented as a superposition of two independent random processes [2, 6]. The first process describes the number of the points, i.e. the network nodes, that have "arrived" at the *observation* region [2, 6]. The second random process represents the uniform random distribution of the network nodes within the observation region [2, 6].

First, the authors discuss the "arrival" situation. A classical Poisson distribution with a discrete probability mass function given by

$$f(k, \lambda) = \frac{\lambda^k e^{-\lambda}}{k!} \tag{15.1}$$

is assumed, where λ is the average number of network nodes which have "arrived" in our compact set $B \in \mathbb{R}^2$, $k \in \{0, 1, 2, \dots\}$ within a unit time interval. In other words, λ is the (average) "arrival rate" (Fig. 15.1).

The second mentioned random process represents the uniform random distribution of the network nodes within the observation region [2, 6]. This means that the average density of network nodes must be the same for any compact set in the Hilbert space $\mathbb{R}^2$. Therefore,

$$\lambda = \frac{\Lambda(B)}{\mu_{\mathrm{L}}(B)}, \tag{15.2}$$

where $B \in \mathbb{R}^2$ is such a compact set in the Hilbert space, $\Lambda(B)$ is the random number of spatially Poisson distributed network nodes within the set B, and $\mu_{\mathrm{L}}(B)$ is a *Lebesgue* measure of the set B [2, 6]. Since the two–dimensional space is considered in this study, the spatial node density has physical unit $1/m^2$.

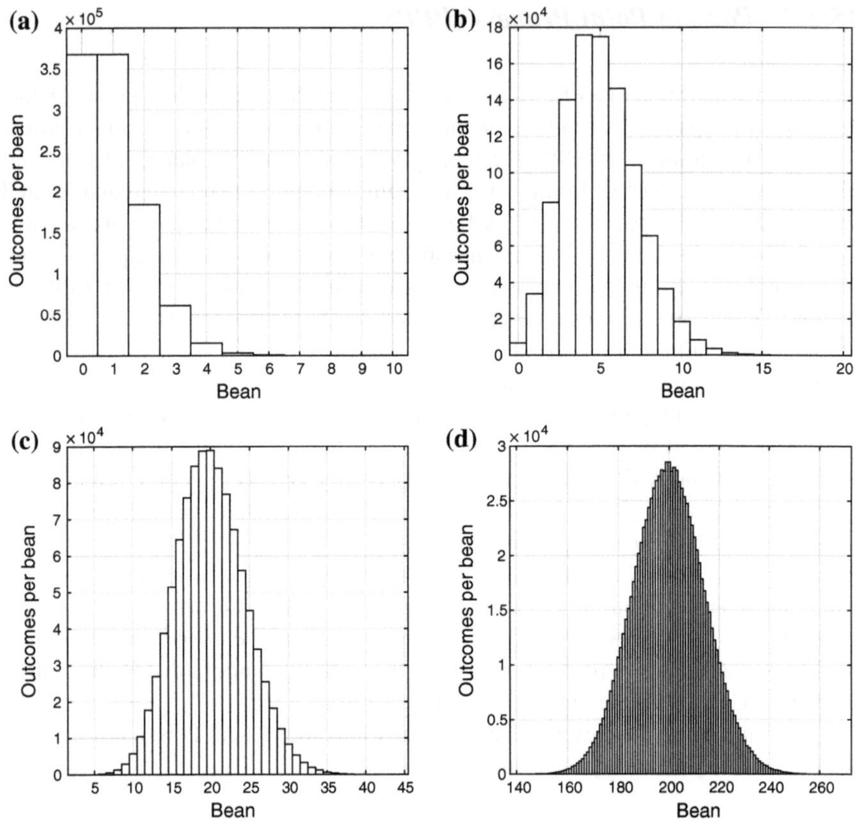

Fig. 15.1 **a** Poisson *Probability Mass Function* (PMF) for $\lambda = 1$, $N_{\text{outcomes}} = 10^6$, **b** Poisson PMF for $\lambda = 5$, $N_{\text{outcomes}} = 10^6$, **c** Poisson PMF for $\lambda = 20$, $N_{\text{outcomes}} = 10^6$, **d** Poisson PMF for $\lambda = 200$, $N_{\text{outcomes}} = 10^6$

15.2.4 Cluster Point Process (CPP)

As already mentioned, the CPP is that random process which describes the spatial distribution of the points, i.e. of the network nodes, when two kinds of points, i.e. of network nodes, exist [2, 6]. These two different kinds of points are the following [2, 6]: the dominant points which are the cluster centers or *immigrants* and the subordinate points distributed near an immigrant or *offsprings*, using the generally accepted terminology taken from biology. The spatial distribution of the immigrants is based on a PPP with the density λ_{imm}, measured in $1/m^2$.

Two general types of offspring distributions can be outlined, namely the Gaussian (normal) distribution with $R_{\text{cluster, normal}}$ being the mean distance between any node and its related cluster center and the uniform distribution within the disk of a given cluster radius $R_{\text{cluster, uniform}}$ [2, 6]. The first kind of the two offspring distributions, i.e. the Gaussian (normal) distribution of the nodes around cluster center, is related to that particular scenario, when there is a dominant either stationary or mobile access point and the subordinate nodes are gathering around this cluster center under the effect of the some physical parameter, for example signal strength [2, 6].

The second kind of offspring distributions, i.e. the uniform distribution of the nodes within a limited area around each immigrant, relates to that scenario, when the network nodes are limited by some environment topology and tend to distribute equidistantly within given space limits [2, 6]. This is e.g. the case when sensors on the human body, wireless devices within a room, mobile wireless users in the building, and mobile station in the urban environment are modelled [2, 6].

Since these two models seem to cover the most relevant scenarios, more exotic offspring distributions are not covered in this study [2, 6].

Within this study the authors assume that every immigrant has the same average (expected) number of offsprings [2, 6]. This mean number of offsprings is denoted by $\lambda_{\text{offspring}}$ [2, 6]. Applying this assumption, the number of network nodes in any cluster in each CPP realization is a Poisson distributed random process with the density parameter $\lambda_{\text{offspring}}$ [2, 6].

The distribution of network nodes' coordinates is not uniform and, in general, the CPP is not homogeneous [2, 6]. But it can be shown, that for the limiting case when the cluster radius approaches infinity, i.e.

$$R_{\text{cluster}} \to \infty\,, \tag{15.3}$$

the overall distribution of the nodes will converge to the superposition of an infinite set of PPP realizations [2, 6]. It follows from [2] that such a superposition is itself a PPP realization, and the overall network node density is given by

$$\lambda_{\text{CPP}} = \lambda_{\text{imm}} \left(\lambda_{\text{offspring}} + 1\right)\,, \tag{15.4}$$

where λ_{CPP} is measured in $1/m^2$ in the two–dimensional case considered through this study. This property that a superposition is itself a PPP realization can be used for verification purposes, since the theoretical bounds for the PPP can be calculated analytically in closed form [4].

15.2.5 Mathematical Model Inputs

The properly selected simulation region defines the mathematical modelling accuracy. Two general properties were taken into account during the selection of the modelling region geometry. Firstly, the region shall be a part of the two–dimensional plane and it shall be large enough to enclose enough network nodes in order to allow a proper statistical analysis. Secondly, the homogeneity of the spatial distribution of the modelled network nodes is assumed, i.e. the properties of the modelled network

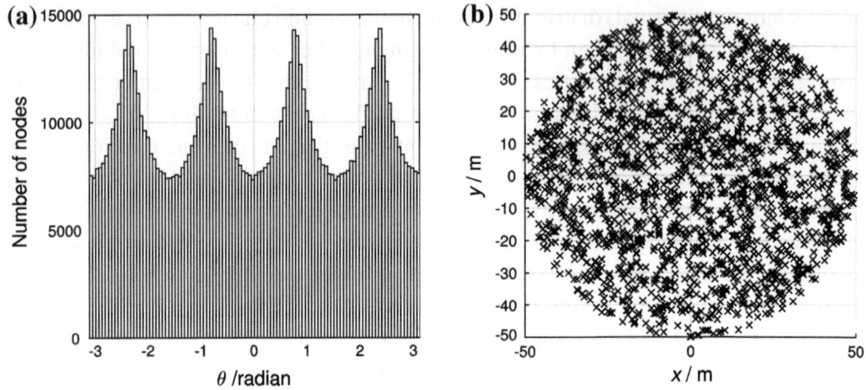

Fig. 15.2 **a** The radial distribution of the nodes for a *homogeneous* PPP, bounded by a two–dimensional ideal square, **b** PPP realization for 2000 nodes over a two–dimensional disk with radius of 50 m

shall not expose any anisotropy or dependence on the orientation of the reference receiver and transmitter pair relative to the coordinates origin and axes.

The first property defines the relatively large centrally–symmetrical two–dimensional object, where the reference receiver is placed into its geometrical center, the second requirements excludes any polygonal figures such as e.g. triangles, squares or hexagons which might by typical for idealized models of networks, having the cellular spatial structure.

Figure 15.2a illustrates the chosen structural effects in the case of an ideal square region. Figure 15.2a shows the distribution of 10^6 network nodes inside an ideal square with respect to the angle of the polar coordinate θ (the coordinate origin is placed into the geometrical center of the simulation region, the area of the simulation region is split into 100 equally spaced sectors). Since the density of network nodes differs for different values of θ, the corresponding interference, simulated over many point process realizations, will also depend on the coordinate θ, and, hence, it is not isotropic. The same histogram can be built for any polygonal simulation region.

However, the only geometrical figure which will yield an isotropic interference pattern is the ideal disk with the reference receiver located at the geometrical center. To model the PPP, the given number of network nodes must be equally distributed inside a simulation region. In the case of the ideal two–dimensional disk the coordinates distribution can be described by *probability density function* (PDF) of two polar coordinates r, θ:

$$f(r,\theta) = f_r(r) f_\theta(\theta) \,. \tag{15.5}$$

In the case of an ideal disk with iid (independent and identically distributed) random variables one yields

$$f_\theta(\theta) = \begin{cases} 1/(2\pi) & \text{if } 0 \le \theta < 2\pi, \\ 0 & \text{otherwise.} \end{cases} \tag{15.6}$$

The PDF $f_r(r)$ can be derived from its *cumulative density function* (CDF). It follows from (15.2) that the mean number of network nodes for a given bounded set can be written as:

$$\langle N \rangle_{\text{nodes}} = \lambda \pi r^2. \tag{15.7}$$

Consider the ideal disk with the overall radius R_d, and the coordinates origin in its geometrical center. The CDF for such system can be written as:

$$F_r(r) = \begin{cases} \langle N \rangle_{\text{nodes}}(r) / \langle N \rangle_{\text{nodes}}(R_\text{d}) & \text{if } 0 \le r \le R_\text{d}, \\ 0 & \text{otherwise}, \end{cases} \tag{15.8}$$

where $\langle N \rangle_{\text{nodes}}(r)$ is the mean number of network nodes inside the disk with the radius r. Taking into account (15.7), we find

$$F_r(r) = \begin{cases} r^2 / R_\text{d}^2 & \text{if } 0 \le r \le R_\text{d}, \\ 0 & \text{otherwise}. \end{cases} \tag{15.9}$$

The PDF $f_r(r)$ of the PPP for the ideal two–dimensional disk can be written as

$$f_r(r) = \frac{\text{d}F_r(r)}{\text{d}r} = \begin{cases} 2r / R_\text{d}^2 & \text{if } 0 \le r \le R_\text{d}, \\ 0 & \text{otherwise}. \end{cases} \tag{15.10}$$

A random variable with the PDF given by (15.10) is required to describe the distribution of the random coordinate r for the PPP on the ideal two–dimensional disk. The random coordinate r can be generated by setting out from a uniform distribution which can be easily obtained in many mathematical packages. Setting out from a simple uniformly iid random variable x with PDF

$$f_x(x) = \begin{cases} 1 & \text{if } 0 \le x \le 1, \\ 0 & \text{otherwise}, \end{cases} \tag{15.11}$$

A transformation must be derived which converts the uniformly iid random variable x into the desired random variable (coordinate) r. To obtain the transformation function $r(x)$ a differential equation has to be solved. Since x is a random variable, the values of the function r being equal to $r(x)$ is also a random variable. The relation between the probability density functions of random variables x and r is described by (15.12):

$$f_r(r)\,\text{d}r = f_x\big(x(r)\big)\text{d}x \quad \text{for } x \ge 0\,. \tag{15.12}$$

In our case, both PDFs are known. Hence, the transformation function $r(x)$ can be easily obtained. Taking into account (15.10) and (15.11),

$$\begin{aligned} 2\int_{u=0}^{r} u\mathrm{d}u &= R_{\mathrm{d}}^2 \int_{v=0}^{x} \mathrm{d}v \\ \Leftrightarrow \quad u^2|_{u=0}^{r} &= R_{\mathrm{d}}^2 u|_{u=0}^{x} \\ \Leftrightarrow \quad r(x) &= R_{\mathrm{d}}\sqrt{x} \quad (\text{for } x \geq 0) \end{aligned} \tag{15.13}$$

follows from (15.12).

A PPP on the two–dimensional disk can be modelled in combined manner. The number of network nodes inside the disk is a random process with a Poisson distribution, the mean number of nodes λ can be calculated by (15.2). Since the number of network nodes inside the simulation region is known, the polar coordinates of the particular network nodes can be generated by the iid random generator and then be transformed according to (15.6) and (15.13).[1] An example is shown in Fig. 15.2b where the disk radius R_{d} is equal to 50 m, and the number of network nodes is equal to 2000.

15.3 Software Modelling for Stochastic Networks

15.3.1 Overview

In this section, we consider the modelling results and how they relate to expected theoretical findings.

15.3.2 Model Convergence Condition

Since the number of calculation resources is limited the estimation of a reasonable number of network nodes for adequate statistics will be given in this subchapter. Skrebtsov et al. [5] shows that, in general, the mean interference for infinite two–dimensional plane converges if the pathloss exponent α is greater than two.

In Fig. 15.3a the relationship between the mean interference and the number of considered network nodes is shown. Figure 15.3a shows that in the case of pathloss exponents α being greater than three the absolute value of the mean interference converges relatively fast, but in the low pathloss exponent case, i.e. for pathloss exponents α between two and three, the number of considered network nodes to approach the theoretical value for infinite number of network nodes is moderately high. Figure 15.3b illustrates this property more precisely. This figure represents a normalized version of the dependence of the modeled interference on the number of network nodes within a given simulation region. It follows that the mean interference converges moderately fast for high values of the pathloss exponent, e.g. $\alpha = 3.8$

[1] This holds because the PDF‘s $f_r(r)$ and $f_x(x)$ are equal to 0 for $r < 0$ and $x < 0$, respectively.

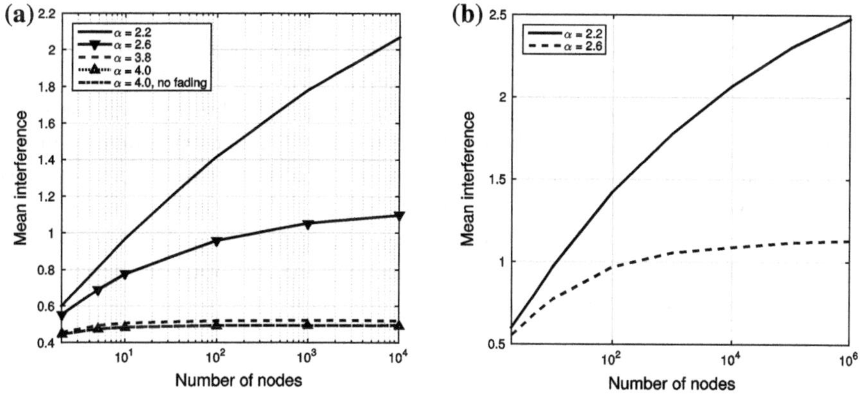

Fig. 15.3 **a** Mean interference, simulated for a two–dimensional disk with respect to the number of network nodes, **b** Mean interference, simulated for a two–dimensional disk with respect to the number of network nodes for low pathloss exponent $\alpha < 3$

and $\alpha = 4.0$: 10^2 network nodes are producing 99 % of the overall interference. The number of network nodes shall thus be larger than 10^4 for lower values of the pathloss exponent ($2 < \alpha \leq 3$).

Figure 15.4a shows the case for a low pathloss exponent $\alpha = 2.6$. Here, a large number of network nodes ($N_{\text{node}} > 10^6$) is recommendable.

A further limitation of the maximum number of network nodes comes from the computation resources limits and forms the application scenario. During this study it was empirically found that the used software development tools and selected algorithms are limiting the maximum number of network nodes to approximately 10^6 network nodes. Normally. the considered uncoordinated ad hoc wireless networks

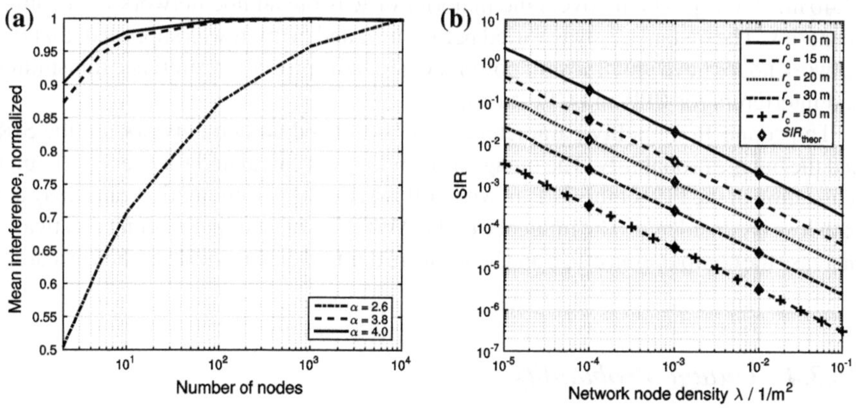

Fig. 15.4 **a** Normalized mean interference with respect to the network nodes number for different pathloss exponents α, **b** Poisson Point Process, *SIR* for different r_c

are limited in some geographical area according to considered application scenario, e.g. rural, urban etc. The number of network nodes in this case is limited by the area of this simulation region and the unit node density λ. The third limiting factor is the self noise produced in the reference receiver. In the case the density of the network nodes is low, the self noise of the reference receiver becomes dominating in the overall signal disturbances, caused by both, interference and noise. In practice, this property governs the maximum achievable SINR (signal to interference and noise ratio) for the low unit node density values, e.g. during this study it was found, that for the unit densities $\lambda < 10^{-6}/\mathrm{m}^2$ the influence of the self noise of the reference receiver becomes dominating in the SINR.

15.3.3 Simulated and Theoretical Bounds Comparison

As already pointed out, the theoretical bounds of the SIR (signal-to-interference ratio) for a homogeneous PPP can be analytically calculated for infinite two–dimensional plane [4, 5]. Although the simulator developed by the authors considers the observation region of a finite size, this size can be made large enough to mimic an infinitely large space.

The following conditions were considered by the simulations: The observation region has the form of an ideal two–dimensional disk with $R_{\mathrm{disk}} \gg r_{\mathrm{c}}$, where r_{c} is a distance between the reference receiver and the reference transmitter. The disk radius R_{disk} was determined for each spatial density in such a way that the observation region always contains 10^5 network nodes. The quantitative values of spatial densities of the disturbing nodes were selected to correspond the real life communication scenarios taken from the interval $\left[10^{-5}/\mathrm{m}^2; 10^{-1}/\mathrm{m}^2\right]$. In the case the interfering network nodes are the nodes of an ideal hexagonal grid, the spatial density represents the distance between neighbouring nodes with adjacent distances taken from the interval [3.40 m; 340 m], which covers the majority of feasible ad hoc network application scenarios. The distance r_{c} was varied between 10 and 50 m. It was supposed, that the network nodes have unit transmit power, i.e. the same power for reference transmitter and interferers for PPP and CPP cases.

The connection quality parameters of the simulated ad hoc network are the SIR and the outage probability. The outage probability was calculated for every particular realization of the PPP or the CPP. In what follows, the signal attenuation caused by the path loss is considered. The influence of the AWGN or of multipath fading drastically increase the computation effort and hence were omitted during the described research.

15.3.4 Outage Probability

In this section, the obtained simulation results for the outage probability are presented. Figure 15.4b depicts the dependence between the SIR and the spatial node density

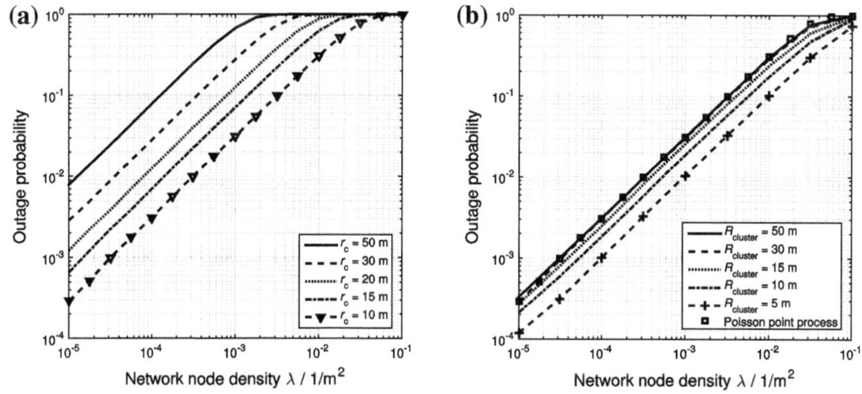

Fig. 15.5 **a** PPP, outage probability for different r_c, **b** CPP, outage probability with respect to spatial node density and fixed number of the offsprings per cluster ($\rho = 20$)

for the different distances between the reference receiver and transmitter for the PPP. The analytically calculated values of the SIR for the wireless network which represent a homogeneous PPP on the infinite two–dimensional plane are also shown in the Fig. 15.4b, represented by diamond symbols and being referred to as SIR_{theor}. The analytical expressions can be found in [5, 7]. The analytically calculated values match the simulation results.

In the Fig. 15.5a, the simulation results for the outage probability in PPP based scenarios are shown, in general, the represented figure is a log scaled CDF, averaged over 10^6 PPP realizations per each value in the plot. An outage is defined as a situation, in which the interference power exceeds a defined quality of service criterion which can be represented by a threshold. In the considered case, the threshold is set to be 10^{-2}. The dependence between the spatial node density and the outage probability for different distances between reference receiver and transmitter is illustrated. The simulation results confirm the intuitive conception that in more tightly populated networks the outage probability is greater.

In the Fig. 15.5b, the simulation results for CPP based scenarios are presented, when a uniformly distributed CPP is assumed. Again, the outage probability is shown, here with respect to spatial node density. The number of offspring nodes per cluster is fixed ($\rho = 20$). In the simulations, the path loss attenuation is considered, and the path loss function is $1/\left(\|\mathbf{r}_{\text{rx}} - \mathbf{r}_{\text{tx}}\|^{\alpha} + 1\right)$, with the attenuation coefficient α being equal to four which is in accordance with commonly accepted channel models; $\mathbf{r}_{\text{rx}}$, $\mathbf{r}_{\text{tx}}$ are the vectors, representing the actual coordinates of the reference receiver and transmitter, correspondingly. In this case, the PPP results can be considered as upper bounds for the SIR and outage probability, and the CPP model verification can be done. The spatial node density for the PPP corresponds to the combined spatial density of CPP which is represented in [5].

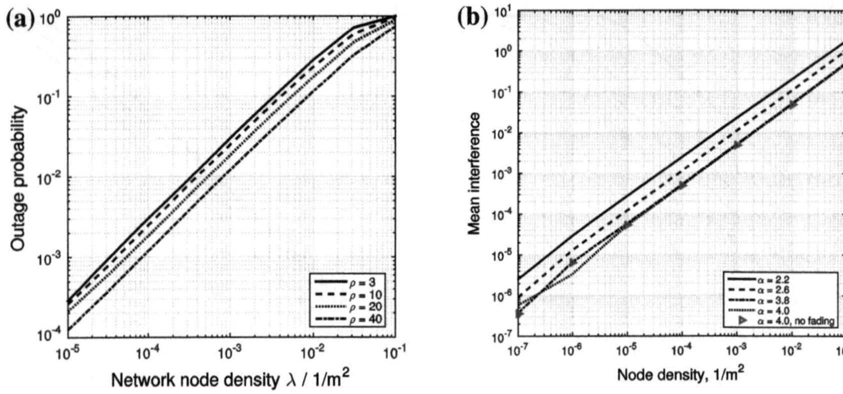

Fig. 15.6 **a** CPP, outage probability with respect to spatial node density and fixed average offspring node number per cluster ($\rho = 3,\ 10,\ 20,\ 40$), **b** Mean interference power with respect to unit node density for $\alpha = 2.2$ (*continuous line*), $\alpha = 2.6$ (*long dashed line*), $\alpha = 3.8$ (*dotted-dashed line*), $\alpha = 4.0$ (*dotted line*) and $\alpha = 4.0$ (*no line, triangular markers*), non–fading case

In the Fig. 15.6a, the simulation results for the outage probability in the CPP are presented with respect to the spatial node density. The number of offspring nodes per cluster is fixed. The cluster population factor ρ, i.e. the average number of the offspring nodes per immigrant node, varies. The equivalent spatial density is calculated by (15.14). The cluster radius is equal to 20 m. The spatial density of the cluster centers (immigrant nodes) was varied accordingly to keep the overall spatial network node density constant for the different values of the cluster population factor ρ. The simulation results illustrated in the Fig. 15.6a show that highly populated clusters with limited linear size produce less interference when the overall spatial node density is kept constant. This result can be explained by the fact that the spatial node density is reciprocal to the square of the mean inter–node distance $\left(\lambda \propto \langle d\rangle_{\text{node}}^{-2}\right)$. The corresponding interference is lower for highly populated clusters under the condition that overall node density is constant, but the absolute difference is not large enough to effectively combat the interference.

15.3.5 *Influence of Fading*

Fading is an important received signal quality degrading factor in wireless data transmission. Hence, the impact of fading on the overall interference is of major interest. The modelling of the fast fading in wireless networks is described in [4].

In the Fig. 15.6b the relationship between the mean interference within the network and the node density is shown. For $\alpha = 4$, both cases with and without fading are considered. In the target range of unit node densities $10^{-5}/\text{m} \leq \lambda \leq 10^{-1}/\text{m}$ the observed behavior is identical. This result is intuitively expected, since the mean

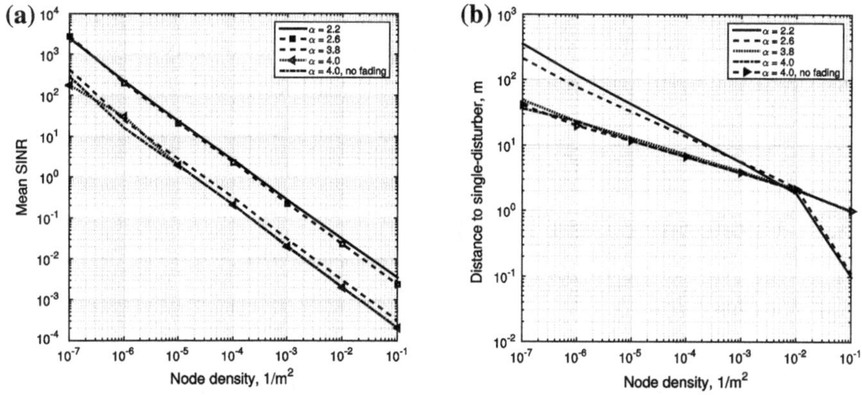

Fig. 15.7 **a** SINR with respect to unit node density for $\alpha = 2.2$ (*continuous line*), $\alpha = 2.6$ (*long dashed line, square markers*), $\alpha = 3.8$ (*long dashed line*), $\alpha = 4.0$ (*dotted line, triangular markers*) and $\alpha = 4.0$ (*dotted-dashed line*), non–fading case, **b** Distance to single equivalent disturbing node with respect to unit node density for $\alpha = 2.2$ (*continuous line*), $\alpha = 2.6$ (*long dashed line*), $\alpha = 3.8$ (*dotted line*), $\alpha = 4.0$ (*dotted-dashed line*) and $\alpha = 4.0$ (*long dashed line, triangular markers*), non–fading case

power is normalized to unit power, same as for the non–fading case, and the mean interference was obtained by averaging over $N = 10^6$ realizations of the PPP.

In the Fig. 15.7a the relationship between the mean SINR and the node density is shown. For $\alpha = 4$, both cases with and without fading are shown. For node density values between $10^{-5}/\text{m} \leq \lambda \leq 10^{-1}/\text{m}$, which are most interesting within the scope of this study, a similar behaviour can be expected.

15.3.6 Equivalent Interference

In some scenarios it is useful to estimate how the useful signal power of the reference transmitter and the overall interference produced by the rest of the network nodes relate to each other. By the assumption that the channel model is the same for every network node within the simulation region the mean interference, averaged over many PPP realizations, can be represented by the equivalent disturbing nodes, their distance to the reference receiver being represented as a function of mean interference with respect to unit node density of the network nodes:

$$r_{\text{eqv}} = h^{-1}\left(\langle \mathcal{I} \rangle\right); \tag{15.14}$$

where $\langle \mathcal{I} \rangle$ is the mean interference and h^{-1} is the inverse function of the channel coefficient. In the Fig. 15.7b the results for different values of the pathloss exponents are shown, the distances to the equivalent disturbing nodes are calculated using the simulation results with respect to unit node density. According to the Fig. 15.7b, the distance to the equivalent single disturbing nodes for high density networks (unit node density $\lambda > 10^{-1}/\text{m}^2$) is less than 1 m for every considered pathloss exponent $\alpha \in (2; 4]$.

15.4 Cooperative Data Transfer

15.4.1 Overview

Cooperative data transfer is an important topic within the scope of achieving the maximum network capacity. In this chapter, we define the conditions on the network and show the statistics for ready–to–relaying nodes with respect to the distribution of the network nodes.

15.4.2 Statistical Approach for Location–Driven "Smart" Relaying

The interference produced by concurrently transmitting network nodes, the additive noise and the fading can drastically limit the communication range within the wireless ad hoc network. Cooperative data transfer is one of the methods to improve the communication range between the network nodes. The most straightforward technique is relaying. Relaying can be relatively simple (amplify and forward) or more sophisticated (decode and forward), single hop and multi hop. More sophisticated cooperative data transfer techniques like network coding or receive cooperative diversity are promising quality improvements but are not considered here. More information on these aspects can be found in [1, 3].

Relaying techniques expose the tradeoff between effective utilization of the available radio resources and the data transfer success probability. In the case that the distance between the receiver and the transmitter is too large, the relaying nodes retransmit the messages being sent by receiver hence producing additional traffic within the network and increasing the overall interference. To find a balance between relaying advantages and disadvantages, a proper relay selection mechanism is required. In the following subchapters, relay selection for stochastically distributed wireless ad hoc networks is proposed and some properties are considered.

Several assumptions are made for the relaying mechanism: Every node collects the information about transactions, the transmitter can mark the message as "relaying desirable" in case the estimated distance to the receiver is greater than a given threshold or the distance to the receiver is unknown to the transmitter. In this study, omnidirectional antennas are considered. Further performance improvements are expected by smart antenna technologies, e.g. by beamforming.

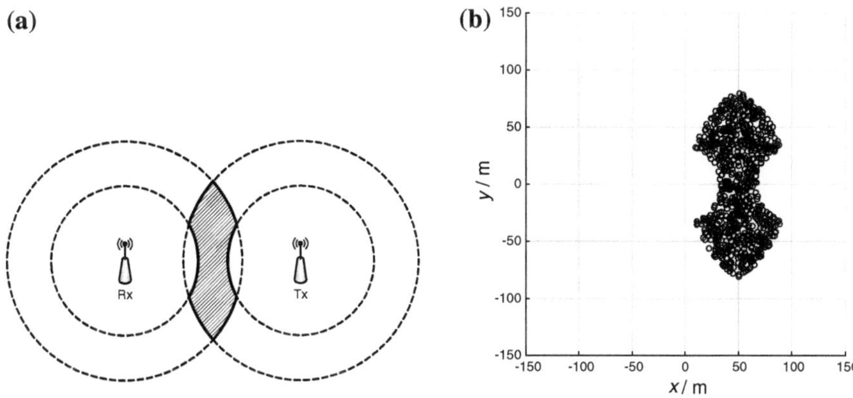

Fig. 15.8 **a** Optimal relay position at the power constraint area intersection, **b** Distribution of the ready–to–relaying network nodes (o-marked points), receiver is at $(0\,\mathrm{m},\ 0\,\mathrm{m})$, transmitter is at $(100\,\mathrm{m},\ 0\,\mathrm{m})$, non–fading case

Hence, the information about the direction is not available to the receiver nodes, the only way to estimate the distance between transmitters and receivers is the measuring of the signal strength at the receiver antenna. Then, using the appropriate pathloss formula, it is possible to estimate the spatial region in which the measured power at the potential relay candidate network nodes fulfills the established constraints.

In the Fig. 15.8a the "optimal" position of the relay candidate node is shown for the ideal case, when the signal power depends only on the distance between the receiver and the transmitter and no fading is present. The areas of the rings bounded by the dashed circles around the receiver and the transmitter, respectively, include those network nodes, for which the estimated signal power lies within the intervals $p_{\mathrm{rx,\,lower\ bound}} < p_{\mathrm{rx,\,est}} < p_{\mathrm{rx,\,upper\ bound}}$, $p_{\mathrm{tx,\,lower\ bound}} < p_{\mathrm{tx,\,est}} < p_{\mathrm{tx,\,upper\ bound}}$. The intersection of the two rings contains those network nodes, for which both power conditions are fulfilled. These network nodes are the best candidates for relaying.

In the Fig. 15.8b the simulated node distribution is shown for the non–fading case. The receiver is placed at $(0\,\mathrm{m},\ 0\,\mathrm{m})$, the transmitter is at $(100\,\mathrm{m},\ 0\,\mathrm{m})$. The constraint for the relay node candidate nodes is identical. In the shown case the optimal relay region is symmetrical with respect to the abscissa and it is equidistant to the receiver and the transmitter. Note that the number of network nodes inside the simulation region was deliberately increased to highlight the area boarders.

15.4.3 Location–Driven Relaying: Influence of Fading

The examples discussed above consider the non–fading wireless channel. In this case the bounds of the area with ready–to–relaying network nodes can be analytically

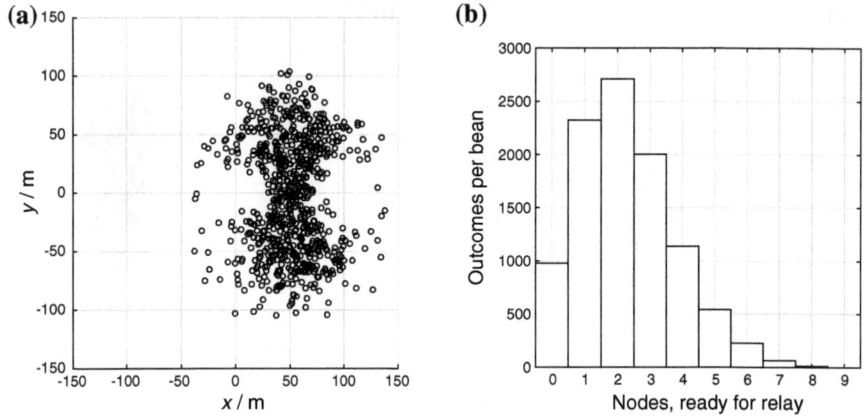

Fig. 15.9 **a** Distribution of the ready–to–relaying network nodes (o-marked points) under the influence of fast Rayleigh fading, receiver is at (0 m, 0 m), transmitter is at (100 m, 0 m), **b** Distribution of ready–for–relay nodes, $0.25\,p_{\mathrm{tx,\,ref}} \leq p_{\mathrm{tx}}$, $p_{\mathrm{rx}} \leq 4.0\,p_{\mathrm{tx,\,ref}}$, fast Rayleigh fading

defined by considering the pathloss [5]. Under real channel conditions, however, more sophisticated signal disturbances are present, e.g. fast frequency–selective fading.

In the Fig. 15.9a the distribution of the ready–to–relaying network nodes is shown. The channel has the same pathloss exponent and power constraints as those for the Fig. 15.8b, the only difference is the presence of fast Rayleigh fading. Due to the Rayleigh fading the ready–to–relaying nodes are no longer bounded by straight geometrical boarders.

If the network is uncoordinated, the network relaying candidate nodes have no possibility to decide, which (if any) of the ready–to–relaying nodes will retransmit the messages sent by the transmitter. The exact number of desirable relay nodes depends on the used relay technique, but in any case too many relay nodes which are ready to transmit simultaneously shall cause collisions and unnecessary power consumption. In order to reduce the number of the relaying nodes it is proposed to configure the ad hoc network based on the properties of the stochastic spatial distribution of the nodes, enabled by the "fine tuning" of the power constraint thresholds to achieve (if possible) the given success probability and desirable statistical distribution of ready–to–relaying nodes.

15.4.4 Simulation Results

In the Fig. 15.9b the distribution of the ready–to–relaying nodes for fast Rayleigh fading is shown. The following power constraints for the distance estimation were used:

- equidistant symmetrical relay node area ($p_{tx}/p_{rx} = 1$);
- $p_{tx,\,low} = p_{rx,\,low} = 0.25 \cdot p_{tx,\,ref}$, where $p_{tx,\,ref}$ is the signal strength of the transmit signal measured at the position of the reference receiver located in the center;
- $p_{tx,\,high} = p_{rx,\,high} = 4.0 \cdot p_{tx,\,ref}$;
- the mean number of *ready–to–relaying* network nodes is $\langle N \rangle_{node} = 2.27$;
- the node density is $\lambda = 10^{-3}/m^2$;
- fast Rayleigh fading.

In the Fig. 15.10a the distribution of the ready–to–relaying nodes for fast Rayleigh fading. Tighter power constraints for the distance estimation were used:

- equidistant symmetrical relay node area ($p_{tx}/p_{rx} = 1$);
- $p_{tx,\,low} = p_{rx,\,low} = 0.33 \cdot p_{tx,\,ref}$;
- $p_{tx,\,high} = p_{rx,\,high} = 3.0 \cdot p_{tx,\,ref}$;

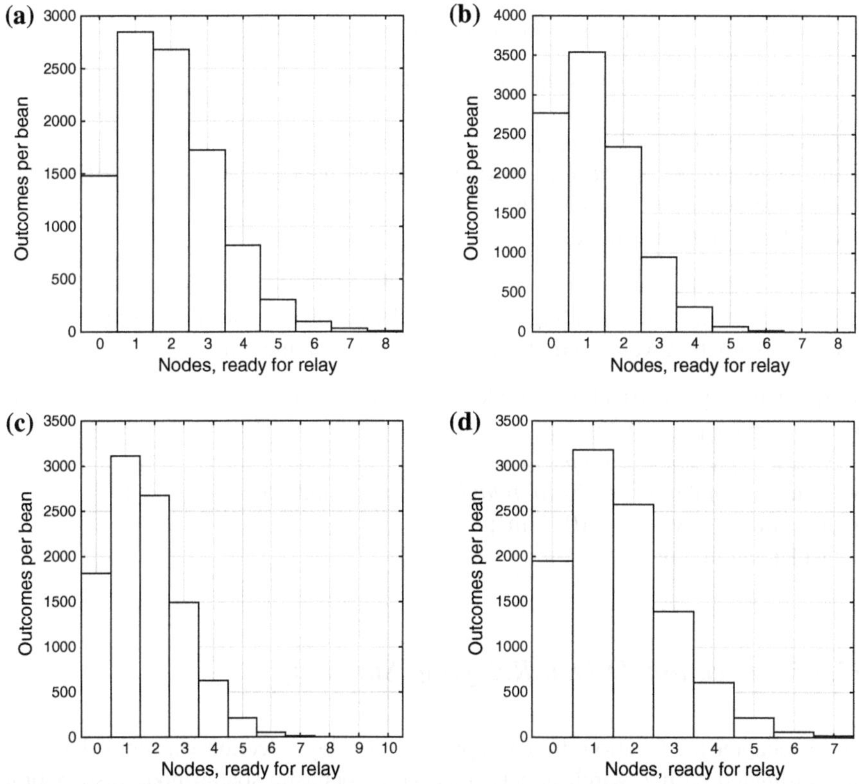

Fig. 15.10 **a** Distribution of ready–for–relay nodes, $0.33\,p_{tx,\,ref} \le p_{tx}\,,\ p_{rx} \le 3.0\,p_{tx,\,ref}$, fast Rayleigh fading, **b** Distribution of ready–for–relay nodes, $0.5\,p_{tx,\,ref} \le p_{tx}\,,\ p_{rx} \le 2.0\,p_{tx,\,ref}$, fast Rayleigh fading, **c** Distribution of ready–for–relay nodes, $0.5\,p_{tx,\,ref} \le p_{tx}\,,\ p_{rx} \le 2.0\,p_{tx,\,ref}$, non–fading channel, **d** Distribution of ready–for–relay nodes, $0.5\,p_{tx,\,ref} \le p_{tx}\,,\ p_{rx} \le 2.0\,p_{tx,\,ref}$, fast Rician fading

- the mean number of *ready–to–relaying* network nodes is $\langle N \rangle_{\text{node}} = 1.92$;
- the node density is $\lambda = 10^{-3}/\text{m}^2$;
- fast Rayleigh fading.

In the Fig. 15.10b the distribution of the ready–to–relaying nodes for fast Rayleigh fading is shown. In comparison to the previous two cases, weakened power constraints for the distance estimation were used:

- equidistant symmetrical relay node area ($p_{\text{tx}}/p_{\text{rx}} = 1$);
- $p_{\text{tx, low}} = p_{\text{rx, low}} = 0.25 \cdot p_{\text{tx, ref}}$;
- $p_{\text{tx, high}} = p_{\text{rx, high}} = 2.0 \cdot p_{\text{tx, ref}}$;
- the mean number of *ready–to–relaying* network nodes is $\langle N \rangle_{\text{node}} = 1.30$;
- the node density is $\lambda = 10^{-3}/\text{m}^2$;
- fast Rayleigh fading.

In the Fig. 15.10c the distribution of the ready–to–relaying nodes of the software modelling for the non–fading case is shown. The following power constraints for the distance estimation were used during this software simulation:

- equidistant symmetrical relay node area ($p_{\text{tx}}/p_{\text{rx}} = 1$);
- $p_{\text{tx, low}} = p_{\text{rx, low}} = 0.25 \cdot p_{\text{tx, ref}}$;
- $p_{\text{tx, high}} = p_{\text{rx, high}} = 2.0 \cdot p_{\text{tx, ref}}$;
- the mean number of *ready–to–relaying* network nodes is $\langle N \rangle_{\text{node}} = 1.70$;
- the node density is $\lambda = 10^{-3}/\text{m}^2$;
- non–fading channel model.

In the Fig. 15.10d the distribution of the ready–to–relaying nodes of the software modelling for fast Rician fading is shown. The following power constraints for the distance estimation were used during this software simulation:

- equidistant symmetrical relay node area ($p_{\text{tx}}/p_{\text{rx}} = 1$);
- $p_{\text{tx, low}} = p_{\text{rx, low}} = 0.25 \cdot p_{\text{tx, ref}}$;
- $p_{\text{tx, high}} = p_{\text{rx, high}} = 2.0 \cdot p_{\text{tx, ref}}$;
- the mean number of *ready–to–relaying* network nodes is $\langle N \rangle_{\text{node}} = 1.64$;
- the node density is $\lambda = 10^{-3}/\text{m}^2$;
- fast Rician fading.

15.4.5 Location–Driven Relaying: Summary

In this chapter the location driven approach for relay selection was considered for uncoordinated ad hoc wireless networks with stochastic spatial node distribution. Some conclusions can be drawn based on the obtained results.

The optimal position for ready–to–relaying network nodes is highly dependent on the selected channel model, especially in the presence of fading. The software modelling has shown that the largest mean number of ready–to–relaying network nodes (averaged over multiple PPP realizations) is observed for the non-fading case.

Rayleigh fading reduces the mean number of ready–to–relaying nodes and, at the same time, these nodes are distributed over a larger area within the simulation region. The mean number of the ready–to–relaying nodes and their spatial distribution lies in between those for the non-fading case and the Rayleigh fading case when Rician fading is considered.

A further important property of the spatial node distribution taken into account during the software modelling is the PMF. As it was already mentioned, in the uncoordinated network it is recommendable to keep the number of ready–to–relaying nodes as low as possible in order to reduce the number of collisions and the extra overhead traffic within the ad hoc network. The software modelling results have shown, that it happens quite often (up to about 20 % of all PPP realizations) that there are no nodes which are fulfilling the relay node conditions. In the case of a desirably low relaying outage rate, the relay selection constraint can be weakened. In this situation, additional effort is required to arrange the relay selection (election) among the relay candidates. In every considered case, the relaying success probability lies between 0.71 and 0.81, i.e. $0.71 < P_{\text{relay}} < 0.81$, for $N_{\text{node}} \in [1\,;\,3]$.

Acknowledgments This work was supported by the German Research Foundation (DFG) within the Priority Program 1397 COIN under Grants JU 2603/19-1 and JU 2603/19-2. Also, the authors gratefully acknowledge the superb collaboration with their colleagues at the Karlsruhe Institute of Technology, especially Professor Dr. Friedrich K. Jondral, Dr. Holger Jäkel and Ralph Tanbourgi. In addition, the typesetting support by Sebastian Rickers is gratefully acknowledged.

References

1. Altieri A, Vega R, Galarza C, Piantanida P (2012) On the balance between cooperation and interference in dense wireless networks. In: 2012 international symposium on wireless communication systems (ISWCS), Paris, France
2. Baccelli F, Blalszczszyn B (2009) Stochastic geometry and wireless networks, volume 1 + 2. Found Trends Netw 3(3–4):249–449
3. Behnad A, Rabiei A, Beaulieu N (2012) Outage probability of dual–hop af opportunistic relaying with directional antennas. In: 2012 IEEE international conference on communications (ICC), Ottawa, Canada
4. Haenggi M, Ganti RK (2008) Interference in large wireless networks. Found Trends Netw 3(2):127–248
5. Skrebtsov A, Bai Z, Bruck GH, Jung P (2013) A novel network simulator based on stochastic spatial models. In: IEEE ICSPCS 2013, Gold Coast, Australia
6. Stoyan D, Kendall WS, Mecke J (1995) Stochastic geometry and its application, 2nd edn. Wiley
7. Tanbourgi R, Elsner JP, Jondral FK (2011) Lowering outage probability in ad hoc networks by nearest neighbour. In: IEEE WCNC 2011, Cancun, Mexixo

Chapter 16
Cooperative Diversity Under Spatial Interference Correlation in Wireless Networks

Ralph Tanbourgi, Holger Jäkel and Friedrich K. Jondral

Abstract Cooperative diversity is a promising technique to increase link reliability and coverage, however, its performance is affected by interference created by concurrently transmitting nodes over the same time-frequency resources. More specifically, interference is spatially correlated across cooperating nodes and this type of correlation may expunge the diversity gains known for the interference-free case. This chapter analyzes this effect using tools from stochastic geometry. First, the diversity order bottleneck for a simple three-node relaying system using selection decode-and-forward is identified and specific conditions under which the diversity order corresponding to the interference-free case can be recovered are found. Second, the concept of cooperative interference cancellation is proposed, which exploits the fact that interference is correlated across cooperative nodes. Possible deployment scenarios, the importance of the correlation of the interference signal, as well as remaining technical challenges are discussed. Finally, a first simulation-based analysis demonstrates that cooperative interference cancellation can increase throughput by around 25 % in a typical 4G network.

16.1 Introduction

The concept of cooperative diversity is experiencing increasing adoption by wireless systems today, driven by its evolution from initially purely information-theoretic research to the design of practical schemes. At the same time, wireless systems are undergoing a fundamental shift due to, e.g., network densification, heterogeneous deployment, limited site-planning, and increased mobility, thereby inevitably

R. Tanbourgi · H. Jäkel (✉) · F.K. Jondral
Karlsruhe Institute of Technology, Karlsruhe, Germany
e-mail: holger.jaekel@kit.edu

R. Tanbourgi
e-mail: ralph.tanbourgi@kit.edu

F.K. Jondral
e-mail: friedrich.jondral@kit.edu

W. Utschick (ed.), *Communications in Interference Limited Networks*,
Signals and Communication Technology, DOI 10.1007/978-3-319-22440-4_16

aggravating the problem of interference. In particular, the interference seen by a mobile user typically appears *random* as the network geometry may rapidly change relative to the location of this user. This "randomness property" generally renders a comprehensive system-level analysis difficult. Moreover, interference is usually correlated over space due to common locations of interferers, which significantly influences the obtainable diversity gains of cooperative diversity schemes.

A promising approach for analyzing interference in wireless networks is *stochastic geometry* [13], which has been proven to provide the required tools to capture the spatial dynamics of the interference. The line of works that apply such tools to problems in wireless networks is considerable, see for instance [4, 13, 25] and the references therein, and some researchers in this field have started considering also the effect of interference correlation. In [8, 19], the spatial and temporal correlation of the interference power was studied for slotted-Aloha based ad hoc networks, assuming a Poisson point process (PPP) for modeling the node locations. The fact that this type of correlation may significantly affect the performance of diversity schemes was demonstrated later in several works, for instance in [10] for selection combining, in [22, 23] for maximal-ratio combining (MRC), and in [12] for retransmission diversity.

In this chapter, we study the effect of interference correlation on cooperative diversity, where cooperating nodes may experience correlated interference. We adopt the modeling and analysis techniques provided by the stochastic geometry framework to understand the diversity behavior of cooperative communications under a fairly realistic interference model and to obtain useful design insights. This chapter is divided into two parts.

In Sect. 16.2, the effect of spatial interference correlation on cooperative diversity is analyzed from a basic viewpoint. For that, we consider a generic three-node configuration with one source, relay, and destination node, surrounded by interferers whose locations follow a PPP. Selection decode-and-forward [16] is assumed as the cooperation protocol with MRC at the destination. This protocol is well-known and can be seen as a *conventional information-centric forwarding* scheme, since only the information-bearing signal is processed and forwarded by the relay. We derive the outage probability of this cooperative transmission, where outage can be due to adverse fading states and/or excessive interference. Using the theoretical results, we then show that the considered scheme cannot provide diversity order gains due to the fact that (i) conventional information-centric relaying cannot alleviate the interference problem at the destination and (ii) interference is spatially correlated across the relay and destination node. However, in the absence of channel fading and with sufficiently large separation between the relay and destination, diversity order gains can be obtained assuming an error-free relay-destination link. Furthermore, the outage probability minimizing relay location is shown to be slightly closer to the destination than to the source. The intuition behind this observation is that occasionally boosting the received power through the relay-destination link outweighs the reliability loss of the source-relay link.

In Sect. 16.3, we study how spatial interference correlation, which was identified as the performance bottleneck for conventional information-centric relaying in

Sect. 16.2, can be exploited in an *interference-aware* cooperative diversity protocol. Roughly speaking, if the relay and the destination observe the same dominant interference, can we benefit from letting the relay forward (part of) the interference to the destination? The destination, being equipped with two copies of the interference signal, could then perform spatial interference cancellation (IC), a powerful technique used in advanced multiple-antenna receivers. Alternatively, parts of the interference signal forwarded by the relay could be used as side information to facilitate IC at the destination node. This interference-aware cooperation approach is commonly known as *interference forwarding* [7] and essentially requires the interference to be correlated across the relay and destination. In this section, we introduce the concept of cooperative IC (CIC), which builds on the latter approach, and discuss possible deployment scenarios and technical challenges associated with CIC. Furthermore, we characterize the correlation coefficient of the interference signal between two cooperating users in the presence of interfering base stations (BSs) in the downlink to understand the influence of the relevant system parameters on the interference correlation. Finally, we consider a practical scenario under fairly realistic model assumptions, where two users in a downlink cellular network cooperate using CIC to cancel dominant out-of-cell interference. Such user cooperation protocols have recently gained practical relevance with the introduction of device-to-device (D2D) communications to cellular networks as foreseen by Third Generation Partnership Program (3GPP) Release 12. A first investigation indicates that throughput gains of around 25 % can be obtained by letting one user amplify and forward the jointly-dominant interference signal to the victim user, who then performs spatial zero-forcing to cancel out the dominant interference.

16.2 Conventional Relaying Under Interference Correlation

Cooperative relaying has been experiencing increasing interest since the seminal work of [5] and is now considered as a core technique in practical systems such as the 3GPP Long Term Evolution (LTE) Rel-10 [14]. However, the performance of cooperative relaying may be strongly affected by interference created by concurrently transmitting devices over the same time-frequency resources. Unfortunately, analyzing the effect of interference has become challenging in spite of increasing mobility, limited site-planning, and heterogeneous deployments [2], rendering the interference observed by a receiver effectively *random*. This section presents an analytical approach to characterize the performance of cooperative relaying in the presence of interference created by randomly located interferers. While the outage performance of cooperative relaying with random interference was already studied for instance in [1, 6], the focus in this section will be on the diversity behavior under spatial interference correlation. We refer the reader to [21] for further details.

16.2.1 System Model

We consider a three-node configuration with a source, a half-duplex relay, and a destination node. The locations of these nodes are denoted by x_S, x_R, and x_D, respectively, and they are assumed arbitrary but fixed. Without loss of generality, the destination node is placed at the origin ($x_D = o \in \mathbb{R}^2$). The block reserved for conveying a packet from the source to the destination is divided into two consecutive time slots. Within the two time slots, selection decode-and-forward (SDF) [16] is used as the relay protocol. In SDF, a packet is first broadcasted by the source node and received by the destination and relay node. The relay then tries to decode the received packet. If the relay succeeds in decoding the packet, it then re-encodes and forwards it to the destination node in the second time slot. Otherwise, the source node re-transmits the packet (repetition coding). At the destination node, the two received copies of the transmitted packet are (linearly) combined using channel state information (CSI) to improve the signal-to-interference ratio. Thereby, the term *selection* in SDF refers to the fact that either the relay or the source transmits in the second time slot, which recovers the dimension loss of conventional decode-and-forward whenever the relay cannot decode the packet sent by the source in the first time slot. This, however, requires a 1-bit feedback from the relay to the source node.

Due to concurrent transmissions by other nodes in the network over the same time-frequency resources, the relay and destination will encounter co-channel interference. We assume that, in one snapshot, the locations of these interfering transmitters follow a stationary planar PPP Φ with density λ. The PPP is a commonly used model for capturing irregular and spatially-dynamic node deployments in various types of wireless networks, see for instance [2, 25]. All transmitted signals undergo a distance-dependent path loss, which for two locations $x_1, x_2 \in \mathbb{R}^2$ is given by $\|x_1 - x_2\|^{\alpha}$ with $\alpha > 2$ being the path loss exponent. Besides path loss, the signals are also subject to channel fading. Hence, the interference from the ith interferer located at $\mathsf{x}_i \in \Phi$ to the destination (relay) node is received with power $\mathsf{g}_{i,\mathrm{D}}\|\mathsf{x}_i\|^{-\alpha}$ ($\mathsf{g}_{i,\mathrm{R}}\|\mathsf{x}_i - x_\mathrm{R}\|^{-\alpha}$) assuming unit transmit power. Similarly, the fading gains of the channels between the source, relay, and destination, are given by $\mathsf{g}_{\mathrm{S,R}}$, $\mathsf{g}_{\mathrm{S,D}}$, and $\mathsf{g}_{\mathrm{R,D}}$. We assume i.i.d. frequency-flat Rayleigh fading, which corresponds to a rich scattering None-Line-of-Sight (NLoS) environment. In this case, $\mathsf{g}_{i,n}$ follows a unit-mean exponential distribution for all $i \in \mathbb{N} \cup \{\mathrm{S}, \mathrm{R}\}$ and $n \in \{\mathrm{R}, \mathrm{D}\}$. The channel is assumed to be static for the duration of the cooperative transmission, i.e., over the two time slots used by SDF, and relay and destination are assumed to have perfect CSI-R. The considered scenario is illustrated in Fig. 16.1.

The sum interference power received at the relay and destination in the two time slots can be expressed as

$$\mathsf{I}_\mathrm{R} := \sum_{\mathsf{x}_i \in \Phi} \mathsf{g}_{i,\mathrm{R}} \|\mathsf{x}_i - x_\mathrm{R}\|^{-\alpha} \tag{16.1}$$

$$\mathsf{I}_\mathrm{D} := \sum_{\mathsf{x}_i \in \Phi} \mathsf{g}_{i,\mathrm{D}} \|\mathsf{x}_i\|^{-\alpha}. \tag{16.2}$$

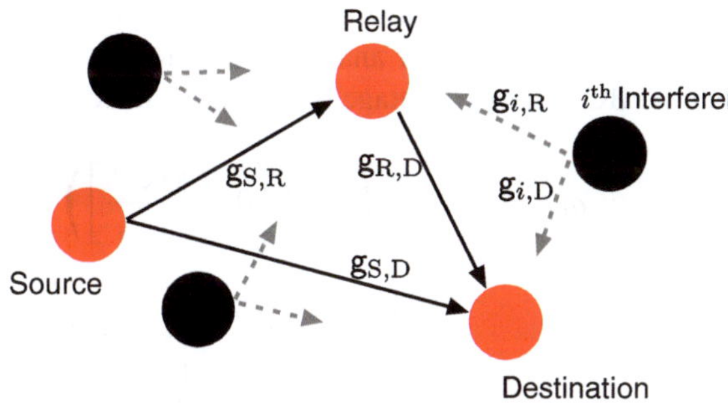

Fig. 16.1 Illustration of the considered cooperative scenario with randomly located interfering transmitters distributed according to a PPP

The signal-to-interference ratio (SIR) at the relay and destination can be written as

$$\mathtt{SIR}_{\mathrm{SR}} = \frac{\mathsf{g}_{\mathrm{S,R}} \|x_{\mathrm{S}} - x_{\mathrm{R}}\|^{-\alpha}}{\mathsf{I}_{\mathrm{R}}}, \tag{16.3}$$

$$\mathtt{SIR}_{\mathrm{SRD}} = \frac{\mathsf{g}_{\mathrm{S,D}} \|x_{\mathrm{S}}\|^{-\alpha} + \mathsf{g}_{\mathrm{R,D}} \|x_{\mathrm{R}}\|^{-\alpha}}{\mathsf{I}_{\mathrm{D}}}, \tag{16.4}$$

and

$$\mathtt{SIR}_{\mathrm{SD}} = \frac{2\mathsf{g}_{\mathrm{S,D}} \|x_{\mathrm{S}}\|^{-\alpha}}{\mathsf{I}_{\mathrm{D}}}. \tag{16.5}$$

The SIR expressions in (16.4) and (16.5) correspond to the cases where the relay succeeds or fails in decoding the packet transmitted by the source, respectively. The factor 2 in (16.5) is due to the fact that the transmitter re-transmits the packet in the second slot. The cooperative transmission may be in outage due to strong interference and/or adverse fading realizations, hence the outage probability is a useful performance metric.

16.2.2 *Optimal Relay Location and Diversity Behavior*

In this section, the optimal relay location and the diversity behavior of SDF are analyzed under the setting described in Sect. 16.2.1. For this purpose, the outage probability will be derived first. The outage probability for SDF can be written as

$$q = \underbrace{\mathbb{P}(\mathtt{SIR}_{\mathrm{SD}} < T,\ \mathtt{SIR}_{\mathrm{SR}} < T)}_{:=q_{\mathrm{BC}}} + \underbrace{\mathbb{P}(\mathtt{SIR}_{\mathrm{SRD}} < T,\ \mathtt{SIR}_{\mathrm{SR}} \geq T)}_{:=q_{\mathrm{MAC}}}, \tag{16.6}$$

where $T > 0$ is a coding and modulation specific threshold, and q_{BC} and q_{MAC} are the outage probability in the broadcast (BC) phase and multiple access channel (MAC) phase, respectively. Analyzing these two expressions separately will turn out to be advantageous in the later analysis. Now, let $\|x_S\| \neq \|x_R\|$.[1] Then, it can be shown that the BC and MAC outage probability can be expressed as [21]

$$q_{BC} = 1 - \exp\left(-\lambda\Omega\left[0, \frac{T\|x_S\|^\alpha}{2r^\alpha}\right]\right) - \exp\left(-\lambda\Omega\left[\frac{T\|x_S - x_R\|^\alpha}{(r^2 + x_R^2 - 2rx_R\cos\phi)^{\frac{\alpha}{2}}}, 0\right]\right) + \exp\left(-\lambda\Omega\left[\frac{T\|x_S - x_R\|^\alpha}{(r^2 + x_R^2 - 2rx_R\cos\phi)^{\frac{\alpha}{2}}}, \frac{T\|x_S\|^\alpha}{2r^\alpha}\right]\right) \quad (16.7)$$

and

$$q_{MAC} = \exp\left(-\lambda\Omega\left[\frac{T\|x_S - x_R\|^\alpha}{(r^2 + x_R^2 - 2rx_R\cos\phi)^{\frac{\alpha}{2}}}, 0\right]\right) - \frac{\|x_R\|^\alpha}{\|x_R\|^\alpha - \|x_S\|^\alpha}\exp\left(-\lambda\Omega\left[\frac{T\|x_S - x_R\|^\alpha}{(r^2 + x_R^2 - 2rx_R\cos\phi)^{\frac{\alpha}{2}}}, \frac{T\|x_S\|^\alpha}{r^\alpha}\right]\right) + \frac{\|x_S\|^\alpha}{\|x_R\|^\alpha - \|x_S\|^\alpha}\exp\left(-\lambda\Omega\left[\frac{T\|x_S - x_R\|^\alpha}{(r^2 + x_R^2 - 2rx_R\cos\phi)^{\frac{\alpha}{2}}}, \frac{T\|x_R\|^\alpha}{r^\alpha}\right]\right) \quad (16.8)$$

where

$$\Omega[f, g] := \int_0^\infty \int_0^\pi 2r\left(1 - \frac{1}{(1 + f(r, \phi))(1 + g(r))}\right) \mathrm{d}\phi\,\mathrm{d}r. \quad (16.9)$$

16.2.2.1 Optimal Relay Location

Using the outage probability expressions in (16.7) and (16.8), we are able to (numerically) find the optimal relay location minimizing the outage probability for SDF. For this, the relay node is assumed to be located on a line connecting the source and destination, see Fig. 16.2.

Figure 16.3 shows the optimal relay position relative to the position of the destination node for different path loss exponents α and interferer densities $\lambda = 10^{-6}$ and $\lambda = 10^{-3}$. The relative position is defined as $\|x_S - x_R\|/\|x_S - x_D\|$. It can be seen that it is strictly better to place the relay closer to the destination for reasonable path loss exponents around $\alpha = 4$, in particular for small outage probabilities (solid line) on the order of 10^{-3}–10^{-2}. For practically non-relevant large outage probabilities on the order of 10^{-1} (dashed line), this observation is weakened as the optimal relay location is closer to the half-line at small path loss exponents around $\alpha = 3$. In conclusion, these observations suggest that there is a slight trend for outage-optimal

[1] The analysis can be carried out for $\|x_S\| = \|x_R\|$ in a similar way, see for instance [1], however, with no additional insights.

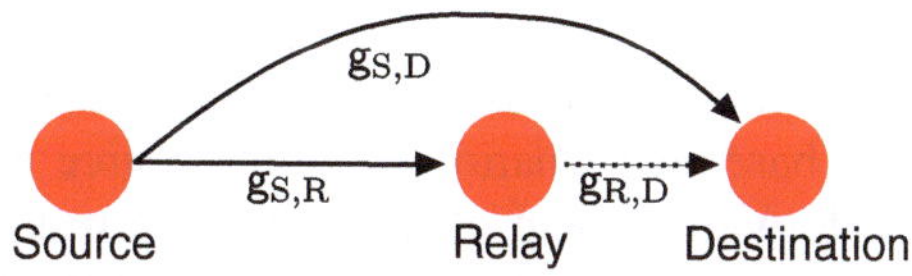

Fig. 16.2 Line configuration of the cooperative three-node relay network

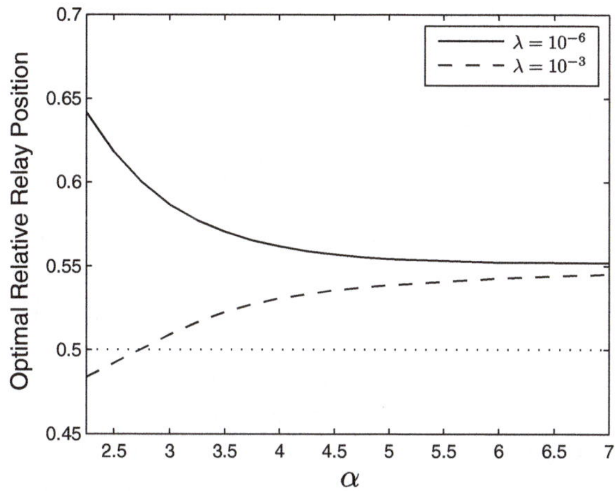

Fig. 16.3 Optimal relay position relative to source-destination distance, i.e., $\|x_S - x_R\|/\|x_S - x_D\|$, versus path loss exponent α. Line configuration assumed. Parameters are: $\|x_S\| = 10$ and $T = 4$

SDF in the sense that receive (Rx) diversity is more beneficial than transmit (Tx) diversity. The intuition behind this result is that the ability to boost the received power through the relay-destination link outweighs the reliability loss of the source-relay link.

16.2.2.2 Diversity Order

The spatial diversity order (SDO) is a useful metric to characterize the diversity gain in decentralized wireless networks. The SDO quantifies the reliability gain (in terms of outage probability slope) as a function of the spatial density of concurrent transmissions λ. The SDO provides an intuitive extension to the conventional diversity order metric, see for instance [16], as it takes into account the spatial resource, which is considered the critical resource in decentralized wireless networks. Furthermore, controlling the density instead of, e.g., transmit power, is a useful technique found in common medium access control protocols such as Aloha (thinning of transmitter set according to medium access probability) or carrier-sense multiple access (spatial inhibition of close-by transmissions).

Definition 16.2.1 (*Spatial Diversity Order*) The SDO is defined as

$$d_\lambda := \lim_{\lambda \to 0} \frac{\log q}{\log \lambda}. \tag{16.10}$$

The subscript λ in d_λ emphasizes that the spatial density of concurrent transmissions is used as the scaling parameter. As an example, consider the following baseline scenario, where the source node communicates with the destination node without help of the relay. The resulting outage probability is known to be $1 - \exp(-\lambda\pi^2 T^{2/\alpha} \|x_S\|^2 \frac{2}{\alpha} \csc(\frac{2}{\alpha}\pi))$ [25]. It can be easily verified that the resulting SDO in this case is $d_\lambda = 1$ as expected, since no diversity can be provided by direct transmission, i.e., without help of the relay. The next result gives the SDO of SDF for the setting described in Sect. 16.2.1 with Rayleigh fading links.

Theorem 16.2.1 (SDO of SDF for Rayleigh fading model) *The SDO of SDF for exponential fading gains (Rayleigh fading) is* $d_\lambda = 1$.

Theorem 16.2.1 shows that simply relaying the packet transmitted by the source does not improve the performance in terms of SDO. This result is in sharp contrast to the interference-free case, where SDF is known to provide a diversity order gain [16]. The reason for this different result is that the relay may indeed provide an array gain in the MAC phase, but it cannot counteract the possibly high interference level at the destination node through forwarding the packet sent by the source. More specifically, in the asymptotic regime, outages are dominated by the occurrence of strong interference rather than adverse fading states, which is due to the heavy-tailed distribution of the interference [11]. Hence, relaying the source packet cannot help alleviating the interference problem at the destination node, which explains why $q_{\mathrm{MAC}} \propto \lambda$ as $\lambda \to 0$.

Furthermore, spatial correlation of the interference across the relay and the destination node renders the BC phase less effective compared to in the interference-free case, which explains why $q_{\mathrm{BC}} \propto \lambda$ as $\lambda \to 0$. In other words, if the destination node is in outage in the first time slot, then it is (more or less) likely that the relay node experiences an outage as well due to the fact that interference is correlated over space. This observation suggests that increasing the relay-destination distance reduces the spatial interference correlation across these two nodes, however, it turns out that the asymptotic slope of $q_{\mathrm{BC}} \propto \lambda$ does not change. It is important to note, though, that the fading distribution co-determines the spatial interference correlation properties, and that abandoning the Rayleigh fading assumption may possibly change the SDO, at least in the BC phase.[2] This will be verified next by focusing on the obtainable SDO in the BC phase when dropping the Rayleigh fading assumption.

The BC phase can be viewed as a distributed single-input multiple-output (SIMO) system, where the destination node has two spatially separated Rx-antennas that are connected through a high-capacity link. Prominent examples of such distributed SIMO systems are for instance radio remote heads (RRH) based network architectures or Coordinated Multi-Point (CoMP) with joint-processing in the uplink [17, 18]. Recall that the Rayleigh fading model applies to rich scattering NLoS environments. On the other extreme, the pure path loss model applies to scenarios with negligible scattering and a dominant Line-of-Sight (LoS) path. Since the above SDO

[2]Obviously, modifying the fading assumption does not change the SDO in the MAC phase due to the inability or the relay to reduce interference at the destination node as explained before.

analysis focused on the Rayleigh fading model, it is reasonable to redo the analysis now assuming the other extreme, namely the pure path loss model. In the pure path loss model, the channel gains are no longer random and now become $\mathsf{g}_{i,n} \equiv 1$ for all $i \in \mathbb{N} \cup \{\mathrm{S}, \mathrm{R}\}$ and $n \in \{\mathrm{R}, \mathrm{D}\}$. Clearly, with unit channel gains, the spatial interference correlation now depends exclusively on the spatial distribution of nodes.

Using the dominant-interferer bounding technique from [25], the outage probability q in the path loss model can be tightly approximated in the asymptotic regime $\lambda \to 0$, yielding

$$q_{\mathrm{BC}} \sim \begin{cases} \lambda |A_{\mathrm{RD}}|, & A_{\mathrm{RD}} \neq \emptyset \quad (16.11a) \\ \lambda^2 |A_{\mathrm{R}}|\,|A_{\mathrm{D}}|, & A_{\mathrm{RD}} = \emptyset, \quad (16.11b) \end{cases}$$

where $A_{\mathrm{R}} := b(x_{\mathrm{R}}, T^{1/\alpha} \|x_{\mathrm{S}} - x_{\mathrm{R}}\|)$, $A_{\mathrm{D}} := b(x_{\mathrm{D}}, (\frac{T}{2})^{1/\alpha} \|x_{\mathrm{S}}\|)$, and $A_{\mathrm{RD}} := A_{\mathrm{R}} \cap A_{\mathrm{D}}$ are the so-called *dominant-interferer regions*, in which the presence of one interferer already suffices to create an outage. Consequently, we obtain the following SDO result for SDF under the pure path loss model.

Theorem 16.2.2 (SDO of SDF in BC phase without fading) *The SDO of SDF in the BC phase in the absence of fading is*

$$d_\lambda = \begin{cases} 1, & \|x_{\mathrm{R}} - x_{\mathrm{D}}\| \leq T^{1/\alpha} \|x_{\mathrm{S}} - x_{\mathrm{R}}\| + (\frac{T}{2})^{1/\alpha} \|x_{\mathrm{S}}\| \quad (16.12a) \\ 2, & \textit{otherwise}. \quad (16.12b) \end{cases}$$

As can be seen by (16.12), the intersection A_{RD} of the two regions A_{R} and A_{D} plays a crucial role for the SDO: if $|A_{\mathrm{RD}}| = 0$, i.e., there is no overlapping of the individual dominant-interferer regions A_{R} and A_{D}, the interference power across the relay and the destination becomes asymptotically uncorrelated as $\lambda \to 0$, thus yielding $d_\lambda = 2$. It is interesting to note that, by looking at (16.12), there exists a hard transition from $d_\lambda = 1$ to $d_\lambda = 2$ in the pure path loss model, which depends only on the geometrical distance between the relay and destination node. Under a more general fading model, e.g., Nakagami-m, which captures the continuum between Rayleigh and no fading, we hence conjecture that the smoothness of the transition will depend on the amount of fading.

The theoretical result from Theorem 16.2.2 is shown in Fig. 16.4, along with simulation results validating the analysis. In particular, it can be seen that the dominant-interferer bounding technique yields a tight approximation to the true q_{BC} in the small outage probability regime. As expected, the slope of q_{BC} does not change ($d_\lambda = 1$) until the relay-destination is sufficiently large such that the respective dominant-interferer regions become non-overlapping ($|A_{\mathrm{RD}}| = 0 \Rightarrow d_\lambda = 2$). In this example, the relay-destination distance has to be at least $\|x_{\mathrm{R}} - x_{\mathrm{D}}\| = 13.72$ for the dominant-interferer regions A_{R} and A_{D} to be disjoint, which can be verified by (16.12a). Note that these regions, and hence their intersection A_{RD} are proportional to $T^{1/\alpha}$. This

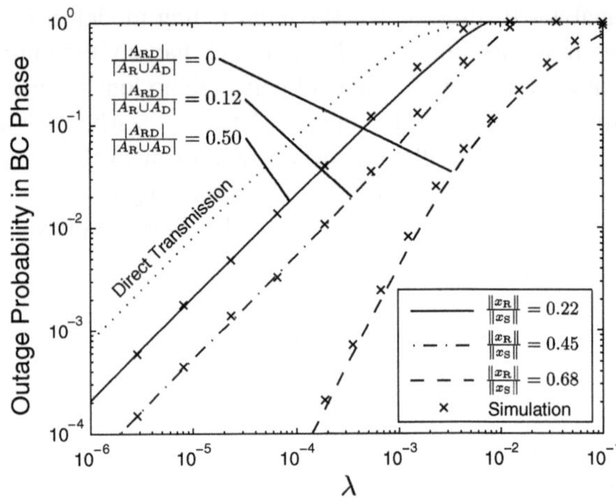

Fig. 16.4 Outage probability q_{BC} in BC phase versus λ for relay locations $x_R = (5, 0)$, $x_R = (10, 0)$, and $x_R = (15, 0)$. Parameters are: $x_S = (18, 0)$, $\alpha = 4$, $T = 0.25$

suggests that small target SIRs, e.g., as in direct-sequence code-division multiple access (CDMA) ad hoc networks [3, 24], are desirable from the viewpoint of the SDO in distributed Rx diversity systems.

16.2.3 Summary

The analysis showed that, although spatially separated, receivers may experience correlated interference due to common locations of interfering transmitters. In this sense, interference correlation is not limited to a few centimeters across Rx antennas, but it propagates through the space surmounting meters. This obviously affects the performance of cooperative diversity schemes; the achievable diversity order in a simple one-relay scenario with Rayleigh fading cannot exceed that of a conventional transmission between two single-antenna nodes. However, the picture changes when considering an Rx diversity system with an error-free relay-destination link. In this case, a diversity order increase is achievable in the absence of channel fading. This is because in the asymptotic regime, outage events are dominated by interfering transmitters within the so-called dominant-interferer regions around the relay and the destination; if these regions are disjoint, the achievable diversity order is increased. For small SIR thresholds T, the interferer regions become disjoint already for small separations between relay and destination. This suggests that signaling schemes that operate at low effective SIR thresholds T, e.g., direct-sequence CDMA, are more desirable when optimizing the diversity order of cooperative Rx diversity systems.

16.3 Cooperative Interference Cancellation

The main conclusion of the last section was that conventional information-centric relaying cannot alleviate the interference problem at the destination and that interference is spatially correlated across close-by receivers. Ideally, a receiver may want to cancel out interference prior to demodulating the desired message or packet. Depending on the allowable complexity, there exist multiple approaches to realize this, for instance using multiple antennas to spatially cancel out interference or (single-antenna) multi-user detection techniques with IC capability. In certain scenarios, however, space limitations may prevent the use of multiple antennas while the required accuracy for estimating the interference signal in order for IC to be beneficial may be too difficult to realize practically. The question that arises now is the following: can we take advantage of the spatial interference correlation across cooperative users to enhance their IC capabilities? Clearly, conventional information-centric relaying cannot improve the IC performance as no information about the interference signal is conveyed from the relay to the destination. In contrast, the concept of *interference forwarding*, where (part of) the interference signal is forwarded to the IC receiver, seems promising. Interference forwarding was first studied in [7] from an information-theoretic perspective, where interference correlation across cooperating users was found essential for this approach to provide gains. Such a cooperation protocol can be seen as cooperative IC (CIC) and is the subject of this section.

In cellular networks, CIC could be realized using D2D as the forwarding link. Since 3GPP D2D is operator-controlled, CIC may be employed by the network as an interference-management tool as follows: a CIC process between a pair of close-by D2D users is triggered by the network whenever this pair experiences dominant interference from the same neighboring BS. Using the interference signal forwarded to the victim user, harmful interference can be canceled out using a conventional IC approach, e.g., zero-forcing spatial IC. The BS may then choose a higher modulation and coding scheme (MCS) for this user to enhance the downlink throughput. Before discussing various system aspects associated with this type of interference-aware cooperation in Sect. 16.3.2, we first present possible deployment scenarios.[3]

16.3.1 Possible Deployment Scenarios

The deployment scenarios explained below are illustrated in Fig. 16.5.

(a) Intra-cell D2D cooperation and no inter-cell coordination: Because interference must be correlated across cooperating receivers for CIC to be beneficial, the cooperative users must be scheduled on the same time-frequency resources in the downlink. Such an allocation is used for instance in multi-user multiple-input

[3]Some parts of Sect. 16.3 are reprinted, with permission, from R. Tanbourgi, H. Jäkel, F.K. Jondral, "Cooperative interference cancellation using device-to-device communications," *IEEE Commun. Magazine*, vol. 52, no. 6, pp. 118–124, June 2014, © 2015 IEEE.

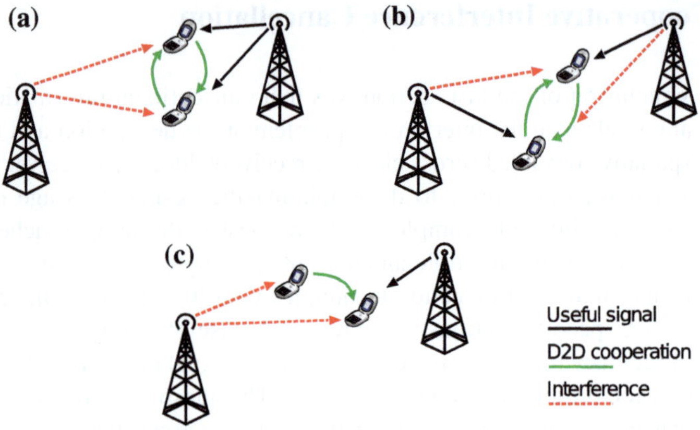

Fig. 16.5 Possible deployment scenarios for D2D-based CIC. **a** Intra-cell D2D Cooperation. **b** Inter-cell D2D Cooperation. **c** One-way D2D Cooperation

multiple-output (MU-MIMO) systems. Here, multiple antennas at the BS are used to serve multiple users of the same cell on the same resources with spatially-precoded orthogonally-multiplexed data streams. At the users, the received signal is free from intra-cell interference. Upon reception, the D2D users may then exchange information regarding the observed inter-cell interference to improve the IC process. Note that a competing technique to suppress this kind of interference is downlink CoMP, where inter-cell interference is pre-canceled. A striking advantage of CIC over downlink CoMP, however, is that no low-latency and high-capacity backhaul is required. Backhaul architectures suitable for CoMP are costly and must be pre-deployed in areas where high mobile traffic is expected with limited possibility for adaptation. In contrast, CIC using D2D is not tied to certain backhaul requirements as it takes place locally between close-by users.

(b) Inter-cell D2D cooperation and limited inter-cell coordination: In cellular systems, cell-edge users may suffer from high inter-cell interference. As already mentioned, CoMP techniques can be employed to solve this problem. However, for the reasons outlined before, the performance of CoMP is usually limited because of imperfect or delayed CSI resulting in non-canceled residual interference. Here, CIC can provide a remedy by additionally post-canceling the residual interference, thereby relaxing the backhaul requirements of CoMP. Furthermore, the required interference correlation across the D2D users can be induced by scheduling two close-by users of different cells on the same radio resources through the existing CoMP backhaul link. Such an intertwining of two IC principles (pre- and post-transmission IC) may help balancing the complexity trade-offs of future cellular networks.

(c) One-way D2D cooperation: In this scenario, a user not scheduled for a downlink session agrees upon cooperating with another close-by user. While in conventional relaying the cooperative user should not forward anything whenever the

channel to the source is in a deep fade, it may now alternatively forward the observed interference. Such an opportunistic relay protocol may then either boost the received useful power or facilitate IC at the victim user.

16.3.2 Challenges and Possible Solutions

We next outline the challenges associated with D2D-based CIC and provide possible approaches for addressing these challenges.

16.3.2.1 Modeling and Analysis Tools

Tools for analyzing cellular networks play an important role for the development of new techniques. Prominent tools include for instance the Wyner-type models and Monte-Carlo simulations. Cellular networks, however, have recently experienced a substantial technology shift, which led to unprecedented issues such as heterogeneity, limited site-planning, irregular deployments and increased mobility, to name just a few. This development has rendered the analysis of cellular networks opaque and user experience difficult to predict. In addition, D2D introduces an ad hoc component, which makes the structure of cellular networks even more complex.

A promising solution to the above challenge is the use of stochastic geometry. In principle, the locations of the BSs and users are modeled by a spatial point process, thereby accounting for irregular deployments, user mobility, and other spatial dynamics. Using such spatial models, key performance metrics such as the signal-to-interference-plus-noise ratio (SINR) or the data rate can be analyzed in a comprehensive way. Owing to the many fundamental results available in the literature, e.g., the Campbell-Mecke Theorem or the Laplace functional for PPPs [20], closed-form expressions for the key performance metrics can be obtained surprisingly often, thereby revealing important performance trends.

For illustration, Fig. 16.6 shows an example of a spatial model for heterogeneous two-tier cellular networks with additional D2D users. The two-tier structure was created by independent superposition of two PPPs. The cell-association regions were obtained using a multiplicatively-weighted Poisson-Voronoi tessellation and users are assumed to be distributed according to a PPP and establish a D2D link if sufficiently close.

Stochastic geometry can also help in better understanding the interference correlation phenomenon, which is crucial for CIC to provide gains. Recent works in this field, e.g., [10], indicate that interference power in decentralized networks is spatially-correlated over several meters. However, due to the cell-association mechanism, interference correlation decays much slower with distance in cellular networks as illustrated in Fig. 16.7. In this figure, the estimated interference correlation between two users in a cellular network is shown, where the BSs locations are modeled by a PPP. For each of the two users, the closest BS is the serving one while interference

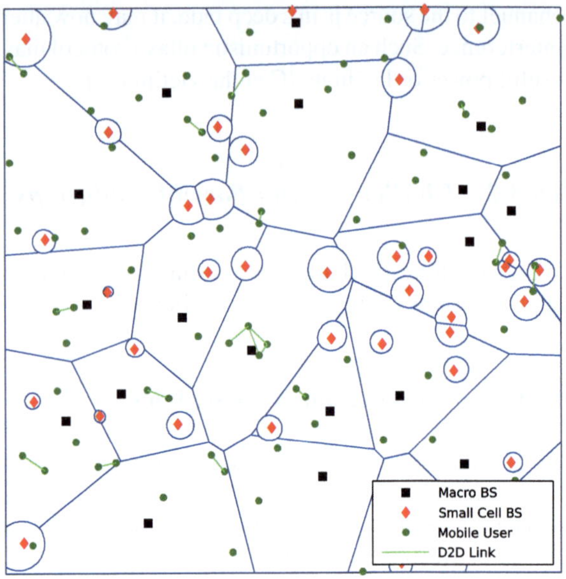

Fig. 16.6 Two-tier cellular network model with D2D links using a Poisson -Voronoi tessellation

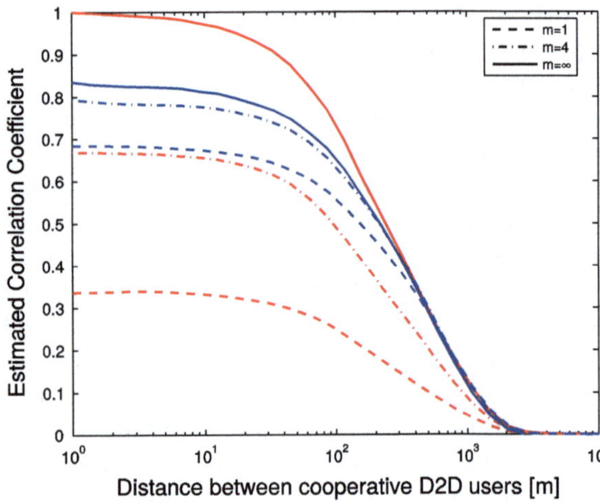

Fig. 16.7 Spatial correlation of the inter-cell interference signal (*blue*) and power (*red*) versus distance between cooperating D2D users. BS locations follow a PPP with density 5 $\mathrm{BS/km^2}$. $\alpha = 4$

is caused by inter-cell BSs. The channel is composed by distance-dependent path loss and Nakagami-m fading with independent and uniformly distributed phases. All BSs transmit complex-valued symbols following a circular-symmetric Gaussian distribution with unit variance. Before correlation, the two signals are aligned with respect to each signal's strongest interferer.

Moreover, it is important to not only look at the correlation of interference powers but also at the correlation of the interference signals. Clearly, correlation of the interference power does not necessarily imply correlation of the interference signals. The interference signal correlation may thereby depend on various system parameters such as path loss exponent and BS density. In the next subsection, we present a first characterization of the interference signal correlation assuming a simple setting.

16.3.2.2 Spatial Correlation of the Interference Signal

While Fig. 16.7 demonstrates, through simulations, that the correlation of the interference signal may span several tens of meters in cellular networks, it is important to understand in which way the correlation depends upon the various system parameters. For this purpose, consider the following scenario: two co-channel users in D2D range located at u_1 and u_2 receive strong interference from a given neighboring BS located at x_1. Taking into account also the interference created by far-off BSs, the total received interference signal at user $n \in \{1, 2\}$ can be expressed as

$$y_n = \mathsf{h}_{1,n} \|x_1 - u_n\|^{-\alpha/2} \mathsf{s}_1 + \sum_{\mathsf{x}_i \in \Phi} \mathsf{h}_{i,n} \|\mathsf{x}_i - u_n\|^{-\alpha/2} \mathsf{s}_i \tag{16.13}$$

where $\mathsf{h}_{i,n}$ is the frequency-flat channel fading gain from the ith interfering BS to node $n \in \{1, 2\}$, α is the path loss exponent, and s_i is the signal transmitted by the ith interfering BS with $\mathbb{E}[\mathsf{s}_i] = 0$ and $\mathbb{E}[|\mathsf{s}_i|^2] = 1$ for all $i \in \mathbb{N}$. For simplicity, we assume that $\mathbb{E}[\mathsf{h}_{i,n}] = 0$ and $\mathbb{E}[|\mathsf{h}_{i,n}|^2] < \infty$ for all $i \in \mathbb{N}$ and $n \in \{1, 2\}$, for instance Rayleigh fading, where $\mathsf{h}_{i,n} \sim \mathcal{CN}(0, 1)$. Note that we do not explicitly model the received useful signal as the primary purpose is to characterize the spatial correlation of the interference signal. For the same reason, we shall assume that the two users can share their observation y_1 and y_2 via an error-free D2D link and estimate the channel to the interfering BS at x_1. The locations $\{\mathsf{x}_i\}$, where $i \in \{2, \ldots\}$, of the remaining interfering BSs that create the residual interference are assumed to follow a stationary point process Φ with density λ. The considered scenario is illustrated in Fig. 16.8.

Without loss of generality, we focus on the first user at u_1 for the subsequent analysis. With the motivation of spatially canceling out the interference from BS x_1, this user will first align the observations y_1, y_2 as follows:

$$y_1 \mapsto y_1' = \mathsf{h}_{1,2} \|x_1 - u_1\|^{\alpha/2} y_1 \tag{16.14}$$

and

$$y_2 \mapsto y_2' = \mathsf{h}_{1,1} \|x_1 - u_2\|^{\alpha/2} y_2 \tag{16.15}$$

In CIC, the user would now perform spatial IC to remove the interference from the neighboring BS at x_1. While the interference from this BS would then be canceled out, the residual interference from the remaining BSs may undesirably increase at

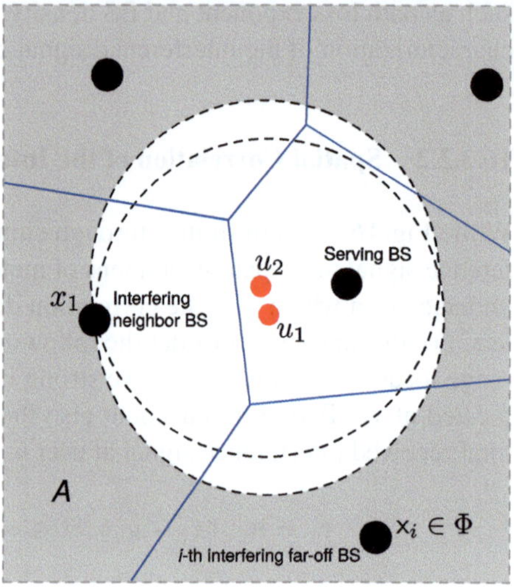

Fig. 16.8 Illustration of the considered scenario. Two D2D users are located at u_1, u_2 and experience interference from a neighboring BS at x_1 and from far-off BSs whose locations $\{\mathsf{x}_i\}$ are modeled by the point process Φ. The shaded region A represents the area in which far-off BSs can be located

the same time. Thereby, the amount of increase strongly depends on the degree of similarity of the residual interference in y_1' and y_2'; clearly, if the similarity is high, much of the residual interference will be canceled out as well, while a low similarity will enhance the residual interference. The correlation coefficient between y_1' and y_2' characterizes this similarity and will be derived next.

The correlation coefficient can be expressed as

$$\rho = \frac{\mathrm{Cov}\left(y_1', y_2'\right)}{\sqrt{\mathrm{Var}\left(y_1'\right)\mathrm{Var}\left(y_2'\right)}}. \tag{16.16}$$

Define $d_{i,n} := \|\mathsf{x}_i - u_n\| / \|x_1 - u_n\|$ for short-hand notation. The covariance can be obtained as

$$\begin{aligned}
\mathrm{Cov}\left(y_1', y_2'\right) &= \mathbb{E}\left[|\mathsf{s}_1|^2|\mathsf{h}_{1,1}|^2|\mathsf{h}_{1,2}|^2\right] + \mathbb{E}\left[\mathsf{h}_{1,2}\mathsf{h}_{1,1}^* \sum_{\mathsf{x}_i,\mathsf{x}_j \in \Phi} \frac{\mathsf{s}_i\mathsf{s}_j^*\mathsf{h}_{i,1}\mathsf{h}_{j,2}^*}{d_{i,1}^{\alpha/2} d_{j,2}^{\alpha/2}}\right] \\
&\quad + \mathbb{E}\left[|\mathsf{h}_{1,1}|^2\mathsf{h}_{1,2} \sum_{\mathsf{x}_i \in \Phi} \frac{\mathsf{s}_1\mathsf{s}_i^*\mathsf{h}_{i,2}^*}{d_{i,2}^{\alpha/2}}\right] + \mathbb{E}\left[|\mathsf{h}_{1,2}|^2\mathsf{h}_{1,1}^* \sum_{\mathsf{x}_i \in \Phi} \frac{\mathsf{s}_1^*\mathsf{s}_i\mathsf{h}_{i,1}}{d_{i,1}^{\alpha/2}}\right] \\
&= \mathbb{E}\left[|\mathsf{h}|^2\right]^2.
\end{aligned} \tag{16.17}$$

Similarly, the variance $\mathrm{Var}\left(y_1'\right)$ can be obtained as

$$\begin{aligned}\mathrm{Var}\left(y_1'\right) &= \mathbb{E}\left[|\mathsf{s}_1|^2|\mathsf{h}_{1,1}|^2|\mathsf{h}_{1,2}|^2\right] + \mathbb{E}\left[|\mathsf{h}_{1,2}|^2 \sum_{\mathsf{x}_i \in \Phi} \frac{|\mathsf{s}_i|^2|\mathsf{h}_{i,1}|^2}{d_{i,1}^{\alpha}}\right] \\ &\quad + \mathbb{E}\left[|\mathsf{h}_{1,2}|^2 \sum_{\mathsf{x}_i \neq \mathsf{x}_j} \frac{\mathsf{s}_i\mathsf{s}_j^*\mathsf{h}_{i,1}\mathsf{h}_{j,1}^*}{d_{i,1}^{\alpha/2}d_{j,1}^{\alpha/2}}\right] \\ &= \mathbb{E}\left[|\mathsf{h}|^2\right]\left(\mathbb{E}\left[|\mathsf{h}|^2\right] + \mathbb{E}\left[\sum_{\mathsf{x}_i \in \Phi} \frac{|\mathsf{s}_i|^2|\mathsf{h}_{i,1}|^2}{d_{i,1}^{\alpha}}\right]\right). \end{aligned} \tag{16.18}$$

Using the Campbell-Mecke Theorem [20], the expectation of the sum in (16.18) can be obtained as

$$\mathbb{E}\left[\sum_{\mathsf{x}_i \in \Phi} \frac{|\mathsf{s}_i|^2|\mathsf{h}_{i,1}|^2}{d_{i,1}^{\alpha}}\right] = \lambda \mathbb{E}\left[|\mathsf{h}|^2\right] \|x_1 - u_1\|^{\alpha} \int_A \frac{1}{\|x - u_1\|^{\alpha}} \,\mathrm{d}x, \tag{16.19}$$

where A is the region in which far-off BSs can lie, see Fig. 16.8. When the inter-user distance $\|u_1 - u_2\|$ is small relative to the distance to the interfering neighbor BS at x_1, (16.19) can be approximated by

$$\begin{aligned}\mathbb{E}\left[\sum_{\mathsf{x}_i \in \Phi} \frac{|\mathsf{s}_i|^2|\mathsf{h}_{i,1}|^2}{d_{i,1}^{\alpha}}\right] &= 2\pi\lambda \mathbb{E}\left[|\mathsf{h}|^2\right] \|x_1 - u_1\|^{\alpha} \int_{\|x_1-u_1\|}^{\infty} r^{-\alpha+1} \,\mathrm{d}r \\ &= \frac{2\pi\lambda}{\alpha - 2} \mathbb{E}\left[|\mathsf{h}|^2\right] \|x_1 - u_1\|^2. \end{aligned} \tag{16.20}$$

Combining (16.17) and (16.20), the correlation coefficient finally becomes

$$\rho = \frac{1}{\sqrt{1 + \frac{2\pi\lambda}{\alpha-2}\|x_1 - u_1\|^2}\sqrt{1 + \frac{2\pi\lambda}{\alpha-2}\|x_1 - u_2\|^2}}. \tag{16.21}$$

Since (16.21) is valid only for small inter-user distances $\|u_1 - u_2\|$, we have $\|x_1 - u_1\| \approx \|x_1 - u_2\|$ in this case, which allows us to rewrite (16.21) in the simple form

$$\rho = \frac{1}{1 + \frac{2\pi\lambda}{\alpha-2}\|x_1 - u_1\|^2}. \tag{16.22}$$

Note that the small-$\|u_1 - u_2\|$ assumption has removed the dependence of the correlation coefficient on the inter-user distance. Nevertheless, (16.22) now allows us to characterize the correlation coefficient as a function of the system parameters for the case of close-by D2D users. First, it can be observed that the correlation coefficient ρ is independent of the channel fading statistics, however, this is mainly due to the zero-mean property of the $\mathsf{h}_{i,n}$. If the fading gains are not zero-mean, then the correlation depends on the fading statistics as well, as illustrated in Fig. 16.7 for Nakagami-m fading. Second, the correlation coefficient increases with the path loss exponent α, which is due to the fact that the interference from far-off BSs becomes weaker. Consequently, the nearest interfering BS x_1 becomes even more dominant. Finally, noting that the expected distance to the nth closest BS is proportional to $\lambda^{-1/2}$ when Φ is a PPP [9], this suggests that the scaling of ρ is independent of the BS density λ.

16.3.2.3 Cooperation Incentives, Resource Allocation and Security

An important question is whether users are keen to provide their resources (e.g., battery power and degrees-of-freedom) to other users without experiencing an obvious immediate benefit. In order to give users reasons to cooperate, incentives have to be introduced. This may be achieved by using methods from game theory and combining them with stochastic geometry for understanding how cooperation achieving behavior depends on and is limited by the spatial configuration of the users. Exploiting the fact that 3GPP D2D is operator-controlled with the possibility of tight control of D2D resources, all CIC higher-level tasks can be performed by the network rather than by the users themselves. This will bring forward cooperation due to several reasons:

- From the user's side, a fair-minded entity takes care of the cooperation-related decisions, increasing the users' confidence in the cooperation process. This way malice and authentication issues are no longer obstructive to real implementation. Security and privacy can additionally be strengthened through encrypting each user's data. In this way, decoding the information bearing data of another user during the cooperation process can be aggravated.
- An operator can boost cooperation by actively promoting it and thereby not relying on decentralized negotiations across the users. Consequently, the number of individual actions are reduced, which translates into power and/or degrees-of-freedom savings.
- BSs have much better overview of communications resources, user requirements, channel states, etc. Especially, managing the resources used for D2D allocation in the BSs could be part of the BS's scheduling algorithm.

16.3.2.4 Implementation and Design Aspects

Four aspects are identified to play a key role for designing and implementing CIC in D2D-assisted cellular networks.

Forwarding Scheme: The way cooperation over the D2D interface is achieved must be compliant with link budget constraints. Depending on the technical constraints, different forwarding strategies are envisaged:

- Amplify-and-Forward (AF): In this scheme, the signal received by the forwarding user is amplified and re-transmitted to the IC user. Importantly, the forwarded signal contains also interference from non-dominant BSs as well as receiver noise, which may limit the IC performance. Because the forwarded signal is a scaled version of the received downlink signal, the same time-frequency resources are used, which renders this scheme less bandwidth-efficient and non-flexible for resource allocation. While for the underlay D2D mode the latter might not be a problem, this is certainly not true for the overlay D2D mode. Due to its simplicity, AF is a prominent candidate for CIC and allows for full-duplex operation in principle.
- Decode-and-Forward (DF): Alternatively, the forwarding user can regenerate the dominant interference signal by decoding it, thereby eliminating residual interference and receiver noise. In case the interference signal must be decoded anyway, such as in scenario b in Sect. 16.3.1, this scheme may be appropriate. Otherwise, the receiver complexity is doubled since additional channel estimation and decoding tasks for the dominant interference signal have to be performed. Compared to AF, DF has the advantage that the decoded interference signal (or symbols) do not have to be forwarded on the same time-frequency resources, which renders this scheme more flexible and compliant with the overlay D2D mode. At the IC user, the forwarded interference signal can be used to zero-force dominant interference in the received downlink signal or as side information to facilitate IC.
- Quantize-and-Forward (QF): In this scheme, the interference signal is first quantized at the forwarding node before it is sent to the IC user. The effect of quantization is twofold: first, the received signal is mapped to a low-space representation which reduces the amount of data to be forwarded; second, residual interference and noise is partially eliminated. Similar to DF, this scheme allows for flexible D2D resource allocation and is therefore suitable for both underlay and overlay D2D mode. Especially for the case of nearby cooperating users, QF is considered to be an efficient forwarding scheme [15].

Synchronization and Estimation: The IC user has to perform synchronization and estimation tasks in order to reject the dominant interference. In contrast to the control of medium access between cooperating D2D devices, the cellular network cannot assist this synchronization process at the IC user. Depending on the forwarding scheme, synchronization can be facilitated by (GPS-based) time-stamping of the received signal before forwarding. As demonstrated by Fig. 16.7, it is already sufficient to align the jointly-strongest interferer in each signal to obtain significant interference correlation. Regarding channel estimation, the fact that reference symbols of neighboring cells are usually orthogonally-multiplexed can be exploited for estimating the channel to the dominant interfering BSs. This may be performed either in the time or in the frequency domain owing to the OFDM signal structure of LTE in the downlink. Hence, the channel estimation and subsequent IC process strongly resembles the conventional zero-forcing multi-antenna receiver principle.

Costs of Cooperation: Besides higher peak data rates and coverage extension, power savings are also anticipated by 3GPP D2D. Therefore, the cooperative methods must work in an energy efficient way in order to not worsen user experience through shorter battery life cycles. The energy used for interference forwarding depends on the bandwidth efficiency of the forwarding scheme as explained above. Furthermore, in the overlay D2D mode, CIC causes a dimension loss since resources are reserved exclusively for the forwarding process. In case this loss is not acceptable, e.g., in full-load situations, the underlay D2D mode may be more appropriate. Besides, by forwarding the interference signal to the IC user, the cooperative user refrains from transmitting/receiving its own data. Hence, CIC is applicable only to asymmetric traffic situations, e.g., CIC during the uplink phase when little or no data must be transmitted to the BS.

Measuring Interference Correlation: An important point to look at is how to determine the interference correlation at the D2D users and when to trigger a CIC process. One possible approach is as follows: the D2D users feedback CSI that allows the serving BS to identify strong interfering BSs. In LTE, this can be accomplished by letting the users measure the received power to each surrounding BS using the cell-specific reference symbols. If two close-by users experience dominant interference from the same interfering BS, this indicates that interference is correlated across these two users. Similarly, interference separability, which is necessary to reliably subtract dominant interference, can be verified in the same way by comparing the interference power from the strongest BS to the one from the second strongest BS or to the residual interference-plus-noise power.

16.3.3 Feasibility Analysis

We next present a first feasibility analysis of CIC in a D2D-assisted OFDMA cellular system similar to LTE. We consider a ring of eight BSs surrounding a pair of two BSs according to the hexagonal grid model with inter-BS distance of 500 m. A D2D user pair separated by a pre-defined distance is placed on the cell edge between the two inner BSs. To account for irregular BS deployments, the grid model is perturbed by randomly shifting the BSs according to a 2D zero-mean Gaussian distribution with standard deviation 50 m. The two D2D users are served by their closest BS, which is likely to be one of the two BSs inside the ring (other realizations are discarded). We assume that the serving BS performs spatial pre-coding to send orthogonal streams to the two D2D users on the same time-frequency resources, see scenario (a) in Sect. 16.3.1. The subsequent analysis and insights, however, apply in principle also to the other two scenarios introduced in Sect. 16.3.1.

16.3.3.1 Considered CIC D2D Scheme

We consider a typical asymmetric traffic scenario, in which both users wish to receive a large amount of data in the downlink, e.g., a video stream or file download. With some resources left for reliable CSI feedback, we assume that the serving BS can identify (i) the victim user (the one with the lower instantaneous SINR) and (ii) the strongest interfering BS for each of the two users. In case the two users experience dominant interference by the same BS, the serving BS then triggers the following CIC procedure: it selects the appropriate MCS for the non-victim user such that this user can reliably decode the desired signal. Furthermore, the serving BS schedules the non-victim user to perform interference forwarding to the victim user during the uplink phase (half-duplex operation), after having decoded and subtracted the desired signal from the received signal. The decision whether or not to trigger a CIC process depends on the degree of interference separability at the victim user. In this analysis, a CIC process is triggered whenever the dominant-interference to residual-interference-plus-noise ratio is above 1.5. We further assume an underlay D2D mode and that AF is used for interference forwarding. At the victim user, the two consecutively received and noise-corrupted signals contain correlated interference due to the jointly-dominant BS. Since reference symbols are orthogonally-multiplexed, the victim user tries to estimate the channels corresponding to the two interference signals from the dominant interfering BS. Then, the channel estimates are used to coherently subtract the dominant interference signal from the received signal containing the desired data. Depending on the channel estimation accuracy (at the users) and the interference separability (verified by the serving BS), the resulting signal has less interference and is then passed through the standard receive chain.

16.3.3.2 First Performance Results

Figure 16.9 shows the simulated performance of CIC for different path loss exponents α in terms of the improvement of the SINR's cumulative distribution function (CDF) and average throughput gain. In this scenario, the D2D user separation was set to 1 m which corresponds to a typical close-proximity situation, e.g., passengers in a bus or customers in a coffee bar. It can be seen that for small α, the interference caused by the strongest BS is less dominant. Consequently, a CIC procedure is triggered less often resulting in an average throughput gain of about 4 % at $\alpha = 3$. With increasing α, dominant interference becomes more separable and hence, CIC becomes more beneficial; however, at some point, receiver noise will start to dominate interference due to a large path loss, thereby expunging the gains of CIC. It can be concluded that the potential gains of CIC are located around 25 % average throughput gain improvement for the system parameters summarized in Table 16.1.

Since interference correlation and D2D link reliability decay with distance, it is important to analyze the resulting performance of CIC when increasing the D2D link distance. The link reliability is affected by the received D2D signal power (which is assumed to be considerably smaller in the underlay D2D mode to protect the uplink)

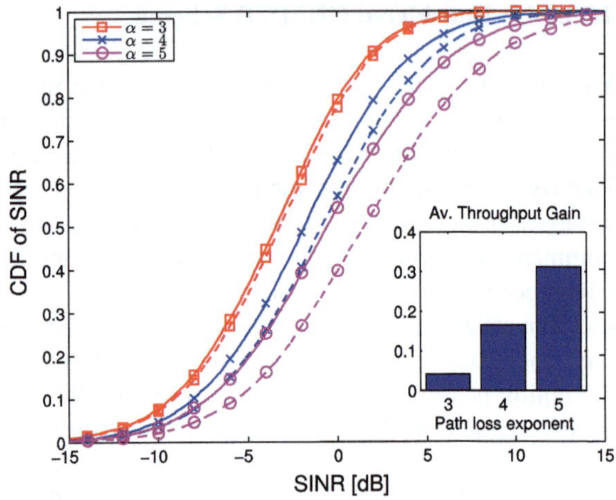

Fig. 16.9 Gains of CIC for a cell-edge scenario. Estimated CDF of the SINR with (*dashed line*) and without CIC (*solid line*) are shown for the victim user for different path loss exponents α. The corresponding average throughput gain is shown in the *lower right* graphic

Table 16.1 System parameters used for simulations

Parameter	Property
Average inter-BS distance	500 m
BS output power	43 dBm
Number of subcarriers	600
Number of OFDM symbols	7
FFT length	1024
Subcarrrier spacing	15 kHz
Reference symbol spacing	7 (frequency), 4 (time)
Thermal noise	−162.2 dBW/carrier
Channel	Extended Pedestrian A Model
Channel estimation	LS + 2D interpolation
Noise figure	5 dB
Maximum user Tx power	23 dBm

as well as by interference by other users transmitting in the uplink. We assume that there is one active user in the other cells (not serving the considered CIC user pair) transmitting over the same time-frequency resources. These interfering users employ power control with full long-term path loss inversion, which corresponds to the worst case in terms of interference to the considered CIC pair. Figure 16.10 shows the average throughput gain of CIC versus the D2D link distance for different D2D Tx powers (0 dBm left, 5 dBm right). In comparison with Fig. 16.7, one can see that, depending on the Tx power allowed on the D2D link, either interference correlation or link reliability limits the performance of CIC in the underlay D2D mode.

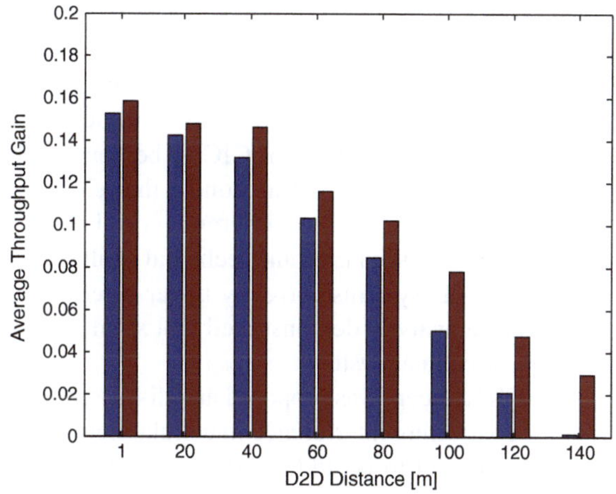

Fig. 16.10 Average throughput gain of CIC versus D2D-user separation for $\alpha = 4$. D2D Tx powers are 0 dBm (*left*) and 5 dBm (*right*)

16.3.4 Summary

We proposed and discussed the concept of CIC using D2D communications, which exploits the spatial interference correlation across close-by users in the downlink. Due to the fact that 3GPP LTE Rel-12 is operator-controlled with the possibility of tight control of D2D resources, we understand this concept as a new network interference-management tool to enhance throughput for users with low reception quality. We presented possible deployment scenarios and discussed several design challenges associated with D2D-based CIC. To each of these challenges, we outlined possible solutions. A first feasibility analysis confirmed that throughput gains of around 25 % can be achieved with CIC.

16.4 Conclusion

The diversity gains of cooperative relaying are well-understood in the noise-limited scenario, but not when interference from other nodes is taken into account. This is mainly due to the fact that interference is correlated across cooperating nodes, thereby affecting the obtainable diversity. Our analysis demonstrated that simply relaying the packet sent by the source as in decode-and-forward and performing MRC at the destination cannot increase the diversity order. However, in poor scattering environments with a dominant LoS path, spatial interference correlation across the relay and destination vanishes at small outage probabilities when the relay-destination distance is sufficiently large. Assuming an error-free relay-destination link, a diversity order increase then becomes possible. This scenario can be viewed as a distributed SIMO system with Rx diversity.

We then took one step further to see how cooperative nodes can exploit the fact that the observed interference is spatially correlated. We showed that interference forwarding is a promising approach and proposed the concept of CIC in the context of D2D communications. Since spatial interference correlation across cooperating nodes is a pre-requirement for CIC to be beneficial, we analyzed the interference correlation properties as a function of the relevant system parameters to obtain a better understanding of the interference correlation phenomenon. We discussed possible deployment scenarios and technical challenges, thereby focusing on operator-controlled D2D systems currently under discussion for 3GPP LTE Rel-12. A first performance analysis demonstrated that significant gains around 25 % are possible in a typical cellular system.

The CIC concept was proposed and discussed as a new interference-management tool for controlling interference in cellular networks. However, some open research problems remain to be addressed in order to understand the net gain of CIC. For instance, the dimension loss due to forwarding (part of) the interference to the victim user must be considered when assessing the overall performance of CIC. Furthermore, CIC (and D2D in general) may increase the interference level in the cellular network at other receivers, depending on the mode of operation. Taking such aspects into account is challenging because this usually requires a large-system analysis.

Acknowledgments This work was supported by the German Research Foundation (DFG) under Grants JO258/21-1 and JO258/21-2. The authors would like to thank their colleagues from KT Duisburg-Essen, namely Andrey Skrebtsov, Guido Bruck, and Peter Jung, for the cooperation within the projects StoCCNets and CREDIT.

References

1. Altieri A, Vega LR, Piantanida P, Galarza CG (2014) Analysis of a cooperative strategy for a large decentralized wireless network. IEEE/ACM Trans Netw 22(4):1039–1051. doi:10.1109/TNET.2013.2269054
2. Andrews JG (2013) Seven ways that HetNets are a cellular paradigm shift. IEEE Commun Mag 51(3):136–144. doi:10.1109/MCOM.2013.6476878
3. Andrews JG, Weber S, Haenggi M (2007) Ad hoc networks: to spread or not to spread? IEEE Commun Mag 45(12):84–91. doi:10.1109/MCOM.2007.4395371
4. Andrews JG, Ganti RK, Haenggi M, Jindal N, Weber S (2010) A primer on spatial modeling and analysis in wireless networks. IEEE Commun Mag 48(11):156–163. doi:10.1109/MCOM.2010.5621983
5. Cover T, Gamal AE (1979) Capacity theorems for the relay channel. IEEE Trans Inf Theory 25(5):572–584. doi:10.1109/TIT.1979.1056084
6. Crismani A, Toumpis S, Schilcher U, Brandner G, Bettstetter C (2014) Cooperative relaying under spatially and temporally correlated interference. IEEE Trans Veh Technol PP(99):1–1. doi:10.1109/TVT.2014.2372633
7. Dabora R, Maric I, Goldsmith A (2008) Interference forwarding in multiuser networks. In: IEEE global telecommunications conference (GLOBECOM), pp 1–5. doi:10.1109/GLOCOM.2008.ECP.192
8. Ganti RK, Haenggi M (2009) Spatial and temporal correlation of the interference in ALOHA ad hoc networks. IEEE Commun Lett 13(9):631–633. doi:10.1109/LCOMM.2009.090837

9. Haenggi M (2005) On distances in uniformly random networks. IEEE Trans Inf Theory 51(10):3584–3586. doi:10.1109/TIT.2005.855610
10. Haenggi M (2012) Diversity loss due to interference correlation. IEEE Commun Lett 16(10):1600–1603
11. Haenggi M, Ganti RK (2008) Interference in large wireless networks. Found Trends Netw 3(2):127–248. doi:10.1561/1300000015
12. Haenggi M, Smarandache R (2013) Diversity polynomials for the analysis of temporal correlations in wireless networks. IEEE Trans Wirel Commun 12(11):5940–5951
13. Haenggi M, Andrews JG, Baccelli F, Dousse O, Franceschetti M (2009) Stochastic geometry and random graphs for the analysis and design of wireless networks. IEEE J Sel Areas Commun 27(7):1029–1046. doi:10.1109/JSAC.2009.090902
14. Hoymann C, Chen W, Montojo J, Golitschek A, Koutsimanis C, Shen X (2012) Relaying operation in 3GPP LTE: challenges and solutions. IEEE Commun Mag 50(2):156–162. doi:10.1109/MCOM.2012.6146495
15. Katz M, Shamai S (2006) Relaying protocols for two colocated users. IEEE Trans Inf Theory 52(6):2329–2344. doi:10.1109/TIT.2005.862090
16. Laneman JN, Tse D, Wornell GW (2004) Cooperative diversity in wireless networks: efficient protocols and outage behavior. IEEE Trans Inf Theory 50(12):3062–3080. doi:10.1109/TIT.2004.838089
17. Lee D, Seo H, Clerckx B, Hardouin E, Mazzarese D, Nagata S, Sayana K (2012) Coordinated multipoint transmission and reception in LTE-advanced: deployment scenarios and operational challenges. IEEE Commun Mag 50(2):148–155. doi:10.1109/MCOM.2012.6146494
18. Sawahashi M, Kishiyama Y, Morimoto A, Nishikawa D, Tanno M (2010) Coordinated multipoint transmission/reception techniques for LTE-advanced [coordinated and distributed MIMO]. IEEE Wirel Commun 17(3):26–34. doi:10.1109/MWC.2010.5490976
19. Schilcher U, Bettstetter C, Brandner G (2012) Temporal correlation of interference in wireless networks with Rayleigh block fading. IEEE Trans Mobile Comput 11(12):2109–2120. doi:10.1109/TMC.2011.244
20. Stoyan D, Kendall W, Mecke J (1995) Stochastic geometry and its applications, 2nd edn. Wiley
21. Tanbourgi R, Jäkel H, Jondral FK (2013) Cooperative relaying in a Poisson field of interferers: a diversity order analysis. In: IEEE international symposium on information Theory, pp 3100–3104. doi:10.1109/ISIT.2013.6620796
22. Tanbourgi R, Dhillon HS, Andrews JG, Jondral FK (2014) Dual-branch MRC receivers under spatial interference correlation and Nakagami fading. IEEE Trans Commun 62(6):1830–1844. doi:10.1109/TCOMM.2014.2321553
23. Tanbourgi R, Dhillon HS, Andrews JG, Jondral FK (2014) Effect of spatial interference correlation on the performance of maximum ratio combining. IEEE Trans Wirel Commun 13(6):3307–3316. doi:10.1109/TWC.2014.041714.131330
24. Weber S, Yang X, Andrews JG, de Veciana G (2005) Transmission capacity of wireless ad hoc networks with outage constraints. IEEE Trans Inf Theory 51(12):4091–4102. doi:10.1109/TIT.2005.858939
25. Weber S, Andrews JG, Jindal N (2010) An overview of the transmission capacity of wireless networks. IEEE Trans Commun 58(12):3593–3604. doi:10.1109/TCOMM.2010.093010.090478

9. Haenggi M (2005) On distances in uniformly random networks. IEEE Trans Inf Theory 51(10):3584–3586. doi:10.1109/TIT.2005.855610
10. Haenggi M (2012) Diversity loss due to interference correlation. IEEE Commun Lett 16(10):1600–1603
11. Haenggi M, Ganti RK (2008) Interference in large wireless networks. Found Trends Netw 3(2):127–248. doi:10.1561/1300000015
[illegible]

Chapter 17
Physical-Layer Key Generation and Reconciliation

**Jon Wallace, Rashid Mehmood, Rajesh Sharma,
Werner Henkel, Oana Graur, Nazia Islam and Alexandra Filip**

Abstract Physical layer security is a technique that makes use of the physical communication channel or medium to provide additional robustness to eavesdroppers and attackers. Key establishment exploiting a multiple-input multiple-output (MIMO) reciprocal wireless channel is proposed and its performance investigated, indicating that secure keys can be rapidly generated between two nodes, even in the presence of close eavesdroppers. The use of reconfigurable antennas is investigated and experimentally proven as a key establishment solution for static channels with limited multipath. Additionally, different options are discussed to handle key-differences due to non-correlated noise together with quantization, either by simply introducing guard intervals or by joint source coding with LDPC codes.

J. Wallace
Lafayette College, 730 High Street, Easton, PA 18042, USA
e-mail: wall@ieee.org

R. Mehmood
Brigham Young University, 459 CB, Provo, UT 84604, USA
e-mail: rashid.mehmood@ieee.org

R. Sharma
Ilmenau University of Technology, Ehrenbergstrasse 29, 98693 Ilmenau, Germany
e-mail: rajesh.sharma@ieee.org

W. Henkel (✉) · O. Graur · N. Islam
Jacobs University Bremen, Campus Ring 1, 28759 Bremen, Germany
e-mail: werner.henkel@ieee.org

O. Graur
e-mail: o.graur@jacobs-university.de

N. Islam
e-mail: n.islam@jacobs-university.de

A. Filip
DLR, Oberpfaffenhofen, 82234 Weßling, Germany
e-mail: alexandra.filip@dlr.de

W. Utschick (ed.), *Communications in Interference Limited Networks*,
Signals and Communication Technology, DOI 10.1007/978-3-319-22440-4_17

17.1 Introduction

Information security is one of the top concerns in today's digitally networked world, where common acts of *cybercrime* include identity theft, credit card fraud, and corporate data breaches. Keeping sensitive information secure from unauthorized parties is becoming increasingly difficult as common access networks like the Internet are used by corporations to share information. Additionally, as wireless technologies become more popular for consumer applications, the potential for eavesdropping and stealing private information is increasing. This is especially true in interference-limited environments, where a high density of users indicates a high potential for eavesdropping and data theft.

Existing security measures typically rely on having a secret key (or a private/public key pair) that is used to encode and decode the information. Although often secure, such methods are vulnerable to brute-force attacks, where large amounts of computing power can be used to learn the keys and decode the sensitive information. There is increasing interest in methods that ensure *perfect secrecy*, or methods of encoding information that are theoretically unbreakable. One such method is the *one-time pad* where identical secret codebooks are possessed by a sender and receiver, and the bits of the common secret key are only used once. The drawback of the one-time pad is the need for a very long key that must be privately distributed to the nodes.

Whereas existing security techniques typically operate on digital data at the data link layer or higher, there is increasing interest in *physical layer* security methods that use special modulation or information about the channel to provide an increased measure of security. For example, when it can be guaranteed that the wireless channel to a legitimate user has higher signal-to-noise (SNR) ratio than eavesdropper channels, it is possible to construct codes that convey information to the intended recipient, while sending a signal that cannot be decoded to eavesdroppers. In contrast to such methods requiring an SNR advantage, key establishment techniques based on wireless channel reciprocity allow long secret keys to be automatically generated, which could be used to provide perfect security as a one-time pad.

This chapter summarizes research performed in the COIN program, whose purpose was to investigate applying security at the physical layer to create an added measure of security in wireless and wired networks.

17.2 Background

This section provides a high-level overview of concepts in cryptography and physical-layer security.

17.2.1 Traditional Cryptography

Security in modern networks is typically accomplished by manipulating the discrete-valued payload data using traditional cryptographic methods [26], a step that is completely separate from modulation and transmission over the physical communication medium. Generally, these methods employ secret keys to encrypt and decrypt information, making reception by eavesdroppers difficult.

It is well understood that the security of any key-based method is directly proportional to the length of the key and inversely proportional to how many message bits are protected with a single key. Security in cryptographic systems is measured in terms of either theoretical or practical security. Theoretical security is the strongest notion of security and indicates how safe the system is, regardless of the processing power and time available to the eavesdropper. The weaker notion of practical security depends on the amount of work that the eavesdropper must perform to break the system using the best possible attack.

A breakthrough in cryptographic systems was the appearance of public-key methods such as RSA, obviating the need to initially transmit a secret key over a secure channel. In such systems, each node has a public and private key. For one node (say Alice) to communicate securely to another node (say Bob), Alice first obtains the public key of Bob over a non-secure channel and encrypts her message. Since the public-private key pair is generated using a so called *one-way function*, the message can only be decoded by using Bob's private key which is kept secret. Although unproven in terms of theoretical security, public-key cryptography appears to be practically secure from known forms of attack. One difficulty with public-key based methods, especially for wireless networks, is the need to distribute public keys to all communicating nodes.

A cryptosystem implementing perfect or *theoretical* security is shown in Fig. 17.1, where $\mathbf{X}$, $\mathbf{Z}$, and $\mathbf{Y}$ represent the plaintext, the secret key, and the cryptogram (the ciphertext), respectively, whereas $\hat{\mathbf{X}}$ and $\hat{\mathbf{Z}}$ denote an enemy's guess of $\mathbf{X}$ and $\mathbf{Z}$. Theoretical security assumes that the enemy has access to $\mathbf{Y}$ but not to $\mathbf{X}$ and $\mathbf{Z}$, and is thus limited to a ciphertext-only attack.

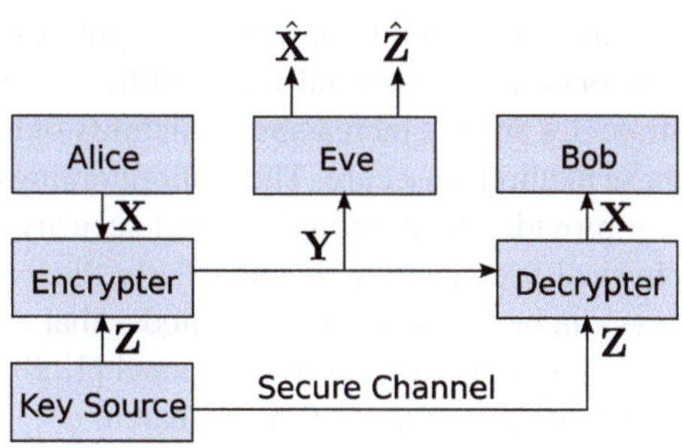

Fig. 17.1 A secret-key cryptosystem illustrating the idea of a one-time pad, which provides perfect or theoretical security. Alice transmits an encoded message to another node Bob, while Eve is an enemy cryptanalyst who tries to steal the message and/or key

According to Shannon, perfect secrecy is achieved if the enemy can do no better estimating $\mathbf{X}$ with knowledge of $\mathbf{Y}$ than he could do without knowing $\mathbf{Y}$. This can be expressed mathematically as [26]

$$H(\mathbf{X}|\mathbf{Y}) = H(\mathbf{X}), \tag{17.1}$$

where $H(\mathbf{X}|\mathbf{Y})$ denotes the conditional entropy of $\mathbf{X}$ given $\mathbf{Y}$.

The simplest implementation of perfect secrecy is the one-time pad. Suppose that the plaintext $\mathbf{X}$, the ciphertext $\mathbf{Y}$, and the key digits $\mathbf{Z}$ are vectors of the same length N that take their values from an L-ary alphabet $\{0, 1, ..., L - 1\}$. Let the key be uniformly chosen at random and used to encipher the plaintext as

$$Y_i = X_i \oplus Z_i, \; i = 1, 2, ..., N, \tag{17.2}$$

where $\oplus$ denotes a modulo-L addition and the subscript i denotes the element index of the vector to which it belongs. The one-time pad provides perfect secrecy no matter what the statistics of $\mathbf{X}$ are. For perfect security, the one-time pad also represents the shortest possible secret key [26].

With a one-time pad, since a wiretapper has no access to the secure channel via which the secret key is transmitted, he or she cannot decrypt the message. Conversely, if the secret key is transmitted via the public channel and the cryptogram via the secure one, he or she still cannot decrypt it. This means there is no logical difference between the secure channel and the public one and hence both channels can be generalized into wire-tap channels, on which Ozarow and Wyner provide an in-depth analysis [40]. They consider the problem, how K data bits can be encoded into N transmitted bits such that the enemy cryptanalyst, upon detecting $\mu < N$ bits, has the maximum possible uncertainty about the original data.

17.2.2 Physical Layer Security

Although existing cryptographic methods provide a good degree of practical security, key distribution for both secret- and public-key cryptography is a burden for wireless networks. Also, especially for public-key methods, the development of new attack strategies and the increasing availability of inexpensive processing power may render these methods insecure. These shortcomings have fueled the search for new methods that provide an additional layer of security by taking advantage of properties of the physical propagation channel. Early theoretical work showed that perfectly secure keys can be established by two nodes that exploit knowledge of the physical channel as well as a public error-free channel [1, 28]. Analysis and practical implementation of this idea is a focus of this research.

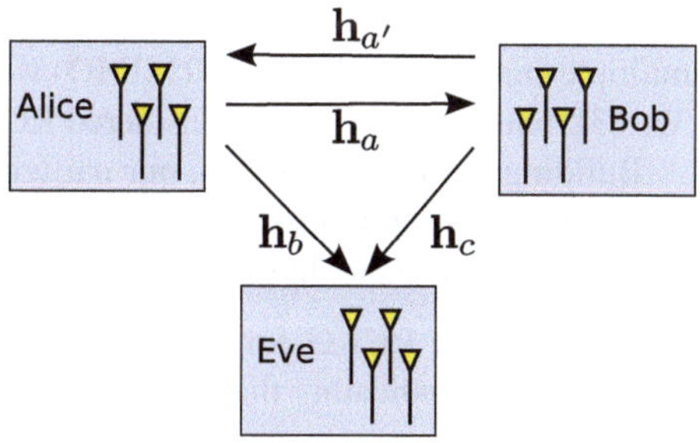

Fig. 17.2 Wireless communications scenario. Alice and Bob are legitimate users and Eve is a potential eavesdropper

One promising method for exploiting the physical layer for improved security is to use a reciprocal wireless channel as a source of common randomness for automatic secret key generation. Figure 17.2 illustrates a generalized system model for describing wireless key establishment methods. Alice and Bob are legitimate users who require secure communication, while Eve is a potential eavesdropper. Vector channels $\mathbf{h}_a = \mathbf{h}_{a'}$ are referred to as the *forward* and *reverse* channels for legitimate communications, which are estimated by Bob and Alice, respectively. Channels $\mathbf{h}_b$ and $\mathbf{h}_c$ convey information to (and are estimated by) Eve. Due to noise or synchronization errors, the nodes have imperfect estimates of the channels, or

$$\hat{\mathbf{h}}_a = \mathbf{h}_a + \boldsymbol{\varepsilon}_2, \quad \hat{\mathbf{h}}_{a'} = \mathbf{h}_{a'} + \boldsymbol{\varepsilon}_1, \quad \hat{\mathbf{h}}_b = \mathbf{h}_b + \boldsymbol{\varepsilon}_3, \quad \hat{\mathbf{h}}_c = \mathbf{h}_c + \boldsymbol{\varepsilon}'_3, \tag{17.3}$$

where $\boldsymbol{\varepsilon}_i$ and $\boldsymbol{\varepsilon}'_i$ are zero-mean complex Gaussian estimation errors.

For a reciprocal channel, $\hat{\mathbf{h}}_a \approx \hat{\mathbf{h}}_{a'}$, Alice and Bob can generate a shared key by simply observing and quantizing $\hat{\mathbf{h}}_a$ and $\hat{\mathbf{h}}_{a'}$. If there is sufficient time-variation of the channels, and $\mathbf{h}_a$ and $\mathbf{h}_{a'}$ vary independently of $\mathbf{h}_b$ and $\mathbf{h}_c$, long keys can automatically be generated that are secure from Eve. By proper choice of the quantization method and/or error control methods, a low probability of key mismatch can be ensured. These keys can then be used as a one-time pad to provide perfect security for highly sensitive information. Generating secret keys in this way is referred to herein as reciprocal channel key generation (RCKG) and was studied in depth in the framework of COIN.

The theoretical foundation of RCKG was established by information theoretic treatments of common randomness and secret-key agreement [1, 28–31]. RCKG is also related to quantum-distributed Gaussian keys [54], and was analyzed for scalar Gaussian channels in [60, 61]. Several practical methods for RCKG appeared at the same time or before the research described in this chapter was begun. In [20] a method is described where the phase of the reciprocal channel is used to randomly rotate the phase of each transmitted data symbol, which is a simple RCKG method with one key symbol per data symbol. Exploiting the delay profile of reciprocal ultrawideband-channels is treated in [58]. In [48], direct quantization of the phase is used for key generation with scalar channels. In [27, 62], a quantization method is experimentally verified that exploits amplitude level crossing to generate keys with low mismatch. Analysis of secure transmission and key agreement techniques for multiple-input multiple-output (MIMO) was also gaining attention [13, 16, 19, 22, 39, 63], but very little work considered RCKG in the context of MIMO.

Building on this foundation, our participation in COIN provided novel analysis, which extended previous results and led to new observations that were not in the previous literature on physical layer security employing RCKG:

- The limits of RCKG for MIMO channels were studied from an *experimental* perspective, indicating that MIMO channels can be used to rapidly generate long and secure keys.
- The measurements facilitated the development of a realistic RCKG protocol that can be applied to real time-varying MIMO channels.
- RCKG employing *reconfigurable* antennas was studied in depth through both simulation and measurement, indicating that highly secure keys can be automatically generated.
- Methods for key generation in *non-Gaussian* channels were developed and analyzed, especially appropriate for reconfigurable antenna techniques.
- A novel beamforming method that is analogous to array synthesis was developed for optimizing security in LOS scenarios.

17.3 Metrics and Simulation of MIMO RCKG

Our initial work in [56, 57] considered using reciprocal fading MIMO channels as the source of common randomness to generate a shared secret key, where we adopted the *intrinsic conditional mutual information* in [28] to define two critical security metrics. The first metric is *available key bits*, which refers to the maximum number of independent key bits that can be generated from each observation of the random channel, or

$$I_{\mathrm{K}} = I(\hat{\mathbf{h}}_a; \hat{\mathbf{h}}_{a'}) = \mathrm{E}\left\{ \log_2 \frac{f(\hat{\mathbf{h}}_a, \hat{\mathbf{h}}_{a'})}{f(\hat{\mathbf{h}}_a) f(\hat{\mathbf{h}}_{a'})} \right\}, \tag{17.4}$$

where $\mathrm{E}\{\cdot\}$ is expectation, and $f(\cdot)$ is a probability density function (PDF).

Given that Eve's channel is not independent of $\hat{\mathbf{h}}_a$ and $\hat{\mathbf{h}}_{a'}$, she may be able to obtain some information about the established key. This is accounted for using the metric *secure key bits* or I_{SK}, which is the number of generated key bits per channel observation that can be secure with respect to Eve, given by

$$I_{\mathrm{SK}} = I(\hat{\mathbf{h}}_a; \hat{\mathbf{h}}_{a'} | \hat{\mathbf{h}}_b, \hat{\mathbf{h}}_c) = \mathrm{E}\left\{ \log_2 \frac{f(\hat{\mathbf{h}}_a, \hat{\mathbf{h}}_{a'} | \hat{\mathbf{h}}_b, \hat{\mathbf{h}}_c)}{f(\hat{\mathbf{h}}_a | \hat{\mathbf{h}}_b, \hat{\mathbf{h}}_c) f(\hat{\mathbf{h}}_{a'} | \hat{\mathbf{h}}_b, \hat{\mathbf{h}}_c)} \right\}. \tag{17.5}$$

When Eve's channels are independent from the Alice-Bob channel, we have $I_{\mathrm{SK}} = I_{\mathrm{K}}$, indicating all available bits are also secure from the eavesdropper. We have also used the metric *vulnerable key bits* which refers to the number of key bits that can potentially be estimated by Eve, which is defined as $I_{\mathrm{VK}} = I_{\mathrm{K}} - I_{\mathrm{SK}}$.

When fading channels exhibit zero-mean joint Gaussian statistics, I_{K} and I_{SK} reduce to fairly simple closed-form expressions, which were given in [56, 57]. When channels are non-Gaussian, obtaining closed-form expressions for I_{K} and I_{SK} is difficult. In [33] we developed a Monte-Carlo method for computing these quantities numerically using expectations.

Our earliest work in the COIN framework involved performing detailed simulations of RCKG using a realistic path-based model from [18] and applying the identified key-generation metrics. The results are detailed in [56, 57] and can be summarized as follows:

- I_{K} is limited by available multipath. For rich multipath, I_{K} scales quadratically with additional antennas, in contrast to the linear growth seen with MIMO channel capacity.
- For single-antenna RCKG, simulated I_{SK} shows that most key bits are secure from an eavesdropper, provided there is some multipath and Eve is at least one wavelength away from legitimate nodes.
- For MIMO RCKG, both richer multipath and wider separation of Eve from Alice/Bob is needed to ensure security.
- High SNR provides very little advantage to eavesdroppers, and array-size advantage is the main concern for security.

17.4 Some First Channel Quantization (CQ) Methods for Key Generation

In the framework of COIN, several techniques for quantizing scalar and MIMO channels were developed, allowing RCKG to be implemented in practice. It was observed that existing methods like [48] were prone to significant key mismatch rate, and more robust quantization methods were needed. We started from simpler procedures to more advanced ones visible from the sequence of sections that follow. The advanced ones are adaptive guard intervals described in Sects. 17.7.1 and 17.7.2 and joint source-coding LDPC-coding schemes covered in Sect. 17.7.

17.4.1 Channel Quantization with Guard Band (CQG)

The first simple quantization method that was developed is reported in [56, 57] and can be seen as a generalization of the CQ methods in [27, 48]. Both amplitude and phase fluctuations are exploited by dividing the space of observable channels into

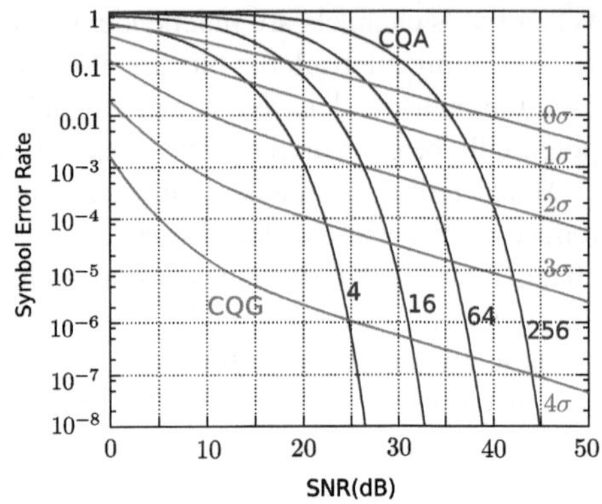

Fig. 17.3 Exact symbol error rate performance of CQ methods for a single unit variance complex channel with varying guard band g for CQG and varying order M for CQA

M equally probable quantization sectors (QSs), and each sector is assigned a unique symbol and bit pattern. At each time instant, Alice and Bob observe the channel, determine the quantization sector the channel is in, and concatenate the appropriate bits to the key. A problem arises when channels are observed near the quantization boundaries, and we suggested guard bands as a solution. An iterative method for finding equal-probability quantization sectors with specified guard band g was given in [56]. Increasing guard bands gradually reduces the efficiency of key generation, but dramatically reduces the symbol mismatch rate. We introduce in Sects. 17.7.1 and 17.7.2 a more advanced vector quantization method based on the *Linde-Buzo-Gray* algorithm [12] and show our results for adaptive guard bands, constructed such that regions with low occurrence probability have narrow guard band width, while more probable regions exhibit a larger guard band width.

Exact symbol error rate and a simple efficiency measure were given in [57], and improved efficiency metrics and exact *bit*-error rates were reported in [49]. Figure 17.3 plots the symbol error rate for CQG with varying SNR and guard band, where σ is the standard deviation of the complex Gaussian channel.

17.4.2 Channel Quantization Alternating (CQA)

The main problem of CQG was the reduced efficiency when large guard bands were used. Channel quantization alternating (CQA) uses a different approach, where four different quantization maps are used for each complex channel dimension, and Alice chooses the map that gives the least probability of mismatch, of which she informs Bob. CQA is similar to CQG once a quantization map has been chosen, but efficiency is not reduced, since channel observations are never discarded. It was found that CQA is very similar to the coset source coding procedure presented in [42] and adopted in [58]. Exact performance of CQA was also derived and reported in [57]. Figure 17.3 shows the symbol error rate and efficiency of CQA, indicating low mismatch rate at a higher efficiency than CQG.

17.4.3 Quantization Methods for Non-Gaussian Channels

Non-Gaussian channels complicate the simple quantization methods (CQA and CQG) developed thus far. We developed one approach in [50], which chooses quantization intervals based on the empirical cumulative distribution function (CDF) of measured channel data. The samples for each record of measurement data are divided into several blocks, where the fading statistics for a single block can be considered stationary. For each block the boundaries for equiprobable sectors in a single real dimension are determined for a desired number of quantization sectors. SER and efficiency graphs for the method can be computed using a Monte-Carlo analysis.

Although the CDF-based quantization overcomes the difficulty of non-Gaussian statistics, a disadvantage of this and other standard CQ methods is the need to send error correction information between Alice and Bob to avoid key mismatch. By transmitting this error control information publicly, the useful randomness of the key is reduced. Instead of quantizing channel snapshots individually, the standard coding procedure assigns codes to *sequences* of channel observations. Using known channel statistics, codebooks can be formed with sufficiently long codes to overcome infrequent mismatches. However, when the channel statistics are non-stationary or unknown, codebook generation is very difficult.

The first kind of coding strategy we published in [50], referred to as *positional coding*, which can be applied to channels with arbitrary fading statistics. The codebook with N_q codes is generated at each node by simultaneously observing N_q sequences of n samples of the reciprocal channel at Alice and Bob. A secret key is then communicated from Alice to Bob by transmitting each code from the codebook in a random order, where the specific order chosen conveys the key. Whereas Alice and Bob have nearly matching codebooks, allowing Bob to decode which key is sent, Eve observes different channels and cannot properly decode. Later in Sect. 17.8, we will give details on how to use capacity-achieving LDPC codes for key reconciliation.

17.5 Measured MIMO RCKG Rates

Direct measurement of I_{K} and I_{SK} for MIMO channels in an indoor scenario was performed and reported in [57], allowing us to compare with simulation results. MIMO measurements were performed with a custom 8×8 channel sounder on the first floor of the Research I building on the Jacobs University Bremen Campus, consisting of classrooms and laboratories.

Several observations were made from the measurement campaign as described below. Figure 17.4 compares I_{K} and I_{SK} for LOS and NLOS scenarios with 4 antennas for Alice, Bob, and Eve and SNR = 15 dB. Line-of-sight (LOS) channels were seen to exhibit about half of the available key bits compared to non-line-of-sight (NLOS) channels. The main reason for this was the removal of the high-power non-fading

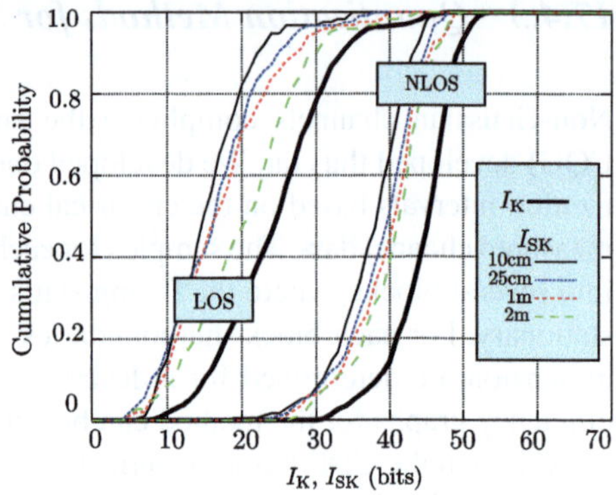

Fig. 17.4 CDFs for I_K and I_{SK} averaged over all LOS and NLOS scenarios for 15 dB SNR and various eavesdropper separations

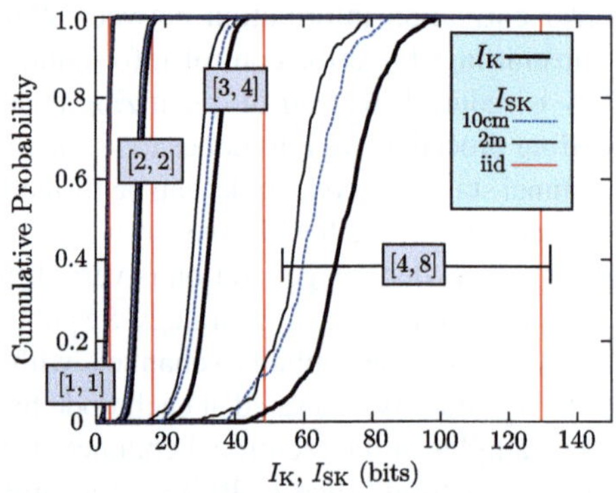

Fig. 17.5 CDFs for I_K and I_{SK} for NLOS scenarios with varying number of antennas. Curves $[N_1, N_2]$ indicate Alice and Eve have N_1 antennas each and Bob has N_2 antennas

component, leading to a lower effective SNR. The results also show that I_{SK} is more sensitive to eavesdropper separation for LOS as compared to NLOS.

The effect of array size for NLOS channels is shown in Fig. 17.5, where curves for (N_1, N_2) indicate N_1 antennas at Alice and Eve and N_2 at Bob. The result shows that for few antennas at Alice and Bob, I_K is near the theoretical maximum (i.i.d. Gaussian channels) and most key bits are secure. However, in contrast to simulations, the gap between ideal I_K and I_{SK} widens for a modest increase in the number of antennas. Also, a significant gap exists between the security metrics of ideal Gaussian i.i.d. curves and measured channels.

Finally, Fig. 17.6 shows the effect of eavesdropper advantage seen in the measurements, highlighting cases where this advantage has the minimum and maximum effect. Eavesdropper advantage has little effect for small arrays that are balanced at Alice and Bob ($N_1 = N_2 = 1$ in this case). For the wider eavesdropper separation of 2 m, having an array size ($N_3 = 4$) and SNR (35 dB) advantage helps Eve negligibly. On the other hand, for unbalanced arrays ($N_1 = 1$, $N_2 = 8$), having more antennas at Eve reduces security significantly, as depicted by the arrows in the figure. As with simulations, having more antennas is much more beneficial to Eve than high SNR.

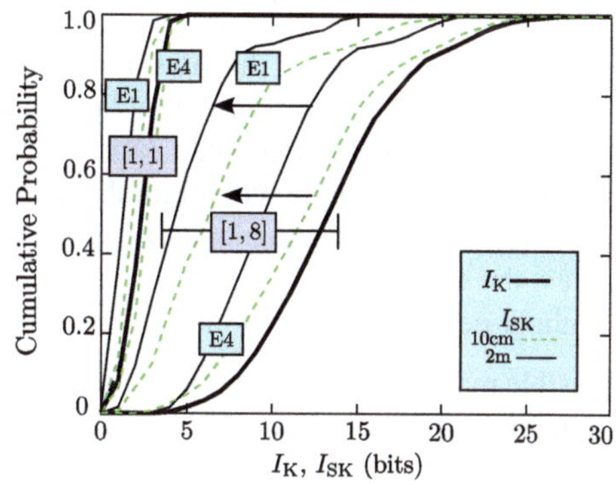

Fig. 17.6 CDFs of I_K and I_{SK} for LOS scenarios with SNR = 15 dB and either 10 cm or 2 m eavesdropper separation. Curves labeled with $[N_1, N_2]$ refer to the case where Alice and Bob have N_1 and N_2 antennas, respectively. For each case, Eve either had 1 antenna (E1) or 4 antennas (E4)

17.6 Key Generation Employing Reconfigurable Antennas

The ability to apply RCKG for fixed antennas is limited when the channel is very slowly fading or static, since the amount of common randomness that can be exploited is limited. In [2] the useful idea of using a *reconfigurable* antenna for key establishment was presented, where random states of an electronically steerable parasitic array (ESPAR) were used to create a random reciprocal channel state at the two communicating nodes, even for a static underlying propagation channel. Although the proof-of-concept demonstrator described in [2] provided evidence that secure keys can be generated this way, many questions remained on the benefits or limits of the approach.

Within the framework of COIN, we explored the effect of reconfigurable antenna complexity on the key generation process [33–36]. Specifically, we studied what *level* and what *type* of complexity is required to ensure not only a high key generation rate (high I_K), but also security from an eavesdropper ($I_{SK} \approx I_K$). The impact of antenna complexity was studied through realistic simulation and direct measurements with a versatile prototype parasitic array having scalable complexity.

17.6.1 Maximum Key Length Versus Antenna Complexity

One of the first questions we addressed was to consider the maximum theoretical number of independent key bits that can be generated from any hypothetical reconfigurable antenna [33]. Consider the scenario where Alice and Bob have reconfigurable antennas and randomly change the states of their antennas to provide a time-varying Alice-Bob channel. This also creates corresponding fluctuations in the Alice-Eve and Bob-Eve channels. Although Eve does not know the *mapping* of key bits to the channels she observes, Eve can record the *sequence* of channels she sees. If the combined complexity of the two reconfigurable antennas is too low, Eve can learn the key by simply trying all possible mappings, which may be much less complex than trying all possible key sequences.

One way to preclude this possibility is to consider how many total secure key bits (N_{bits}) must be generated during static channel conditions. By making the combined reconfigurable antenna complexity large enough, such that the number of possible mappings to search is larger than $2^{N_{\text{bits}}}$, a reduced-complexity brute-force attack is not possible. Given a single reconfigurable antenna at one of the communicating nodes with N_{RE} reconfigurable elements, where each element can assume N_{RS} possible states, the total number of antenna states is $N_{\text{RS}}^{N_{\text{RE}}}$. For a quantization order of N_q symbols per channel observation, each channel sample that Eve observes has N_q possible mappings. Thus, the total combination of mappings to check is $N_q^{(N_{\text{RS}}^{N_{\text{RE}}})}$ and we require

$$N_{\text{RS}}^{N_{\text{RE}}} \log_2 N_q \geq N_{\text{bits}} \tag{17.6}$$

to avoid a reduced-complexity brute-force attack. For example, to be able to securely generate 10^5 key bits using a reconfigurable antenna that uses two-state switches $N_{\text{RS}} = 2$, we require $N_{\text{RE}} \approx 16$ elements. Key lengths like 10^5 seem long in the context of existing cryptosystems, but the number is quite modest if the key is used as a one-time pad. Noting the rapid increase in the left-hand side of (17.6) with increasing N_{RS}, an interesting possibility is to use reconfigurable elements that have virtually continuous states (such as a varactor connected to an analog noise source). Such an approach appears to preclude the reduced-order brute-force attack altogether.

17.6.2 *Normality of Reconfigurable Antenna Channels*

Ideally, channel states created by random control of a reconfigurable antenna should conform to a complex Gaussian distribution with the maximum variance, thus providing the maximum key generation rate. The ability of parasitic arrays to achieve this was studied in [33]. The largest deviation from an ideal complex Gaussian distribution was seen in the channel phase, as shown in Fig. 17.7, which compares CDFs of the phase for reconfigurable antenna channels with the ideal uniform CDF. The labels LR and HR refer to low reconfigurability ($N_{\text{RE}} = 4$) and high reconfigurability

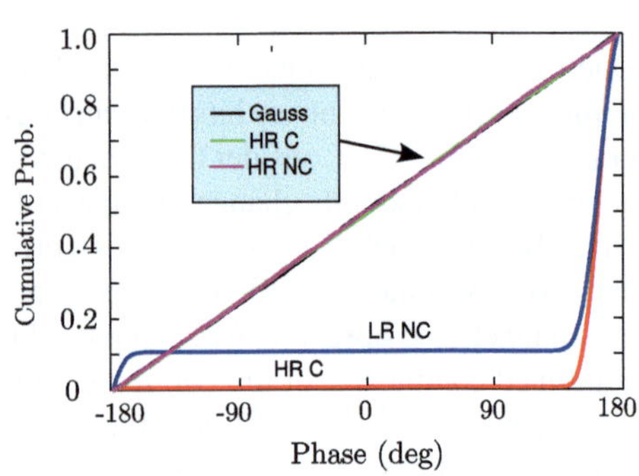

Fig. 17.7 Phase distribution of the channel obtained using a reconfigurable parasitic array for a static, LOS propagation channel

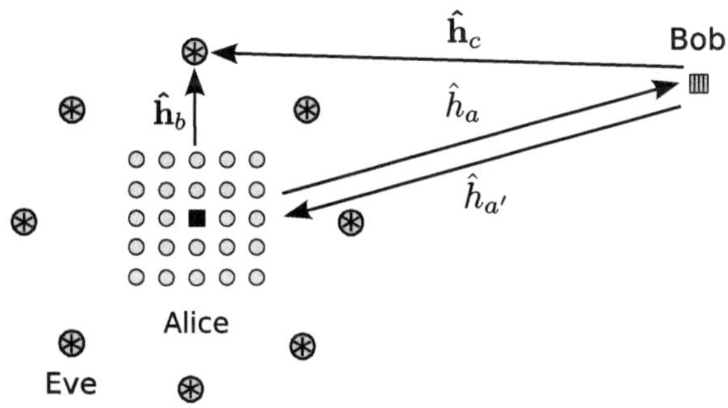

Fig. 17.8 Antenna arrangement in the security analysis of reconfigurable antennas, where Bob has a single antenna (*hatched square*), Alice has a reconfigurable array with one feed (*black square*) and programmable elements (*empty circles*), and Eve has an array surrounding Alice (*asterisks*)

($N_{\mathrm{RE}} = 64$), and C and NC refer to control or no control of the reflection coefficient (matching) at the antenna feed. As can be seen, the statistics are nearly uniform for high complexity and reflection control, which is not the case for a low complexity antenna.

17.6.3 Measured Key Generation Versus Antenna Complexity

The security of key generation employing reconfigurable antennas was studied in [36] through direct measurement, where the topology in Fig. 17.8 was assumed. Here, Alice is equipped with a reconfigurable array having a single active element, while Bob only has a single antenna. Eve has an active array that surrounds the reconfigurable array, allowing Eve to gain the most information about the key generation process. Different levels of complexity of the reconfigurable antenna were considered by considering different numbers and arrangements of the reconfigurable elements in the square grid depicted in Fig. 17.8.

Figure 17.9 summarizes key generation statistics averaged over four indoor measurement locations. The curves show that secure key bits drop in a nearly linear fashion with increased antennas at Eve (N_{E}). Here, uniform assignment of the reconfigurable states was assumed, and for this choice, using simple switches $N_{\mathrm{RS}} = 2$ gave the highest secure key rate. Also, approximately $N_{\mathrm{RE}} = 16$ reconfigurable elements are needed to achieve peak security.

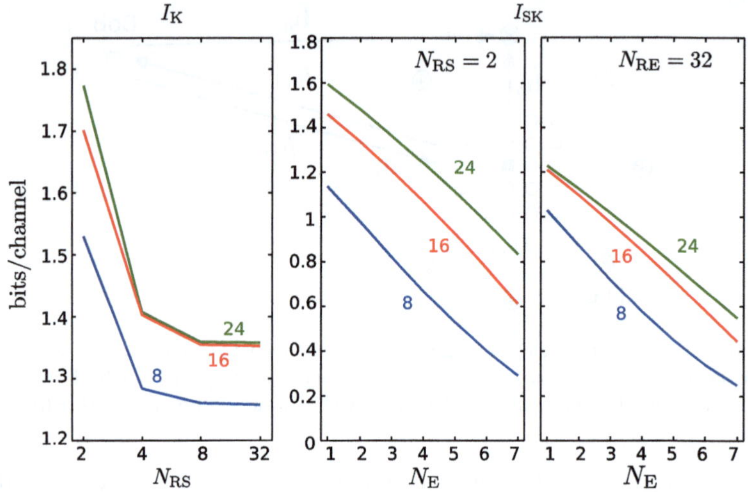

Fig. 17.9 Measured I_K and I_{SK} for the reconfigurable array as a function of N_E for different values of N_{RE} and N_{RS}. Each curve has the N_{RE} value indicated with a number next to the plot

17.6.4 Secure Array Synthesis

Line-of-sight (LOS) channels are arguably the most challenging for employing physical layer security, since there is no multipath to provide common randomness for key generation. Section 17.6 explored the idea of using a reconfigurable antenna to synthetically generate random channels, where the states of antenna elements were controlled in an ad-hoc way. An important outstanding question is how a reconfigurable or adaptive array can be *optimally* used in LOS scenarios to convey information (such as a key) to a desired node, and avoid sending that information to eavesdroppers.

One method that can be employed in line-of-sight (LOS) channels is standard array synthesis, which solves the problem of finding the antenna weights required to send the signal in certain directions, while avoiding transmission in other directions. In the context of security, it is desirable to synthesize a *signal pattern* that sends the information-bearing signal to a desired node, while limiting transmission of that information in other directions. Security can be further enhanced by synthesizing one or more *noise patterns*, whose purpose is to send artificial noise, i.e., jam potential eavesdroppers. Although array synthesis is a very mature topic, the problem of optimal synthesis of *joint* signal and noise patterns for optimal security had not been solved.

We posed the problem of optimal secure array synthesis using the system model shown in Fig. 17.10 [37, 38]. A LOS channel is assumed, where Alice has an array and wants to convey a secret message to Bob. Eve is a potential eavesdropper that resides outside of an *exclusion sector* that surrounds Bob. In some situations it may be possible to ensure that an exclusion sector is free of eavesdroppers, such as when physical access to the space is not possible. If, however, an eavesdropper is able to

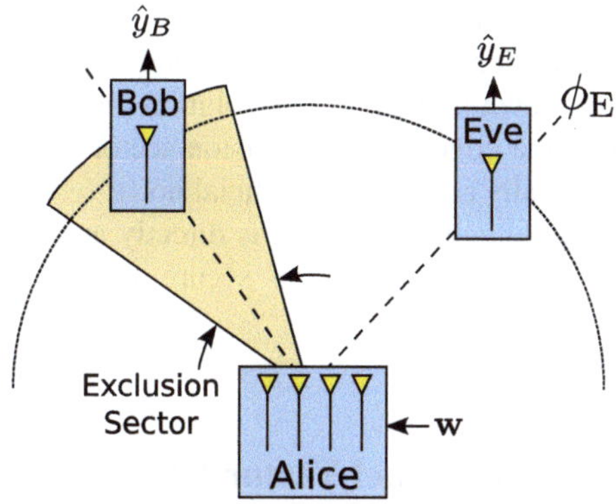

Fig. 17.10 System model for the secure array synthesis problem

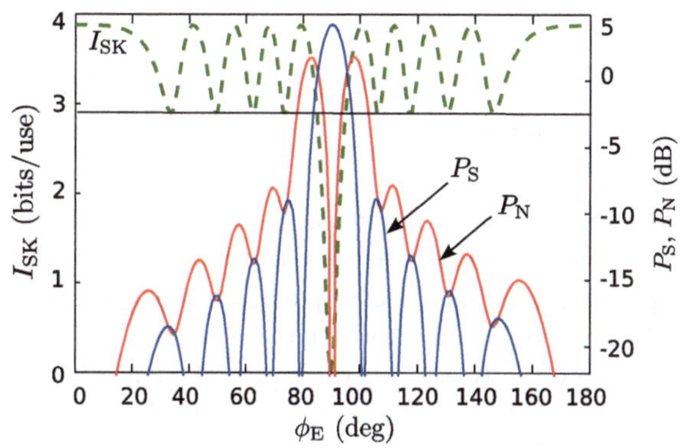

Fig. 17.11 Radiated signal (P_S) and noise (P_N) for the optimal SDP solution for a patch ULA at Alice with $N_T = 10$ elements and a 10° exclusion sector. The *horizontal line* is the minimum I_{SK} outside of the exclusion sector

penetrate the exclusion sector, it should be understood that physical layer security can no longer provide protection and security must rely on higher-level protocols. The goal is to choose the transmit signaling to convey as much secure information to Bob for all possible angles of Eve outside the exclusion sector.

We posed this problem mathematically by defining the secrecy capacity as

$$C_S = \max_{p(\mathbf{w})} \min_{\phi_E} I(\mathbf{w}; \hat{y}_B | \hat{y}_E), \tag{17.7}$$

where $p(\mathbf{w})$ is the PDF of random weights applied to the array, ϕ_E is Eve's angle, and $I(\mathbf{w}; \hat{y}_B | \hat{y}_E)$ is the conditional mutual information of the transmit weights and Bob's receive signal, indicating how much information is conveyed from Alice to Bob that is hidden from Eve. Assuming a joint Gaussian distribution for the transmit weights, it was found that this problem is convex and can be solved using semi-definite programming (SDP).

A representative result is shown in Fig. 17.11 for a patch antenna array with 10 antennas and an exclusion sector of 10°. Here, the optimal transmit covariance is found using SDP and used for key generation. The resulting $I_{SK}(\phi_E)$ for different Eve angles is plotted, indicating an equi-ripple response outside of the specified exclusion sector, where the minimum value is the true I_{SK}. Additionally, the transmit pattern can be decomposed into noise and signal patterns, denoted P_N and P_S, respectively. As shown in the plot, the solution finds the joint signal and noise patterns to ensure that the edge of the exclusion sector as well as directions of signal sidelobes are optimally covered by artificial noise.

Although the method is directly applicable for key establishment, further work is needed to extend this secure array synthesis technique for other transmission applications.

17.7 Channel Vector Quantization for Key Generation and Key Reconciliation Techniques

17.7.1 *Channel Quantization—The Linde-Buzo-Gray Algorithm*

For quantization one might use the density-based Lloyd-Max quantizer [24, 32] or, in case of measured distributions, its training based counterpart, the Linde-Buzo-Gray algorithm [12]. Such vector quantizers would ideally yield equally distributed intervals (codebook values), which is desired for the generation of a good key sequence. (We are currently working on improved quantization schemes strictly ensuring equal probability). The exact steps of the Linde-Buzo-Gray algorithm are given in more detail in [10, 12, 23]. Given a sequence of 2-dimensional channel samples, the real and imaginary parts of the CSI measurements are quantized jointly, obtaining the final codebook of length N_q, $\mathbf{A} = \{\mathbf{a}_1, \mathbf{a}_2, \ldots, \mathbf{a}_{N_q}\}$, and the corresponding quantization regions, $\mathcal{R}_i$, for each codebook vector $\mathbf{a}_i$.

The algorithm starts from a single point which is the average of all measured values. This is split into two points with some spacing in between, followed by some optimization for the new points and their corresponding Voronoi regions. Just like in the Lloyd-Max algorithm, Voronoi regions are resulting from half-distance splitting and codebook entries are conditional means inside Voronoi regions.

For a line-of-sight (LOS) channel, Figs. 17.12 and 17.14 show the channel histograms obtained from a simulated antenna configuration with one feed element and 2 and 4, respectively, additional reconfigurable elements.

In Figs. 17.13 and 17.15, we illustrate the quantization regions delivered by the Linde-Buzo-Gray algorithm for codebook sizes $N_q = 16$ and $N_q = 32$, respectively, for the channel densities shown in Figs. 17.12 and 17.14.

While for 2 and 4 reconfigurable elements the statistics are non-symmetric, non-Gaussian, for a sufficiently large number of reconfigurable elements, such as 24, the central limit theorem holds and the channel density obtained approaches an ideal circularly symmetric Gaussian, as shown in Fig. 17.16. For 24 reconfigurable elements, the quantization results are plotted in Fig. 17.17. The Gaussian case leads to more regular quantization regions.

17.7.2 *Channel Quantization with Adaptive Guard Bands*

As previously explained in Sect. 17.4.1, when a point from the channel distribution is very close to a quantization threshold, due to the effect of noise, or mismatches coming from synchronization offsets or differences in the circuitry, it might cross the quantization boundary, ending up in being quantized to different regions on both sides. Thus, by specifically designing guard bands around the quantization thresholds, any point that will end up within one of them will be discarded when generating the encryption key. Unlike the guard band methods described in previous subsections, we now consider the possibility of introducing *adaptive* guard bands, such that, at regions with small occurrence probability, the guard band width is narrower than at regions with high probability.

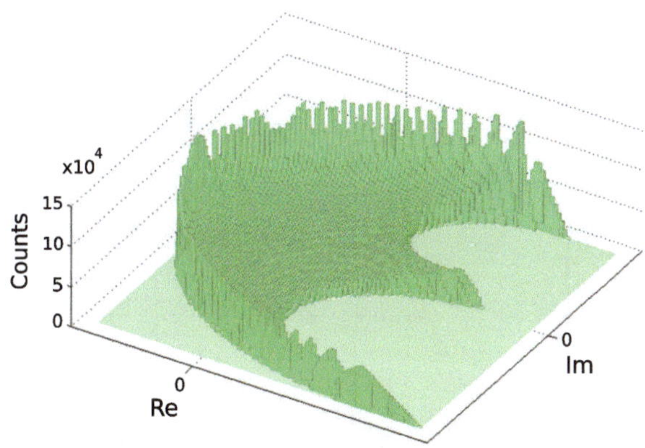

Fig. 17.12 Channel measurements—2 reconfigurable antenna elements with equally distributed channel states

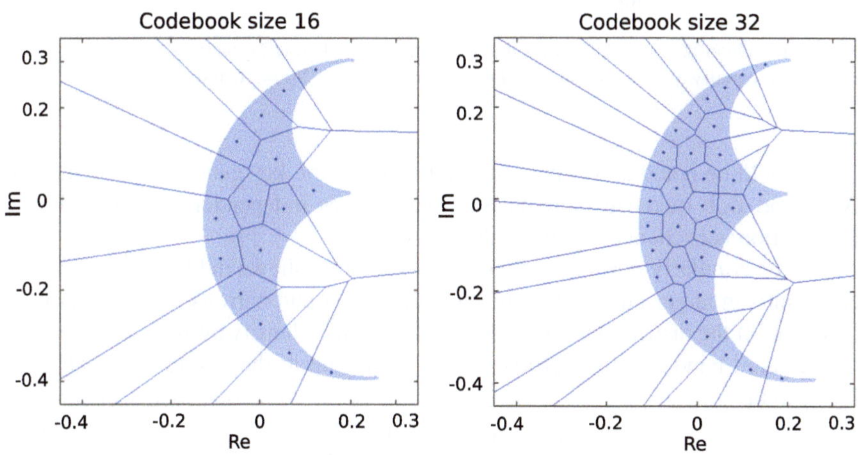

Fig. 17.13 Quantization results for a line-of-sight (LOS) channel; 2 reconfigurable antenna elements with equally distributed random states; codebook sizes $N_q = 16$ and $N_q = 32$

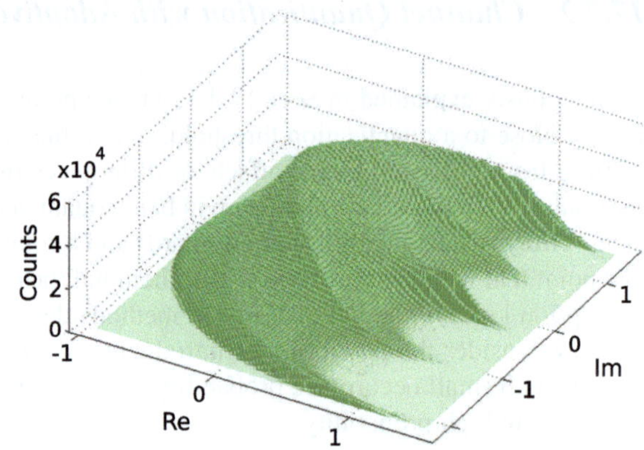

Fig. 17.14 Channel measurement —4 reconfigurable antenna elements with equally distributed channel states

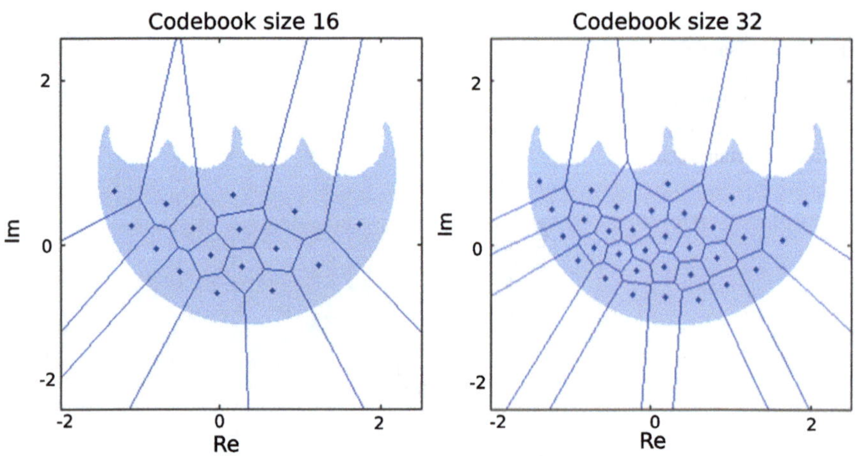

Fig. 17.15 Quantization results for a line-of-sight (LOS) channel; 4 reconfigurable antenna elements with equally distributed random states; codebook sizes $N_q = 16$ and $N_q = 32$

It should be noted at this point that for this specific key generation scheme both Alice and Bob quantize their measurement values independently and no further information is exchanged between the users.

For the circularly symmetric Gaussian distribution shown in Fig. 17.16, this strategy results in larger guard bands towards the center and narrower ones towards the outer part of the distribution. If we denote the channel density by $p(c)$, and the width of the guard band by g, the probability of traversing to the wrong region, P_{trav}, is found to be

$$P_{trav} \propto p(c)\frac{1}{2}\text{erfc}\frac{g}{\sqrt{2\sigma_n^2}}\,, \tag{17.8}$$

Fig. 17.16 Channel measurement—24 reconfigurable antenna elements with equally distributed channel states

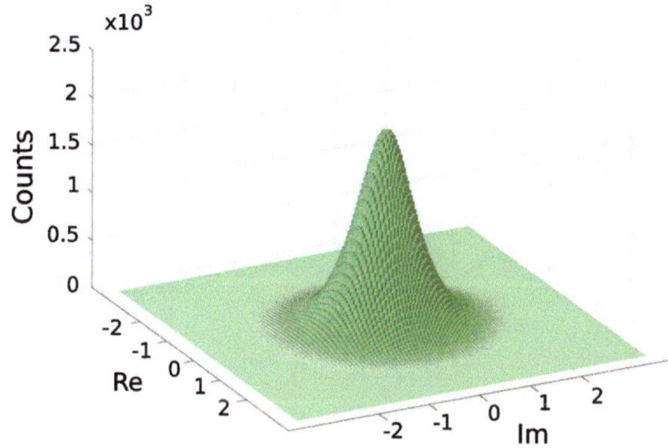

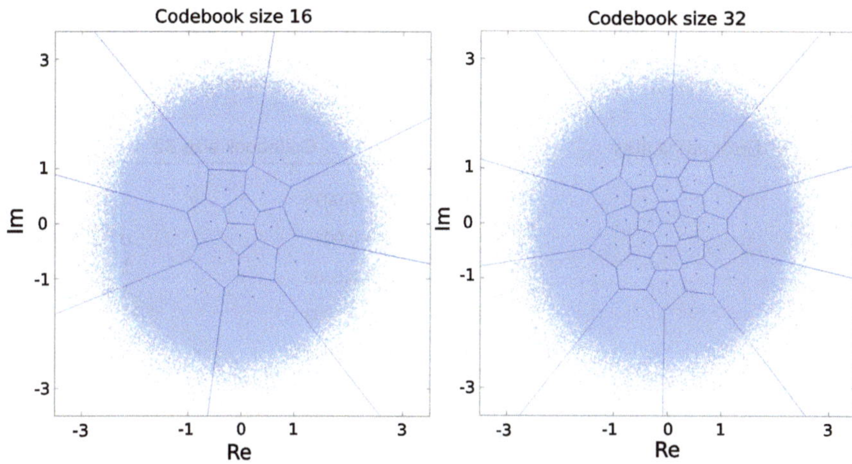

Fig. 17.17 Quantization results for a line-of-sight (LOS) channel; 24 reconfigurable antenna elements with equally distributed random states; codebook size $N_q = 16$ and $N_q = 32$

where σ_n^2 denotes the AWGN noise variance. Solving for the guard band width[1] g, this can be derived as

$$g = \sqrt{2\sigma_n^2}\text{erfc}^{-1}[k/p(c)]\,, \tag{17.9}$$

with $k = uP_{trav}$, where u is just a constant to be modified in order to obtain different widths. It now becomes evident that the width of the guard bands depends on the channel density distribution and on the noise variance. Note that a smaller k implies a larger guard band width. Figure 17.18 illustrates the adaptive guard band construction for a codebook size $N_q = 32$, for two extreme values of k. For the left plot, a small value of k results in very large guard band regions. This, while minimizing the

[1] We ignore the effect of the number of nearest neighbors, but it can be argued that this effect is negligible.

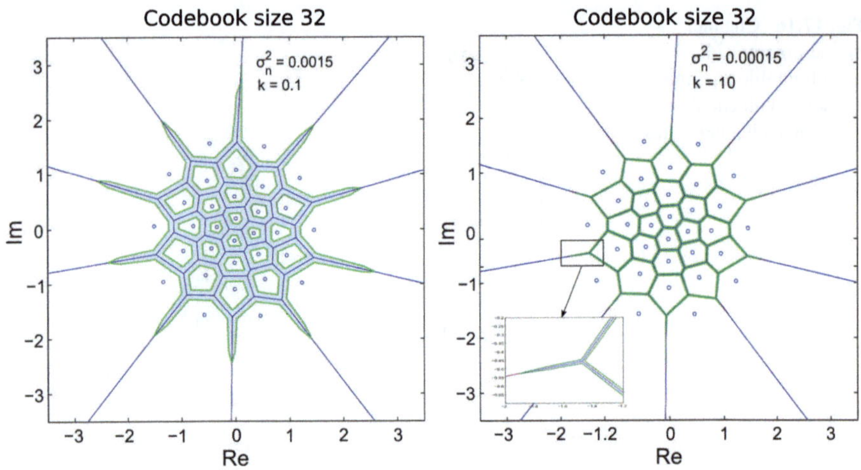

Fig. 17.18 *Left* small k, large guard bands; *Right* large k, narrow guard bands; $N_q = 32$

Fig. 17.19 Error probability

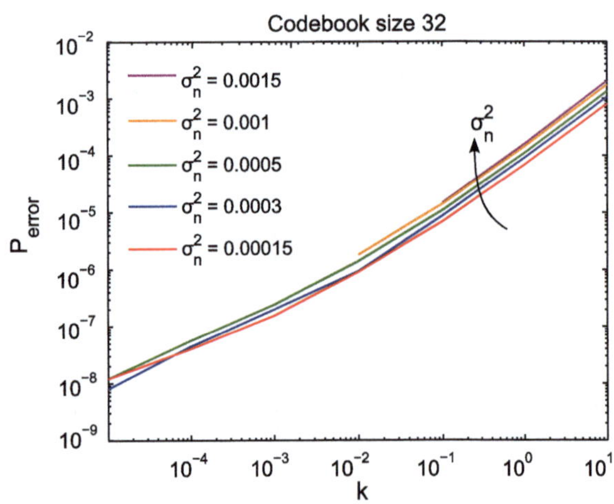

probability of quantizing a point to the wrong region, also significantly reduces the number of available bits for key generation, since it discards a significant fraction of measurement values. The second scenario, shown on the right, produces very narrow guard bands, close to the quantization boundaries. Unlike the first case, now more key bits are available, but also the probability of key mismatches increases.

Performance simulation results for different noise variances, as well as for different widths have been given in [11]. For convenience, we show in Figs. 17.19 and 17.20 the probability of error, i.e., quantizing to the wrong region, and the *efficiency*, as functions of k and noise variance σ_n^2, both for a codebook size of $N_q = 32$. The probability to obtain a valid key, or efficiency, is defined as one minus the probability to be in a guard interval or in a different Voronoi region.

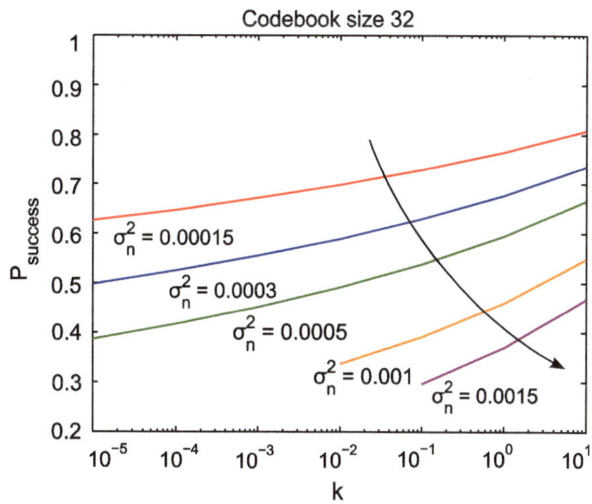

Fig. 17.20 Efficiency

A reduction in k results in a lower error probability as well as a lower efficiency. There is a trade-off between high efficiency and small error probability, depending on the width of the guard bands. As an alternative, we consider a channel-coding based method in Sect. 17.8, which does not reduce efficiency by guard bands, however, requires code redundancy instead.

17.8 Key Reconciliation Based on LDPC Codes

Since the overall efficiency of the previous guard band schemes was significantly affected by discarding points close to the quantization thresholds, as discussed in Sect. 17.7.2, a different approach could permit both legitimate users to use all measurement values for their key generation, thus directly increasing efficiency, while using channel coding for correcting the keys to the same value, resulting also in a minimization of the key mismatch probability.

One natural implementation of this reconciliation problem makes use of Slepian-Wolf coding, which is a kind of lossless compression with side information or joint source coding. Both Alice and Bob obtain measurements of the reciprocal channel, leading to the generation of two correlated keys, however, not always identical due to, e.g., non-correlated noise on both sides. Further exchange of information is thus required between the legitimate users, once the measurement phase has been completed. This, however, needs to be performed in such a way that secrecy is not compromised, implying that neither Alice nor Bob send the actual CSI information to each other. Nevertheless, parities or syndromes will be transmitted, which need to be further protected by privacy amplification [30], which is not further discussed in here.

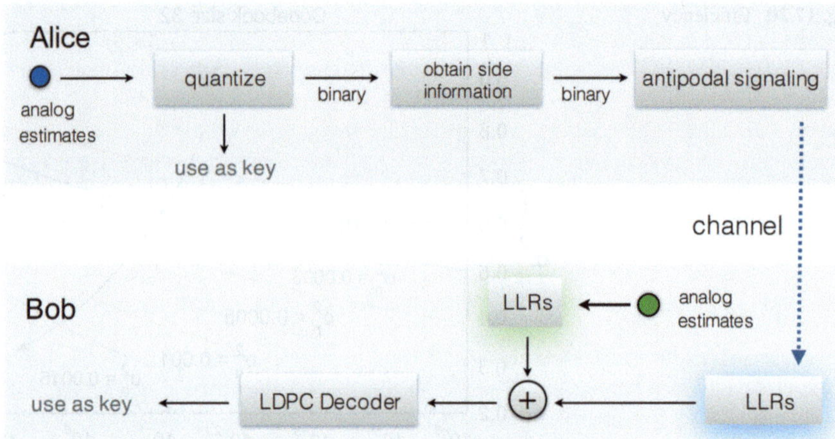

Fig. 17.21 System model—key reconciliation based on LDPC codes

17.8.1 System Model

Following the notations introduced in Sect. 17.2.2 and the illustration (binary case) in Fig. 17.21, Alice measures the reverse channel, obtaining estimate vector $\hat{\mathbf{h}}_a$, quantizes her values and computes the side information vector **s** to be sent to Bob. The key obtained by Alice after quantization is assumed to be correct, forcing Bob to go through additional steps in order to ensure he finally has an identical key. The side information computed by Alice is further sent over the physical channel in the shape of an antipodal signal (BPSK). Bob obtains his own forward channel estimates $\hat{\mathbf{h}}_b$, along with the possibly corrupted side information sent by Alice, and proceeds to combining them in order to obtain the exact same key Alice generated after her quantization step.

Low Density Parity Check (LDPC) codes are used to obtain the reconciled key on Bob's side. The input of the LDPC decoder consists of two sets of log-likelihood ratios (LLRs), one for the *information symbols*, obtained from Bob's own channel estimates $\hat{\mathbf{h}}_b$, and one set for the *parity* or *syndrome symbols*, sent by Alice. For better performance, *soft decoding* is implemented, i.e., no quantization of the measured analog values is performed on Bob's side.

17.8.2 Slepian-Wolf Coding

Slepian-Wolf key reconciliation exploits the correlation of two jointly distributed random variables arising from the channel reciprocity. The minimum number of bits,

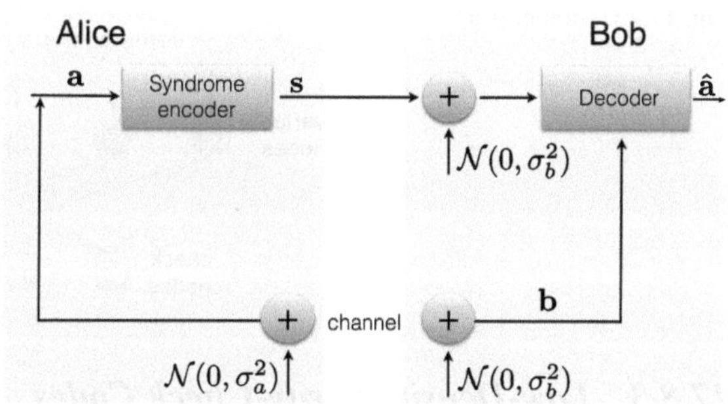

Fig. 17.22 Slepian-Wolf coding—syndrome approach

M_S, that need to be exchanged during reconciliation is given by the Slepian-Wolf lower bound [51], as follows

$$M_S \geq M_p = K \cdot H(\hat{h}_a | \hat{h}_b) , \tag{17.10}$$

where K stands for the length of the information bits sequence and H denotes the entropy. There are two possible approaches for Slepian-Wolf coding, both generating the required side information to be sent to Bob for correct decoding, namely the *parity approach* and the *syndrome approach.* We will limit our description here to the syndrome approach only, since both construction methods yield similar results and their choice has negligible effect on the overall performance of the reconciliation scheme. The terms *parity symbols* and *syndrome symbols* will be used interchangeably through the current section to refer to the side information sent from Alice to Bob.

For the **syndrome approach**, Alice compresses her key $\mathbf{a}$ of length K to a syndrome $\mathbf{s} = [s_1, s_2, \ldots s_{M_S}]$ and sends it over the physical channel to Bob, as illustrated in Fig. 17.22. The syndrome vector is computed according to

$$\mathbf{s} = \mathbf{Ha}, \tag{17.11}$$

where $\mathbf{H}$ denotes the $M_S \times K$ sparse parity-check matrix of the LDPC code. Regardless of the method used to generate the parities, BPSK signaling can, e.g., be used for transmission over the physical channel.

It should be noted, however, that the side information is subject to being corrupted by additional AWGN noise when transmitted over the physical channel, and this should be accounted for during code design. In [52], the transmission channel is assumed to be noiseless and ideal side information is assumed to be available at Bob, not corrupted by additional AWGN. This is, however, not accurate, and the Slepian-Wolf lower bound on the total number of bits that need to be further exchanged is not tight, since it does not account for the extra redundancy introduced by the code protecting the parities.

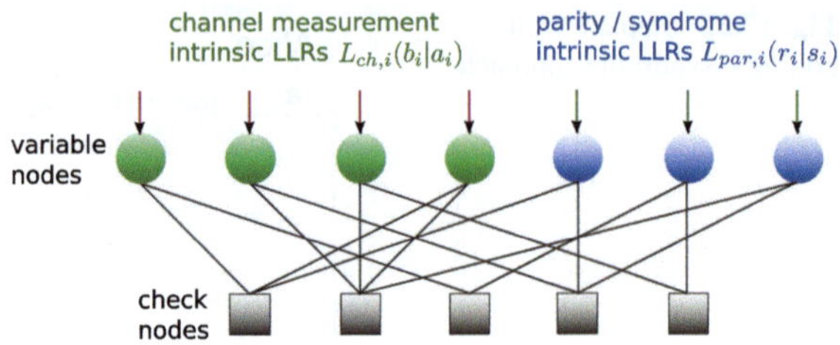

Fig. 17.23 Tanner graph

17.8.3 Low-Density Parity-Check Codes

Low-density parity-check codes (LDPC) are known to reach capacity when the codeword length goes to infinity, but are also very close to capacity at more moderate lengths of a few thousand [43]. More specifically, irregular codes show better performances. In the presence of bursty errors, irregular non-binary LDPC codes are the prime choice [7, 25].

An irregular LDPC code is defined by the degree distribution of the variable nodes and check nodes in the Tanner graph representation of its parity check matrix. Figure 17.23 shows an exemplary Tanner graph, here with the specialty that the intrinsic information inputs are coming from two different channels, one resulting from channel measurements (probing), the other from actually transmitted parity symbols, experiencing, e.g., only additive white Gaussian noise.

Belief-Propagation (BP) is used as an efficient message passing decoding algorithm. For BP decoding, messages are passed iteratively along the connecting edges from variable nodes to check nodes and vice versa. Each node on the variable-node and check-node side passes on messages at each iteration along every edge after combining the messages which arrive at that node along the other edges.

One typically defines the following two polynomials for degree distribution representation.

- Polynomial associated to variable nodes:

$$\lambda(x) = \sum_{i=2}^{t_{cmax}} \lambda_i x^{i-1} , \tag{17.12}$$

  where λ_i is the proportion of edges of the graph connected to variable nodes of degree i, and t_{cmax} is the maximum number of edges linked to a variable node.
- Polynomial associated to check nodes:

$$\rho(x) = \sum_{j=2}^{t_{rmax}} \rho_j x^{j-1} , \tag{17.13}$$

  where ρ_j is the proportion of edges of the graph connected to check nodes of degree j, and t_{rmax} is the maximum number of edges linked to a check node.

The two distributions are not independent, but linked by the code rate

$$R = 1 - \frac{M}{N} = 1 - \frac{M/E}{N/E} = 1 - \frac{\sum_{j=2}^{t_{rmax}} \rho_j / j}{\sum_{i=2}^{t_{cmax}} \lambda_i / i}, \tag{17.14}$$

where N is the codeword length, $M = M_p(1+\beta)$ is the overall number of parity symbols, β is the fraction of additionally required parities additionally to the ones given by the Slepian-Wolf bound, E is the total number of edges. Note that the sums compute the average over $1/j$ and $1/i$, where j and i denote the check and variable node degrees, respectively. These averages are, of course, equal to M/E and N/E, respectively.

Since we have to distinguish two different channel properties, the descriptive tool would be Richardson and Urbanke's multi-edge-type description [44, 45], which allows to distinguish different edge types and different channel input distributions. The first allows, e.g., the definition of unequal error protecting codes, which is not our concern. However, we do have two different input distributions and hence need to distinguish two variable node types, giving way to two variable node degree distributions $\lambda^{(1)}(x)$ and $\lambda^{(2)}(x)$.

17.8.3.1 Log-Likelihood Ratio Computation

In general, LDPC decoding can be performed either in the probability domain or in the log-likelihood domain, through the use of log-likelihood ratios (LLR). Several studies have shown that an LLR representation poses some advantages over the probability domain representation, such as increased resilience against quantization and finite precision errors [55]. The LLR domain also has the advantage of transforming operations such as multiplications and divisions into additions and subtractions, thus reducing the required hardware complexity. While the LLR computation for the parity bits is trivial, since they just experience a standard AWGN channel with variance σ_b^2, the calculation of the LLRs for Bob's information bits is much more problematic, due to the fact that Bob's decoding of the information bits is subject to what Alice quantized to.

LLR for Parity Bits

In general, for a transmitted BPSK symbol $s_{x,i}$, and received $r_{x,i}$, the Gaussian relation

$$p_{par,i}(r_{x,i}|s_{x,i}) = \frac{1}{\sqrt{2\pi}\sigma_b} \cdot e^{-(r_{x,i}-s_{x,i})^2/2\sigma_b^2} \tag{17.15}$$

leads to the LLR

$$L_{par,i}(r_{x,i}) = \ln \frac{p_{par,i}(r_{x,i}|s_{x,i} = +1)}{p_{par,i}(r_{x,i}|s_{x,i} = -1)} = \ln \frac{e^{+2r_{x,i}/2\sigma_b^2}}{e^{-2r_{x,i}/2\sigma_b^2}} = \frac{2}{\sigma_b^2} \cdot r_{x,i}$$
$$= \frac{2}{\sigma_b^2} \cdot (s_{x,i} + n_{x,i}) \,, \qquad (17.16)$$

where uppercase P is used to denote probability, while lowercase p denotes probability densities. The mean and variance of $L_{par,i}(r_{x,i})$ are given by

$$\mu_L = 2/\sigma_b^2 \cdot s_{x,i} \quad \text{and} \quad \sigma_L^2 = \left[\frac{2}{\sigma_b^2}\right]^2 \cdot \sigma_b^2 = 4/\sigma_b^2 \,, \qquad (17.17)$$

and considering $s_{x,i} = +1$, we obtain the relation between mean and variance

$$\mu_L = \sigma_L^2/2 \,. \qquad (17.18)$$

The density of the log-likelihood ratio (LLR) of a Gaussian is a special case of a so-called *consistent* density, which has an exponential symmetry given by

$$p(x) = e^x \cdot p(-x), \forall x \in \mathbb{R} \,. \qquad (17.19)$$

The LLR of a Gaussian has the convenient property that mean and variance are not independent.

LLR for Information Bits

The undisturbed complex value $c = c_x + jc_y$ from the channel distribution is disturbed by AWGN to $a = a_x + ja_y$ on Alice's side before quantization, and to $b = b_x + jb_y$ on Bob's side. For a normalized zero-mean complex Gaussian channel,[2] such as the one obtained by using 24 reconfigurable antenna elements, the channel density distribution, $p(c)$, simplifies to

$$p(c) = \frac{1}{2\pi\sigma_{ch_x}\sigma_{ch_y}} e^{-\frac{(c_x-\mu_{ch_x})^2}{2\sigma_{chx}^2} - \frac{(c_y-\mu_{ch_y})^2}{2\sigma_{chy}^2}} = \frac{1}{2\pi\sigma_{ch}^2} e^{-\frac{(c_x^2+c_y^2)}{2\sigma_{ch}^2}} \,. \qquad (17.20)$$

The disturbance at Bob, $p(b|c)$, is then found to be

$$p(b|c) = \frac{1}{2\pi\sigma_b^2} e^{-\frac{(b_x-c_x)^2+(b_y-c_y)^2}{2\sigma_b^2}} \,. \qquad (17.21)$$

Similarly, given the ideal channel measurement c, the disturbance at Alice, $p(a|c)$, is found to be

---

[2]Real and imaginary components not correlated.

$$p(a|c) = \frac{1}{2\pi\sigma_a^2} e^{-\frac{(a_x - c_x)^2 + (a_y - c_y)^2}{2\sigma_a^2}}, \tag{17.22}$$

where σ_a^2 and σ_b^2 are the variances of the independent AWGN noise present at Alice and Bob, respectively. Again, the means of the real and imaginary noise components, at both Alice and Bob, are assumed to be zero. It follows from Bayes' rule that

$$p(c|b) = \frac{1}{2\pi\sigma_b^2} e^{-\frac{|c-b|^2}{2\sigma_b^2}} \cdot \frac{p(c)}{p(b)},$$

$$p(c|a) = \frac{1}{2\pi\sigma_a^2} e^{-\frac{|c-a|^2}{2\sigma_a^2}} \cdot \frac{p(c)}{p(a)},$$

where $p(a)$ and $p(b)$ are given by

$$p(a) = \frac{1}{2\pi(\sigma_{ch}^2 + \sigma_a^2)} e^{-\frac{(a_x^2 + a_y^2)}{2(\sigma_{ch}^2 + \sigma_a^2)}}, \tag{17.23}$$

$$p(b) = \frac{1}{2\pi(\sigma_{ch}^2 + \sigma_b^2)} e^{-\frac{(b_x^2 + b_y^2)}{2(\sigma_{ch}^2 + \sigma_b^2)}}. \tag{17.24}$$

On Bob's side, having measured a value b, we need to determine the probability of what Alice may have quantized to, i.e., $P(a \in \mathcal{R}_i|b)$ and $P(a \notin \mathcal{R}_i|b)$.

$$P(a \in \mathcal{R}_i|b) = \int_{\mathcal{R}_i} \int_{\mathcal{R}} p(a|c) \cdot p(c|b)\, dc\, da, \tag{17.25}$$

$$P(a \notin \mathcal{R}_i|b) = \int_{\mathcal{R}\setminus\mathcal{R}_i} \int_{\mathcal{R}} p(a|c) \cdot p(c|b)\, dc\, da. \tag{17.26}$$

Note that c and a are, of course, complex and hence, the corresponding integrals are two-dimensional. Exploiting Bayes' rule again, it follows that

$$p(b|a \in \mathcal{R}_i) = P(a \in \mathcal{R}_i|b) \cdot \frac{p(b)}{P(a \in \mathcal{R}_i)}, \tag{17.27}$$

$$p(b|a \notin \mathcal{R}_i) = P(a \notin \mathcal{R}_i|b) \cdot \frac{p(b)}{P(a \notin \mathcal{R}_i)}. \tag{17.28}$$

Notation $\mathcal{R}_i$ stands for an individual quantization region.

The LLR ratios for the information symbols, assuming a zero-mean circularly symmetric Gaussian channel density, and N_q equiprobable arbitrary quantization regions, can be numerically computed according to (17.29),

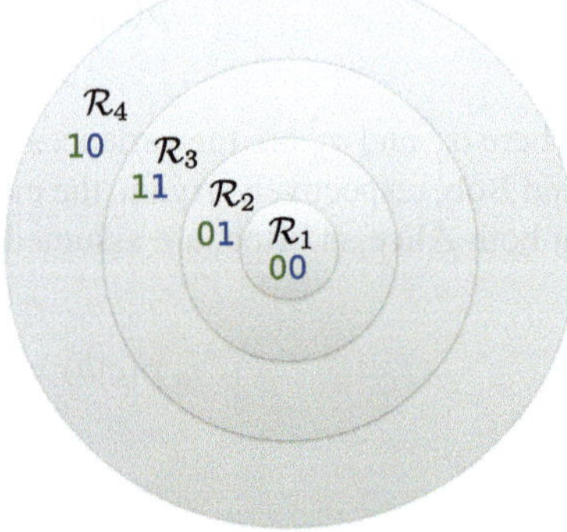

Fig. 17.24 Four concentric quantization regions

$$L_{ch}(b) = \ln \frac{p(b|a \in \mathcal{R}_i)}{p(b|a \notin \mathcal{R}_i)} = \ln \frac{(N_q - 1)P(a \in \mathcal{R}_i|b)}{P(a \notin \mathcal{R}_i|b)} ;$$

$$L_{ch}(b) = \ln \frac{(N_q - 1) \int\limits_{-\infty}^{\infty} \int\limits_{-\infty}^{\infty} \int\limits_{\mathcal{R}_{i_y}} \int\limits_{\mathcal{R}_{i_x}} \mathcal{I}\, da_x\, da_y\, dc_x\, dc_y}{\sum\limits_{\mathcal{R}_k, k \neq i} \int\limits_{-\infty}^{\infty} \int\limits_{-\infty}^{\infty} \int\limits_{\mathcal{R}_{k_y}} \int\limits_{\mathcal{R}_{k_x}} \mathcal{I}\, da_x\, da_y\, dc_x\, dc_y} , \tag{17.29}$$

with

$$\mathcal{I} = e^{-\frac{(a_x - c_x)^2 + (a_y - c_y)^2}{2\sigma_a^2} - \frac{(c_x - b_x)^2 + (c_y - b_y)^2}{2\sigma_b^2} - \frac{(c_x^2 + c_y^2)}{2\sigma_{ch}^2} + \frac{(b_x^2 + b_y^2)}{2(\sigma_{ch}^2 + \sigma_b^2)}} .$$

$(\mathcal{R}_{i_x}, \mathcal{R}_{i_y})$ denotes the set of coordinates for all points within region $\mathcal{R}_i$.

The complexity of Eq. (17.29) can be drastically reduced by exploiting the circular symmetry of the Gaussian channel distribution using polar coordinates, and designing concentric circular quantization regions, such as the ones shown in Fig. 17.24. For the two-bit quantization previously shown, at the input of the LDPC decoder it is necessary to switch between two different sets of channel intrinsic LLRs, each corresponding to one bit. This is inherent in the fact that the integration limits change accordingly to the bit for which the LLR is calculated. For the first bit in this example, in the numerator of (17.29), we integrate over the regions where the first bit (in green) has a value of 0, $\mathcal{R}_1$ and $\mathcal{R}_2$, while in the denominator we consider the regions where the first bit has a value of 1, $\mathcal{R}_3$ and $\mathcal{R}_4$, respectively. The LLR in (17.29) is a function of b and can be computed in advance and stored, for a wide range of b values, or (b_x, b_y), in order to speed up the LDPC decoder. Figures 17.25 and 17.26 show the results. Further details can be found in [14, 15, 17]. For a binary codebook, $N_q = 2$, the LLR at variable node i simplifies to

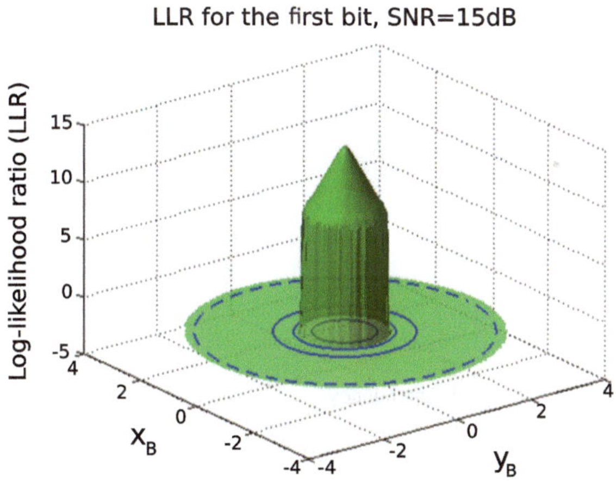

Fig. 17.25 Log-likelihood ratio for the first bit, SNR = 15 dB

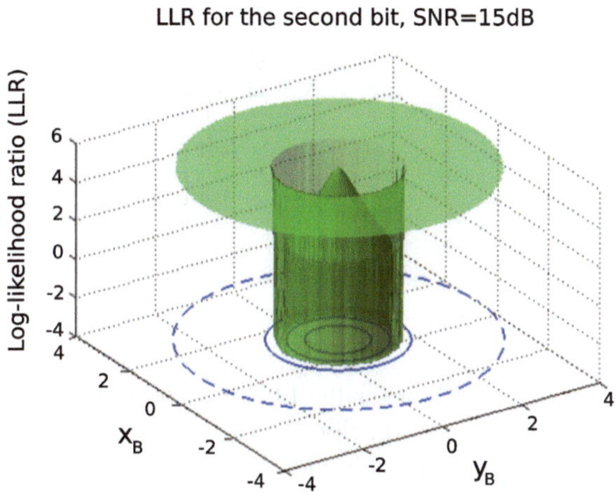

Fig. 17.26 Log-likelihood ratio for the second bit, SNR = 15 dB

$$L_{ch,i}(b_{x,i}) = \ln \frac{\int_{-\infty}^{0} \int_{-\infty}^{+\infty} e^{-\frac{(a_x - c_x)^2}{2\sigma_a^2}} \cdot e^{-\frac{(c_x - b_x)^2}{2\sigma_b^2}} \cdot e^{-\frac{c_x^2}{2\sigma_{ch}^2}} \, dc_x \, da_x}{\int_{0}^{+\infty} \int_{-\infty}^{+\infty} e^{-\frac{(a_x - c_x)^2}{2\sigma_a^2}} \cdot e^{-\frac{(c_x - b_x)^2}{2\sigma_b^2}} \cdot e^{-\frac{c_x^2}{2\sigma_{ch}^2}} \, dc_x \, da_x} \,. \tag{17.30}$$

The PDF of the log-likelihood ratio $L_{ch,i}$ is shown in Fig. 17.27. It is not a consistent density, i.e., mean and variance are not in a fixed relation for different SNRs.

Fig. 17.27 PDF of the log-likelihood ratio $L_{ch,i}$ of the measurement key bits

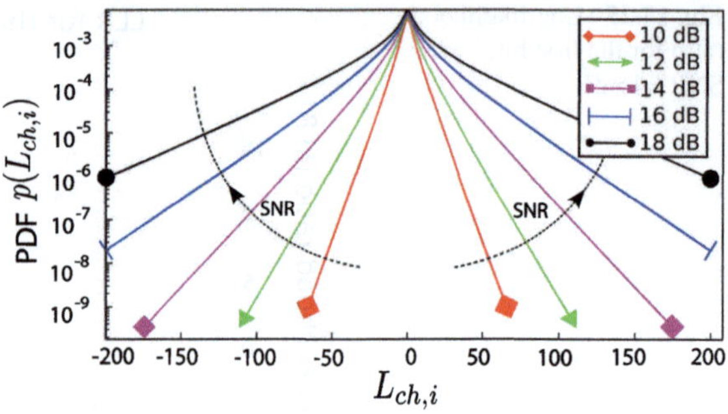

Fig. 17.28 LLR dependency relation $\mathcal{L}_{ch}(b_x)$

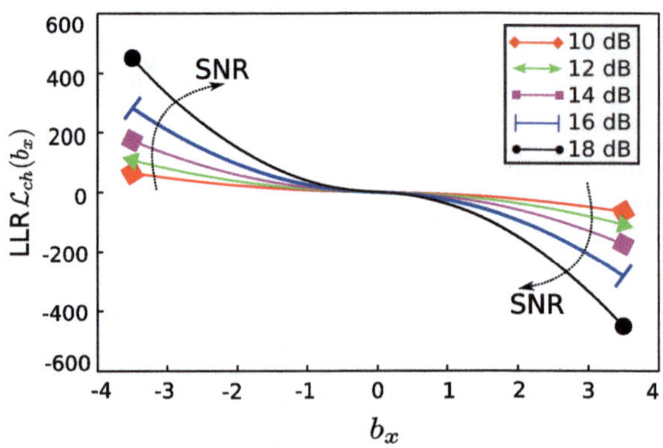

In Fig. 17.28, $\mathcal{L}_{ch}(b_{x,i}|a_{x,i})$ is shown for different SNR[3] values. We switched to a calligraphic font for the LLR to indicate that we see it as a functional relation for density transformation. $\mathcal{L}_{ch}(b_x)$ is an odd symmetric function of b_x, the analog received value at Bob's end. We also observe the expected increase with the SNR. This function was also used to obtain Fig. 17.27 by applying the following transform rule

$$p(L_{ch,i}|a_{x,i} = \pm 1) = \sum_{b_{x,i}:\mathcal{L}_{ch}^{-1}(L_{ch,i})=b_{x,i}} \frac{p_{b_{x,i}|a_{x,i}=\pm 1}(b_{x,i})}{|\frac{d}{d(b_x)}\mathcal{L}_{ch}(b_{x,i}|a_{x,i})|} . \tag{17.31}$$

Message Passing and Density Evolution

Let us first be reminded of the combination of information that is treated at a variable node. We consider a general channel input vector **x** and a channel output vector **y** and its components x_i and y_i, respectively. The log-likelihood ratio in a MAP sense addressing for a single component x_i given a received vector **y** is then

[3] SNR$= 10\log_{10}\frac{\sigma_{ch}^2}{\sigma_{a/b}^2}$, $\sigma_a^2 = \sigma_b^2$.

$$\ln \frac{P(x_i=+1|\mathbf{y})}{P(x_i=-1|\mathbf{y})} = \ln \frac{p(\mathbf{y}|x_i=+1)}{p(\mathbf{y}|x_i=-1)} + \ln \frac{P(x_i=+1)}{P(x_i=-1)}$$
$$= \underbrace{\ln \frac{p(y_i|x_i=+1)}{p(y_i|x_i=-1)}}_{L_{\text{intrinsic}}} + \underbrace{\ln \frac{p(\mathbf{y}_{\setminus i}|x_i=+1)}{p(\mathbf{y}_{\setminus i}|x_i=-1)}}_{L_{\text{extrinsic}}} + \underbrace{\ln \frac{P(x_i=+1)}{P(x_i=-1)}}_{L_{\text{a-priori}}} . \quad (17.32)$$

Assuming equally distributed data (which is also required for a good key sequence), the processing of log-likelihood ratios at the variable node, during iteration l, is given by

$$L_{v_i c_j}^{(l)} = L_{ch,i} + \sum_{k \neq j} L_{c_k v_i}^{(l-1)}, \quad (17.33)$$

where indices vc and cv indicate the directions variable-to-check node and check-to-variable node, respectively. The operations at the check-node side are nonlinear and given by [6]

$$L_{c_i v_j}^{(l)} = 2 \tanh^{-1} \left(\prod_{k \neq j} \tanh \left(\frac{L_{v_k c_j}^{(l)}}{2} \right) \right) . \quad (17.34)$$

The two sides of the Tanner graph—variable and check nodes—represent individual repetition and single-parity-check codes.

The sum of log-likelihood ratios on the variable-node side according to (17.33) means the convolution of their corresponding densities. In here, we simplify the treatment by postulating all densities in the message passing to be consistent Gaussian, except for the intrinsic ones originating from the channel measurements. The parity densities are consistent, anyhow. One should however note that this assumption is far from correct during the first decoding iterations, especially, in the very first, when LLRs at the variable-node side are not modified by the variable nodes at all and just forwarded to the check nodes and also from there, only non-consistent densities will be returned. Nevertheless, at later iterations, the central limit theorem will ensure that all other densities are approximately consistent Gaussians and we hence limited the exact treatment of the densities to the variable-node side, realizing the sum of messages from the check nodes by simply adding means (variances) and convolving the resulting densities with the incoming intrinsic ones at the measurement variable nodes, meaning

$$p(L_{v_i c_j}^{(l)}|a_{x,i} = \pm 1) = p(L^{*(l-1)}_{cv}|a_{x,i} = \pm 1) \star p(L_{ch,i}|a_{x,i} = \pm 1) . \quad (17.35)$$

In here, L^* represents the already performed combination of all LLRs from the check-nodes incident to the current variable node, except for the outgoing edge. The convolution of those components is realized by a factor of degree minus one times the mean (variance).

To investigate the convergence, the mutual information is computed from the densities according to (17.36) and for the check-node treatment, we assume a consistent density making use of the relation between mutual informations and means (variances).

$$x_{v_i c_j}^{(l)} = \int_{\mathcal{R}} p(L_{v_i c_j}^{(l)} | a_x = +1) \log_2 \left(\frac{2p(L_{v_i c_j}^{(l)} | a_x = +1)}{p(L_{v_i c_j}^{(l)} | a_x = +1) + p(L_{v_i c_j}^{(l)} | a_x = -1)} \right) \mathrm{d}L_{v_i c_j}. \tag{17.36}$$

In case of a consistent $\mathcal{N}(m, 2m)$ Gaussian variable τ, this is typically expressed by the so-called J-function [6, 8, 47].

$$J(m) := 1 - \frac{1}{\sqrt{4\pi m}} \int_{\mathcal{R}} e^{-\frac{(\tau - m)^2}{4m}} \log_2(1 + e^{-\tau}) \, \mathrm{d}\tau = x_\tau. \tag{17.37}$$

The check-node function will be described just as in the standard message passing decoding and only back at the variable-node side, the actual convolution operation is performed. In the later listed linear program, the check-to-variable node equation is shown, but, due to space limitations, we do not provide the derivation in here, but rather refer to [6, 46].

An exact density evolution on both sides, especially for the first iterations, would, of course, consider all incoming densities at the variable-node side to be non-consistent, leading to manifold convolutions (products of characteristic functions). On the check-node side, a logarithmic reformulation will require 2-dimensional convolutions (product of 2D characteristic functions). This will exceed the intended tutorial character of this chapter. Anyhow, the described simplifications still lead to a viable realization.

Linear Optimization Algorithm and Results

A linear program for finding the optimized degree distributions is listed in Algorithm **1** without being able to describe all the individual constraints in here. One will nevertheless recognize the central convergence steps starting from (11.42) and describing the evolution of the mutual information including the described convolutive operations. The superscript (j) is used to distinguish the two classes of variable nodes, the values 1 and 2 of which refer to variable and parity nodes, respectively. The proportion distribution constraints are discussed in [9]. Equations (11.47) and (11.49) follow from [8].

Note the check-node side update step (11.49). We notice, within the summation, the J^{-1} function is applied to $(1 - x_{vc})$, i.e., addresses the mutual information on an outgoing edge of a variable node. As already mentioned, we assume consistency to be able to use the J^{-1} function, the inverse of (17.37).

Algorithm 1 Linear programming algorithm

Optimize

$$\min_{\beta\in\mathbb{R}^+}(1+\beta)\,, \tag{11.38}$$

subject to

1. Proportion distribution constraints [9]

1.1

$$\sum_{j=1}^{2}\sum_{i=2}^{d_{\text{vmax}_j}}\lambda_i^{(j)}=1 \tag{11.39}$$

1.2

$$\sum_{i=2}^{d_{\text{vmax}_2}}\frac{\lambda_i^{(2)}}{i}=\sum_{i=2}^{d_{\text{cmax}}}\frac{\rho_i}{i} \tag{11.40}$$

1.3

$$\frac{M_p}{N}(1+\beta)\sum_{i=2}^{d_{\text{vmax}_1}}\frac{\lambda_i^{(1)}}{i}=\sum_{i=2}^{d_{\text{cmax}}}\frac{\rho_i}{i}\,,\quad \beta\geq 0 \tag{11.41}$$

2. Convergence condition

$$F\left(\boldsymbol{\lambda},\boldsymbol{\rho},x_{vc}^{(l)}\right)>x_{vc}^{(l-1)}\ \text{ with }, \tag{11.42}$$

$$x_{vc_{j=1}}^{(l)}=0; \tag{11.43}$$

for $i=2:d_{\text{vmax}_j}\ ,\quad j=1$: information bits,

$$p(L_{vc}^{(l)}|a_x=+1)=\\ p_{(L_{cv}^{l-1}|a_x=+1)}((i-1)m)\star p(L_{\text{intrinsic}}|a_x=+1)\,, \tag{11.44}$$

$$p(L_{vc}^{(l)}|a_x=-1)=\\ p_{(L_{cv}^{l-1}|a_x=-1)}((i-1)m)\star p(L_{\text{intrinsic}}|a_x=-1)\,, \tag{11.45}$$

m = mean of the Gaussian densities.

$$x_{vc_{j=1}}^{(l)}=x_{vc_{j=1}}^{(l)}+\lambda_i^{(j=1)}\int_{\mathcal{R}}p(L_{vc}^{(l)}|a_x=+1)\ldots\\ \log_2\left(\frac{2p(L_{vc}^{(l)}|a_x=+1)}{p(L_{vc}^{(l)}|a_x=+1)+p(L_{vc}^{(l)}|a_x=-1)}\right)\mathrm{d}L_{vc}\,. \tag{11.46}$$

end for.

$$x_{vc_{j=2}}^{(l)}=\sum_{i=2}^{d_{\text{vmax}_{j=2}}}\lambda_i^{(j=2)}J\left(\frac{2}{\sigma_b^2}+(i-1)\,J^{-1}\left(x_{cv}^{(l-1)}\right)\right), \tag{11.47}$$

$$x_{vc}^{(l)}=\sum_{j=1}^{2}x_{vc_j}^{(l)} \tag{11.48}$$

$$x_{cv}^{(l-1)}=1-\sum_{h=2}^{d_{\text{cmax}}}\rho_h J\left((h-1)\,J^{-1}\left(1-x_{vc}^{(l-1)}\right)\right). \tag{11.49}$$

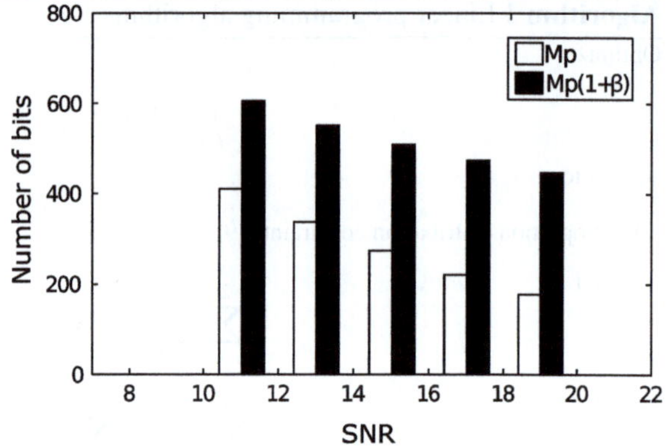

Fig. 17.29 Redundancy requirements delivered by the linear program

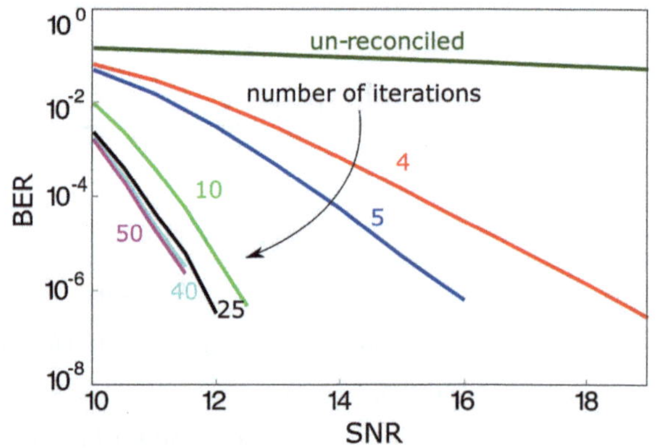

Fig. 17.30 BER results

We provide the results of the linear optimization with 50 iterations. The routine minimizes $1+\beta$, where β is the fraction of correction parity bits required in addition to the ones given by the conditional entropy $H(\hat{h}_a|\hat{h}_b)$ in Eq. (17.10) already required for reconciliation for error-free transmission of the parities. As check-node degree distribution, we used $\rho(x) = 0.98x^9 + 0.02x^{10}$. The maximum variable node degrees for the two classes were chosen as $d_{vmax_1} = 15$ and $d_{vmax_2} = 15$. The length of the information word is $n = 2^{10} = 1024$. $M_p(1+\beta)$ is the total required redundancy.

Some results are presented in Fig. 17.29. As expected, with increasing SNR, the required redundancy decreases.

BER Simulation

Figure 17.30 shows the BER against SNR plot for an LDPC code with degree distributions optimized at 11 dB.

17.9 Some Remarks on Wiretap Channels

The reader might have interpreted the chapter more in relation to Wyner's wiretap channel [59]. These works, however, make strongly use of the so-called *secrecy capacity*, i.e., the difference in capacities (SNRs) between the legitimate channel between Alice and Bob and the wiretapper channels Alice–Eve or Bob–Eve. One makes use of those differences such that the eavesdropper will be close to error probabilities of 0.5, whereas the legitimate receiver is still fine. Papers in that direction use some kind of coset code or multi-level constructions [3–5], might also use LDPC [53] or Polar codes [21], without trying to cover all the references that are addressing the wiretap channel. However, this is not so much related to key generation and key reconciliation by measuring the physical reciprocal channel. The wiretap channel is, however, a very suitable concept for the transmitted parities from the Slepian-Wolf coding. The measurement channel is usually considered to be perfectly secure, however, also there some leakage [41] might be considered, which is outside the scope of this contribution.

17.10 Summary

This chapter has summarized research in the framework of COIN whose goal was to extend existing methods of physical-layer key establishment and to analyze such methods from both a theoretical and experimental perspective. The specific focus was methods based on quantization of a reciprocal wireless channel, where more sophisticated quantization and reconciliation approaches were developed to provide higher efficiency and lower key mismatch rate. The use of reconfigurable antennas for key generation was also studied in depth to understand the potential benefits and limitations of this approach. Experimental campaigns to characterize key generation using time-varying channels and reconfigurable antennas were performed, indicating that long and secure keys can be generated in practice using the developed methodologies.

Due to uncorrelated noise (and non-ideal circuitry) quantization methods might lead to key mismatch. We investigated two approaches for key reconciliation, one based on guard intervals instead of quantization thresholds, the other based on Slepian-Wolf coding with additional redundancy, realized by a multi-edge type LDPC code. For quantization, we have mainly focused on the Linde-Buzo-Gray algorithm, although it is not yet perfect in the intended equal distribution of quantization regions and hence equal distributed i.i.d. key bits. We showed how log-likelihood ratios (LLRs) and their distributions are determined in the binary and non-binary case that describe the input to the LDPC decoder. Those channel measurement LLRs are non-Gaussian and non-consistent. We designed an optimized LDPC code using the non-consistent density inside the density evolution. Results were shown for both the guard-band and the coding approaches.

Acknowledgments This work was supported by the German Research Foundation (DFG) under Grants WA 2735/5-1/2 and HE 3654/11-1/2.

References

1. Ahlswede R, Csiszar I (1993) Common randomness in information theory and cryptography. I. Secret sharing. IEEE Trans Inf Theory 39:1121–1132
2. Aono T, Higuchi K, Ohira T, Komiyama B, Sasaoka H (2005) Wireless secret key generation exploiting reactance-domain scalar response of multipath fading channels. IEEE Trans Antennas Propag 53:3776–3784
3. Bloch M, Barros J (2011) Physical-layer security: from information theory to security engineering. Cambridge University Press, Cambridge
4. Bloch M, Thangaraj A, McLaughlin S, Merolla JM (2006a) LDPC-based Gaussian key reconciliation. In: Information theory workshop, 2006. ITW '06 Punta del Este. IEEE, pp 116–120
5. Bloch MR, Thangaraj A, McLaughlin SW, Merolla JM (2006b) LDPC-based secret key agreement over the Gaussian wiretap channel. Proceedings of IEEE international symposium on information theory. Seattle, USA, pp 1179–1183
6. Chung SY, Richardson T, Urbanke R (2001) Analysis of sum-product decoding of low-density parity-check codes using a Gaussian approximation. IEEE Trans Inf Theory 47(2):657–670
7. Davey M, MacKay D (1998) Low-density parity check codes over GF(q). IEEE Commun Lett 2(6):165–167
8. von Deetzen N, Sandberg S (2010) On the UEP capabilities of several LDPC construction algorithms. IEEE Trans Commun 58(11):3041–3046
9. Etesami J, Henkel W (2012) LDPC code construction for wireless physical-layer key reconciliation. In: 1st IEEE international conference on communications in China (ICCC), pp 208–213
10. Filip A, Mehmood R, Wallace J, Henkel W (2013) Physical-layer key generation supported by RECAP antenna structures. In: 9th international ITG conference on systems, communication and coding (SCC), pp 1–6
11. Filip A, Mehmood R, Wallace J, Henkel W (2014) Variable guard band construction to support key reconciliation. In: IEEE international conference on acoustics, speech and signal processing (ICASSP), pp 8173–8177
12. Gersho A, Gray RM (1991) Vector quantization and signal compression. Kluwer Academic Publishers, Norwell
13. Goel S, Negi R (2008) Guaranteeing secrecy using artificial noise. IEEE Trans Wirel Commun 7:2180–2189
14. Graur O, Islam N, Filip A, Henkel W (2015a) Quantization aspects in LDPC key reconciliation for physical layer security. In: 10th international ITG conference on systems, communications and coding SCC 2015, pp 1–6
15. Graur O, Islam N, Filip A, Henkel W (2016) Quantization and LLR computation for physical layer security. In: Proceedings of the International Zürich Seminar on Communications (IZS)
16. Hero AO III (2003) Secure space-time communication. IEEE Trans Inf Theory 49:3235–3249
17. Islam N, Graur O, Henkel W, Filip A (2015) LDPC code design aspects for physical-layer key reconciliation. In: IEEE Global Communications Conference (Globecom) 2015
18. Jensen MA, Wallace JW (2004) A review of antennas and propagation for MIMO wireless communications. IEEE Trans Antennas Propag 52:2810–2824
19. Kim H, Villasenor JD (2008) Secure MIMO communications in a system with equal numbers of transmit and receive antennas. IEEE Commun Lett 12:386–388
20. Koorapaty H, Hassan A, Chennakeshu S (2000) Secure information transmission for mobile radio. IEEE Commun Lett 4:52–55

21. Koyluoglu O, El Gamal H (2012) Polar coding for secure transmission and key agreement. IEEE Trans Inf Forensics Secur 7(5):1472–1483
22. Li X, Ratazzi EP (2005) MIMO transmissions with information-theoretic secrecy for secret-key agreement in wireless networks. In: Proceedings of 2005 IEEE Military Communication Conference (MILCOM'05), Atlantic City, NJ, vol 3, pp 1353–1359
23. Linde Y, Buzo A, Gray R (1980) An algorithm for vector quantizer design. IEEE Trans Commun 28(1):84–95
24. Lloyd S (1982) Least squares quantization in PCM. IEEE Trans Inf Theory 28(2):129–137
25. MacKay D, Wilson S, Davey M (1999) Comparison of constructions of irregular Gallager codes. IEEE Trans Commun 47(10):1449–1454
26. Massey J (1988) An introduction to contemporary cryptology. Proc IEEE 76:533–549
27. Mathur S, Trappe W, Mandayam N, Ye C, Reznik A (2008) Radio-telepathy: extracting a secret key from an unauthenticated wireless channel. In: Proceedings of 14th ACM international conference on mobile computing and networking, San Francisco, CA, pp 128–139
28. Maurer U (1993) Secret key agreement by public discussion from common information. IEEE Trans Inf Theory 39:733–742
29. Maurer U, Wolf S (2003a) Secret-key agreement over unauthenticated public channels–part I: definitions and a completeness result. IEEE Trans Inf Theory 49:822–831
30. Maurer U, Wolf S (2003b) Secret-key agreement over unauthenticated public channels–part III: privacy amplification. IEEE Trans Inf Theory 49:839–851
31. Maurer U, Wolf S (2003c) Secret-key agreement over unauthenticated public channels–part II: the simulatability condition. IEEE Trans Inf Theory 49:832–838
32. Max J (1960) Quantizing for minimum distortion. IEEE Trans Inf Theory 6(1):7–12
33. Mehmood R, Wallace JW (2011) Wireless security enhancement using parasitic reconfigurable aperture antennas. Proceedings of the European conference on antennas and propagation. Italy, Rome, pp 2761–2765
34. Mehmood R, Wallace JW (2012) Experimental assessment of secret key generation using parasitic reconfigurable aperture antennas. Proceedings of the European conference on antennas and propagation. Czech Republic, Prague, pp 1151–1155
35. Mehmood R, Wallace JW (2012b) MIMO capacity enhancement using parasitic reconfigurable aperture antennas (RECAPs). IEEE Trans Antennas Propag 60:665–673
36. Mehmood R, Wallace JW, Jensen MA (2014a) Key establishment employing reconfigurable antennas: impact of antenna complexity. IEEE Trans Wirel Commun 13:6300–6310
37. Mehmood R, Wallace JW, Jensen MA (2014) Optimal array patterns for encryption key establishment in LOS channels. Proceedings of the IEEE antennas and propagation society international symposium. Memphis, TN, pp 478–479
38. Mehmood R, Wallace JW, Jensen MA (2014) Secure array synthesis. IEEE Trans Antennas Propag Submitted
39. Mohammadi MS (2009) MIMO minimum leakage-physically secure wireless data transmission. Proceedings of the international conference application of information and communication technologies. Baku, Azerbaijan, pp 1–5
40. Ozarow LH, Wyner AD (1984) Wire-tap channel II. AT&T Bell Lab Tech J 63:2135–2157
41. Pierrot A, Chou R, Bloch M (2013) Experimental aspects of secret key generation in indoor wireless environments. In: IEEE 14th workshop on signal processing advances in wireless communications (SPAWC), pp 669–673
42. Pradhan SS, Ramchandran K (2003) Distributed source coding using syndromes (DISCUS): design and construction. IEEE Trans Inf Theory 49:626–643
43. Richardson T, Urbanke R (2001) The capacity of low-density parity-check codes under message-passing decoding. IEEE Trans Inf Theory 47(2):599–618
44. Richardson T, Urbanke R (2004) Multi-edge type LDPC codes
45. Richardson T, Urbanke R (2008) Modern coding theory. Cambridge University Press
46. Richardson T, Shokrollahi M, Urbanke R (2001) Design of capacity-approaching irregular low-density parity-check codes. IEEE Trans Inf Theory 47(2):619–637

47. Sandberg S, von Deetzen N (2010) Design of bandwidth-efficient unequal error protection LDPC codes. IEEE Trans Commun 58(3):802–811
48. Sayeed A, Perrig A (2008) Secure wireless communications: secret keys through multipath. Proceedings of the 2008 IEEE international conference acoustics, speech, and signal processing. Las Vegas, NV, pp 3013–3016
49. Sharma RK, Wallace JW (2010) Bit error rate and efficiency analysis of wireless reciprocal channel key generation. Proceedings of the IEEE conference on wireless information technology and systems. Honolulu, HI, pp 1–4
50. Sharma RK, Wallace JW (2011) Physical layer key generation methods for arbitrary fading channels. In: Proceedings of the 2011 IEEE antennas and propagation society international symposium, pp 1368–1371
51. Slepian D, Wolf J (1973) Noiseless coding of correlated information sources. IEEE Trans Inf Theory 19(4):471–480
52. Sun X, Wu X, Zhao C, Jiang M, Xu W (2010) Slepian-Wolf coding for reconciliation of physical layer secret keys. In: Wireless Communications and Networking Conference (WCNC), 2010 IEEE, pp 1–6
53. Thangaraj A, Dihidar S, Calderbank A, McLaughlin S, Merolla JM (2007) Applications of LDPC codes to the wiretap channel. IEEE Trans Inf Theory 53(8):2933–2945
54. Van Assche G, Cardinal J, Cerf NJ (2004) Reconciliation of a quantum-distributed Gaussian key. IEEE Trans Inf Theory 50:394–400
55. Voicila A, Declercq D, Verdier F, Fossorier M, Urard P (2010) Low-complexity decoding for non-binary LDPC codes in high order fields. IEEE Trans Commun 58(5):1365–1375
56. Wallace J (2009) Secure physical layer key generation schemes: performance and information theoretic limits. Proceedings of the IEEE international conference on communications. Dresden, Germany, pp 1–4
57. Wallace JW, Sharma RK (2010) Automatic secret keys from reciprocal MIMO wireless channels: measurement and analysis. IEEE Trans Inf Forensics Secur 5:381–392
58. Wilson R, Tse D, Scholtz RA (2007) Channel identification: secret sharing using reciprocity in ultrawideband channels. IEEE Trans Inf Forensics Secur 2:364–375
59. Wyner A (1975) The wire-tap channel. Bell Syst Tech J 54(8):1355–1387
60. Ye C, Reznik A, Shah Y (2006) Extracting secrecy from jointly Gaussian random variables. Proceedings of IEEE international symposium on information theory. Seattle, WA, pp 2593–2597
61. Ye C, Reznik A, Sternberg G, Shah Y (2007) On the secrecy capabilities of ITU channels. In: Proceedings of the 66th IEEE vehicular technology conference, Baltimore, MD
62. Ye C, Mathur S, Reznik A, Shah Y, Trappe W, Mandayam NB (2010) Information-theoretically secret key generation for fading wireless channels. IEEE Trans Inf Forensics Secur 5:240–254
63. Zhou X, Kyritsi P, Eggers P, Fitzek F (2007) The medium is the message: secure communication via waveform coding in MIMO systems. Proceedings of the IEEE 65th vehicular technology conference. Dublin, Ireland, pp 491–495

Chapter 18
Physical Layer Cooperation in One-Way Relaying Systems

Meng Wu, Frank Ludwig, Dirk Wübben, Armin Dekorsy, Karl-Dirk Kammeyer and Steffen Paul

Abstract The application of cooperative strategies in wireless networks offers several advantages like improved performance, increased system throughput, and coverage extension. To this end, the source-destination link is expanded by additional transceivers in the link, e.g., mobile terminals, relays, or sensor nodes. By clustering several nodes, a virtual antenna array (VAA) is constructed, such that coding and signal processing approaches known from multiple antenna systems can be applied in principle to these distributed systems. Depending on the allowed amount of cooperation between the nodes within the VAA, different distributed reception and transmission schemes can be implemented. This chapter aims at the investigation of relaying networks with distributed single-antenna relays, adapted distributed signal processing strategies, and different levels of cooperation within the VAA. The focus also lies on implementation aspects in the physical layer and corresponding energy consumption considerations. The latter is especially important for wireless sensor networks with battery power restrictions per node. Therefore, the total energy consumption at the VAA for signal transmission as well as transceiver circuit and baseband processing is analyzed.

M. Wu (✉) · D. Wübben · A. Dekorsy · K.-D. Kammeyer
Department of Communications Engineering, University of Bremen,
28359 Bremen, Germany
e-mail: wu@ant.uni-bremen.de

D. Wübben
e-mail: wuebben@ant.uni-bremen.de

A. Dekorsy
e-mail: dekorsy@ant.uni-bremen.de

K.-D. Kammeyer
e-mail: kammeyer@ant.uni-bremen.de

F. Ludwig · S. Paul
Department of Communication Electronics, University of Bremen,
28359 Bremen, Germany
e-mail: ludwig@me.uni-bremen.de

S. Paul
e-mail: steffen.paul@me.uni-bremen.de

W. Utschick (ed.), *Communications in Interference Limited Networks*,
Signals and Communication Technology, DOI 10.1007/978-3-319-22440-4_18

18.1 Introduction

The concept of cooperation attracts increasing interests in wireless communications that combats fading effectively. This can be achieved by, e.g., introducing relaying nodes to expand the source-destination link with improved performance. To this end, different relaying schemes were presented such as amplify-forward (AF) and decode-forward (DF) [10]. Employing AF, the relay simply forwards an amplified version of its receive signal, whereas the source message is decoded and re-encoded before re-transmission in DF. Furthermore, when multiple relays are available to support the transmission from source to destination, the relays can be clustered into a virtual antenna array (VAA) [5], where coding and signal processing approaches known for multiple antenna systems can be applied. In [9] a distributed relaying network with single-antenna relays applying DF was investigated. Therein, only the relays decoding the source message correctly transmit to the destination using orthogonal space-time block code (OSTBC). It was shown that the diversity degree depends on the number of cooperating relays, but not on the number of relays with correct decoding result.

In this chapter, coded orthogonal frequency division multiplexing (OFDM) is utilized for relaying communications to combat multi-path fading. Firstly, a single-relay network is considered, where OFDM-based relaying schemes are presented and compared. Secondly, the network is extended to a multiple-relay scenario. Therein, inter-relay cooperation (IRC) strategies are developed that allow message exchange between relays. Depending on the decoding status at the VAA, different types and amounts of exchanged information are adjusted accordingly. The compromise between the improved decoding performance and the required time overhead is investigated in the throughput analysis. Finally, the total energy consumption at the relays for not only signal transmission but also RF circuit and baseband processing is taken into account for analysis.

18.2 Single-Relay Network

A triplet relaying network with a single relay R supporting the one-way transmission from source S to destination D is shown in Fig. 18.1. The system operates in half-duplex mode with presence of the source-destination link. In the first phase, the

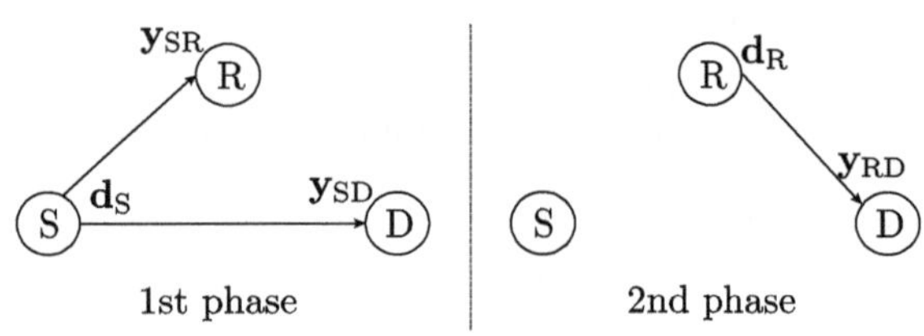

Fig. 18.1 A single-relay system consisting of one source S, one destination D and one relay R. The direct transmission from S to D is present

information word $\mathbf{b}_S$ at S is encoded to a codeword $\mathbf{c}_S$. This codeword is interleaved and mapped to the symbol vector $\mathbf{d}_S$, which is then broadcasted to both R and D. Subsequently, R processes its receive signal vector $\mathbf{y}_{SR}$ and forwards a relay message $\mathbf{d}_R$ to D in the second phase. Finally, D estimates the source message using the receive signal vector $\mathbf{y}_{SD}$ from S and $\mathbf{y}_{RD}$ from R. Since OFDM is applied, this results in a varying link quality per subcarrier. Correspondingly, the conventional relaying schemes in [10] should be adjusted to OFDM transmission and may be modified to yield improved performance [16], as presented in the sequel.

18.2.1 Amplify-Forward

For amplify-forward (AF), the relay retransmits an amplified version of the receive signal $\mathbf{y}_{SR}$ constrained by its power budget P_R. Concerning OFDM systems, the amplification can be done either in frequency domain or in time domain, as illustrated in the following.

AF with constant power (AF-CP) When the amplification is performed in frequency domain, the same power is assigned to each subcarrier. Denoting $P_{R,k}$ the power on subcarrier k yields $P_{R,k} = P_R/N_K$ with N_K being the number of subcarriers and $k = 0, 1, \ldots, N_K - 1$. To meet this power constraint per subcarrier, the amplification factor on the kth subcarrier can be calculated as

$$\beta_k = \sqrt{\frac{P_{R,k}}{|h_{SR,k}|^2 P_{S,k} + \sigma_n^2}} . \tag{18.1}$$

Here, $h_{SR,k}$ denotes the channel coefficient on subcarrier k of the SR link and σ_n^2 is the noise variance. Furthermore, the power P_S at S is equally assigned over all subcarriers as $P_{S,k} = P_S/N_K$. The processing chain at R for AF-CP is shown in Fig. 18.2. At D the receive signals from S and R are maximum ratio combined (MRC) to achieve the maximum signal-to-noise-ratio (SNR). It is noted that fast Fourier transform (FFT) and inverse FFT (IFFT) as extra baseband processing effort are required at R due to signal amplification in frequency domain.

AF with constant gain (AF-CG) In contrast to AF-CP, the amplification can also be done in time domain with an equal gain on all subcarriers. This is achieved by applying a common amplification factor β to all subcarriers with β defined as

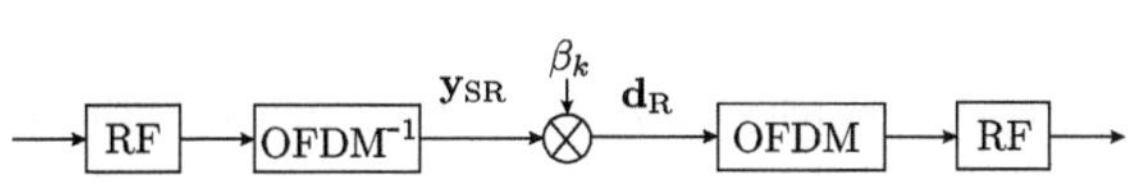

Fig. 18.2 Processing chain at the relay for AF-CP. The amplification is done in frequency domain with different β_k per subcarrier

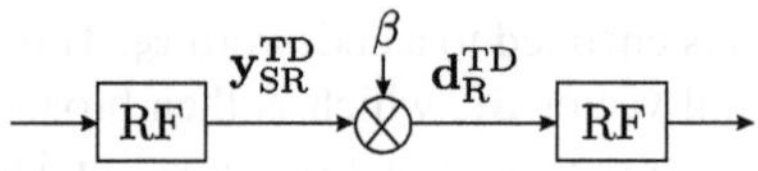

Fig. 18.3 Processing chain at the relay for AF-CG. The amplification is done in time domain with a common β over all subcarriers

$$\beta = \sqrt{\frac{P_{\mathrm{R}}}{\sum_{k=0}^{N_{\mathrm{K}}-1} |h_{\mathrm{SR},k}|^2 P_{\mathrm{S},k} + N_{\mathrm{K}} \sigma_n^2}}\,. \quad (18.2)$$

The processing chain at R for AF-CG is shown in Fig. 18.3 with the time domain signals denoted by superscript 'TD'. The advantage of using AF-CG over AF-CP is that the signal processing at the relay remains in time domain. Thus, the effort for implementing FFT and IFFT can be saved. This aspect is elaborated later on as the energy consumption for not only transmission, but also RF circuit and baseband processing will be taken into account for practical reasons.

18.2.2 Decode-Forward

When AF is applied, the coding gain is ignored at the relay, which can be exploited by decode-forward (DF). For DF, the receive signal $\mathbf{y}_{\mathrm{SR}}$ at R is demodulated and decoded to estimate the source message $\mathbf{b}_{\mathrm{S}}$. Note that a soft-input soft-output decoder is employed with the log-likelihood ratio (LLR) vector $\mathbf{L}_{\mathrm{b}}$ representing the decoder output. Afterwards, the same channel code as at the source is used to encode the estimated information bits $\hat{\mathbf{b}}_{\mathrm{S}}$, yielding the code bits which are then interleaved and mapped to an OFDM symbol, as shown in Fig. 18.4. Assuming perfect decoding at the relay, MRC is used at D to optimally combine the receive signals from S and R. However, if decoding errors occur at the relay, these errors will be propagated to D which jeopardizes the performance of MRC, especially when the RD link is much stronger than the SR link.

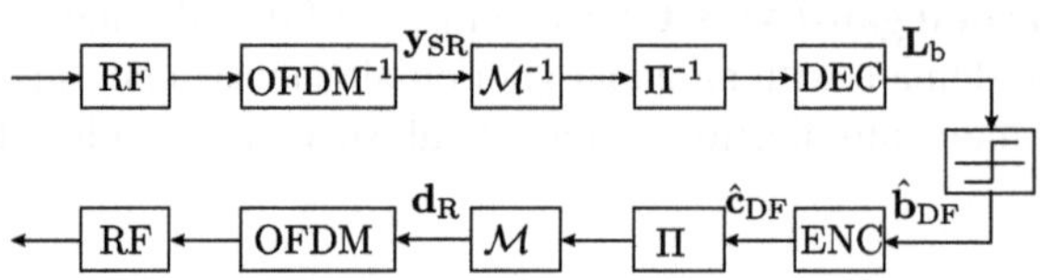

Fig. 18.4 Processing chain at the relay for DF. The estimated information bits $\hat{\mathbf{b}}_{\mathrm{DF}}$ are re-encoded to yield the estimated code bits $\hat{\mathbf{c}}_{\mathrm{DF}}$

18.2.3 Decode-Quantize-Forward

In order to mitigate the impact of error propagation in DF, decode-quantize-forward (DQF) [16] can be applied by exploiting the reliability knowledge of the SR link at D, e.g., SNR of the SR link denoted as $\mathrm{SNR_{SR}}$. In contrast to DF, DQF directly takes the estimates of the code bits from the decoder at R. The corresponding processing chain for DQF is presented in Fig. 18.5 with $\mathbf{L}_\mathrm{c}$ denoting the LLR vector of the whole codeword at the decoder output.The principle of DQF is based on the relationship between code bit errors (CBEs) at R and $\mathrm{SNR_{SR}}$, as shown in Fig. 18.6. Therein, the CBEs and the channel gain $|h_{\mathrm{SR},k}|$, or equivalently, $\mathrm{SNR}_{\mathrm{SR},k} = P_{\mathrm{S},k}|h_{\mathrm{SR},k}|^2/\sigma_n^2$, are presented for both DF (after re-encoding) and DQF with $N_\mathrm{K} = 256$ subcarriers and a fixed arbitrary channel realization. For these simulations, BPSK and an LDPC code are employed. It can be observed that the CBEs for DF are uniformly distributed over all subcarriers, i.e., there is no explicit relationship between $\mathrm{SNR_{SR}}$ and the CBEs. This is because for an erroneous information sequence, the errors will be distributed over the whole codeword after re-encoding due to the applied LDPC encoder structure. Thus, it is difficult to exploit the reliability information of the SR link at D for DF. For DQF, it can be observed that less CBEs occur at high $\mathrm{SNR_{SR}}$ whereas erroneous decisions of the code bits are more likely for low $\mathrm{SNR_{SR}}$. This

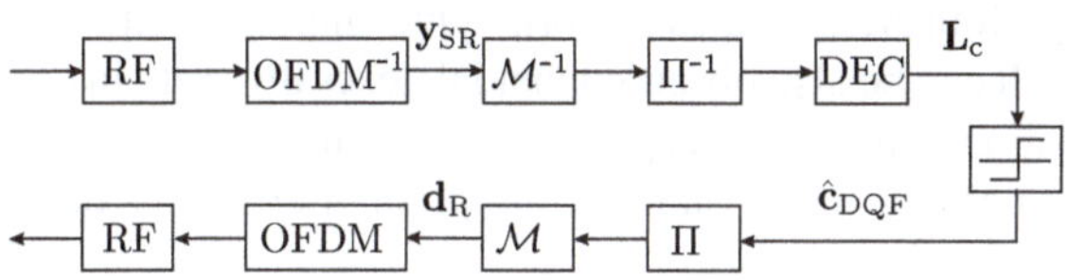

Fig. 18.5 Processing chain at the relay for DQF. The estimated code bits $\hat{\mathbf{c}}_\mathrm{DQF}$ are directly generated from the decoder without re-encoding

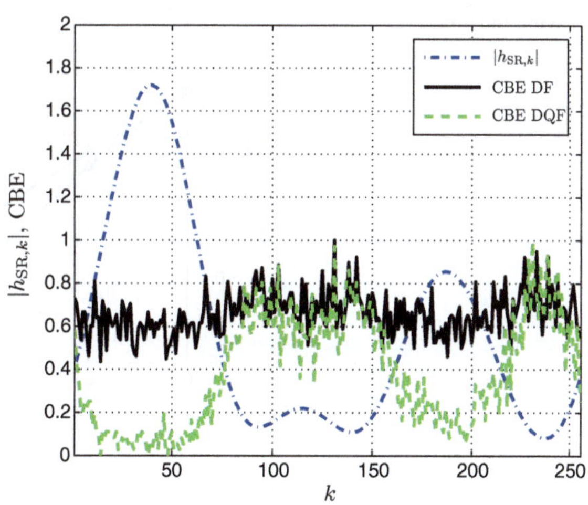

Fig. 18.6 Normalized CBEs and channel gain $|h_{\mathrm{SR},k}|$ of DF and DQF per subcarrier for one channel realization with $N_\mathrm{H} = 5$ taps in time domain and $N_\mathrm{K} = 256$ subcarriers

characteristic motivates a modified MRC (mMRC) at D that reduces the effect of decoding errors at R on the overall performance [16]. Based on the assumption of applying a zero forcing (ZF) equalizer at R, mMRC is constructed in such a way that the subcarrier with a lower $\mathrm{SNR_{SR}}$ from the relay experiences a greater suppression at D. Correspondingly, the influence of error propagation can be mitigated.

18.2.4 Performance Evaluation

For link-level simulations of a coded OFDM-based relaying system with $N_{\mathrm{K}} = 256$ subcarriers, all links are assumed to be block Rayleigh fading with $N_{\mathrm{H}} = 5$ channel taps and path-loss exponent $\alpha = 4$. The relay is located in the middle between the source and the destination. The modulation alphabet is 16-QAM for the relaying schemes and 4-QAM for the direct transmission (DT). Thus, a fair comparison is achieved due to the half-duplex constraint for relaying. An LDPC code of rate $R_{\mathrm{C}} = 0.5$ is applied to encode each OFDM frame individually. Furthermore, the power is equally assigned to S and R, i.e., $P_{\mathrm{S}} = P_{\mathrm{R}}$. Optimal power allocation schemes for P_{S} and P_{R} can be found in [12, 13] but are not applied here.

Figure 18.7 shows the BER performance for DT, AF-CP, AF-CG, DF, DQF with common MRC (cMRC) as well as mMRC. The relaying approaches outperform the DT significantly. Both AF schemes perform quite similarly, with slight gain for AF-CP over AF-CG. It can be also observed that DF loses diversity in contrast to AF due to error propagation caused by decoding errors at the relay. Furthermore, DQF in combination with cMRC based on the assumption of error-free decoding at R leads

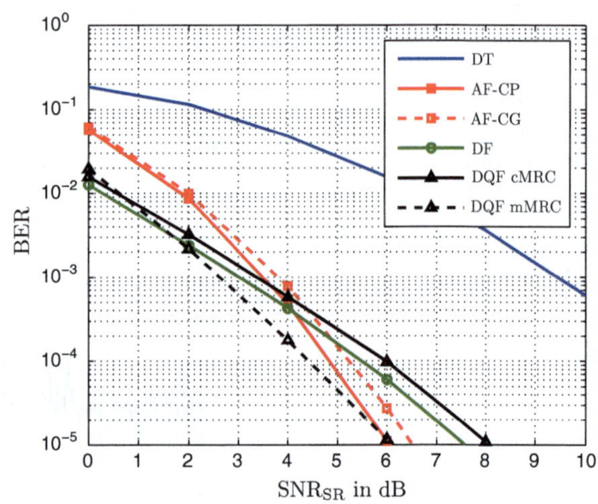

Fig. 18.7 BER performance for DT and a relaying network using different relaying schemes

to even worse performance compared to DF. However, when mMRC is employed at D, DQF achieves the best performance in the investigated SNR region, as the effect of error propagation is limited by mMRC.

18.3 Multiple-Relay Network

In this section, the single-relay network considered above is extended to a multiple-relay scenario, i.e., M single-antenna relays constituting a VAA, as shown in Fig. 18.8. In case of one relay equipped with multiple antennas, many multiple-input multiple-output (MIMO) techniques are available to achieve spatial diversity gain. For example, MRC and generalized selective combining (GSC) can be applied to exploit receive diversity. If the channel state information (CSI) is not available at the transmitter side, space-time codes (STCs) are common candidates to exploit transmit diversity. However, the application of these MIMO schemes needs to be reconsidered if the antennas are geographically distributed in the VAA. In this context, the transceivers in the VAA can cooperate with each other to approach a conventional MIMO system, which requires information exchange among the relays for both transmission and reception at the VAA. This aspect is addressed in the sequel.

18.3.1 Inter-Relay Cooperation

In order to facilitate information exchange within the VAA, inter-relay cooperation (IRC) is introduced that allows dedicated transmissions among the relays. To this end, a time division operating mode is proposed as shown in Fig. 18.9. Therein, the time duration for both transmissions from S to VAA and from VAA to D is T_{OFDM}. IRC occupies an independent time slot with duration T_{IRC}.

Employing DF as the relaying scheme, the relays decode the source message individually upon reception. Assuming that a cyclic redundancy check (CRC) code is

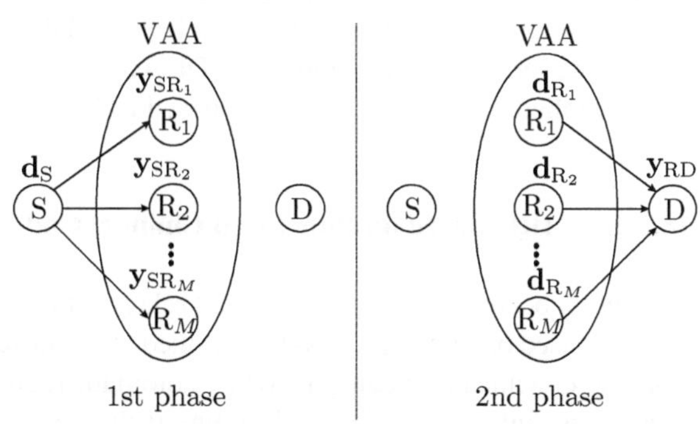

Fig. 18.8 A multiple-relay network consisting of one source S, one destination D and M relays R_m. The direct transmission from S to D is not present

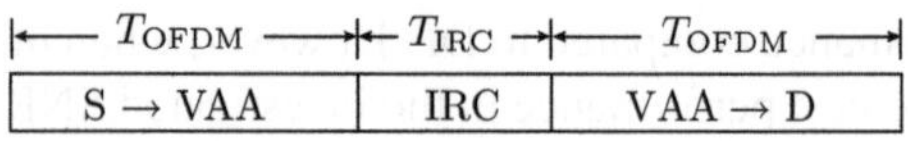

Fig. 18.9 Time line for relaying transmission with IRC

used at the source, each relay R_m is aware of its own decoding status. The relays now inform each other about their decoding status by broadcasting a one-bit acknowledgment (ACK) or negative acknowledgement (NAK) denoted as CRC bit. Furthermore, a fixed network structure is assumed, thus each relay R_m knows its own index m. This facilitates the relays to send the CRC bits in a pre-fixed order without extra overhead, e.g., R_1 sends first and after receiving the CRC bit from R_m, R_{m+1} sends its CRC bit. Consequently, all relays are aware of the set $\mathcal{D}$ containing all relays that decode the source message correctly. Depending on the cardinality $|\mathcal{D}|$ three events can be distinguished as follows.

- **Event E_1: $|\mathcal{D}| = M \rightarrow$ all relays decode correctly**
  The exchange of auxiliary information among the relays is not necessary since all relays have already decoded the source message correctly. Therefore, M relays participate in the transmission to D using, e.g., distributed orthogonal space-time block code (D-OSTBC).
- **Event E_2: $1 \leq |\mathcal{D}| < M \rightarrow$ some relays decode correctly**
  An IRC scheme based on punctured channel codes is applied for re-decoding at the relays with decoding errors [14] as illustrated in Sect. 18.3.1.1. These $(M - |\mathcal{D}|)$ relays exchange their CRC bits again after re-decoding.
- **Event E_3: $|\mathcal{D}| = 0 \rightarrow$ no relay decodes correctly**
  An improved IRC scheme based on receive signal exchange is applied for re-decoding at the relays with decoding errors [15] as illustrated in Sect. 18.3.1.2. All relays have to exchange their CRC bits again after re-decoding.

After IRC, the relays achieving successful re-decoding transmit to D using D-OSTBC. Specifically, each involved relay transmits its corresponding layer in the OSTBC matrix [9]. Note that the source-destination link is not present in the multiple-relay scenario for simplicity, in other words, the destination is only able to achieve information from the relays. Therefore, if all relays still have decoding errors after IRC, the relays may, e.g., switch to AF because DF probably causes error propagation. However, as a simple approach, DF with all relays participating in D-OSTBC is still applied in this case for the proposed IRC scheme.

18.3.1.1 IRC Based on Punctured Channel Codes

If event E_2 occurs, i.e., some but not all relays achieve successful decoding, the IRC scheme based on punctured codes is applied [14]. Basically, this scheme is motivated from incremental redundancy in hybrid automatic request (HARQ), which makes use of auxiliary information for re-decoding if the initial decoding fails [3]. As shown

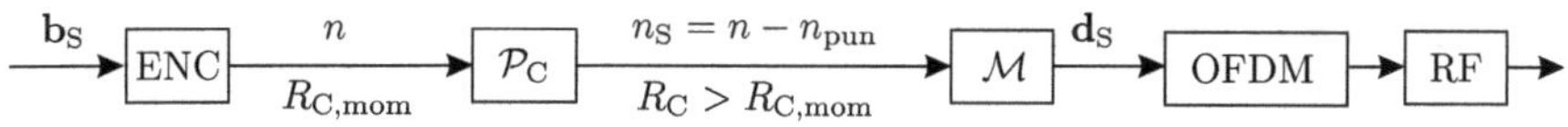

Fig. 18.10 Processing chain at source S by using a punctured channel code with n_{pun} punctured bits

in Fig. 18.10 for the processing chain at S, the information bits are encoded to a codeword of length n using a channel code with mother rate $R_{C,mom}$. Afterwards, n_{pun} bits are punctured, resulting in a shortened codeword of length $n_S = n - n_{pun}$ and code rate $R_C > R_{C,mom}$. The punctured codeword is further interleaved and modulated for transmission to the relays. It is noted that the interleaver is omitted here and in the sequel for simplicity.

After local decoding and exchanging the CRC bits, one of the relays with correct decoding result $R_\mu \in \mathcal{D}$, e.g., the one with the smallest index $\mu = \min_{R_m \in \mathcal{D}} m$, re-generates and broadcasts $n_{IRC} \leq n_{pun}$ punctured bits to the other relays. As shown in Fig. 18.11, the upper dashed block depicts the processing chain at relay R_μ. After successful decoding, all or part of the punctured bits from the mother codeword are generated. Note that less punctured bits exchange leads to smaller amount of overhead but results in less performance gain. This trade-off is to be observed in the simulation results later on. The punctured bits are then modulated using a higher modulation alphabet $\mathcal{M}_{IRC}$ and transmitted over a single-carrier inter-relay (IR) channel. Since the relays within the VAA are assumed to be close to each other, this results in a robust IR transmission. Subsequently, as shown in the lower dashed block in Fig. 18.11, each relay R_j that has decoding errors re-decodes the source message with the help of the punctured bits received from the IR channel. In order to inform the other relays about the re-decoding status, these relays broadcast the CRC bits again to determine the new set $\mathcal{D}'$, where $\mathcal{D}' \supseteq \mathcal{D}$ holds. This indicates that this IRC scheme may increase the number of relays with correctly decoded source message. Finally, only the relays in set $\mathcal{D}'$ transmit to D using D-OSTBC.

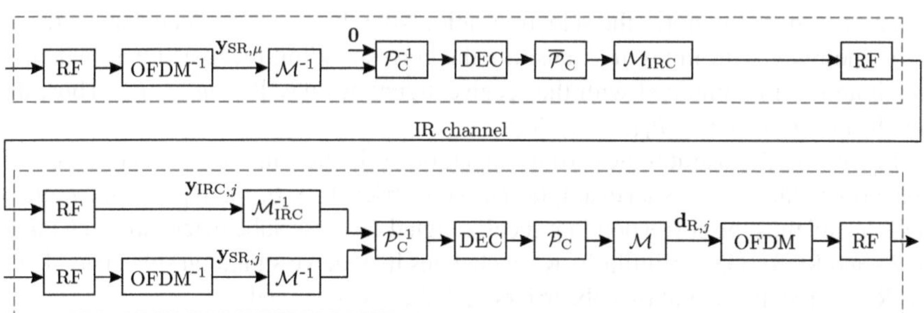

Fig. 18.11 Processing chain at relay R_μ (*upper dashed block*) and one of the relays R_j that fails successful decoding (*lower dashed block*)

18.3.1.2 IRC Based on Receive Signal Exchange

The IRC scheme based on exchanging punctured bits does not improve the performance in E_3, i.e., when all relays encounter imperfect decoding. In this case, other types of information rather than hard decided bits should be exchanged, e.g., the receive signal. Motivated by a single-input multiple-output (SIMO) system using signal combining strategies to exploit receive diversity, an IRC scheme based on receive signal exchange is applied in E_3 [15]. Therein, two cooperation schemes are distinguished as follows.

Full cooperation The whole received OFDM frames at the relays are shared with each other for MRC, resulting in $\Phi_{\text{IRC}} = MN_{\text{K}}$ complex symbols to be exchanged in the IRC phase. This leads to a dramatic decrease of throughput but is still investigated to serve as a benchmark and termed full cooperation.

Partial cooperation In order to decrease the overhead, only part of the receive signal is exchanged within the VAA, which is termed partial cooperation. In [8, 18], GSC was proposed that combines the receive signal at some branches instead of all of them in a SIMO system. This combining strategy can be applied at the VAA based on partial cooperation with decreased amount of exchanged information compared to full cooperation. To determine which part of the receive signal is exchanged, a subcarrier wise relay index μ_k is firstly defined as

$$\mu_k = \arg\max_m \{|h_{m,k}|^2\}\,, \tag{18.3}$$

which corresponds to the strongest SNR over all relays. This results in a SNR_{SR} related vector

$$\hat{\mathbf{h}} = [|h_{\mu_1,1}|^2, |h_{\mu_2,2}|^2, ..., |h_{\mu_{N_{\text{K}}},N_{\text{K}}}|^2]\,. \tag{18.4}$$

Sorting $\hat{\mathbf{h}}$ in a decreasing order yields the sorted vector $\hat{\mathbf{h}}_{\text{s}}$. For partial cooperation, only the subcarriers with respect to the first $\lceil \rho N_{\text{K}} \rceil$ terms in $\hat{\mathbf{h}}_{\text{s}}$, or in other words, the part of subcarriers with the highest SNRs, are selected for receive signal exchange, where $0 < \rho \leq 1$ denotes the selection ratio. Note that for the selected subcarriers, only one relay with index μ_k transmits the corresponding receive signal, which is maximum ratio combined with the receive signal at relay R_m, $m \neq \mu_k$. Thus, the overhead is reduced to $\Phi_{\text{IRC}} = \lceil \rho N_{\text{K}} \rceil$.

Figure 18.12 illustrates the partial cooperation scheme with $M = 3$ relays, $N_{\text{K}} = 8$ subcarriers and $\rho = 1$ as an example. The subcarriers $0, 1, 6, 7$ at R_1 have the highest SNR_{SR} and thus the corresponding receive signal on these subcarriers are exchanged to R_2 and R_3 for GSC. Similarly, R_2 exchanges its receive signal on subcarrier 4 and for R_3 the receive signal on subcarriers 2, 3, 5 are exchanged.

Exchange of CSI For both full and partial cooperation schemes, the relays need to share the CSI of the SR links for subcarrier wise relay selection as well as message combining. This requires extra overhead and should be considered adequately. To this end, each relay has to exchange N_{H} complex values for the time-domain channel

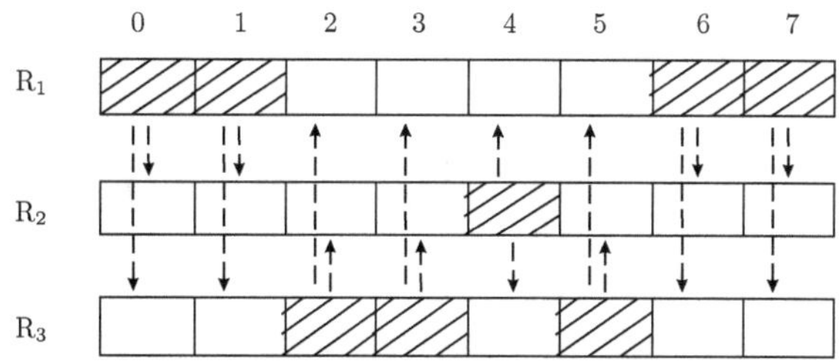

Fig. 18.12 An example for partial cooperation with $M = 3$ relays, $N_K = 8$ subcarriers and $\rho = 1$. The hatched subcarriers with the highest SNR_{SR} over all relays are selected for receive signal exchange

taps within the VAA, leading to an overhead of $\Phi_{CH} = MN_H$. Assuming that all channels are block fading and stay invariant for N_{inv} OFDM frames, during which these channel coefficients need to be exchanged once, the data rate is decreased by a factor δ given as

$$\delta = \frac{N_{inv}\Phi_{IRC}}{N_{inv}\Phi_{IRC} + \Phi_{CH}} \tag{18.5}$$

for IRC in E_3. Clearly, this effect becomes negligible with growing N_{inv}. Note that considering the energy consumption for baseband processing, $M(M-1)$ extra FFT efforts are required to obtain the channel coefficients in frequency domain.

18.3.2 Throughput Analysis

The dedicated time slot assigned to IRC decreases the system throughput. Specifically, when punctured bits are exchanged in E_2, an M_{IRC}-QAM modulation alphabet is used, resulting in a time duration $T_{pun} = \lceil \frac{n_{IRC}}{M_{IRC}} \rceil T_s$ with T_s denoting the symbol duration in single-carrier transmission. On the other hand, the receive signal and the channel coefficients are exchanged in E_3 by AF or digital transmission of the quantized signal [15]. As an example, AF is adopted here with a time duration $T_{rec} = \Phi_{IRC} T_s/\rho$. Furthermore, event E_3 is decomposed into three sub-events as

$$E_4 = E_3 \cap (E_3 \to E_1) \tag{18.6a}$$

$$E_5 = E_3 \cap (E_3 \to E_2) \tag{18.6b}$$

$$E_6 = E_3 \cap (E_3 \to E_3) \ , \tag{18.6c}$$

where $E_3 \to E_i$, $i = 1, 2, 3$, represents the event transfer that E_3 turns to E_i after exchanging the receive signal and re-decoding. Noticeably, if $E_3 \to E_2$ occurs, i.e., at least one but not all relays decode correctly, the punctured bits are further exchanged among the relays. Since the presented IRC scheme is adaptive, the total time required for IRC T_{IRC} depends on different events as

$$T_{\mathrm{IRC}} = \begin{cases} MT_s & \text{if } \mathrm{E}_1 \\ (2M - |\mathcal{D}|)\, T_s + T_{\mathrm{pun}} & \text{if } \mathrm{E}_2 \\ 2MT_s + T_{\mathrm{rec}} & \text{if } \mathrm{E}_4, \mathrm{E}_5 \\ \left(3M - |\mathcal{D}''|\right) T_s + T_{\mathrm{rec}} + T_{\mathrm{pun}} & \text{if } \mathrm{E}_6\,. \end{cases} \tag{18.7}$$

Here, $\mathcal{D}''$ represents the set containing all relays with correct decoding results after receive signal exchange in E_6. Note that each relay requires one individual symbol duration T_s to broadcast its CRC bit, resulting in a time duration MT_s in total to determine the set $\mathcal{D}$ in the VAA. In contrast, only the relays with decoding errors have to update and send their decoding status after re-decoding.

In order to achieve a fair comparison between the schemes with and without IRC, the throughput of the overall system is analyzed that takes into account the extra time consumption T_{IRC}. Since T_{IRC} depends on different events as shown in (18.7), the averaged throughput of the overall system η is calculated by

$$\eta = \sum_{\substack{j=1 \\ j \neq 3}}^{6} p_j \frac{N_{\mathrm{K}} \log_2 M_{\mathrm{S}} R_{\mathrm{C}}}{2T_{\mathrm{OFDM}} + T_{\mathrm{IRC},j}} \left(1 - \mathrm{FER}_{\mathrm{D}}\right) , \tag{18.8}$$

where p_j and $T_{\mathrm{IRC},j}$ denote the event probability and time duration for IRC when E_j occurs, respectively. Each subcarrier is modulated by M_{S}-QAM for OFDM transmission. The frame error rate (FER) at the destination is denoted as $\mathrm{FER}_{\mathrm{D}}$. Both p_j and $\mathrm{FER}_{\mathrm{D}}$ can be achieved numerically by simulations. The throughput η defined in (18.8) represents the number of correct information bits received at D per unit time such that it captures the impact of the extended time duration T_{IRC} in case of IRC.

18.3.3 Performance Evaluation

A distributed relaying system with $M = 4$ relays using DF is considered. The constituted VAA joints the direct line between S and D with $d_{\mathrm{SR}} = d_{\mathrm{RD}}$. All channels are assumed to be block Rayleigh fading with $N_{\mathrm{H}} = 5$ multi-path taps and path-loss exponent $\alpha = 4$. The channel coefficients are assumed to be invariant during $N_{\mathrm{inv}} = 10$ OFDM frames. One OFDM frame occupies $N_{\mathrm{K}} = 256$ subcarriers with 16-QAM modulation. IRC is only subject to AWGN disturbance with $\mathrm{SNR}_{\mathrm{IRC}} = 30$dB at the receiver side due to the assumption of closely located relays. 256-QAM modulation is used when the punctured bits are exchanged among the relays.

For the transmission from VAA to D, rate matching is applied to compensate the rate loss caused by OSTBC. Denoting R_{C} the original channel code rate and R_{ST} the OSTBC rate, the channel code rate for the RD link transmission is adjusted to R'_{C} with $R_{\mathrm{C}} = R_{\mathrm{ST}} R'_{\mathrm{C}}$. Depending on the events, the VAA employs an OSTBC of rate $R_{\mathrm{ST}} = 3/4$ [11] when 3 or 4 relays transmit and $R_{\mathrm{ST}} = 1$ (Alamouti [1]) when 2 relays transmit. Therefore, the channel code rate for the RD link transmission is adapted to $R_{C,RD} = 2/3$ in case of using the 3/4-OSTBC for rate matching. For IRC, S uses rate-compatible punctured convolutional (RCPC) codes [7] with a mother code rate $R_{C,mom} = 1/3$. The constraint length is set to $L_C = 4$ and the corresponding generator polynomial is $(13, 15, 11)_8$. By puncturing $n_{pun} = 512$ bits, the effective code rate $R_C = 1/2$ is achieved for transmission.

In Fig. 18.13, the probabilities for different number of relays with correct decoding results before re-recoding $|\mathcal{D}|$ and after re-decoding $|\mathcal{D}'|$ are presented. It can be observed in (a), that event E_1 ($|\mathcal{D}| = 4$) occurs more likely in the high SNR region and vice versa for E_3 ($|\mathcal{D}| = 0$). Note that E_2 is the superposition of $|\mathcal{D}| \in \{1, 2, 3\}$ and occurs more likely in the medium SNR region. Figure 18.13b, c show these probabilities after re-decoding when considering IRC only for E_2 with $n_{IRC} = 128$ and $n_{IRC} = 512$, respectively. In this case, the curve for $|\mathcal{D}'| = 0$ remains unchanged since E_3 is ignored. Furthermore, the curves for 1, 2 and 3 relays without decoding errors after re-decoding totally vanish when all punctured bits $n_{IRC} = 512$ are exchanged. By exchanging receive signal when E_3 occurs, the probability for all relays still in error after re-decoding gets smaller, as can be observed in Fig. 18.13d. It is also shown that exchanging larger amount of receive signal leads to improved re-decoding quality at the relays.

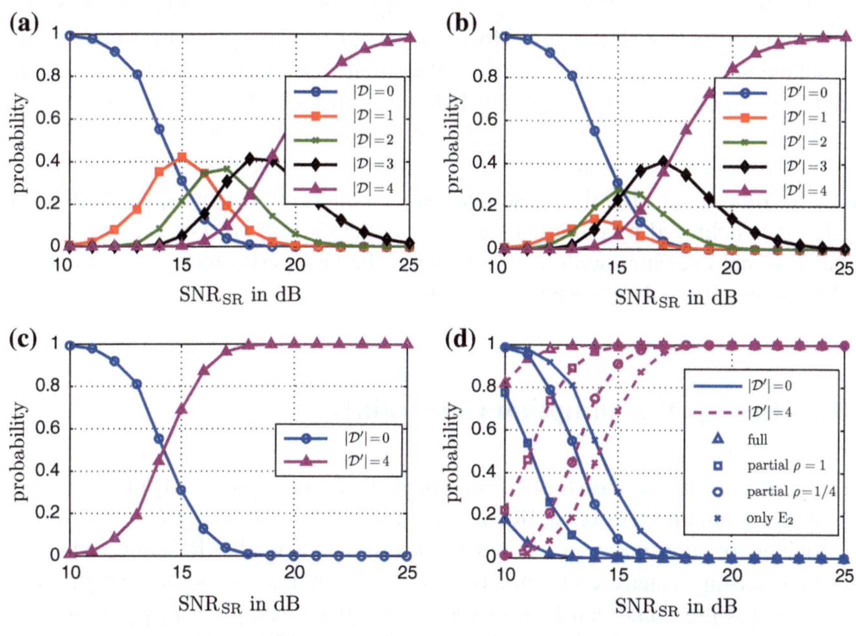

Fig. 18.13 Probabilities for different number of relays with correct decoding results **a** before re-decoding, **b** and **c** after re-decoding when only considering IRC for E_2 with $n_{IRC} = 128$ and $n_{IRC} = 512$, respectively, and **d** after re-decoding when considering IRC for both E_2 and E_3 with $n_{IRC} = 512$

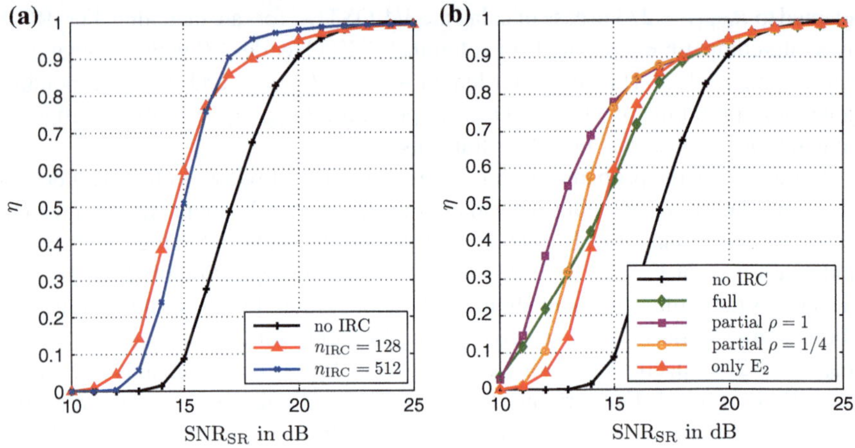

Fig. 18.14 System throughput η when **a** only considering IRC for E_2 and **b** considering IRC for both E_2 and E_3 with $n_{IRC} = 512$

Figure 18.14 shows the throughput performance for different cooperation schemes. When only considering IRC for E_2, exchanging more punctured bits leads to higher throughput in the low SNR region. However, exchanging less punctured bits becomes advantageous as SNR_{SR} increases since the smaller amount of overhead gets dominant in this case. When considering IRC for both E_2 and E_3, exchanging receive signal when all relays are in error further improves the throughput in general, especially in the low SNR region. Due to large amount of overhead, using full cooperation in E_3 becomes even slightly worse than only considering E_2 for higher SNRs. The trade-off is achieved by employing partial cooperation in E_3. For example, partial receive signal exchange with $\rho = 1$ achieves the best performance over the whole SNR region among the investigated scenarios.

18.4 Energy Consumption Constraints

The extra time consumption for IRC was taken into account previously in the throughput analysis. On the other hand, IRC also requires additional energy cost. To this end, the total energy consumption for signal transmission as well as RF circuit and baseband processing is analyzed for practical concerns. In this section, the energy model in [14, 15] is presented that focuses on the energy consumption E_{relay} at all relays for transmitting one OFDM frame from S to D via the VVA, which can be split into two parts as

$$E_{relay} = E_{SRD} + E_{IRC} \,. \tag{18.9}$$

Here, E_{SRD} denotes the energy required to support the source-relay-destination transmission whereas E_{IRC} is the energy dedicated to IRC. Furthermore, both parts can be decomposed into the energy components for signal transmission, RF circuit and baseband processing, yielding

$$E_{\mathrm{SRD}} = E_{\mathrm{SRD,signal}} + E_{\mathrm{SRD,RF}} + E_{\mathrm{SRD,base}} \tag{18.10a}$$

$$E_{\mathrm{IRC}} = E_{\mathrm{IRC,signal}} + E_{\mathrm{IRC,RF}} + E_{\mathrm{IRC,base}} \,. \tag{18.10b}$$

Each component of the energy consumption will be elaborated in the following subsections, respectively.

18.4.1 Transmit Signal Energy

The energy consumption for signal transmission from VAA to D is related to the transmit power P_{R} at all relays. Correspondingly, it depends on SNR of the RD link $\mathrm{SNR}_{\mathrm{RD}} = P_{\mathrm{R}}/(d_{\mathrm{RD}}^{\alpha}\sigma_n^2)$ with d_{RD} denoting the distance from VAA to D. On the other hand, the relays are assumed to be close to each other within the VAA, i.e., $d_{\mathrm{IRC}} \ll d_{\mathrm{RD}}$ with d_{IRC} denoting the distance between the relays. This yields the proportional relation

$$\frac{P_{\mathrm{IRC}}}{P_{\mathrm{R}}} \propto \left(\frac{d_{\mathrm{IRC}}}{d_{\mathrm{RD}}}\right)^{\alpha} \ll 1 \,, \tag{18.11}$$

where P_{IRC} represents the transmit power for IRC. Thus, compared to the signal energy consumption $E_{\mathrm{SRD,signal}}$ for the RD transmission, the energy for signal transmission $E_{\mathrm{IRC,signal}}$ in the IRC phase can be reasonably neglected.

18.4.2 RF Circuit Energy

For a transceiver circuit, there are three modes, namely, active, sleep and wake-up mode. By only considering the active mode that consumes the most energy, a power model for RF circuit in the active mode specified for sensor networks was developed in [4], which is adjusted and used here. Furthermore, the circuit energy of most RF parts is assumed to be fixed in general. The power consumption in one transmitter P_{tx} and one receiver P_{rx} is the summation of those fixed components, e.g., mixer, low noise amplifier (LNA) and intermediate frequency amplifier (IFA), etc., as listed in Table 18.1. However, the power consumption of the power amplifier P_{PA} is proportional to the transmit signal power as defined in [6]. As a result, the RF circuit energy for the RD transmission can be written as

$$E_{\mathrm{SRD,RF}} = (M\,P_{\mathrm{rx}} + M_{\mathrm{act}}\,P_{\mathrm{tx}} + P_{\mathrm{PA}}) \cdot T_{\mathrm{OFDM}} \,. \tag{18.12}$$

Table 18.1 RF parameters for sensor networks in [4]

Parameters	Values (mW)
Circuit power of mixer	$P_{mix} = 30.3$
Circuit power of low noise amplifier	$P_{LNA} = 20$
Circuit power of frequency synthesizer	$P_{syn} = 50$
Circuit power of intermediate frequency amplifier	$P_{IFA} = 3$
Circuit power of filter	$P_{fil} = 2.5$

Here, M_{act} is number of active relays participating in the RD transmission. When IRC is not applied, M_{act} is always equal to the number of all relays M. In contrast, M_{act} varies and has to be adjusted according to different events considering IRC. Specifically, in case of event E_1, E_4 and E_6, M_{act} is still equal to M; otherwise, M_{act} is equal to the number of relays re-decoding the source message correctly.

To evaluate the RF circuit energy consumption $E_{IRC,RF}$ for IRC, the overhead of exchanging the CRC bits is neglected since each relay only needs to exchange one bit per OFDM frame. If the punctured bits are exchanged in E_2 and E_5, the corresponding RF energy consumption is calculated as

$$E_{IRC,RF,\,pun} = \left(P_{tx} + \left(M - |\mathcal{D}^*|\right) P_{rx}\right) \cdot T_{pun} \tag{18.13}$$

because only the relays decoding erroneously need to receive the punctured bits. Here, $\mathcal{D}^* = \mathcal{D}$ for event E_2 and $\mathcal{D}^* = \mathcal{D}''$ for event E_5. When the receive signal is exchanged in E_3, it is equivalent to assume that one relay transmits all the signals being selected and $(M - 1)$ relays receive. This leads to the RF energy cost

$$E_{IRC,RF,rec} = \left(P_{tx} + (M-1) P_{rx}\right) \cdot T_{rec}\ . \tag{18.14}$$

Consequently, the RF energy consumption $E_{IRC,RF}$ for IRC is calculated by averaging $E_{IRC,RF,pun}$ and $E_{IRC,RF,rec}$ depending on the involved events as

$$E_{IRC,RF} = \left(\Pr\{E_2\} + \Pr\{E_5\}\right) E_{IRC,RF,pun} + \Pr\{E_3\}\, E_{IRC,RF,rec}\ . \tag{18.15}$$

18.4.3 Baseband Processing Energy

The energy consumed by the baseband processing is estimated at CMOS level, where one CMOS gate consumes static energy (subthreshold leakage current, gate leakage current, etc.) and dynamic energy (switching power and short-circuit current). It was shown in [2], that the switching component is dominant in a "well-designed" circuit. Therefore, the static energy can be neglected, leading to the energy for one CMOS gate (here an inverter is considered) that only consists of the dynamic part as

$$E_{\mathrm{CMOS}} = \xi C_{\mathrm{L}} V_{\mathrm{dd}}^2 \ . \tag{18.16}$$

Here, C_{L} denotes the load capacity of one CMOS gate, $0 \le \xi \le 1$ denotes the effective switching factor whereas V_{dd} is the supply voltage. Based on this energy model for one CMOS gate, the baseband processing energy for the SRD transmission and IRC are calculated by counting the number of involved CMOS gates N_{SRD} and N_{IRC}, respectively, which yields

$$E_{\mathrm{SRD,base}} = N_{\mathrm{SRD}} E_{\mathrm{CMOS}} \tag{18.17a}$$

$$E_{\mathrm{IRC,base}} = N_{\mathrm{IRC}} E_{\mathrm{CMOS}} \ . \tag{18.17b}$$

The baseband processing energy captures IFFT, FFT, interleaver, de-interleaver, memory units and the decoding block using the Viterbi algorithm. Further details on energy models for baseband processing can be found in [17].

18.4.4 *Performance Evaluation*

In Fig. 18.15 the energy consumed by the RF circuit $E_{\mathrm{RF}} = E_{\mathrm{SRD,RF}} + E_{\mathrm{IRC,RF}}$ and baseband processing $E_{\mathrm{base}} = E_{\mathrm{SRD,base}} + E_{\mathrm{IRC,base}}$ is shown for full cooperation and partial cooperation in E_3. Note that all $n_{\mathrm{pun}} = 512$ punctured bits are exchanged in E_2 and E_5 as an example. It can be observed that this energy cost depends on $\mathrm{SNR}_{\mathrm{SR}}$, specifically, it decreases with growing $\mathrm{SNR}_{\mathrm{SR}}$. This is due to the fact that exchanging receive signals and punctured bits in IRC occurs more frequently at lower $\mathrm{SNR}_{\mathrm{SR}}$. However, IRC is rarely required at high $\mathrm{SNR}_{\mathrm{SR}}$. Furthermore, full cooperation consumes much more RF circuit and baseband energy because larger

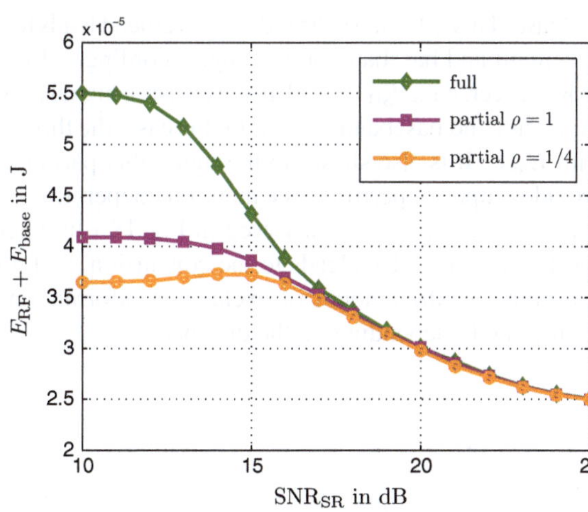

Fig. 18.15 RF circuit plus baseband energy consumption at the relays for full and partial cooperation

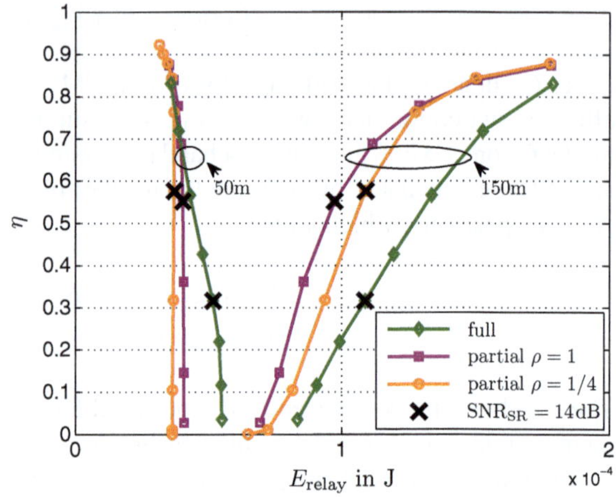

Fig. 18.16 System throughput η over energy consumption E_{relay} when considering IRC for both E_2 and E_3 with $n_{\mathrm{IRC}} = 512$, the SD distance is set to $d_{\mathrm{SD}} = 50\,\mathrm{m}$ and $d_{\mathrm{SD}} = 150\,\mathrm{m}$, respectively

amount of receive signals have to be exchanged. Less energy consumption can be achieved by partial cooperation with smaller ρ.

The throughput shown in Fig. 18.14b is re-considered with respect to the total energy consumption E_{relay} at the relays in Fig. 18.16. When the SD distance is set to $d_{\mathrm{SD}} = 50\,\mathrm{m}$, the energy for RF circuit and baseband processing dominates the total energy consumption with negligible transmit signal energy. Note that $\mathrm{SNR}_{\mathrm{SR}}$ increases upwards with the adjacent markers corresponding to a 1dB step on each curve. Moreover, both the throughput η and the energy E_{relay} depend on $\mathrm{SNR}_{\mathrm{SR}}$, as already observed in Figs. 18.14 and 18.15, respectively. With growing $\mathrm{SNR}_{\mathrm{SR}}$, the throughput η increases but the total energy E_{relay} decreases because the transmit signal energy can be neglected. As a result, the throughput even increases with less energy consumption for small SD distances. Furthermore, it is observed that a smaller ρ, e.g., $\rho = 1/4$, for partial cooperation achieves better performance compared to $\rho = 1$ in this case. This is because the reduced overhead leads to less energy consumption in the RF circuit and baseband processing, according to Fig. 18.15. When the SD distance grows larger, e.g., $d_{\mathrm{SD}} = 150\,\mathrm{m}$, the transmit signal power becomes comparable to the RF circuit and baseband energy. In this case, the throughput increases monotonically with E_{relay}. It is also shown in the figure that partial cooperation with $\rho = 1/4$ still provides superior performance than full cooperation even at low $\mathrm{SNR}_{\mathrm{SR}}$. In contrast, opposite performance is observed in Fig. 18.14b when ignoring the RF circuit and baseband energy. This leads to the conclusion, that with respect to the total energy consumption, the amount of exchanged information for IRC needs to be chosen appropriately depending on the distance.

18.5 Conclusion

In this chapter, one-way relaying networks are considered. Assuming coded OFDM transmission, several relaying schemes such as amplify-forward with constant power (AF-CP), amplify-forward with constant gain (AF-CG), decode-forward (DF) and decode-quantize-forward (DQF) are investigated in a single-relay scenario. It is shown that DQF with a modified MRC achieves the best performance. Extending the network to a multiple-relay scenario, inter-relay cooperation (IRC) schemes allowing information exchange between the DF-based relays are proposed. Depend on the decoding status at the relays, punctured bits or receive signals are exchanged leading to improved decoding performance. The overhead required for IRC is incorporated by resorting to a throughput analysis. Moreover, the total energy consumption at the relays for signal transmission as well as RF circuit and baseband processing is analyzed, indicating that the source-destination distance plays an important role in choosing proper amount of exchanged information for IRC in practical systems.

Acknowledgments This work was supported in part by the German Research Foundation (DFG) under grant Wu 499/7 and Pa 438/4 within the priority program "Communication in Interference Limited Networks (COIN)", SPP 1397.

References

1. Alamouti S (1998) A simple transmit diversity technique for wireless communications. IEEE J Sel Areas Commun 16(8):1451–1458
2. Chandrakasan A, Sheng S, Brodersen R (1992) Low-power CMOS digital design. IEEE J Solid-State Circuits 27(4):473–484
3. Cheng JF (2006) Coding performance of hybrid ARQ schemes. IEEE Trans Commun 54(6):1017–1029
4. Cui S, Goldsmith A, Bahai A (2003) Energy-constrained modulation optimization for coded systems. In: IEEE global telecommunications conference (GLOBECOM'03), San Francisco, CA, USA
5. Dohler M (2003) Virtual antenna arrays. Ph.D. thesis, University of London
6. Dohler M, Li Y (2010) Cooperative communications: hardware. Channel & PHY Addison-Wesley
7. Hagenauer J (1988) Rate-compatible punctured convolutional codes (RCPC Codes) and their applications. IEEE Trans Commun 36(4):389–400
8. Kim S, Kim Y, Simon M (2004) Generalized selection combining based on the log-likelihood ratio. IEEE Trans Commun 52(4):521–524
9. Laneman J, Wornell G (2003) Distributed space-time-coded protocols for exploiting cooperative diversity in wireless networks. IEEE Trans Inf Theory 49(10):2415–2425
10. Laneman J, Tse D, Wornell G (2004) Cooperative diversity in wireless networks: efficient protocols and outage behavior. IEEE Trans Inf Theory 50(12):3062–3080
11. Tarokh V, Jafarkhani H, Calderbank A (1999) Space-time block codes from orthogonal designs. IEEE Trans Inf Theory 45(5):1456–1467
12. Wu M, Wübben D, Dekorsy A (2011) BER-based power allocation for amplify-and-forward and decode-and-forward relaying systems. In: 15th international ITG workshop on smart antennas (WSA'11), Aachen, Germany

13. Wu M, Wübben D, Dekorsy A (2011) BER-based power allocation for decode-and-forward relaying with M-QAM constellations. In: 7th international wireless communications and mobile computing conference (IWCMC'11), Istanbul, Turkey
14. Wu M, Xue W, Wübben D, Dekorsy A, Paul S (2011) Energy-aware design of inter-relay cooperation for distributed relaying networks. In: 8th international symposium on wireless communication systems (ISWCS'11), Aachen, Germany
15. Wu M, Xue W, Wübben D, Dekorsy A, Paul S (2012) An improved inter-relay cooperation scheme for distributed relaying networks. In: 16th international ITG workshop on smart antennas (WSA'12), Dresden, Germany
16. Wübben D, Wu M (2011) Decode-quantize-forward for OFDM-based relaying systems. In: IEEE 73rd vehicular technology conference (VTC'11-Spring), Budapest, Hungary
17. Xue W, Paul S (2012) Model for Energy optimization of baseband architectures in wireless communications. In: 16th international ITG workshop on smart antennas (WSA'12), Dresden, Germany
18. Yue L (2000) Analysis of generalized selection combining techniques. In: IEEE 51st vehicular technology conference (VTC'00-Spring), Tokyo, Japan

Chapter 19
Physical Layer Cooperation in Two-Way Relaying Systems

Meng Wu, Frank Ludwig, Dirk Wübben, Armin Dekorsy and Steffen Paul

Abstract To combat the loss of spectral efficiency caused by the half-duplex constraint in one-way relaying systems, two-way relaying communications is considered using physical-layer network coding. Based on the principle of network coding, two sources intend to exchange information with each other via a relay in two phases. Specifically, both sources transmit to the relay simultaneously and the relay broadcasts a joint message back to the sources. In order to estimate the network coded message from the superimposed receive signal at the relay, different detection and decoding schemes are designed and analyzed in this chapter. For practical concerns, hardware imperfection such as carrier frequency offset mismatch is taken into account. This impact on the presented detection and decoding schemes is studied and corresponding approaches are introduced to mitigate the impairments. Furthermore, the network is extended to a distributed relaying scenario, where adaptive broadcast transmission strategies depending on the decoding status at the relays are investigated. To confirm the theoretical results, a selection of two-way relaying schemes are implemented on real-time hardware platforms and are shown to be practical in a real transmission.

M. Wu (✉) · D. Wübben · A. Dekorsy
Department of Communications Engineering, University of Bremen,
28359 Bremen, Germany
e-mail: wu@ant.uni-bremen.de

D. Wübben
e-mail: wuebben@ant.uni-bremen.de

A. Dekorsy
e-mail: dekorsy@ant.uni-bremen.de

F. Ludwig · S. Paul
Department of Communication Electronics, University of Bremen,
28359 Bremen, Germany
e-mail: ludwig@me.uni-bremen.de

S. Paul
e-mail: steffen.paul@me.uni-bremen.de

W. Utschick (ed.), *Communications in Interference Limited Networks*,
Signals and Communication Technology, DOI 10.1007/978-3-319-22440-4_19

19.1 Introduction

In two-way relaying systems, two sources A and B exchange information with each other supported by a relay R. To achieve this goal, several relaying protocols are available. For example, orthogonal transmissions are employed in a four-phase protocol, which is a straightforward extension from one-way relaying communications as discussed in the Chap. 18. In a three-phase protocol, both sources transmit to the relay successively while the relay broadcasts a network coded message back to the sources in the third time slot [6]. The required transmission phases can be further reduced to two by letting both sources transmit to the relay simultaneously, resulting in a two-phase protocol using physical-layer network coding (PLNC) [26]. In this case, a network coded message is directly estimated from the superimposed receive signal at the relay and broadcasted to the sources. In this chapter, we focus on the two-phase protocol and investigate different PLNC detection and decoding schemes to estimate a bitwise XOR based network coded message at the relay. These schemes are analyzed with respect to mutual information according to [12] and the error performance is examined by link level simulations as well as measurements based on hardware platforms. Moreover, practical issues such as carrier frequency offset (CFO) mismatch on PLNC [23] is investigated, and corresponding approaches to alleviate the impact are presented. Finally, the network is extended to a distributed two-way relaying scenario, where a modified adaptive broadcast transmission scheme is proposed that outperforms the traditional one [27]. Note that the investigations presented in this chapter will be re-considered with waveform designs in the Chap. 20.

19.2 Physical-Layer Network Coding

A two-phase two-way relaying network using PLNC is shown in Fig. 19.1. In the multiple-access (MA) phase, the information words $\mathbf{b}_{\mathrm{A}}$ and $\mathbf{b}_{\mathrm{B}}$ at both sources are encoded to source codewords $\mathbf{c}_{\mathrm{A}}$ and $\mathbf{c}_{\mathrm{B}}$, which are then mapped to symbol vectors $\mathbf{d}_{\mathrm{A}}$ and $\mathbf{d}_{\mathrm{B}}$ by M_{S}-QAM modulation and broadcasted to R simultaneously. Based on decoded-forward (DF), R estimates an XOR based network coded message $\mathbf{c}_{\mathrm{R}}$ termed as relay codeword from the receive signal $\mathbf{y}_{\mathrm{R}}$ and broadcasts it to the sources

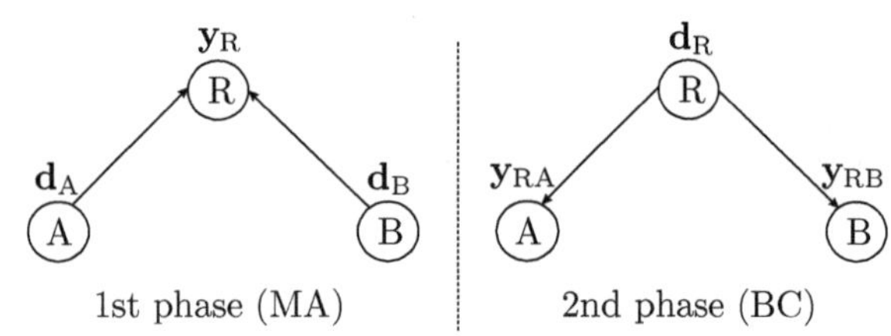

Fig. 19.1 A two-phase two-way relaying system consisting of two sources *A* and *B* that exchange information with each other via one relay *R* using PLNC

in the broadcast (BC) phase. Since the sources know what they transmitted in the previous time slot, each source estimates the relay codeword and determines its desired message by network decoding using this a-priori information.

19.2.1 APP-Based Detection and Decoding Schemes

The key issue of PLNC is to estimate the relay codeword $\mathbf{c}_\mathrm{R}$ from the receive signal $\mathbf{y}_\mathrm{R}$ at R. To this end, different detection and decoding schemes are presented in [20, 21, 25] based on the a-posteriori probability (APP) obtained from $\mathbf{y}_\mathrm{R}$. Assuming the application of orthogonal frequency division multiplexing (OFDM), the superimposed receive signal on subcarrier k, $k = 0, 1, \ldots, N_\mathrm{K} - 1$, at the relay can be written as

$$y_{\mathrm{R},k} = h_{\mathrm{AR},k} d_{\mathrm{A},k} + h_{\mathrm{BR},k} d_{\mathrm{B},k} + n_{\mathrm{R},k} \,. \tag{19.1}$$

Here, $h_{\mathrm{AR},k}$ and $h_{\mathrm{BR},k}$ denote the kth channel coefficients in frequency domain on both links in the MA phase and $n_{\mathrm{R},k}$ represents complex Gaussian noise. Omitting the subcarrier index k and denoting $d_\mathrm{AB} = (d_\mathrm{A}, d_\mathrm{B})$ the pair of transmit symbols from both sources, the subcarrier wise APP that d_AB is transmitted conditioned on receiving y_R is given by

$$\Pr\{d_\mathrm{AB}|y_\mathrm{R}\} = \frac{p\{y_\mathrm{R}|d_\mathrm{AB}\}\Pr\{d_\mathrm{AB}\}}{\Pr\{y_\mathrm{R}\}} = \frac{p\{y_\mathrm{R}|d_\mathrm{AB}\}}{\sum_{\forall d_\mathrm{AB}} p\{y_\mathrm{R}|d_\mathrm{AB}\}} \,. \tag{19.2}$$

For some fixed channel realizations h_AR and h_BR in one MA transmission, the probability density of y_R based on the symbol pair d_AB in (19.2) reads

$$p\{y_\mathrm{R}|d_\mathrm{AB}\} = \frac{1}{\pi\sigma_n^2} \exp\left\{-\frac{|y_\mathrm{R} - h_\mathrm{AR}d_\mathrm{A} - h_\mathrm{BR}d_\mathrm{B}|^2}{\sigma_n^2}\right\} \tag{19.3}$$

with σ_n^2 denoting the noise variance. Moreover, equal a-priori probabilities are assumed for the transmit signals at both sources. Based on the APPs, different detection and decoding schemes at R are presented to estimate $\mathbf{c}_\mathrm{R}$ from $\mathbf{y}_\mathrm{R}$ in the sequel.

19.2.1.1 Separated Channel Decoding

In order to generate the network coded signal at R, the individual source messages can be firstly estimated, which essentially corresponds to a traditional MA transmission [3]. To this end, separated channel decoding (SCD) is performed to achieve the estimates $\hat{\mathbf{c}}_\mathrm{A}$ and $\hat{\mathbf{c}}_\mathrm{B}$ explicitly from $\mathbf{y}_\mathrm{R}$. Subsequently, the relay codeword is calculated by $\mathbf{c}_\mathrm{R} = \hat{\mathbf{c}}_\mathrm{A} \oplus \hat{\mathbf{c}}_\mathrm{B}$. Using the APPs defined in (19.2), the log-likelihood ratio (LLR) L_A of each code bit c_A from source A yields

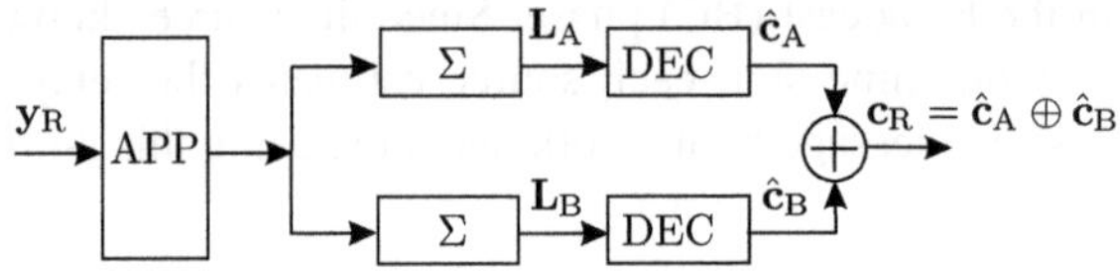

Fig. 19.2 Block diagram for P-SCD. $\mathbf{c}_A$ and $\mathbf{c}_B$ are decoded individually and then network coded

$$L_A = \log_2 \frac{\Pr\{c_A = 0|y_R\}}{\Pr\{c_A = 1|y_R\}} = \log_2 \frac{\sum_{d_{AB} \in \Omega_A^0} \Pr\{d_{AB}|y_R\}}{\sum_{d_{AB} \in \Omega_A^1} \Pr\{d_{AB}|y_R\}}, \qquad (19.4)$$

where the sets Ω_A^0 and Ω_A^1 contain all the symbol pairs d_{AB} with the involved bit c_A equal to 0 and 1, respectively. The LLR L_B for code bit c_B can be calculated similar to (19.4). After determining the LLRs, channel decoding is applied to separately calculate the estimates $\hat{\mathbf{c}}_A$ and $\hat{\mathbf{c}}_B$, which are then network coded and broadcasted in the BC phase. The corresponding block diagram at R is shown in Fig. 19.2.

Parallel SCD Versus Successive SCD The SCD scheme presented above decodes the source messages $\mathbf{c}_A$ and $\mathbf{c}_B$ in parallel and thus is termed P-SCD. Alternatively, the decoding results from the stronger link in the MA channel can be subtracted from the receive signal and the other codeword is estimated from this interference reduced signal. Such a successive SCD scheme is termed S-SCD. For example, assuming that the AR link has a higher signal to noise ratio (SNR) compared to the BR link without loss of generality, $\hat{\mathbf{c}}_A$ is firstly decoded using the LLR vector $\mathbf{L}_A$. Subsequently, the signal contribution from $\mathbf{d}_A$ is subtracted from $\mathbf{y}_R$, after which $\hat{\mathbf{c}}_B$ is estimated in succession. Note that S-SCD can be performed iteratively.

For investigation of the performance upper-bound of both P-SCD and S-SCD, a mutual information (MI) based analysis is applied. To this end, the subcarrier wise MI $C_{S,A,k}$ between the individual code bit $c_A = \xi \in \{0, 1\}$ and the receive signal y_R for decoding the source message from A is determined as [12]

$$\begin{aligned} C_{S,A} &= I(c_A; y_R) \\ &= \sum_{c_A=\xi} \int_{-\infty}^{\infty} \Pr\{c_A = \xi, y_R\} \log_2 \frac{\Pr\{c_A = \xi, y_R\}}{\Pr\{c_A = \xi\} \Pr\{y_R\}} \mathrm{d}y_R \\ &= \frac{1}{M_S^2} \sum_{c_A=\xi} \int_{-\infty}^{\infty} \sum_{d_{AB} \in \Omega_A^\xi} p\{y_R|d_{AB}\} \log_2 \frac{\sum_{d_{AB} \in \Omega_A^\xi} p\{y_R|d_{AB}\}}{\Pr\{c_A = \xi\} \sum_{\forall d_{AB}} p\{y_R|d_{AB}\}} \mathrm{d}y_R \end{aligned} \qquad (19.5)$$

with $\Pr\{c_A = \xi\} = \frac{1}{2}$ and the probability density of y_R defined in (19.3). Note that the index k is omitted for simplicity. For the above MI calculation, the signal from source B is treated as non-Gaussian interference due to fixed channel coefficients in one MA transmission. Since there is no closed-form solution for the integral in (19.5), it has to be solved numerically. Similarly, the MI $C_{S,B}$ between c_B and y_R can be obtained. Furthermore, it is assumed that each OFDM symbol is independently encoded. Therefore, the performance of P-SCD to estimate the network coded signal for one OFDM symbol is constrained by

$$C_{\mathrm{S}}^{\mathrm{P}} = \min\left\{ \frac{1}{N_{\mathrm{K}}} \sum_{k=0}^{N_{\mathrm{K}}-1} C_{\mathrm{S,A},k}, \frac{1}{N_{\mathrm{K}}} \sum_{k=0}^{N_{\mathrm{K}}-1} C_{\mathrm{S,B},k} \right\} . \tag{19.6}$$

Here, the subcarrier wise MI, e.g., $C_{\mathrm{S,A},k}$ defined in (19.5) is averaged over all subcarriers. The minimum MI from parallely decoding the two individual source messages is dominant since both $\mathbf{c}_{\mathrm{A}}$ and $\mathbf{c}_{\mathrm{B}}$ have to be estimated correctly in order to obtain an error-free network coded message. Contrarily, the performance upper-bound for S-SCD is given by the MI

$$C_{\mathrm{S}}^{\mathrm{S}} = \min\left\{ \max\left\{ \frac{1}{N_{\mathrm{K}}} \sum_{k=0}^{N_{\mathrm{K}}-1} C_{\mathrm{S,A},k}, \frac{1}{N_{\mathrm{K}}} \sum_{k=0}^{N_{\mathrm{K}}-1} C_{\mathrm{S,B},k} \right\}, \frac{1}{N_{\mathrm{K}}} \sum_{k=0}^{N_{\mathrm{K}}-1} C_{\mathrm{SU},k} \right\} , \tag{19.7}$$

where taking the maximum MI from both initial individual decoding guarantees successful decoding of the message from the stronger uplink. Subsequently, successive decoding regarding the interference free signal is performed essentially in a single-user (SU) scenario. Assuming again that the AR link has a higher SNR over the BR link, the subcarrier wise MI C_{SU} between c_{B} and the interference free signal can be calculated as

$$C_{\mathrm{SU}} = I\left(c_{\mathrm{B}}; y_{\mathrm{R}} | d_{\mathrm{A}} = 0\right) . \tag{19.8}$$

Similar to P-SCD, the minimum of these two terms is taken, leading to the MI with respect to the network coded message for S-SCD.

19.2.1.2 Joint Channel Decoding and Physical-Layer Network Coding

For PLNC the relay is not interested in the individual messages from the sources but only broadcasts the network coded signal in the BC phase. In this context, the relay codeword $\mathbf{c}_{\mathrm{R}} = \mathbf{c}_{\mathrm{A}\oplus\mathrm{B}}$ can be directly estimated from $\mathbf{y}_{\mathrm{R}}$ without the knowledge of the individual source messages $\mathbf{c}_{\mathrm{A}}$ and $\mathbf{c}_{\mathrm{B}}$. Since channel decoding and network coding is performed concurrently, this scheme is termed joint channel decoding and physical-layer network coding (JCNC) [24, 25]. Assuming that the same linear channel code is used at both sources, the XORed packet is also a valid codeword of that code. Therefore, the LLR value $L_{\mathrm{A}\oplus\mathrm{B}}$ for $c_{\mathrm{A}\oplus\mathrm{B}} = c_{\mathrm{A}} \oplus c_{\mathrm{B}}$ can be calculated as

$$L_{\mathrm{A}\oplus\mathrm{B}} = \log_2 \frac{\Pr\left\{c_{\mathrm{A}\oplus\mathrm{B}} = 0 | y_{\mathrm{R}}\right\}}{\Pr\left\{c_{\mathrm{A}\oplus\mathrm{B}} = 1 | y_{\mathrm{R}}\right\}} = \log_2 \frac{\sum_{d_{\mathrm{AB}} \in \Psi^0} \Pr\left\{d_{\mathrm{AB}} | y_{\mathrm{R}}\right\}}{\sum_{d_{\mathrm{AB}} \in \Psi^1} \Pr\left\{d_{\mathrm{AB}} | y_{\mathrm{R}}\right\}} . \tag{19.9}$$

Here, Ψ^0 and Ψ^1 contain all the symbol pairs with $c_{\mathrm{A}\oplus\mathrm{B}} = 0$ and $c_{\mathrm{A}\oplus\mathrm{B}} = 1$, respectively. Subsequently, the XORed message can be decoded directly using $L_{\mathrm{A}\oplus\mathrm{B}}$.

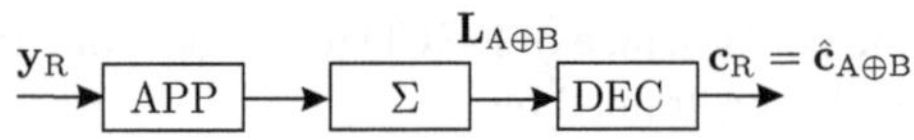

Fig. 19.3 Block diagram for JCNC. $\mathbf{c}_R$ is directly estimated from $\mathbf{y}_R$

The corresponding block diagram at R is shown in Fig. 19.3. Moreover, omitting index k, the subcarrier wise MI between the XORed bit $c_{A\oplus B}$ and the receive signal y_R for JCNC is given by [12]

$$
\begin{aligned}
C_J &= I\left(c_{A\oplus B}; y_R\right) \\
&= \sum_{c_{A\oplus B}=\xi} \int_{-\infty}^{\infty} \Pr\left\{c_{A\oplus B}=\xi, y_R\right\} \log_2 \frac{\Pr\left\{c_{A\oplus B}=\xi, y_R\right\}}{\Pr\left\{c_{A\oplus B}=\xi\right\}\Pr\left\{y_R\right\}} \mathrm{d}y_R \\
&= \frac{1}{M_S^2} \sum_{c_{A\oplus B}=\xi} \int_{-\infty}^{\infty} \sum_{d_{AB}\in\Psi^\xi} p\left\{y_R | d_{AB}\right\} \log_2 \frac{\sum_{d_{AB}\in\Psi^\xi} p\left\{y_R | d_{AB}\right\}}{\Pr\{c_{A\oplus B}=\xi\} \sum_{\forall d_{AB}} p\{y_R | d_{AB}\}} \mathrm{d}y_R
\end{aligned}
\tag{19.10}
$$

with $\Pr\{c_{A\oplus B}=\xi\}=\frac{1}{2}$. Note that (19.10) can be interpreted as the MI of transmitting the network coded signal $c_{A\oplus B}$ to the relay via a virtual channel. For example, this implies that $c_{A\oplus B}=0$ is mapped to the noise-free receive signal $\pm(h_A+h_B)$ and $c_{A\oplus B}=1$ is mapped to $\pm(h_A-h_B)$ for BPSK. Consequently, the MI regarding each OFDM transmission involving one codeword is obtained by averaging the subcarrier wise MI $C_{J,k}$ in (19.10) over all subcarriers as $\frac{1}{N_K}\sum_k C_{J,k}$.

19.2.1.3 Generalized Joint Channel Decoding and Physical-Layer Network Coding

The separate decoding of both source codewords $\mathbf{c}_A$ and $\mathbf{c}_B$ by SCD and the joint decoding of the XORed codeword $\mathbf{c}_{A\oplus B}$ from the receive signal $\mathbf{y}_R$ by JCNC operate in the binary bit-level. Alternatively, the decoding process at the relay can be performed in the non-binary field $\mathbb{F}_{M_S^2}$. Compared to SCD and JCNC, this scheme takes into account the channel codes from both sources simultaneously and thus is able to fully exploit the available information contained in the superimposed receive signal at the relay, which is termed generalized joint channel decoding and physical-layer network coding (G-JCNC) [20, 21]. Since the decoding is performed in the non-binary field $\mathbb{F}_{M_S^2}$, higher decoding complexity is expected for G-JCNC.

In [20, 21] a modified sum-product algorithm (SPA) based on the investigations in [4, 9] at the relay is employed considering the application of the same LDPC code at both sources for the G-JCNC scheme. Specifically, instead of the LLRs, the APPs are directly fed to the non-binary channel decoder in $\mathbb{F}_{M_S^2}$ and updated iteratively within the modified SPA. Therefore, the sum of APPs shown in (19.5) for SCD and (19.10) for JCNC is excluded. Subsequently, the APPs at the decoder output are mapped to

Fig. 19.4 Block diagram for G-JCNC. The decoder operates in $\mathbb{F}_{M_S^2}$

the relay codeword by PLNC. In Fig. 19.4 the corresponding block diagram at R is shown. Furthermore, omitting index k, the subcarrier wise MI between the transmit symbol pair d_{AB} at the sources and the receive signal y_R at the relay with respect to G-JCNC is calculated as [12]

$$\begin{aligned} C'_G &= I\left(d_{AB}; y_R\right) \\ &= \sum_{\forall d_{AB}} \int_{-\infty}^{\infty} \Pr\left\{d_{AB}, y_R\right\} \log_2 \frac{\Pr\left\{d_{AB}, y_R\right\}}{\Pr\left\{d_{AB}\right\} \Pr\left\{y_R\right\}} \mathrm{d}y_R \\ &= \frac{1}{M_S^2} \sum_{\forall d_{AB}} \int_{-\infty}^{\infty} \Pr\left\{y_R | d_{AB}\right\} \log_2 \frac{M_S^2 \Pr\left\{y_R | d_{AB}\right\}}{\sum_{\forall d_{AB}} \Pr\left\{y_R | d_{AB}\right\}} \mathrm{d}y_R \,. \end{aligned} \quad (19.11)$$

Here, C'_G essentially corresponds to the sum-rate in the MA phase. In order to achieve a fair comparison with SCD and JCNC, the code rate at the sources has to be smaller than $C_G = C'_G/2$ to recover the individual source messages successfully at R assuming a symmetric MA channel. Furthermore, the subcarrier wise MI C_G in (19.11) has to be averaged over all subcarriers as $\frac{1}{N_K} \sum_k C_{G,k}$ for OFDM transmissions.

19.2.2 Performance Evaluation

A two-way relaying network is considered, where the relay in the middle of the sources. Firstly, MI of the APP-based schemes are compared in multi-path fading channels with $N_H = 5$ i.i.d. taps using QPSK, as shown in Fig. 19.5. One OFDM

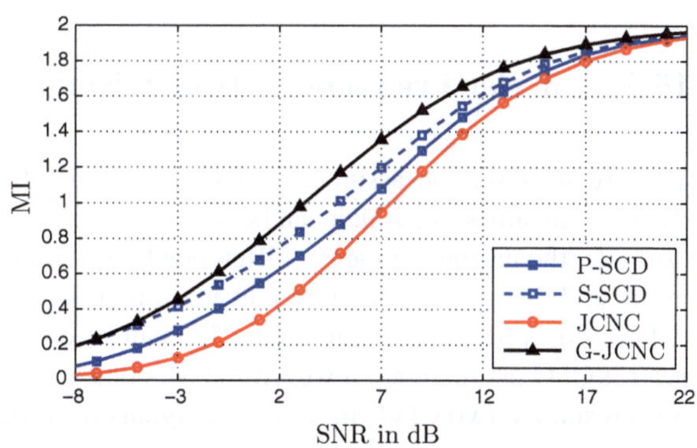

Fig. 19.5 MI of the APP-based schemes in multi-path fading channels using QPSK

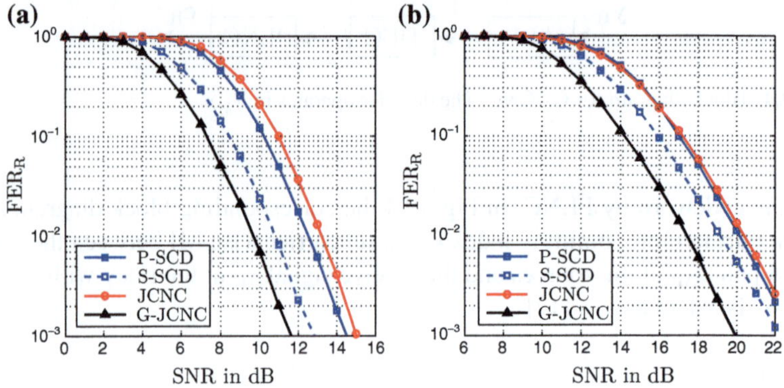

Fig. 19.6 FER performance of the APP-based schemes in multi-path fading channels with QPSK at R, **a** for $R_C = 0.5$ and **b** for $R_C = 0.875$

symbol contains $N_K = 1024$ subcarriers. It can be observed in the figure, that SCD and G-JCNC outperform JCNC significantly, especially in the low SNR region. Furthermore, G-JCNC achieves the best performance while S-SCD is superior to P-SCD.

In Fig. 19.6 the frame error rate (FER) of the network coded message for the APP-based schemes at R is presented in the multi-path fading channel. When an LDPC code of medium rate $R_C = 0.5$ is applied, P-SCD performs approximately 1 dB better than JCNC. S-SCD improves the performance by 2 dB over P-SCD and is 1 dB worse than G-JCNC. In a high code rate scenario, e.g., $R_C = 0.875$, JCNC almost approaches P-SCD, whereas G-JCNC still achieves tremendous gains. These results correspond to the MI observations in Fig. 19.5.

Actually, the performance of the APP-based schemes is strongly dependent on the relative phase difference of the two incoming signals from the sources at the relay. Therefore, an optimal phase control strategy is presented in [17]. Moreover, the impact of introducing multiple antennas at the relay is elaborated in [18]. These aspects are not discussed in this chapter.

19.3 Carrier Frequency Offset Mismatch

The previous section assumes perfect synchronization in time and frequency for the MA transmission, which can be destroyed by several practical impairments. For example, the carrier frequency difference between the local oscillators at the transmitter and the receiver due to hardware imprecision causes carrier frequency offset (CFO). For a SU transmission, the CFO can be completely compensated assuming perfect CFO estimation. However, the MA phase of two-way relaying communications results in two CFOs between the signals from both sources to the relay. These

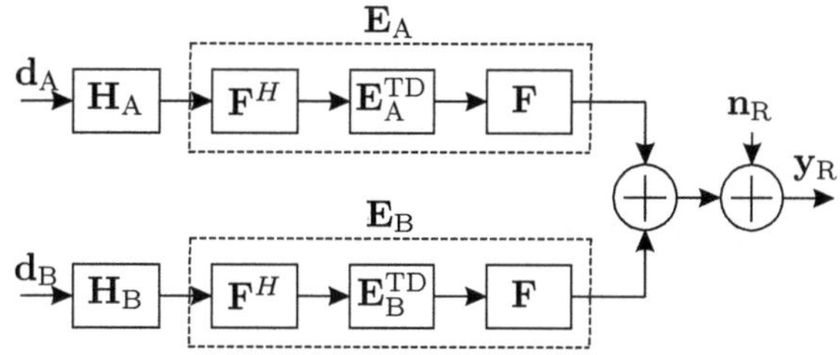

Fig. 19.7 Block diagram of the MA phase with CFO impact for one whole OFDM symbol vector

two CFOs, if not matched, cannot be totally compensated even though both of them are perfectly estimated. Such a CFO mismatch leads to performance degradation and corresponding approaches are studied in [19] to mitigate the impact, as discussed in the sequel.

For CFO-related analysis, one whole OFDM symbol vector is considered instead of the subcarrier wise description in (19.1) to formulate the MA transmission. To this end, the $N_\text{K} \times 1$ receive signal vector $\mathbf{y}_\text{R}$ in frequency domain is given by

$$\mathbf{y}_\text{R} = \mathbf{H}_\text{AR}\mathbf{d}_\text{A} + \mathbf{H}_\text{BR}\mathbf{d}_\text{B} + \mathbf{n}_\text{R} \tag{19.12}$$

when CFO in both links are absent. Here, $\mathbf{H}_\text{AR}$ and $\mathbf{H}_\text{BR}$ denote the $N_\text{K} \times N_\text{K}$ diagonal channel matrices, where each entry of both matrices represents a subcarrier wise channel coefficient $h_{\text{AR},k}$ or $h_{\text{BR},k}$. With the presence of CFOs, the channels in the MA phase are distorted. Applying the CFO model presented in [8, 14], the system equation in matrix representation is re-formulated as

$$\mathbf{y}_\text{R} = \mathbf{V}_\text{A}\mathbf{d}_\text{A} + \mathbf{V}_\text{B}\mathbf{d}_\text{B} + \mathbf{n}_\text{R} \tag{19.13a}$$

$$= \underbrace{\boldsymbol{\Lambda}_\text{A}\mathbf{d}_\text{A} + \boldsymbol{\Lambda}_\text{B}\mathbf{d}_\text{B}}_{\text{desired signal}} + \underbrace{\overline{\boldsymbol{\Lambda}}_\text{A}\mathbf{d}_\text{A} + \overline{\boldsymbol{\Lambda}}_\text{B}\mathbf{d}_\text{B}}_{\text{ICI}} + \mathbf{n}_\text{R} \tag{19.13b}$$

with the corresponding block diagram shown in Fig. 19.7. Therein, the CFO between each source, e.g., source A, and relay R is covered by $\mathbf{E}_\text{A} = \mathbf{F}\mathbf{E}_\text{A}^\text{TD}\mathbf{F}^H$ with dimension $N_\text{K} \times N_\text{K}$, yielding the equivalent channel matrix $\mathbf{V}_\text{A} = \mathbf{E}_\text{A}\mathbf{H}_\text{AR}$. The $N_\text{K} \times N_\text{K}$ DFT matrix is denoted by $\mathbf{F}$, which is unitary and thus $\mathbf{F}\mathbf{F}^H$ is equal to the identity matrix $\mathbf{I}$. The $N_\text{K} \times N_\text{K}$ CFO distortion matrix $\mathbf{E}_\text{A}^\text{TD}$ is diagonal and defined in time domain as

$$\mathbf{E}_\text{A}^\text{TD} = \text{diag}\left\{1 \quad \text{e}^{\frac{j2\pi\varepsilon_\text{A}}{N_\text{K}}} \quad \dots \quad \text{e}^{\frac{j2\pi\varepsilon_\text{A}(N_\text{K}-1)}{N_\text{K}}}\right\}. \tag{19.14}$$

Here, ε_A denotes the CFO from the AR link normalized to subcarrier spacing, which is assumed to be perfectly known at the relay. Note that without CFO, i.e., $\varepsilon_\text{A} = 0$, $\mathbf{E}_\text{A}$ is reduced to an identity matrix, which indicates that the orthogonality of the channel matrix $\mathbf{H}_\text{AR}$ is preserved. However, this orthogonality is destroyed by the non-diagonal matrix $\mathbf{E}_\text{A}$ if $\varepsilon_\text{A} \neq 0$. In this case, the equivalent channel matrix $\mathbf{V}_\text{A} = \boldsymbol{\Lambda}_\text{A} + \overline{\boldsymbol{\Lambda}}_\text{A}$ is composed of the diagonal matrix $\boldsymbol{\Lambda}_\text{A}$ corresponding to the desired signal and the non-diagonal interference matrix $\overline{\boldsymbol{\Lambda}}_{\mathrm{A}}$ resulting in inter-carrier interference (ICI). The CFO influence to the BR link can be deduced similarly.

19.3.1 CFO Compensation

For a SU transmission, the CFO ε between the transmitter and the receiver can be compensated in time domain by element wise multiplying the kth time domain signal with $\mathrm{e}^{-\frac{j2\pi\varepsilon k}{N_{\mathrm{K}}}}$. However, nulling both CFOs at R for the MA transmission in (19.13) is impossible if $\varepsilon_{\mathrm{A}} \neq \varepsilon_{\mathrm{B}}$. In this case, complete compensation of one CFO, e.g., ε_{A} from the AR link, results in $\varepsilon'_{\mathrm{A}} = 0$ and $\varepsilon'_{\mathrm{B}} = \varepsilon_{\mathrm{B}} - \varepsilon_{\mathrm{A}} \neq 0$, where $\varepsilon'_{\mathrm{A}}$ and $\varepsilon'_{\mathrm{B}}$ denote the remaining CFOs after compensation. Alternatively, an average CFO compensation strategy by $-\frac{\varepsilon_{\mathrm{A}}+\varepsilon_{\mathrm{B}}}{2}$ presented in [10] can be applied. Such an average compensation yields both remaining CFOs as

$$\varepsilon'_{\mathrm{A}} = \varepsilon_{\mathrm{A}} - \frac{\varepsilon_{\mathrm{A}}+\varepsilon_{\mathrm{B}}}{2} = \frac{\varepsilon_{\mathrm{A}}-\varepsilon_{\mathrm{B}}}{2} \tag{19.15a}$$

$$\varepsilon'_{\mathrm{B}} = \varepsilon_{\mathrm{B}} - \frac{\varepsilon_{\mathrm{A}}+\varepsilon_{\mathrm{B}}}{2} = \frac{\varepsilon_{\mathrm{B}}-\varepsilon_{\mathrm{A}}}{2}\,. \tag{19.15b}$$

Applying the average CFO compensation strategy, the performance degradation due to CFO mismatch can still be significant when the initial relative CFO $|\varepsilon_{\mathrm{A}} - \varepsilon_{\mathrm{B}}|$ is large. To this end, a more sophisticated scheme based on the space-alternating generalized expectation-maximization (SAGE) algorithm [5] is presented in [23], which is beyond the scope of this chapter for ease of practical implementation.

19.3.2 Inter-Carrier Interference Cancellation

After CFO compensation in time domain, soft ICI cancellation (ICIC) [16] in frequency domain can be performed to further mitigate the impairment caused by the CFO mismatch. Note that soft ICIC leads to superior performance compared to hard ICIC, and thus is applied here. Basically, the ICI components shown in (19.13b) are reconstructed using the estimated individual messages and are subtracted from the receive signal. Subsequently, the decoding is performed again with respect to the ICI reduced signal. This ICIC process can be performed in an iterative manner. Note that since JCNC only produces the relay codeword $\mathbf{c}_{\mathrm{R}}$ without explicit estimation of $\mathbf{c}_{\mathrm{A}}$ and $\mathbf{c}_{\mathrm{B}}$, ICIC is not applicable for this scheme. On the other hand, the individual source messages are also estimated for SCD and G-JCNC besides $\mathbf{c}_{\mathrm{R}}$, which can be used to perform ICIC.

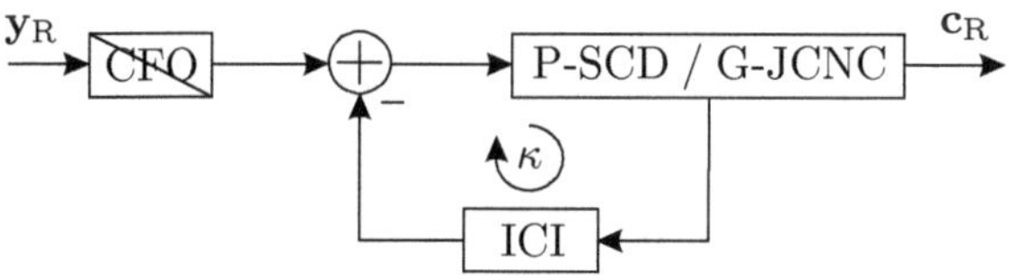

Fig. 19.8 Block diagram of P-SCD and G-JCNC with CFO compensation and ICIC at the relay. The ICIC process can be performed iteratively

19.3.2.1 ICIC for P-SCD and G-JCNC

Considering P-SCD and G-JCNC, both of which estimate the individual source messages simultaneously, the above depicted ICIC process can be represented as

$$\tilde{\mathbf{y}}_\mathrm{R} = \mathbf{y}_\mathrm{R} - \overline{\mathbf{\Lambda}}_\mathrm{A}\tilde{\mathbf{d}}_\mathrm{A} - \overline{\mathbf{\Lambda}}_\mathrm{B}\tilde{\mathbf{d}}_\mathrm{B} \tag{19.16a}$$

$$= \underbrace{\mathbf{\Lambda}_\mathrm{A}\mathbf{d}_\mathrm{A} + \mathbf{\Lambda}_\mathrm{B}\mathbf{d}_\mathrm{B} + \mathbf{n}_\mathrm{R}}_{\text{ICI-free signal}} + \underbrace{\overline{\mathbf{\Lambda}}_\mathrm{A}(\mathbf{d}_\mathrm{A} - \tilde{\mathbf{d}}_\mathrm{A}) + \overline{\mathbf{\Lambda}}_\mathrm{B}(\mathbf{d}_\mathrm{B} - \tilde{\mathbf{d}}_\mathrm{B})}_{\text{remaining ICI}} , \tag{19.16b}$$

where $\tilde{\mathbf{d}}_\mathrm{A}$ and $\tilde{\mathbf{d}}_\mathrm{B}$ denote the soft estimates of the source symbol vectors at R from the previous detection step. $\tilde{\mathbf{y}}_\mathrm{R}$ denotes the interference-reduced signal. For ease of representation, let (19.13) denote here the system equation after the initial average CFO compensation. Inserting (19.13b) into (19.16a), it can be observed that the ICI reduced signal contains the ICI-free signal without CFO mismatch and the remaining ICI as shown in (19.16b). This ICIC process can be performed iteratively to further improve the performance, as shown in Fig. 19.8 for the corresponding block diagram with CFO compensation and ICIC. As the iteration number κ increases, less amount of remaining ICI leads to reduced impairment.

19.3.2.2 ICIC for S-SCD

For S-SCD with CFO compensation and ICIC, the corresponding block diagram is presented in Fig. 19.9. After initial average CFO compensation, the codeword from the stronger link, e.g., $\hat{\mathbf{c}}_\mathrm{A}$, is estimated. Subsequently, the soft ICI as well as the desired signal with respect to $\tilde{\mathbf{d}}_\mathrm{A}$ is subtracted from the receive signal as

$$\tilde{\mathbf{y}}_\mathrm{R} = \mathbf{y}_\mathrm{R} - (\mathbf{\Lambda}_\mathrm{A} + \overline{\mathbf{\Lambda}}_\mathrm{A})\tilde{\mathbf{d}}_\mathrm{A} \tag{19.17a}$$

$$= \underbrace{(\mathbf{\Lambda}_\mathrm{B} + \overline{\mathbf{\Lambda}}_\mathrm{B})\mathbf{d}_\mathrm{B}}_{\text{signal from B}} + \mathbf{n}_\mathrm{R} + \underbrace{(\mathbf{\Lambda}_\mathrm{A} + \overline{\mathbf{\Lambda}}_\mathrm{A})(\mathbf{d}_\mathrm{A} - \tilde{\mathbf{d}}_\mathrm{A})}_{\text{remaining interference from A}} . \tag{19.17b}$$

Assuming correct decoding of the source message from A, i.e., $\tilde{\mathbf{d}}_\mathrm{A} = \mathbf{d}_\mathrm{A}$, both the signal and the ICI components contributed from $\mathbf{d}_\mathrm{A}$ are completely removed. Since the resulting interference reduced signal contains only the signal contribution from

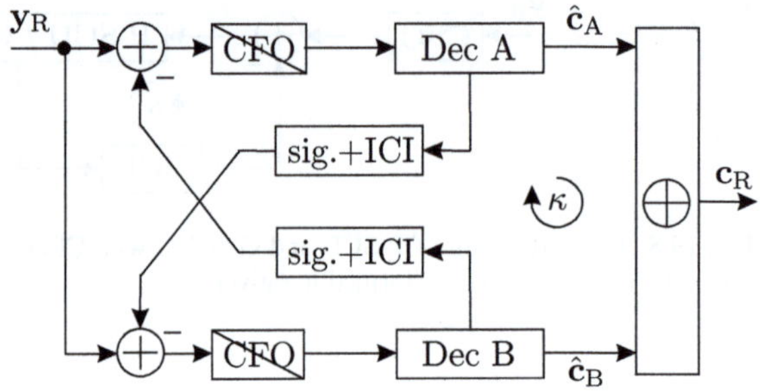

Fig. 19.9 Block diagram of S-SCD with CFO compensation and ICIC at the relay. The ICIC process together with successive decoding can be performed iteratively

$\mathbf{d}_B$, the corresponding CFO leading to the ICI term $\overline{\mathbf{\Lambda}}_B$ can be completely compensated as in a SU transmission. Afterwards, $\hat{\mathbf{c}}_B$ is estimated from this interference free signal. Such an ICIC process in combination with successive decoding also can be performed iteratively for further performance improvements.

19.3.3 Performance Evaluation

Figure 19.10 shows FER_R at the relay for P-SCD. Average CFO compensation is performed in all cases. It can be observed in Fig. 19.10a, that the CFO mismatch results in catastrophic performance without ICIC for larger relative CFO $|\varepsilon_A - \varepsilon_B| = 0.4$. However, ICIC improves the performance significantly with increasing number of iterations, which converges at the 4th iteration. In case of smaller relative CFO $|\varepsilon_A - \varepsilon_B| = 0.2$ as shown in Fig. 19.10b, the presented ICIC strategy with 1 iteration approaches the ideal case without CFO already, indicating that further iterations for ICIC are not necessary in this case.

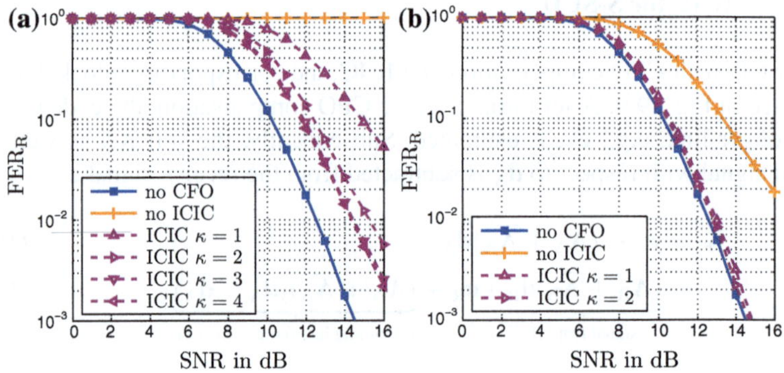

Fig. 19.10 FER performance of P-SCD with ICIC, **a** for $|\varepsilon_A - \varepsilon_B| = 0.4$ and **b** for $|\varepsilon_A - \varepsilon_B| = 0.2$

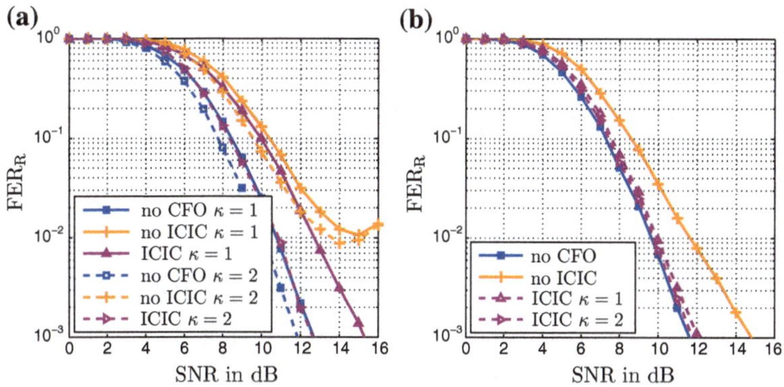

Fig. 19.11 FER performance of **a** S-SCD and **b** G-JCNC with ICIC for $|\varepsilon_A - \varepsilon_B| = 0.2$

The FER performance for S-SCD and G-JCNC with $|\varepsilon_A - \varepsilon_B| = 0.2$ are shown in Fig. 19.11. Note that the iterations regard to both ICIC and successive decoding for S-SCD. In this case, a significant performance loss is observed Fig. 19.11a without ICIC, which improves the FER dramatically, especially at the 2nd iteration. Comparing Fig. 19.11b with Fig. 19.10b, G-JCNC is more robust against CFO mismatch than P-SCD without ICIC. Furthermore, ICIC with 1 iteration already leads to significantly improved performance by approaching the ideal case with minor SNR loss.

19.4 Hardware Implementation of PLNC

For the practical application of PLNC in a real-time transmission, a hardware demonstrator of the two-way relaying network has been developed. Therefore, the relay is implemented on a multiprocessor board consisting of several field-programmable gate arrays (FPGAs) and powerful digital signal processors (DSPs). For synchronization and channel estimation purposes a radio frame has to be defined for the real wireless transmission. Furthermore, radio frequency (RF) and OFDM parameters have to be defined.

19.4.1 Radio Frame Structure

For the real-time transmission it is necessary to arrange the OFDM symbols in a radio frame which includes synchronization signals and pilot symbols. In this work the setup of the radio-frame is inspired by the established LTE standard [1]. Every frame consists of 120 OFDM symbols, where 6 OFDM symbols are arranged as a slot and two slots represent one subframe. In this manner, the whole frame is decomposed

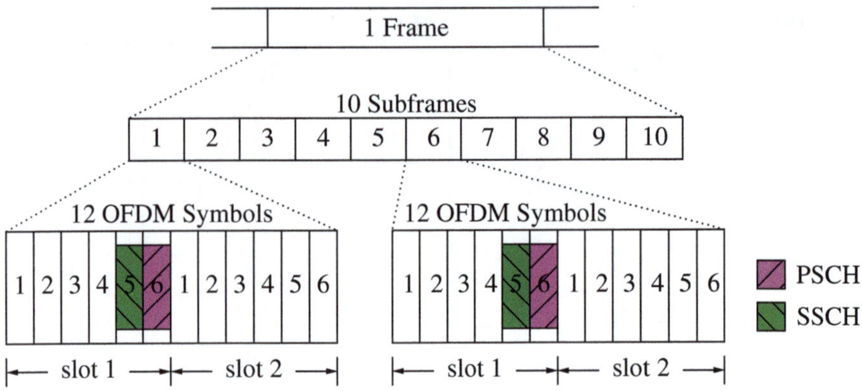

Fig. 19.12 Schematic illustration of the frame structure

into 10 subframes as illustrated in Fig. 19.12. Similarly, 12 subcarriers are arranged to a physical resource block (PRB). The synchronization signals are embedded in the first and the sixth subframe at the end of their first slot. Two different synchronization signals are used namely, a primary (PSCH) and a secondary (SSCH) synchronization signal. The PSCH is located in the sixth OFDM symbol of the first and sixth subframe and consists of a Zadoff-Chu-sequence. Zadoff-Chu-sequences are complex-valued sequences with constant magnitude and correlation characteristics in time and frequency domain that make them perfectly suited for synchronization tasks [2]. The secondary synchronization signal consists of two interleaved maximum length sequences (m-sequences) and is arranged in the fifth OFDM symbol vector of the first and the sixth subframe. Both synchronization signals are grouped symmetrically around the DC-carrier on 62 subcarriers whereas the remaining subcarriers and the DC-carrier are set to zero.

In order to estimate the channel effects and to compensate them in the relay, cell specific reference signals are added in the frames of both sources A and B. Therefore, QPSK modulated symbols are used for these pilots. A lattice-type pilot arrangement is achieved by inserting the pilot tones along the time and frequency axes with a fixed period. These scattered pilot tones in both frequency and time can be used to interpolate the channel estimates in the time-frequency grid. As the transmitted signals of source A and B are superimposed in the multiple access phase, the scattered pilots of A and B are arranged in an orthogonal pilot pattern. An illustration of the time-frequency grid of the first subframes transmitted from sources A and B is shown in Fig. 19.13. With orthogonal mapping of the reference signals, the effort of channel estimation can be kept low because the pilots will not be superimposed by interference from the other source.

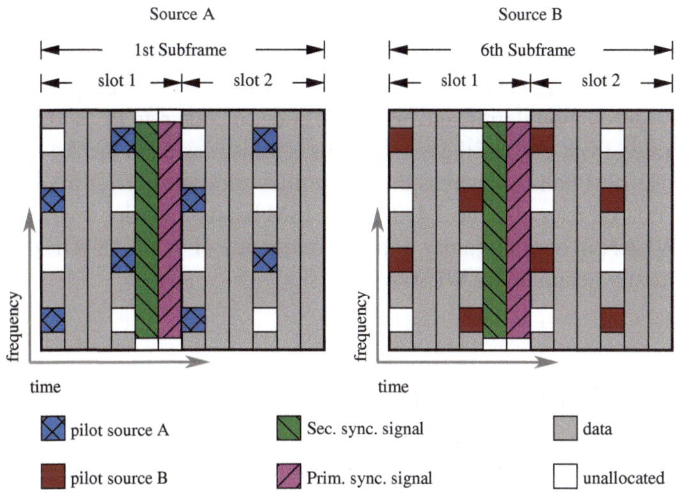

Fig. 19.13 Orthogonal pilot pattern and synchronization signals of source A and B in the time-frequency grid

19.4.2 OFDM Parameters

Besides the radio frame structure, OFDM parameters have to be defined for the real-time transmission. As aforementioned, the structure of the radio frame is inspired by the LTE standard, accordingly the OFDM parameters are similar to LTE. The fundamental parameters used in this work by the hardware (HW) demonstrator are shown in Table 19.1 and compared to the 20 MHz bandwidth mode of the LTE standard. The FFT size is set to $N_{\text{FFT}} = 2048$ and both sources allocate 6 PRBs with data, the other subcarriers are unused. Furthermore, the cyclic prefix (CP) length is set to

Table 19.1 OFDM and RF transmission parameters of the HW demonstrator in this work and the 20 MHz bandwidth LTE mode

Parameter	LTE	HW Demonstrator
Bandwidth	20 MHz	≈20 MHz
FFT size N_{FFT}	2048	2048
CP length	512	512
Frame duration	10 ms	7.7 ms
PRBs	100	6
Carrier frequency f_{c}	–	2.484 GHz
Sampling frequency f_{s}	30.72 MHz	40 MHz
Subcarrier spacing Δf_{sc}	15 kHz	≈20 kHz
Modulation	QPSK, 16-/64-QAM	QPSK
Channel code	Turbo code	LDPC code, $R_{\text{C}} = 0.5$

512. The embedded oscillator of the HW demonstrator generates the sampling clock of the analog to digital converters (ADC), which is $f_s = 40\,\text{MHz}$. This results in a bandwidth of 20 MHz, a frame duration of 7.7 ms and a subcarrier spacing of 20 kHz. Moreover, QPSK modulation and an LDPC code of code rate $R_C = 0.5$ are applied. The integrated front-ends of the relay supports transmission in the 2.4– 2.5 GHz and 5.2–5.8 GHz ISM bands, whereas the transmitter supports carrier frequencies up to 3.2 GHz. WLAN uses also the 2.4–2.5 GHz ISM bands. In order to get less conflicts with the WLAN in our laboratory a carrier frequency of $f_c = 2.484\,\text{GHz}$ was chosen, which is hardly occupied by WLAN.

19.4.3 Hardware Demonstrator

The general set-up of the test-bed for the real-time transmission is shown in Fig. 19.14. The baseband processing of sources A and B are implemented on one Rohde & Schwarz Baseband Signal Generator and Fading Simulator (R&S AMU200A [13]). The RF part is implemented on two identical unsynchronized Rohde & Schwarz Vector Signal Generator (R&S SMBV100A [13]). The advantage of the usage of high precision laboratory equipment for the sources A and B is that the transmission will be nearly ideal. Imperfections on the transmitter side, like the CFO, can be adjusted manually. The focus of the demonstrator is the relay R which is implemented on a Sundance multiprocessor system [15]. It consists of the SMT148 carrier board, a SMT911 MIMO transceiver card, a SMT351 FPGA module and a SMT362 processor module with two DSPs and one FPGA. The front-ends on the SMT911 card basically consist of the radio frequency (RF) transceiver chips MAX2829 [11] and analog to digital as well as digital to analog converters (ADC/DAC). The baseband processing is partitioned to several FPGAs and powerful DSPs on the SMT362 card. With the Virtex 4 and the Universal Serial Bus (USB) module on the SMT148 carrier board the system communicate with a PC. In our HW demonstrator the Host PC is used to

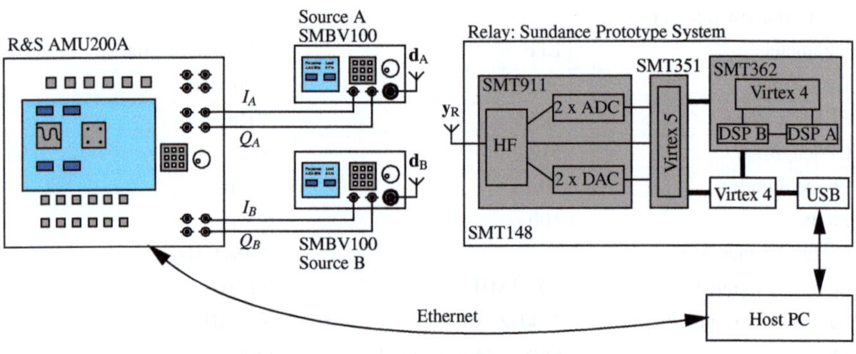

Fig. 19.14 General setup of the test-bed for the real-time transmission of PLNC

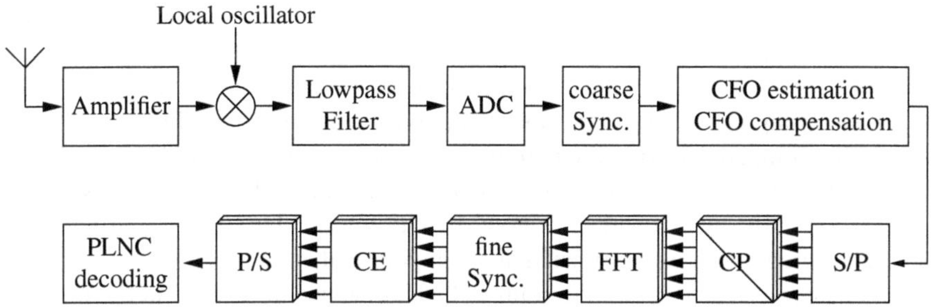

Fig. 19.15 System block diagram of the receiver at relay R

generate the messages of the sources A and B which are transmitted over Ethernet to the signal generator. Furthermore, the PC is used for the evaluation, specifically the comparison of the transmit data to the received data at the relay.

In Fig. 19.15 the receive path at the relay R is illustrated in a block diagram. Initially, the antenna receives the continuous analog signal which is amplified to match the signal power of the receiver input limits. Next, a mixer converts the signal down to the baseband domain.The mixer is controlled by the local oscillator of the SMT911 board. After generating the baseband signal, it is filtered by a lowpass filter and sampled by an ADC. The sampling clock is produced by another local oscillator on the SMT911, the sampling frequency is $f_s = 40\,\text{MHz}$. Subsequently, the CP search of the superimposed OFDM signals of sources A and B is accomplished in the coarse synchronization by the minimum mean squared error (MMSE) criterion [2] for a joint symbol and carrier frequency synchronization. With the detected CP and its repetition the CFO is estimated in time domain by calculating the phase difference between the CP and the corresponding part in the OFDM symbol vector. In the following, the average CFO is compensated and the received serial data is converted into parallel OFDM symbol vectors. Next, the CP is removed and OFDM symbols are transferred to frequency domain by FFT. For fine synchronization, the synchronization signals PSCH and SSCH will be detected by a cross-correlation in order to determine the beginning and end of the received frame. With the channel equalizer the channel effects will be negated, therefore a pilot based linear channel estimation is performed. Furthermore, the pilots have to be removed and the remaining QPSK symbols are transferred in a serial data sequence. Finally, the PLNC decoding is performed by using the APP-based schemes presented in Sect. 19.2.1.

19.4.4 Measurement

The described HW demonstrator is used in a nearly symmetric two-way relaying network. The amplification of the received signal is set to a constant value. Therefore, it can be assumed that the noise power in the receiver is nearly constant. On the transmitter side the baseband noise will be scaled with the RF signal amplification.

As this work focuses on the general application of PLNC in real-time transmission and on the impact of the CFO, all other imperfections were not estimated, analysed or compensated. In general, the other impairments are quantization, sampling clock offset, sampling clock jitter, IQ-imbalance, non-linear power amplifiers and phase noise [7]. Compared to the simulation results before, these non-ideal conditions decrease the bit error rate (BER) performance. In addition, the channel is estimated by a simple least square channel estimation with linear interpolation. Furthermore, only the average CFO was estimated and compensated.

In our measurement setup, the sources had a distance of 40 cm and the distance of both sources to the relay was about 450 cm. Therefore, Fig. 19.16a shows the measured $\mathrm{BER_R}$ at the relay R for $\Delta f_c = f_{c\mathrm{A}} - f_{c\mathrm{B}} \approx 0\,\mathrm{Hz}$, where $f_{c\mathrm{A}}$ and $f_{c\mathrm{B}}$ denotes the carrier frequencies of source A and B, respectively. The transmitter level denotes the RF output power in dBm of both R&S vector signal generators and is equal for both sources. In the case of $\Delta f_c = 0\,\mathrm{Hz}$, the carrier frequency of both signal generators are identical. It can be seen, that G-JCNC outperforms JCNC and P-SCD and that JCNC is little bit better than P-SCD. In Fig. 19.16b, c the measured $\mathrm{BER_R}$ for unequal carrier frequencies are presented. Therefore, in Fig. 19.16b $\Delta f_c = 400\,\mathrm{Hz}$ corresponds to $|\varepsilon_\mathrm{A} - \varepsilon_\mathrm{B}| = 0.02$ and in Fig. 19.16c $\Delta f_c = 800\,\mathrm{Hz}$ corresponds to $|\varepsilon_\mathrm{A} - \varepsilon_\mathrm{B}| = 0.04$. For the small differential carrier frequency of $\Delta f_c = 400\,\mathrm{Hz}$ the performance of P-SCD and JCNC is hardly decreased compared to the case of identical carrier frequencies. G-JCNC loses performance for very small transmit power. The ICIC for P-SCD and G-JCNC increases the performance insignificant. Furthermore, G-JCNC outperforms JCNC and PSCD in all scenarios. For $\Delta f_c = 800\,\mathrm{Hz}$ the BER performance without ICIC is insufficient for P-SCD and JCNC, instead the ICIC increases the performance significantly, also for G-JCNC.

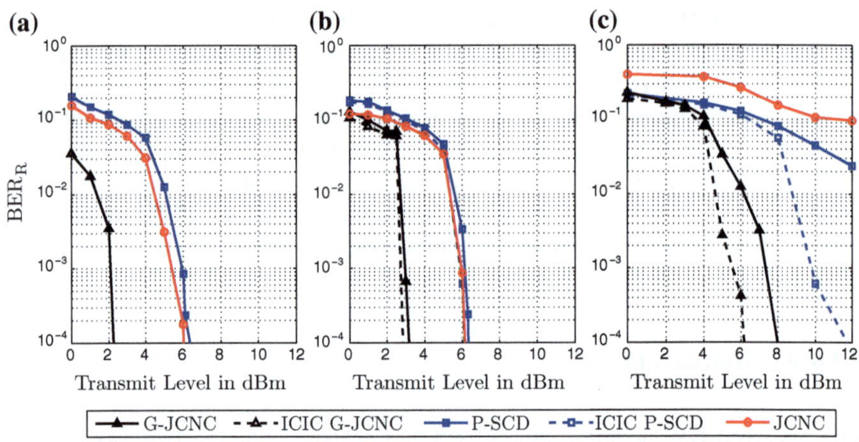

Fig. 19.16 Measured BER of the XORed packet at relay R with average CFO compensation. **a** Source A and B with same carrier frequency $f_{c\mathrm{A}} - f_{c\mathrm{B}} \approx 0$ Hz. **b** Differential carrier frequency of $\Delta f_c = |f_{c\mathrm{A}} - f_{c\mathrm{B}}| = 400\,\mathrm{Hz}$ corresponds to $|\varepsilon_\mathrm{A} - \varepsilon_\mathrm{B}| = 0.02$. **c** Differential carrier frequency of $\Delta f_c = |f_{c\mathrm{A}} - f_{c\mathrm{B}}| = 800\,\mathrm{Hz}$ corresponds to $|\varepsilon_\mathrm{A} - \varepsilon_\mathrm{B}| = 0.04$

In the measurement results the performance of the different decoding schemes diverges from the simulation results presented before. JCNC is for $\Delta f_c = 0\,\text{Hz}$ and $\Delta f_c = 400\,\text{Hz}$ a little bit better than P-SCD. On the other hand, at $\Delta f_c = 800\,\text{Hz}$ JCNC inferior to P-SCD. Therefore, several reasons exist. One of the most significant reason is the effective channel. In the simulation multi-path fading channels are assumed. The transmit channels in our laboratory are more like AWGN channels with some phase shift. The type of the effective channel have a huge impact on the mutual information. The MI for JCNC is the highest at larger SNR in AWGN channels and also the phase difference between the channels has an impact to the MI, as analysed in [17]. There was shown, that in AWGN channels the MI for JCNC strongly depends on the phase difference at higher SNR. Depending on the phase difference the MI is twice larger for JCNC as for SDC and vice versa. Here, we have not adjusted the phase difference, in order to get a realistic scenario where both sources are unsynchronized and do not cooperate. JCNC benefits from the accidental phase difference between the channels in the measurements for Fig. 19.16a, b. Additionally, the simulation benefits from a perfect channel knowledge. Furthermore, the real-world imperfections and the superposition of other RF signals in the laboratory affects to the measurement results. The impact of quantization, sampling clock offset, sampling clock jitter, IQ-imbalance, non-linear power amplifiers, phase noise and other non-idealities are not recognized in the simulation results. But for all these impairments, the efficiency of the PLNC schemes were confirmed in the real world transmission.

19.5 Extension to Two-Way Distributed Relaying

In this section, the two-phase two-way relaying system depicted in Fig. 19.1 is extended to distributed relaying with M relays. As shown in Fig. 19.17 for the system model, each relay R_m, $m = 1, 2, \ldots, M$ estimates the network coded message from its receive signal in the MA phase. Subsequently, the relays broadcast to the sources simultaneously in the BC phase. In the sequel, two adaptive broadcast transmission strategies presented in [22] are examined.

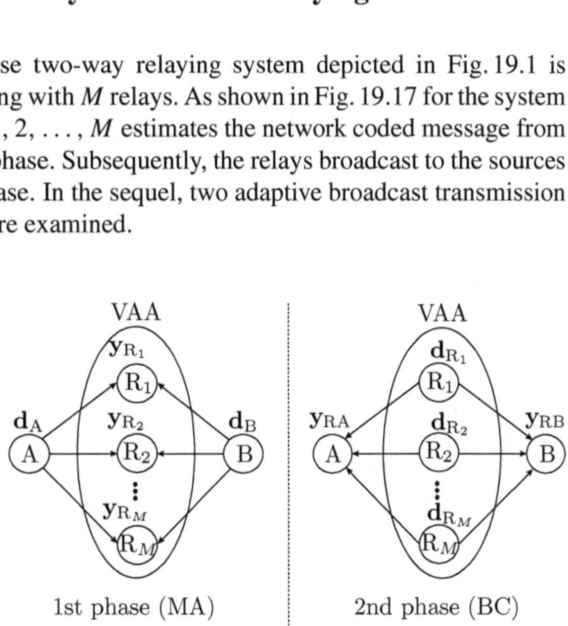

Fig. 19.17 A two-phase two-way distributed relaying system consisting of two sources A and B that exchange information with each other via M relays R_m

19.5.1 Traditional Adaptive Broadcast Transmission

Assuming the application of the same cyclic redundancy check (CRC) code in both sources, each relay R_m can determine its decoding status. In order to protect the BC phase from decoding errors propagated from relays, adaptive relaying has been proposed where only relays with correct network coded message $\mathbf{c}_R$ participate in the BC transmission using distributed orthogonal space-time block code (D-OSTBC) [27]. Specifically, denoting $\mathcal{D}$ the set that contains the relays decoding the source messages correctly in the MA phase, relays $R_m \in \mathcal{D}$ transmit their corresponding layer of the D-OSTBC to exploit spatial diversity. If no relay is able to decode correctly, i.e., $\mathcal{D} = \emptyset$, a *MA outage* occurs resulting in an end-to-end (e2e) outage.

19.5.2 Modified Adaptive Broadcast Transmission

Recall that SCD and G-JCNC estimate not only the relay message $\mathbf{c}_R$ but also the individual source messages $\mathbf{c}_A$ and $\mathbf{c}_B$. Thus, it might happen that some relays only decode one source message correctly while the network coded message is erroneous. This case is ignored in the traditional adaptive BC transmission scheme leading to an e2e outage. By incorporating this case a modified adaptive scheme is presented. Collecting the relays that decode $\mathbf{c}_A$ correctly in $\mathcal{D}_A$ and that decode $\mathbf{c}_B$ correctly in $\mathcal{D}_B$, this modified scheme is described as follows.

- If $\mathcal{D} \neq \emptyset$, the relays $R_k \in \mathcal{D}$ transmit the relay message using D-OSTBC as in the traditional scheme.
- If $\mathcal{D} = \emptyset$, $\mathcal{D}_A \neq \emptyset$ and $\mathcal{D}_B \neq \emptyset$, the relays $R_i \in \mathcal{D}_A$ and $R_j \in \mathcal{D}_B$ transmit the correctly estimated source message $\mathbf{d}_A$ and $\mathbf{d}_B$ using their respective D-OSTBCs simultaneously in the BC phase. Upon reception, each source subtracts the self-interference term with respect to the own information and performs D-OSTBC detection to estimate the message from the counterpart.
- Otherwise, the system is in outage.

Signaling Overhead In practice, each relay $R_m \in \mathcal{D}$ transmits its own layer of the applied D-OSTBC whereas the other relays switch into reception mode. Therefore, the relays not in $\mathcal{D}$ are able to overhear the signal transmitted by $R_m \in \mathcal{D}$ and recognize that at least one relay has decoded the relay message correctly. On the other hand, if one relay does not recognize any transmission during a specified time slot termed idle time, it knows that no relay estimated $\mathbf{c}_R$ correctly. In this case, the relay, if in $\mathcal{D}_A$ or $\mathcal{D}_B$, forwards the individual message $\mathbf{d}_A$ or $\mathbf{d}_B$ using its own layer of the applied D-OSTBC. Otherwise, the system is in outage. By such an implicit CRC exchange between the relays termed *CRC over the air*, only an idle time is required as extra signaling overhead, the impact of which is discussed in [22].

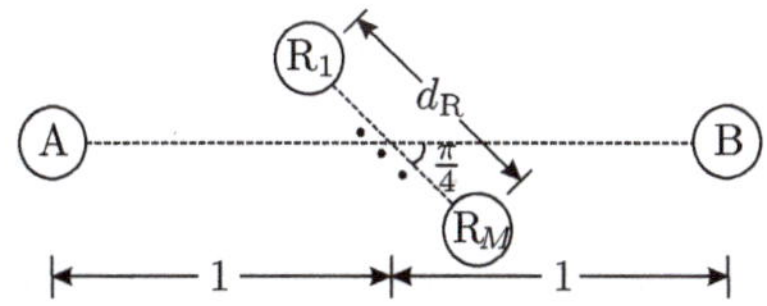

Fig. 19.18 A distributed two-way relaying network with M linearly positioned relays. The *two lines* connecting the sources and the relays respectively form an angle of $\pi/4$

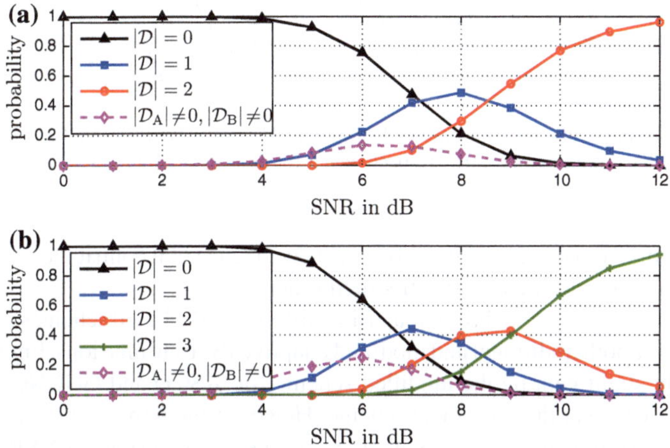

Fig. 19.19 Probabilities for different number of relays decoding the XORed message correctly applying P-SCD, **a** for $M = 2$ and **b** for $M = 3$ relays

19.5.3 Performance Evaluation

A distributed two-way relaying network is shown in Fig. 19.18, where M relays are equally distributed on a line $R_1 - R_M$ with d_R denoting the distance between R_1 and R_M. Therefore, the distance between neighboring relays yields $d_R/(M-1)$. Furthermore, the angle of the two lines connecting the sources A − B and the relays $R_1 - R_M$ respectively is set to $\pi/4$. Note that $d_R = 0$ corresponds to a symmetric network. In this case, all relays are assumed to be located at the same position but still the involved channels are uncorrelated. The path-loss factor is set to $\alpha = 4$. D-OSTBC with rate matching is employed as in the Chap. 18.

The probabilities that different number of relays decode the XORed message correctly are numerically calculated and shown in Fig. 19.19 for a symmetric network. P-SCD is assumed as an example with $M = 2$ or $M = 3$ distributed relays. As can be observed, the probabilities that no relay ($|\mathcal{D}| = 0$) and all relays ($|\mathcal{D}| = M$) succeed to decode the network coded message change from 1 to 0 and from 0 to 1 with growing SNR, respectively. When $|\mathcal{D}| = 0$, the event probability for both $\mathcal{D}_A$ and $\mathcal{D}_B$ being not empty increased with more relays. Since the gain of the modified

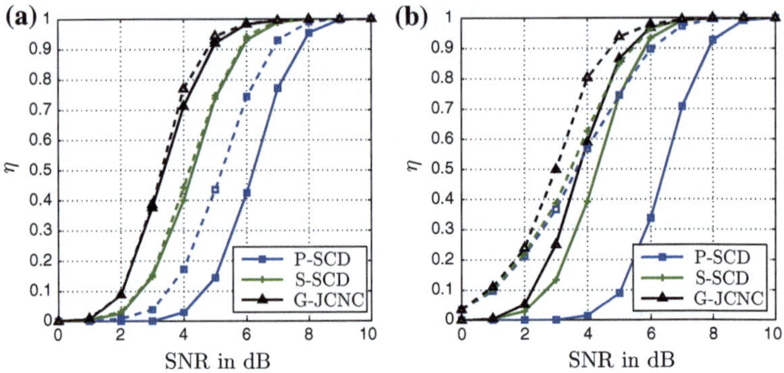

Fig. 19.20 Throughput η for APP-based schemes with $M = 4$. *Solid lines* and *dashed lines* are for the traditional and modified schemes, respectively, **a** for $d_R = 0$ and **b** for $d_R = 0.4$

scheme stems from this event, more frequent occurrence of it contributes to enhanced performance improvement of the modified scheme.

The normalized system throughput using different APP-based schemes is shown in Fig. 19.20 for both traditional and modified adaptive BC transmissions. Figure 19.20a shows only slight gain by the modified scheme for S-SCD and G-JCNC in a symmetric network over the traditional scheme. However, the improvement is enhanced greatly in an asymmetric network with $d_R = 0.4$, especially in the low SNR region, as depicted in Fig. 19.20b. This is because the asymmetric topology leads to increased chances that the relays decode only one source message correctly. Note that the overhead for idle time is ignored here but is incorporated in the throughput analysis in [22].

19.6 Conclusion

In this chapter, two-phase two-way relaying networks using physical-layer network coding (PLNC) are considered. In order to estimate the network coded message from the receive signal, several a-posteriori probability (APP) based detection and decoding schemes are demonstrated and compared with respect to mutual information and frame error rate. For practical concerns, the impact of carrier frequency offset (CFO) mismatch in the multiple-access (MA) transmission is investigated. The resulting performance degradation can be mitigated by the presented CFO compensation and inter-carrier interference (ICI) cancellation strategies. When the system is extended to distributed two-way relaying, a modified adaptive broadcast transmission scheme is proposed, which outperforms its traditional counterpart especially in asymmetric networks. Furthermore, the APP-based PLNC schemes are implemented on real-time hardware platforms in a laboratory environment.

Acknowledgments This work was supported in part by the German Research Foundation (DFG) under grant Wu 499/8-1 and Pa 438/6-1 within the priority program "Communication in Interference Limited Networks (COIN)", SPP 1397.

References

1. 3GPP (2013) Technical Specification Group Radio Access Network; Evolved Universal Terrestrial Radio Access (E-UTRA); Physical channels and modulation (Release 12). http://www.3gpp.org/ftp/Specs/html-info/36211.htm
2. Chiueh TD, Tsai PY, Lai IW (2012) Baseband receiver design for wireless MIMO-OFDM communications. Wiley
3. Cover T, Thomas J (1991) Elements of information theory. Wiley series in telecommunications
4. Davey M, Mackay D (1998) Low density parity check codes over GF(q). IEEE Commun Lett 2(6):165–167
5. Fessler J, Hero A (1994) Space-alternating generalized expectation-maximization algorithm. IEEE Trans Signal Process 42(10):2664–2672
6. Fragouli C, Boudec J, Widmer J (2006) Network coding: an instant primer. ACM SIGCOMM Comput Commun Rev 36(1):63–68
7. Horlin F, Bourdoux A (2008) Digital compensation for analog front-ends: a new approach to wireless transceiver design. Wiley
8. Kotzsch V, Holfeld J, Fettweis G (2009) Joint detection and CFO compensation in asynchronous multi-User MIMO OFDM systems. In: IEEE 69th vehicular technology conference (VTC-Spring'09), Barcelona, Spain
9. Kschischang F, Frey B, Loeliger H (2001) Factor graphs and the sum-product algorithm. IEEE Trans Inf Theory 47(2):498–519
10. Lu L, Wang T, Liew S, Zhang S (2012) Implementation of physical-layer network coding. In: IEEE international conference on communications (ICC'12), Ottawa, Canada
11. Maxim Integrated (2015) MAX2828, MAX2829—Single-/Dual-Band 802.11a/b/g World-Band Transceiver ICs. http://datasheets.maxim-integrated.com/en/ds/MAX2828-MAX2829.pdf
12. Pfletschinger S (2011) A practical physical-layer network coding scheme for the uplink of the two-way relay channel. In: 45th asilomar conference on signals, systems and computers (ASILOMAR'11), Monterey, CA, USA
13. Rohde and Schwarz (2015). http://www.rohde-schwarz.de
14. Schellmann M, Jungnickel V (2009) Multiple CFOs in OFDM-SDMA uplink: interference analysis and compensation. EURASIP J Wireless Commun Networking
15. Sundance Multiprocessor Technology (2015). http://www.sundance.com
16. Wang X, Poor H (1999) Iterative (Turbo) soft interference cancellation and decoding for coded CDMA. IEEE Trans Commun 47(7):1046–1061
17. Wu M, Wübben D, Dekorsy A (2013) Mutual information based analysis for physical-layer network coding with optimal phase control. In: 9th international ITG conference on systems, communications and coding (SCC'13), Munich, Germany
18. Wu M, Wübben D, Dekorsy A (2013) Physical-layer network coding in coded of dm systems with multiple-antenna relay. In: IEEE 77th vehicular technology conference (VTC'13-Spring), Dresden, Germany
19. Wu M, Ludwig F, Woltering M, Wübben D, Dekorsy A, Paul S (2014) Analysis and implementation for physical-layer network coding with carrier frequency offset. In: International ITG workshop on smart antennas (WSA'14), Erlangen, Germany
20. Wübben D (2010) Joint Channel decoding and physical-layer network coding in two-way QPSK Relay systems by a generalized sum-product algorithm. In: 7th international symposium on wireless communication systems (ISWCS'10), York, United Kingdom

21. Wübben D, Lang Y (2010) Generalized sum-product algorithm for joint channel decoding and physical-layer network coding in two-way relay systems. In: IEEE global communications conference (GLOBECOM'10), Miami, FL, USA
22. Wübben D, Wu M, Dekorsy A (2014) Adaptive broadcast transmission in distributed two-way relaying networks. In: European signal processing conference (EUSIPCO'14), Lisbon, Portugal
23. Xia X, Xu K, Xu Y (2012) Asynchronous physical-layer network coding scheme for two-way OFDM relay. http://arxiv.org/abs/1204.2692v1
24. Zhan A, He C (2008) Joint design of channel coding and physical network coding for wireless network. In: International conference on neural networks and signal processing, Zhejiang, China
25. Zhang S, Liew S (2009) Channel coding and decoding in a relay system operated with physical-layer network coding. IEEE J Sel Areas Commun 27(5):788–796
26. Zhang S, Liew S, Lam P (2006) Hot topic: physical layer network coding. In: International conference on mobile computing and networking (MobiCom'06), Log Angeles, CA, USA
27. Zhu K, Burr A (2011) Relay selection aided distributed space-time block code for two-way relay channel with physical-layer network coding. In: IEEE 73rd vehicular technology conference (VTC'11-Spring), Budapest, Hungary

Chapter 20
Generalized Multi-Carrier Waveforms in Two-Way Relay Systems

Matthias Woltering, Stephan Schedler, Dirk Wübben, Armin Dekorsy and Volker Kühn

Abstract In this chapter, two-way relaying networks using Physical-Layer Network Coding (PLNC) are considered with practical constraints. In a Multiple Access (MA) phase, two users transmit messages simultaneously on the same physical resources. The relay constructs a relay message based on the superimposed receive signal of both users. This message is broadcasted to the users which are able to extract the desired data from the other user. However, as the relay is not able to resolve shifts in time and frequency for the individual channels, the transmission scheme needs to be adjusted, if perfect synchronization cannot be ensured. This chapter deals with the proper selection of waveforms for multi-carrier systems in the Two-Way-Relay Channel (TWRC) using PLNC. We address Generalized Frequency Division Multiplexing (GFDM) compared to Orthogonal Frequency Division Multi-plexing (OFDM) under practical constraints like Carrier Frequency Offsets (CFOs) and Timing Offsets (TOs) within double dispersive channels. For OFDM rectangular transmit filters are used, which have a broad spectrum shape and are sensitive to shifts, especially in frequency domain. GFDM applies Gaussian waveforms that are inherently less susceptible to synchronization offsets than theirs rectangular counterparts, but additional interference is introduced even in a perfectly synchronized and flat fading channel. Thus, interferences caused by the channel or by the

M. Woltering (✉) · D. Wübben · A. Dekorsy
Department of Communications Engineering, University of Bremen,
Otto-Hahn-Allee NW1, 28359 Bremen, Germany
e-mail: woltering@ant.uni-bremen.de

D. Wübben
e-mail: wuebben@ant.uni-bremen.de

A. Dekorsy
e-mail: dekorsys@ant.uni-bremen.de

S. Schedler · V. Kühn
Institute of Communications Engineering, University of Rostock,
Richard-Wagner-Str. 31 (Haus 8), 18119 Rostock (Warnemünde), Germany
e-mail: stephan.schedler@uni-rostock.de

V. Kühn
e-mail: volker.kuehn@uni-rostock.de

W. Utschick (ed.), *Communications in Interference Limited Networks*,
Signals and Communication Technology, DOI 10.1007/978-3-319-22440-4_20

non-orthogonal Gaussian impulse shape are treated by an equalizer at the relay. The simulation results show that the Gaussian waveform outperforms the conventional OFDM system if the nodes are not perfectly synchronized.

20.1 Introduction

Modern communication systems have evolved to networks supporting an increasing number of services. A key technology is the principle of cooperative communication. It offers transmission techniques to enhance the coverage for applications with required Quality of Service (QoS) constraints [2, 7, 17]. Besides cooperation among sources and destinations, the utilization of relay stations has gained significant interest [2, 13]. Intermediate relay nodes can reduce the path loss significantly and offer spatial diversity. In a Two-Way-Relay Channel (TWRC) the spectral efficiency can be improved significantly by Physical-Layer Network Coding (PLNC), where the data of two users is simultaneously transmitted to an assisting relay in a Multiple Access (MA) phase. In contrast to separately handling the users, the relay constructs a common message.

As a consequence, the relay R in the system model depicted in Fig. 20.1 uses the superposition of signals of both users A and B by applying PLNC detection techniques [18, 32, 35] which utilize the superposition of the two users in the detection block to construct a relay message as also introduced in the Chap. 19. Subsequently, in the Broadcast phase, a network coded signal is transmitted by the relay [20, 32, 35, 36]. As the user terminals are aware of their own messages, they can extract the message of the other user after detecting the relay message. In the TWRC applying Orthogonal Frequency Division Multiplexing (OFDM), the superposition of the two messages at the relay contains additional interference caused by a double dispersive channel as well as impairments like Carrier Frequency Offsets (CFOs) and Timing Offsets (TOs), which result in Inter-Symbol Interference (ISI) and Inter-Carrier Interference (ICI). In contrast to Point-to-Point (P2P) transmissions, the relay cannot eliminate synchronization offsets between the users as it receives the superposition of both transmitted signals in the MA phase, i.e., the transmission system should be as robust as possible towards synchronization offsets. Since OFDM is widely used as a multi-carrier transmission scheme in current mobile radio standards, the combination of OFDM and PLNC has been applied in several publications, e.g., [8–10, 14, 25, 31, 33]. However, as OFDM implements rectangular transmit and receive filters with its $\sin(\omega)/\omega$ shape in frequency domain it has some considerable drawbacks like high out-of-band radiation, high Peak to Average Power Ratio (PAPR), and relatively large sensitivity to CFOs [11, 19]. In [30] it was shown that OFDM in TWRC suffer from CFO causing ICI.

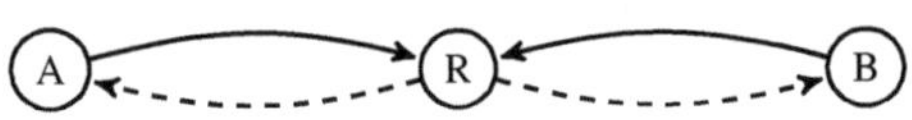

Fig. 20.1 A TWRC with MA phase (*solid lines*) and BC phase (*dashed lines*)

Besides OFDM many alternative multi-carrier schemes are known, using different transmit and receive filters compared to the rectangular one in OFDM. The term Filter Bank Multi-Carrier (FBMC) in general is used for general filter-bank multi-carrier schemes. However, FBMC is mostly used in context with Offset-QAM (OQAM) which results again in an orthogonal multi-carrier scheme like in [5]. It uses sub-carrier wise filtering with well-localized filters. Thus, in most cases, FBMC refers to Offset-QAM/Filter Bank Multi-Carrier (OQAM/FBMC) or OQAM/OFDM, but unfortunately, the term FBMC might also be used for multi-carrier applying any kind of filter-banks. Another scheme is Universal Filtered Multi-Carrier (UFMC). It implements parallel OFDM system, which are filtered block-wise to have a better spectral shape [26].

More general schemes, which introduce non-orthogonal multi-carrier transmissions are given in [3, 6]. Overviews of non-orthogonal waveforms in mobile applications are given in [22, 34]. The authors in [3, 6] both use the term Generalized Frequency Division Multiplexing (GFDM), but with a different meaning. The paper [6] deals with a block-based realization of a non-orthogonal multi-carrier transmission scheme similar to OFDM. Whereas, thesis [3] gives a general view on GFDM. According to the work of Du [3], we will use GFDM as a term for a multi-carrier scheme applying general waveforms.

An OFDM system is perfectly orthogonal in the ideal case, i.e., no interference is introduced by the channel. On the other hand, a Gaussian transmit/receive filter introduces ISI and ICI even in this ideal case as it does not satisfy the first Nyquist criterion [22]. However, the Gaussian transmit/receive filter slopes decay fast in time and frequency domain such that interference is mainly limited to adjacent time-frequency points and it allows for a more flexible spectrum shaping [5, 22]. By this means, the spectral efficiency, the influence of doubly dispersive channels, or sensitivity to synchronization offsets can be improved as shown in [12]. Nevertheless, this comes at the cost of additional interference and an increased computational complexity for equalization even in the ideal case, but for the general case the complexity for OFDM increases, sharply.

As the superposition of two users under realistic channel conditions always introduces CFOs and TOs and therefore ISI and ICI, we propose to use GFDM for a TWRC with a relay performing PLNC to be more robust against these effects. Subsequently, we will investigate the robustness of Gaussian transmit/receive filters in a PLNC system. PLNC detection techniques as introduced in the Chap. 19 to construct the relay message are extended to non-orthogonal multi-carrier schemes. The impact of the non-orthogonal waveforms and the doubly dispersive channels are treated by an linear equalizer and furthermore by a soft Interference Cancellation (sIC) feedback of already decided data. We will show that the application of a Gaussian waveform and a subsequent equalizer leads to a system that is less sensitive to synchronization errors than OFDM. To evaluate the overall system performance we will provide simulation results for a PLNC system.

20.2 Generalized Frequency Division Multiplexing for PLNC

Within this section we will introduce Generalized Frequency Division Multiplexing for the TWRC. To distinguish different transmission schemes we will stick to the following nomenclature.

GFDM	This is the most general transmission scheme. In this work we term multi-carrier schemes GFDM always using Gaussian transmit/receive filters.
plain-OFDM	By using rectangular transmit/receive filters GFDM change to plain OFDM, if no guard interval, like a Cyclic Prefix (CP) is used.
CP-OFDM	Similar to plain OFDM. GFDM will change to CP-OFDM if besides the rectangular filters also a CP is added as guard interval.

We assume a PLNC system where all nodes are equipped with a single antenna. In the MA phase both users A and B transmit simultaneously their *user messages* to the relay node R in the MA-phase. Based on the receive signal at the relay, the relay constructs a common *relay message*, which is transmitted in the BC phase. After the detection of this relay message at each user, they are able to remove their own messages from the relay message to extract the messages of the other user. Subsequently, we will focus on the MA phase of the PLNC scheme only, as this is the critical phase where synchronization offsets might have severe influence.

20.2.1 System Model

A block diagram of the MA phase is depicted in Fig. 20.2. We assume that the relay is positioned in the middle of users A and B, such that the individual channels H_{A} and H_{B} have identical statistics. Binary sequences $\mathbf{b}_{\mathrm{A}}$ and $\mathbf{b}_{\mathrm{B}}$ of user A and user B are encoded by a linear Forward Error Correcting (FEC) code $\mathcal{C}$ with code rate R_{C}. The coded binary sequences $\mathbf{c}_{\mathrm{A}} = \mathcal{C}(\mathbf{b}_{\mathrm{A}})$ and $\mathbf{c}_{\mathrm{B}} = \mathcal{C}(\mathbf{b}_{\mathrm{B}})$ are modulated to symbol sequences $d_{\mathrm{A}}^{(k,\ell)}$ and $d_{\mathrm{B}}^{(k,\ell)}$. These symbols are arranged in matrices $\mathbf{D}_{\mathrm{A}}$ and $\mathbf{D}_{\mathrm{B}}$ of dimension $N_{\mathrm{K}} \times N_{\mathrm{L}}$. N_{K} and N_{L} denoting the number of symbols per frame in frequency and time dimension, respectively. A single lattice point in the rectangular

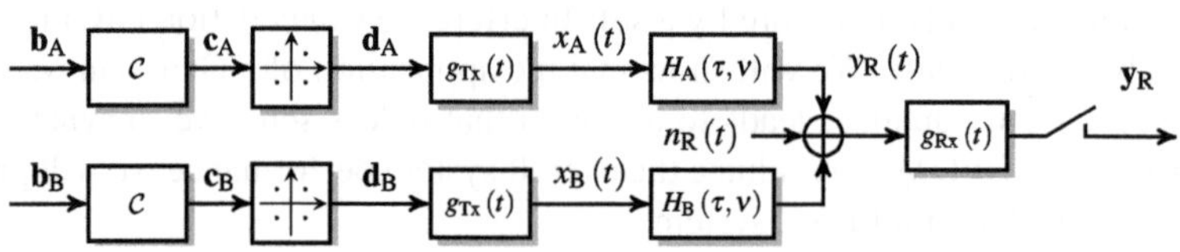

Fig. 20.2 Block diagram of the MA phase of a TWRC system with GFDM

time-frequency grid is denoted (k, ℓ), with sub-carrier index $0 \leq k \leq N_K - 1$ and time slot $0 \leq \ell \leq N_L - 1$. For each symbol $d_i^{(k,\ell)}$ with $i \in \{A, B\}$, the transmit filter $g_{Tx}(t)$ is shifted to the corresponding lattice point (k, ℓ) with

$$g_{Tx}^{(k,\ell)}(t) = g_{Tx}(t - \ell T) e^{j2\pi k F t}, \tag{20.1}$$

where t is the continuous time variable. F denotes the sub-carrier spacing and T the temporal spacing between symbols in the time-frequency grid. The corresponding matched filter at the receiver is given by

$$g_{Rx}^{(k',\ell')}(t) = g_{Tx}^{*}\left(-t - \ell' T\right) \cdot e^{-j2\pi (k' F)(-t)}. \tag{20.2}$$

Here, (k', ℓ') represent the lattice points in the time-frequency grid at the receiver. Both transmit signals will pass individual channels $H_i(\tau, \nu)$ assuming a tapped delay line channel model. The delay-Doppler function is given by the Fourier transform $H_i(\tau, \nu) \bullet\!\!-\!\!\circ\, h_i(\tau, t)$ w.r.t. t and is modeled as

$$H_i(\tau, \nu) = \sum_{\mu=0}^{N_h-1} h_i^{(\mu)} \delta(\tau - \tau_i^{(\mu)} T - \Delta\tau_i T) \delta(\nu - \nu_i^{(\mu)} F - \Delta\nu_i F), \tag{20.3}$$

where $h_i^{(\mu)}$, $\tau_i^{(\mu)}$, $\nu_i^{(\mu)}$ are the complex channel coefficient, relative time delay and relative Doppler shift of tap $\mu = 0, \ldots, N_h - 1$. Additionally, relative CFO $\Delta\nu_i$ and relative TO $\Delta\tau_i$ can occur for both users individually. For example, the offsets might have been caused by imperfect synchronization of nodes.

The relay receives the superposition of both users messages and circular Additive White Gaussian Noise (AWGN) $n_R(t) \sim \mathcal{CN}(0, P_N)$. The receive signal after matched filtering and sampling at the relay at the lattice point (k', ℓ') is given by [24]

$$y_R^{(k',\ell')} = \sum_{k=0}^{N_K-1} \sum_{\ell=0}^{N_L-1} \upsilon_A^{(k',\ell'),(k,\ell)} d_A^{(k,\ell)} + \sum_{k=0}^{N_K-1} \sum_{\ell=0}^{N_L-1} \upsilon_B^{(k',\ell'),(k,\ell)} d_B^{(k,\ell)} + n_R^{(k',\ell')}. \tag{20.4}$$

The elements $y_R^{(k',\ell')}$ can be described by a corresponding matrix $\mathbf{Y}_R$. The coefficients [24]

$$\begin{aligned}\upsilon_i^{(k',\ell'),(k,\ell)} = &\iint H_i(\tau, \nu) e^{-2j\pi\left(kF\tau + \left(F(k'-k) - \nu\right)\left(\frac{1}{2}\left((\ell'+\ell)T + \tau\right)\right)\right)} \\ &\times A^{*}\left(T\left(\ell - \ell'\right) + \tau, F\left(k - k'\right) + \nu\right) \, d\nu d\tau \end{aligned} \tag{20.5}$$

combine the impact of the channel $H_i(\tau, \nu)$ and transmit and receive filters for the corresponding transmit signals $d_i^{(k,\ell)}$ on the receive signal $y_R^{(k',\ell')}$. The coefficients

$\upsilon_i^{(k',\ell'),(k,\ell)}$ in (20.5) are elements of the equivalent effective channel matrix $\mathbf{V}_i$, which will be introduced in the Sect. 20.2.3. The ambiguity function [3]

$$A(\tau,\nu) = \int g_{\mathrm{Rx}}\left(t+\frac{\tau}{2}\right) g_{\mathrm{Tx}}^*\left(t-\frac{\tau}{2}\right) e^{-j2\pi\nu t} dt \tag{20.6}$$

describes these influences on a specific time and frequency point given filter $g_{\mathrm{Tx}}(\cdot)$ and $g_{\mathrm{Rx}}(\cdot)$.

20.2.2 *Waveforms*

In order to investigate the impact of doubly dispersive channels in a TWRC two basic multi-carrier schemes will be compared in this chapter, namely a classical scheme with rectangular transmit/receive filter as well as a generalized scheme with Gaussian transmit/receive filter [3].

20.2.2.1 Rectangular Waveform

For the rectangular waveform the transmit and receive filters are given by

$$g_{\mathrm{Tx}}(t) = \begin{cases} 1/\sqrt{T} & \text{for } |t| \le \frac{T}{2} \\ 0 & \text{else} \end{cases}, \qquad g_{\mathrm{Rx}}(t) = \begin{cases} 1/\sqrt{T} & \text{for } |t| \le \frac{T-T_{\mathrm{G}}}{2} \\ 0 & \text{else} \end{cases}, \tag{20.7}$$

where T_{G} is the duration of the guard interval. For an AWGN channel, the coefficients in a perfectly synchronized OFDM system ($\Delta\nu = 0$, $\Delta\tau = 0$) become $\upsilon_i^{(k',\ell'),(k,\ell)} = 1$ for the desired signal part $(k,\ell) = (k',\ell')$ and $\upsilon^{(k',\ell'),(k,\ell)} = 0$ for all interference terms $(k,\ell) \neq (k',\ell')$. Hence, this scheme is called orthogonal. The ambiguity function of the rectangular pulse shape with guard interval can be computed by solving (20.6) and results in [3]

$$A(\tau,\nu) = \begin{cases} -\frac{1}{\pi T\nu}\sin\left(\pi\nu\left(T_{\mathrm{G}}/2-\tau-T\right)\right)e^{+j\pi\frac{T_{\mathrm{G}}}{2}} & \text{for } -T+\frac{T_{\mathrm{G}}}{2} \le \tau \le -\frac{T_{\mathrm{G}}}{2} \\ -\frac{1}{\pi T\nu}\sin\left(\pi\nu\left(T_{\mathrm{G}}-T\right)\right)e^{-j\pi\tau} & \text{for } -\frac{T_{\mathrm{G}}}{2} \le \tau \le +\frac{T_{\mathrm{G}}}{2} \\ -\frac{1}{\pi T\nu}\sin\left(\pi\nu\left(T_{\mathrm{G}}/2+\tau-T\right)\right)e^{-j\pi\frac{T_{\mathrm{G}}}{2}} & \text{for } +\frac{T_{\mathrm{G}}}{2} \le \tau \le T-\frac{T_{\mathrm{G}}}{2} \\ 0 & \text{else.} \end{cases} \tag{20.8}$$

In the sequel, we assume a guard interval used like a CP. Thus, this scheme is called *CP-OFDM*. Note, if the transmit and the receive filter are equal $g_{\mathrm{Tx}}(t) = g_{\mathrm{Rx}}(t)$, we call this scheme *plain-OFDM*, because no guard interval is used.

20.2.2.2 Gaussian Waveform

Contrary to the rectangular waveform, multi-carrier systems using Gaussian waveforms are non-orthogonal, as the first Nyquist criterion is not fulfilled. Thus, even if the system is perfectly synchronized, the coefficients $\upsilon_i^{(k',\ell'),(k,\ell)}$ are not zero for interference terms $(k,\ell) \neq (k',\ell')$, i.e., the system has to deal with ICI and ISI. The filters for a Gaussian impulse shape with configuration factor α are given by [22]

$$g_{\mathrm{Tx}}(t) = g_{\mathrm{Rx}}(t) = (2\alpha)^{\frac{1}{4}} e^{-\pi\alpha t^2}. \tag{20.9}$$

Note that the Fourier transform of (20.9) is again the Gaussian function, where parameter α influences the localization in time and frequency domain. For $\alpha = 1$ the Gaussian waveform has identical responses in time and frequency [22], and most interference is restricted to direct neighbors in the time-frequency grid. The related ambiguity function for the Gaussian waveform is given by [3]

$$A(\tau,\nu) = e^{-\frac{\pi\left(\nu^2+\alpha^2\tau^2\right)}{2\alpha}}. \tag{20.10}$$

For an AWGN channel and perfect synchronization ($\Delta\nu = 0$, $\Delta\tau = 0$), the equivalent channel coefficients with Gaussian transmit/receive filters become [24]

$$\upsilon_{i,\mathrm{Gaussian}}^{(k',\ell'),(k,\ell)} = e^{-\frac{1}{2}\pi\left((k'-k)^2F^2+(\ell'-\ell)^2T^2\right)} \cdot e^{-j\pi(k'-k)F(\ell'+\ell)T}. \tag{20.11}$$

20.2.3 Matrix Representation

With the vec $\{\mathbf{A}\}$ operator which stacks the columns of a matrix $\mathbf{A}$ into a vector the sampled receive signals $y_{\mathrm{R}}^{(k',\ell')}$ in frame $\mathbf{Y}_{\mathrm{R}}$ are stacked in vector $\mathbf{y}_{\mathrm{R}} = \mathrm{vec}\{\mathbf{Y}_{\mathrm{R}}\}$

$$\mathbf{y}_{\mathrm{R}} = \mathbf{V}_{\mathrm{A}} \cdot \mathbf{d}_{\mathrm{A}} + \mathbf{V}_{\mathrm{B}} \cdot \mathbf{d}_{\mathrm{B}} + \mathbf{n}_{\mathrm{R}}, \tag{20.12}$$

where the data symbols $d_i^{(k,\ell)}$ of frames $\mathbf{D}_{\mathrm{A}}$ and $\mathbf{D}_{\mathrm{B}}$ are rearranged in vectors $\mathbf{d}_{\mathrm{A}} = \mathrm{vec}\{\mathbf{D}_{\mathrm{A}}\}$ and $\mathbf{d}_{\mathrm{B}} = \mathrm{vec}\{\mathbf{D}_{\mathrm{B}}\}$ of length $N_{\mathrm{L}} \cdot N_{\mathrm{K}}$. The effective channel matrices $\mathbf{V}_i$ consist of coefficients $\upsilon_i^{(k',\ell'),(k,\ell)}$ defined in (20.5) connected with

$$[\mathbf{V}_i]_{(k'+\ell' N_{\mathrm{K}},k+\ell N_{\mathrm{K}})} = \upsilon_i^{(k',\ell'),(k,\ell)}. \tag{20.13}$$

Figure 20.3 depicts the absolute values of matrix $\mathbf{V}_i$ for a rectangular and a Gaussian transmit/receive filter, considering a flat fading channel with a CFO of $\Delta\nu_i = 0.2$ and without CFO. For simplicity, a frame of size $N_{\mathrm{L}} \times N_{\mathrm{K}} = 7 \times 7$ is shown. As the Gaussian filter is well localized, the influence to adjacent lattice

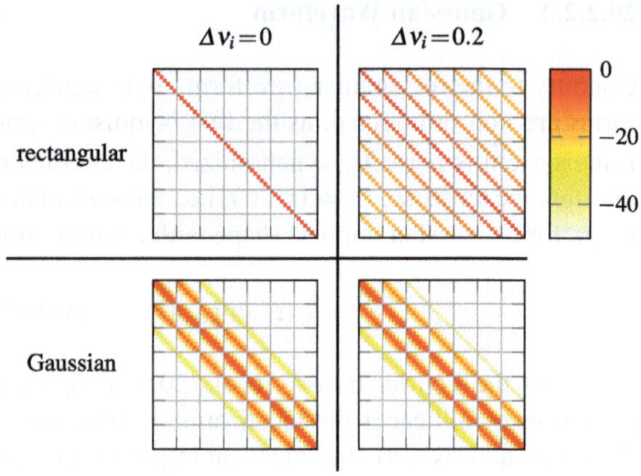

Fig. 20.3 Absolute values of the elements in matrix $\mathbf{V}_i$ in logarithmic scale for (**a**) rectangular waveform and (**b**) Gaussian waveform. Flat channel with and without influence of additional CFO

points is small. In contrast, the rectangular waveform introduces no influences on the off-diagonals, therefore it is orthogonal in the ideal case. However, a CFO has huge impact in case of a rectangular waveform, whereas the influence on the Gaussian waveform is smaller.

Without loss of generality, the effective channel matrices $\mathbf{V}_i = \mathbf{\Lambda}_i + \overline{\mathbf{\Lambda}}_i$ can be separated further into a diagonal matrix $\mathbf{\Lambda}_i = \operatorname{diag}\{\mathbf{V}_i\}$ and an off-diagonal matrix $\overline{\mathbf{\Lambda}}_i = \mathbf{V}_i - \mathbf{\Lambda}_i$ corresponding to the desired part and the interference terms of the transmission to the relay, respectively. Thus the signal $\mathbf{y}_\mathrm{R}$ can be written as

$$\mathbf{y}_\mathrm{R} = \underbrace{\mathbf{\Lambda}_\mathrm{A} \cdot \mathbf{d}_\mathrm{A} + \mathbf{\Lambda}_\mathrm{B} \cdot \mathbf{d}_\mathrm{B}}_{\text{desired part}} + \underbrace{\overline{\mathbf{\Lambda}}_\mathrm{A} \cdot \mathbf{d}_\mathrm{A} + \overline{\mathbf{\Lambda}}_\mathrm{B} \cdot \mathbf{d}_\mathrm{B}}_{\text{interference part}} + \underbrace{\mathbf{n}_\mathrm{R}}_{\text{noise part}} . \quad (20.14)$$

Based on (20.14) the received power for the desired user signal $i \in \{\mathrm{A}, \mathrm{B}\}$ on time-frequency point (k', ℓ') can be calculated with

$$P_{i,\mathrm{S}}^{(k',\ell')} = \mathrm{E}\left|[\mathbf{\Lambda}_i \cdot \mathbf{d}_i]_{(k',\ell')}\right|^2. \quad (20.15)$$

Furthermore, the interference powers caused by the users i can be evaluated for each receive time-frequency point

$$P_{i,\mathrm{I}}^{(k',\ell')} = \mathrm{E}\left|\left[\overline{\mathbf{\Lambda}}_i \cdot \mathbf{d}_i\right]_{(k',\ell')}\right|^2. \quad (20.16)$$

20.2.4 *Relaying Strategies*

The achievable performance depends significantly on the applied modulation as well as the actual detection strategy at the relay. Subsequently, we will distinguish Separate

Channel Decoding (SCD), Joint Channel decoding and Network Coding (JCNC), and Generalized Joint Channel decoding and Network Coding (G-JCNC) schemes that have been introduced already in the Chap. 19 for discrete input alphabets. Within this section we will introduce mutual information for these strategies assuming Gaussian input alphabets, rather than discrete alphabets.

20.2.4.1 Separate Channel Decoding (SCD)

In the SCD scheme, the signals of the users are detected individually, treating the symbol d_{B} of user B as interference for the signal d_{A} of user A and vice versa. If the interference of user B is assumed to be normal distributed, the mutual information $I\left(d_{\mathrm{A}};\, y_{\mathrm{R}}\right)$ of the A → R link is given by P2P channel capacity, where the noise power is substituted by sum over interference power and noise power. For Gaussian input alphabets, it is given by

$$I\left(d_{\mathrm{A}}^{(k',\ell')};\, y_{\mathrm{R}}^{(k',\ell')}\right) = \log_2\left(1 + \frac{P_{\mathrm{A,S}}^{(k',\ell')}}{P_{\mathrm{B,S}}^{(k',\ell')} + P_{\mathrm{A,I}}^{(k',\ell')} + P_{\mathrm{B,I}}^{(k',\ell')} + P_{\mathrm{N}}}\right). \tag{20.17}$$

In order to generate the BC signal correctly, both user messages have to be decoded without errors. Thus, the mutual information of SCD is given by the worse link for each time-frequency point, i.e.,

$$C_{\mathrm{S}} = \sum_{(k',\ell')} \min\left\{I\left(d_{\mathrm{A}}^{(k',\ell')};\, y_{\mathrm{R}}^{(k',\ell')}\right),\, I\left(d_{\mathrm{B}}^{(k',\ell')};\, y_{\mathrm{R}}^{(k',\ell')}\right)\right\} \tag{20.18}$$

$$= \sum_{(k',\ell')} \log_2\left(1 + \min\left\{\frac{P_{\mathrm{A,S}}^{(k',\ell')}}{P_{\mathrm{B,S}}^{(k',\ell')}+P_{\mathrm{A,I}}^{(k',\ell')}+P_{\mathrm{B,I}}^{(k',\ell')}+P_{\mathrm{N}}},\, \frac{P_{\mathrm{B,S}}^{(k',\ell')}}{P_{\mathrm{A,S}}^{(k',\ell')}+P_{\mathrm{A,I}}^{(k',\ell')}+P_{\mathrm{B,I}}^{(k',\ell')}+P_{\mathrm{N}}}\right\}\right). \tag{20.19}$$

Note that the SCD scheme might be improved if decoding is done serially [29]. In that case, the signal decoded first, can be subtracted from the receive signal, i.e., there is no superposition of two user signals in the second decoding step. However, the mutual information depends on the decoding order.

20.2.4.2 Joint Channel Decoding and Network Coding (JCNC)

Due to the PLNC strategy, the relay is not necessarily interested in the individual signals, but in the relay message, which is an XOR signal that will be transmitted in the BC phase. Thus, the relay node does not need to detect d_{A} and d_{B} individually. It is sufficient to directly estimate the XORed codeword based on the receive signal $y_{\mathrm{R}}^{(k',\ell')}$. However, with Gaussian input, an XOR combination of the signals transmitted by user A and B is not defined. Liew et al. argue in [37] that achievable rates are nonetheless limited by the worse user, i.e.,

$$C_{\mathrm{J}} = \sum_{(k',\ell')} \log_2 \left(1 + \frac{\min\left\{ P_{\mathrm{A,S}}^{(k',\ell')}, P_{\mathrm{B,S}}^{(k',\ell')} \right\}}{P_{\mathrm{A,I}}^{(k',\ell')} + P_{\mathrm{B,I}}^{(k',\ell')} + P_{\mathrm{N}}} \right) \tag{20.20}$$

and it was shown that this bound is tight in low as well as high Signal to Noise Ratio (SNR) regions.

20.2.4.3 Generalized Joint Channel Decoding and Network Coding (G-JCNC)

Instead of detecting the source signals separately with subsequent XOR combination (SCD), the individual codewords might be decoded jointly within a Generalized Sum-Product Algorithm [31, 32]. To fully exploit the available information about the superimposed receive signal as well as the coding gain, decoding is done with respect to an extended Galois field. The achievable rate for the MA phase of the GFDM system with Gaussian input is given by

$$C_{\mathrm{G}} = \frac{1}{2} \sum_{(k',\ell')} \log_2 \left(1 + \frac{P_{\mathrm{A,S}}^{(k',\ell')} + P_{\mathrm{B,S}}^{(k',\ell')}}{P_{\mathrm{A,I}}^{(k',\ell')} + P_{\mathrm{B,I}}^{(k',\ell')} + P_{\mathrm{N}}} \right), \tag{20.21}$$

where factor $1/2$ comes from subsequent mapping to the XORed relay codeword. Note that (20.21) is just an upper bound on the sum rate of the MA phase in a PLNC system assuming uncorrelated source messages, no direct link, and no prioritization of users [20, 21]. First, the computed sum rate might not be feasible, as individual rates are not necessarily in the rate region of the MA-channel. Second, we have assumed that channels H_{A} and H_{B} have identical statistics such that the MA phase is the bottleneck in the PLNC transmission scheme. If average channel qualities of users A and B differ significantly, the computed sum rate C_{G} is not necessarily meaningful in the PLNC context, as the corresponding mutual information cannot be broadcasted over the worse link in the second transmission phase.

20.3 Interference Handling in a Two-Way-Relay Channel

In the TWRC a superposition of two user signals is received at the relay. As the detection and decoding block can utilize this superposition in the PLNC detection process, the aim at the equalizer output is to reduce the influence of the interference (ISI and ICI) introduced by the channel and the waveforms in order to provide a signal, depending on the desired signals of user A and B. As described in [16], the influence of non-orthogonal filters and doubly dispersive channels are highly complex and the optimal sequence estimator is a Viterbi algorithm working on a state diagram including all possible states in time and frequency. Subsequently, we focus on linear equalizers that have low computational effort. Additionally, we introduce an iterative structure in the decoding/detection process which utilize already decided data, to further boost the performance.

20.3.1 Linear Equalizer

First, we focus on a Linear Equalizer (LE) taking only adjacent symbols in the time-frequency grid into account. Let us define a window operator

$$\mathcal{W}_{N_\mathrm{N}}^{(k',\ell')}\{\mathbf{A}\} = \mathrm{vec}\left\{[\mathbf{A}]_{(k'-N_\mathrm{N}:k'+N_\mathrm{N})(\ell'-N_\mathrm{N}:\ell'+N_\mathrm{N})}\right\} \tag{20.22}$$

that selects N_N neighboring elements in a window of size $(2N_\mathrm{N}+1)\times(2N_\mathrm{N}+1)$ around a lattice point (k',ℓ') of frame $\mathbf{A}$ [28]. Without loss of generality, we assume that the window is square. The operator can be extended easily to a window that uses different number of neighbors in time and frequency direction. By applying the window operator in (20.22) to the receive signal in (20.12), the signal

$$\mathbf{y}_{\mathcal{W}}^{(k',\ell')} = \mathcal{W}_{N_\mathrm{N}}^{(k',\ell')}\{\mathbf{Y}_\mathrm{R}\}\,. \tag{20.23}$$

is generated, which selects a subset of $(2N_\mathrm{N}+1)^2$ neighboring signals around a specific lattice point (k',ℓ') as well as the desired signal for the LE. The restriction to neighboring symbols is motivated by the auto-ambiguity function of the Gaussian transmit/receive filter. Contrary to the ambiguity function of a rectangular transmit/receive filter, the Gaussian waveform is well concentrated in time and frequency, i.e., the interference terms are limited to adjacent lattice points. The model is depicted in Fig. 20.4.

Here, the windowed receive signal $\mathbf{y}_{\mathcal{W}}^{(k',\ell')}$ is used as equalizer input and similar to (20.12) it can be written as

$$\mathbf{y}_{\mathcal{W}}^{(k',\ell')} = \tilde{\mathbf{V}}_\mathrm{A}\cdot\mathbf{d}_\mathrm{A} + \tilde{\mathbf{V}}_\mathrm{B}\cdot\mathbf{d}_\mathrm{B} + \mathbf{n}_{\mathcal{W}}, \tag{20.24}$$

where $\tilde{\mathbf{V}}_\mathrm{A}$ and $\tilde{\mathbf{V}}_\mathrm{B}$ are matrices matching the data $\mathbf{d}_i$ to the receive signal $\mathbf{y}_{\mathcal{W}}^{(k',\ell')}$ and $\mathbf{n}_{\mathcal{W}}$ is the windowed noise vector. The size of these matrices is $(2N_\mathrm{N}+1)^2\times(N_\mathrm{L}\cdot N_\mathrm{K})$. The output of the LE for the symbol at (k',ℓ') is given by

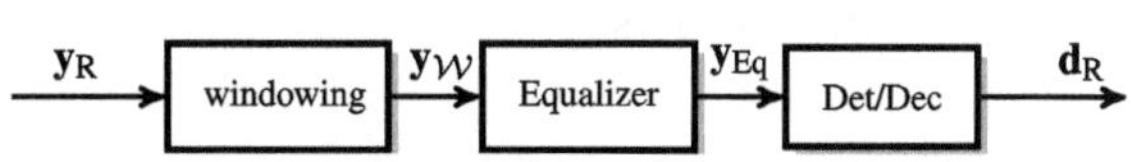

Fig. 20.4 Linear equalizer with windowing

$$y_{\mathrm{Eq}}^{(k',\ell')} = \left(\mathbf{z}^{(k',\ell')}\right)^{\mathrm{T}} \mathbf{y}_{\mathcal{W}}^{(k',\ell')}, \tag{20.25}$$

where vector $\mathbf{z}^{(k',\ell')}$ contains the coefficients of the LE. Note that the equalizer is calculated for each time-frequency point (k', ℓ') separately and the equalizer output is stacked into vector $\mathbf{y}_{\mathrm{Eq}}$. In contrast to single user P2P transmissions, where the receiver wants to estimate the signal of each user separately, in a TWRC applying PLNC, it is sufficient to estimate the superposition $d_{\mathrm{AB}}^{(k',\ell')} = d_{\mathrm{A}}^{(k',\ell')} + d_{\mathrm{B}}^{(k',\ell')}$. Thus, the equalizer coefficients to estimate the superposition are determined assuming Minimum Mean Square Error (MMSE) criterion, by solving the optimization problem

$$\mathbf{z}_{\mathrm{MMSE}}^{(k',\ell')} = \underset{\mathbf{z}^{(k',\ell')}}{\arg\min}\ \mathrm{E}\left\{\left|y_{\mathrm{Eq}}^{(k',\ell')} - d_{\mathrm{AB}}^{(k',\ell')}\right|^2\right\}. \tag{20.26}$$

The solution is given by [1]

$$\left(\mathbf{z}_{\mathrm{MMSE}}^{(k',\ell')}\right)^{\mathrm{T}} = \mathbf{e}^{\mathrm{T}}\left(\tilde{\mathbf{V}}_{\mathrm{A}}^{\mathrm{H}} + \tilde{\mathbf{V}}_{\mathrm{B}}^{\mathrm{H}}\right)\left(\tilde{\mathbf{V}}_{\mathrm{A}}\,\tilde{\mathbf{V}}_{\mathrm{A}}^{\mathrm{H}} + \tilde{\mathbf{V}}_{\mathrm{B}}\,\tilde{\mathbf{V}}_{\mathrm{B}}^{\mathrm{H}} + \frac{P_N}{P_S}\,\mathbf{I}\right)^{-1}, \tag{20.27}$$

where vector $\mathbf{e} = [0, \ldots, 0, 1, 0, \ldots, 0]^{\mathrm{T}}$ selects the row that corresponds to the desired symbol at (k', ℓ'). Note, by using no window operator the calculation of the equalizer coefficients will take all elements $N_{\mathrm{K}} N_{\mathrm{L}} \times N_{\mathrm{K}} N_{\mathrm{L}}$ of a frame. The calculation is done for all elements at once $\mathbf{y}_{\mathrm{Eq}} = \mathbf{Z}\,\mathrm{vec}\,\{\mathbf{Y}_{\mathrm{R}}\}$.

20.3.1.1 Complexity of Linear Equalizer

As the complexity is mainly influenced by the inversion of matrices we focus here on the complexity of inversions in the LE exploiting the full receive signal (20.12) and the reduced signal (20.24) by windowing. Considering a smaller number of neighbors reduces the complexity. The impact of the interference is directly connected to the window size. If N_{N} adjacent neighbors are considered by the equalizer, the influence caused by the $(2N_{\mathrm{N}} + 1)^2$ symbols around lattice point (k', ℓ') will be considered. The interference of lattice points that are more than $N_{\mathrm{N}} + 1$ symbols away, is not considered by the equalizer. In (20.27) an inversion of a matrix with dimension $(2N_{\mathrm{N}}+1)^2 \times (2N_{\mathrm{N}}+1)^2$ has to be applied for each symbol at (k', ℓ'). The larger N_{N}, the smaller the impact of the residual interference. For a frame size of $N_{\mathrm{Fr}} = N_{\mathrm{L}} \cdot N_{\mathrm{K}}$ symbols, the complexity of a full inversion in (20.27) is $\mathcal{C}_{\mathrm{full}} \sim \mathcal{O}(N_{\mathrm{Fr}}^3)$. Whereas a reduced equalizer requires N_{Fr} inversions of a matrix with size $(2N_{\mathrm{N}} + 1)^2$ which is $\mathcal{C}_{\mathrm{reduced}} \sim \mathcal{O}(N_{\mathrm{Fr}} \cdot ((2N_{\mathrm{N}} + 1)^2)^3)$. Note, that the windowed equalizer scales with the number of neighbors.

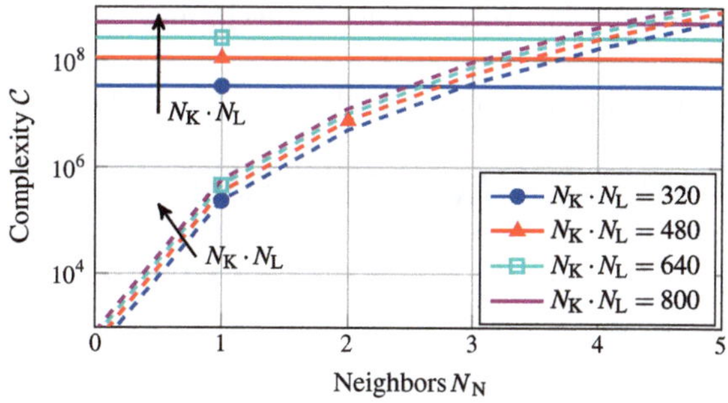

Fig. 20.5 Complexity of a full inversion (—), and inversion of reduced size matrix (- - -) for different frame sizes

Figure 20.5 shows the complexity of applying a LE which uses the full signal (20.12) and the windowed version (20.24) considering only some neighbors. As it can be observed, taking only a few neighbors N_N the complexity is reduced. If more neighbors are taken into account the complexity increases strongly. For a frame of size $N_{Fr} = 320$ elements, complexity can be reduced only if $N_N \leq 3$.

20.3.2 TWRC with Soft Interference Cancellation

In order to improve the estimation performance, we subtract the estimated interference of the previous iteration from the receive signal $\mathbf{y}_R$ before equalizing. Therefore, the equalizer is extended with a feedback path [27], where the interference terms of already estimated data symbols $\tilde{\mathbf{d}}_A$ and $\tilde{\mathbf{d}}_B$ from the PLNC detection block are fed back in a soft manner. A block diagram of sIC is shown in Fig. 20.6.

The property of providing soft symbols is essential for iterative detection of TWRC systems [30]. The popular JCNC scheme, which directly estimates the relay codeword without providing separate estimates for the source codewords $\mathbf{c}_A$ and $\mathbf{c}_B$, can not be used with sIC [35].

Note that in the iterative sIC structure the window operator introduced in (20.22) is used again to reduce the complexity in the equalizing step, but it is omitted here for the sake of simplicity.

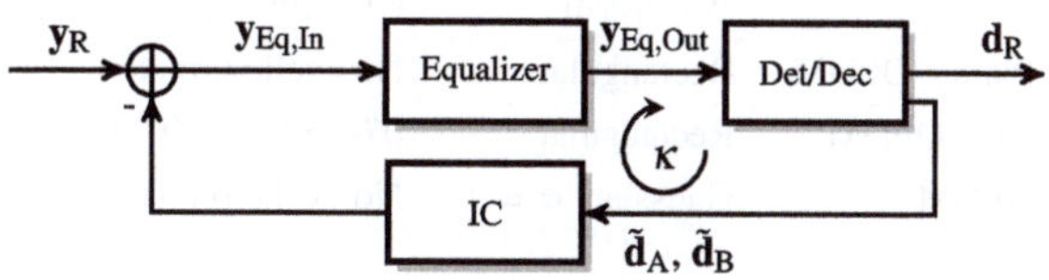

Fig. 20.6 Equalizer with soft interference cancellation

20.3.2.1 Soft Interference Cancellation

In the κth iteration the signal using (20.14) at the equalizer input is given by

$$\begin{aligned}\mathbf{y}_{\text{Eq,In}}^{(\kappa)} &= \mathbf{y}_{\text{R}} - \left(\overline{\mathbf{\Lambda}}_{\text{A}} \cdot \tilde{\mathbf{d}}_{\text{A}}^{(\kappa-1)} + \overline{\mathbf{\Lambda}}_{\text{B}} \cdot \tilde{\mathbf{d}}_{\text{B}}^{(\kappa-1)}\right) \\ &= \mathbf{\Lambda}_{\text{A}} \cdot \mathbf{d}_{\text{A}} + \mathbf{\Lambda}_{\text{B}} \cdot \mathbf{d}_{\text{B}} + \overline{\mathbf{\Lambda}}_{\text{A}} \left(\mathbf{d}_{\text{A}} - \tilde{\mathbf{d}}_{\text{A}}^{(\kappa-1)}\right) + \overline{\mathbf{\Lambda}}_{\text{B}} \left(\mathbf{d}_{\text{B}} - \tilde{\mathbf{d}}_{\text{B}}^{(\kappa-1)}\right) + \mathbf{n}_{\text{R}} ,\end{aligned} \tag{20.28}$$

where $\tilde{\mathbf{d}}_{\text{A}}^{(\kappa-1)}$ and $\tilde{\mathbf{d}}_{\text{B}}^{(\kappa-1)}$ denote the estimates for the transmit signals achieved in the previous iteration $\kappa - 1$. Note that for the first iteration the feedback symbols are initialized by $\tilde{\mathbf{d}}_i^{(0)} = 0$ yielding the LE (20.27). Assuming the estimates of $\tilde{\mathbf{d}}_i^{(\kappa-1)}$ are tight, the residual interference terms get close to zero. Hence, only the diagonal parts of the effective channel matrix $\mathbf{V}_i$ will remain. After N_{I} iterations the detector provides the estimated relay codeword $\hat{\mathbf{c}}_{\text{R}}$ which is mapped to the relay symbol vector $\mathbf{d}_{\text{R}}$.

20.3.3 Performance Results

Here we will compare different multi-carrier configurations given in Table 20.1.

In the performance analysis the envelope of the complex channel gains $h_i^{(\mu)}$ are Rayleigh distributed. The relative time delay $\tau_i^{(\mu)}$ are random variables with uniform distribution in the interval $[0, \tau_{\max}]$, where parameter $\tau_{\max}$ is the normalized maximum offset. Note, that $\tau_i^{(\mu)}$ is given as a relative value regarding the symbol duration T, which will be important for the CP-OFDM case. The relative Doppler shift $\nu_i^{(\mu)}$ are also uniformly distributed within the interval $[-\nu_{\max}, \nu_{\max}]$ with $\Delta\nu_{\text{B}} = -\Delta\nu_{\text{A}}$ [30], where parameter $\nu_{\max}$ is the normalized maximum, again. At the relay, the full knowledge of the channels as well as the offsets and the Doppler and delay spread is assumed.

Table 20.1 Multi-carrier configuration

	Waveform	CP with T_{CP}	Equalizer
Plain—OFDM	Rectangular	No cyclic prefix	LE & sIC
CP—OFDM	Rectangular	$0T$, $\approx 0.1T$, $0.2T$, or $0.3T$	1-tap Equalizer
GFDM	Gaussian, $\alpha = 1$	No cyclic prefix	LE & sIC

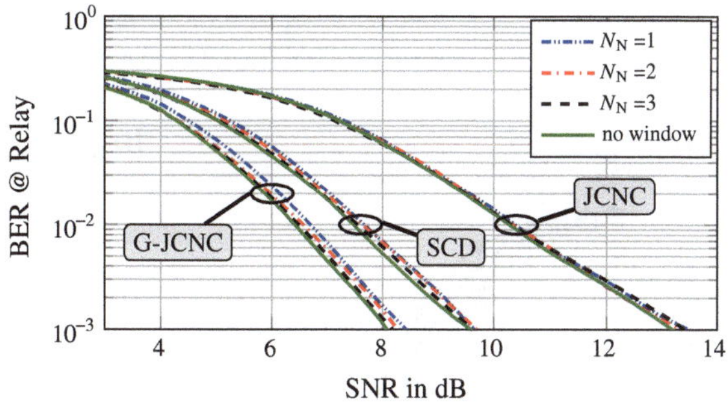

Fig. 20.7 BER performance at the relay of SCD, JCNC and G-JCNC using Gaussian transmit and receive filter, including the effect of delay and Doppler spread

20.3.3.1 Influences of the Number of Neighbors N_N for the Reduced LE

Here the performance of the LE regarding different number of neighbors N_N in (20.23) is investigated for the different PLNC detection techniques introduced in Sect. 20.2.4. Figure 20.7 shows the Bit Error Rate (BER) performance regarding the XORed bits. Here, delay and Doppler spreads are limited to $\nu_{max} = 0.08$ and $\tau_{max} = 0.08$. It can be observed that G-JCNC outperforms the different relaying strategies. The BER performance regarding different N_N is rather small. Also the performance of the reduced scheme in comparison to the case of using no window achieves approximately the same performance. This motivates to use only a small window size in the sequel with a lower complexity, while achieving approximate the full performance.

20.3.3.2 Investigations for Frequency Selective Channels

Figure 20.8 shows the BER performance at the relay regarding different delay spreads exemplary shown with G-JCNC as an example. In Fig. 20.8a the multi-carrier schemes are compared in a frequency flat fading channel $N_h = 1$, showing only the influences of the waveforms. Thus, $\overline{\boldsymbol{\Lambda}}_i$ has only zero elements in case of OFDM, whereas the Gaussian waveform will have interference terms $\overline{\boldsymbol{\Lambda}}_i \neq \mathbf{0}$. As expected, the best performance of the orthogonal schemes is achieved by OFDM with no CP, because the channel introduces no ISI/ICI and a simple one-tap equalizer is sufficient to equalize the channel impact. Larger CPs lead to an SNR loss $\gamma_{CP} = 10\log_{10}(T/(T + T_{CP}))$ in comparison to the plain-OFDM, which are $\gamma_{CP} \approx -0.39$ dB, -0.75 dB, and -1.18 dB for the corresponding CP lengths. The plain-OFDM applying LE also achieves the same performance, but at higher costs, due to an equalizer with higher complexity. By using sIC the performance is not improved, because no interference was introduced by the channel. The GFDM scheme with Gaussian waveform has a worse performance due to the additional interference. However, by utilizing the sIC structure with $\kappa = 3$ iterations, the system gains in performance and it is similar to CP-OFDM with a CP of length $T_{CP} = 0.3T$.

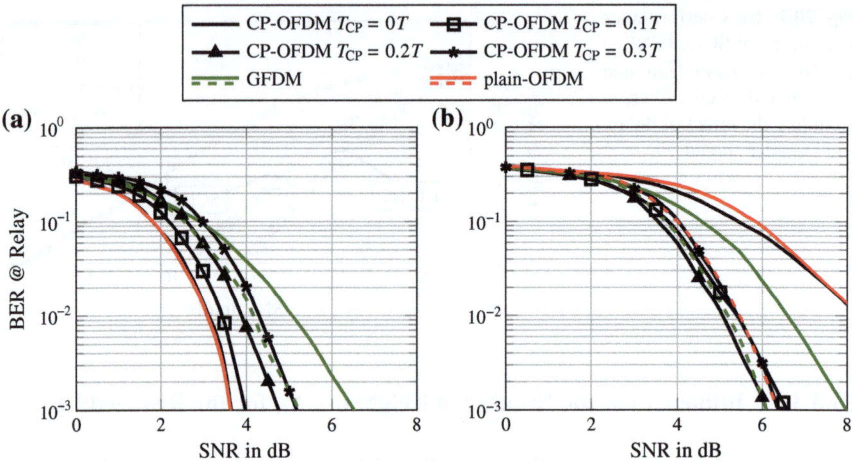

Fig. 20.8 BER performance at the relay regarding the different multi-carrier schemes. Here only delay spreads are considered. Results are given for the linear equalizer (—) and sIC with $\kappa = 3$ iterations (- - -). Exemplary: G-JCNC is used as PLNC detection scheme. **a** One tap channel. **b** Delay spread $\tau_{max} = 0.2$

Figure 20.8b shows the performance in a frequency selective channel ($\tau_{max} = 0.2$) introducing ISI. The performance of CP-OFDM with no CP has a severe degradation due to the impact of ISI. The CP-OFDM with CP length of $T_{CP} = 0.2T$ can deal with this interference, because the maximum delay does not exceed the CP length. For a CP of $T_{CP} = 0.3T$ and $T_{CP} = 0.1T$ the performance is slightly worse, due to the SNR loss of the CP and the violation the CP length, respectively. The schemes using sIC will achieve approximately the same performance. In the sequel, we will always use the CP-OFDM with a CP length of $T_{CP} = 0.2T$ as benchmark.

20.3.3.3 Analysis of Different Number of Iterations and Additional CFOs

In contrast to the previous scenario, Fig. 20.9 shows the performance with additional CFO $\Delta\nu_A = 0.2$ and $\Delta\nu_B = -0.2$ introduced at the transmitter. As stated in [15, 30] an average CFO compensation is always possible and will achieve a good performance. Therefore, we only focus on symmetric CFOs, i.e., $|\Delta\nu_A| = |\Delta\nu_B|$. Using the LE (i.e., no iteration $\kappa = 0$), the CFOs cause a huge performance loss in the BER of CP-OFDM even by applying a large CP. Here the GFDM outperforms OFDM. This effect comes by the rectangular waveform and its large spreading over the frame as illustrated in Sect. 20.2.3. By applying $\kappa = 1$ iteration, the performance achieves a huge gain in both multi-carrier schemes. Further iterations only achieves minor gains. In particular GFDM always outperforms the OFDM schemes.

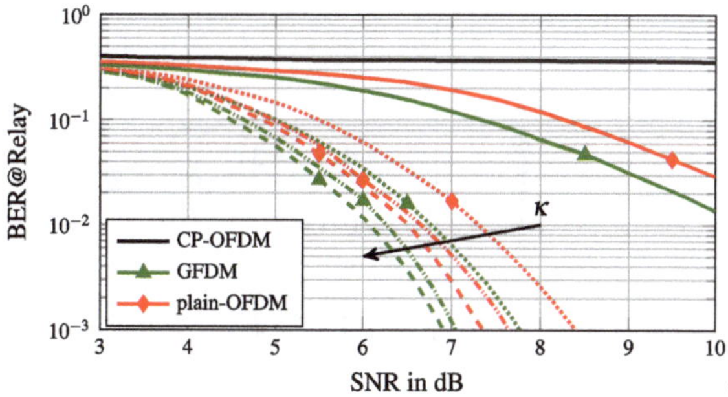

Fig. 20.9 Investigations on iterations using channels with high additional CFOs of $\Delta\nu_A = 0.2$ and $\Delta\nu_B = -0.2$ with LE(—), $\kappa = 1$ iteration (.....), $\kappa = 2$ iterations (– ·· – ·· -), or $\kappa = 3$ iterations (- - -)

20.3.3.4 Robustness in Doubly Dispersive Channels

The channel parameters for Fig. 20.10 include a general Doppler spread of maximum $\nu_{\max} = \pm 0.2$ and a delay spread of $\tau_{\max} = 0.2$. Here, no additional CFOs are assumed. In this figure, the performance results of SCD are also shown. Regarding the LE, G-JCNC outperforms SCD by approx. 1dB at BER of 10^{-2} due to exploiting the full channel coding gain in the detection process. The gains of G-JCNC to SCD are around 0.5 dB at BER of 10^{-2} using sIC. In this scenario, the BERs also increase considerably for CP-OFDM. We can observe that applying sIC is more robust against spreading in time and frequency. Especially, the multi-carrier scheme GFDM with its good localized Gaussian waveform outperforms OFDM applying rectangular waveform.

Further investigation regarding the robustness against different delay $\tau_{\max}$ and Doppler spreads $\nu_{\max}$ are shown in Fig. 20.11 for the G-JCNC scheme. Here the BER performance at a fixed noise variance of SNR = 5dB is considered. For OFDM with no CP in Fig. 20.11a with a one-tap equalizer the performance is highly sensitive against offsets in time and frequency direction. Figure 20.11b shows the improvement by using a CP (here $0.3T$), which increases the robustness against ISI. The system is clearly more robust against time spreads, but if the time spread τ_{max} exceeds the CP T_{CP}, huge degradations in the BER performance are achieved, due to the simple equalizer structure. The improvement by an iterative equalizer in OFDM with sIC is illustrated in Fig. 20.11c. This scheme is more robust against delay spreads, but has a severe influence with the presence of Doppler shifts, due to the large spreading of the rectangular filter in frequency domain. GFDM applying Gaussian waveforms with good localization properties is given in Fig. 20.11d. It is more robust with spreads in time and frequency direction. This could be expected due to the isotropic property of the Gaussian waveform. Thus, GFDM with Gaussian waveforms is well suited for TWRC using PLNC and outperforms the other schemes under practical constraints.

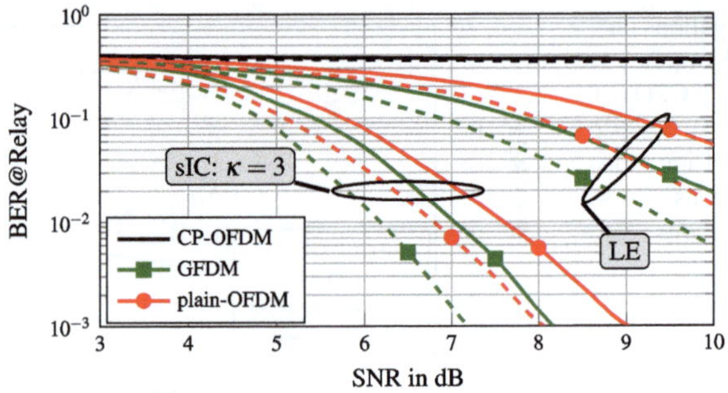

Fig. 20.10 Influences of Doppler and delay spread. $\tau_{\text{max}} = 0.2$ and $\nu_{\text{max}} = \pm 0.2$ and a delay spread $\tau_{\text{max}} = 0.2$. Results are depicted for G-JCNC (- - -) and SCD (—)

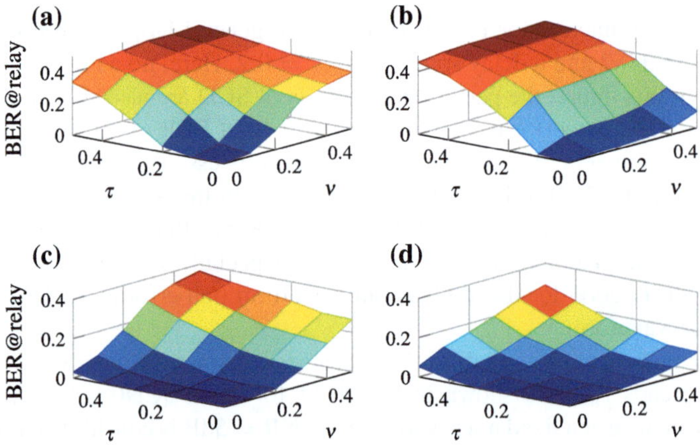

Fig. 20.11 BER performance at the relay at a noise variance SNR = 5 dB using G-JCNC detection with different τ_{max} and ν_{max} **a** CP-OFDM, $T_{\text{CP}} = 0T$, **b** CP-OFDM, $T_{\text{CP}} = 0.3T$, **c** plain-OFDM, $k = 3$, **d** GFDM, $k = 3$

20.4 Mutual Information

To evaluate the performance of the transmission system meaningful QIFs are required. So far, we have presented BER results, only. However, the achievable mutual information is also well suited to evaluate the quality of the receive signal at the relay. It was shown in [18, 29] that mutual information is a good measure to approximate the decoder performance in a practical TWRC. Therefore, we will present mutual information based results for dispersive channels in this section.

Within this section the GFDM system with Gaussian waveform is compared with a CP-OFDM system in the TWRC, where the individual channels $H_{\text{A}}(\tau, \nu)$, $H_{\text{B}}(\tau, \nu)$ are modeled by the COST207-Typical-Urban channel model [4]. The duration of the OFDM symbol was chosen to be $T = 71.3\,\mu\text{s}\,(= 1/F)$, where guard interval is $T_G = 4.7\,\mu s$ ($\approx 6.5\,\%$). The guard was chosen such that the CP-OFDM system can handle time dispersion of the COST207 channel model. For a fair comparison the sampling rate and sub-carrier spacing of the basic GFDM system were chosen such that the same factor TF is achieved as in CP-OFDM. The TOs $\Delta\tau_i$ are random variables with uniform distribution in interval $[0, \Delta\tau_{max}]$, where parameter $\Delta\tau_{max}$ is the normalized maximum offset. The CFOs $\Delta\nu_i$ are also uniformly distributed within the interval $[-\Delta\nu_{max}, \Delta\nu_{max}]$ with $\Delta\nu_B = -\Delta\nu_A$ [30], where parameter $\Delta\nu_{max}$ is the normalized maximum, again.

20.4.1 Performance Without Equalizer

Figure 20.12 compares the mutual information of CP-OFDM and GFDM systems without equalizer for different relaying strategies according to definitions (20.18), (20.20), and (20.21). The left plot shows the achievable mutual information without synchronization offsets, where the other two plots visualize the influence of disturbances. In general, SCD and JCNC perform similarly in the low SNR region, where the system performance is limited by the noise. Compared to SCD, JCNC with Gaussian input alphabets is advantageous in the high SNR region as the performance does not suffer from the interference caused by the second user. The G-JCNC is superior in the whole SNR range.

Due to the non-orthogonal waveform, the performance of GFDM without equalizer (dashed lines) is considerably worse than the CP-OFDM system (solid lines), as ISI as well as ICI from neighboring lattice points disturb the receive signal at the relay. In the high SNR region, the GFDM results are thus limited by the Signal to Interference Ratio (SIR). On the other hand, GFDM results are almost independent of the introduced synchronization offsets, i.e., the system is very robust against additional disturbances. In contrast, OFDM is not interference limited, in general. If the system is synchronized perfectly, the achievable mutual information is much larger.

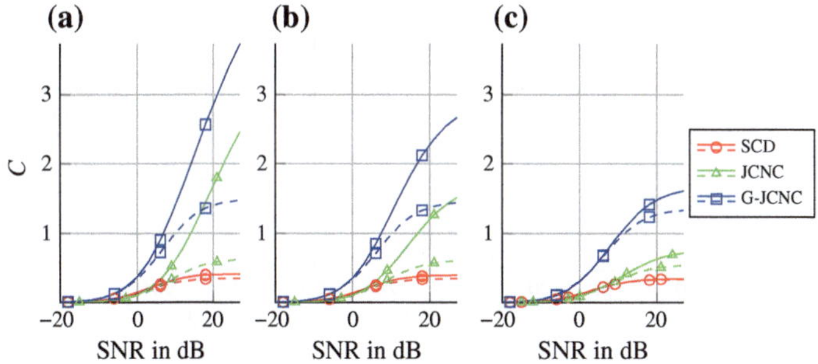

Fig. 20.12 Average mutual information of the MA phase in the TWRC for COST207-Typical-Urban channel model. CP-OFDM (—) and GFDM **without** equalizer (- - -) are compared. **a** $\tau_{max} = \Delta\nu_{max} = 0$. **b** $\tau_{max} = \Delta\nu_{max} = 0.1$. **c** $\tau_{max} = \Delta\nu_{max} = 0.2$

However, if the system is disturbed by synchronization offsets, the performance degrades considerably.

Another point, not shown in the figures, is that the degradation of the CP-OFDM system differs significantly for synchronization offsets in time or frequency. As long as the TO can be compensated by the guard interval, there are no losses in the CP-OFDM system. If the TO is longer than the guard interval, there is a severe degradation. However, these losses are nonetheless smaller than the gain of the Gaussian system due to the increased robustness. If the increased robustness of the GFDM system without equalizer has to pay off, considerable disturbances are required.

20.4.2 Performance for Unknown Synchronization Offsets

The basic motivation to use Gaussian waveforms was to design a system that is robust towards synchronization offsets. To analyze the robustness of the system we will consider the offsets to be unknown at the relay, i.e., the relay cannot adjust the equalizer to TOs and CFOs. Figure 20.13 shows the mutual information of the MA phase for COST207-TU channel model assuming that synchronization offsets are not known at the relay. The equalizer used in the GFDM system is a MMSE equalizer with $N_{\mathrm{N}} = 2$.

Even if only a small number of neighboring lattice points is considered, the interference caused by Gaussian waveform can be eliminated efficiently as significant interference terms are covered by window the $\mathcal{W}_{N_{\mathrm{N}}}^{(k',\ell')}$ with $N_{\mathrm{N}} = 2$. Compared to the results without equalizer in Sect. 20.4.1, GFDM is no longer limited by interference caused by the non-orthogonal waveform. Figure 20.13 shows also that GFDM with equalizer is more sensitive to synchronization errors. However, the gain of the equalizer is larger than the loss due to synchronization offsets. Furthermore, sensitivity to offsets is still lower than in the OFDM system. Nonetheless, TOs and

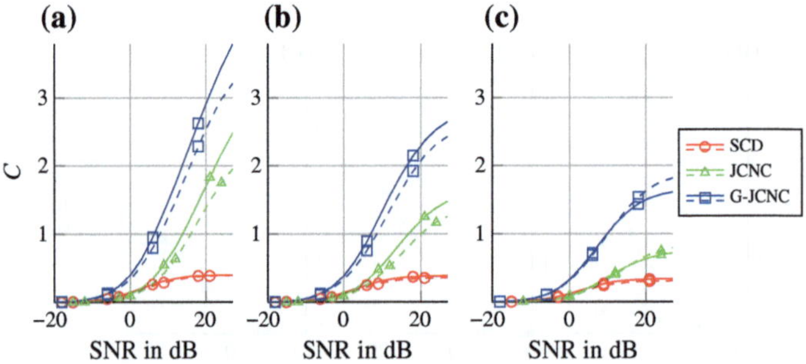

Fig. 20.13 Average mutual information of the MA phase in the TWRC for COST207-Typical-Urban channel model. CP-OFDM (—) and GFDM **with** MMSE equalizer (- - -) are compared assuming that synchronization offsets are unknown at the relay. **a** $\tau_{\max} = \Delta\nu_{\max} = 0$. **b** $\tau_{\max} = \Delta\nu_{\max} = 0.1$. **c** $\tau_{\max} = \Delta\nu_{\max} = 0.2$

CFOs degrade achievable mutual information considerably and synchronization offsets have to be rather large (>20 %) to allow GFDM a gain over CP-OFDM. Note that the degradation is caused by the fact that TOs and CFOs are assumed not to be known at the relay. The computed equalizer coefficients are therefore not optimal.

20.4.3 Influence of Lattice Spacing

The results in the previous section indicate that the application of GFDM in the TWRC to improve the robustness regarding synchronization offsets hardly pays off if linear equalizers are used. However, GFDM systems have an additional advantage that has not been exploited so far. Contrary to OFDM systems, where sampling positions are fixed by the roots of the transmit filter, lattice spacing is not necessarily fixed in GFDM. The choice of the lattice spacing is a trade-off [23].

Let Δ be the relative lattice spacing, such that symbol duration T and sub-carrier spacing F of GFDM are equivalent to the CP-OFDM system for $\Delta = 1$. ($\Delta = 1$ is also equivalent to the setup in the previous section). Spacings $\Delta < 1$ correspond to shrinking the time-frequency-grid, i.e., increasing spectral efficiency, and $\Delta > 1$ correspond to stretching, i.e., decreasing spectral efficiency. Note that Δ does not change the shape of the waveform, but the position of the lattice points in the time-frequency grid is modified. Figure 20.14 depicts the ambiguity function of the Gaussian waveform at lattice point $\ast$ as well as the positions of neighboring lattice points $+$ in a time-frequency grid for different lattice spacings Δ. Employing the Gaussian waveforms does always cause interference between the lattice points. Thus, improving spectral efficiency by shrinking the lattice comes at the expense of an increased interference level. Altering T or F in an OFDM system requires a redesign of the filters in order to maintain orthogonality of the waveforms. Furthermore, increasing spectral efficiency by decreasing T as well as F would cause a loss in orthogonality and severe interference occurs due to the slow decay of the sinc-

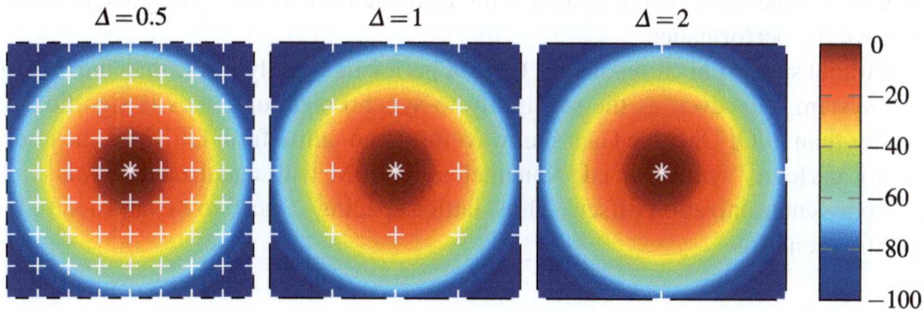

Fig. 20.14 Ambiguity function of the Gaussian waveform versus time and frequency, where white crosses correspond to the position of neighboring lattice points

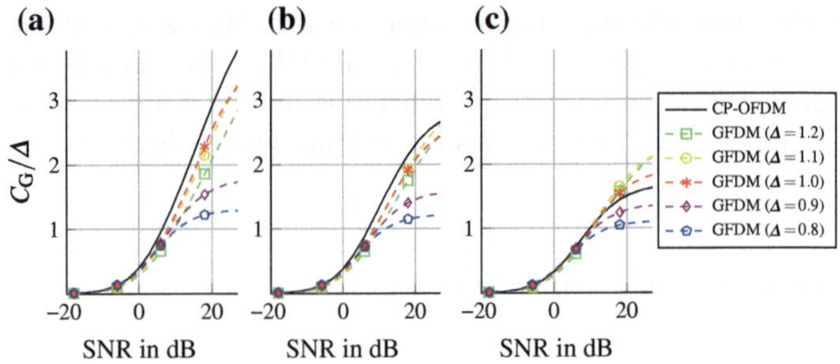

Fig. 20.15 Normalized average mutual information of G-JCNC for COST207-Typical-Urban channel model. CP-OFDM and GFDM **with** MMSE equalizer are plotted for different lattice spacings $\varDelta$ assuming that synchronization offsets are unknown at the relay. **a** $\tau_{\max} = \varDelta\nu_{\max} = 0$. **b** $\tau_{\max} = \varDelta\nu_{\max} = 0.1$. **c** $\tau_{\max} = \varDelta\nu_{\max} = 0.2$

function. Contrary, significant interference is still limited to adjacent lattice points in the GFDM system as long as $\varDelta$ does not become too small.

Figure 20.15 shows the normalized mutual information **with** MMSE equalizer in the GFDM system for different lattice spacings $\varDelta = \{0.8, \ldots, 1.2\}$ as well as the mutual information of the OFDM system. Note that the transmit energy need to be normalized for the different lattice spacings. Again, synchronization offsets are assumed to be unknown at the relay. The performance of the GFDM system can be improved significantly by adjusting lattice spacing $\varDelta$, where the optimal choice of $\varDelta$ depends on the actual SNR. However, the achievable improvements by modifying $\varDelta$ are limited, as the choice of $\varDelta$ is a trade-off between spectral efficiency and interference from neighboring lattice points. On the one hand, spectral efficiency increases if the lattice spacing decreases. On the other hand, the amount of interference to neighboring lattice points increases as well. In the low SNR region, the GFDM performance is limited by the noise rather than the interference, i.e., shrinking the time-frequency-grid can improve system performance. Here, additional interference is irrelevant compared to the noise. In the high SNR region, the reverse strategy is beneficial. As the system is interference limited, increasing distances can improve the performance.

Without synchronization offsets, GFDM performs slightly worse than the orthogonal system. However, as the sensitivity to synchronization errors is significantly smaller than in OFDM, the Gaussian waveform can outperform OFDM if the nodes are not perfectly synchronized. Nonetheless, GFDM with equalizer does also suffer from synchronization offsets. Thus, lattice spacing $\varDelta$ need to be adjusted to the specific scenario.

20.5 Conclusions

In this chapter, Generalized Frequency Division Multiplexing (GFDM) systems with Gaussian filters are applied to the Two-Way-Relay Channel (TWRC) with PLNC to reduce the impact of time and frequency impairments. In general, the BER performance as well as achievable mutual information for Gaussian input alphabets was evaluated for three different detection/decoding schemes when imperfect synchronization leads to time and frequency offsets. To treat the interference caused by the Gaussian transmit and receive filters, a MMSE equalizer has been added to the GFDM system. In order to reduce the remaining interference after linear equalization further, we developed an iterative TWRC detection scheme. The performance results indicate, that the GFDM system with Gaussian waveforms is more robust against offsets than OFDM with rectangular waveforms. Furthermore, it was shown that the performance of GFDM can be improved further if the lattice spacing is adjusted dynamically.

Acknowledgments This work was supported in part by the German Research Foundation (DFG) under Grants KU 1221/18-1 and WU 499/10-1 within the priority program "Communication in Interference Limited Networks (COIN)", SPP 1397.

References

1. Bhat U, Duman T (2012) Decoding strategies for physical-layer network coding over frequency selective channels. In: Wireless communications and networking conference (WCNC), IEEE. France, Paris
2. Dohler M, Li Y (2010) Cooperative communications: hardware. Channel & PHY, Wiley, Chichester, U.K
3. Du J (2008) Pulse shape adaptation and channel estimation in generalised frequency division multiplexing systems. Licentiate thesis in electronics and computer systems, KTH, Stockholm, Sweden
4. Failli M (1989) Digital land mobile radio communications COST 207. Technical report, Commission of the European comunities
5. Farhang-Boroujeny B (2011) OFDM versus filter bank multicarrier. IEEE Signal Process Mag **28**(3):92–112. doi:10.1109/MSP.2011.940267
6. Fettweis G, Krondorf M, Bittner S (2009) GFDM–generalized frequency division multiplexing. In: IEEE 69th vehicular technology conference, (VTC Spring 2009), Barcelona, Spain. doi:10.1109/VETECS.2009.5073571
7. Fitzek FH, Pedersen MV, Heide J, Médard M (2010) Network coding: applications and implementations on mobile devices. In: 5th ACM workshop on performance monitoring and measurement of heterogeneous wireless and wired networks (PM2HW2N '10), ACM, New York, NY, USA. doi:10.1145/1868612.1868628
8. Han B, Wang W, Peng M (2012) Optimal resource allocation for network coding in multiple two-way relay OFDM systems. In: IEEE wireless communications and networking conference (WCNC 12), Paris, France
9. Ho CK, Zhang R, Liang YC (2008) Two-way relaying over OFDM: optimized tone permutation and power allocation. In: IEEE international conference on communications (ICC'08), Bejing, China. doi:10.1109/ICC.2008.734

10. Jitvanichphaibool K, Zhang R, Liang YC (2008) Optimal resource allocation for two-way relay-assisted OFDMA. In: IEEE global telecommunications conference (GLOBECOM 2008), New Orleans, LO, USA. doi:10.1109/GLOCOM.2008.ECP.791
11. Kim YH, Song I, Kim HG, Chang T, Kim HM (1999) Performance analysis of a coded OFDM system in time-varying multipath rayleigh fading channels. IEEE Trans Veh Technol **48**(5):1610–1615. doi:10.1109/25.790538
12. Kozek W, Molisch A (1998) Nonorthogonal pulseshapes for multicarrier communications in doubly dispersive channels. IEEE J Sel Areas Commun **16**(8). doi:10.1109/49.730463
13. Laneman J, Tse D, Wornell GW (2004) Cooperative diversity in wireless networks: efficient protocols and outage behavior. IEEE Trans Inf Theory **50**(12):3062–3080. doi:10.1109/TIT.2004.838089
14. Li F, Zhu G, Wang D (2011) Optimal power allocation for two-way relaying over OFDM using physical-layer network coding. J China Univ Posts Telecommun **18**(1):9–15. doi:10.1016/S1005-8885(10)60021-0
15. Lu L, Wang T, Liew SC, Zhang S (2013) Implementation of physical-layer network coding. Phys Commun 6. doi:10.1016/j.phycom.2012.02.008
16. Matheus K, Knoche K, Feuersanger M, Kammeyer KD (1998) Two-dimensional (recursive) channel equalization for multicarrier systems with soft impulse shaping (MCSIS). In: Global telecommunications conference (GLOBECOM 1998), vol 2. doi:10.1109/GLOCOM.1998.776871
17. Pabst R, Walke BH, Schultz D, Herhold P, Yanikomeroglu H, Mukherjee S, Viswanathan H, Lott M, Zirwas W, Dohler M, Aghvami H, Falconer D, Fettweis G (2004) Relay-based deployment concepts for wireless and mobile broadband radio. IEEE Commun Mag **42**(9):80–89. doi:10.1109/MCOM.2004.1336724
18. Pfletschinger S (2011) A practical physical-layer network coding scheme for the uplink of the two-way relay channel. In: Conference record of the forty fifth asilomar conference on signals, systems and computers (ASILOMAR). doi:10.1109/ACSSC.2011.6190375
19. Pollet T, Van Bladel M, Moeneclaey M (1995) BER sensitivity of OFDM systems to carrier frequency offset and wiener phase noise. IEEE Trans Commun **43**(2):191–193. doi:10.1109/26.380034
20. Popovski P, Yomo H (2007) Physical network coding in two-way wireless relay channels. In: IEEE international conference on communications (ICC'07), Glasgow, UK. doi:10.1109/ICC.2007.121
21. Rankov B, Wittneben A (2006) Achievable rate regions for the two-way relay channel. In: Information theory, 2006 IEEE international symposium on, pp 1668–1672. doi:10.1109/ISIT.2006.261638
22. Sahin A, Guvenc I, Arslan H (2014) A survey on multicarrier communications: prototype filters, lattice structures, and implementation aspects. IEEE Commun Surveys Tuts **16**(3):1312–1338. doi:10.1109/SURV.2013.121213.00263
23. Schedler S, Kühn V (2015) Influence of lattice spacing in disturbed generalized frequency division multiplexing systems. In: IEEE 81th vehicular technology conference (VTC2015-Spring), vol 81. Glasgow, Great Britain
24. Schedler S, Woltering M, Wübben D, Kühn V, Dekorsy A (2014) Investigation on gaussian waveforms to improve robustness in physical layer network coding. In: 18th international OFDM workshop 2014 (InOWo'14), Essen, Germany
25. Shin H, Lee J (2011) Joint resource allocation for multiuser two-way OFDMA relay networks with proportional fairness. In: IEEE vehicular technology conference (VTC Fall), San Francisco, CA, USA
26. Vakilian V, Wild T, Schaich F, ten Brink S (2013) Universal-filtered multi-carrier technique for wireless systems beyond LTE. In: IEEE global telecommunications conference (GLOBECOM), Atlanta, GA, USA
27. Woltering M, Wübben D, Dekorsy A (2015a) Physical layer network coding with gaussian waveforms using soft interference cancellation. In: IEEE 81th vehicular technology conference (VTC2015-Spring), vol 81. Glasgow, Great Britain

28. Woltering M, Wübben D, Dekorsy A, Schedler S, Kühn V (2015b) Physical layer network coding using gaussian waveforms: a link level performance analysis. In: 10th international ITG conference on systems, communications and coding (SCC 2015), Hamburg, Germany
29. Wu M, Wübben D, Dekorsy A (2013) Mutual information based analysis for physical-layer network coding with optimal phase control. In: 9th international ITG conference on systems, communications and coding (SCC' 13), Munich, Germany
30. Wu M, Ludwig F, Woltering M, Wübben D, Dekorsy A, Paul S (2014) Analysis and implementation for physical-layer network coding with carrier frequency offset. In: International ITG workshop on smart antennas (WSA2014), Erlangen, Germany
31. Wübben D (2010) Joint channel decoding and physical-layer network coding in two-way QPSK Relay systems by a generalized sum-product algorithm. In: 7th international symposium on wireless communication systems (ISWCS 2010), York, UK
32. Wübben D, Lang Y (2010) Generalized sum-product algorithm for joint channel decoding and physical-layer network coding in two-way relay systems. In (2010) IEEE global telecommunications conference (GLOBECOM). IEEE, Miami, Fl, USA. doi:10.1109/GLOCOM.2010.5683819
33. Wübben D, Wu M, Dekorsy A (2014) Physical-layer network coding with multiple-antenna relays. In: MIMO processing for 4G and beyond: fundamentals and evolution, CRC Press
34. Wunder G, Jung P, Kasparick M, Wild T, Schaich F, Chen Y, ten Brink S, Gaspar I, Michailow N, Festag A, Mendes L, Cassiau N, Ktnas D, Dryjanski M, Pietrzyk S, Eged B, Vago P, Wiedmann F (2014) 5GNOW: non-orthogonal, asynchronous waveforms for future mobile applications. IEEE Commun Mag **52**(2). doi:10.1109/MCOM.2014.6736758
35. Zhang S, Liew S, Lam P (2006a) Hot topic: physical-layer network coding. In: 12th annual international conference on mobile computing and networking (MobiCom 06), Los Angeles, CA, USA
36. Zhang S, Liew S, Lam P (2006b) On the synchronization of physical-layer network coding. In: IEEE information theory workshop (ITW'06), Chengdu, China, pp 404–408. doi:10.1109/ITW2.2006.323830
37. Zhang S, Liew SC, Wang H, Lin X (2009) Capacity of two-way relay channel. access networks, 4th international conference, accessnets (2009) Hong Kong, China, 1–3 Nov 2009. Revised selected papers, Springer, pp 219–231

Chapter 21
Linear Multi-Cell Precoding for Throughput Optimization Considering Outage

Xin Zhang, Richard Fritzsche, Andreas Festag and Gerhard Fettweis

Abstract Linear precoding provides the potential to boost the data throughput in a cellular multiple-input-multiple-output downlink system. It exploits spatial diversity and enables multiple users to be served on the same time-frequency resources. In a previous study we presented the maximization of the weighted sum data rate in multi-cell networks considering a transmit power constraint per base station with imperfect channel state information (CSI) at the transmitter. Beside imperfections in the precoding matrix design, impaired CSI can also cause outage if the actual channel does not support the data rate assigned for transmission. As the new 5th generation wireless communication era is dawning, future cellular networks are expected to be more resilient. In this chapter, we investigate the aspect of potential outages with respect to the joint optimization of the precoding matrix and the transmission rate. In this context, we target at the maximization of throughput considering both imperfect CSI at the transmitter and potential outages. We quantify the performance by means of Monte Carlo simulations.

21.1 Introduction

To support the high demands on ubiquitous and data intensive applications, an increasing amount of information needs to be exchanged every day. A significant improvement in the system capacity is one of the main targets for future cellular

X. Zhang (✉) · R. Fritzsche · A. Festag · G. Fettweis
Vodafone Chair Mobile Communications Systems,
Technische Universität Dresden, München, Germany
e-mail: xin.zhang@ifn.et.tu-dresden.de

R. Fritzsche
e-mail: Richard.Fritzsche@tu-dresden.de

A. Festag
e-mail: Andreas.Festag@tu-dresden.de

G. Fettweis
e-mail: Gerhard.Fettweis@vodafone-chair.com

W. Utschick (ed.), *Communications in Interference Limited Networks*,
Signals and Communication Technology, DOI 10.1007/978-3-319-22440-4_21

communication systems [35]. One way to achieve this goal is to increase the spectral efficiency by employing multi-antenna systems, such as Multi-Input-Multi-Output (MIMO), which exploits spatial diversity by deploying multiple antennas at the Base Station (BS) and User Equipment (UE) [25]. The 4th generation of cellular networks makes use of multi-user MIMO techniques, where the BS serves multiple UEs on the same radio resource [16]. In the future cellular communication systems, interference between cells is one of the major limiting factors. To handle inter-cell interference, the antennas of multiple BSs can be combined to form a virtual super BS. Such cooperative provisioning of data transmission from multiple BSs to multiple UEs is typically referred to as Coordinated Multi-Point (CoMP) [22, 37]. CoMP allows the mobile users to experience a consistent performance and quality whether they are close to the center of a cell or at its outer edges [19]. In this regard, using the technique in the downlink, data of multiple users is jointly precoded, e.g., at a Central Node (CN) and transmitted from the cooperating BSs in a coherent fashion.

In general, precoding can be classified into two main types, namely, linear and non-linear. With non-linear dirty paper coding [6], the capacity of the multi-user MIMO broadcast channel can be achieved in theory [3]. However, practical implementations, such as Tomlinson-Harashima precoding [17, 26], require much higher computational complexity compared to linear precoding approaches. Due to its low complexity, linear precoding methods are considered in this work.

Precoding requires that Channel State Information (CSI) of all jointly served UEs being available at the BS. Several of the existing studies in this area are based on the assumption that CSI is precisely available at the transmitter side, see [21, 30, 34] for single-cell and [31] for multi-cell scenarios. However, substantial performance degradation can be expected if CSI is only imperfectly available at the BS side [20]. The imperfection can be caused by noisy pilot reception, quantization noise and delays for feedback transmission.

One option to combat performance losses is to increase the accuracy of CSI, e.g., by optimizing the feedback link from the UE to the BS [4, 13]. In setups with cooperative BSs, the sharing of CSI among the BSs is crucial. This is of special interest when the precoding is performed in a distributed manner [9]. In this case different backhaul delays need to be compensated [14, 15].

When imperfections in the channel information are present, robust precoding techniques give the opportunity to mitigate performance losses by integrating knowledge of the error statistics into the precoding design [7, 29]. Robust precoding for the multi-cell broadcast channel have been addressed in [10, 33].

In order to design the precoding matrix, different optimization functions have been considered. While the previously mentioned studies focused on Mean Square Error (MSE) related objectives, the maximization of the achievable data rate is of special interest in order to evaluate the potential of the communication link. The method in [36] addressed the maximization of the Weighted Sum Rate (WSR) for perfect CSI by introducing an additional zero forcing constraint. The authors in [5] discovered that the WSR maximization problem can be solved by means of the weighted sum MSE minimization. The proposed solution was intended for single BS setups. Precoding for cooperating BSs have been addressed by [23], where the transmit power is restricted per group of antennas. A solution for the WSR maximization, which is robust against impairments in the CSI, has been presented in [27]. An extension was given in [12], where the solution took into account that the CSI error variance can be different for each BS-UE link. Further extensions to distributed precoding were addressed in [11, 24].

Beside misalignments in the precoding matrix, an imperfect CSI also causes uncertainty in the assignment of the transmission rate. Since the actual achievable rate is not precisely known at the BS, the transmission rate might not be supported by the channel, resulting in potential outages [32]. For optimizing the throughput of a communication link, the outage probability of the transmitted data block need to be taken into account. In multi-antenna systems the analytical assessment of the outage probability is a complex task. Algorithms for single-user MIMO have been presented in [28], while for multi-user MIMO systems no solution is known.

In this work, we address the problem of throughput maximization for the multi-cell broadcast channel, considering that transmissions can be in outage. For this purpose the objective targets at the joint optimization of the precoding matrix and the transmission rate. In order to tackle the complex problem of joint optimization, both—the precoding matrix and the transmission rate—are optimized separately in an alternating fashion. The lack of a closed form expression for the outage probability is overcome by making use of the concentration inequality [18] to translate the problem into a WSR maximization problem with adapted weighting factors.

The remainder of this chapter is organized as follows: The system model and problem formulation will be introduced in Sect. 21.2. The proposed alternating optimization algorithm will be presented in Sect. 21.3, including an analysis of the relation between the outage probability and the assigned data rate. A performance evaluation of the proposed algorithm and a comparison with reference algorithm will be provided in Sect. 21.4. Finally, conclusions will be drawn in Sect. 21.5.

Notation: Conjugate, transposition and conjugate transposition is denoted with $(\cdot)^*$, $(\cdot)^T$ and $(\cdot)^H$, respectively. The trace of a matrix is written as $\mathrm{tr}(\cdot)$, $\det(\cdot)$ denotes determinant. The operator $\mathrm{dg}(\cdot)$ replaces each non-diagonal element of a matrix with zero, while $\mathrm{diag}(\cdot)$ places the elements of a vector on the diagonal of a matrix. Similarly, $\mathrm{blkdiag}(\cdot)$ puts matrices on the diagonal of a block diagonal matrix. Expectation is denoted with $\mathbb{E}\{\cdot\}$ and probability by $\mathbb{P}\{\cdot\}$. $\mathbb{C}$ donates the set of complex numbers and $\mathcal{N}_{\mathbb{C}}(\mathbf{m}, \mathbf{C})$ refers to a multi-variate complex normal distribution with mean vector $\mathbf{m}$ and covariance matrix $\mathbf{C}$.

21.2 System Model and Problem Formulation

In this section, we first introduce a macroscopic picture of the considered cooperative cellular system and the involved entities. We then zoom into a more detailed view of the data transmission in a cooperative multiuser-MIMO system, including linear joint precoding of multiple BSs. We further model the effect of imperfections in the channel knowledge and define metrics for performance evaluation, including achievable data rate, outage probability and throughput. Finally, we specify the optimization problem considered in this work.

21.2.1 System Setup

Figure 21.1 illustrates a high level view of a cellular system setup for cooperative multi-cell transmission in the downlink.

The cooperation cluster is comprised of M collaborating BSs, which jointly serve K UEs on the same radio resource. In this work, all BSs are located in a single dimensional (1-D) setup. The BSs are connected to a CN and exchange user data and CSI over a backhaul network. Outside the cooperation cluster M_{OC} BSs are assumed to be located symmetrically w.r.t. the 1-D setup of the cooperation cluster. Those BSs generate interference to the users located within the cooperation cluster, as illustrated in Fig. 21.1.

The mean gain of the channel between BS m and UE k is given as

$$\lambda_{k,m} = \beta d_{k,m}^{-\alpha}, \tag{21.1}$$

where α is the path loss exponent, $d_{k,m}$ denotes the distance between UE k and BS m, and β is the adjusting coefficient account for shadow fading effects.

Interference from outer-cluster BSs, also referred to as inter-cluster interference (ICI), is assumed to be Gaussian distributed noise. For the 1-D models, where all BSs are located on a straight line, the variance for inter-cluster interference received at UE k is given as

$$\sigma_{\mathrm{I},k}^{2} = \eta\gamma\beta \sum_{i=1}^{M_{OC}/2} [(id_I + d_{k,1}^{-\alpha}) + (id_I + d_{k,M})^{-\alpha}], \tag{21.2}$$

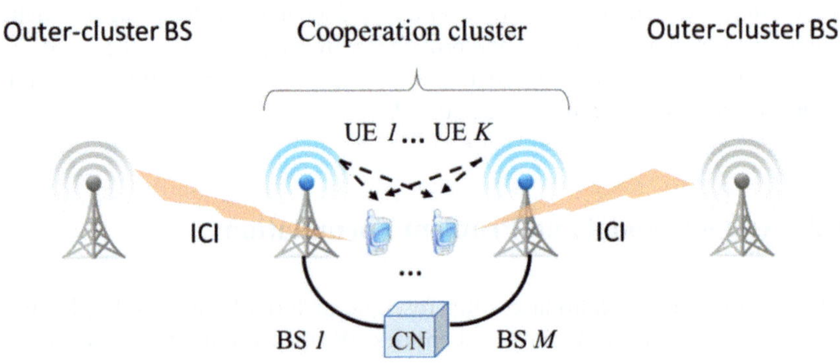

Fig. 21.1 A cellular system setup for cooperative multi-cell transmission in the downlink

where η is a parameter to indicate the isolation of a cluster, which is obtained in real 3-D dimensional setups by, e.g., antenna tilting. γ denotes the maximum transmit power at each BS and d_I indicates the inter-site distance.

21.2.2 Data Transmission

Figure 21.2 describes the network MIMO transmission from the cooperative BSs to the jointly served UEs. It is assumed that the precoding matrix is computed at the CN, based on centrally available CSI. Each BS m obtains its specific parts of the matrix as well as the data of all users via the backhaul network. The backhaul network is assumed to not applying compression to the pre-coded data. At the BSs, the user data is precoded and transmitted to the UEs. Note that, due to the unconstrained backhaul, this option is logically equivalent to precoding the data at the CN and forwarding the results to the respective BSs.

In this setup, each BS is equipped with B_m antennas, where $m = 1, \ldots, M$, and each UE is equipped with U_k antennas, where $k = 1, \ldots, K$. The data intended for user k is denoted as $\mathbf{d}_k$.

The data of all K users is combined in a single vector $\mathbf{d} = [\mathbf{d}_1^T, \ldots, \mathbf{d}_K^T]^T$, which is input to each BS. All elements of $\mathbf{d} \sim \mathcal{N}_{\mathbb{C}}(\mathbf{0}, \mathbf{I})$ are assumed to be independent and identically distributed according to a complex Gaussian distribution with zero mean and unit variance. The overall precoding is performed by multiplying the user data vector with the precoding matrix

$$\mathbf{B} = [\bar{\mathbf{B}}_1, \ldots, \bar{\mathbf{B}}_K] = [\mathbf{B}_1^T, \ldots, \mathbf{B}_M^T]^T , \tag{21.3}$$

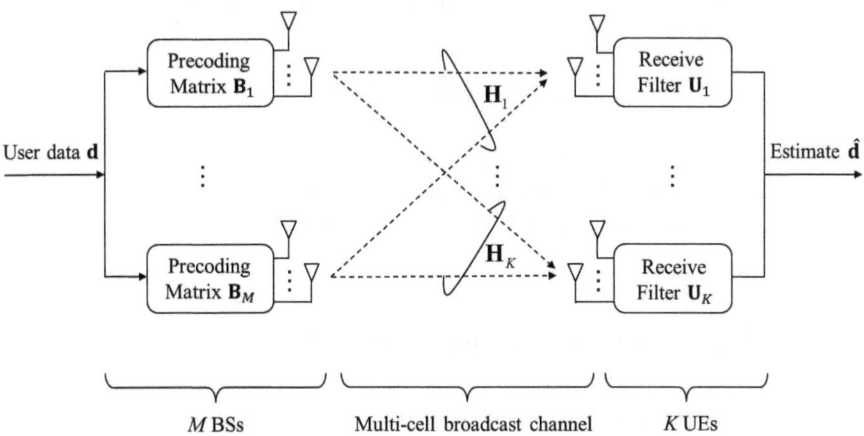

Fig. 21.2 Model of a cooperative multiuser-MIMO system

where $\mathbf{B}_m \in \mathbb{C}^{[B_m \times U]}$ denotes the portion that is applied at the mth BS. In order to meet the per-BS transmit power constraints, the precoding matrix needs to satisfy $\mathrm{tr}(\mathbf{B}_m \mathbf{B}_m^H) \leq \gamma_m, \forall m$. On the other hand, $\bar{\mathbf{B}}_k \in \mathbb{C}^{[B \times U_k]}$ denotes the portion of the precoding matrix that is applied to precode the data intended for the kth UE.

The precoded data is coherently transmitted from all M BSs to all K UEs over a multi-cell broadcast channel, which is denoted by $\mathbf{H} \in C^{[U \times B]}$, where $B = \sum_{m=1}^{M} B_m$ and $U = \sum_{k=1}^{K} U_k$ are the overall number of receive antennas at the UE and transmit antennas at the BS side, respectively. Each element in the matrix $\mathbf{H}$ corresponds to the channel state of a link between a particular transmit and receive antenna pair. In this work, block-static fading is assumed. The channel state during a transmission block is assumed to remain constant and subsequent blocks are not correlated. According to Rayleigh fading each link between BS m and UE k is complex Gaussian distributed with zero mean and variance $\lambda_{k,m}$. The channel matrix can be split up into

$$\mathbf{H} = [\mathbf{H}_1^T, \ldots, \mathbf{H}_K^T]^T , \tag{21.4}$$

where the sub-matrix $\mathbf{H}_k \in C^{[U_k \times B]}$ corresponds to the channel among all B BS antennas and the antennas of UE k (see Fig. 21.2).

The data received at the UEs is impaired by interference from outer cluster BSs and thermal noise, which are both assumed to follow a Gaussian distribution. The variance of the thermal noise is $\sigma_{\bar{n}}^2$. The total noise variance received at each UE k is $\mathbf{n}_k \sim \mathcal{N}_{\mathbb{C}}(0, \sigma_{\mathrm{n},k}^2)$ where $\sigma_{\mathrm{n},k}^2 = \sigma_{\bar{n}}^2 + \sigma_{\mathrm{I},k}^2$.

The signals, which are received at all K UEs are comprised in vector

$$\mathbf{y} = [\mathbf{y}_1^T, \ldots, \mathbf{y}_K^T]^T = \mathbf{HBd} + \mathbf{n} , \tag{21.5}$$

where $\mathbf{n} = [\mathbf{n}_1^T, \ldots, \mathbf{n}_K^T]^T$ is the overall noise vector. At each UE k the received signals are equalized by a linear receive filter $\mathbf{U}_k$. The processed symbols after equalization are combined in vector

$$\hat{\mathbf{d}} = \mathbf{U}(\mathbf{HBd} + \mathbf{n}) = \mathbf{Uy} , \tag{21.6}$$

where the matrix $\mathbf{U} = \mathrm{blkdiag}(\mathbf{U}_1, \ldots, \mathbf{U}_K)$ represents the combined receive filter. The overall vector of estimated data symbols can be obtained by stacking the equalized data symbols $\hat{\mathbf{d}}_k$ of all K UEs into a single vector.

21.2.3 Imperfect Channel State Information

For aligning the precoding matrix to the instantaneous channel state, CSI need to be available at the BS side. However, in practice the transmitted channel information can be impaired by several effects, e.g., noisy pilot reception, quantization or feedback delays for frequency division duplex systems. Based on the assumption that CSI is obtained via minimum MSE (MMSE) estimation, imperfections in the CSI can be modeled by substituting the actual channel to

$$\mathbf{H} = \hat{\mathbf{H}} + \mathbf{E}\,, \tag{21.7}$$

where $\hat{\mathbf{H}}$ denotes the CSI matrix, which is uncorrelated with the Gaussian error matrix $\mathbf{E} = [\mathbf{E}_1^T, \ldots, \mathbf{E}_K^T]^T$ with $\mathbf{E}_k = [\mathbf{E}_{k,1}, \ldots, \mathbf{E}_{k,M}]$ and $\text{vec}(\mathbf{E}_{k,m}) \sim \mathcal{N}_{\mathbb{C}}(\mathbf{0}, \varepsilon_{k,m}\mathbf{I})$. In our study, we assume the error variance $\varepsilon_{k,m}$ of a given BS-UE link is potentially different from the other links due to independent mean channel gains $\lambda_{k,m}$. Still, the elements of $\mathbf{E}_{k,m}$ have equal variance due to the same statistics in all sub-links of a particular BS-UE link. Sublink here refers to the connection between a particular transmit and receive antenna pair of one link.

21.2.4 Performance Evaluation Metrics

The performance of the system defined in (21.5) can be evaluated by the achievable rate, which corresponds to the maximum data rate a user can obtain when the precoding matrix $\mathbf{B}$ is applied at the BSs. The achievable rate of UE k is denoted by

$$R_k = \log\det(\mathbf{I} + \mathbf{H}_k\bar{\mathbf{B}}_k\bar{\mathbf{B}}_k^H\mathbf{H}_k^H\mathbf{C}_k^{-1})\,, \tag{21.8}$$

where $\mathbf{H}_k\bar{\mathbf{B}}_k\bar{\mathbf{B}}_k^H\mathbf{H}_k^H$ refers to the meaningful portion of the signal received at user k, while $\mathbf{C}_k = \sum_{l\neq k}\mathbf{H}_k\bar{\mathbf{B}}_l\bar{\mathbf{B}}_l^H\mathbf{H}_k^H + \sigma_{\text{n},k}^2\mathbf{I}$ indicates noise and interference to this user. As can be seen from (21.8), the achievable rate is subject to the instantaneous channel state and the choice of the precoding matrix $\mathbf{B}$.

When CSI is perfectly available at the BS side, the achievable rate of each user is precisely known. In order to achieve it accurately, this exact data rate needs to be assigned for transmission. In contrast, when CSI is impaired, the achievable user rate is not known at the BS side and the assigned transmission rate can only be determined based on imperfect channel estimates. In the case when the assigned transmission rate exceeds the actually achievable rate, the correct decoding of the transmitted block cannot be guaranteed. As a result, outage can occur.

Assuming a particular estimate $\hat{\mathbf{H}}$ is known, from the BS perspective, the actual channel is a random variable according to (21.7). Due to this uncertainty, the probability that a data block, which is transmitted at a data rate $\bar{R}_k$, cannot be decoded at UE k is

$$p_{\text{out},k} = \mathbb{P}\{R_k < \bar{R}_k\}\,. \tag{21.9}$$

The rate that is successfully transmitted on average to UE k is

$$T_k = (1 - p_{\text{out},k}) \cdot \mathbb{E}\{\bar{R}_k\}\,, \tag{21.10}$$

which refers to the throughput of UE k. In addition, the sum user throughput is denoted by $T = \sum_{k=1}^{K} T_k$.

21.2.5 Problem Formulation

Outage probability is an indication of how reliable a link is on average. Inspired by the latest developments for 5G cellular networks, which has considerably higher requirements on data throughput and network reliability than the current 4G networks [8], we are interested in investigating the aspect of outages with respect to the design of the precoding matrix.

By combination of the user specific transmission rates into a single vector $\bar{\mathbf{r}} = [\bar{R}_1, \ldots, \bar{R}_K]^T$, the following formulation of the problem is reached:

$$\begin{aligned}[\mathbf{B}^\star, \bar{\mathbf{r}}^\star] = \; & \arg\max_{\mathbf{B},\bar{\mathbf{r}}_k} \textstyle\sum_{k=1}^{K} (1 - p_{\text{out},k}) \bar{R}_k \,, \\ & \text{s.t. } \operatorname{tr}(\mathbf{B}_m \mathbf{B}_m^H) < \gamma_m \quad \forall m \end{aligned} \tag{21.11}$$

where $\mathbf{B}^\star$ and $\bar{\mathbf{r}}^\star = [\bar{R}_1^\star, \ldots, \bar{R}_K^\star]^T$ denotes the optimized precoding matrix and the optimized transmission rate vector, where $\bar{R}_k^\star$ is the optimized rate for UE k. The function in (21.11) targets the maximization of the sum user throughput with per-BS transmit power constraints when CSI is imperfect and outage is considered.

In general, outage occurs when the actual channel does not support the assigned transmission rate. In order to avoid outage, the assigned rate can be decreased. However, such reduction also affects the overall throughput. More precisely, for a given precoding matrix, the throughput is a quasi-concave function of the assigned transmission rate. Hence, the optimal value can be found by, e.g., an approximation procedure known as Newton's method [2].

The joint optimization of the precoding matrix and the transmission rates is not a trivial task due to several challenges. First of all, the transmission rate is not fixed. Even if the transmission rate were fixed, no closed form expression for the outage probability is known when precoding is applied. In this work we present a solution which translates the joint optimization into an alternating algorithm. An approximation is used to find a suboptimal solution for the precoder optimization problem. More details on the solution are given in the next section.

21.3 Optimization Algorithm

In this section, we present a suboptimal solution for the throughput maximization problem in (21.11). It is known that joint optimization of two variables can be translated to alternating optimization [2]. This basically results in an iterative algorithm, where every iteration consists of two main steps. First, one variable is assumed fixed, while the other one is optimized. Secondly, the optimized variable is assumed fixed, while the first one is optimized. After both variables have been updated a new iteration is executed. If both individual optimizations lead to an improvement of the overall metric, convergence is achieved. It is known that with alternating optimization a local optimum can be obtained, while global optimality cannot be guaranteed. In the following, the derivation of the proposed optimization approach is presented.

21.3.1 Optimizing the Transmission Rate

Based on the approach of splitting up the joint optimization problem in (21.11), the optimization of the transmission rate for a given channel estimate and precoding matrix can be written as

$$\begin{aligned} \bar{\mathbf{r}}^{\star} = \arg\max_{\bar{\mathbf{r}}} \ & \textstyle\sum_{k=1}^{K}(1 - \mathbb{P}\{R_k(\mathbf{B}) < \bar{R}_k\})\bar{R}_k\,, \\ \text{s.t.} \ & \bar{R}_k \geq 0 \quad \forall k. \end{aligned} \tag{21.12}$$

The achievable rate in (21.12) is obtained by including (21.7) in (21.8) and the result into (21.9), where the probability is calculated w.r.t. the error $\mathbf{E}$.

Figure 21.3 shows the outage probability plotted against a function of the normalized transmission rate for different realizations of the CSI matrix. The normalization is w.r.t. the expected achievable rate $\mathbb{E}\{\sum_{k=1}^{K} R_k\}$, where the expectation is also w.r.t. the error. It can be observed, that for a given channel estimate and precoding matrix, the outage probability is a monotonically increasing function of the assigned transmission rate. As mentioned before, the throughput T_k is a quasi-concave function of the average assigned transmission rate $\bar{R}_k$ for each UE k. In order to obtain a value of the outage probability for a particular transmission rate, we use Monte Carlo simulations over the error. The result is illustrated in Fig. 21.4, where the obtained throughput is plotted against a function of the normalized assigned transmission

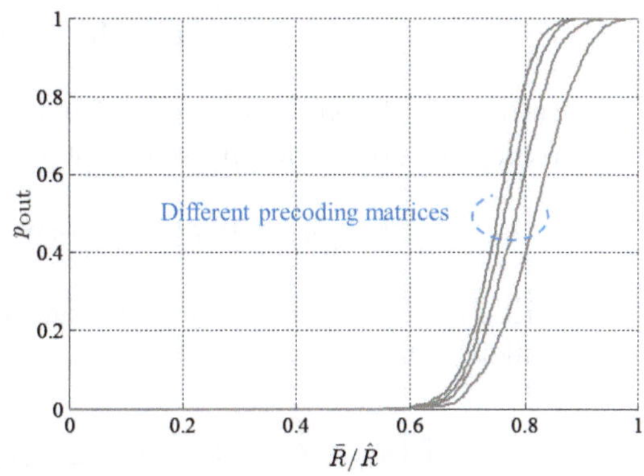

Fig. 21.3 Outage probability versus a function of the assigned transmission rate normalized w.r.t the estimated rate

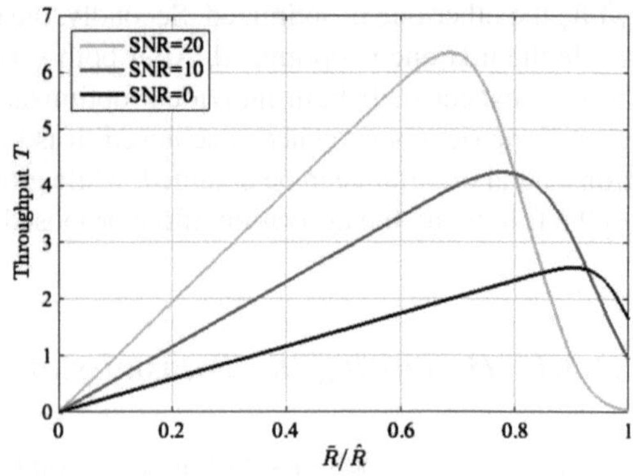

Fig. 21.4 Throughput considering outage versus a function of the assigned transmission rate normalized w.r.t. the estimated rate

rate. It can be observed that it is optimal to reduce the transmission rate to 70–90 % w.r.t. the expected rate. Furthermore, it can be seen, that on average the backing-off increases with the SNR.

21.3.2 Optimizing the Precoding Matrix

In order to find the precoding matrix **B**, which maximizes the throughput for a given transmission rate vector $\bar{\mathbf{r}} = [\bar{R}_1, \ldots, \bar{R}_K]$, the following optimization problem can be formulated:

$$\begin{aligned} \mathbf{B}^\star = \arg\max_{\mathbf{B}} & \textstyle\sum_{k=1}^{K} (1 - \mathbb{P}\{R_k(\mathbf{B}) < \bar{R}_k\})\bar{R}_k \,, \\ & \text{s.t. } \operatorname{tr}(\mathbf{B}_m \mathbf{B}_m^H) < \gamma_m \quad \forall m. \end{aligned} \tag{21.13}$$

Since, the transmission rates $\bar{R}_k$ are fixed, (21.13) reduces to

$$\begin{aligned} \mathbf{B}^\star = \arg\max_{\mathbf{B}} & \textstyle\sum_{k=1}^{K} -\mathbb{P}\{R_k(\mathbf{B}) < \bar{R}_k\}\bar{R}_k \,, \\ & \text{s.t. } \operatorname{tr}(\mathbf{B}_m \mathbf{B}_m^H) < \gamma_m \quad \forall m. \end{aligned} \tag{21.14}$$

In order to handle the probability expression in (21.14), we make use of the concentration inequality [18], which gives the relation

$$\mathbb{P}\{x \geq a\} = \mathbb{P}\{f(x) \geq f(a)\} \leq \mathbb{E}\{f(x)\}/f(a) \,, \tag{21.15}$$

for a strictly increasing and non-negative function $f(x)$. According to (21.14), we do the following transformation

$$\mathbb{P}\{R_k < \bar{R}_k\} = 1 - \mathbb{P}\{R_k \geq \bar{R}_k\} \geq 1 - \frac{E\{f(R_k)\}}{f(\bar{R}_k)} \,. \tag{21.16}$$

We now need to do a restriction on the function $f(x)$, in order to keep tractability of the optimization. Although more strictly increasing and non-negative functions exist (21.14), we find that only the linear function of the form $f(x) = a + bx$, with $a, b \in \mathbb{R}_0^+$, provides a solvable solution. On that basis the objective function is transformed into:

$$\mathbf{B}^{\star} = \arg\max_{\mathbf{B}} \ \Sigma_{k=1}^{K} \nu_k \mathbb{E}\{R_k\}, \quad \text{s.t. } \operatorname{tr}(\mathbf{B}_m \mathbf{B}_m^H) < \gamma_m \quad \forall m \tag{21.17}$$

with the user specific coefficients

$$\nu_k = b \cdot \bar{R}_k / (a + b \cdot \bar{R}_k). \tag{21.18}$$

The problem in (21.17) is of the same form as the WSR maximization problem, which has already been addressed in our previous work [12]. The weighting factors ν_k are determined by the assigned transmission rates as well as the choice of the coefficients a, b. How to solve the optimization problem in (21.17) is described in the following.

The achievable rate expression in (21.8) can be rewritten to

$$R_k = -\log\det(\mathbf{M}_k), \tag{21.19}$$

with the MMSE covariance matrix

$$\mathbf{M}_k = (\mathbf{I} + \bar{\mathbf{B}}_k^H \mathbf{H}_k^H \mathbf{C}_k^{-1} \mathbf{H}_k \bar{\mathbf{B}}_k)^{-1}. \tag{21.20}$$

Based on (21.19), the expected achievable rate in (21.17) can be lower bounded to

$$\mathbb{E}\{R_k\} = \mathbb{E}\{-\log\det(\mathbf{M}_k)\} \geq -\log\det(\mathbb{E}\{\mathbf{M}_k\}), \tag{21.21}$$

by applying Jensen's inequality [27], which states that for any random variable X and convex function φ, $\varphi\mathbb{E}(x) \leq \mathbb{E}(\varphi(x))$ holds.

It was shown in [5, 12] that the original WSR maximization problem can be translated into a Weighted Sum MMSE (WMMSE) minimization problem, where precoding matrix and receive filter need to be optimized jointly. A solution of this optimization problem can be found by alternatingly optimizing the receive filter matrix $\mathbf{U}$ and the precoding matrix $\mathbf{B}$, resulting in the iterative algorithm, which is stated in Algorithm 1 (see [12]). By this approach a local optimum can be found, similar to the alternating optimization for the overall throughput.

With a given precoding matrix, first the receive filters are optimized w.r.t. the MMSE between $\mathbf{d}$ and $\hat{\mathbf{d}}$ in step (a) by computing

$$\mathbf{U}_k^{\text{MMSE}} = \mathbf{B}_k^H \hat{\mathbf{H}}_k^H (\hat{\mathbf{H}}_k \mathbf{B} \mathbf{B}^H \hat{\mathbf{H}}_k^H + (\sigma_{n,k}^2 + \zeta_k)\mathbf{I})^{-1} \tag{21.22}$$

for each UE k with $\zeta_k = [\varepsilon_{k,1}\mathbf{1}_{1\times B_1}, \ldots, \varepsilon_{k,M}\mathbf{1}_{1\times B_M}] \cdot \mathrm{diag}(\mathbf{B}\mathbf{B}^H)$. Note that (21.22) takes into account that the CSI is only imperfectly available. For initialization of the precoding matrix the Wiener filter is used. A more detailed analysis is given in, e.g., [5]. Based on the same precoding matrix the weighting matrices $\mathbf{W}_k = \nu_k(\bar{\mathbf{M}}_k^{MMSE})^{-1}$ are calculated for each UE k in step (b). The MMSE matrix

$$\bar{\mathbf{M}}_k^{\mathrm{MMSE}} = (\mathbf{I} + \mathbf{B}_k^H\hat{\mathbf{H}}_k^H\bar{\mathbf{C}}_k^{-1}\hat{\mathbf{H}}_k\mathbf{B}_k)^{-1} \tag{21.23}$$

Algorithm 1: Weighted Sum Rate Maximization according to [12]

Input: $\hat{\mathbf{H}}$, σ_n^2, γ_m, ε
Output:$\mathbf{B}$
Set iteration index $i = 0$
Init: $\mathbf{B}^i = \mathbf{B}^{\mathrm{init}}$
Repeat
  Update $i = i + 1$
  (a) Update of the receive filter $\mathbf{U}_k^i|\mathbf{B}^{i-1}$ $\forall k$ with (21.22)
  (b) Update of the weighting matrix $\mathbf{W}_k^i|\mathbf{B}^{i-1}$ $\forall k$ with (21.23)
  (c) Update of the precoding matrix $\mathbf{B}^i|\mathbf{U}^i, \mathbf{W}^i$ with (21.25)
Until convergence

with the interference matrix

$$\bar{\mathbf{C}}_k = (\sigma_n^2 + \zeta_k)\mathbf{I} + \sum_{l=1, l\neq k}^{K} \hat{\mathbf{H}}_k\mathbf{B}_l\mathbf{B}_l^H\hat{\mathbf{H}}_k^H \tag{21.24}$$

results from substituting the channel in (21.6) with (21.7) and calculating the MSE covariance matrix between $\mathbf{d}$ and $\hat{\mathbf{d}}$ by applying the MMSE receive filters in (21.22).

Based on the weighting matrices, the precoding matrix can be found by solving the optimization problem

$$\begin{aligned}\mathbf{B}^\star = \arg\min_{\mathbf{B}} & \textstyle\sum_{k=1}^K \mathrm{tr}(\mathbf{W}_k\bar{\mathbf{M}}_k) \\ \text{s.t. } & \mathrm{tr}(\mathbf{B}_m\mathbf{B}_m^H) < \gamma_m \quad \forall m.\end{aligned}$$

A solution for the optimization problem in (21.25) is given by the transmit Wiener filter approach, which minimizes the average weighted sum MSE. It is obtained by the following expression

$$\mathbf{B} = b\tilde{\mathbf{B}} = b\left(\hat{\mathbf{H}}^H\mathbf{U}^H\mathbf{W}\mathbf{U}\hat{\mathbf{H}} + \mathrm{dg}(\mathbf{D}) + \frac{\sigma_k^2\mathrm{tr}(\mathbf{W}\mathbf{U}\mathbf{U}^H)}{\gamma}\mathbf{I}\right)^{-1}\hat{\mathbf{H}}^H\mathbf{U}^H\mathbf{W}, \tag{21.25}$$

where $\mathbf{D} = \mathbf{G}^T \text{diag}(\mathbf{U}^H \mathbf{W} \mathbf{U}) \mathbf{1}_{1 \times U}$ is the regularization matrix, which accounts for CSI imperfections. Furthermore, $\mathbf{W} = \text{blkdiag}(\mathbf{W}_1, \ldots, \mathbf{W}_K)$ denotes the overall weighting matrix and the extended error variance matrix $\mathbf{G} = [\mathbf{G}_1^T, \ldots, \mathbf{G}_K^T]^T$ with $\mathbf{G}_k = [\mathbf{G}_{k,1}, \ldots, \mathbf{G}_{k,M}]$, which consists of the link-wise error variances $\mathbf{G}_{k,m} = \varepsilon_{k,m} \cdot \mathbf{1}_{U_k \times B_m}, \forall k, m$. In order to satisfy the per-BS transmit power constraints, the scaling factor

$$b = \max_m \sqrt{\gamma_m / tr(\hat{\mathbf{B}}_m \hat{\mathbf{B}}_m^H)} \tag{21.26}$$

is applied, with the unscaled Wiener filter matrix $\hat{\mathbf{B}} = [\hat{\mathbf{B}}_1, \ldots, \mathbf{B}_M]^T$.

Algorithm 2: Proposed throughput maximization

Input: $\hat{\mathbf{H}}$, σ_n^2, ρ_m, ε
Output:$\mathbf{B}$, $\bar{R}_k$
Set iteration index $i = 0$
Init: $\mathbf{B}^i$=$\mathbf{B}^{\text{init}}$ according to Algorithm 1, $\nu_k = 1, \forall k$
Repeat
  Update $i = i + 1$
  (a) Compute the data rate $\bar{R}_k$, $\forall k$ by solving (21.12)
  (b) Update weights ν_k, $\forall k$ with (21.18)
  (c) Calculate $\mathbf{B}^i$ with ν_k according to Algorithm 1
Until convergence

21.3.3 Overall Optimization Algorithm

Now we combine the findings of Sects. 21.3.1 and 21.3.2 to the final algorithm. As stated before, we translate the joint optimization of the precoding matrix and the transmission rate into an alternating optimization algorithm, which is illustrated in Algorithm 2.

In this regard, first the precoding matrix is initialized according to Algorithm 1 without taking care of potential outages. On that basis the transmission rate is optimized in step (a), in order to maximize the throughput as discussed in Sect. 21.3.1. Therefore, the quasi-concave optimization problem in (21.12) can be solved by means of the Newton method [2], while the expectation is determined by Monte Carlo simulations.

Based on the optimized rate vector $\bar{\mathbf{r}}$ the user weights are updated in step (b) according to (21.18). The weights are used as an input for the precoding optimization in step (c), which is done according to Algorithm 1.

In order to obtain the user weights in step (b) of the Algorithm 2 the coefficients a and b need to be determined. In general, it can be seen that for $a = 0$, the weights result in $\nu_k = 1, \forall k$ and step (c) is not affected by the results of step (a) and (b).

Consequently, any further iteration lead to the same result as obtained after step (a). We chose the coefficients to be $a = b = 1$. More details on the choice of the coefficients as well as results on the overall performance are given in the next section.

21.4 Performance Evaluation

This section will present the results of the performance evaluation for the proposed algorithm. Since our solution is based on the method in [12], we will first present the results of the reference methods in an environment with imperfect CSI and no outage. Next we move on to compare the results of our proposed algorithm with all reference methods when outage is considered. Furthermore, we include the effect of inter-cluster interference. Lastly, we analyze the parametrization of the approximation function w.r.t. the performance of our proposed method.

In all simulations, we assume the number of BSs is two, each with two transmit antennas; the number of UEs is also two and each UE is equipped with two receive antennas. Radio propagation is modeled according to an urban macro-cell scenario as defined by 3GPP [1], where the inter-site distance d_I is 500 m, $\beta = -144.5$ dB, and $\alpha = 3.5$. The two UEs are located at the cell edge. According to the 1-D setup, introduced in Sect. 21.2, the distance to the BSs is $d_{k,m} = d_I/2, \forall k, m$. The variance of the CSI imperfection is $\varepsilon_{k,m} = 0.1 \cdot \lambda, \forall k, m$.

All algorithms were executed in a Monte Carlo simulation with 10,000 channel realizations. For our proposed algorithm, 100 outer iterations and 100 inner iterations were computed. Outer iterations refer to the alternating computation of precoding matrix and transmission rate, i.e., index i in Algorithm 2. Inner iterations are related to the calculation of the precoding matrix as given in [12]. In all figures,"P-CSI" stands for perfect CSI and "I-CSI" for imperfect CSI. "Outage" indicates that outage is considered and "ICI" means that inter-cluster interference is included.

Figure 21.5 shows the data rate performance of our reference methods when CSI is impaired but the transmission is assumed to be perfect, i.e., no outage can occur.

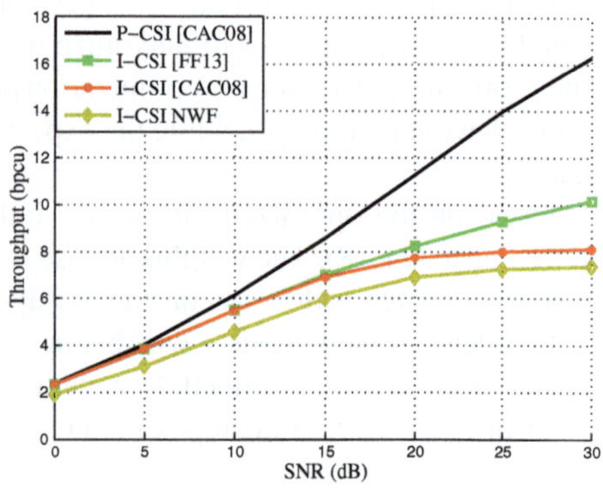

Fig. 21.5 Data rate in *bpcu* against SNR without inter-cluster interference for various algorithms

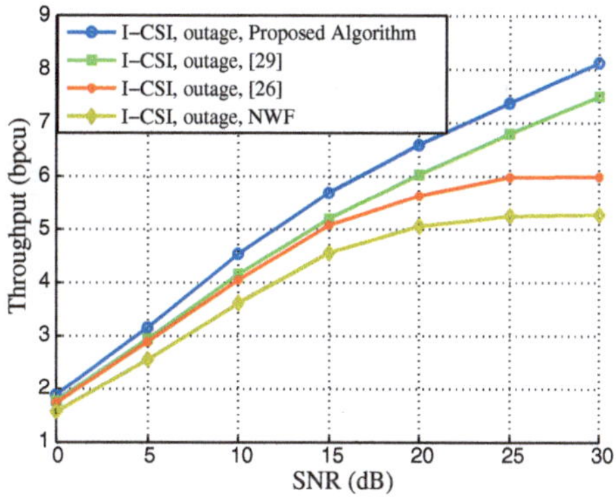

Fig. 21.6 Throughput in *bpcu* against SNR considering outage *without* inter-cluster interference for various algorithms

An upper bound is shown for comparison, which is obtained from [5] under perfect CSI. It is represented by a black line. Considering that CSI is only imperfectly available for computing the precoding matrix, a substantial performance loss can be expected; corresponding results are plotted in red. The green line refers to the robust precoding method in [12], which is able to partially compensate the degradation in data rate, especially in the high SNR regime. It can be observed that the robust precoding method yields 10 bpcu (*bits per channel use*) at a SNR of 30 dB, which is around 2 bpcu better than the method in [5]. In addition, the yellow line refers to the Normalized Wiener Filter (NWF), which is clearly outperformed by the algorithms mentioned before.

Figure 21.6 shows the performance of our proposed solution (Algorithm 2) and the three reference methods presented in Fig. 21.5. This figure illustrates the throughput when the transmission can be potentially in outage. Because the three reference methods did not consider outage in their respective designs, they would yield poor results when data is transmitted at the rate obtained by assuming perfect CSI. To allow for a fair comparison, we assume all these three methods are aware of potential outage, and data is transmitted at the optimized transmission rates $\bar{\mathbf{r}}^{\star}$. The blue line represents the performance of our proposed method. The green, red, and yellow lines represent the reference methods, respectively. From the figure, we observe that our proposed algorithm shows performance gains compared to [12] for the whole SNR regime, while the gains slightly increase with the SNR. At SNR = 30 dB, the gain compared to [12] is 0.5 bpcu and compared to [5] it is 2 bpcu. The improvement of our proposed algorithm is due to the alternating optimization of precoding matrix and transmission rate, where in each iteration the throughput is increased.

Figure 21.7 presents the throughput as a function of the SNR considering inter-cluster interference. From the figure, we observe that—when ICI is present—the performance of all methods is mainly reduced at high SNR, while the relations between the presented algorithms are similar to the case without ICI (Fig. 21.6). At SNR = 30 dB our proposed algorithm obtains 6.9 bpcu, while [12] achieves 6.4 bpcu. These results are roughly 15 % degraded compared to Fig. 21.6.

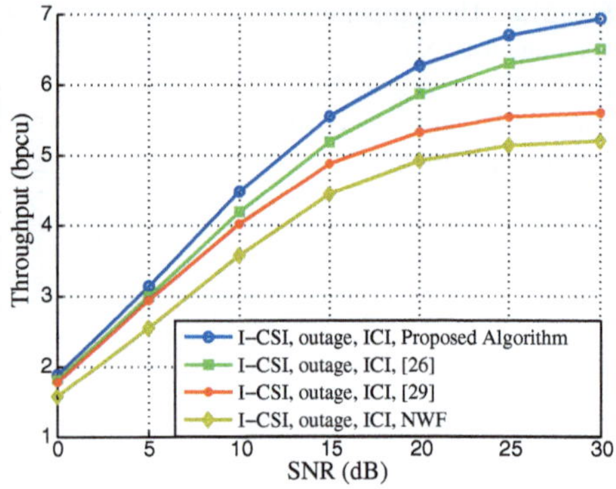

Fig. 21.7 Throughput in *bpcu* against SNR considering outage *with* inter-cluster interference for various algorithms

21.4.1 Parametrization of the Approximation Function

In the proposed method, we applied an approximation, which is based on the concentration inequality, see (21.15). The approximation results in coefficients, which are used as input to the precoding matrix computation. According to (21.18) they are defined as

$$\nu_k = \frac{b \cdot \bar{R}_k^\star}{a + b \cdot \bar{R}_k^\star}. \tag{21.27}$$

We further investigate the impact of the coefficients w.r.t. the values of a and b. We analyze two cases: (a) when $a \gg b$, and (b) when $b \gg a$.

Figure 21.8 plots the throughput as a function of the SNR considering outage without inter-cluster interference. Two different combination of a and b values are chosen: (a) a = 1,000, b = 1, and (b) a = 1, b = 1,000. We observe that there is no difference in throughput for all SNR regions for the considered combination of a and b values. The reason for this observation is due to the strictness of the concentration inequality approximation in (21.15). If an approximation is very strict about its para-

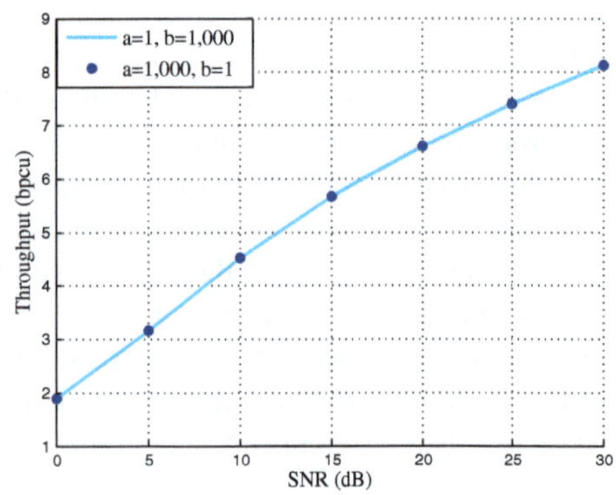

Fig. 21.8 The impact of the value of a and b on the throughput in *bpcu* versus SNR

meters, any slight changes in them will result in changes in the output. According to the concentration inequality [18], the approximation is valid for all types of probability distributions. Based on the observation of the simulation results, we can draw the conclusion that this approximation is relatively relaxed. However, the strictness and convergence behavior of this approximation is beyond the scope of this work.

21.5 Conclusion

In this work, we investigated linear precoding in a network MIMO downlink system, when channel state information (CSI) is only imperfectly available at the BS side and transmissions can be in outage. We target at the maximization of the overall throughput by jointly optimizing the precoding matrix and the transmission rate. We proposed an iterative optimization strategy that alternately computes the precoding filter and the transmission rate. In order to optimize the precoding matrix with a given transmission rate, we made use of the concentration inequality. We evaluated our proposed algorithm by means of Monte Carlo simulations, showed that gains in the throughput performance can be achieved, and compared the proposed algorithm with state-of-the-art solutions.

Acknowledgments The authors would like to thank Prof. Eduard Jorswieck and Dr. Pin-Hsun Lin from Technische Universität Dresden for their valuable discussions.
This work was supported by the German Science Foundation (DFG) within the priority program COIN under project grant Fe 423/12-3.
The computations were performed on a computing cluster at the Center for Information Services and High Performance Computing (ZIH) at TU Dresden.

References

1. 3GPP (2009) 3rd Generation partnership project (3GPP) Technical specification group radio access network (2010) Further advancements for E-UTRA physical layer aspects (Release 9). Technical report 3GPP TR 36.814
2. Boyd S, Vandenberghe L (2004) Convex optimization, 1st edn. Cambridge University Press
3. Caire G, Shamai S (2003) On the achievable throughput of a multiantenna Gaussian broadcast channel. IEEE Trans Inf Theory **49**(7):1691–1706
4. Castro P, Joham M, Castedo L, Utschick W (2007) Optimized CSI feedback for robust THP design. In: Proceedings of the Asilomar conference on signals, systems and computers (ACSSC '07). Pacific Grove, CA, USA, pp 1956–1960
5. Christensen SS, Agarwal R, Carvalho E, Cioffi JM (2008) Weighted sum-rate maximization using weighted MMSE for MIMO-BC beamforming design. IEEE Trans Wirel Commun **7**(12):4792–4799
6. Costa MHM (1983) Writing on dirty paper (Corresp.). IEEE Trans Inf Theory **29**(3):439–441
7. Dietrich FA, Hunger R, Joham M, Utschick W (2003) Robust transmit Wiener filter for time division duplex systems. In: Proceedings IEEE international symposium on signal processing and information technology (ISSPIT'03), Darmstadt, Germany, pp 415–418

8. Fettweis GP, Alamouti S (2014) 5G—Personal mobile internet beyond what cellular did to telephony. IEEE Commun Mag **52**(2):140–145
9. Fritzsche R, Fettweis GP (2011) CSI distribution for joint processing in cooperative cellular networks. In: Proceedings IEEE vehicular technology conference (VTC'11-Fall), San Francisco, CA, USA, pp 1–5
10. Fritzsche R, Fettweis GP (2012) Robust precoding with general power constraints considering unbounded channel uncertainty. In: Proceedings of the IEEE international symposium on wireless communication systems (ISWCS'12), France, Paris, pp 161–165
11. Fritzsche R, Fettweis GP (2013) Distributed robust sum rate maximization in cooperative cellular networks. In: Proceedings IEEE CoNET workshop at IEEE international conference on communications (ICC '13), Budapest, Hungary, pp 291–295
12. Fritzsche R, Fettweis GP (2013) Robust sum rate maximization in the multi-cell MU MIMO downlink. In: Proceedings of the IEEE wireless communication and networking conference (WCNC), Shang Hai, China, pp 3180–3184
13. Fritzsche R, Fettweis GP (2014) Optimized quantization resolution for limited CSI feedback with outage. In: Proceedings international symposium on wireless communications systems (ISWCS'14), Barcelona, Spain, pp 584–588
14. Fritzsche R, Ohlmer E, Fettweis GP (2013) Where to predict the channel for cooperative multicell transmission over correlated subcarriers? In: Proceedings IEEE international conference on acoustics, speech, and signal processing (ICASSP'13), Vancouver, Canada, pp 5051–5055
15. Fritzsche R, Ohlmer E, Fettweis GP (2013) Where to predict the channel in cooperative cellular networks with backhaul delay? In: Proceedings international ITG conference on systems, communications and coding (SCC'13), Munich, Germany, pp 1–6
16. Gesbert D, Kountouris M, Heath RW, Chan-Byoung C, Salzer T (2007) Shifting the MIMO paradigm. IEEE Signal Process Mag **24**(5):36–46
17. Habendorf R, Fettweis GP (2006) Nonlinear predistortion for OFDM SDMA systems. In: Proceedings IEEE workshop on signal processing advances in wireless communications (SPAWC'06), Cannes, France, pp 1–5
18. Hoeffding W (1963) Probability inequalities for sums of bounded random variables. J Am Stat Assoc **58**(301):13–30
19. Irmer R, Droste H, Marsch P, Grieger M, Fettweis G, Brueck S, Mayer HP, Thiele L, Jungnickel V (2011) Coordinated multipoint: concepts, performance, and field trial results. IEEE Commun Mag **49**(2):102–111
20. Jindal N (2005) MIMO broadcast channels with finite rate feedback. In: Proceedings IEEE global telecommunications conference (GLOBECOM'05), vol 3. St. Louis, MO, USA
21. Joham M, Kusume K, Gzara MH, Utschick W, Nossek JA (2002) Transmit wiener filter for the downlink of TDDDS-CDMA systems. In: Proceedings IEEE seventh international symposium on spread spectrum techniques and applications (ISSSTA'02), vol 1. Prague, Czech, pp 9–13
22. Karakayali MK, Foschini GJ, Valenzuela RA (2006) Network coordination for spectrally efficient communications in cellular systems. IEEE Trans Wireless Commun **13**(4):56–61
23. Kaviani S, Simeone O, Krzymien W, Shamai S (2012) Linear precoding and equalization for network MIMO with partial cooperation. IEEE Trans Veh Technol **61**(5):2083–2096
24. de Kerret P, Fritzsche R, Gesbert D, Salim U (2014) Robust precoding for network MIMO with hierarchical CSIT. In: Proceedings international symposium on wireless communications systems (ISWCS'14), Barcelona, Spain, pp 987–991
25. Keusgen W (2003) On limits of wireless communications when using multiple dual-polarized antennas. Wireless Pers Commun **1**:204–210
26. Kusume K, Joham M, Utschick W, Bauch G (2005) Efficient Tomlinson-Harashima precoding for spatial multiplexing on flat MIMO channel. In: Proceedings IEEE international conference on communications (ICC '95), Seoul Korea, pp 2021–2025
27. Negro F, Ghauri I, Slock D (2012) Sum rate maximization in the noisy MIMO interfering broadcast channel with partial CSIT via expected weighted MSE. In: Proceedings of the IEEE international symposium on wireless communication systems (ISWCS'12), France, Paris, pp 576–580

28. Ohlmer E, Fettweis GP (2011) Rate adaptation for time variant MIMO channels with linear receivers. In: Proceedings of the IEEE vehicular technology conference (VTC'11 Fall), Budapest, Hungary, pp 1–5
29. Shenouda MB, Davidson TN (2006) Robust linear precoding for incertain MISO broadcast channels. In: Proceedings IEEE international conference on acoustics, speech and signal processing (ICASSP '06), Toulouse, France, p 4
30. Shi S, Schubert M, Boche H (2008) Downlink MMSE transceiver optimization for multiuser MIMO systems: MMSE balancing. IEEE Trans Signal Process **56**(8):3702–3712
31. Shuying S, Schubert M, Vucic N, Boche H (2008) MMSE optimization with per-base-station power constraints for network MIMO systems. In: Proceedings IEEE international conference on communications (ICC '08), Beijing, China, pp 4106–4110
32. Tse D, Viswanath P (2008) Fundamentals of wireless communications, 1st edn. Cambridge University Press
33. Vucic N, Boche H, Shuying S (2009) Robust transceiver optimization in downlink multiuser MIMO systems. IEEE Trans Signal Process **57**(9):3576–3587
34. Wiesel A, Eldar Y, Shamai S (2006) Linear precoding via conic optimization for fixed MIMO receivers. IEEE Trans Signal Process **54**(1):161–176
35. Wunder G, Jung P, Kasparick M, Wild T, Schaich F, Yejian C, Brink S, Gaspar I, Michailow N, Festag A, Mendes L, Cassiau N, Ktenas D, Dryjanski M, Pietrzyk S, Eged B, Vago P, Wiedmann F (2014) 5GNOW: Non-orthogonal, asynchronous waveforms for future mobile applications. IEEE Commun Mag **52**(2):97–105
36. Yoo T, Goldsmith A (2006) On the optimality of multiantenna broadcast scheduling using zero-forcing beamforming. IEEE J Sel Areas Commun **24**(3):528–541
37. Zhang H, Dai H (2004) Cochannel interference mitigation and cooperative processing in downlink multicell multiuser MIMO networks. EURASIP J Wireless Commun Netw Spec Multiuser MIMO Netw (2):222–235

28. [illegible] (2011) Rate adaptation for time variant MIMO channels with [illegible] feedback. In: Proceedings of the IEEE vehicular technology conference (VTC [illegible]), Budapest, Hungary, pp 1–5
29. [illegible] (2010) Robust linear precoding for multi-user MIMO [illegible] channels. In: Proceedings IEEE international conference on acoustics, speech and signal processing (ICASSP) [illegible]
[illegible]

MIX
Papier aus verantwortungsvollen Quellen
Paper from responsible sources
FSC® C105338

If you have any concerns about our products,
you can contact us on
ProductSafety@springernature.com

In case Publisher is established outside the EU,
the EU authorized representative is:
Springer Nature Customer Service Center GmbH
Europaplatz 3, 69115 Heidelberg, Germany

Printed by Libri Plureos GmbH
in Hamburg, Germany